AGARD Conference Proceedings
Number Thirty Eight

New Experimental Techniques in Propulsion and Energetics Research

The Advisory Group for
Aerospace Research and
Development of NATO

1970

Editors

David Andrews

Industrial and Marine Gas Turbine Division
Rolls Royce Ltd
and

Jean Surugue

Director Scientifique, Energie et Propulsion
ONERA

Printed and Published by
Technivision Services
Slough, Bucks England
A Division of Engelhard Hanovia International Ltd.

International Standard Book Number 0.85102.024.0

Library of Congress Catalog Card Number 73-82415

Contents

Préface

Les méthodes expérimentales appliquées à l'étude de la combustion ont été présentées dans un manuel AGARD publié il y a sept ans. Depuis cette époque, les méthodes décrites dans ce livre ont été développées et de nouvelles méthodes, fort importantes ont été découvertes. Le Groupe de Travail AGARD sur la Propulsion et l'Energétique a donc jugé opportun de tenir une Réunion Technique à Munich du 11 au 15 Septembre 1967 afin de discuter ces nouvelles techniques expérimentales et le présent document est la version finale des Comptes Rendus de cette réunion.

La plupart de ces nouvelles techniques ont été développées pour l'analyse d'écoulements à grande vitesse - supersoniques et hypersoniques - tels par exemple que ceux qui ont lieu dans les statoréacteurs hypersoniques.

La recherche sur la combustion qui met en oeuvre la cinétique chimique a largement profité du développement de techniques utilisant le tube à choc. L'étude des réactions de combustion, qui n'ont que des durées très faibles, présente un grande intérêt. La comparaison des techniques de tube à choc avec celles qui ont été développées dans l'étude des régimes permanents doit être faite. Les deux techniques donnent des informations incomplètes, mais elles semblent complémentaires.

Les méthodes de visualisation si fructueuses dans tous les domaines de l'aérodynamique semblent en revanche difficilement applicables aux problèmes de combustion où les faibles variations de pression n'induisent que de faibles variations de densité, donc un contraste optique insuffisant. Les progrès obtenus dans certains laboratoires de recherche ont donné l'espoir que des techniques optiques améliorées pourraient servir à l'investigation des zones de combustion; les méthodes utilisant le laser en sont un exemple.

La combustion dans un courant gazeux pose toujours le problème de la description du développement dans le temps des réactions chimiques prenant place dans le sein du fluide. Si ce problème n'a pas encore été résolu, c'est dû au fait que les méthodes de prélèvement et d'échantillonnage sont encore incertaines et les méthodes d'analyses imparfaites. Pour ces dernières, il faut insister sur le développement de l'emploi du spectromètre de masse dans l'industrie. La simplicité de ces instruments, l'amélioration de leur temps de réponse et de leur sensibilité ont considérablement augmenté la précision des mesures effectuées.

Cependant, le domaine dans lequel le plus de techniques nouvelles sont apparues est sans conteste celui de la recherche sur les plasmas. Et même, s'il n'est pas tout à fait permis de considérer les gaz issus de la combustion comme des plasmas, ils sont néanmoins suffisamment ionisés pour que les techniques initialement conçues pour le diagnostic et la description des plasmas puissent leur être transposées. Les études sur la combustion n'ont pas été privées du bénéfice de ces techniques: quelques unes ont été adaptées avec un succès certain.

Tels sont les développements décrits dans cet ouvrage. Nous avons eu quelques difficultés pour le rendre clair et concis et nous espèrons qu'il sera utile pour les Chercheurs et Ingénieurs travaillant dans le domaine de la combustion.

David Andrews
Jean Surugue

Preface

Experimental methods for the study of combustion were the subject of an AGARD handbook published seven years ago. Since that time, the methods then described have been further developed and important new methods have been discovered. The Propulsion and Energetics Panel of AGARD therefore considered it timely to hold a Technical Meeting in Munich from the 11th to the 15th September 1967 in order to discuss these new experimental techniques, and this present volume is an edited version of the Proceedings.

Most of the new techniques have been developed for the diagnosis of high velocity flow - both supersonic and hypersonic,- such as occurs,for instance, in hypersonic ramjets.

Combustion research,which brings into prominence questions of chemical kinetics, has been stimulated by the development of techniques employing shock tubes. Great interest is being shown in the study of combustion reactions taking place in very short intervals of time. The comparison of these techniques with those that have been developed for studying steady-state conditions is the question of the day. Both give incomplete information, but they seem to be complementary.

Visualisation methods,which are so valuable in a general way in aerodynamics, are difficult to apply to the problems now in question because the relatively small changes in pressure level cause small changes in density and hence weak optical contrasts. Progress made as a result of the work of several research laboratories has raised hopes of improvements in techniques for measuring conditions in combustion zones; one in particular, of which much is expected, in the use of lasers.

Combustion occuring in gas streams always poses the problem of describing the time history of the chemical reactions taking place. That this has not been resolved already is due in part to the uncertain validity of sampling methods and in part to the analytical methods that have to be employed. On the latter point, however, there is justification for drawing attention to the growth in the industrial use of mass spectrographs. Their simplicity and the improvements made in their speed and sensitivity have greatly increased the accuracy of the measurements obtained.

The field in which the greatest novelty is seen however is without doubt in plasma research. And even if it is not permissible to speak of combustion gases as plasmas, they are nevertheless sufficiently ionised for techniques originally evolved for the diagnosis and description of plasmas to be used for them too. Indeed combustion studies have not been deprived of the benefit of these techniques : some have been transposed already with satisfactory results.

These then are the developments described in this volume.

We have been at some pains to make the volume lucid and concise, and hope that it will be useful to Scientists and Engineers working in this field.

David Andrews
Jean Surugue

Part 1

Shock Tubes and Related Techniques

1-1

New Experimental Techniques for Kinetic Studies in Shock Tubes

K. L. WRAY.
Avco Everett Research Laboratory, Everett, Massachusetts, U. S. A.

Summary

Since about 1960 the number of techniques available to the experimentalist to investigate kinetic processes in shock tubes has grown significantly. It is the purpose of this chapter to review the advantages and limitations of these new techniques and to illustrate their applications. Older methods which have already been reviewed in detail elsewhere are excluded. Those that are discussed here can conveniently be divided into two categories: 1) diagnostic techniques applicable to monitoring time histories behind shock waves; and 2) techniques which may be used in the preparation of nonequilibrium distributions, either in the unshocked gas or behind the shock. In the first category are such methods as infrared emission and absorption, electron absorption and scattering, the hook method, which uses a spectrometer crossed with an interferometer, mass spectrometry, laser interferometry, and some new concepts in single pulse shock tubes. In the second category are techniques such as electric discharges, flash photolysis, and the production of free radicals by shocking unstable molecules. Also included in this category is the two phase shock tube.

Sommaire

Le nombre de techniques expérimentales mises à la disposition des chercheurs pour étudier la cinétique des réactions a considérablement augmenté depuis 1960. Dans le présent exposé les avantages et les limitations de ces nouvelles techniques sont analysés et leurs applications sont présentées. On n'indique pas les méthodes plus anciennes qui ont déjà été analysées en détail. Les méthodes décrites ici peuvent être classées en deux catégories: 1) techniques de diagnostic pouvant être utilisées pour la commande de l'histoire du phénomène en aval de l'onde de choc; 2) techniques que l'on peut utiliser pour la préparation de distributions hors équilibre dans le gaz frais ou en aval du choc. La première catégorie comprend des méthodes telles que l'émission et l'absorption infra-rouge, l'absorption et la dispersion d'électrons, la méthode dite du crochet (elle utilise un spectromètre et un interféromètre croisés), la spectrométrie de masse, l'interférométrie laser et quelques nouvelles techniques utilisables dans les tubes à choc à simple impulsion. Dans la seconde, on trouve des techniques telles que les décharges électriques, la photolyse par éclair et la production de radicaux libres par chocs sur des molécules instables. On comprend dans cette catégorie le tube à choc à deux phases.

Introduction

In this Chapter 'kinetic processes' are considered in a broad sense so as to include not only the classical chemical processes but also energy relaxation processes such as translational and vibrational relaxation, collisional excitation and de-excitation of electronic states, and chemiluminescent processes including those involving electrons. Techniques strictly applicable to monitoring electron concentration, such as microwaves, Langmuir probes, and various coil techniques, are deliberately not considered.

It is not the purpose of this Chapter to review older techniques that have already appeared in review articles and books, though the distinctions drawn might, in a few cases, be considered rather arbitrary. Deliberately excluded from this review are such well established methods as ultraviolet and visible absorption and emission, line reversal, conventional single pulse shock tubes, X-ray densitometry, conventional interferometry and schlieren, pressure pickups, and heat transfer gages. These methods are reviewed in references (1) to (6). Reference (3) has a good chapter on measurement of temperature which, in particular, covers the spectral line reversal method in detail; (4) has a good discussion of the single pulse technique; (6) is a reference book for those starting to do chemical kinetics in shock tubes.

The new experimental techniques discussed in this Chapter can conveniently be divided into two categories: 1) diagnostic techniques applicable to monitoring time histories behind shock waves; this would include densities, temperatures, and concentrations; and 2) techniques that may be used in the preparation of nonequilibrium distributions, either in the unshocked gas or behind the shock.

New Diagnostic Techniques

Low Density Shock Tubes

In 1960 Lin and co-workers (7) designed and built a 24 in. diameter low density shock tube at the Avco Everett Research Laboratory. The original purpose for this apparatus was to study the rate of ionization in shock heated air at temperatures sufficiently high so that the kinetics could not be resolved at initial shock tube pressures above 1 torr, which is about the lower limit for conventional 1.5 in. shock tubes. The low density shock tube was constructed of stainless steel, with a low pressure section having an inside diameter of 24 in. and a length of 54 ft. The driver which was only 5 in. inside diameter and 6 ft. long was coupled to the low pressure section via a 4 ft. long transition section, located immediately downstream of the diaphragm. Shocks could be initiated in the conventional manner either by pressure or combustion breaks. The shock tube was found to operate satisfactorily down to initial pressures (P_1) of $10\,\mu$. In these initial studies (7) (8) Lin measured the curvature of the shock front.

In his ionization experiments (9) Lin inserted the microwave probe several inches into the shock tube to avoid the boundary layer flow near the tube walls. To minimize aerodynamic disturbances to the flow due to the presence of the microwave probe, the probe, a waveguide, was terminated with a knife edge plate at its end.

The knife edge technique has invariably been used in association with this facility. For example, Wray (10) used it to study the coupling of the O_2-Ar rates of dissociation and vibrational relaxation. Incident shocks covered the temperature range between 5000 and 18,000°K; P_1 was varied from 0.25 to 10 torr. UV absorption at 1470 Å was used to monitor the O_2 concentration as a function of time. The microwave technique was used to simultaneously monitor the electron concentration to ensure that electrons were not produced during the chemical relaxation of the oxygen.

Using four heat transfer gages, mounted 90° apart around the circumference of the shock tube, Wray showed that under the conditions of the experiment, tilt was insignificant. Lin (7) had found the shock front curvature (defined as the distance along the center line from the shock front to the plane formed by the intersection of the shock front and the shock tube walls) to be independent of the shock speed and to vary inversely as the square root of the initial pressure. His data for curvature as a function of initial pressure, P_1, is fitted by the following equation (10):

$$\delta \text{ (mm)} = 1.75.P_1^{-\frac{1}{2}} \text{ (torr}^{-\frac{1}{2}}) \qquad \text{(Eq. 1-1-1)}$$

At the lowest P_1 used in these experiments, shock front curvature would amount to 3.5 mm. The optical slits used for all the experiments were 1 mm wide. Hence curvature represented a serious loss in resolution.

To avoid this difficulty the optical path was reduced to 12 in. by inserting 1 in. diameter, 6 in. long pipes into opposite sides of the shock tube (Fig. 1-1-1). Each of these pipes was terminated inside the shock tube by a flat plate 2.5 in. high and 1.5 in. wide, the edges of which were tapered at an angle of 9° 44' to form knife edges. The width of the plate was arrived at by making the leading edge project twice the standoff distance of the bow shock formed in front of the pipe, and the height was such that the Mach wave formed at the upper and lower corners of the leading edge would not cross the window located at the center of the plate. The center of each knife edge plate contained a flush calcium fluoride window, behind which was located the slit.

Assuming the shock front surface to be spherical, the amount of curvature between two parallel planes, the knife edges, placed within the shock tube decreases as the square of the ratio of the distance between the two planes to the shock tube diameter. Hence for the runs at 0.25 torr the shock front curvature was effectively reduced to 0.87 mm. Furthermore, this technique reduces the problem of O-atom recombination in the cold, dense boundary layer, since on these plates the boundary layer grows only to a steady state thickness rather than continuously as on the shock tube wall.

In subsequent work, using the low density facility, Wray and Freeman (11) studied the shock front structure in oxygen at high Mach numbers, again using UV absorption at 1470 Å. This work was carried out in pure O_2 at a P_1 of 15 and 30 μ, the Mach number range covered was 4 to 21. The knife edge technique made the experiment possible. In this work the optical system extended into the shock tube 10 in. on each side leaving only a 4 in. optical path between plates. At $P_1 = 15\mu$ the shock front curvature across the 24 in. tube is equal to 14 mm, so with knife edges placed 4 in. apart the curvature across the optical path was reduced by a factor of 36 or $\delta = 0.4$ mm. With 0.5 mm slits, the total resolution was 0.9 mm. Since at 15 μ the upstream mean free path in O_2 is 3.6 mm, this experiment had a resolution equal to 0.25 ambient mean free paths.

Infrared Emission and Absorption

The infrared emission technique for monitoring the concentration of heteronuclear diatomic molecules and polyatomic molecules has proved to be extremely valuable in recent years. The technique has also been used to monitor the time history of the vibrational energy behind incident and reflected shock waves. Indeed the first use of the IR emission technique in shock tube work was by Windsor, Davidson, and Taylor (12) in 1957, who used a lead sulphide infrared detector with a rise time of 30 μ sec to follow CO vibrational relaxation, monitoring its first overtone at a wavelength of 2.35 μ. A germanium filter was used to isolate a band pass between 2.0 and 2.8 μ. Both incident and reflected shock waves in a pyrex shock tube were employed. Some crude vibrational relaxation times were measured and the workers clearly demonstrated the important effect of trace amounts of H_2O.

Since the work of Windsor *et al* (12), the IR technique has been used successfully to follow the time history of a variety of species. Included in these studies are the

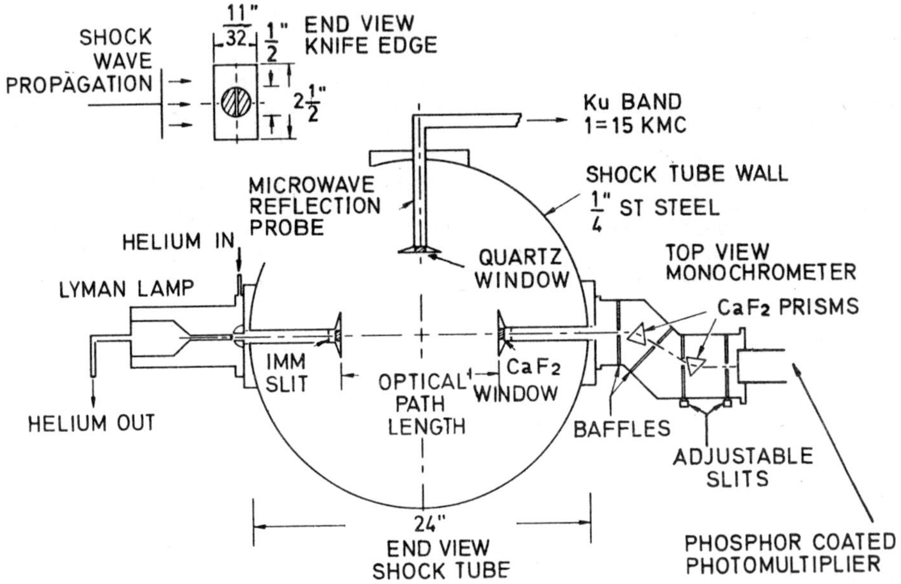

Fig. 1-1-1 Schematic diagram of the 24 in. diameter low density shock tube showing the use of knife edges on a UV absorption and microwave reflection experiment.

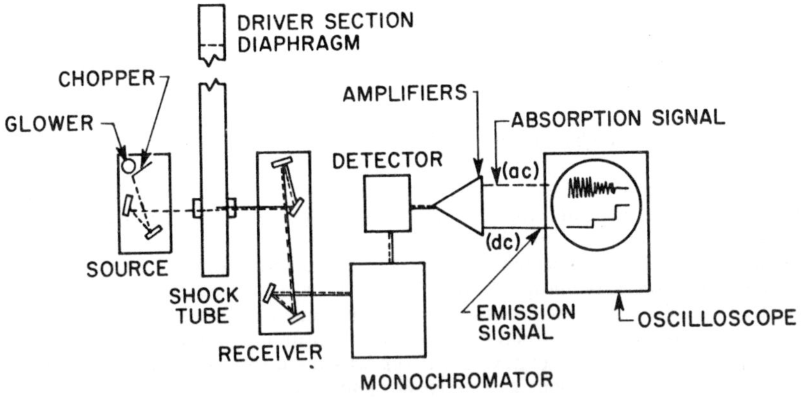

Fig. 1-1-2 Schematic diagram showing the apparatus used to simultaneously measure IR absorption and emission on the shock tube.

thermal decomposition of NH_3, (13, 14), CO_2 (15-17), HF (18), HCl (19, 20), DCl (20), and N_2O (21). The IR technique has also been used along the original lines of following the vibrational relaxation by Taylor *et al* (22) in which work the vibration-vibration coupling in gas mixtures such as NO-CO, NO-N_2, and CO_2-N_2 were studied. In this work the shock tube was supplied with a test section constructed of lucite, a material which the authors claim has a very low reflectivity in the IR, thus eliminating reflection problems. Some of their work was carried out in the Avco Everett 24 in. diameter shock tube using the knife edge technique.

Although measurements of radiation *per se* are not to be dwelt on in this chapter, we might mention here studies carried out by Taylor (23) on the emission continuum from air and nitrogen. He worked at wavelengths out to 8μ, taking intensity measurements as a function of wavelength point by point. Recently Camm *et al* (24) have constructed a high speed scanning infrared spectrometer that can be synchronized with a pulse source and triggered within 15 μsec. This was an f/12 Ebert type monochromator which had a scanning rate of 200 Å/μsec, a range of 2 to 6μ, and a spectral resolution of 700 Å. The scanning was accomplished by spinning a mirror with a magnetic field, this being produced by a current pulse through a coil wrapped around the mirror. When pulsed, the mirror accelerated for about 15μsec and then coasted at constant speed. The mirror position could be manually set and the instrument had a built in wavelength-defining signal. A ribbon type InSb detector was used. This device was employed in making continuum measurements behind reflected shocks.

Wilson (25) has used free-free radiation from ions at 6 μ, where there is no line interference, to monitor electron production behind very strong shocks in air. He used a 6 in. diameter, arc driven shock tube to get shock speeds between 9 and 12.6 mm/μ sec. Even at the low initial pressure used in this experiment (50 μ Hg), microwave and coil techniques were not capable of the necessary spatial resolution. Wilson used gold and copper doped germanium detectors which, with the aid of knife edges, enables a spatial resolution of 1 mm and a rise time of 0.1 μ sec. In this way, he was able to extend the high temperature limit of the ionization measurements in air significantly.

Recently the IR technique of simultaneously monitoring emission and absorption to obtain equilibrium temperatures has been successful. The work of Lauver *et al* (26) and Penzias *et al* (27) are very similar and both groups have obtained CO_2 temperatures behind incident and reflected shock waves in good agreement with theory. The temperatures that were covered in these measurements are in the range 1000 to 4000°K. Both CO_2-Ar and CO_2-N_2 mixtures were used. This technique involves the simultaneous measurement of the spectral radiance N_λ and the spectral absorptivity which is, of course, equal to the spectral emissivity ϵ_λ. By Kirchhoff's law:

$$\frac{N_\lambda(T)}{\epsilon_\lambda(T)} = N_\lambda(bb)(T), \qquad \text{(Eq. 1-1-2)}$$

where N_λ (bb) (T) is the spectral radiance of a black body, and which, of course, immediately yields a temperature.

The details of the technique are best realized by examining figure 1-1-2 which shows schematically the experimental setup. There is no need for the continuum source used to make the emissivity measurement to be as bright as the shock tube source itself, since the absorption and emission signals are separated using a narrow band amplifier set to pass only the modulated chopping frequency. In the work of Penzias (27) the light was chopped at 165 kc, while Lauver *et al* (26) used 80 kc. Although this technique has not yet been applied to measuring chemical

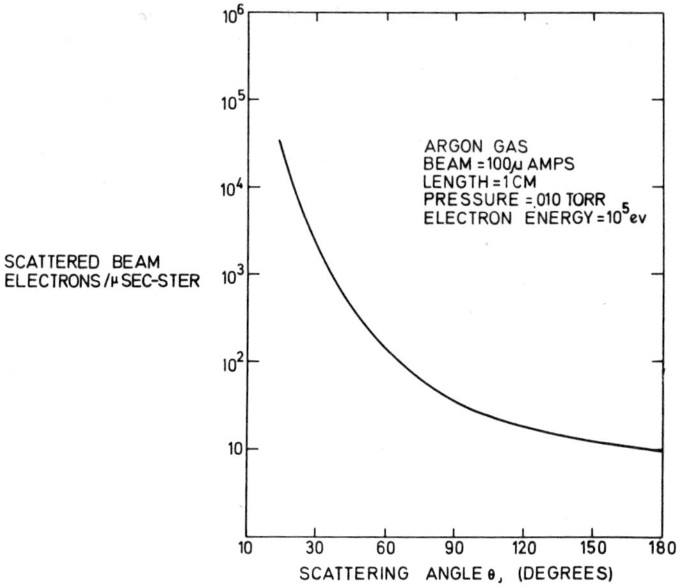

Fig. 1-1-3 Theoretically calculated scattered electron beam intensity as a function of the scattering angle.

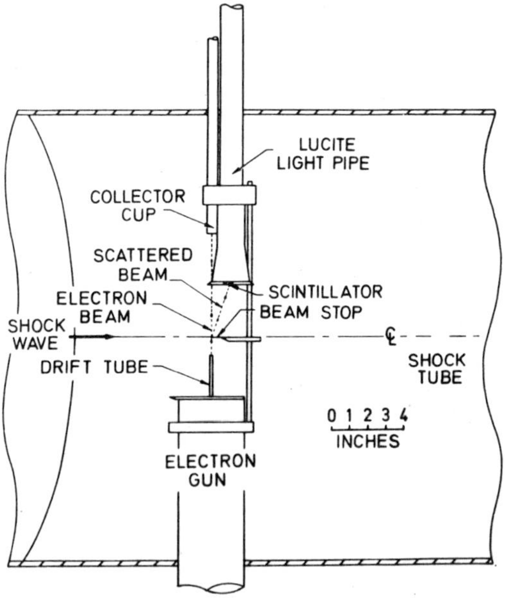

Fig. 1-1-4 Schematic diagram showing the arrangement of the electron beam scattering experiment in the 24 in. diameter shock tube.

kinetics in shock tubes, it certainly has potential to do that, apparently with great precision. The advantages such a technique has over the line reversal technique are: a) there is no need to introduce a tracer species such as Na or Cr; and b) it is applicable over a wide range of temperatures, there being no limit set by the brightness temperature of the source.

Electron Beam Scattering

The technique of measuring densities behind shock waves employing the absorption of an electron beam originated with Ballard and Venable (28), and has been reviewed elsewhere (6). Although electron beam absorption has recently been used (29, 30) in large diameter shock tubes running at low initial pressures to get shock front density profiles in Ar and He over the Mach number range of 1.65 to 9.5, it is felt that the technique that utilizes the scattered electrons has more future in shock tube chemical kinetic applications.

The electron beam scattering technique was developed by Camac (31), originally to study shock front thicknesses but has since been applied to more conventional chemical problems in the shock tube. In this technique the electrons scattered through small angles are monitored. In the original work (31, 32) Camac measured the shock thicknesses of argon and nitrogen at a P_1 of 15 to 60 μ.

The equation for Rutherford scattering of electrons gives the scattered beam intensity in electrons/sec-ster as:

$$I_s = 7.8 \times 10^{-16} I_0 n Z^2 \ell \left(\frac{1 + \frac{V}{V + 10^6}}{V \sin^2 \theta/2} \right)^2 \qquad \text{(Eq. 1-1-3)}$$

where I_0 = electron/sec in the incident beam, n = the number density of particles of atomic number, Z, ℓ = the path length, V = the electron energy, and θ = the scattering angle. Camac used 90 keV electrons and a primary beam having an intensity of 100 μ amps. Equation 1-1-3 is plotted in figure 1-1-3 for typical operating conditions; note the strong forward scattering in this phenomenon. For angles equal to or less than 20° a 1 % density determination can be made in times of the order of 1μ sec (i.e., $S/N \geq \sqrt{10^4}$). Clearly this is a major design criterion.

The effects of shock front curvature - which are present at the low densities employed in this study - were minimized by utilizing knife edges on the electron gun and detector as shown in figure 1-1-4. Also seen in figure 1-1-4 is the beam stop that defined the scattering region. Because of the strong angular dependence (Fig. 1-1-3), most of the counts came from the region near the beam stop. A drift tube passed the beam from the electron gun to within 1 in. of the scattering region. A collector cup was used to measure the intensity from the primary beam. In these experiments the electron beam was turned on before the shock arrived at the measuring station, but after the recording oscilloscope had been triggered. Hence, the signal due to scattering by the P_1 gas was recorded and represented a calibration that was available for each run.

Although Camac did his original work using a solid scintillator, it had been shown by Duff (33) that these tended to acquire a charge within the material producing a nonlinear response. Duff showed that liquid scintillators did not have this spurious effect. In subsequent work Camac (17) used a liquid scintillator.

Applying the electron beam and IR techniques to CO_2 relaxation, Camac (17) measured the translational-rotational, vibrational, and chemical relaxation of CO_2 from

Mach 5 to 25 in the Avco Everett 24 in. shock tube; P_1 ranged from 0.015 torr to 0.12 torr. The scattered electron beam yielded the density variation, while the IR measurement at 4.3μ yielded the vibrational energy of the ν_3 mode. The IR system utilized an InSb detector with a rise time of 0.5 μ sec. Knife edges were employed utilizing an optical path length of 12 in. The optical axis of the infrared set-up was 0.5 in. upstream from and perpendicular to the electron beam apparatus which was essentially as described above.

In a series of papers Muntz and co-workers (34-36) have described the development of a technique in which an electron beam is used to excite ground state nitrogen molecules directly to the N_2^+ (B) state which gives rise to the N_2^+ (1-) band system. For high electron energies, optical selection rules are obeyed and there is no interaction of the electron with the angular momentum of the nuclei. At low densities the quenching collisions are negligible, making the intensity directly proportional to the density. Under these conditions comparison of the data with calibration data gives directly a point measurement of the density. The point size is given by the beam width and the length of the beam actually observed. This work was based on earlier work by Schumacher and Gadamer (37) who passed a thin electron beam across the viewing field of a low density air wind tunnel.

Muntz (34) used a 17.5 kV, 250 μ amp, 1.65 mm diameter electron beam. The nitrogen pressure was 0.33 torr, but the electron gun was differentially pumped down to 7×10^{-5} torr. Spectral plates were taken of the (0, 0), (0, 1), and (1, 0) bands of the N_2^+ (1-) system, and rotational temperatures were deduced by plotting rotational line relative intensities in the appropriate spectroscopic manner.

After making some detailed measurements in static systems, the beam experiment was mounted 2.5 mm downstream of an exit nozzle on the center line of the flow. The radiative lifetime of the N_2^+ (1-) band system is 6.5×10^{-8} sec, thus the radiation is localized even in high speed flows. In these low density flow experiments they were able to evaluate rotational temperatures to $\pm$ 2%, vibrational temperatures to $\pm$ 10%, and the density to $\pm$ 8%.

In (35) and (36) Muntz and co-workers have applied this technique to measuring the density in hypersonic shock tunnel flows and the density and temperature in laminar hypersonic near wakes behind 10° cones.

In (38) Muntz and Marsden reviewed the earlier work and included some new work on the factors affecting the resolution of the electron beam excitation technique. They also gave a detailed analysis of some electron beam excited spectra obtained in He, N_2, NO, CO_2, O_2, and Ar.

Muntz (39) used the electron excitation technique in a free jet experiment to measure the velocity distribution. The basic principle involved here was to measure the Doppler profile of the atomic lines in emission. He used the helium line at 5015.67 Å. The technique was first tried out in static tests in He at 0.1 torr. Temperatures in the neighborhood of room temperature were measured within 1%. In the jet expansion experiment the beam passed through the center line of the He jet but could be moved to any position along the direction of flow. The beam was used to: a) provide flow visualization for photographic studies of the jet; b) measure the temperature ahead of and behind the standing shock wave; c) obtain the relative density profiles as functions of distance along the jet; and d) obtain some velocity measurements by observing the blue shift looking directly upstream. The results of all these measurements were in good agreement with theory and were capable of extremely high precision.

Recently, Robben and Talbot (40) applied the electron beam fluorescence method to the study of shock thicknesses in a low density wind tunnel operated with He, Ar, and N_2 from Mach 1.5 to 17.4. In further work (41, 42) they measured the rotational temperature (via the N_2^+ (1-), 0, 0 band analysis described by Muntz (34)) in the tunnel employing various nozzles and actually measured the rotational temperature profile through the shock.

It is to be noted that the work described in (34 to 42), which are reviewed in the above paragraphs, is not shock tube work, although Muntz (39) suggests that the electron beam excitation technique should have useful applications in shock tube work. Indeed, recently Camac (43) has applied the work of Muntz and co-workers to a chemical kinetic study in a shock tube. He has used the electron beam to excite N_2^+(1-) system radiation behind incident shocks in pure N_2 at a P_1 of 0.05 to 0.5 torr. The process under study is $N_2^+(B) + N_2(X) \rightarrow N_2^+(A \text{ or } X) + N_2(X)$. To date the process has been studied by him at room temperature and at 2650°K behind the incident shock. The apparatus used by Camac in these recent studies is essentially the same as that which was used in his electron beam scattering experiments (31, 32) with the addition of a PM-filter combination monitoring the radiation at 3914 Å, i.e., the N_2^+(1-) 0, 0 band. The radiation was observed at right angles to the electron beam direction. In this experiment the electron beam was turned on shortly before the shock wave arrived at the test station and the radiation was seen to rise as the shock wave made its transit across the optical slit. The rise in radiation is due to the compression across the shock front but is somewhat less than a factor of ρ_2/ρ_1 due to the increased collisional quenching of the B state. This is the effect that is studied.

Mass Spectrometry

In carrying out kinetic experiments the ideal diagnostic tool would allow the experimenter to have a continuous record of all the chemical species during the course of the experiment. The mass spectrometer comes close to satisfying this ideal. In 1961 Bradley and Kistiakowsky (44) coupled a time-of-flight mass spectrometer to a shock tube. As we shall see, the question with this experimental technique is whether the gas is being cleanly sampled from the free stream or whether a finite amount of the sample comes from the boundary layer gas, or from gas that has contacted the inlet surface.

In order to test the technique, these workers chose to measure the thermal decomposition of N_2O over the temperature range 1780 to 2000°K. This reaction has a known temperature dependence and the mechanism involves a free radical; namely, atomic oxygen. P_1 was varied between 2 and 10 torr, and the gas was analyzed at intervals of 50 or 100 μ sec behind the reflected shock. The gas was sampled through a 5 mil diameter hole in a 1 mil thick gold foil window in the end plate. The pressure behind the reflected shock was equal to 0.4 atm so that the 5 mil diameter hole was about 100 mean free paths across. The gas was sampled for times of the order of 0.5 msec during which time a cold boundary layer grew out from the end plate, the thickness of which was calculated to be 40 mils at the end of the test time. It was estimated by the authors that only 3% of the sampled gas was taken in from the boundary layer.

The ion source was, of course, differentially pumped, the pressure during the run being of the order of 10^{-8} atm. The mass spectrometer was designed for minimum collisions of molecules with the walls. It was estimated by the authors that less than 10% of the ions that were detected arose from molecules that had struck the walls. The mass analysis showed the presence of N_2O, N_2, O_2, and O, in agreement with predictions.

In a subsequent paper, Bradley and Kistiakowsky (45) studied the polymerization of acetylene (C_2H_2) over the temperature range 1800 to 2700°K in C_2H_2-Ar mixtures, and the oxidation of C_2H_2 between 950 and 1100°K in C_2H_2-O_2-Ar mixtures. They found polymers of approximate masses 51, 75, and 97 probably corresponding to molecules C_4H_4, C_6H_4 or C_6H_6 and $C_8H_?$.

In 1964, Kistiakowsky and Michael (46) modified the Bendix time-of-flight mass spectrometer to allow detection of ions. They studied the oxidation of methane and acetylene and found the first ion produced to be $C_3H_3^+$.

In 1965, the apparatus that had been used by Bradley and Kistiakowsky was modified by Dove and Moulton (48) so that it was a considerably more advanced instrument than the old one. They increased the pumping speed in the ionization region and modified the electronics so as to increase the repetition rate of the mass analysis to 20 μ sec. Some of the background noise caused by chemi-ions produced in the shock heated gas was eliminated by rejecting these ions. Also the pinhole sampling technique was modified to the extent that the sample was drawn through a small cone of base diameter 1.1 mm and height 0.3 mm projecting into the shock tube; at its apex was a 0.08 mm diameter hole.

In work similar to that of Bradley and Kistiakowsky, Diesen and Felmlee (49) studied the thermal dissociation of Cl_2 over the temperature range 1700 to 3200°K. They used 0.5 %Cl_2-0.5 %Kr-99% Ar mixtures, P_1 was varied between 1 and 10 torr. The krypton was added as an internal pressure standard. In their work the mass spectral analysis was carried out in 25μ s, with a time resolution on individual mass peaks of better than 1μ sec. The sampling leak in the end plate was a small (2 mil) diverging nozzle; they used a thin (30 mil) end plate, which was located close (80 mil) to the electron beam of the spectrometer to minimize effects of back scattering and background build-up. The effect of the boundary layer at the end plate was difficult to assess; however, the data are in reasonable agreement with literature values.

In a subsequent investigation, Diesen (50) studied the thermal decomposition of hydrazine (N_2H_4) in dilute mixtures with Ar over the temperature range 1200 to 2500°K. In this work the mass spectral analysis gave the time history of the species N_2H_4, NH_3, N_2, H_2, and NH_2. These species accounted for all the mass within the experimental error of 10 %, clearly indicating the power of this technique.

In subsequent work Diesen studied the thermal dissociation of fluorine (51) and the decomposition of NF_2 (52).

In a recently published report, Felmlee, Petrella, and Diesen (47) describe an experiment in which the time-of-flight mass spectrometer is used to sample shock heated NO_2 that has also been subjected to flash photolysis subsequent to shock heating. The flash photolysis of NO_2 is known to proceed in the following way:

$$NO_2 + h\nu \longrightarrow NO\ (^2\pi) + O\ (^3P) \qquad (R1)$$

$$O + NO_2 \xrightarrow{k_2} O_2 + NO. \qquad (R2)$$

The quantum yield for (R1) is of order unity, and the rate of (R2) has been measured near room temperature and is known to be fast.

The photolysis was accomplished by means of a semi-torus shaped argon flash lamp which fitted around the end of the pyrex shock tube. The initial mixture was 0.33% NO_2 - 0.17% Kr in Ne. The flash duration was 8 μ sec during which time 10^{18} photons/cm^2 were emitted in the wavelength interval 2000 to 3500 Å; 15-25% of the NO_2 was decomposed in the flash. Mass spectra showed NO, NO_2, O_2 and O, although data on O were rather poor due to fragmentation of the other species. At the reflected shock temperature (1035°K) the rate constant k_2 was found to have twice the value that it did at room temperature; this agreed well with the calculated value using the temperature dependence also obtained near room temperature.

This report (47) also gave an excellent treatment of sampling based on considering the leak in the shock tube end plate as a source of a freely expanding jet. Ion molecule reactions are found to be of most concern. For reactions of this type:

$$R^+ + R \xrightarrow{k_3} C^+ + D \ , \qquad \text{(R3)}$$

the following relationship between the parent ion R^+ and the false ion C^+ is derived:

$$\frac{(C^+)}{(R^+)} \approx 0.15 \left(\frac{D}{X_e}\right)^2 \frac{t_P}{2} (R)_0 \ k_3 \ , \qquad \text{(Eq. 1-1-4)}$$

where D is the hole diameter, X_e is the distance from the orifice to the electron beam, t_P is the pulse time, $(R)_0$ is the concentration of reactant behind the reflected shock wave, and k_3 is the rate constant of (R3) ($k_3 \lesssim 1 \times 10^{12}$ liters/mole-sec). On the basis of this equation the authors have questioned the significance of some observations made under certain conditions using the mass spectrometer/shock tube technique. The sampling technique of Kistiakowsky and co-workers and Diesen and co-workers has been questioned by numerous workers. Marsters, Bauer and Resler (53) have made some admittedly crude calculations of the effects of the end plate boundary layer on these sampling techniques, and concluded that the boundary layer must completely dominate the sample. They abandoned the technique of sampling through the end plate and instead employed a technique which utilizes just the incident shock as shown schematically in figure 1-1-5. As can be seen, an inner sleeve cuts out the central portion of the shock wave. After reaching the end of the sleeve the gas behind the incident shock undergoes a free expansion. The first nozzle which has a diameter of 2mm produces a Prandtl-Meyer expansion and is designed in such a way that it will have an attached shock associated with it, but there will be no shock across the nozzle so that the state of the gas is undisturbed. Somewhere within the second nozzle (diameter = 8mm) the flow goes free molecular and, hence, thereafter there are essentially no collisions.

The experiment was designed for studies over a temperature range 800 to 3000°K. Although there have not been any extensive data published utilizing this sampling technique in conjunction with the time-of-flight mass spectrometer, Prof. Resler (54) has indicated that the aerodynamics seems to be quite clean and that the sampling procedure is unquestionably sound. If further work bears this out, undoubtedly the mass spectrometer will become an extremely powerful and trustworthy dianostic instrument to be used in chemical kinetic studies on the shock tube.

The mass spectrometric experiments described so far have all been carried out with time-of-flight instruments. Recently Gutman *et al* (55, 56) described the use of a quadrupole mass filter instrument on a shock tube. These instruments do not have the great advantage of being able to scan over the entire mass range, as do the time-of-flight instruments, and the sensitivity of both type mass spectrometers seems to be comparable. However, the quadrupole type does enable one to monitor

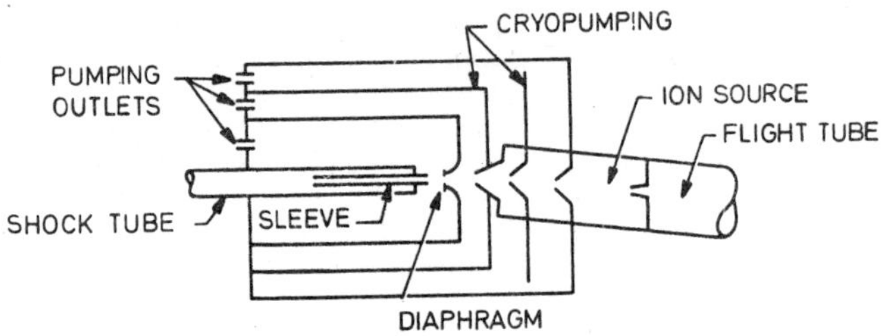

Fig. 1-1-5 Highly schematic diagram showing the sampling technique for a time-of-flight mass spectrometer employing a free jet expansion of the gas behind the incident shock wave.

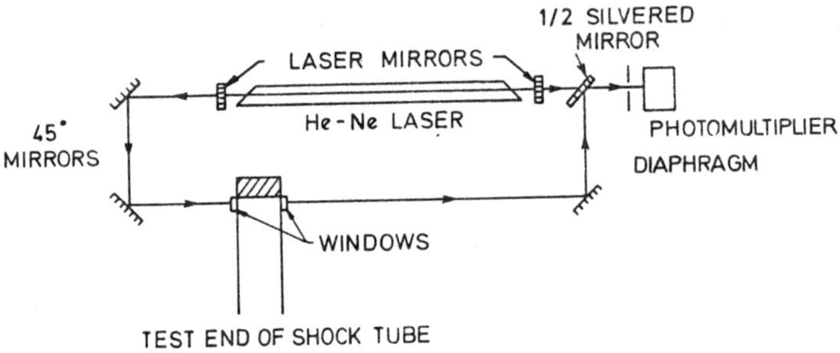

Fig. 1-1-6 Schematic diagram showing the incorporation of a He-Ne laser into an arm of an interferometer for employment on the shock tube.

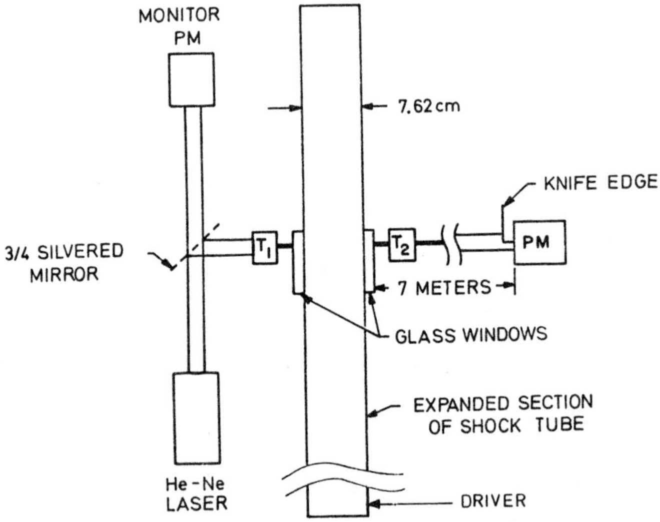

Fig. 1-1-7 Schematic diagram showing the employment of a He-Ne laser in a high sensitivity schlieren apparatus coupled to a shock tube.

continuously a single mass peak. The sampling technique used in this work was similar to that used by Diesen. Shock waves were made into 1, 2, and 4% N_2O in Ar mixtures at temperatures between 1800 and 3500°K behind the reflected shock. The species N_2O, O_2, N_2, NO, and O were monitored. Note that the time history of only one of these species could be monitored for each shock, hence, five times as many runs had to be made in this study than if a time-of-flight mass spectrometer had been used. Hay (57) has done a similar study on the decomposition of COS in which COS, CO, S, S_2, CS, and SO were monitored.

Single Pulse Shock Tubes

The first single pulse shock tube resulted from the work of Glick, Squire, and Herzberg (58). There has been a number of excellent chemical kinetic studies made in single pulse shock tubes, and the technique has been reviewed in several places (4) (6). We give here only some recent modifications of the technique.

Lifshitz, Bauer, and Resler (59) have added a dump tank 3.5 in. downstream from the diaphragm. Rapid cooling of the heated gas behind the reflected shock is achieved by the generation of the expansion wave due to the intersection of the reflected shock with the contact surface. Further cooling is accomplished by the expansion wave generated in the driver section when it reflects off the driver end plate and eventually reaches the test section. In order to achieve the best cooling rate the tube can be tuned, so that the reflected shock wave meets the gas interface and the reflected expansion wave coming from the driver section at approximately the same point. The tuning is accomplished by varying the length of the driver section with a screw fed piston. The detrimental effects of reflected waves, which under ordinary circumstances would reheat the sample, is eliminated in this experiment by the presence of the dump tank. In the initial work on the cis-trans-isomerization of butene-2 (59) samples were removed from the shock tube and analyzed by vapor phase chromatography in the usual manner of single pulse shock tube work.

In subsequent work (60) mixtures of cyanogen (C_2N_2) and argon were heated to temperatures between 1900 and 2300°K behind the reflected shock producing CN radicals. Following the arrival of the expansion wave, the association of the radicals was monitored by absorption spectroscopy at 3883 Å, the 0,0 band head of the CN violet system. The optical station was located 7 mm from the end wall.

This same apparatus was used by Tsang (61) for measurement of the relative rates of decomposition of two compounds in the same reflected shock. The rate of decomposition of one of these compounds, namely isopropyl bromide, was well known and was used as a standard. The molecules under study were tert-butyl chloride and tert-butyl bromide. Mixtures of each of these compounds with isopropyl bromide were run in the shock tube and their relative rates of decomposition were evaluated. Analysis was made by gas chromatographic techniques which gave the percent reaction for each component of the mixture. This method of measuring relative rates eliminates the uncertainties involved in the single pulse technique of the detailed temperature time history and has great promise for measuring accurately the chemical kinetic rate constants in the shock tube.

In another type modification of the single pulse technique, Jacobs *et al* (62) used an expansion chamber located a few tube diameters from the end of the shock tube. This expansion chamber had a cylindrical diaphragm separating it from the shock tube proper. The pressure behind the incident shock as it passed this diaphragm was insufficient to rupture it. After reflection, the wave once again passed the diaphragm, the pressure now being large enough to rupture the diaphragm at which point a centered expansion wave rapidly cooled the gas behind the reflected shock.

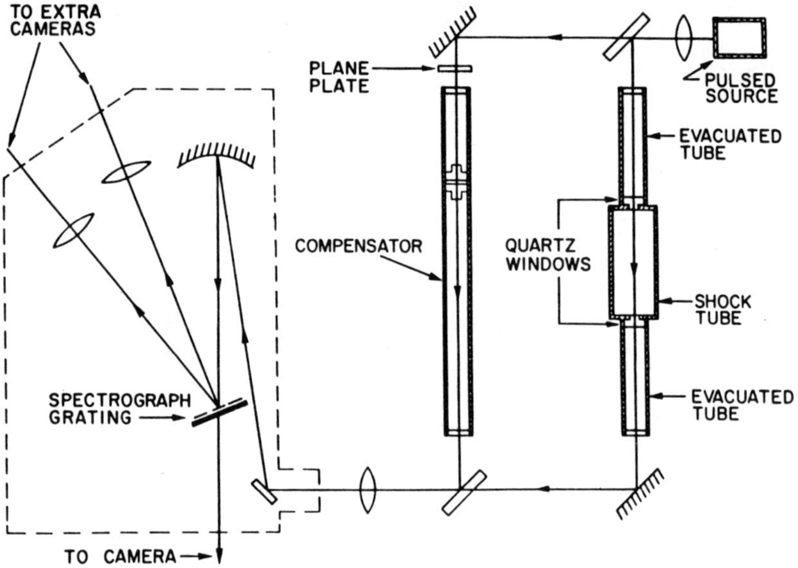

Fig. 1-1-8(a) Schematic diagram showing the crossing of an interferometer and spectrometer (the hook technique) in shock tube work.

(b) Fringe pattern in the vicinity of an absorbing line without the plane parallel plate.

(c) Fringe pattern in the vicinity of an absorbing line with the plane parallel plate showing the hooks.

The apparatus was designed to be able to investigate three body atomic recombination rates at temperatures and pressures close to those in rocket chambers and nozzles. For a centered expansion wave the rate of change of temperature with time is inversely proportional to the distance from the point at which the measurement is being made to the origin of the wave. Hence, it is clear that one wants to locate the observation station very near the auxiliary diaphragm, although one must be certain that the wave is one dimensional by the time it reaches the observation point. For the 3 in. diameter shock tube used by Jacobs *et al*, this distance was approximately 6 in. For their experimental setup these workers claim a temperature drop from 4000 to 3040°K in 0.3 msec or 3×10^6 degrees per second. They made observations 1 in. from the end wall on 1 % Cl_2-99% A mixtures following Cl_2 in absorption, and simultaneously monitored the pressure history with transducers as a function of time.

In recent work Brabbs and Belles (63) used a single pulse technique to measure the rate constant for the three body recombination $CO + O + M \longrightarrow CO_2 + M$. They shocked dilute CO_2-A mixtures to temperatures between 2800 and 3600°K behind the reflected shock. Subsequent to the arrival of the expansion wave, they monitored the pressure and the intensity of the CO + O recombination continuum 6 mm from the endplate.

Although it is a little out of place, we should mention here for the sake of completeness the work of Wilson (64) in which the three body recombination rate of atomic oxygen was measured at 2700°K. Here the O_2 was dissociated behind relatively strong shocks to produce the O-atoms. The shock wave then passed between two air foils at an appropriate angle to cause a Prandtl-Meyer expansion sufficiently rapid to freeze the chemistry through the expansion. The central portion of this expanded flow was then cut out with a constant area channel and optical measurements of the O_2 concentration using UV absorption at 2283 Å were made 11 cm downstream from the initial expansion point.

Although substantial progress has been made in the single pulse type shock tube and its various modifications, there are still serious problems with knowing the temperature time history of the cooling process. Recombination processes are of great interest to the chemical kineticist and the shock tube is the most appropriate way of preparing a sample of free radicals under known conditions. It is clear that more progress is highly desirable in the understanding and design of techniques which allow these free radicals to be rapidly cooled and monitored during their recombination.

Laser Interferometry and Schlieren

The schlieren and interferometer techniques are well established diagnostic methods that have been applied to shock tube studies for many years. The invention of the laser with its properties of coherent radiation, narrow band width, and high intensity has added new possibilities to these old techniques.

Thornton and co-workers (65) (66) have incorporated a laser into an arm of an interferometer and have used it to measure the pressure jump at the shock front and subsequent electron formation in argon shocks at a temperature of 10 800°K. From the fringe shift they calculated an electron density of 2.8×10^{16} electrons/cm^3, which is in good agreement with theory. The apparatus is shown in figure 1-1-6. A helium-neon laser putting out light at 0.63μ was used. The laser was used with one plane and one spherical mirror adjusted so that but one laser mode was observed. It provided a source of high enough intensity to override the plasma source in the shock tube. As can be seen in the figure, the laser was used in one arm of the Mach-Zehnder type interferometer. However, the beam was not split in the usual manner: instead, since laser radiation is coherent, the light coming from both ends of the laser was used in the two arms of the instrument. Because of the extremely narrow beam angle of the laser, long path interferometry could be accomplished. A similar instrument is described by Besse and Kelley (67) who used it to measure the degree of ionization in a Mach 18 shock in argon as a function of time.

Also employing a He-Ne laser, Kiefer and Lutz (68) (69) have built a quantitative schlieren system of high sensitivity. In three excellent papers, these authors describe the utilization of the laser schlieren device for the study of the vibrational relaxation of deuterium (69) and hydrogen (70), and the vibrational relaxation of oxygen by oxygen atoms (71). This technique is similar to the integrated schlieren system of Resler and Scheibe (72), but is about four times more sensitive.

The apparatus is shown schematically in figure 1-1-7. The laser beam was demagnified at T_1 by a factor of 6 before going through the glass windows on the shock tube. The beam diameter at this point was equal to 0.8 mm. The beam was then focused onto the knife edge located directly in front of the photomultiplier, which was 7 m from the shock tube. The laser beam itself was monitored simultaneously in order to make base line corrections. The laser put out 5 mwatts of total power with a beam divergence of 2.7×10^{-4} radians. The demagnification at T_1 increased the beam divergence by a factor of 6 so that the increase in the beam diameter between telescopes on either side of the shock tube was not negligible. The geometry employed represented a best compromise.

The experiments on vibrational relaxation of deuterium (69) were carried out in a 3 in. shock tube in D_2-Ar mixtures, P_1 was varied from 5 to 220 torr and the temperature range covered was 1000 to 3000°K. For fractional changes in light intensity of 15 percent or less, the deflection detector was linear to 2 percent. The experimental resolution was limited by shock curvature at the tube walls. The minimum detectable density gradient (with Ar) was less than 5×10^{-5} g/l/mm, which corresponded to a fractional signal change of less than 10^{-3}. This would correspond to 0.002 fringe/mm in an experiment carried out with an interferometer using the same diameter shock tube. The maximum density gradient that would still lie in the 2 percent region was equal to 1×10^{-2} g/l/mm.

The Hook Method

In 1962 Dunaev *et al* (73) published a paper entitled 'The Application of Rozhdestvenskii's Hook Method in the Investigation of Gas Dynamic Processes in Shock Tubes.' Although I don't believe there have been any subsequent papers, it is clear that the hook technique is an extremely powerful one and might profitably be more frequently applied to shock tube investigations. The hook method is some 55 years old. It was originated by Rozhdestvenskii, who, over the years, has utilized the method in many spectroscopic investigations. In fact, the method seems to have been utilized extensively in Russia, where it is well known. It has been applied in many instances to the measurement of oscillator strengths of vapors in ovens and electric discharges. The review by Marlow (74) is primarily a mathematical treatment of the theory, where pertinent equations describing the hook phenomenon are developed for single lines, multiple lines, and various cases of broadened lines. Penkin's excellent review (75) leans more toward a description of the experimental aspects. A brief outline of the hook method can be found in (76).

Basically the technique crosses an interferometer with a spectrometer. A highly schematic diagram of Dunaev's (73) apparatus is shown in figure 1-1-8a. If no optical path difference existed between the arms of the interferometer, i.e., if there were a vacuum in the shock tube and no extra plane parallel plate, then in the focal plane of the spectrometer, the zero order fringe would be horizontal. When the dispersing medium is put into the shock tube, the zero order fringe traces out the dispersion curve, i.e., Sellmeier's formula for dispersion in the neighborhood of an absorbing line. A photograph (75) taken under such conditions, but not of shock tube work, is shown in figure 1-1-8b. In principle, such data could yield the product of the concentration times the oscillator strength, but such an analysis is neither easily done nor very accurate. By introducing the extra plane parallel plate, the fringes near the absorbing line take on a hook shape as can be seen in the photograph (75) shown in figure 1-1-8c. The hooks form at the wavelength where the plate effect is equal but opposite to that of the dispersing vapor in the test cell. The product of the concentration of the absorbing atoms and the oscillator strength, Nf, can be readily obtained from a photograph such as is shown in figure 1-1-8c by means of the following relationship:

$$Nf = \frac{\pi K}{\ell r_0 \lambda_0^3} (\lambda_2 - \lambda_1)^2 , \qquad \text{(Eq. 1-1-5)}$$

where ℓ is the optical path length in the cell, r_0 is the classical electron radius, λ_0 is the wavelength of the absorbing line, and $(\lambda_2 - \lambda_1)$, obtained directly from the picture, is the separation of the hooks in a given fringe in wavelength space. K is a calibration factor for the apparatus which is equal to $P\lambda$ where P is the number of fringes per unit wavelength measured at wavelength λ which should be in the vicinity of λ_0.

Dunaev ***et al*** (73) used the hook method to measure densities, temperatures, and some new f-numbers for mercury. In their work they used a shock tube of rectangular cross section 3.8 x 7.6 cm. The entire downstream section of the shock tube was heated to 250°C to get the desired pressure of mercury. Mach numbers between 6 and 11.5 were obtained. The spectrometer, with a dispersion of 5Å per mm, had three cameras covering the range 2500 to 5800 Å. The continuum light source was pulsed and had a lifetime of about 3 to 5 μs. To prevent convection currents coming off the hot windows from disturbing the optical system, these workers used evacuated side tubes over the windows (Fig. 1-1-8a). From the hook at 2537 Å the number of mercury atoms in the ground state was determined using the known f-number for that transition. The concentration of excited mercury atoms was obtained from the hooks near the lines of the mercury triplet at λ = 4047, 4358, and 5461 Å. From the relative number densities of the ground state atoms before and after the shock, they were able to compute the density ratio, and from the ratio of excited to ground state atoms they got the temperature ratio.

It would seem that these workers chose a particularly hard experiment to first try the hook technique on the shock tube, namely working with mercury in a heated low pressure section. However, their results were certainly very encouraging. The Rozhdestvenskii hook method would seem to have a future in following the concentration of atoms and especially the excited atoms in chemical reactions behind shock waves.

New Techniques for Sample Preparation

Electric Discharges

The use of an electric discharge to prepare an initially nonequilibrium distribution in a shock tube was developed in 1963 by Wray and Teare (77) (78) and in 1965 by
c Hartunian and co-workers (79) (80). The two techniques are substantially different, the first using a single condensed discharge over the entire downstream end of the

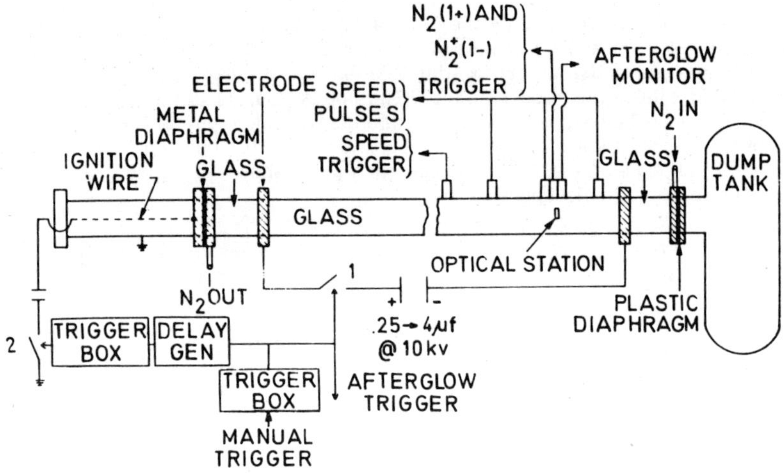

Fig. 1-1-9 Schematic diagram of the condensed electrical discharge technique coupled to the shock tube.

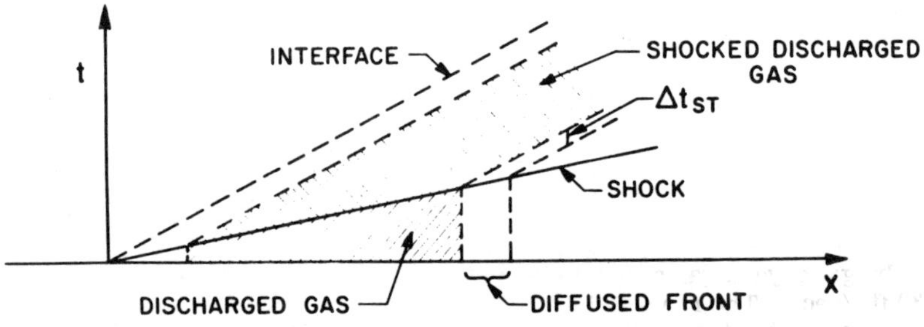

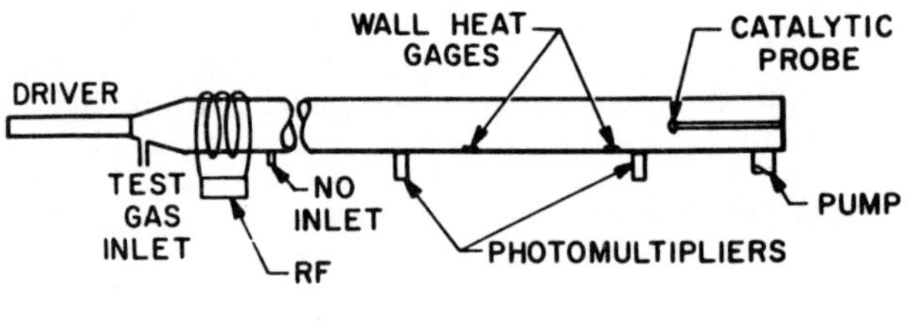

Fig. 1-1-10 Schematic diagram of the RF glow discharge shock tube along with an x-t diagram.

shock tube, the latter using a continuous RF discharge over a small region of the shock tube just downstream of the diaphragm station.

Wray (77) (78) studied the excitation of the $N_2(1+)$ and N_2^+ (1-) systems in shock heated N-N_2 mixtures. The electric discharge was used to make the nitrogen atoms before the shock wave arrived. The apparatus is shown schematically in figure 1-1-9. The shock tube used in this work had a stainless steel driver section and the low pressure side was a 10 ft. long Corning glass pipe, 1.5 in. in diameter. Stainless steel ring electrodes whose inside diameters were also 1.5 in. were placed at both ends of this 10 ft. length. The driver and dump tank were insulated from the ring electrodes with 6 in. lengths of Corning pipe.

After the nitrogen flow was started through the test section and the driver was loaded with the combustion mixture necessary to produce the desired shock strength, the run was initiated by triggering switch 1 which allowed the capacitor bank across the electrodes on the test section of the shock tube to discharge through the N_2 gas. At the same instant a time delay generator was activated, which, after a preset time of about 50 msec, fired a trigger box that activated switch 2, igniting the combustion mixture. Within a few milliseconds the diaphragm ruptured and a shock wave moved into the test gas.

Runs were made covering the temperature range 7000 to 18 600°K, the temperature before vibrational relaxation, at a P_1 of 1 torr. The ratio of atoms to molecules was $N/N_2 = 0$ (no discharge), 0.025, 0.07, and 0.27. The $N_2(1+)$ system, monitored in a broad radiation band between 6600 and 8000 Å, and the $N_2^+(1-)$ system, monitored at 4272 Å, were followed as functions of time with photomultipliers behind the incident shock. Comparison of the experimental radiation slopes with theory yielded rate constants for the production of triplet state nitrogen molecules in N-N_2 collisions.

The motivation for the work on the so-called glow discharge shock tube (79) came from the need for a technique to produce a known step function of atoms capable of producing fluxes of the order of 10^{19} particles/sec to catalytic probes with rise times $\leqslant$ 0.5 msec. Such a device was needed to properly calibrate catalytic probes which were to be used in hypersonic shock tunnel tests and to identify the source of some time dependent heat transfer measurements made with catalytic gages in a shock tube.

Figure 1-1-10 shows a schematic diagram of the glow discharge shock tube along with an x-t diagram. The test section of the shock tube was made of 4 in. diameter pyrex tubing; oxygen gas at 0.6 torr pressure was flowed through the apparatus at 10 to 30 ft./sec. The run was initiated by suddenly turning on the RF energy which partially dissociated the gas just downstream of the diaphragm station. When the step function of dissociated gas was within about 1 ft. of the catalytic probe, the diaphragm was ruptured, moving a shock out into the predissociated gas. The shocks were sufficiently weak so that no thermal dissociation occurred, the shock wave serving simply to compress and accelerate the atoms produced in the discharge. With the slow flow it took of the order of 100 msec for the front to cross the probe, behind the shock it took only 0.5 msec (Δt_{ST} in figure 1-1-10). In their initial experiments they established that for a silver oxide catalytic probe the surface kinetics are such as to allow rapid response to changes in the atom flux.

In subsequent work Hartunian *et al* (80) used the glow discharge shock tube to study chemiluminescent reactions at elevated temperatures. In this work the radiation coming from the reactions $CO + O \rightarrow CO_2 + h\nu$ and $NO + O \rightarrow NO_2 + h\nu$ was monitored before and after shock heating and compression. The same experimental set-

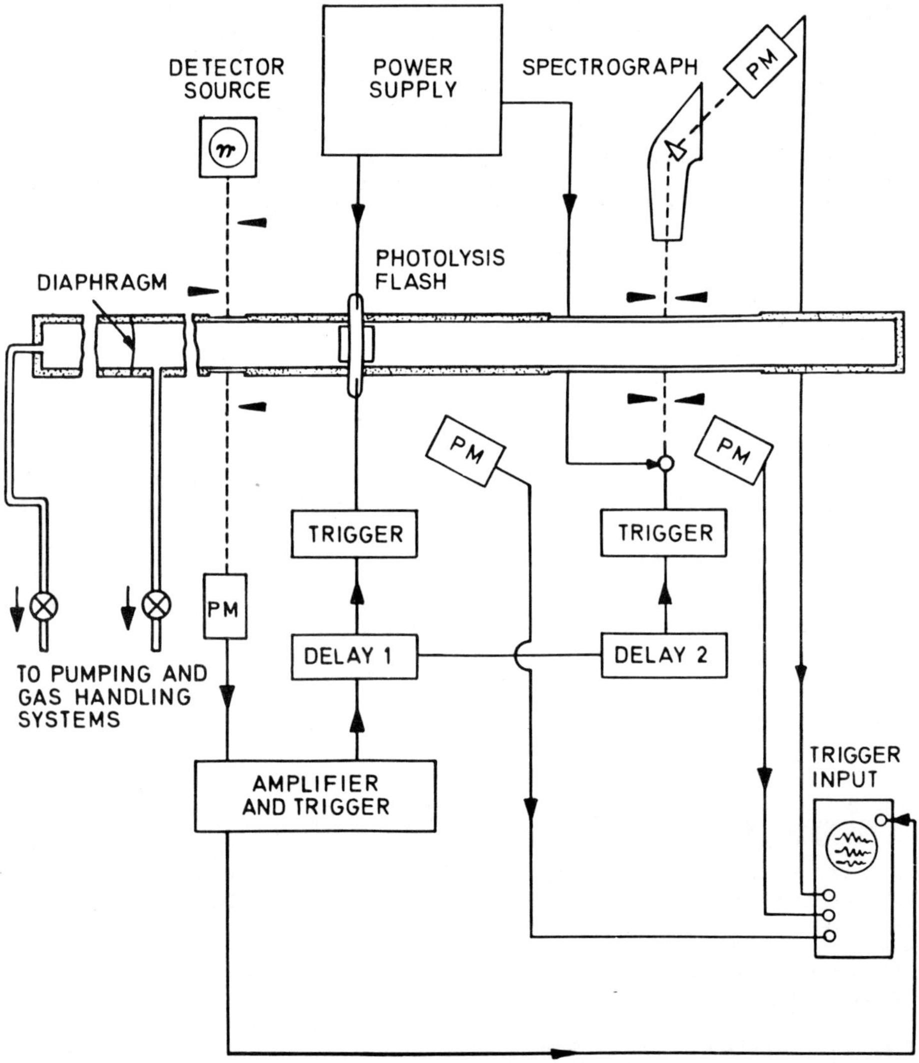

Fig. 1-1-11 Schematic diagram showing the coupling of the flash photolysis technique to a shock tube.

up was used as described above for the earlier investigation. The flow of O_2 with the RF discharge was established and NO or CO was mixed into the dissociated stream 4 ft. downstream from the discharge. The pertinent radiation was monitored with a collimated photomultiplier at an optical station located 15 ft. downstream of the discharge. The initial pressure in the shock tube was 1 torr of O_2 and the Mach number range covered was 1.2 to 4.5. From the oscilloscope records of the photomultiplier output, the rate constant at the elevated temperature can immediately be evaluated from the relation:

$$\frac{I_2}{I_1} = \frac{k(T_2)}{k(T_1)}\left(\frac{\rho_2}{\rho_1}\right)^2 \qquad \text{(Eq. 1-1-6)}$$

From a series of such measurements, the authors have evaluated the temperature dependence for the above two chemiluminescent reactions.

Flash Photolysis

The flash photolysis technique has been used for many years to produce large concentrations of free radicals. Many room temperature recombination experiments have been carried out using this method. In 1960 Burns and Hornig (81) combined the flash photolysis method with a shock tube in order to study bromine recombination at elevated temperatures. They used a 3 in. diameter pyrex shock tube and worked with 0.6% Br_2-Ar mixtures at a P_1 of 114 torr. The photolysis was accomplished with 8 xenon-filled quartz lamps arranged around the tube inside a reflecting container. The lamps all discharged within 1 μ sec of each other with a pulse width at half height of 7μ sec. Typically the flash photolysis dissociated approximately 20 percent of the bromine into atoms, the shock wave then compressed and heated the mixture to 950°K. The Br_2 concentration was monitored downstream from the flash photolysis region employing a photomultiplier filter combination at 4420 Å. Very few runs were made in this experiment, but the technique holds extreme promise.

Bradley and Tuffnell (82) also used a flash photolysis technique coupled to a shock tube. In ordinary flash photolysis it is often necessary to use flash durations larger than 100 μ sec in order to get adequate quantum efficiencies. This obviously leads to relatively poor time resolution. Also, it is difficult to get information leading to the temperature dependence of reaction rates in ordinary flash photolysis experiments. In coupling the photolysis to a shock tube, the rapid flow continuously removes the photolyzed gas from the region of the source.

In this work (82) the gas was photolyzed subsequent to the arrival of the shock wave. The gases studied were $COCl_2$ and $(COCl)_2$, both of which yield COCl upon photolysis. The apparatus is shown in figure 1-1-11. The shock tube employed was of square cross section 8 x 16 cm, and was made out of a copper wave guide. Two rectangular shaped optical quality quartz windows mounted on opposite sides of the shock tube admit the radiation from the U shaped photolysis flash lamp which had a duration of 95μ sec. Both the photolytic flash and the spectroscopic flash were monitored as a function of time with photomultipliers. The spectroscopic flash, which served as the light source for the absorption measurements, had a duration of 11 μ sec; the absorption spectrum was recorded on film from 2500 to 5000 Å. The spectrometer could also be used as a monochromator by putting an adjustable slit in the focal plane. Extremely weak shocks were used in this study, the temperature being 480°K, but the gas flow velocity behind the shock was 0.39 mm/μsec which was the reason for coupling the flash photolysis experiment to the shock tube.

As already mentioned in Section 2.4, Diesen and co-workers (47) have used a flash photolysis technique in conjunction with a mass spectrometer on a shock tube.

The Ozone Technique

In order to measure the three body recombination of oxygen atoms by argon catalysts at elevated temperatures, Wray (83) in 1963, shocked dilute ozone (O_3)-Ar mixtures in a conventional 1.5 in. diameter stainless steel shock tube. The compositions of the mixtures were 0.25, 0.5 and 1.0 %O_3 in Ar at initial pressures of 800, 400 and 200 torr, respectively. The purpose of the ozone was to produce an excess of atomic oxygen behind the shock.

The thermal dissociation of ozone has been investigated by shock tube techniques and other means. The mechanism is as follows:

$$O_3 + M \xrightarrow{k_4} O + O_2 + M \qquad \text{(R4)}$$

$$O_3 + O \xrightarrow{k_5} 2O_2 \qquad \text{(R5)}$$

At high concentrations of M and at relatively high temperatures, (R4) can be made sufficiently fast so that (R5) plays no role. Under these conditions if the temperature is not too high, there will be an overabundance of O-atoms which must subsequently recombine by the reaction:

$$O + O + M \longrightarrow O_2 + M \, . \qquad \text{(R6)}$$

The O_2 formed via (R6) was monitored behind the shock wave by absorption of 1270 Å radiation. The temperature range covered in the experiment was 1340 to 2920°K. The low temperature limit is set by the overbearance of (R5) and the high temperature limit by the reverse of (R6). Over this temperature range, (R4)(and (R5)) is extremely fast compared to (R6). The O_3 disappears in small fractions of a microsecond for the conditions of this experiment. The concentration of oxygen atoms (O) produced at the shock front is given by (83):

$$2\,\frac{k_4}{k_5}\,\ln\left[1 - \frac{k_5}{k_4}\,\frac{(O)}{(M)}\right] + \frac{(O)}{(M)} + \frac{(O_3)_i}{(M)} = 0, \qquad \text{(Eq. 1-1-7)}$$

where $(O_3)_i$ and (M) are the concentrations of ozone and catalysts before any reaction occurs.

The rate constants, k_4 and k_5, were measured by Jones and Davidson (84) in a shock tube experiment covering the temperature range 769 to 910°K. Wray's work (83) yielded the oxygen concentration just behind the shock front. When these measurements were compared with equation 1-1-7, using the rate constants of Jones and Davidson, there was a clear disagreement, the data indicating a value for k_5/k_4 as much as ten times larger than that given by extrapolation of the Jones and Davidson rates.

The ozone technique was used by Kiefer and Lutz (85) to measure the recombination of O-atoms by O_2 at high temperature. The technique was similar to that just described except that the O_3 and O_2 concentration was varied between 10 and 30 percent, the rest being made up of either argon or krypton. The diagnostic technique used by Kiefer and Lutz was the x-ray densitometer method which has been reviewed elsewhere (5) (6). In this work a discrepancy of about a factor of 5 was also found in the ratio k_5/k_4.

These same authors (71) used the ozone technique to measure the catalytic efficiency of O in vibrationally relaxing O_2. The diagnostic technique used here was the laser schlieren described elsewhere in this review paper. The mixtures shocked in this case were 0.3-1.2% O_3 in O_2. These data also indicated that the extrapolated Davidson rates were off by approximately a factor of 10.

Recently Fishburne and Edse (86) have employed the ozone technique to measure reactions between atomic oxygen and nitrous oxide over the temperature range 1700 to 2300°K. The mixtures shocked were 2% N_2O, 0.5-2 %O_3, the rest being O_2. The diagnostic technique employed was to follow the infrared emission from N_2O. Therefore, in this work the authors did not have a measurement of the amount of atomic oxygen produced at the shock front. They assumed this to be correctly given by equation 1-1-7 using the rate constants of (84). Their data are all analyzed on the basis of this assumption, making their conclusions very suspect.

Because of the great importance of reactions involving atomic oxygen, it is anticipated that the ozone technique will be employed in a great many studies in the future. The mechanism by which ozone decomposes in the shock front, and precise measurements of the reaction rates for the processes involved, are of paramount importance.

Two-phase Shock Tubes

Ever since emission spectroscopy experiments were conducted in shock tubes, radiation from impurities (including dust) has been a problem. Recent workers have attempted to introduce dust into a shock tube in a controlled manner. In 1957 Nicholls and Parkinson (87) used the shock tube to excite spectra of astrophysical interest under controlled temperature conditions. A few grams of the powdered material were placed near the end plate of a shock tube and spectroscopic plates were taken through a quartz window mounted in the end plate. The driven gas was Ar at P_1 between 3 and 60 torr. The temperatures behind the incident shock were 2100 to 5500°K and behind the reflected shock 4800 to 12 000°K. Such materials as MgO, Al_2O_3, CaO, TiO, Cr_2O_3, SrO, BaO, WO_3, Fe_2O_3, and a sample of the 1939 Dresden meteorite were subjected to shock heating.

Nicholls, Parkinson and co-workers (87-91) have investigated many different spectra on the shock tube. The powders are usually put in the shock tube on kleenex tissue or mylar tape, or sometimes just spread out on a steel platform mounted in the center of the shock tube. The references given here are only a partial list - (90) is a review of seven years of work in the field.

In 1961 Berry *et al* (92) studied the UV absorption spectra of RbF, RbBr, CsF and CsCl. The shock tube employed was of square cross section, 8.5 cm on a side. The optical path involved 4 traverses across the shock tube. The samples were prepared by depositing solutions on cellulose tissue or thin perforated aluminum foil and allowing them to dry; from 0.1 to 1 gram of material was deposited on the support. The salt impregnated tissues or foil were then mounted across the shock tube at a location 90 cm upstream from quartz windows. The downstream section of the shock tube was filled with Ar at a P_1 between 15 and 25 torr.

For both CsCl and RbCl sharp absorption edges at 3424 Å and 3325 Å were identified, this energy difference corresponding very closely to the spin orbit splitting of Cl^- with the larger energy giving a precise measurement of the electron affinity of Cl. The process observed is, of course, $Cl^- + h\nu \rightarrow Cl + e$. The presence of the alkali metal in these experiments is necessary to produce sufficiently high

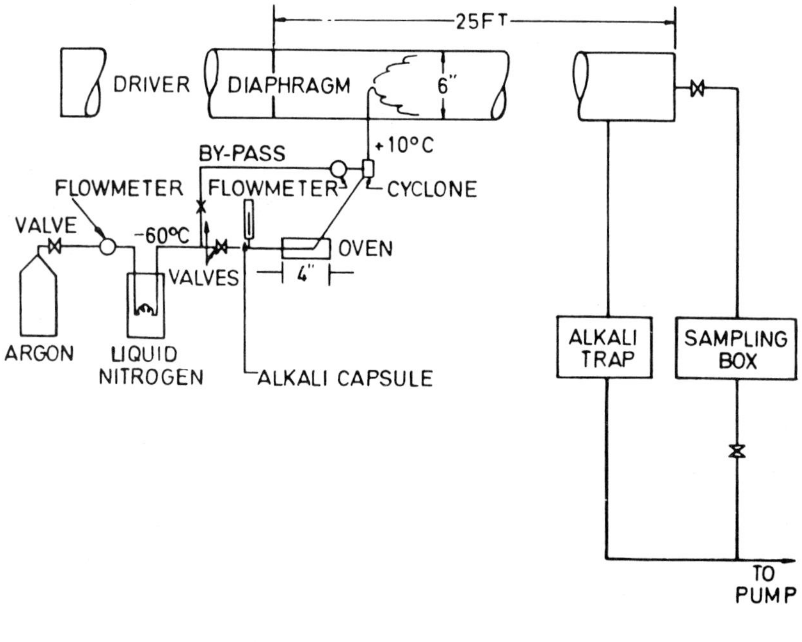

Fig. 1-1-12 Schematic diagram showing the technique for producing a cesium aerosol and introducing it into the shock tube.

electron concentrations and thence sufficiently high ion concentrations so that these can be observed in absorption.

In subsequent work Berry and co-workers (93) observed the absorption spectra of Cl^-, Br^-, and I^- in the vapors of their rubidium and cesium salts. Furthermore, in this work the ion densities were determined from the width of the alkali metal atomic lines. These concentrations in conjunction with absolute intensity measurements at various wavelengths yielded cross sections for the processes $X^- + h\nu \rightarrow X + e$. Using stronger shocks to vaporize and dissociate the fluoride salts, the same kind of information was obtained for the fluoride ion (94). Similar work has been done in emission, yielding results in good agreement with the absorption measurements (95).

Recently Berry *et al* (96) have studied the process $O + e \rightarrow O^- + h\nu$ in emission around threshold at 7800 to 8800 Å by shocking the vapors produced from K_2O_2 and Rb_2O.

A group at Heliodyne Corporation has been working on a dusty shock tube facility for several years, they call this SAPAG, which stands for shock accelerated particles and gases (97-99). The principal effort here has been to learn how to distribute a dust uniformly throughout a shock tube. Most of the work done was with teflon dust, although other materials also were investigated.

Vertical and horizontal glass and steel shock tubes were tried: ultimately a horizontal metal shock tube was used because it was found that glass attracted the highly desiccated injected teflon powder by electrostatic charge. The 'burning' criterion within the test time requires that the particles be of submicron size. With such large surface areas involved, handling the dust becomes a major problem. Teflon dust is made by grinding the plastic at liquid nitrogen temperatures where it is quite brittle. Subsequent to grinding, the particles must be cleaned up by outgassing in a dry box. Many injection schemes were tried; the best one seemed to be one in which the outgassed dust was loaded directly into a transfer chamber in the dry box, put into an injection tank, premixed with high pressure gas, and then expanded into the driven section of the shock tube.

Powder distributions in the shock tube were determined from weight measurements. Aluminum foil strips were placed in the tube for several hours during which time the dust settled out on the strips: these were then carefully removed and weighed. The distribution was found to be fairly uniform about 1 m away from the injector. Typical mixtures in the shock tube would be a P_1 of 40 torr plus 1 mole per cent teflon powder.

UV absorption measurements were made behind incident shocks looking at the CF_2 radical. Calculations indicated that the burnup times were short compared to the test time, and furthermore, that the CF_2 radical would go to local equilibrium rapidly. These experiments indicated that a uniform test slug in the dusty shock tube had been accomplished.

Using the Nicholls and Parkinson technique of spreading a sample on mylar film, Tite (100) looked at shock heated metal powders in an oxidizing atmosphere. The samples were small spheres 38 to 147 μ in diameter. In particular, he looked at the blue-green AlO system with time resolution. The purpose of this investigation was to try to distinguish whether the primary means by which the radius of the particle decreases is by a combustion or an ablation type process. It was concluded that the process is primarily one of combustion.

Many workers have tried to design and construct a hot shock tube so as to be able to study materials whose vapor pressures at room temperature are too low to allow their use. However this has never been a very productive endeavor. The problem of keeping a uniform temperature over the entire shock tube is indeed overwhelming, although several people have tried to study alkali metal ionization processes in such a tube.

Louis (101) avoided this problem by filling his shock tube with an aerosol of cesium. He examined the problem of uniform suspension and evaporation of the particles in the shock front. His work showed that a uniform slug of test gas can be obtained a short distance behind the shock. Use of the pure alkali metal eliminates the problem of dissociation of ionic compounds and the presence of anions. To make the settling out of the aerosol particles negligible, an area/volume ratio much greater than unity and a high flow velocity of the aerosol down the tube are needed.

Stokes' law corrected for slip - since the particles involved here are smaller than the mean free path - gives the settling velocity as:

$$V_S = \frac{2}{9} \frac{\rho_L r^2 g}{\mu} (1 + A\ell/r), \qquad \text{(Eq. 1-1-8)}$$

where ρ_L is the density of the particles of radius r, μ is the viscosity, A is Milliken's constant, and ℓ is the mean free path. Using this equation, one finds that settling is negligible for particles a few tenths of a micron in diameter when the flow velocity down the tube is $\geqslant$ 1 m/sec.

Figure 1-1-12 is a schematic diagram of Louis' apparatus. Argon is cooled through a liquid nitrogen trap to minus 60°C. It is then blown over a pool of liquid cesium in an oven. The partial pressure of cesium, and hence its concentration in the aerosol, up to a few weight percent, is governed by the temperature of the liquid cesium. After the aerosol is formed, it goes next to a cyclone separator where it is mixed with more argon and particles greater than 1 micron are centrifuged out of the stream.

The apparatus seems to function particularly well. Electromicrograms yield particle diameters less than 0.3μ and light scattering experiments show the aerosol to be uniformly distributed in the shock tube. Theoretical calculations of particle evaporation times indicate them to be much smaller than the test time, which is of the order of 3 x 10^{-3} seconds. Furthermore, diffusion times between particles is of the order of a microsecond, hence a uniform slug of test gas is rapidly formed behind the shock. Meaningful information on chemical kinetic rate has been obtained with this apparatus.

Conclusion

This review of the new techniques available for monitoring time histories in the shock tube has shown that in recent years several methods have been developed which enable the kineticist to obtain detailed information. In addition, various techniques have been outlined which enable the kineticist to prepare novel initial conditions in the shock tube and so investigate atomic and free radical reactions at high temperatures.

Sufficient details, such as the pressure and temperature ranges covered by the various workers, have been mentioned in the hope that this paper will prove useful to investigators who are looking for the 'right technique' to answer a particular kinetics question. The extensive bibliography included here can then be utilized to obtain 'state of the art' information on the chosen technique.

1. **Palmer, H.,** 'The Shock Tube as a Tool in Fuel and Combustion Research,' J. Inst. Fuel 34, Sept., 1961, 359-367.

2. **Jost, W., Just, T.,** 'Schnelle Gasreaktionen in Stosswellen und Detonationen,' Angewandte Chemie 74, Dec. 1962, 919-941.

3. **Gaydon, A., Hurle, I.,** 'The Shock Tube in High-Temperature Chemical Physics,' Reinhold Pub. Corp., New York, 1963.

4. **Bauer, S.,** 'Chemical Kinetics in Shock Tubes,' Science 141, 3584, Sept., 1963, 867-879.

5. **Bradley, J.,** 'Chemical Applications of the Shock Tube,' Royal Inst. of Chemistry, London, Lecture Series 6, 1963, 1-29.

6. **Greene, E., Toennies, J.,** 'Chemical Reactions in Shock Waves,' Edward Arnold (Publishers) Ltd., London, 1964.

7. **Lin, S., Fyfe, W.,** 'Low-Density Shock Tube for Chemical Kinetic Studies,' Phys. Fluids 4, Feb., 1961, 238-249.

8. **Lin, S.,** 'Low-Density Shock Tube Studies of Reaction Rates Related to the High Altitude Hypersonic Flight Problem,' Rarefied Gas Dynamics, ed. L. Talbot, Academic, New York, 1961, 623-642.

9. **Lin, S., Neal, R., Fyfe, W.,** 'Rate of Ionization Behind Shock Waves in Air I. Experimental Results,' Phys. Fluids 5, Dec. 1962, 1633-1648.

10. **Wray, K.,** 'Shock-Tube Study of the Coupling of the O_2-Ar Rates of Dissociation and Vibrational Relaxation,' J. Chem. Phys. 37, 6, Sept., 1962, 1254.

11. **Wray, K., Freeman, T.,** 'Shock Front Structure in O_2 at High Mach Numbers,' J. Chem. Phys. 40, 10, May 1964, 2785-2789.

12. **Windsor, M., Davidson, N., Taylor, R.,** 'Measurement of the Vibrational Relaxation Time of CO behind a Shock Wave by Infrared Emission,' J. Chem. Phys. 21, 1, July 1957, 315-6.

13. **Jacobs, T.,** 'Shock-Tube Measurements of the Homogeneous Rate of Decomposition of NH_3 in NH_3-Ar Mixtures,' Paper presented at the Eighth Int. Symp. on Combustion, California Inst. of Technology, Pasadena, 1960. Pub. by The Williams and Wilkins Co., Baltimore, 1962, 151-4.

14. **Jacobs, T.,** 'Further Shock-Tube Studies by Infrared Emission of the Decomposition of Ammonia,' J. Phys. Chem. 67, March 1963, 665-7.

15. **Davies, W.,** 'Carbon Dioxide Dissociation at 3500° to 6000°K,' J. Chem. Phys. 41, 6, Sept., 1964, 1846-1852.

16. **Davies, W.,** 'Carbon Dioxide Dissociation at 6000° to 11,000°K,' J. Chem. Phys. 43, 8, Oct., 1965, 2809-2818.

17. **Camac, M.,** 'CO_2 Relaxation Processes in Shock Waves,' Fundamental Phenomena in Hypersonic Flow, ed. J. Gordon Hall, Cornell Univ. Press, Ithaca, New York, 1966, 195-215.

18. **Jacobs, T., Giedt, R., Cohen, N.,** 'Kinetics of Decomposition of HF in Shock Waves,' J. Chem. Phys. 43, 10, Nov., 1965, 3688-3693.

19. Fishburne, E., 'Gaseous Reaction Rates at High Temperatures. II. The Dissociation of Hydrogen Chloride,' J. Chem. Phys. 45, 11, Dec., 1966.

20. **Jacobs, T., Cohen, N., Giedt, R.,** 'Kinetics of Hydrogen Halides in Shock Waves: HCl and DCl,' J. Chem. Phys. 46, 5, March 1967, 1958-1968.

21. **Fishburne, E., Edse, R.,** 'Shock-Tube Study of Nitrous Oxide Decomposition,' J. Chem. Phys. 41, 5, Sept., 1964, 1297-1304.

22. **Taylor, R., Camac, M., Feinberg, R.,** 'Measurements of Vibration-Vibration Coupling in Gas Mixtures,' Paper presented at Eleventh Combustion Symp. Berkeley, California, Aug., 1966.

23. **Taylor, R.,** 'Continuum Infrared Radiation from High-Temperature Air and Nitrogen,' J. Chem. Phys. 39, 9, Nov., 1963, 2354-2360.

24. **Camm, J., Taylor, R., Lynch, R.,** 'Synchronized High-Speed Scanning Infrared Spectrometer,' J. Appl. Optics, 6, 5, May 1967, 885-891.

25. **Wilson, J.,** 'Ionization Rate of Air behind High-Speed Shock Waves,' Phys. Fluids 9, 10, Oct., 1966.

26. **Lauver, M., Hall, J., Belles, F.,** 'Shocked Gas Temperature Measurements by Infrared Monochromatic Radiation Pyrometry,' Proc. of the 5th Int. Shock Tube Symp. U.S. Naval Ordnance Lab., White Oak, Silver Spring, Maryland, May 1966, 1913-1921.

27. **Penzias, G., Dolin, S., Kruegle, H.,** 'Determination of Shock Temperatures from Simultaneous Infrared Emission and Absorption Measurements,' Proc. of the 5th Int. Shock Tube Symp. U.S. Naval Ordnance Lab., White Oak, Silver Spring, Maryland, May 1966, 171-182.

28. **Ballard, H., Venable, D.,** 'Shock-Front-Thickness Measurements by an Electron Beam Technique,' Phys. Fluids, 1, 3, May-June, 1958, 225-229.

29. **Grunow-Schultz, F., Frohn, A.,** 'Density Distribution in Shock Waves Traveling in Rarefied Monatomic Gases,' Proc. of the Fourth Symp. on Rarefied Gas Dynamics, Vol. I, ed. J.H. de Leeuw, Academic Press, New York-London, 1965, 250-264.

30. **Russell, D.,** 'Electron-Beam Measurements of Shock-Wave Thickness,' Proc. of the Fourth Symp. on Rarefied Gas Dynamics, Vol.I, ed. J.H. de Leeuw, Academic Press, New York-London, 1965, 265-276.

31. **Camac, M.,** 'Argon and Nitrogen Shock Thicknesses,' Paper pres. at AIAA Aerospace Sciences Meeting, New York, Jan., 1964.

32. **Camac, M.,** 'Argon Shock Structure,' Proc. of the Fourth Symp on Rarefied Gas Dynamics, Vol.I, ed. J.H. de Leeuw, Academic Press, New York-London, 1965, 240-9.

33. **Duff, R.,** 'Shock-Tube Performance at Low Initial Pressure,' Phys. Fluids, 2, 2, March-April 1959, 207-216.

34. **Muntz, E.,** 'Static Temperature Measurements in a Flowing Gas,' Phys. Fluids 5, 1, Jan., 1962, 80-90.

35. **Zempel, R., Mallin, J., Muntz, E.,** 'An Electron Beam Excitation Technique for the Measurement of Low Densities in Hypersonic Shock Tunnel Flows,' G.E. 63SD707, Aug., 1963.

36. **Softley, E., Muntz, E., Zempel, R.,** 'Experimental Determination of Pressure, Temperature and Density in Some Laminar Hypersonic Near Wakes,' G.E. R64SD35, May 1964.

37. **Schumacher, B., Gadamer, E.,** 'Electron Beam Fluorescence Probe for Measuring the Local Gas Density in a Wide Field of Observation,' Can. J. Phys. 36, 1958, 659-671.

38. **Muntz, E., Marsden, D.,** 'Electron Excitation Applied to the Experimental Investigation of Rarefied Gas Flows,' Proc. of the Third Symp. on Rarefied Gas Dynamics, Vol.II, ed. J.A. Laurmann, Academic Press, New York-London, 1963, 495-526.

39. **Muntz, E.,** 'The Direct Measurement of Velocity Distribution Functions,' Proc. of the Fourth Symp. on Rarefied Gas Dynamics, Vol. II, ed. J.H. de Leeuw, Academic Press, New York-London, 1966, 128-150.

40. **Robben, F., Talbot, L.,** 'Measurement of Shock Wave Thickness by the Electron Beam Fluorescence Method,' Phys. Fluids 9, 4, April 1966, 633.

41. **Robben, F., Talbot, L.,** 'Measurements of Rotational Temperatures in a Low Density Wind Tunnel,' Phys. Fluids 9, 4, April 1966, 644-652.

42. **Robben, F., Talbot, L.,** 'Experimental Study of the Rotational Distribution Function of Nitrogen in a Shock Wave,' Phys. Fluids 9, 4, April 1966.

43. **Camac, M.,** 'Gas-Dynamic Measurements with Electron Beams,' Paper pres. at A.P.S., Division of Fluid Dynamics, 1966 Div. Meeting, Stanford Univ., California, Nov., 1966 (Abstract). Also M. Camac, private communication, May 1967.

44. **Bradley, J., Kistiakowsky, G.,** 'Shock Wave Studies by Mass Spectrometry. I. Thermal Decomposition of Nitrous Oxide,' J. Chem. Phys. 35, 1, July 1961.

45. **Bradley, J., Kistiakowsky, G.,** 'Shock Wave Studies by Mass Spectrometry. II. Polymerization and Oxidation of Acetylene,' J. Chem. Phys. 35, 1, July 1961.

46. **Kistiakowsky, G., Michael, J.,** 'Mechanism of Chemi-Ionization in Hydrocarbon Oxidations,' J. Chem. Phys. 40, 5, March 1964.

47. **Felmlee, W., Petrella, R., Diesen, R.,** 'Mass Spectral Studies of Kinetics behind Shock Waves Direct Sampling and Flash Photolysis of NO_2,' The Dow Chemical Co., Midland, Michigan, Tech. Rep. SL 175311a, May 1966.

48. **Dove, J., Moulton, D.,** 'Shock Wave Studies by Mass Spectrometry. III.

Description of Apparatus; Date on the Oxidation of Acetylene and of Methane,' Proc. Roy. Soc. 283, 1393, Jan., 1965, 216-228.

49. **Diesen, R., Felmlee, W.,** 'Mass Spectral Studies of Kinetics behind Shock Waves, I. Thermal Dissociation of Chlorine,' J. Chem. Phys. 39, 9, Nov., 1963, 2115-2120.

50. **Diesen, R.,** 'Mass Spectral Studies of Kinetics behind Shock Waves. II. Thermal Decomposition of Hydrazine,' J. Chem. Phys. 39, 9, Nov., 1963.

51. **Diesen, R.,** 'Mass Spectral Studies of Kinetics behind Shock Waves. III. Thermal Dissociation of Fluorine,' J. Chem. Phys. 44, 10, May 1966.

52. **Diesen, R.,** 'Comment on NF_2 Decomposition behind Shock Waves,' J. Chem. Phys. 45, 2, July 1966, 759-760.

53. **Marsters, G., Bauer, S., Resler, E.,** 'Optimized Geometry for Coupling a Mass Spectrometer to a Shock Tube,' Proc. 5th Int. Shock Tube Symp., U.S. Naval Ordnance Lab., White Oak, Silver Spring, Maryland, May 1966.

54. **Resler, E.,** Private Communication, May 1967.

55. **Gutman, D., Hay, A., Belford, R.,** 'Shock Wave Studies with a Quadrupole Mass Filter. I. Experimental Apparatus: Its Design and Performance,' J. Phys. Chem. 70, 6, June 1966, 1786-1792.

56. **Gutman, D., Belford, R., Hay, A., Pancirov, R.,** 'Shock Wave Studies with a Quadrupole Mass Filter. II. The Thermal Decomposition of Nitrous Oxide,' J. Phys. Chem. 70, 6, June 1966, 1793-1800.

57. **Hay, A.,** 'High Temperature Gas Kinetic Study of Carbonyl Sulfide Decomposition Performed with a Shock Tube and Quadrupole Mass-Filter,' Res. Group in Shock Tube Kinetics, Noyes Chem. Lab., Dept. of Chem. and Chem. Eng., Univ. of Illinois, Res. Rep. No.3, Feb., 1966.

58. **Glick, H., Squire, W., Herzberg, A.,** 'Shock Tube Technique for the Study of High Temperature Gas Phase Reactions,' Fifth Symp. on Combustion, Reinhold Pub. Corp. New York, 1955, 393-402.

59. **Lifshitz, A., Bauer, S., Resler, E.,** 'Studies with a Single-Pulse Shock Tube. I. The Cis-Trans Isomerization of Butene-2,' J. Chem. Phys. 38, 9, May 1963, 2056-2063.

60. **Bauer, S., Watt, W.,** 'Association of CN in an Expansion Wave,' J. Chem. Phys. 44, 4, Feb., 1965, 1327-1333.

61. **Tsang, W.,** 'Comparative Rate Measurements with a Single-Pulse Shock Tube,' J. Chem. Phys. 40, 4, Feb., 1964, 1171-2.

62. **Jacobs, T., Hartunian, R., Giedt, R., Wilkins, R.,** 'Direct Measurements of Recombination Rates in a Shock Tube,' Phys. Fluids 6, 7, July 1963, 972-4.

63. **Brabbs, T., Belles, F.,** 'Recombination of Carbon Monoxide and Atomic Oxygen at High Temperatures,' Paper pres. Eleventh Symp. on Combustion, Berkeley, California, Aug., 1966.

64. **Wilson, J.,** 'An Experiment to Measure the Recombination Rate of Oxygen,'

J. Fluid. Mech. 15, 4, April 1963, 497-512.

65. **Thornton, E.**, 'Incorporation of a Laser into the Arm of an Interferometer for Measurement of Transient Phase Changes,' J. Appl. Phys. 36, 11, Nov., 1965, 3539-3541.

66. **Deuel, R., Kirchner, L., Thornton, E.**, 'The Laser Interferometer as a Diagnostic Tool in Shock-Tube Experiments,' Appl. Phys. Letters 8, 3, Feb., 1966, 59-60.

67. **Besse, A., Kelley, J.**, 'Interferometer for Shock Tube,' Rev. Sci. Inst. 37, 11, Nov., 1966, 1497-1499.

68. **Kiefer, J., Lutz, R.**, 'Simple Quantitative Schlieren Technique of High Sensitivity for Shock Tube Densitometry,' Phys. Fluids 8, 7, July 1965

69. **Kiefer, J., Lutz, R.**, 'Vibrational Relaxation of Deuterium by a Quantitative Schlieren Method,' J. Chem. Phys, 44, 2, Jan., 1966, 658-667.

70. **Kiefer, J., Lutz., R.**, 'Vibrational Relaxation of Hydrogen,' J. Chem. Phys., 44, 2, Jan., 1966, 668-672.

71. **Kiefer, J., Lutz, R.**, 'The Effect of Oxygen Atoms on the Vibrational Relaxation of Oxygen,' Paper pres. Eleventh Symp. on Combustion, Berkeley, California, Aug., 1966.

72. **Resler, E., Scheibe, M.**, 'Instrument to Study Relaxation Rates behind Shock Waves,' J. Acoust. Soc. Am. 27, Sept., 1955, 932-938.

73. **Dunaev, Y., Tumakaev, G., Shukhtin, A.**, 'The Application of Rozhdestvenskii's Hook Method in the Investigation of Gasdynamic Processes in Shock Tubes,' Sov. Phys.-Tech. Phys. 6, 9, March 1962, 815-820.

74. **Marlow, W.**, 'The Hakenmethode, A Review Article,' Lockheed Missiles and Space Co., Sunnyvale, California, LMSC A034392, 1964, 41-94.

75. **Penkin, N.**, 'The Determination of Oscillator Strengths in Atomic Spectra,' J. Quant. Spectrosc. Radiat. Trans. 4, 1, 1964. This article is in Russian but an English version exists (J. Tech. Harvard College Observatory).

76. **Foster, E.**, 'The Measurement of Oscillator Strengths in Atomic Spectra,' Rep. on Progress in Physics, Vol.27, Inst. of Physics and the Physical Soc. London, 1964, 489-551.

77. **Wray, K., Teare, J.**, 'Excitation Mechanism for the $N_2(1+)$ Band System in Shock-Heated Nitrogen,' Proc. IIIrd Int. Conf. on the Physics of Electronic and Atomic Collisions, ed. M.R.C. McDowell, North-Holland Pub., Co., Amsterdam, 1964, 1123-1131.

78. **Wray, K.**, 'Excitation Studies on the $N_2(1+)$ and $N_2^+(1-)$ Systems in Shock-Heated N-N_2 Mixtures,' J. Chem. Phys. 44, 2, Jan., 1966, 623-632.

79. **Thompson, W., Hartunian, R.**, 'Catalytic Probe Response to High Atom Flux in a Glow Discharge Shock Tube,' AIAA J. 3, 4, Jan-June 1965, 790-1.

80. **Hartunian, R., Thompson, W., Hewitt, E.**, 'Glow-Discharge Shock Tube

for Studying Chemiluminescent, Surface-Catalytic, and Gas-Phase Reaction Rates; Temperature Dependence of NO-O and CO-O Chemiluminescence,' J. Chem. Phys. 44, 5, March 1966, 1765-9.

81. **Burns, G., Hornig, D.,** 'A Combined Flash Photolysis and Shock Wave Method for the Study of Bromine Atom Recombination over a Wide Temperature Range,' Can. J. Chem. 38, 1960, 1702-1713.

82. **Bradley, J., Tuffnell, R.,** 'A Combined Flash-Photolysis Shock-Tube Technique for the Study of Fast Photochemical Reactions,' Proc. Roy. Soc. 280, 1381, Feb., 1964, 198-210.

83. **Wray, K.,** 'Shock-Tube Study of the Recombination of O Atoms by Ar Catalysts at High Temperatures,' J. Chem. Phys. 38, 7, April 1963, 1518.

84. **Jones, W., Davidson, N.,** 'The Thermal Decomposition of Ozone in a Shock Tube,' J. Am. Chem. Soc. 84, Aug., 1962, 2868-2878.

85. **Kiefer, J., Lutz, W.,** 'Recombination of Oxygen Atoms at High Temperatures as Measured by Shock-Tube Densitometry,' J. Chem. Phys. 42, 5, March 1965, 1709-1714.

86. **Fishburne, E., Edse, R.,** 'Reaction between Atomoc Oxygen and Nitrous Oxide,' J. Chem. Phys. 44, 2, Jan., 1966, 515-520.

87. **Nicholls, R., Parkinson, W.,** 'Shock Excitation of Atomic and Molecular Spectra,' J. Chem. Phys. 26, 2, Feb., 1957, 423-424.

88. **Nicholls, R., Parkinson, W.,** 'Shock-Induced Physical and Chemical Surface Changes on Oxide Powders,' Can. J. Phys. 36, 5, May 1958, 625-7.

89. **Parkinson, W.,** 'Time Resolved Absorption Studies in a Shock Tube: A New Band System of BaO,' Proc. Phys. Soc. 78, 5, Nov., 1961, 705-9.

90. **Nicholls, R., Parkinson, W., Reeves, E.,** 'The Spectroscopy of Shock-Excited Powdered Solids,' Appl. Opt. 2, 9, Sept., 1963, 919-930.

91. **Parkinson, W., Reeves, E.,** 'The Spectrum of Titanium Nitride,' Can. J. Phys. 41, 4, April 1963, 702-4.

92. **Berry, R., Reimann, C., Spokes, G.,** 'Absorption Spectrum of Gaseous Cl^- and Electron Affinity of Chlorine,' J. Chem. Phys. 35, Dec., 1961.

93. **Berry, R., Reimann, C., Spokes, G.,** 'Absorption Spectra of Gaseous Halide Ions and Halogen Electron Affinities: Chlorine, Bromine, and Iodine.' J. Chem. Phys. 37, 10, Nov., 1962, 2278-2290.

94. **Berry, R., Reimann, C.,** 'Absorption Spectrum of Gaseous F^- and Electron Affinities of the Halogen Atoms,' J. Chem. Phys. 38, 7, April 1963.

95. **Berry, R., David, C.,** 'Radiative Capture of Electrons by Halogen Atoms,' Proc. IIIrd Int. Conf. on the Physics of Electronic and Atomic Collisions, ed. M.R.C. McDowell, North-Holland Pub. Co., Amsterdam, 1964, 543.

96. **Berry, R., Mackie, J., Taylor, R., Lynch, R.,** 'Spin-Orbit Coupling and Electron-Affinity Determinations from Radiative Capture of Electrons by Oxygen Atoms,' J. Chem. Phys. 43, 9, Nov., 1965, 3067-3074.

97. **Hickman, R.,** 'Progress on SAPAG Facility, II,' Heliodyne Corp., Los Angeles, California, Research Note 12, Dec., 1963.

98. **Hickman, R., Ory, H., Tannenbaum, I.,** 'Progress on SAPAG Facility, III,' Heliodyne Corp. Los Angeles, California, Res. Note 15, June 1964.

99. **Holden, R., Hooker, W.,** 'The Preparation and Injection of Teflon/Gas Dispersions for Shock Tube Investigations,' Heliodyne Corp. Los Angeles, California, Res. Note 24, June 1965.

100. **Tyte, D.,** 'Interaction of Metallic Powders with a Shock Wave through an Oxidizing Atmosphere,' J. Appl. Phys. 37, 2, Feb., 1966, 802-6.

101. **Louis, J.,** 'An Alkali Metal Vapor Shock Tube and the Measurements of the Conductivity and Electron Concentration of Cesium in Argon,' Proc. 5th Int. Shock Tube Symp. U.S. Naval Ordnance Lab., White Oak, Silver Spring, Maryland, May 1966, 485-500.

d

Commentary on Chapter 1-1

K.G. Sulzmann

Dep.of the Aerospace and Mechanical Engineering Sciences, and Institute for Radiation Physics and Aerodynamics, University of California, San Diego, La Jolla, California.

Dr. Wray has undertaken the formidable task of presenting an instructive chapter on recent techniques for shock tube kinetic studies. In the opinion of this reviewer, he is one of the few most suited men for this undertaking, because Dr. Wray has participated in and contributed to many of the applications of shock tube techniques to kinetic studies, and he has the privilege of working with one of the organizations where research with shock tubes is done on a large and, for many workers in the field, overriding scale.

Since Dr. Wray has specialized his chapter in a well defined way, excluding many techniques which had been established until about ten years ago, I might add to his list of general references the recent book by Zeldovich and Raizer (1), this excellent treatise contains a rather unbiased listing of all important references in the field up to about 1963, including the AGARDograph No. 41, while at the same time this book provides the reader with an extensive background on the physics of shock waves and high-temperature hydrodynamic phenomena which is essential for the successful use of present methods and the development of future techniques.

Also, since schlieren techniques are of considerable advantage for density measurements related to relaxation phenomena, and since these techniques will presumably find even wider applications in the future because of their increased sensitivity when used with laser light sources, I would like to point out a recent paper by DeBoer (3) on modifications and improvements of the system by Resler and Scheibe, referred to by Dr. Wray; in this article the optical design parameters for integrating schlieren systems are discussed in great detail and a simple method for calibrating the density scale is given.

Furthermore, with reference to the fast scanning spectrometers and the interference techniques pointed out by Dr. Wray, I would like to mention an interesting interferometric method for high speed scanning of spectral lines which has been developed by Cooper and Burgess (4) (5), although this method has not yet been applied to shock tube studies, it appears very promising to this reviewer for the determination of time resolved electron and neutral densities by measurements of line shifts and line shapes. The technique involves essentially a fast scanning Fabri-Pérot or Fizeau interferometer with a time resolution and a repetition time for individual line scans of better than 10^{-6} seconds. Furthermore, the general area of optical dispersion, the interesting Hook Method mentioned by Dr. Wray, and related techniques are also discussed in any early review article by Korff and Breit (6).

In connection with sampling techniques for chemical reaction studies, I might also add to Dr. Wray's list of recent techniques a method developed and used by Lauer and co-workers (7) for the continuous sampling of shock tube reaction products which is based upon the idea of the multiple shock machine (8) as examplified by the Cornell Aeronautical Laboratory superheater (3). For this method essentially the concept of single pulse shock tubes is combined with continuous operation by arranging a small and rotating shock tube in such a way that a repetitive cycle can be maintained for reactant loading, shock heating, product expansions and product removal; the apparatus has been used in conjunction with gas chromatographic procedures for the quantitative analysis of trace constituents in the pyrolysis of hydrocarbons with the advantage, that product to reactant mole fractions of less than 1 percent could be determined as a function of temperature in argon diluted mixtures containing only 1 percent of hydrocarbons.

Because the accurate and reproducible determination of species concentrations is of great concern in chemical kinetic studies, I might also add that with the infrared techniques, mentioned by Dr. Wray, very good concentration measurements can be performed if sufficient care is taken in carrying out the necessary calibration procedures. For example, by using the mechanical light chopper method, first employed by Penner and collaborators, (10) for intermittent emission and transmission measurements, this reviewer (11) was able to demonstrate for a particular application that CO_2 concentrations on the order of 1 mole percent can easily be determined behind incident shock waves with an accuracy of about 5 percent.

But rather than add further material to the already extensive list of references given by Dr. Wray, I want to close with the general remark that the need for methods which allow the determination of kinetic data with sufficient precision, as well as with sufficient reproducibility, cannot be overemphasized. This does not only apply to diagnostic techniques for time resolved measurements but it applies equally to the preparation of reactant mixtures and their initial state, so that the rate of increase in precision of shock tube kinetic data will keep pace with the rate of increase in sophisticated techniques required for the understanding of detailed kinetic mechanisms.

References

1. **Zeldovich, Ya. B., Raizer, P.,** 'Physics of Shock Waves and High Temperature Hydrodynamic Phenomena,' eds. W. D. Hayes and R. F. Probstein, Vol. 1 (Vol. 2 in preparation), Academic Press, New York, 1966.

2. **Ferri, A.,** (ed), 'Fundamental Data Obtained from Shock Tube Experiments,' AGARDograph No. 41, Pergamon Press, New York, 1961.

3. **DeBoer, P.C.T.,** 'Optics of the Integrating Schlieren Instrument,' Rev. Sci. Instr., 36, 1965, 1125-1142.

4. **Cooper, J., Greig, J.R.,** 'Rapid Scanning of Spectral Line Profiles using an Oscillating Fabry-Pérot Interferometer,' J. Sci. Instrum. 40, 1963, 433.

5. **Burgess, D.D., Cooper, J.,** 'An Interferometer designed for the Detection of Small Shifts in Wavelength of Spectral Lines emitted by Plasmas, and for Measuring the Ratio of such Shifts to the Line Width,' J. Sci. Instrum, 44, 1965, 829-833.

6. **Korf, S.A., Breit, G.,** 'Optical Dispersion,' Rev. Mod. Phys. 4, 1932.

7. **Lauer, J.L., Berchtold, M., Shang, J.Y.,** 'Continuous Shock Wave Reactor for Chemical Production and Reaction Studies,' Chem. Eng. Sci. 22(2), 1967, 209-215.

8. **Seipple, C.,** U.S. Patent No. 2,399,394. 1946.

9. **Herzberg, A., Glick, H.S., Squire, W.,** U.S. Patent No. 2,832,666 (1958) **Glick, H.S., Herzberg, A., Squire, W., Weatherston, R.,** U.S. Patent No. 2,902,337. 1959.

10. **Hooker, W.W., Lapp, M., Weber, D., Penner, S.S.,** 'Multiple-Path

Techniques for the Determination of Physico Chemical Data Behind Shock Fronts,' J. Chem. Phys. 25, 1087, 1956.

11. **Sulzmann, K.G.P.,** 'High Temperature Shock Tube CO_2 - Transmission Measurements at 4.25μ,' J.Q.S.R.T. 4, 1963, 375-413.

1-2

The Role of Shock Tubes in Opacity Measurements

W. WURSTER
Aerodynamic Research Department Cornell Aeronautical Lab., Inc. Buffalo, N.Y. U.S.A.

Summary

Quantitative spectral studies of the radiation from molecules, atoms and electrons will be discussed. These studies are chosen to demonstrate the usefulness of the shock tube in obtaining the fundamental radiative constants for these systems. Instrumentation unique to such studies, from the infrared to the vacuum ultraviolet, will be described. Critical problems associated with spectral purity and species identification will be stressed. Finally, the discussion will include the extension of these techniques to the study of radiation from normally solid species.

Sommaire

On discute d'études spectrographiques quantitatives du rayonnement de molécules, d'atomes et d'électrons. Ces études ont été choisies pour démontrer l'intérêt de l'utilisation du tube à choc pour obtenir des constantes fondamentales du rayonnement de ce système. On décrit l'instrumentation particulière à ces études, depuis l'infrarouge jusqu' à l'ultraviolet.

Des problèmes critiques sont soulevés à propos de la pureté spectrale et de l'identification des espèces. La discussion comprendra l'extension de ces techniques à l'étude du rayonnement d'espèces normalement à l'état solide.

Introduction

Experimental research in spectroscopy can be divided into two broad categories. One is the measurement of the wavelengths of the radiation arising from chemical species excited in any conceivable manner, and with little regard to the populations of the excited states involved. Through the production of energy level schemes for radiating systems, this work has contributed to the development and present status of quantum mechanics, and to our knowledge of atomic and molecular structure.

The second category consists of research in which absolute intensity measurements of the radiation are performed, and in which full utilization is made of the vast body of published wavelength data and of the established energy level diagrams. The objective of the research in this area is to determine the fundamental constants that relate energy level populations to the resultant emission intensities. These constants are the experimentally determined inputs to the well-known formulae that describe the optical properties of matter, such as emissivity, reflectivity and opacity. In practice, it is from a measurement of these properties, such as the opacity, of a carefully defined system, that the desired fundamental constants are obtained. Thus by the term 'opacity measurements' is meant measurements designed to yield these constants, often referred to as the transition probabilities or f-numbers for particular radiating systems.

These studies depend on the measurement of radiation from species excited under conditions in which thermodynamic equilibrium is established and to which a meaningful temperature can be ascribed. The shock tube can be utilized to produce such conditions, and this chapter shows its use in basic studies to determine the optical properties of radiating systems.

The principal features of shock tube spectroscopy can be illustrated by discussing typical experiments in which were measured the discrete line radiation of atoms, the band spectra of molecules and the continua from free-free and free-bound electron transitions. A variety of instruments has been used by the researchers in this field, covering a wavelength range that extends from 700 Å in the vacuum ultraviolet, to 5 microns in the infrared. The design of the instrumentation is generally dictated by the shock tube operating parameters, and it is this correlation that will be stressed. While the shock tube offers the capability for very unique excitation conditions for gases, it has several inherent limitations that will also be discussed.

General Discussion

In recent years several books have been published, describing shock tubes, their operation and their application to problems of physics and chemistry (1) (2) (3) (4). The discussion here relates just to properties of importance to radiation measurements and is in effect a supplement up-dating a review by S. S. Penner (46).

Shock Tubes

There exists today a great variety of shock tubes, mostly being essentially two unequal lengths of tubing, separated by a diaphragm. In operation, the shorter- or driver-section is filled with a gas under high pressure, and the other comprises the test- or driven-section, in which is placed the gas to be processed by the shock waves. When the diaphragm is ruptured, the high pressure driver gas expands into the driven section with sufficient velocity to form a shock wave ahead of it, which heats, compresses and accelerates the test gas. At a given station along the shock tube, first the incident shock wave passes, then the heated test gas, and finally the expanding driver gas. At a station near the end of a closed tube, the reflected shock wave moving back upstream can also be utilized to further process and heat the oncoming test gas. In such cases the experiment is usually considered terminated when this reflected shock interacts with the oncoming driver gas.

Driver section diameters range between 1 inch and 8 inches, and some are designed to accommodate pressures as high as 30 000 lb/in^2. The diaphragms range from very thin mylar to copper or steel plates 1/4 inch thick. The driven sections of shock tubes also vary greatly, with diameters ranging between 1 inch and 24 inches, and lengths exceeding 100 feet. Cross sectional geometries of circular, square and rectangular design have been used. Many tubes are designed with electric-arc heating of driver gases, and recently a novel imploding-wall system has been employed to create extremely fast shock waves (5).

The major fraction of shock tube research today is in the field of kinetics where departures from conventional shock tube design are frequently needed. For example, in experiments where the structure of shock waves, or the chemical rate processes in gases, is under investigation it is often desirable to decrease collision rates by lowering the overall pressure levels of the experiment. However, the low pressure limit is dictated by the growth of the boundary layer on the tube walls, which adversely affects the state of the test gas. It has been shown (5) that larger tube diameters alleviate this problem, and hence, for such studies, a family of large diameter shock tubes, up to 24 inches (7) (8) has been developed in a number of laboratories.

Another general comment that relates to design differences between kinetic studies and equilibrium radiation studies concerns the required purity of the test gases. For the radiation experiments, the most adverse effect of impurities arises when the radiation from unwanted species overlaps the desired data. This problem can generally be surmounted by better wavelength resolution, so that spectral discrimination can be employed against impurity radiation. On the other hand, for kinetic studies, the effects of contaminants can enter the data directly. For example, a few parts per million of oxygen can significantly affect the dissociation rate measurement of hydrogen (9). For reasons such as these, a number of high-purity shock tubes have been built. They feature, for example, assembly without O-rings, being bakable to high temperatures, and capable of evacuation to better than 10^{-7} torr with leak rates as low as 10^{-3} microns per minute.

Such stringent criteria do not normally apply to shock tubes for equilibrium radiation measurements. The most critical parameter for such measurements is the knowledge of the thermodynamic state of the test gas that has been processed by the shock waves. This applies to either the incident shock wave alone, or to the combined effects of the incident and reflected shock waves in a closed tube.

The most sensitive characteristic of the gas is its temperature. For a gas in equilibrium at elevated temperatures, use is made of the Boltzmann distribution for the determination of the populations of the excited states that contribute to the measured radiation intensity. Since this dependence is exponential, it is greatly affected by the temperature. For example, at 6000°K, and for radiation from an energy level at 60 000 cm^{-1} (such as from a nitrogen band system) an error of only 0.5 percent in the temperature gives rise to an error in the population of 10 percent. This error applies directly to the final transition probability for that state. The value of the temperature and density of the processed test gas is generally calculated, making use of the measurement of the incident shock velocity, the initial pressure in the test section, and the thermochemical properties of the test gas involved. In some cases, use is made of line-reversal techniques which provide an independent experimental measurement of the temperature. This complicates the instrumentation considerably, but is warranted by the importance of the temperature in such experiments. It thus becomes clear that quantitative shock tube spectroscopy relies very heavily upon the instrumentation required to determine the state of the gas and to verify that equilibrium conditions prevail while the data are being recorded.

When studies are performed in the gas behind the incident shock wave, the test time begins as the shock wave passes the viewing port, and ends when the interface appears, marking the transition to driver-gas flow. Because the shock wave has a higher velocity than the interface, the test time increases with the distance along the driven tube length. The tube length is limited, however, by increasing shock wave attenuation, which results in enthalpy gradients in the region of the test gas between the shock and the interface. Also, the expansion wave, reflected from the end of the driver section, will eventually overtake the test gas and destroy the flow.

In a shock tube 30 feet long, calculated test times in air under ideal-gas assumptions range from about 1 msec with a Mach 4 shock to 200μ sec with a Mach 12 shock. The equilibrium temperatures are 1200° and 5600°K, respectively. These figures show the increased need for fast-response instrumentation for high temperature work. In practice, because of the non-ideal nature of interface mixing, the test time is usually reduced by 1/3 to 1/2, which decreases the time still more. For this reason, high temperature studies are usually conducted behind the reflected shock wave. Testing times behind the reflected shock are comparable with those behind incident shocks, while affording the advantage of higher temperatures and densities for a given shock strength. The incident shock temperatures of 1200° and 5600°K are increased to 2400° and 8500°K, respectively. In addition, the

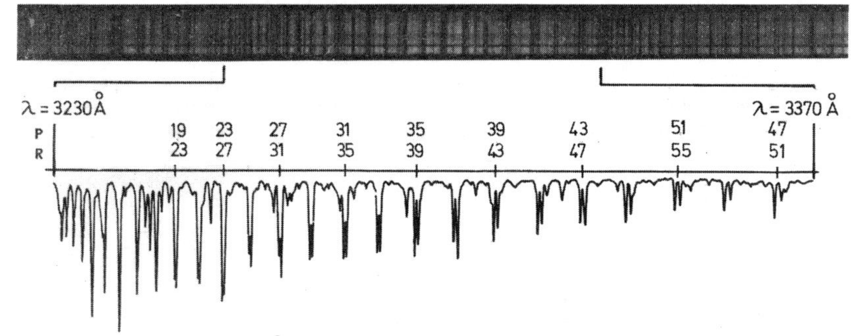

Fig. 1-2-1 Absorption spectrum of the (0, 13) band of the Schumann-Runge system of Oxygen. T=3410° K, $\rho/\rho_o = 3.6$

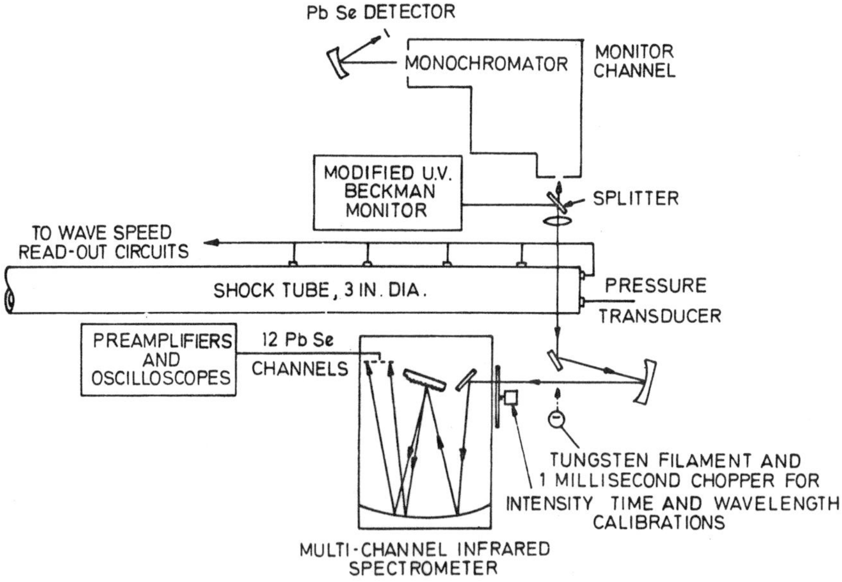

Fig. 1-2-2 Experimental apparatus for infrared emission measurements.

heated gas behind the reflected shock is ideally at rest with respect to the shock tube.

Optical Instrumentation

In principle, the purpose of the optical instrumentation is to yield quantitative intensity data as a function of wavelength, over the wavelength interval subtended by the radiating system being studied. In practice, several factors must be taken into account in instrument design, depending upon the particular experiment being conducted. Most difficulties are probably attributable to the low light levels. Taking into account the available intensity and the short testing times, there are only a limited number of photons available for counting and analysis.

Usually, when poor signal-to-noise levels are encountered, wavelength resolution has to be sacrificed. That this should be a last resort in any situation cannot be too strongly emphasized. It is through the recognition and identification of spectral features that the experimenter can most readily verify that the recorded intensities which constitute his data correspond only to transitions in the species under study. Impurity radiations in various degrees are invariably present, and if chemically or mechanically unavoidable, must be accounted for in the data analysis.

Both photographic and photoelectric (or photoconductive) types of radiation registration have been utilized in shock tube spectroscopy. Clearly, the maximum amount of spectral detail can be obtained through the use of photographic plates. However, in emission work, they are useful only for atomic lines and some strong molecular bands that form bandheads. This follows naturally from requiring high temperatures to obtain more radiation at a given emissivity, whereas most molecules dissociate appreciably at temperatures between 4000 and 9000°K. For absorption studies, the limit is determined by the lamp used as a background source. Today, flashlamps are available (10) (11) that emit continuum spectra at extremely high brightness temperatures, and with fair reproducibility, permitting reliable values of nonabsorbed signals to be obtained. However, the photographic plate is an integrating detector, so in absorption as well as in emission work, care must be taken to avoid plate exposure to unwanted radiation from the shock tube. Several high speed shutters have been developed to meet this need (12) (13).

Photoelectric detection, with its inherently higher sensitivity, is the diagnostic most commonly used in shock tube research. Furthermore, its output can be obtained as a function of time, making this method indispensable for kinetic analyses. Even for equilibrium studies, this feature is beneficial, since it permits assessment as to whether equilibrium (or at least steady-state) conditions have been obtained in the experiment. Indeed, the determination of useful testing times in shock tubes is most critically measured in this manner.

The price for such convenience is lower wavelength resolution. Because of physical size, a choice must be made between a multichannel system, and a rapid-scanning single-channel system. Both have been used to advantage. In the multi-channel approach, the radiation appearing at the focal plane of a dispersing system is divided and directed to several detectors. Units using as many as 12 channels have been employed with shock tube work (14). The wavelength interval subtended by each detector depends on the dispersion of the spectrograph and the dimensions of the aperture at the focal plane.

Better resolution can be obtained using a rapid-scanning technique (15), one of which will be described later. By sweeping the desired portion of the spectrum across a fixed exit slit, it is converted into a time-varying potential, that can be displayed and recorded by oscillographic means. The spectral resolution in this case depends on dispersion, slit size and scan rate. This technique also requires the assumption, or preferably, the experimental check that all properties of the radiating gas sample

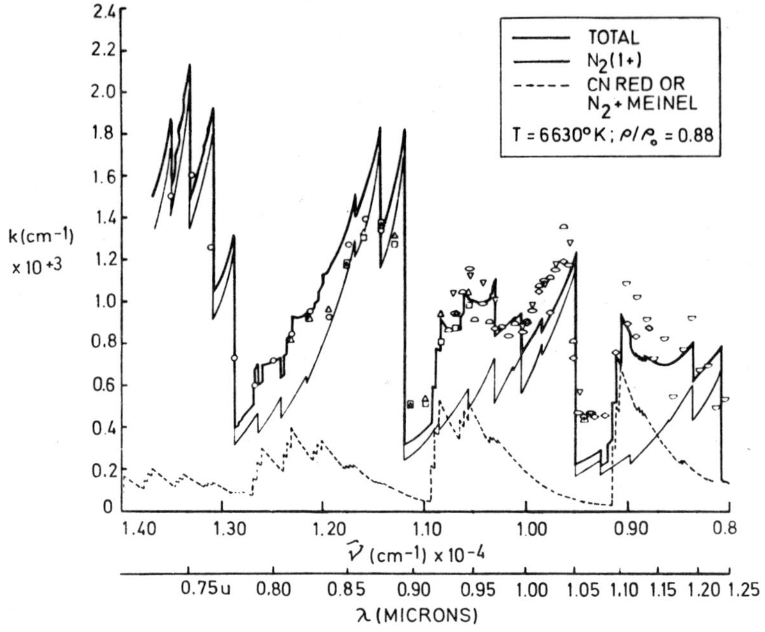

Fig. 1-2-3 Radiation from nitrogen.

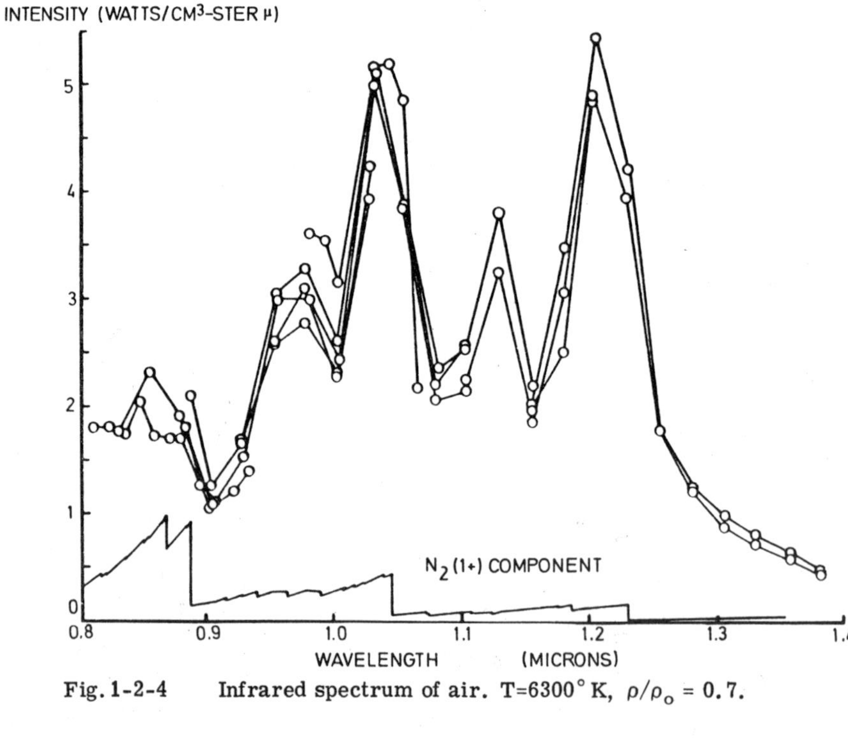

Fig. 1-2-4 Infrared spectrum of air. T=6300° K, $\rho/\rho_o = 0.7$.

remain constant over the scanning time interval. In addition, the time constant for the actual detector must be shorter than in the multichannel case, in order to adequately record spectral variations of the scan within the available shock tube test time.

Filters are extremely convenient to define wavelength intervals, because they can be coupled directly to a detector. On the other hand, they offer no flexibility in the choice of wavelength (except for the small change resulting from tilting interference filters). This is a handicap when good filters with well-defined transmission passbands are specified, because they are expensive which essentially precludes their use for survey work. It is generally true that filters are at best only suitable after spectrally resolved exploratory measurements have been made, in which the desired radiating species has been identified and shown to be free of unwanted impurity or background radiation. Even then, their use involves the convolution of their transmission curves with the spectral intensities under measurement. Quantatitive results obtained with the use of filters must generally be very critically appraised.

The following experiments, chosen to illustrate the general features already discussed, highlight those experimental aspects that are relevant to the measurement of different types of spectra over a wide range of wavelengths. Where applicable, they have been chosen from research performed at the Cornell Aeronautical Laboratory, with which the author is most familiar.

Molecular Band Spectra

Interest in the radiative properties of pure and contaminated air was a strong stimulus for initiating much of the shock tube research that has been done on molecular spectra. Such species as N_2, O_2, NO, N_2^+, CN, C_2, OH and NH have been extensively studied in a number of laboratories (14), (16-28) and the reports (29) (30) tabulate the radiative properties of air over broad ranges of temperature and density.

Figure 1-2-1 shows an example of the spectral detail that can be obtained under favorable circumstances with the use of photographic plates. It shows a portion of the band spectrum of the O_2 Schumann-Runge system in the near ultraviolet (18). The spectrum was obtained in an absorption experiment, wherein the continuum radiation from a flashlamp was passed through shock heated O_2 and into a large quartz spectrograph. The O_2 had been processed to the revealed conditions by the incident and reflected shock wave, and the flashlamp was triggered to radiate during the test-time interval. The intensity distribution of the rotational lines in this band is shown below the spectrum. The high wavelength resolution in this case permitted the spectral line shapes, and hence the absorption of each line to be measured. The opacities for some 18 bands were obtained by making use of the absorption coefficients for each line of these band distributions and the transition probability for the O_2 Schumann-Runge system was obtained.

When photographic methods are not feasible, recourse must be had to electronic means of detection, introducing problems of lower wavelength resolution, and correspondingly less knowledge of the spectral detail. The first positive band system in N_2 is illustrative of such a case (14). This band system extends throughout the near infrared to about 1.5 microns. To measure the spectrum, a 12-channel infrared spectrometer was constructed, with a resolution compatible with the known vibrational structure of the band system. Its deployment was typical of many shock tube-spectrometer systems, and is shown for illustrative purposes in figure 1-2-2. The entire spectrum was covered by moving the 12 wavelength intervals along the spectrum for successive shock tube tests. The results are shown in figure 1-2-3. It was found that the N_2 spectral distribution could not be made to fit the measurements in the 0.95 and 1.1μ regions, even though the prominent N_2 band sequence heads at 0.76, 0.88 and 1.06μ were well matched. The data were explained by postulating the presence of a second radiating source. Both CN, a common shock tube impurity, and N_2^+ were known to have band systems in this region. From the fit of the combined spectrum to the data, there is little doubt that the second component is real. The transition probability for the N_2(1+) system was determined by the scale factor required to fit the N_2 component to the predicted spectrum. It was subsequently shown that the radiation of the second component does not scale properly with N_2^+ concentration, and it was concluded that it arises from about 10 ppm of CN. The advantage of multichannel viewing with good dispersion is evident, since failure to discriminate against the contaminant corresponds to a factor of two in the intensity at the CN bandhead positions.

As demonstrated in this example, the analysis of many molecular band studies are made by comparing the recorded intensities arising from many discrete lines in a given wavelength bandpass with intensity distributions of a "synthetic" spectrum. These distributions are generated by using the known spectroscopic constants for the molecule excited to the conditions which apply to the measurement. The arbitrary scale factor which matches the amplitudes of the observed and theoretical spectrum is related to the desired f-number. Line-by-line analyses are tedious and costly, and increasing use is being made by approximate methods, where molecular band distributions are developed from just-overlapped line models, or smeared models (30, 50-52). These calculations are more tractable and have been shown to compare favorably with more exact treatments.

When the spectroscopic constants for the radiating system are unknown, analysis is more difficult and opacites cannot be determined precisely. Consider figure 1-2-4, which shows the results of spectral measurements in air (21), using the same instrumentation. In this case the source of the radiation was unknown. It was clear that the radiation in this wavelength region in shock-heated air was about five times that which could be attributed to N_2 alone. By making a series of measurements in which the N_2-O_2 ratio was varied, and by measuring the temperature dependence of the prominent peaks in the spectrum, it was shown that the radiation could be explained by considering transitions among excited electronic states in NO. Radiation from these states had not previously been considered in air radiation analyses. Because of inadequate wavelength resolution, assignment to particular states was not possible and the analysis required for definitive opacity determinations could not be performed. Hence, the NO-hypothesis was quite controversial (15) (31), and has only recently gained acceptance. Dr. K. Wray, using an arc-heated airstream is now mapping this NO spectrum in detail, from which transition probability assignments can then be made.

Atomic Line Radiation

Quantitative measurement of line radiation presents yet another difficulty, which again can be traced to low light intensities. The analysis of data for optically thin radiating layers does not require a detailed knowledge of spectral line shapes and widths. It is just necessary that the emissivity of the gas at the line centers is small ($\epsilon << 1$), self-absorption is negligible and every volume of the gas can contribute equally to the recorded intensity. To permit this assumption to be made, the power is limited by the gas temperature and an emissivity of order 10^{-1}. The signal/noise ratio that can be accepted in a given experiment is thus determined by the efficacy of the instrumentation in gathering, dispersing and recording the radiation. In molecular spectral recording, the radiation from many lines in a small wavelength interval can be added to increase the signal.

Thus, to bring about these increased signal strengths, most experiments on line radiation require a radiating particle density corresponding to an optically thick sample. In this case the analysis for the line strength is no longer straight forward, since the details of the spectral line shape now play an important role. A curve of growth analysis is usually used (32), which relates the recorded intensity to the density of radiators and the line strength, through a parameter, which is the ratio of the width of the collision-induced line to that of the Doppler line. Unless this parameter can be measured, the collision width must be calculated, and may introduce a sizeable error in the final determination of the transition probability.

At the Universities of Michigan and Maryland, line-strength or f-number measurements on atomic lines using the shock tube on Ne, C, O, N, A and Cv have been made. Wilkerson (33) has measured the strengths of many lines of chromium atom and ion. He recorded photographically many emission lines of these species from a mixture of shocked neon with a trace of chromium carbonyl. A range of temperatures and Cr concentrations were used in the experiments, and the data were reduced by the thick-gas curve-of-growth analysis.

The f-numbers for some 21 lines of the chromium ion were measured by Shackleford (53). By using a window in the end wall of the shock tube, and viewing the growing volume of test gas behind the reflected shock, the curve-of-growth of the radiation could be observed directly. A linear rise in the intensity was taken to correspond to optically thin radiation.

It should be mentioned that most studies have been of species that are pure gases, or that are in the form of volatile compounds that can readily be put into the shock tube in gaseous form. This facilitates calculation of the concentration of radiating atoms. However, the Harvard Group (34) studying the auto-ionized line strengths of aluminum in connexion with astrophysical problems, has used a mixture of aluminum and calcium sulphides in a known ratio, finely ground, placed on cellulose tissue suspended in the shock tube, and subjected to excitation by the shock waves. Absorption spectra were recorded, using an intense flashlamp (10), and the atomic lines were then analyzed by densitometric techniques. The aluminum concentration was deduced from measurements on the accompanying calcium lines, whose line strengths were known.

The same group recently reported (35) a study of iron and chromium line strengths, in which correction for the line emission from the shock-heated gas was made in the data analysis.

Continuum Radiations

Interest in superorbital reentry has promoted research to measure the radiation arising from electron reactions. At temperatures exceeding 10 000°K, diatomic molecules are almost entirely dissociated, and appreciable ionization prevails. The radiation from such a gaseous sample is composed of atomic and ionic line radiation, and the continua associated with free-free and free-bound electron transitions. Even though the testing times are short ($<100\ \mu$ sec), the shock tube is virtually the only technique that produces equilibrium thermal excitation of the species under study.

To study the infrared bremsstrahlung radiation from electrons scattered by neutrals and molecules, Camm, Taylor and Lynch (15) have developed a novel scanning spectrometer, designed specifically for use with shock tubes. It is a grating instrument, with a mirror placed in front of the focal plane. The mirror is rotated by electromagnetic pulses and, by sweeping the spectrum across the detector, permits scanning at the rate of about $1\ \mu$ in $50\ \mu$s. The principal advantage of the instrument is the output display of a continuous spectrum, which permits the continuum to be separated from overlapping lines and bands.

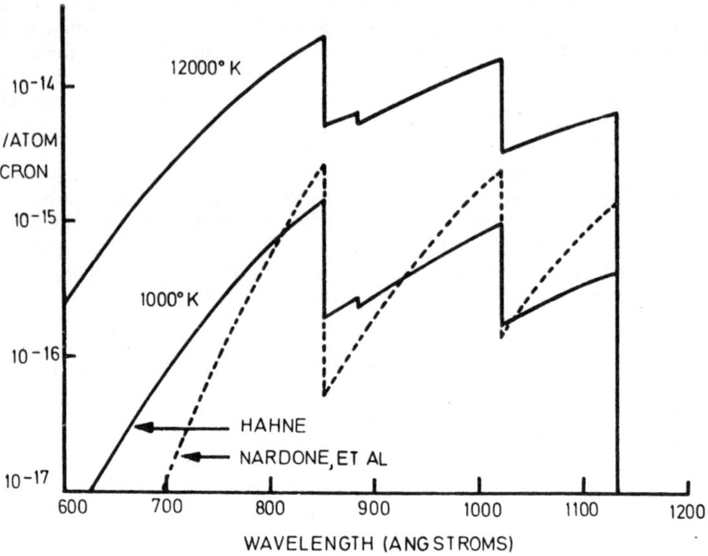

Fig. 1-2-5 $N^+ + e$ Recombination continuum.

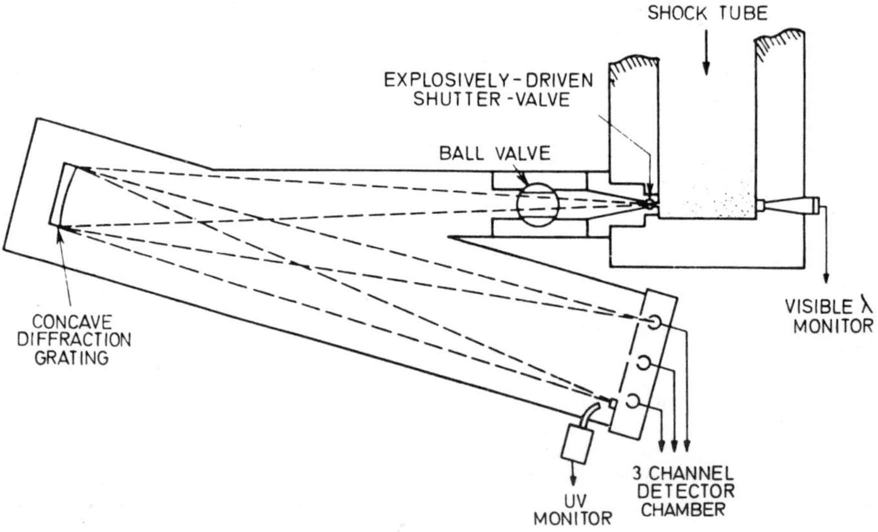

Fig. 1-2-6 Vacuum ultraviolet radiation experiment.

Fig. 1-2-7 Spectrum of $Al-0_2$-Argon aerosol. T=4000° K.

The coupling of image intensifiers with spectrometers has advantages for shock tube research in that they can be readily synchronized with the motion of the shock wave, have the high light gain characteristics of photoelectric devices and yet provide the continuous spectral coverage given by photographic plates (36).

A very strong continuum arises in the vacuum ultraviolet region of the spectrum when electrons recombine radiatively with nitrogen and oxygen ions. In such cases the continuum has a sawtooth profile, with a leading edge at the wavelength corresponding to the ionization energy, followed by a monotonic decrease to shorter wavelengths. Figure 1-2-5 shows a theoretical prediction of the spectrum arising from the $N^+ + e$ recombination in air. The figure also demonstrates the spread in the predictions (37) (38). The superposition of sawtooth profiles results from the fact that recombination transitions can take place to various levels of the neutral atom from several states of the ion.

It is obvious that a measurement of the intensity of the continuum places no great demands on spectral resolution. Fairly broad wavelength intervals can be used, and require only the assumption or knowledge that no strong line radiation is present in that interval. On the other hand, since the continua are generally featureless, species identification on this basis is almost impossible. In this case, however, the intensity jumps across the leading edges of the different levels may be utilized.

However, no window materials exist that transmit radiation below 1100 Ångstroms, and as most of this radiation lies in that range, the detector must be coupled directly, without windows, to the radiating gas sample.

Recent shock-tube experiments (39-42) measuring total wavelength-integrated radiation in air, illustrate problems of quantitative spectroscopy in the vacuum ultraviolet. Hoshizaki and coworkers (39) examined the radiation behind a reflected shock wave in air at temperatures as high as 17 000°K. This high temperature was obtained by using an electric arc-heated driver gas, capable of driving a Mach 30 shock wave into 200μ pressure of air in the test section of the shock tube. Just ahead of the reflecting wall, a splitter plate was used to core out a portion of the shock tube flow, and apertures in the plate permitted the detecting gages to view the shock heated air. The thin film, heat transfer gages of blackened platinum, were isolated in cavities, all contained within the shock tube. The test time was about 15μ s terminated when the shock wave that was formed in the cavity struck the gages. A virtue of this configuration was the opportunity it gave for varying the optical pathlength. Provisions were also made to insert various windows in the optical path. The radiation recorded with and without windows was compared with the available theoretical predictions of the radiation.

A different approach to the same problem was taken by Gruszczynski and Warren (40). They permitted the shock wave and gas-flow in a shock tube to impinge upon a gage mounted on a sting in the tube. The measurement made was of the radiation from the stagnation region of the bow shock in front of the gage. The gage itself, called a cavity gage, was a small enclosed cylinder, with a slot that permitted radiation to enter and strike the inner platinum surface, which served as a fast-response resistance thermometer. In operation, the gage was sealed off from the test gas by a small latex membrane, and filled with a transparent gas. Prior to shock impingement, the membrane was broken, so that during the test, no air would expand into the gage cavity. Such expansion would cool the air, rendering it strongly absorbing, and thereby decrease the recorded intensity signal. An arc-heated shock tube was used, with Mach numbers as high as 25 driven into 300μ of initial air pressure. Again, the wavelength-integrated data, with and without windows, were compared with available radiation predictions. Because the measurements were broad band, the details of the various contributions of the continua and lines to the radiation could not be resolved, and hence extrapolation to other thermodynamic conditions was difficult (41) (42).

At CAL we have begun to measure the wavelength distribution of the continuum from the $N^+ + e$ recombination. A schematic diagram of the apparatus used is presented in figure 1-2-6. Here is shown a vacuum grating spectrometer with three detectors in the focal plane, permitting the radiation in three 30 Å intervals to be measured. The detectors are bare photomultiplier tubes constructed of stainless steel. The spectrometer is separated from a small aperture in the shock tube, which serves as the entrance slit, by a hollow plunger. This plunger acts as a shutter-valve, and is explosively driven across the shock tube aperture, permitting the detectors to view the shock-heated gases during a 50 μs interval of the test time. The shock tube in this case is a conventional pressure driven tube, and the temperatures between 9000 and 14 000°K are achieved by using neon as a diluent with 1 to 5 percent nitrogen.

The plunger being hollow, this configuration permits the effects of vacuum or varying pressures using different "window" gases to be assessed. One finding to date has been that the use of any gas - even one that is transparent - that restricts the flow of test gas into the spectrometer tends to block or absorb radiation. This seems to indicate that the thin layer of cool gas in the boundary layer absorbs radiation originating in the shocked-gas volume. By evacuating the plunger and draining off some of the boundary layer, higher radiation levels have been obtained. This phase of experiment is still in progress. So far, the radiation intensity discontinuity at the photoelectric edge in nitrogen has been identified, and the continuum intensities for optically thick nitrogen concentrations have been measured and found to scale properly with the black body radiation as a function of temperature. Measurement with optically thin concentrations, from which the opacity can be determined, will be made when the boundary layer behavior is better understood.

Spectra of Normally Solid Species

Questions of stellar abundances of the elements in astrophysical studies, and the interaction of ablative material with hypersonic flow fields in re-entry studies, have both given rise to a new phase of shock tube research. This deals with quantitative spectroscopic measurements of radiation emitted by species that normally occur as solids. Nicholls *et al* (43) reported a series of experiments on a large number of compounds. The technique used was to place powders on plastic tape suspended in the shock tube. Under shock processing, the powder became partially vaporized and the product species became thermally excited. At temperatures between 3000 and 6000°K, many molecular spectra were obtained, while at 5000 to 8000°K, the atomic line radiation was dominant, due to the dissociation of the diatomic species. These studies resulted in a bibliography of spectra for these species, but the experiments were of a qualitative nature and were not designed to yield opacities or transition probabilities for the bands observed. This was due to inability to determine the concentrations of the radiating species. A recent report by Drake *et al* (44) described a similar but quantitative study of one of the BeO molecular band systems. The concentration of BeO in that case was estimated by thermodynamic calculations, in which the energy required for powder vaporization and excitation was deduced from the difference between the ideal and measured temperatures. The resultant emissivity of the band was stated to within a factor of 2 to 3.

At CAL we have developed (45) a different approach which we anticipate will permit good quantitative results to be obtained for metallic species. In this case, a wire of suitable material is exploded in a small chamber containing a controlled atmosphere, thereby forming a metallic aerosol. The aerosol is then expanded into the test section of the shock tube and is subsequently vaporized and excited by the shock waves. The aerosol particles are very small (~ 0.05μ), ensuring complete vaporization and an equilibrium gaseous mixture. In the present study of the blue-green band system of AlO, the concentration of AlO is obtained by a measurement of the Al atoms in the equilibrium mixture containing Al, AlO, O_2 and Argon. So far the technique has proved very promising, yielding clean spectra of the atomic lines and band spectra (Fig. 1-2-7). The bands of AlO and the doublets of Al have also been monitored photoelectrically by detectors placed behind the focal plane of the spectrometer, and have yielded data from which a preliminary transition probability of this band system was obtained.

The method is quite general and it is planned to extend these studies to other metallic oxides. It is limited by the available f-number data for the atoms used as diagnostics, and by the thermochemical data needed to obtain species concentrations in the equilibrium mixture.

Conclusion

A general discussion has been presented on the use of shock tubes for the determination of the basic radiative properties of excited chemical species. It was hoped to portray the broad scope of such research, and by a description of several experiments, to promote an appreciation of some of the specific problems in this field. Several aspects of the experiments, as, for example, the absolute intensity calibration of the detectors have not been included, but can be found in the references to the individual publications. Quantitative shock tube spectroscopy, not only for opacity measurements, but also for chemical-kinetic studies, is a very active research area. The shock tube, by virtue of its unique capabilities, is providing fundamental data needed to solve some of the technological problems of today and tomorrow.

Acknowledgment

The preparation of this Chapter was supported by the Cornell Aeronautical Laboratory on internal funding.

References

1. **Gaydon, A.G., Hurle, I.R.,** 'The Shock Tube in High-Temperature Chemical Physics,' New York, Reinhold, 1963.

2. **Greene, E.F., Toennies, J.P.,** 'Chemical Reactions in Shock Waves, New York, Academic Press, 1964.

3. **Bradley, J.N.,** 'Shock Waves in Chemistry and Physics,' London, Methuen; New York, Wiley, 1962.

4. **Zeldovich, I.B., Raizer, N.P.,** 'Physics of Shock Waves and High Temperature Hydrodynamic Phenomena,' Vol. 1, Academic Press, New York and London, 1967.

5. **Gill, S.P.,** 'Theory and Operation of a Jet-Augmented Implosive Shock Tube,' Proc. Sixth Int. Shock Tube Symp., Freiburg, Germany, April, 1967.

6. **Mirels, H.,** Phys. Fluids, Vol. 6., 1963, 1201.

7. **Liepmann, H.W., Roshko, A., Coles, D., Sturtevant, B.,** Rev. Sci. Instr. Vol. 33., 6, 1962.

8. **Camac, M.,** 'Rarified Gas Dynamics,' Vol. 1, Ed. J.H. de Leeuw, Academic Press, New York, 1965, 240-9.

9. **Myerson, A.L., Watt, W.S., Joseph, P.J.,** 'Atom Formation Rates Behind Shock Waves in Hydrogen and the Effect of Added Oxygen,' Cornell Aero. Lab. Rept. No. AD-1689-A-7, Nov. 1966. submitted to J. Chem. Phys.

10. **Wheaton, J.E.G.,** Appl. Optics. Vol. 3, 1964, 1247.

11. **Goldstein, R., Mastrup, F.N.,** J. Opt. Soc., Vol. 56, 1966, 765.

12. **Wurster, W.H.,** Rev. Sci. Instr., Vol. 28, 1957, 1093.

13. **Camm, J.C.,** Rev. Sci. Instr., Vol. 31, 1960, 278.

14. **Wurster, W.H.,** J. Chem. Phys., Vol. 36, 1962, 2111.

15. **Camm, J.C., Taylor, R.L., Lynch, R.,** Appl. Optics Vol. 6, 1967, 885.

16. **Wray, K.L., Connolly, T.J.,** J.Q.S.R.T., Vol. 5, 1965, 111.

17. **Dronov, A.P., Sobolev, N.N., Faizullov, F.S.,** trans. Optics and Spectr. Vol. 21, 1966, 301.

18. **Treanor, C.E., Wurster, W.H.,** J. Chem. Phys. Vol. 32, 1960, 758.

19. **Keck, J.C., Camm, J., Kivel, B., Wentink, T.,** Ann. Phys. Vol. 7, 1959.

20. **Daiber, J.W., Williams, M.J.,** J.Q.S.R.T., Vol. 1, 1961, 135.

21. **Wurster, W.H., Treanor, C.E., Thompson, H.M.,** J. Chem. Phys. Vol. 37, 1962, 2560.

22. **Camac, M., Feinberg, R.M.,** J.Q.S.R.T. Vol. 7, 1967, 581.

23. **Allen, R.A., Camm, J.C., Keck, J.C.,** J.Q.S.R.T. Vol. 1, 1961, 269.

24. **Buttrey, D.E., Gibson, J.B.,** Air Force Special Weapons Center, AFSWC RTD - TDR 63-3047, 1963.

25. **Fairbairn, A.R.,** J.Q.S.R.T. Vol. 6, 1966, 325.

26. **Kudryavtsev, E.M., Gippius, E.F., Derbeneva, S.S., Pechenov, A.N., Sobolev, N.N.,** trans. High Temperature, Vol. 1, 1963, 338.

27. **Watson, R.,** J.Q.S.R.T., Vol. 4, 1964, 1.

28. **Harrington, J.A., Modica, A.P., Libby, D.R.,** J.Q.S.R.T., Vol. 6, 1966.

29. **Gilmore, F.R.,** 'The Contribution of Generally-Neglected Band Systems and Continua to the Absorption Coefficient of High-Temperature Air,' Rand Corp. Memo., RM-4335-PR, Oct. 1964.

30. **Allen, R.A.,** 'Air Radiation Tables: Spectral Distribution Functions for Molecular Band Systems,' AVCO-Everett Res. Lab. Res. Rep. 236, April 1966.

31. **Wurster, W.H., Marrone, P.V.,** J.Q.S.R.T., Vol. 7, 1967, 591.

32. **Penner, S.S.,** 'Quantitative Molecular Spectroscopy and Gas Emissivities,' Addison-Wesley, Reading, Mass.: London; 1959.

33. **Wilkerson, T.,** 'The Use of the Shock Tube as a Spectroscopic Source with an Application to the Measurement of the gf-Values for Lines of Neutral and Singly Ionized Chromium.' Univ. of Michigan Tech Note, Physics Dept., AFOSR 1151, 1961.

34. **Garton, W.R.S., Parkinson, W.H., Reeves, E.M.,** Astro-phys. J. Vol. 140, 1964, 1269.

35. **Huber, M., Tobey, F.L.,** 'Fe I, Cr I and Cr II gf-values from Shock Tube Measurements,' Harvard Coll. Obs. Shock Tube Spectroscopy Lab., Sci. Rep. No. 16, 1967.

36. **Liu, I.D.,** Appl. Optics, Vol.6, 1967, 1195.

37. **Hahne, G.E.,** 'The Vacuum Ultraviolet Radiation from N^+ and O^+ Electron Recombination in High-Temperature Air,' NASA Tech. Note TN-D-2794, June 1965.

38. **Nardone, M.C., Breene, R.G., Zeldin, S.S., Riethof, T.R.,** 'Radiance of Species in High Temperature Air,' Gen. Electric Space Sciences Lab. Rep. R-63-SD3, June 1963.

39. **Hoshizaki, H., Wood, A.D., Andrews, J.C., Wilson, K.H.,** Radiant Energy Transfer Measurements in Air, NASA CR-585, Sept., 1966.

40. **Gruszczynski, J.S., Warren, W.R.,** AIAA Journ. Vol.5, 1967, 517.

41. **Nerem, R.M.,** AIAA Journ. Vol.4, 1966, 1485.

42. **Allen, R.A., Textoris, A., Wilson, J.,** J.Q.S.R.T., Vol.5, 1965, 95.

43. **Nicholls, R.W., Parkinson, W.H., Reeves, E.M.,** Appl. Optics, Vol.2, 1963, 919.

44. **Drake, G.W.F., Tyte, D.C., Nicholls, R.W.,** J.Q.S.R.T., Vol.7, 1967.

45. **Wurster, W.H.,** 'Shock Tube Spectroscopy of Ablative Species,' Proc. Sixth Int. Shock Tube Symp., Freiburg, Germany, April 1967.

46. AGARDograph No. 41 Fundamental Data Obtained from Shock Tube Experiments (Ed. A. Ferri) Pergamon Press London 1961.

47. **Krindach, N.I., Kudryavtsev, R.M., Sobolev, N.N., Tunitskii, L.N., and Faizullov, F.S.,** trans., Optics and Spectr., Vols. 14, 15 1963.

48. **Sviridov, A.G., Sobolev, N.N., Sutovski, V.M.,** J.Q.S.R.T. Vol.5, 1965.

49. **Sviridov, A.G., Sobolev, N.N., Novgorodov, M.Z.,** J.Q.S.R.T. Vol.6, 1966.

50. **Golden, S.A.,** J.Q.S.R.T., Vol. 7, 1967, p. 225.

51. **Patch, R., Shackleford, W. and Penner, S.S.** J.Q.S.R.T. Vol. 3, 1963,

52. **Golden, S.A.,** J.Q.S.R.T. Vol. 2, 1962, p. 201.

53. **Shackleford, W.L.,** J.Q.S.R.T., Vol. 5, 1965, 303.

Commentary on Chapter 1-2

K.G. Sulzmann

Within the last 20 years, the shock tube has become an indispensable tool for the determination of quantitative spectroscopic data of high temperature systems. In view of the many books and review articles on this subject, Dr. Wurster in his chapter has elected generally to demonstrate the merits of shock tube techniques for opacity studies; at the same time he has outlined some of the inherent problems encountered in shock tube spectroscopy with the help of several examples.

Although by no means exhaustive, additional references for books and general review articles on the subject are given at the end of this review (1) (2) the chosen examples can be considered as typical for the application of well known spectroscopic techniques to shock tube research. Since the emphasis of the paper is on the general usefulness of the shock tube as a tool primarily for the production of thermally excited radiating and absorbing systems for quantitative spectroscopic analysis, this reviewer feels it is worthwhile to amplify some of the general points mentioned by Dr. Wurster; at the same time, I would like to take the liberty and add a few remarks about some additional problems related to precision opacity measurements in shock tubes.

Dr. Wurster has rightly pointed out, that for a meaningful analysis of spectroscopic data in terms of transition probabilities the thermodynamic state of the system under investigation has to be known. This knowledge has to include the relevant species concentrations, the population numbers or the temperature in the system as well as the spacial distribution of these parameter along the line of sight. Evidently, thermodynamic equilibrium, including uniformity of the state parameters along the line of sight, constitutes the preferred situation for quantitative measurements because of the inherent simplifications in the data analysis under this condition. I want to stress the point, that this condition can neither be assumed *ad hoc* nor can it be inferred from the existence of steady state absorption and emission signals alone. Rather, the validity of the assumptions concerning the uniform distribution of radiators and absorbers and the existence of partial or complete equilibrium, including the existence of a meaningful temperature, requires experimental justifications.

In this connection, the combination of plate spectroscopy with photoelectric measurements, as mentioned, by Dr. Wurster, is of considerable help in the regions of the photographic spectrum. Additional measurements of excitation temperatures can also be performed without doing plate spectroscopy, and the unwary investigator may be referred to the literature on this subject, e.g., references 6 through 9 at the end of this review. Furthermore, knowledge of the kinetic mechanisms occurring in the high temperature system under study is a valuable and often necessary prerequisite for meaningful opacity measurements related to life times, particularly in systems where collision limiting can occur; here, knowledge of the kinetics will aid in the selection of experimental conditions under which a Boltzmann distribution is established by collisional excitation and de-excitation over the relevant radiating and absorbing states. As an example, I might mention the work by Skackelford on measurements of gf-values for singly ionized chromium (7) and also refer to review literature on non-equilibrium phenomena and relaxation processes in shock waves (8) not mentioned in the chapter.

Furthermore, I think it is worthwhile to stress the importance of impurity control mentioned by Dr. Wurster; at the same time I want to emphasize the equally important problem of reproducible test sample preparation for precision opacity measurements on homogeneous and heterogeneous systems. Since for each system under investigation usually different techniques are required, I might refer for a typical example to the procedure developed by R. Watson for water vapor studies (9) or to the methods developed for studies of other heterogeneous systems which are reported on during this meeting. Since prolonged flushing of the shock tube with the relevant test mixture is often the only way to ensure the presence of the wanted mixture composition, it seems possible, that many of the techniques which have been developed for the preparation of uniform flows of homogeneous and heterogeneous test mixtures for flame studies (10) may be attractive for future shock tube applications. Finally, in this connection, I want to emphasize the need for, and the critical importance of, sufficient reliability checks on the procedures used for and on the result obtained by shock tube techniques for absolute opacity measurements.

References

1. **Ladenburg, R. W., Lewis, B., Pease, R.N., Taylor, H.S.,** (eds.) 'Physical Measurements in Gas Dynamics and Combustion,' High Speed Aerodynamics and Jet Propulsion, Vol. IX, Princeton Univ. Press, 1954.

2. **Griem, H.K.,** 'Plasma Spectroscopy,' McGraw-Hill, New York, 1964.

3. **Gaydon, A.G.,** 'Temperature Measurements and Relaxation Processes in Shock Tubes,' Proc. Int. Symp. on Fundamental Phenomena in Hypersonic Flows, J. Gordon Hall, ed. Cornell Univ. Press, Ithaca, New York, 1965.

4. **Penner, S.S.,** 'Spectroscopic Methods of Temperature Measurements,' Temperature, Its Measurements and Control in Science and Technology, Vol. III, Part 1, Reinhold Publ. Corp., New York, 1962.

5. **Tourin, R.H.,** ibid., Vol. III, part 2, Ch. 43.

6. **Lochte-Holtgreven, W.,** 'Production and Measurement of High Temperatures,' Rep. Progr. Phys. 21, 1958, 312-383.

7. **Shackleford, W.L.,** 'Measurements of gf-Values for Singly Ionized Chromium using the Reflected Wave Region of a Shock Tube,' Ph. D. Thesis, Calif. Inst. of Techn. 1964, and J.Q.S.R.T. 5, 1964, 303-312.

8. **Losev, S.A., Osipov, A.I.,** 'The Study of Nonequilibrium Phenomena in Shock Waves,' Soviet Phys.-Usp. (English Transl.) 4, 1962, 525.

9. **Watson, R.,** 'Procedure for Reliable Preparation of Shock Tube Test Gas Mixtures Containing Water Vapor,' Rev. Sci. Instr. 33, 1962, 1113-4.

10. **Mavrodineanu, R., Boiteux, H.,** 'Flame Spectroscopy,' John Wiley and Sons, Inc., New York, 1965.

*

1-3

Shock Tube Studies on Two-Phase Systems

W. J. HOOKER, A. L. MORSELL and R. WATSON
Heliodyne Corporation, Van Nuys, California.

Summary

A shock tube facility is described which employs a two-phase system of solid particles and gases as the primary driven fluid. The techniques for producing and uniformly suspending the particulate matter in the driven section of the shock tube are discussed, and comparisons between calculated and measured particle vaporization times behind a shock wave are presented. A description is given of a new concept in optical instrumentation which permits high resolution, time resolved absorption spectra to be obtained for a shock tube flow, which greatly reduces the amount of experimentation required to define the reaction kinetics of a complex chemical system. The application of the two-phase shock tube to studies of hypervelocity impact phenomena is described, and experimental data is presented.

Sommaire

On décrit un tube à choc utilisé pour l'étude d'un système à deux phases, gaz et particules solides, comme fluide actif. La technique utilisée pour la production et la suspension des particules dans la section de travail du tube est analysée et on présente une comparaison des temps de vaporisation des particules en aval d'une onde de choc obtenus par mesure et par calcul. On présente un nouveau montage optique permettant une résolution très fine dans le temps pour les études de spectre d'absorption dans les tubes à choc, et qui réduit considérablement le nombre d' essais nécessaires pour définir la cinétique des réactions chimiques dans une tuyère complexe. L'application du tube à choc à deux phases aux études d'impacts hypersoniques est décrite avec présentation de données expérimentales.

Introduction

Motivation for Two-Phase Shock Tube Development

The powder injection shock tube described in the following sections was originally developed to extend shock tube technique to cover a large number of chemical systems. The chemical species that can be studied with a conventional shock tube are those that are gases or vapours at about room temperature, thus excluding a large class of compounds, including the products of decomposition of solids such as are found in heatshields and solid propellants. The two-phase shock tube overcomes this restriction.

Particle Dynamics in a Two-Phase System

The primary consideration governing the choice of solid particle size for any given set of shock tube conditions is the degree to which the particles are required to vaporize during the available testing time. If equilibrium gas phase measurements are required, the particles must vaporize in a time less than the experimental

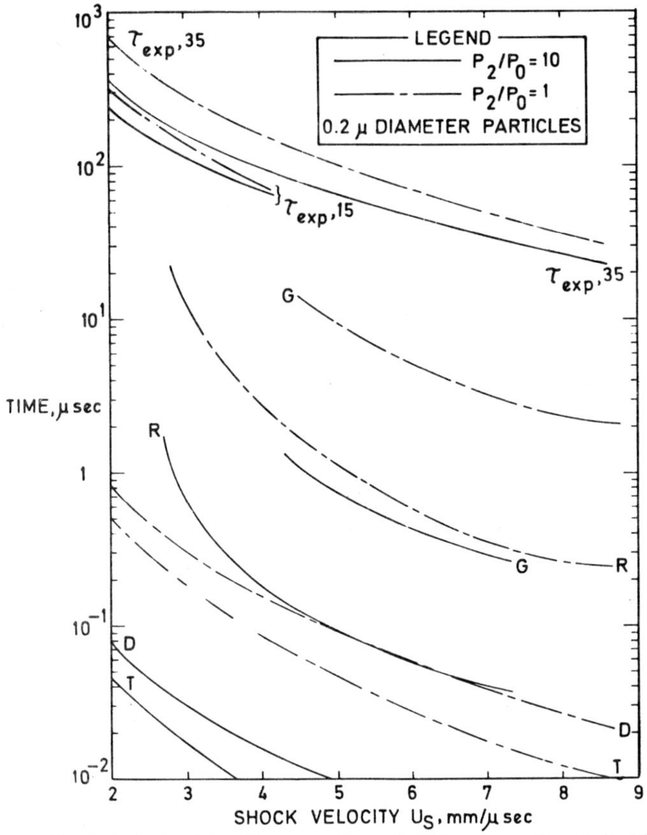

Fig. 1-3-1 Shock tube testing times and particle burn-up times in air for various materials versus shock velocity. T-Teflon; D-Delrin; R-Phenolic Refrasil; G-Graphite;

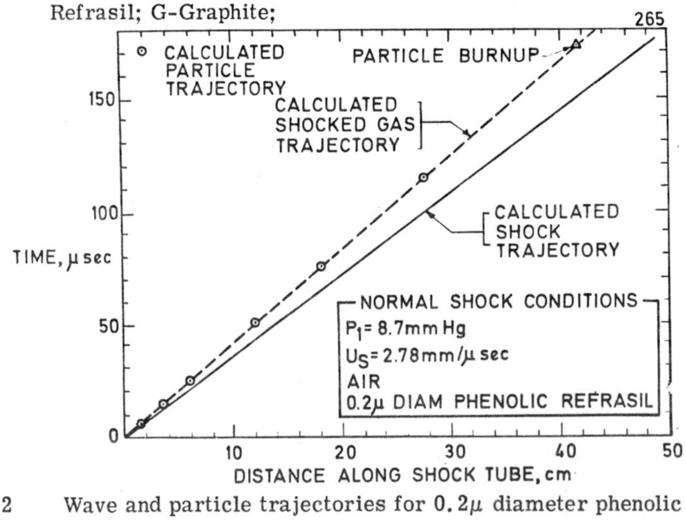

Fig. 1-3-2 Wave and particle trajectories for 0.2μ diameter phenolic refrasil powder.

testing time in order to produce a gaseous mixture near the end of the available testing time with predictable chemical composition. If gas phase reactions between the carrier gas and the particulate products of decomposition are to be studied, the particle vaporization time must be much less than the sum of the experimental testing time and the observed reaction times. Conversely, if the properties of the particulate matter in the shock-heated mixture are to be studied, the particle vaporization time must be greater than the experimental testing time.

The particle vaporization times shown in figure 1-3-1 were calculated on the assumption of free molecular energy and momentum transfer to the particles from pure, equilibrium air. The basic formulations for these calculations followed the works of Ashley (1) and Sauer (2). In all cases, a thermal accommodation coefficient of unity was assumed: this has been verified experimentally (3) for teflon. What is clearly seen from figure 1-3-1 is that all materials, even those with heats of vaporization as large as that of graphite, vaporize in times that are very short compared to available experimental testing times. Indeed, the calculated vaporization times are short compared to the large majority of observed chemical reaction times. The vaporization times for particles with different initial diameters scale linearly with size (3).

The curves of figure 1-3-2 show that submicron size particles are accelerated to the local shocked gas velocity virtually instantaneously. The momentum and energy transfer processes assumed in these calculations were the same as those used in deriving the curves of figure 1-3-1.

When the flow behind the incident shock wave becomes a continuum with respect to the particle diameter, the particle acceleration can be computed on the basis of pressure drag forces. For strong shock waves in air, the time for a non-vaporizing particle to reach 90 percent of shocked gas velocity can be read off from the curves of figure 1-3-3 (following (3)).

Powder Injection Shock Tube Facility

Mechanical Operation

The powder injection system used now is represented in figure 1-3-4. To prevent agglomeration, the powder is processed under moisture-free conditions, and the demountable powder chamber, shown on top of the injection tank, is loaded with a measured amount of powder in a dry box. The powder chamber is separated from the injection tank by a small electrically-controlled valve. Larger electrically-controlled valves separate the injection tank from the shock tube and the dummy volume from the shock tube. The shock tube dump tank is isolated from the shock tube by an aluminum foil diaphragm. In order to avoid interference with the powder injection process, the diaphragm is contoured to match the inside of the tube. The powder chamber is just large enough to contain a few grams of powder, the injection tank has a volume about equal to that of the shock tube, and the dummy volume is approximately 6 times the volume of the shock tube.

Just before injection the shock tube and dummy volume are evacuated, the injection tank is pressurized to several cm Hg, and the powder chamber is pressurized to 1,000 psig. The valve between the dummy volume and the shock tube is open; the other valves are closed. Upon initiation of an electronically-timed automatic firing sequence, the powder chamber valve is opened and the powder is blown violently into the injection tank. About one second later, long enough to achieve reasonable mixing of the powder with the gas in the tank, the valve connecting the injection tank with the shock tube is opened and the contents of the injection tank rush into the driven section of the shock tube and into the dummy volume tank. The injection tank valve and the dummy volume valve are closed immediately after injection; and as

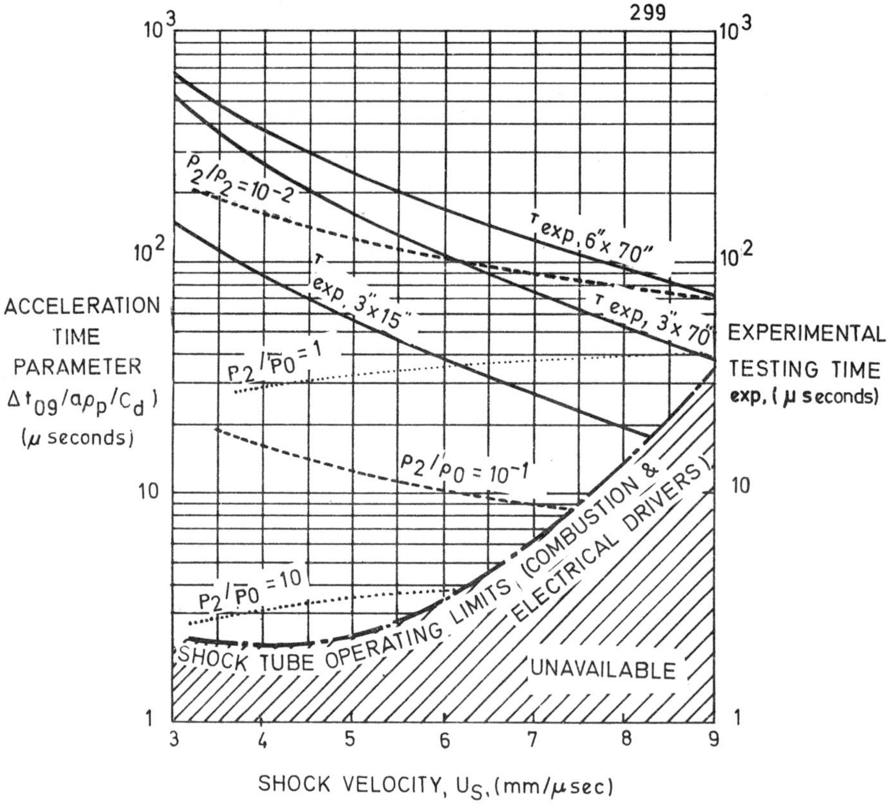

Fig. 1-3-3 Particle Acceleration and Shock Tube Testing Times versus Shock Velocity.

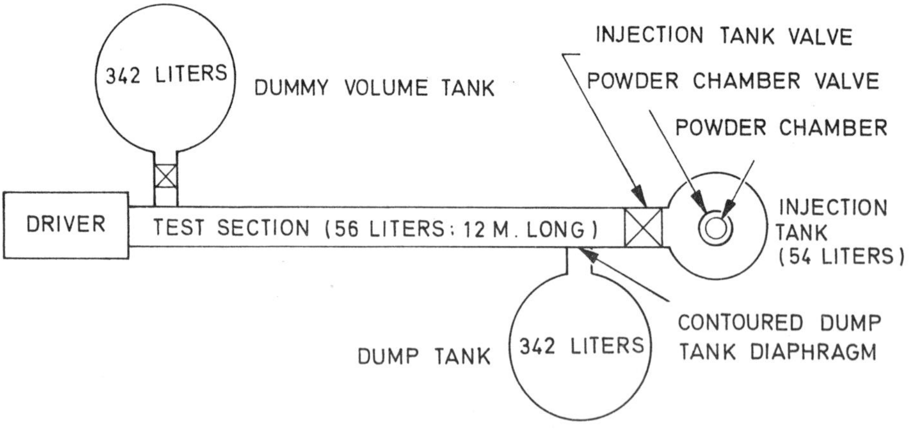

Fig. 1-3-4 Schematic diagram of the powder injection shock tube.

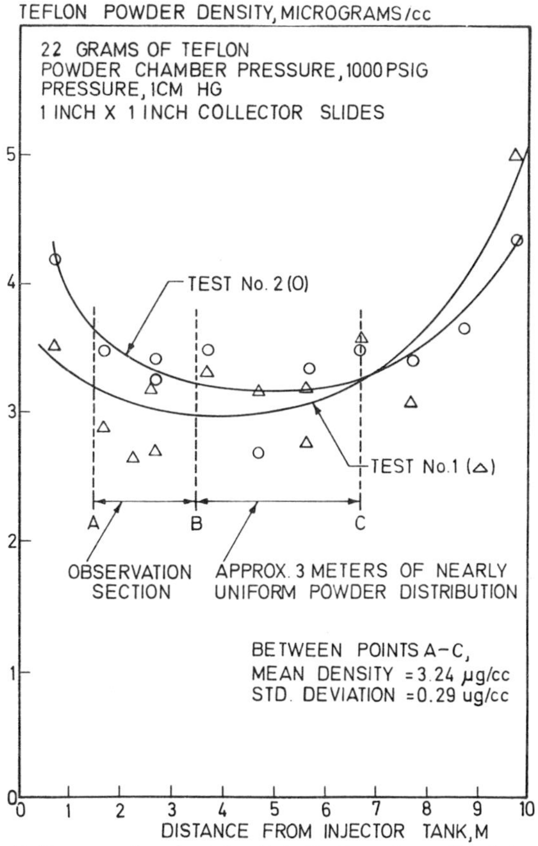

Fig. 1-3-5 Teflon powder density distribution along the driver section, obtained by direct weight measurements.

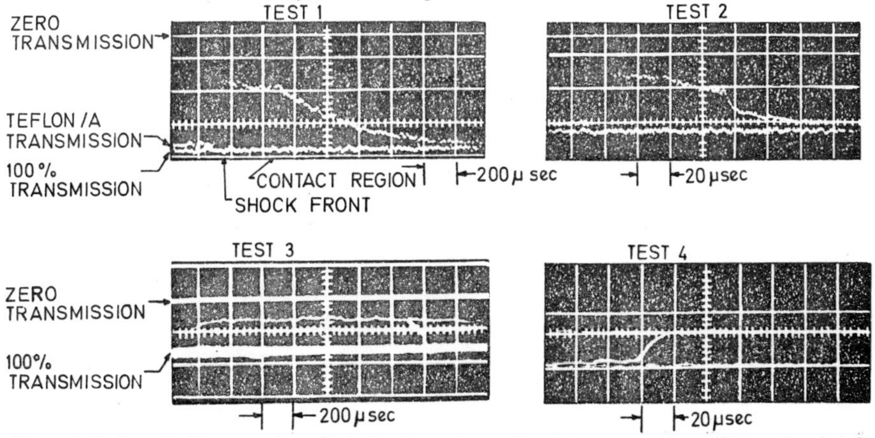

Fig. 1-3-6 Teflon powder distribution along the driven section of the shock tube as determined by CF_2 absorption at 2536A in argon in four shock tube tests. The gain settings are the same for all four records; tests 1 and 2 are at slow sweep speeds while tests 3 and 4 are fast.

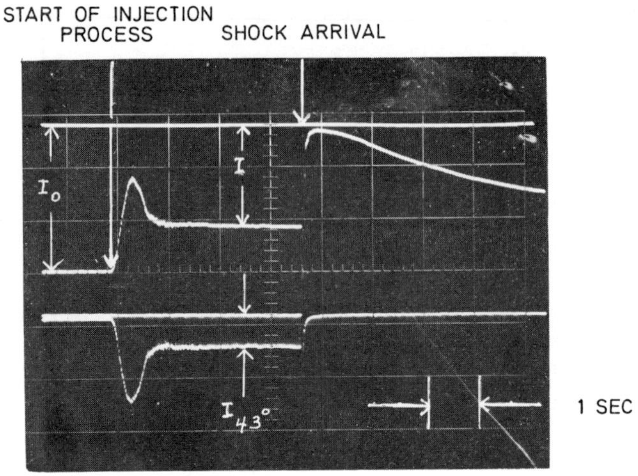

Fig. 1-3-7 Transmitted and scattered light signals from a suspension of carbon particles in air.

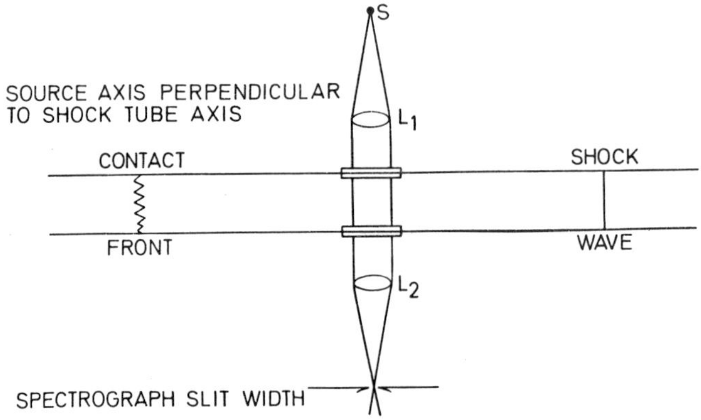

Fig. 1-3-8 Schematic of the optical system for a conventional spectrographic absorption measurement in a shock tube.

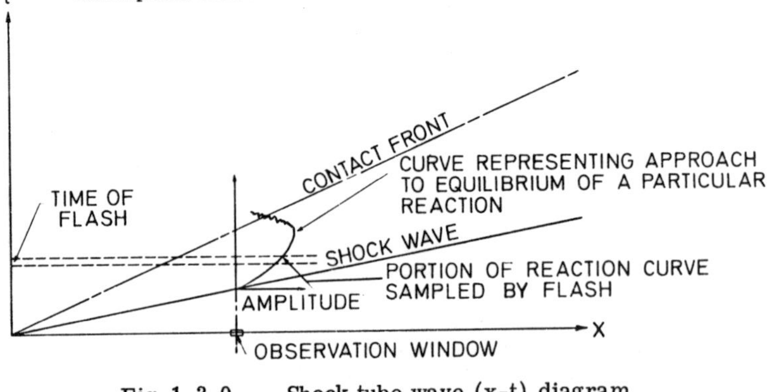

Fig. 1-3-9 Shock tube wave (x-t) diagram

soon as they are closed, the shock tube is fired.

Powder Distribution by Direct Weighing

The amount of powder suspended in the shock tube and the uniformity of the distribution of the powder have been measured by several means, the most direct of which are weight measurements of the quantity of powder that settles on small squares of stainless steel foil arranged at one-meter intervals along the inside of the shock tube. Because the weight differences are very small, and because absorption of water from the atmosphere can cause large weighing errors, one of the shock tubes has been coupled to a dry box and the measurements are made without ever exposing the powder samples to the laboratory room atmosphere. Figure 1-3-5 shows the results of two such tests.

The lines on the figure at 1.5 and 6.5 meters from the injector bound a region for which a nearly uniform powder distribution is desirable. In this region the standard deviation of the weighings from the mean of 3.24 $\mu g/cm^3$ is 0.29 $\mu g/cm^3$. Powder to the left of this region is never seen at an observation point and powder on the right gets turbulently mixed with driven gas at the contact front.

Powder Distribution by Gas Phase Optical Absorption

Another technique for measuring the powder distribution makes use of the fact that all the powder/gas mixture upstream of an observation station in a shock tube is swept by the station in a shock-heated condition when the shock tube is fired. If the powder decomposes rapidly to form a stable gas-phase compound, a measurement of spectral absorption from this compound as a function of time will give a measure of the original powder distribution along the length of the tube. This measurement has been made on the teflon powder, which decomposes to the relatively stable compound CF_2(4) when heated in an inert argon diluent. Figure 1-3-6 shows oscilloscope traces of CF_2 absorption at 2536 Å for four shock tube tests in which all parameters were maintained constant in order to deduce reproducibility of injection; both slow and fast sweep traces are shown to indicate the overall duration and uniformity of absorption. The 10 microsecond rise time on the fast sweep traces is the time required for the shock wave to traverse the observation port. The initial driven argon pressure for these tests was maintained at 1 cm Hg, and the injected teflon powder was maintained at approximately 4 mole percent. In each case, zero and one-hundred percent transmission lines were put onto the film shortly before the teflon powder/argon mixture was injected. The transmission through the teflon/argon mixture prior to shock arrival does not coincide with the one-hundred percent transmission line because of the loss of light by scattering from the teflon powder initially suspended in the test section. It is seen from figure 1-3-6 that the powder concentration is quite uniform in the useful test gas length upstream of the observation port.

Powder Distribution by Light Scattering

A third technique for measuring injected powder uniformity is the light scattering method. Monitoring scattered light at a given station along the shock tube driven section provides a measure of injection reproducibility, as well as the settling rate of powder at that station. In (5), Deirmendjian shows by numerical computation that the intensity of radiation scattered approximately 40° from the forward direction is a direct measure of the total number of particles in the scattering volume, independent of their size distribution. We have made use of this phenomenon in the powder injection shock tube experiments, and an oscilloscope record of such a measurement is shown in figure 1-3-7.

In figure 1-3-7 time increases from left to right. The top trace corresponds to the

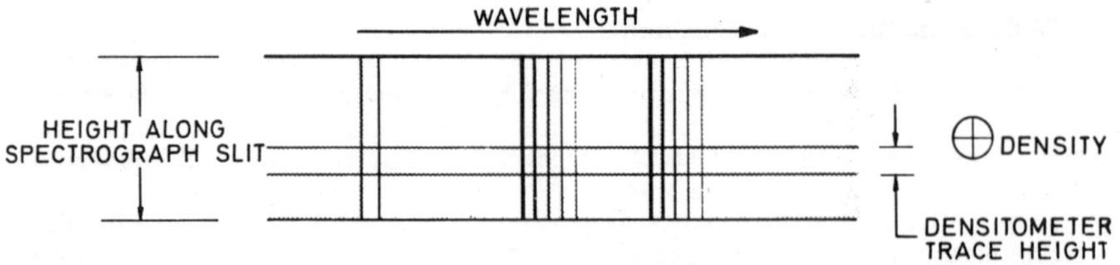

Fig. 1-3-10 Schematic of the plate record for a conventional spectrographic absorption measurement in a shock tube.

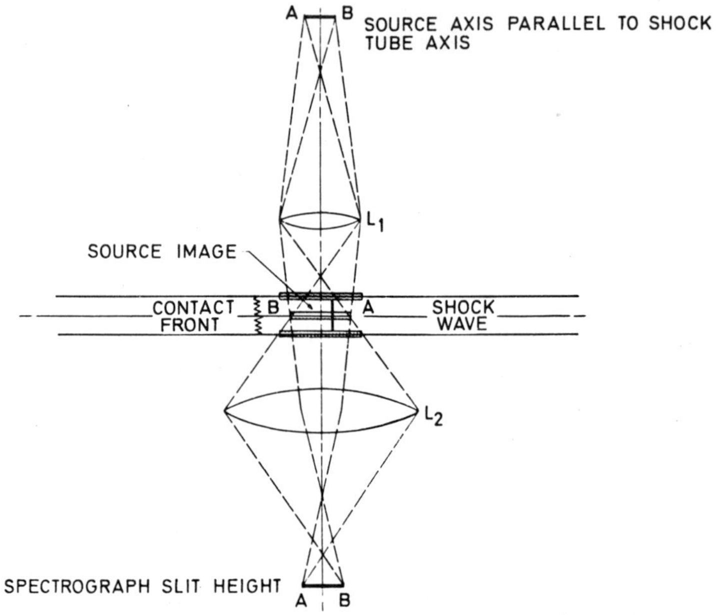

Fig. 1-3-11 Schematic of the optical system for the TRASE (Time Resolved Absorption Spectra Experiment) spectrographic absorption measurement in a shock tube.

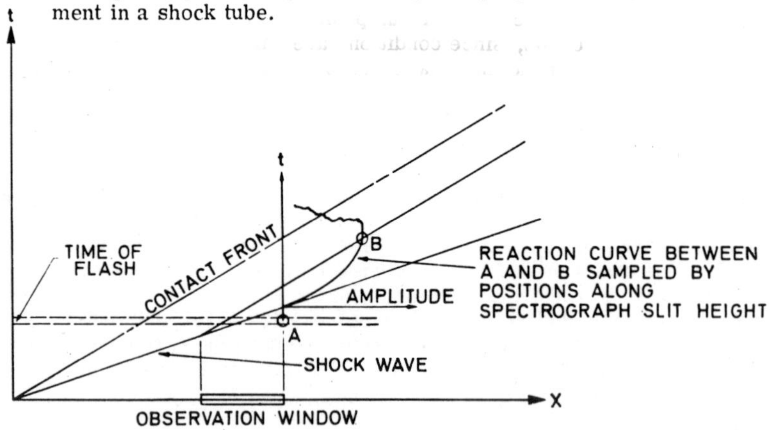

Fig. 1-3-12 Shock Tube wave (x-t) diagram for the TRASE system.

intensity of light transmitted across a diameter of the shock tube, while the bottom trace corresponds to the intensity of light scattered at 43° from the forward direction. I_0 is the intensity of the transmitted light in vacuo, and I is the net intensity transmitted after the powder has been injected; $I_{43°}$ is the intensity scattered at 43°. After an initial transient of approximately one second, associated with the opening and closing of the injector valves, the I and $I_{43°}$ intensities are seen to be quite steady for several seconds, indicating that a quiescent powder/gas dispersion of uniform distribution has been achieved. The sharp change in signals at the time indicated by 'shock arrival' is due to water vapor behind the contact front from the combustion driver. On the 1 sec/cm time scale shown, the actual experimental testing time is immeasureably small, being on the order of 100 μ s.

Summarizing the material of this section, it is seen that the techniques required for producing a uniform and reproducible powder/gas dispersion in the driven section of a shock tube have been developed satisfactorily.

Trase

(Time Resolved Absorption Spectra Experiment)

The optical technique discussed in this section was developed to make it possible to obtain reaction rate data on complex chemical systems in a shock tube. It is similar to a technique used in flame studies.

A schematic diagram presented in figure 1-3-8 shows the optical arrangement for a conventional flash absorption apparatus on a shock tube. The wave diagram of figure 1-3-9 shows that the plate absorption record obtained defines the state of the gas at a pre-selected position behind the incident shock wave. If the degree of completion of any particular chemical reaction in the system under study is measured as shown in figure 1-3-9, then the complete reaction curve can be obtained only by a series of reproducible experiments with varying exposure delay times. This, in general, is a prohibitive procedure due to the large number of test runs required, and the difficulties involved in providing reproducible conditions. In fact, this procedure is generally only useful for obtaining absorption spectra for systems in chemical equilibrium.

Shown in figure 1-3-10 is a schematic diagram of a spectroscopic plate record obtained using the optical arrangement illustrated in figure 1-3-8. In a typical plate absorption spectrum, the height of the densitometer slit which scans the plate to produce a density/wavelength plot, is small compared to the actual spectrograph slit image height. Since traces at various positions along the image height produce the same absorption records, since conditions are uniform across a cross-section of a shock tube, the bulk of the information recorded is redundant and, hence, wasted. The TRASE (Time Resolved Absorption Spectra Experiment) apparatus was designed to make better use of the information handling capability of a spectrograph plate record.

The optical arrangement of the TRASE apparatus is shown in figure 1-3-11, and a wave diagram for this experiment is shown in figure 1-3-12. The feature central to the operation of the TRASE equipment is the direct conjugation between positions along the shock tube axis and positions along the spectrograph slit height. This produces a one-to-one relationship between positions along the spectrograph plate record obtained and distance behind the shock wave in the shock tube. Since position behind the shock wave is directly proportional to time, the TRASE spectrograph plate is a record of absorption strength versus wavelength and chemical reaction time behind the shock wave.

A schematic diagram of a TRASE spectrograph plate is shown in figure 1-3-13.

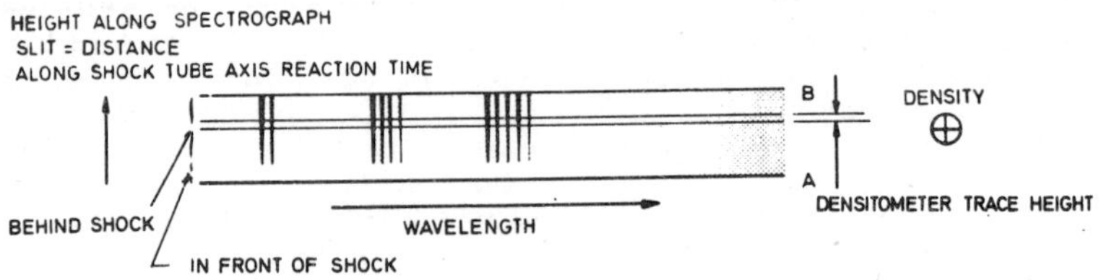

Fig. 1-3-13 Schematic of the plate record for the TRASE spectrographic absorption measurement in a shock tube.

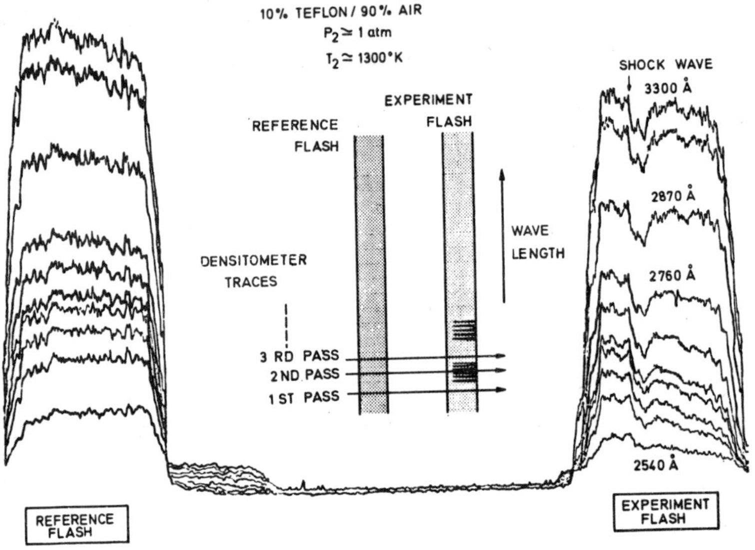

Fig. 1-3-14 Densitometer traces of a TRASE plate record obtained with a teflon/air mixture in the two-phase shock tube.

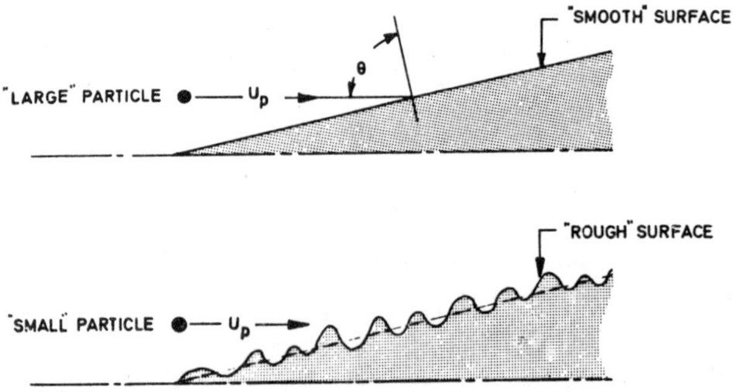

Fig. 1-3-15 Impact of a particle onto a "smooth" and a "rough" surface at a velocity U_p and angle of incidence, θ.

Densitometry of the portions of the record corresponding to position in front of the shock wave produces a record of incident flash source intensity (I_0) versus wavelength. The ratio of the net transmitted intensity (I) to I_0 at any wavelength vs time produces a record of the concentration of absorbers responsible for the observed spectral absorption, i.e., the reaction rate curve. Densitometry at all wavelengths corresponding to known spectral transitions produces a comprehensive reaction rate profile for the chemical system studied.

The densitometer record from an experiment involving the vaporization of teflon powder in air is presented in figure 1-3-14. As shown in the inset to figure 1-3-14, the densitometer traces were made at fixed wavelength intervals for both the Reference Flash, (obtained with a vacuum in the shock tube prior to firing) and the Experiment Flash. The wavelength region presented in figure 1-3-14 corresponds to the interval over which there is measurable absorption from the CF_2 radical (4) (6). The principal features of this record are the following:

(a) at each wavelength the plate density of the Reference Flash is greater than the Experiment Flash due both to the scattering of light by the teflon powder, and to gas phase molecular absorption.

(b) the dip in the density records at all wavelengths in the Experiment Flash just behind the position indicated by 'Shock Wave' is due to the increased scattering of light by the teflon powder that is compressed behind the shock wave; the total vaporization time of the teflon particles can be derived from the length, (time interval) of the dip;

(c) the slope of the density traces behind the shock wave in the approximate spectral interval of 2540 to 2760 Å indicates the rate of disappearance of the CF_2 radical due to oxidation; if the CF_2 oxidation reactions had been sufficiently fast to have achieved completion within the physical distance behind the shock wave viewed in this measurement, the overall rate history of CF_2 disappearance would have been determined;

(d) the plate density behind the shock wave, i.e., the net optical transmission, increases at longer wavelengths where CF_2 absorption is no longer important, and eventually surpasses the density in front of the shock wave;

(e) the monotonic increase in plate density with wavelength is due to the combination of spectrograph plate response (Eastman 103-0 plate) and flash source radiant energy distribution (Suntron X-6 Xe lamp with 6/e emission width of approximately 6 microseconds).

The plate record of figure 1-3-14 is only of qualitative usefulness insofar as specific reaction rate information is concerned. However a large aperture system currently under design at our laboratory, combined with a shorter duration flash source, also under development at our laboratory, will provide a powerful new tool for complex reaction rate studies.

Two-Phase Shock Tube Hypervelocity Impact Experiments

In the very extensive literature (7 through 15) on hypervelocity impact phenomena, it is demonstrated that good correlation is obtained between the mass of target crater removed and the normal component of kinetic energy of the impacting projectile; the cosine-squared, or Newtonian, approximation. The proportionality constant is dependent on the target and projectile materials, but has been found to be independent of projectile size for projectile diameters ranging from several inches down to 1/64 inch (400 microns). Performing hypervelocity impact tests with projectiles in the micron-size range would thus provide an important extension

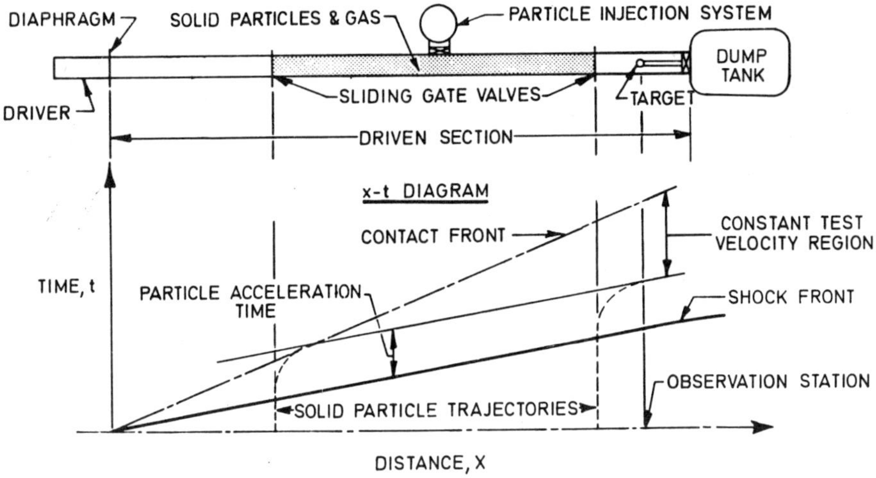

Fig. 1-3-16 Schematic of particle injection shock tube, and associated distance-time (x-t) diagram.

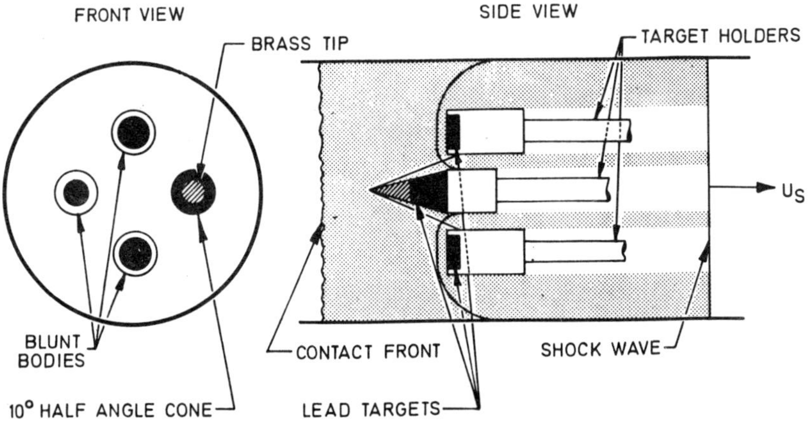

Fig. 1-3-17 Schematic of the targets used in the shock tube particle impact tests.

Fig. 1-3-18 Target used in the shock tube impact tests.

of the available literature data.

Further consideration at this laboratory of the mechanics of interaction of micron-size projectiles with material surfaces revealed that the scale of the surface roughness of the target, when compared to the projectile dimension, could be important. This is illustrated schematically in figure 1-3-15. All of the hypervelocity impact data reported in the open literature were obtained under test conditions similar to those shown in figure 1-3-15 (a). Under these circumstances, the mean target surface inclination is clearly defined, and a meaningful impact angle θ, is obtained. However when the dimensions of the impacting projectile are reduced to the micron-size range, test conditions similar to those shown in figure 1-3-15 (b) can be expected. In this latter situation the impact angle is not uniquely determined, and the Newtonian scaling relationship can be expected to become invalid. Thus the ratio of surface roughness to projectile dimension can be expected to be an important scaling parameter.

Shown in figure 1-3-16 is a schematic diagram of a powder injection shock tube facility for performing hypervelocity impact tests. The shock tube used for the tests to be described later in this chapter did not have the sliding gate valves shown in figure 1-3-16: instead, the entire driven section of the shock tube was filled with the injected particles. A schematic diagram of the target rake used for these studies is shown in figure 1-3-17. The three blunt targets and the conical target were arranged so that the bow shock waves generated by the supersonic flow behind the primary shock wave did not interfere with one another. The tip of the conical target was of a material different from that of the remainder of the target so as to eliminate any effects of near normal impact near the nose of the target. The technique used for determining hypervelocity impact damage in these experiments was to measure the total weight loss from the impacts. A view of the actual targets used in these experiments is shown in figure 1-3-18.

A preliminary series of shock tube measurements has been made by us to determine the angular dependence of mass removal by hypervelocity impact, and the influence of surface roughness on this mass removal. The particles used were 5 micron diameter Alundum. The shock tube conditions chosen for these experiments were an initial pressure of 1 cm Hg and a shock velocity of approximately 4 mm/μsec. Prior to each experiment, the four lead targets were carefully weighed and then mounted in the holder. After the experiment, the lead models were re-weighed, and the mass difference noted. Since it was not possible, in this initial series of measurements, to determine the actual number density of material particles in the shocked flow, it was decided to compute the mass loss from an impact on the conical model by normalizing it to the mass loss on the blunt models, as is shown below.

If, as is shown in figure 1-3-17 we let:

L = the length of the column of particles between the shock front and the contact front,

n_p = the particle number density, and

A_p = the projected, or swept, area of the target,

then the total number of particles, N_T, striking the target during the testing time is given by:

$$N_T = A_p Ln_p \qquad \text{(Eq. 1-3-1)}$$

Therefore, the mass removed, M_R, per impact is given by:

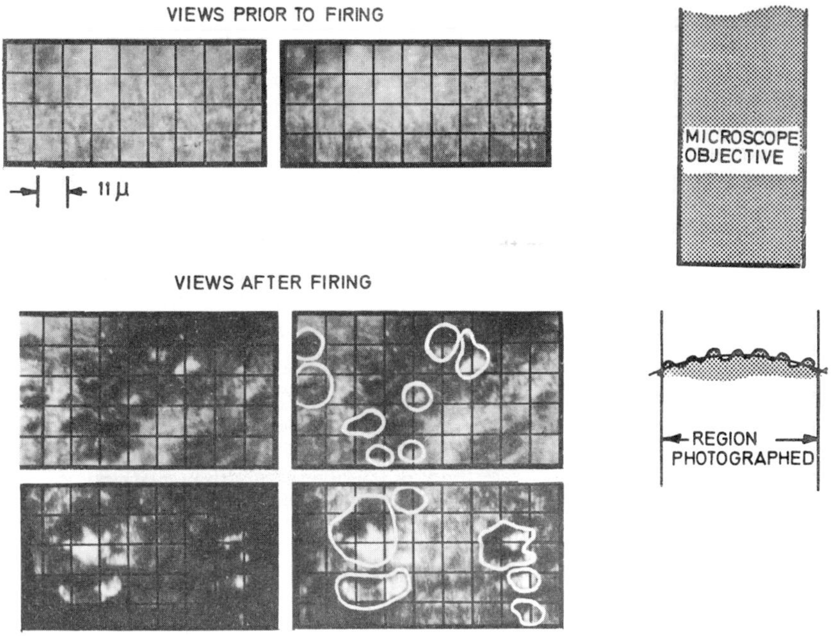

Fig. 1-3-19 Photomicrographs of the conical lead models used in the particle impact shock tube tests.

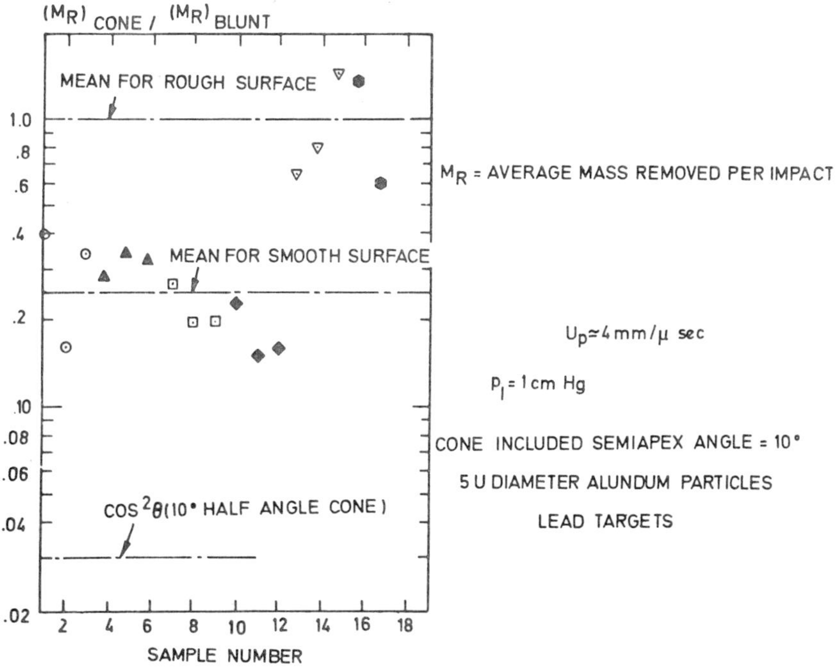

Fig. 1-3-20 Ratio of the mass removed per conical impact, $(M_R)_{cone}$, to the mass removed per blunt impact, $(M_R)_{blunt}$, versus sample number.

$$M_R = \frac{M_T}{N_T} = \frac{M_T}{A_p\, L\, n_p}\,, \qquad \text{(Eq. 1-3-2)}$$

where M_T is the total mass removed from the target.

During any one shock tube experiment, L and n_p are fixed, and the mass removed per impact on the conical surface divided by the similar quantity for the blunt surface is given by the ratio:

$$\frac{(M_R)_{cone}}{(M_R)_{blunt}} = \frac{(M_T)_{cone}}{(M_T)_{blunt}} \times \frac{(A_p)_{blunt}}{(A_p)_{cone}} \qquad \text{(Eq. 1-3-3)}$$

Using the data reduction method of equation 1-3-3, three independent data points were obtained for each experiment. One series of experiments was made with models whose surfaces were relatively smooth when compared to the dimensions of the particles used,the conical models were cast in a polished mold while the blunt models were trimmed with a razor blade. Photomicrographs of two areas of the models used in this series are shown at the top of figure 1-3-19. Another series of experiments was run, using models that had previously been exposed to the high velocity particle flow in the shock tube. The surfaces of these models were relatively rough, or 'sand-blasted', by comparison with the particle dimensions, as is shown by the two "fuzzy" photographs in the lower left of Figure 1-3-19. At the right of these photographs the same photographs are repeated, but with the edges of the craters outlined so as to improve the visualization of the result. The photographs are fuzzy because the region photographed is slightly out of focus due to the curvature of the surface being viewed.

In figure 1-3-20 we show the results of these two series of experiments. Plotted is the mass removed per conical impact normalized by the mass removed by a blunt impact, versus the individual sample number. It is seen that the data group about two points. Those corresponding to the previously mentioned smooth surface fall at a ratio of approximately 0.25, while those corresponding to the rough surface center about unity. For comparison purposes the fraction predicted based on the cosine-squared Newtonian law, namely 0.03, is shown.

What is immediately evident from figure 1-3-20 is that under no conditions is the mass loss from a conical impact within an order of magnitude as small as that predicted based on the cosine-squared law and, indeed, for the roughened surface the mass loss from a conical impact is virtually the same as that from a normal impact. This is a direct result of the fact that when the surface roughness dimensions are comparable to, or greater than, that of the dimensions of the impacting particles, every impact is nearly a normal incidence impact. In fact, an argument can be made that, if small proturberances are merely sheared by an impacting particle, rather than vaporized, the fraction shown on the ordinate of figure 1-3-20 can exceed unity.

References

1. **Ashley, H.,** 'Applications of the Theory of Free-Molecule Flow to Aeronautics', J. Aero. Sci. II, 95, 1949.

2. **Sauer, F.M.,** 'Convective Heat Transfer From Spheres in a Free-Molecule Flow', J. Aero. Sci. 18, 353. 1951.

* 3. **Hooker, W.J., Watson, R., Morsell, A.L.,** 'Measurements with Powdered Solids in Shock Tubes: I. Vaporization Rates of Small Particles;

II. Hypervelocity Impact Tests with Large Particles', Paper No. F4 pres. at the American Physical Soc. Sixth Int. Shock Tube Symp., Freiburg, Germany, April 1967.

4. **Modica, A.P., LaGraff, J.E.,** 'Decomposition and Oxidation of C_2F_4 Behind Shock Waves', J. Chem. Phys. 43, 3383, 1965; **Modica, A.P.,** 'Kinetics and Equilibria of the Difluorocarbene Radical Decomposition Behind Shock Waves', J. Chem. Phys. 44, 1585, 1966.

5. **Deirmendjian, D.,** 'Scattering and Polarization Properties of Water Clouds and Hazes in the Visible and Infrared', Appl. Optics 3, 187, 1964.

6. **Hooker, W.J., Sellers, R.P. Jr.,** 'The Absorption Spectra of Shock-Heated Teflon/Argon and Teflon/Nitrogen Mixtures', Heliodyne Corp., RN 21, June 1965.

7. 'Study of the Phenomena of Hypervelocity Impact', Summary Rep. GMDRL: TR63-216, June 1963.

8. **Posever, F.C.,** *et al* 'Investigation of Structural Implications of Meteroid Impact,' Tech. Doc. Rep. April 1962 - June 1964 (NAA) July 1964.

9. **Hermann, W., Jones, A.H.,** 'Survey of Hypervelocity Impact Information,' ASRL Rep. No. 99-1, MIT. 1961.

10. **Charters, A.C., Locke, G.S.J.,** 'A Preliminary Investigation of High-Speed Impacts: The Penetration of Small Spheres into Thick Copper Targets', NACA RM A-58 B-26, 1958.

11. **Summers, J.L.,** 'Investigation of High Speed Impact: Regions of Impact and Impact at Oblique Angles, NASA TN D-94. 1959.

12. **Whipple, F.L.,** 'Meteoritic Phenomena and Meteorites, Physics and Medicine of the Upper Atmosphere'. Ed. White, C.S., Benson, O.O., Univ. of New Mexico Press. 1952.

13. **Bjork, R.L.,** 'Effects of a Meteoroid Impact on Steel and Aluminum in Space.' P-1662, Rand Corp. Dec. 1958.

14. **Bjork, R.L.,** 'Meteroid vs Space Vehicles', ARS J 31, 803. 1961.

15. **Lieblein, S., Clough, N., McMillan, A.R.,** Hypervelocity Impact Damage Characteristics in Armoured Space Radiator Tubes', (GMDRL), 1 Sept. 1964.

Acknowledgement

The work reported herein was supported in part by the Advanced Research Projects Agency, whose help we gratefully acknowledge.

List of Symbols

a	particle radius, microns
A_p	target projected area, cm^2
C_d	particle drag coefficient
D	delrin
G	graphite
I	transmitted light intensity, w - cm^{-2}
I_0	incident light intensity, w - cm^{-2}
$I_{43°}$	scattered light intensity at 43° from the forward direction, w - cm^2
L	separation between the shock and contact front, cm
M_R	target mass removed per impact, g
M_T	total target mass removal, g
n_p	particle number density behind the shock wave, cm^{-3}
N_T	total number of particle impacts
P_0	reference pressure = 1 atm
P_1	pressure in front of the shock wave, mm Hg
P_2	pressure behind the incident shock wave, atm
R	phenolic refrasil
T	teflon
t	time, μ sec
U_p	particle velocity, mm - μsec^{-1}
U_s	shock velocity, mm - μsec^{-1}
x	distance along the shock tube axis, cm
$\Delta t_{0.9}$	time for a particle to be accelerated to 90% of shocked gas velocity, μ sec
μ	micron = 10^{-4} cm
μg	microgram = 10^{-6} g
μsec	microsecond = 10^{-6} second
ρ_o	standard air density at STP, $1.29 \times 10^{-3} g\text{-}cm^{-3}$
ρ_2	gas density behind the shock wave, $g\text{-}cm^{-3}$
ρ_p	particle bulk density, $g\text{-}cm^{-3}$
θ	angle of incidence between an impacting particle and the surface normal, degrees
$\tau_{exp,\ 15}$	experimental shock tube testing times at distances of 15 ft. and 35 ft.,
$\tau_{exp,\ 35}$	respectively, from the diaphragm (similarly for other distances).

Subscripts

blunt	quantity for a blunt (flat) surface
cone	quantity for a conical surface

Commentary on Chapter 1-3

J. Valensi

Les auteurs doivent être félicités du souci qu'ils ont eu de contrôler par des méthodes différentes les conditions des essais et, en particulier, l'homogénéité des suspensions. Ils doivent être aussi félicités pour le souci qu'ils ont eu d'adapter pour les tubes à choc, de nouvelles techniques de mesure.

Je ne retiendrai de cet excellent mémoire que quelques points:

Suspension des particules solides dans un gaz porteur

La méthode de production de suspensions dans le gaz porteur décrite par les auteurs est une méthode essentiellement discontinue. On peut la rapprocher des méthodes déjà suggérées par d'autres auteurs, par exemple par Lau en 1964, pour ensemencer un gaz afin d'obtenir une valeur déterminée de la conductivité électrique. Ces dernières méthodes cependant, se trouvent restreintes au cas de substances à l'état de vapeur à la température ambiante, alors que la méthode qui vient d'être décrite possède le très grand mérite de s'appliquer à une classe importante de matériaux tels que ceux qui entrent dans la composition des boucliers thermiques des capsules spatiales ou dans celle des poudres solides dont la vaporisation se produit seulement à haute température.

Dans la méthode de Lau que je décrirai brièvement, un jet de gaz est projeté contre la surface libre d'une substance liquide, de tension de vapeur notable à la température ambiante. Une certaine quantité de vapeur est ainsi entraînée avec le gaz dans une chambre comportant un orifice fermé par une vanne à ouverture rapide. Lorsque la quantité de vapeur nécessaire a été introduite dans la chambre, la vanne est ouverte et le brouillard se répand dans le tube à choc, à la pression voulue. Or, les mesures ont montré que la conductivité du gaz ainsi ensemencé dépendait largement de l'intervalle de temps séparant l'ouverture de la vanne de la rupture de la membrane du tube à choc. Le phénomène doit être attribué au dépôt des particules sur les parois du tube par gravité.

Par opposition, Louis, chez A.V.C.O., utilise un flux continu de gaz ensemencé. Il s'agit d'un courant d'argon froid qui lèche d'abord la surface libre d'une certaine masse de métal alcalin à l'état liquide, placée dans un creuset, puis qui traverse le tube d'essais, le débit étant entretenu par une pompe à palettes.

Il semble que cette méthode permette d'atteindre une très bonne homogénéité de la suspension car, à aucun instant, le flux gazeux ne se trouve interrompu et la vitesse de chute des particules microscopiques de métal condensées dans le courant d'argon froid est beaucoup plus faible que la vitesse de ce courant.

Or, dans le mémoire, l'intervalle de temps séparant l'instant où la vanne distributrice de poudre est ouverte de celui où la membrane est rompue n'est pas précisé, et, les résultats des mesures de distribution de densité effectuées par les auteurs, par pesées directes, présentent une dispersion notable.

Dans ces conditions, je voudrais demander au Dr. Hooker s'il pense que sa méthode pourrait être encore perfectionnée et si, en particulier, il pourrait envisager une méthode continue, pour le remplissage du tube d'essais.

Temps d'évaporation des particules

En raison de leur masse, un certain temps est nécessaire pour que les particules soient mises en vitesse à l'aval du choc, tandis qu'elles subissent l'évaporation. Le

Dr. Hooker a montré que le temps d'évaporation pouvait être très court par rapport à la durée utile de la rafale. Par conséquent, une colonne homogène peut être obtenue à faible distance de l'aval du choc, étant admis que la suspension est uniforme. Cependant, si des études de cinétique chimique doivent être réalisées derrière le choc primaire, il y a lieu de prendre en considération la cinétique du processus d'évaporation, tandis que la vitesse des particules croît en tendant à devenir égale à celle du gaz. Les atomes et les molécules qui s'évaporent de la surface des particules peuvent, par le jeu des collisions, atteindre des températures supérieures à la température de translation du gaz. D'autre part, la concentration des substances chimiques au voisinage immédiat des particules peut être beaucoup plus élevée que ne l'indique la pression partielle de la vapeur dans le tube d'essais.

Pour ces deux raisons, la cinétique du processus d'évaporation peut être une cause de non équilibre. Il ne faut d'ailleurs pas perdre de vue que la durée utile de la rafale peut être très courte, en particulier lorsqu'on emploie des chocs très intenses.

Conditions des essais

Des informations précises, qui manquent dans le mémoire, sur les conditions des essais seraient très utiles: nombre de Mach du choc, pressions initiales, nature du gaz mené, mode de fonctionnement du tube à choc. Il ne faut d'ailleurs pas perdre de vue que la température du gaz, aussi bien que sa densité, sont déterminées par les conditions initiales et par le nombre de Mach du choc, donc par le rapport de la pression initiale du gaz menant à la pression initiale du gaz mené.

Indépendamment donc de tout effet de gaz réel, on voit que certains groupes de valeurs de la pression, de la température et de la densité volumique derrière le choc, peuvent être impossibles à réaliser simultanément. Il serait donc intéressant de trouver dans le mémoire une discussion sur les conditions optimales que l'on peut réaliser en vue d'une étude déterminée.

Si, maintenant, on prend en considération les effets de gaz réels, on voit immédiatement que, si l'on fait des études de cinétique chimique, l'état du gaz mené peut jouer un rôle important, suivant qu'il est dissocié ou non. Si le gaz utilisé est facilement ionisable, comme l'argon par exemple, alors les pressions et températures requises pour l'étude peuvent être telles que ce gaz soit fortement ionisé.

Simulation d'impacts aux hypervitesses

Le mémoire fait état d'un résultat nouveau et très intéressant, qui concerne les cas où les particules sont de dimensions comparables à celles des aspérités à la surface du modèle.

Cependant, il faut remarquer que le tube à choc ne saurait, en aucun cas, permettre la simulation des impacts de météorites dans l'espace. En effet, la pression dans le tube à choc est relativement élevée, or le comportement mécanique d'un métal dans le vide de l'espace diffère beaucoup de celui sous pression, même faible. D'autre part, en tube à choc, une onde de choc se forme devant l'obstacle puisqu'il y a écoulement gazeux; une telle onde n'existe évidemment pas dans le vide spatial.

Enfin, la vitesse des particules que l'on peut obtenir dans le tube à choc, qui est au plus égale à la vitesse de l'écoulement gazeux, ne dépassera guère 3 km/s alors que la vitesse relative des météorites est de l'ordre de 30 km/s.

Reply to Commentary

W.J. Hooker

1. The elapsed time between the opening of the injection tank valve, and the initiation of shock propagation, is typically several seconds. This is shown for one set of tests in figure 1-3-7 of the chapter. Unfortunately the time scale viz., 1 sec/cm, was left off, although the time is quoted in the text.

2. The dispersion in the measured weight distribution shown in figure 1-3-5 corresponds to a standard deviation of approximately 10 percent, which is actually quite good for the microgram quantities of powder involved. The gas phase absorption measurements of figure 1-3-6 and the light scattering measurements of figure 1-3-7, indicate that the dispersion is generally less than indicated by the direct weighings.

3. We have considered several continuous flow powder injector devices, but have not devoted sufficient effort to produce a working model. However, in one version of the injection tank, a small fan was installed to provide continuous agitation, and it was found that the injected powder tended to quickly collect along the periphery of the tank. In the light of this, we have emphasized the 'batch' approach.

4. We have used a light gas driver (H_2, He), and a combustion driver ($H_2 + O_2$ + He) for the powder shock tube. The bulk of our experimentation has been with several mass percent of powder in air and argon at initial pressures ranging from 0.5 to 10 cm Hg; 1 mm Hg initial pressure is the lowest at which reasonable powder suspension has been achieved. Shock velocities in the range of 1.5 to 5 mm/μ sec are regularly generated.

In this connection, it should be noted that the temperature-density operating map attainable by a conventional shock tube is the same for the powder shock tube, with modification for the powder heat of vaporization.

In closing, much more development work could be devoted to the improvement of the powder shock tube technique. However, the present state of development is sufficient to perform meaningful experiments, unattainable by conventional shock tube procedure.

1-4

Combustion Research in a Shock Tunnel

J. SWITHENBANK and R. J. PARSONS
University of Sheffield, UK

Summary

The flight corridor for scramjet operation extends to about Mach 20 at altitudes up to 50,000 metres. Simulation of the combustion chamber conditions for research and development therefore requires air at extremely high enthalpy and pressure. The tailored interface shock tunnel is an economic test facility which can attain the required conditions, despite the short testing time. The supersonic combustion test chamber may be directly connected to the shock tunnel nozzle. Starting of the flow in the test section is accompanied by a transient shock whose strength may be reduced by pre-evacuation. Fuel (usually hydrogen) is injected into the test section and the mixing, reaction and aerodynamic processes are investigated.

In addition to conventional pressure and shock speed instrumentation, the feasibility of the measurement of gas velocity by electromagnetic induction has been studied and shown to be a promising technique at the higher flight speeds. The electrical conductivity of the gas limits the range of applicability and radio frequency probes have been used to indicate its magnitude. High speed gas sampling valves may be used to measure the fuel concentration in the mixing region, however care must be taken with probe design. The proven value of flow visualization techniques may be attained by high speed cine photography.

Sommaire

Le corridor de vol des statoréacteurs supersoniques s'étend presque vers des Mach de l'ordre de 20 à des altitudes allant jusqu'à 50 km. La simulation des conditions de fonctionnement des foyers nécessite donc la fourniture de l'air à des enthalpies et pressions très élevées pour permettre les recherches et les mises au point de ces engins. Le tube à choc à tuyère profilée est un moyen d'essai économique et peut fournir les conditions désirées malgré la brièveté de son fonctionnement.

Le foyer de combustion supersonique à essayer est directement placé à la sortie de la tuyère de la soufflerie à tube à choc. Le démarrage de l'écoulement dans la section d'essais est accompagné par le passage d'une onde de choc instationnaire dont l'intensité peut être réduite par vidage préalable de la chambre. Le combustible (en général de l'hydrogène) est injecté dans la section d'essais et on étudie les processus de mélange, la réaction chimique et l'écoulement des gaz.

En plus de l'instrumentation classique pour la mesure de la pression et de la vitesse de choc, on a également étudié la possibilité d'utilisation de l'induction électromagnétique pour la mesure directe de la vitesse du gaz. Cette technique semble spécialement intéressante aux vitesses élevées. La faible conductivité électrique du gaz limite les possibilités d'emploi de cette technique et on utilise des sondes HF pour déterminer la valeur de cette conductivité.

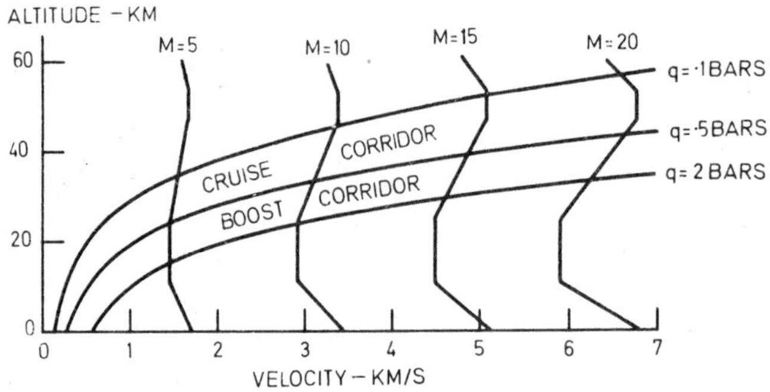

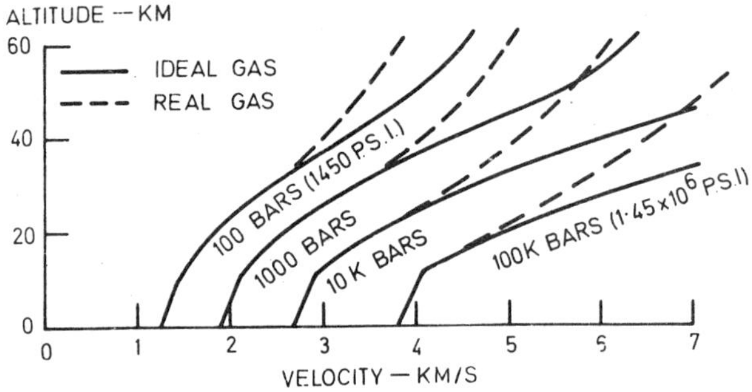

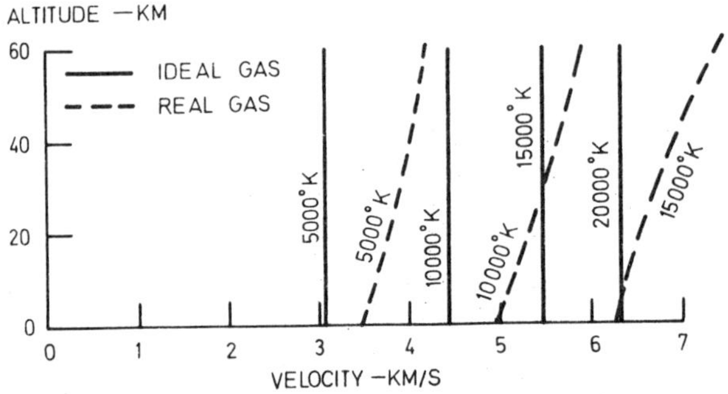

Fig. 1-4-1
Flight stagnation conditions

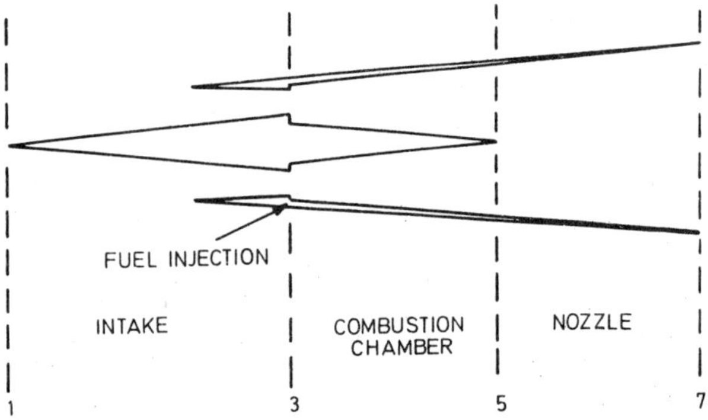

Fig. 1-4-2
Engine stations

On peut également utiliser des vannes à ouverture rapide pour prélever des échantillons de gaz et pour mesurer la concentration en combustible dans la zone de mélange, cependant la forme de la sonde joue un grand rôle dans cette manipulation.

La cinématographie à grande vitesse donne également d'intéressants résultats, car elle permet la visualisation de l'écoulement.

Introduction

At the present time, the scramjet (supersonic combustion ramjet) appears to be the most promising power plant for hypersonic propulsion. Extensive performance calculations (1) (2) have been carried out based on hypothetical efficiency of the intake, combustor, nozzle and vehicle.

However experimental investigation of these components is still at an early stage. Although a complete understanding of aerodynamics and chemical kinetics would permit confident predictions of component behavior, such an objective is too vast, and it is necessary to confine attention to the actual conditions likely to be encountered. Experimental research programmes must therefore be closely associated with the hypersonic mission to be achieved if the accumulation of irrelevant data is to be avoided.

Figure 1-4-1 (a) shows the flight corridor of a scramjet powered vehicle, since this illustrates the range of stagnation pressure and stagnation temperature to be considered. These parameters are plotted on figures 1-4-1 (b) and (c) for both ideal air ($\gamma = 1.4$) and real air (equilibrium), and it can be seen that real gas effects must be included in this speed range when stagnation conditions are required. For cruise missions a maximum speed of about Mach 15 appears likely (3), whilst for boost acceleration missions, speeds up to Mach 20 are conceivable. As shown in figure 1-4-1 (a), cruise vehicles would also operate at lower dynamic pressure (higher altitude) than boost vehicles, and flight stagnation pressures and temperatures up to 5000 bars, 8000°K. and 10^5 bars, 14000°K. for the cruise and boost missions respectively.

Combustion Chamber Conditions

In this particular study interest is centered on the combustion chamber conditions rather than the ambient conditions, and an air stream corresponding to that at the exit of an intake is therefore required for experimental research. Practical intakes do not diffuse the air with complete efficiency and the intake efficiency decreases with the amount of diffusion required. In general, two parameters are required to specify the amount of diffusion and efficiency of diffusion, and in this study velocity ratio V_3/V_1 and diffusion factor M_1/M_3 will be used to define the former, whilst kinetic energy efficiency η_D and process efficiency K_D will be used to define the latter. Reference stations through the engine are illustrated on figure 1-4-2. Although the velocity ratio V_3/V_1 and kinetic energy efficiency η_D are the more obvious intake performance parameters, the Mach number ratio M_1/M_3 and process efficiency K_D remain more nearly constant for scramjet engines as flight Mach number and amount of diffusion are varied. As shown in (3) and (4) the optimum value of M_1/M_3 is approximately 3 and the likely value of K_D is approximately 0.9 in the hypersonic speed range. The static pressure p_3 and temperature t_3 at the entry to a scramjet combustor therefore depend on the flight Mach number M_1, dynamic pressure q, diffusion factor M_1/M_3 and process efficiency K_D of the intake. It will be shown that the temperature t_3 is normally below 2000°K. and therefore ideal gas relations can be used to illustrate the relationship between these parameters with little loss in accuracy.

Considering first the static temperature t_3, it is apparent that for ideal gas this is

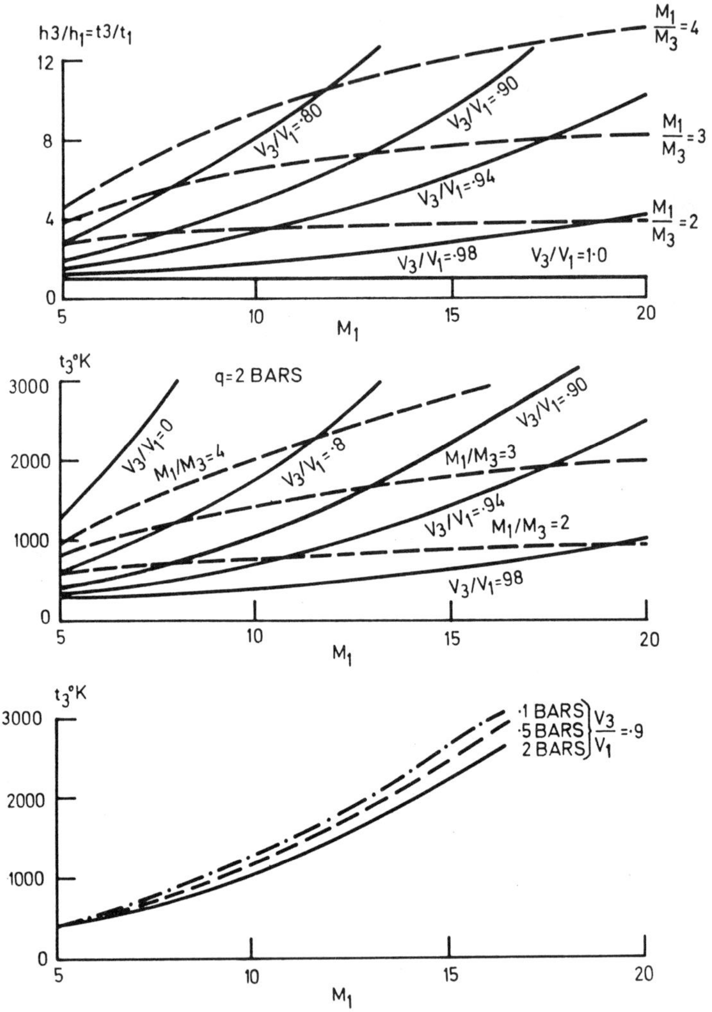

Fig. 1-4-3   Combustion chamber temperatures

independent of the intake efficiency, thus:

$$\frac{t_3}{t_1} \approx \frac{h_3}{h_1} = 1 + \frac{V_1^2}{2Jh_1}\left\{1 - \frac{V_3^2}{V_1^2}\right\} \qquad \text{(Eq. 1-4-1)}$$

or in terms of the diffusion factor:

$$\frac{t_3}{t_1} = \left\{\frac{1}{M_1^2} + \frac{\gamma - 1}{2}\left(\frac{M_1}{M_3}\right)^2\right\} \Big/ \left\{\frac{1}{M_1^2} + \frac{\gamma - 1}{2}\right\}$$

(Eq. 1-4-2)

These functions are plotted on figure 1-4-3 (a) for the hypersonic speed range, and it can be seen that the velocity in the combustion chamber is only a few percent below its free stream value, especially at the higher hypersonic speeds. Since the standard ambient temperature t_1 is defined by the ARDC atmosphere, the value of the combustion chamber entry temperature t_3 can be plotted as a function of the flight Mach number M_1 for any given value of dynamic pressure. In figure 1-4-3 (b), t_3 is plotted against M_1 for q = 2 bars, and a range of intake diffusion factors. Taking $M_1/M_3 \sim 3$ it is apparent that the combustion chamber entry temperature lies between 800°K. and 2000°K. in this speed range, the higher temperatures being associated with the higher flight speeds. The small effect of dynamic pressure on t_3 is illustrated on figure 1-4-3 (c) for q = 0.1, 0.5 and 2 bars.

Turning now to the pressure in the combustion chamber, this is given by:

$$\frac{p_3}{p_1} = \left[\frac{1 + \frac{V_1^2}{2Jh_1}\left(1 - \frac{V_3^2}{V_1^2}\right)}{1 + \frac{V_1^2}{2Jh_1}(1 - \eta_D)}\right]^{\frac{\gamma}{\gamma - 1}} \qquad \text{(Eq. 1-4-3)}$$

or in terms of the diffusion factor and process efficiency:
(where $\eta_D = K_D + (1 - K_D)(V_3/V_1)^2$, see (5)).

$$\frac{p_3}{p_1} = \left[1 - K_D\left\{1 - \frac{M_3^2}{M_1^2}\right\} \Big/ \left\{1 + \frac{2}{(\gamma - 1) M_1^2}\right\}\right]^{\frac{-\gamma}{\gamma - 1}} \qquad \text{(Eq. 1-4-4)}$$

This latter function is plotted in figure 1-4-4 (a) for the relevant range of K_D and M_1/M_3, and it can be seen that the static pressure is normally increased by about two orders of magnitude in the intake. Again the Mach number/altitude relationship given by particular values of the dynamic pressure can be used to obtain the absolute value of the parameter - in this case combustor entrance pressure. Taking q = 0.1, 0.5 and 2 bars at $K_D = 0.9$ and $M_1/M_3 = 3$, the characteristics plotted in figure 1-4-4 (b) are obtained and it is apparent that the p_3 range of interest lies between 0.1 and 5 bars.

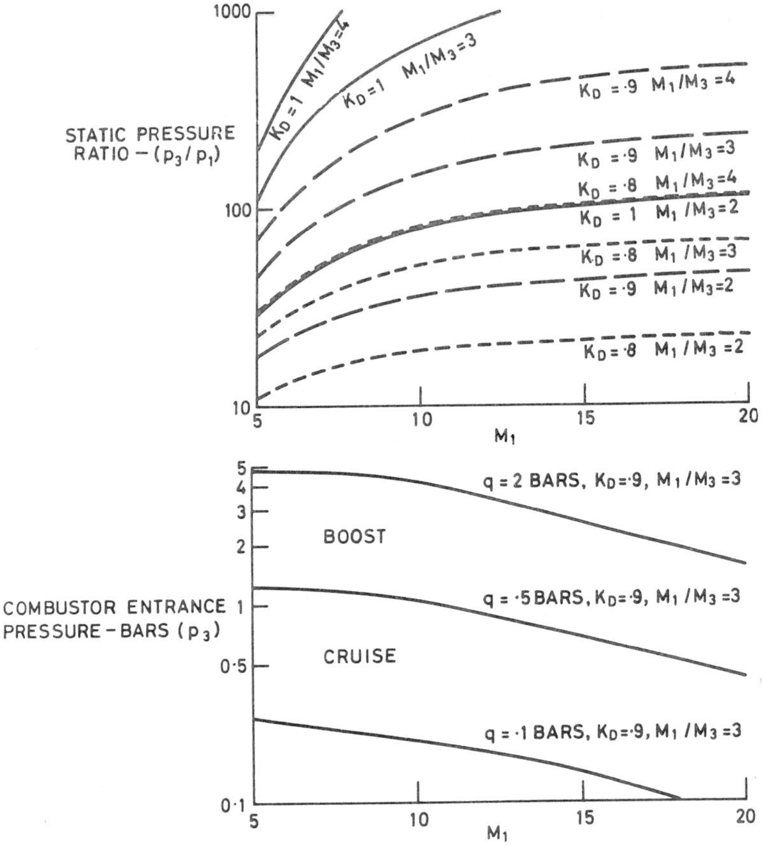

Fig. 1-4-4 Combustion chamber pressures

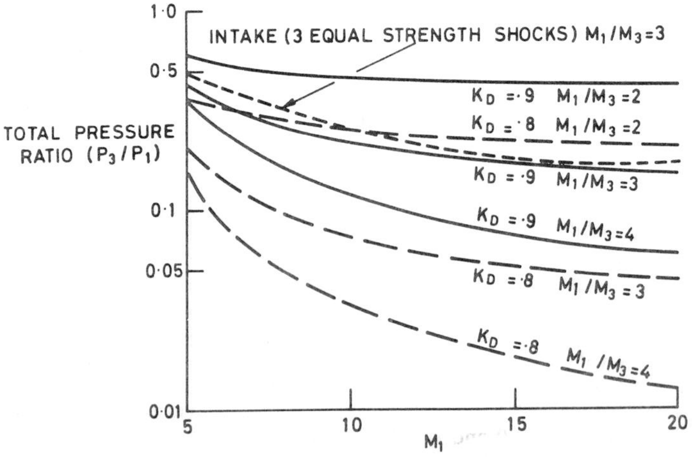

Fig. 1-4-5 Intake pressure recovery

In a flight scramjet these pressures, temperatures and Mach numbers at entry to the combustor are derived from the flight stagnation conditions by non-isentropic diffusion in the intake. In the ground test facility, isentropic expansion from a stagnation reservoir can be obtained in a Laval nozzle designed for the appropriate Mach number. Thus the ground test facility only needs to provide a stagnation pressure P_3 that is considerably lower than P_1, however the stagnation enthalpy must be the same in both systems for complete simulation. Although real gas effects become more significant as the stagnation conditions are approached, the magnitude of the reduction in stagnation pressure in the intake can be conveniently illustrated by ideal gas relations thus:

$$\frac{P_3}{P_1} = \left[1 + (1 - K_D) \left\{ \frac{h_3}{h_1} - 1 \right\} \right]^{\frac{-\gamma}{\gamma - 1}}$$

$$= \left[1 + (1 - K_D) \left(1 - \frac{M_3^2}{M_1^2} \right) \Big/ \left\{ \frac{2}{(\gamma - 1) M_1^2} + \frac{M_3^2}{M_1^2} \right\} \right]^{\frac{-\gamma}{\gamma - 1}} \qquad \text{(Eq. 1-4-5)}$$

This function is plotted in figure 1-4-5 for a range of K_D and M_1/M_3, and it is shown that the stagnation pressure in the ground facility is about an order of magnitude less than the equivalent flight case. As can be seen by reference to figures 1-4-1 (b) and 1-4-1 (c), this still represents extremely high pressures, well in excess of 1 kilobar at the higher hypersonic speeds. However, this reduction in stagnation pressure is a welcome feature, and means that it is much more practicable to carry out realistic 'connected' tests of a scramjet combustor than tests of an engine complete with intake.

Scaling Parameters

The above discussion shows that complete simulation of all gas dynamic parameters in a scramjet combustor is an arduous task for any facility, and we must consider which scaling parameters are likely to be most significant so that some of the requirements may be relaxed. The important phenomena to be studied are the turbulent mixing of fuel and air, with simultaneous chemical reaction, and the interaction of the resulting pressure gradients with the duct geometry:

a) The mixing is governed by the velocity and density ratios, Reynolds numbers, and the boundary layer at the fuel jet entry.

b) The chemical reactions are governed by the local pressure, temperature, concentration and mixedness of the fuel and air throughout the reaction zone.

c) The interaction between area changes in the duct and the heat addition process can be very large, especially at transonic flow conditions and near combustion limits.

To study all these factors simultaneously, it is necessary that the pressures, temperatures, velocities and dimensions of the fuel and air streams should be correctly simulated. Of course facilities which study only one factor e.g. mixing, chemical kinetics, or aerodynamics, have been used but we now wish to study the integrated effect, and little compromise is possible.

Fortunately sufficient is now known about supersonic combustors that the critical regions can be identified and it is found that the major problems can be studied without maximum test facility capability, i.e. 10^5 bars at 14 000°K. Considering first the higher hypersonic speeds, Ferri *et al* (6) showed that the combustion time is sufficiently short (less than 100 μs) that the system is controlled by mix processes. The pressure and temperature can then be reduced with little loss in accuracy since the mixing depends predominantly on velocities and densities. In addition, at these speeds a large part of the stagnation energy is associated with the velocity, and a small reduction in velocity greatly reduces the required stagnation conditions. It is therefore concluded that for flight speeds between Mach 10 and Mach 20, connected combustion testing can be adequately covered by stagnation temperatures of 5000°K. and stagnation pressures of 200 bars. If however an intake is to be included, as in complete engine tests, then the Mach Number in the intake is important and the reduction in temperature must be carefully matched to the reduction in velocity so that the Mach number is unchanged. In this case at M_1 = 20 a combustion chamber temperature of 1500°K. would correspond to a stagnation temperature of ~ 8000°K. real gas, which is much more readily attainable than 14 000°K. required for true simulation. The corresponding stagnation pressure is approximately 10^5 bars at $M_1 = 20$, however this can be reduced if higher stagnation temperatures are available. This follows because the overall combustion time is a function of both pressure and temperature, and a reduction in combustor static pressure can be compensated for by an increase in static temperature.

Consider next the intermediate flight speeds between Mach 7 and Mach 10, at which the supersonic combustion is usually controlled by the chemical kinetics. In this case the Mach number in the combustor is not a critical parameter, whilst temperature, pressure, mixture strength and residence time are important. The stagnation conditions can therefore be reduced significantly provided the mixing is simulated by retaining the correct velocity relationship between the two streams (7). A stagnation temperature of ~ 2200°K. is adequate to study this region.

Finally consider the lower flight speeds below Mach 7 when the combustion chamber is operating transonically. Dobrowolski has shown (8) that the critical condition of thermal choking, which is $M_5 = 1$ in a constant area duct, occurs at other Mach numbers in non-constant area ducts. It is essential therefore that the Mach number is correctly simulated if this limit is approached. This phenomenon is particularly important below a flight Mach number of 7 - which is also the region at which the combustor static temperature is too low for spontaneous ignition of the fuel, and some form of recirculation or piloting (9) is necessary to stabilize the flame. The spread of flame from a pilot is only slightly affected by the static temperature of the main stream, hence true temperature simulation is not too critical in this region. Indeed some transonic combustion tests have been carried out with no air preheating (10).

Facilities

From the foregoing it is apparent that supersonic combustors for the different flight regimes can be conveniently studied with different facilities. Conventional supersonic wind tunnel systems modified to handle the potentially explosive exhaust, with vitiation or heat exchanger preheating up to say 1000°K., can be used for studies up to $M_1 = 7$.

Ceramic preheaters e.g. pebble beds or arc heaters can give temperatures above 2200°K. and are therefore useful to cover the intermediate range between $M_1 = 7$ and $M_1 = 10$.

The higher speed range from Mach 10 to Mach 20 requires minimum stagnation conditions of 5000°K at 200 bars for connected tests, and up to 8000°K. at 10^5 bars for complete engine tests. These pressures and temperatures are beyond those attainable by any present day continuous ground test facility and intermittent facilities must be considered. It would however be most valuable if the arc heated wind tunnel could be developed to this capability, although free flight tests using a rocket boosted test vehicle will offer a cheaper alternative in the immediate future.

Two types of intermittent facility can attain the required conditions - the hot shot tunnel and the reflected shock tunnel. Current hot shot tunnels are expensive to build since they need considerable electrical energy storage, and also suffer from the disadvantage that the oxygen in the air is depleted during heating. It is anticipated that these disadvantages can be largely overcome.

However the reflected shock tunnel appears to be an attractive alternative at the present time. This study will therefore be concerned with the application of a 15 cms (6") shock tunnel built at Sheffield University, to the study of supersonic combustion for scramjet operation between Mach 10 and Mach 20. A general view of this facility is given in figure 1-4-17.

Although connected test techniques are being used at present with this facility, complete engine tests will eventually be essential. This arises because the velocity profiles at the exit of a real scramjet intake are likely to be very non-uniform, since different streamlines are processed by vastly different shock and boundary layer conditions. It may be possible to extend the applicability of connected testing by the design of nozzles which give deliberately non-uniform flow, thus duplicating the intake diffuser exit flow determined independently on a hypersonic wind tunnel.

The Shock Tunnel

Many references, e.g. (11)detail the design of shock tunnels, and only the points relevant to supersonic combustor testing will be discussed here.

The main disadvantage of the shock tunnel arises from the short testing time of a few milliseconds. The important point is that this time should be long compared to the important combustor processes. In the flight speed range being considered, figures 1-4-3 and 1-4-4 show that the intake velocity ratio V_3/V_1 is about 0.9, hence the air velocity in the combustor is greater than $\sim$ 3000 m/s. In (3) it is shown that the fuel injection velocity will be about half the air velocity, i.e. $\sim$1500 m/s, hence all the gases will flow through a combustor of 1 metre length in less than 0.75 milliseconds. Testing times in excess of one millisecond are therefore adequate to study the combustion process.

The tailored interface mode of operation of the reflected shock tunnel gives the longest running time for a given length of driven tube, and the capability of such a facility is shown on figures 1-4-6 (a) and 1-4-6 (b) as a function of primary shock Mach number (12). It can be seen that air stagnation temperatures of $\sim$ 5000°K. and 8000°K. required for direct connected and complete engine tests, can be produced with cold hydrogen, hot hydrogen and hot helium driver gases. The corresponding testing time varies from 0.3 to 0.15 m.sec/metre and hence a driven tube length in excess of 7 metres is required. A great excess in length can result in shock attenuation problems: the supersonic combustion research shock tunnel at Sheffield University has a length of 8.7 metres.

With cold hydrogen driving, the diaphragm is simply burst by pressure. The hot helium driving follows the technique developed by General Electric (13), in which a mixture of 70 percent helium and stoichiometric hydrogen: oxygen, is ignited by eight high energy discharge igniters, each 4 μF at 1500 volts. The combustion results in a smooth six to seven fold pressure rise over a period of about 20 m.sec., at which time the diaphragm opens. The required driver pressure is usually about

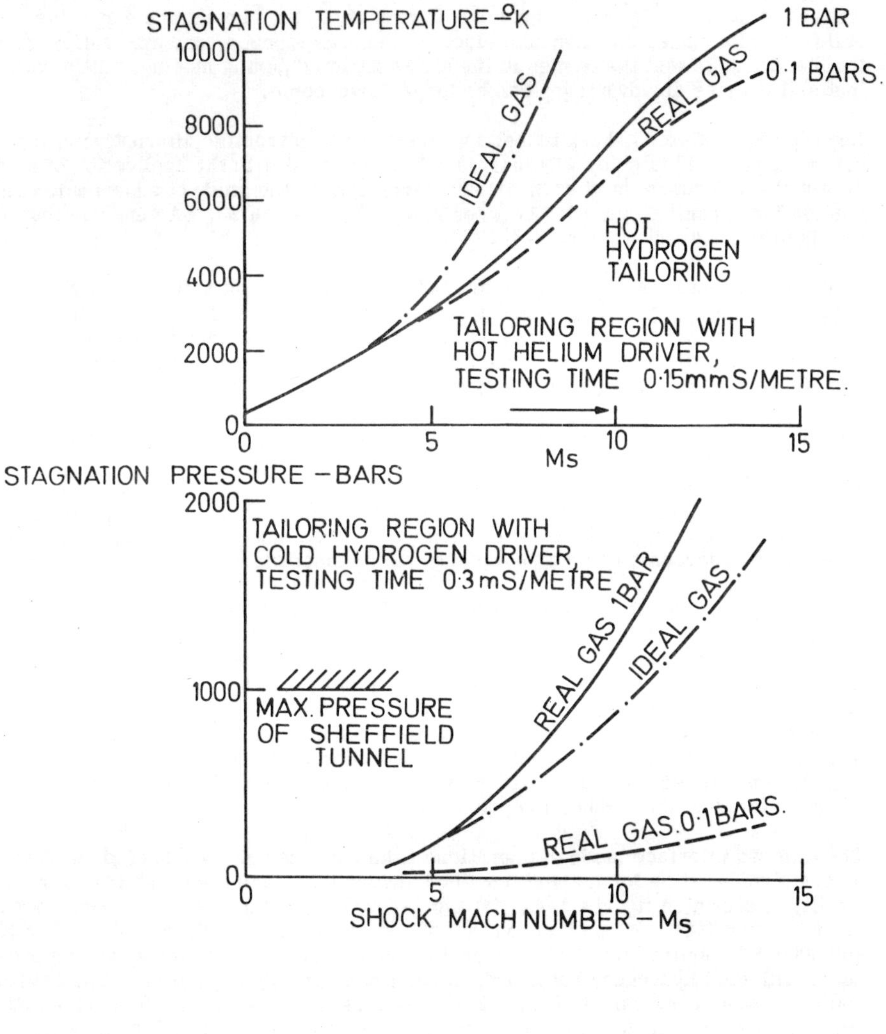

Fig. 1-4-6 **Shock tunnel stagnation conditions**

20 percent greater than the final air stagnation pressure. A detailed description of the facility design and calibration is given by Wood (12).

It is interesting to note that hot hydrogen driving would be very suitable for scramjet testing; and in view of the high cost of helium, a rapid electrical hydrogen heating technique, in which the driver tube is cooled by thermal inertia, could be well worth developing. The alternative of combustion heating of hydrogen, burning some hydrogen with oxygen, can result in undesirable detonations, and indeed less than 70 percent helium mixtures have produced very rough combustion in our facility.

In addition to the familiar limitation of testing time due to the arrival at the nozzle of the tail or reflected head of the expansion from the diaphragm, recent studies have shown that contact surface interactions in the driven tube may also curtail the useful running period. This process has been analysed (14) where it is shown that two phenomena are responsible, namely contact surface instability and shock bifurcation caused by shock interaction with the boundary layer. This latter process results in annular jets of driver gas penetrating the shock heated gas in the stagnation region and diluting the test gas flowing through the nozzle. The onset of these two processes usually occurs at similar shock Mach numbers, and calculated critical shock Mach numbers for the two driver systems in current use at Sheffield are shown in figure 1-4-7 (a). Fortunately these limits are above the tailoring Mach numbers but experiments are in hand to investigate this problem and will be reported in due course.

Since the stagnation temperatures are so high, the air is dissociated a few percent, and the possibility of the flow freezing in the nozzle must be considered. It is well known that the flow freezes in hypersonic nozzles with an area ratio of about 10^3, however in this study area ratios are approximately 10 and the test section static pressures and temperatures are still comparatively high. The freezing point in a hypersonic nozzle has been analysed by Bray (15) based on results calculated at Cornell Aeronautical Laboratory. Data from this reference are plotted in figure 1-4-7 (b) together with the combustion chamber conditions corresponding to a range of flight Mach numbers, dynamic pressures and diffusion factors. It can be seen that the flow is frozen for all conditions, hence the translational temperature will be lower than would be obtained with equilibrium flow. However it will be noted that the freezing occurs only at temperatures just above those being considered, so that the effects of freezing will be slight. In general it is expected that freezing effects will increase the oxygen atom level above its equilibrium value at entry to the combustion zone.

Test Section

Flow Starting

The design of a connected combustion test section poses some interesting problems when compared to the hypersonic shock tunnel since the ratios of test section area to throat area are much smaller, and the test section static pressure is usually much greater. In fact with test section static pressures of about 1 bar as shown above, it might appear that the conventional evacuated dump tank could be dispensed with, and the exhaust discharged direct to atmosphere. Indeed our facility is not fitted with a dump tank, and free jet tests at $M_T = 3$ have been successfully conducted with a robust $M = 3$ nozzle. However when a 1 metre long combustion chamber is fitted, the tunnel starting conditions must be considered since high pressures can be built up during the starting process. This phenomenon is considered in (16) and (17) and in this case can be interpreted as follows.

On bursting of the nozzle diaphragm, a shock wave travelling at M^* is initiated at the throat and advances through the nozzle gas followed successively by the contact

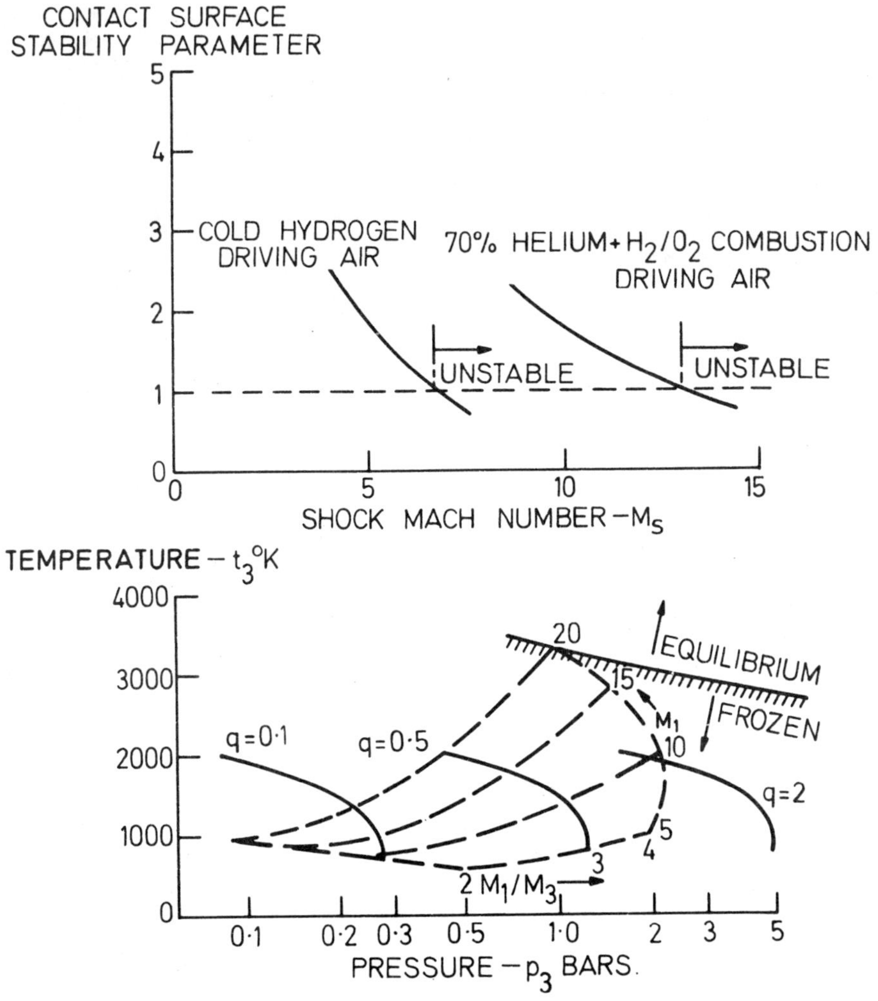

Fig. 1-4-7   Contact surface instability

surface, a backward facing shock and air at the required test conditions. The pressure between the two shocks (p_y) is determined by the initial pressure in the nozzle (p_N) and the shock Mach number, which decreases from M_s^* to M_{ST} as it passes down the nozzle.

For any given stagnation pressure and nozzle area ratio, there is a maximum initial nozzle pressure that will restrict this starting pressure to the required level. Any further decrease in p_N will give a decrease in p_y even though it results in an increase in the starting shock Mach number. As a second requirement, in order that the starting process should be rapid, there exists a minimum value of M_s^* which is a function of the nozzle area ratio (15). This is interpreted in terms of the combustor test Mach number (M_T) on figure 1-4-8 (a) and it can be seen that $M_{s\,min}^*$ increases as the test Mach number increases. The corresponding minimum starting pressure ratio (P_s/p_N) to give this Mach number is plotted against the test Mach number in figure 1-4-8 (b), thus the nozzle section should be evacuated to give at least this pressure ratio to ensure rapid starting. At this minimum level, the starting shock Mach number in the test section is given by:

$$M_{ST} = \left\{ 1 + \frac{(M_s^{*\,2} - 1)}{(A_T/A^*)^{\cdot 4}} \right\}^{\frac{1}{2}} \qquad \text{(Eq. 1-4-6)}$$

This value is plotted on figure 1-4-8 (a) and it can be seen that the minimum value of M_{ST} is approximately Mach 5 throughout the interesting range of test section Mach numbers. The corresponding static pressure rise (p_y/p_N) through this shock is approximately 30. Referring again to figure 1-4-8 (b) it can be seen that the running pressure ratio is approximately an order of magnitude less than the minimum starting pressure ratio, so that if the test section is evacuated to about one tenth to one thirtieth of the test static pressure ($p_T = p_3$), then the starting process will be rapid and the overpressure during the starting shock will be reduced to negligible proportions. Since test section pressures of about 1 bar are usually required, the connected test shock tunnel configuration does not require such high vacuum as the conventional hypersonic shock tunnel, however some evacuation is essential if the test section is not to be damaged by the starting shock.

Our shock tunnel is therefore fitted with a blast tube 6 metres long and 0.25 metres diameter exhausting to atmosphere through a Melinex diaphragm. This tube is normally evacuated to about 0.1 bars, and tests have proved this arrangement to be satisfactory - necessary even, since earlier omission of the vacuum resulted in failure of the test section!

Test Section Design

The scramjet combustor test section is very similar to a conventional supersonic wind tunnel, having a nozzle to accelerate the flow, followed by the test section proper. Due to the high static pressure it is not usually necessary to include a diffuser unless the scramjet exhaust nozzle is included with the combustion chamber, and even in this case it would be better to use a large evacuated dump tank. At present we do not include the exhaust nozzle, and the combustor is simply a variable area duct about 1 metre long. The length is determined by the combustor mixing and kinetics that we wish to study, and in common with other supersonic wind tunnels the length determines the minimum acceptable cross-sectional area due to friction considerations.

For a scramjet combustor the boundary layers are almost invariably turbulent (3), whilst in the test section the Reynolds numbers are $\sim 10^6$ and the boundary layer is

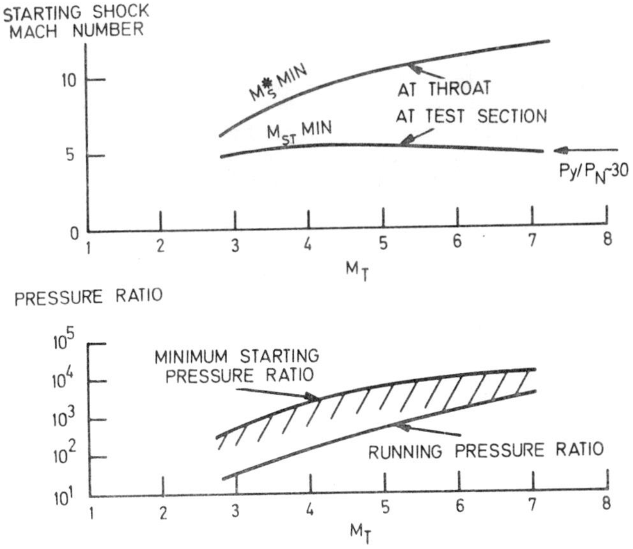

Fig. 1-4-8 Tunnel starting

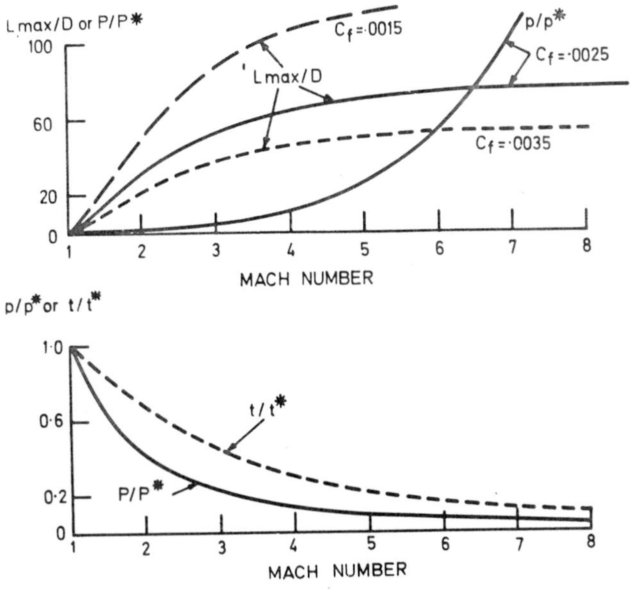

Fig. 1-4-9 Friction effects

marginally turbulent (transition will occur at a Reynolds number of about 5 x 10^6). The corresponding mean skin friction coefficient can be taken as ~ 0.0025, and figure 1-4-9 (19) shows that the flow in a constant area duct will then decrease from Mach 4 to Mach 3 over a length equal to 11 duct diameters with a corresponding reduction of stagnation pressure to 25 percent of its initial value. It is therefore essential that the length/diameter ratio of the test section should be as small as possible, especially at the higher test Mach numbers, and a maximum value of 10 is suggested. Since the combustor length is one metre, the minimum useful diameter is approximately 10 cms. at the exit of the shock tunnel nozzle. The nozzle area ratio then determines the throat area as $\sim$ 6.5 cm^2., and since the shock tube driver to throat area ratio should be about 25, the corresponding minimum diameter of the shock tube is $\sim$ 0.145 metres i.e. approximately 6 inches. When these dimensions are associated with the stagnation pressure capability of $> 10^3$ bars as described above, the result is a large shock tunnel compared to many University shock tubes, but is nevertheless the minimum practicable size for this particular field of work.

Returning to the friction considerations in a constant area tube, the large effect of friction coefficient is also illustrated in figure 1-4-9 (a). Although a value of 0.0025 was assumed above, the value varies from about 0.0015 to 0.0035 for laminar and turbulent boundary layers respectively (19), and the length required for a given change in Mach number is inversely proportional to C_f. It may therefore be possible to take advantage of this difference between the boundary layers in flight engines and those in connected model tests by scaling the diameter down in inverse proportion to the change in friction coefficient, i.e. by a factor of approximately 3. The change in duct static pressure and temperature resulting from the friction is illustrated in figure 1-4-9 (b), and it can be seen that a factor of two can be readily induced. The large effect of such a change on the combustion kinetics will be appreciated, as will the fact that measurements of wall static pressure can be interpreted as due to either heat release or friction effects. Experimental results can not therefore be analysed in terms of one dimensional inviscid flow, and detailed knowledge of the flow field is necessary.

Although a constant area test section has been discussed above for simplicity, practical combustors will usually vary in area to produce, for example, combustion at constant pressure, constant Mach number, etc. The required variation of pressure with flow area is often specified by the Crocco relation (8) (20) :

$$\frac{p}{p_3} = \left\{ \frac{A}{A_3} \right\}^{\epsilon/(1-\epsilon)} \qquad \text{(Eq. 1-4-7)}$$

and the area of the combustor should be variable along its length to constitute a versatile research apparatus. This variation is most readily accomplished by the use of a two dimensional test section with flexible walls supported on screw jacks. If the side walls are transparent, conventional optical techniques can be used for flow visualization and temperature measurement. The two dimensional geometry is also suitable for the essential flow profile measurements, as discussed above. A diagram of a basic Mach 4 test section is given in figure 1-4-10, together with a photograph of the nozzle section in figure 1-4-18.

Fuel Injection

The supersonic combustion process can be investigated by using either small disturbance methods or near stoichiometric fuel injection. In the former, the principle of equivalence between heat, mass and volume sources can be invoked to show that a small fuel jet, even with combustion, will not greatly alter the flow pattern in

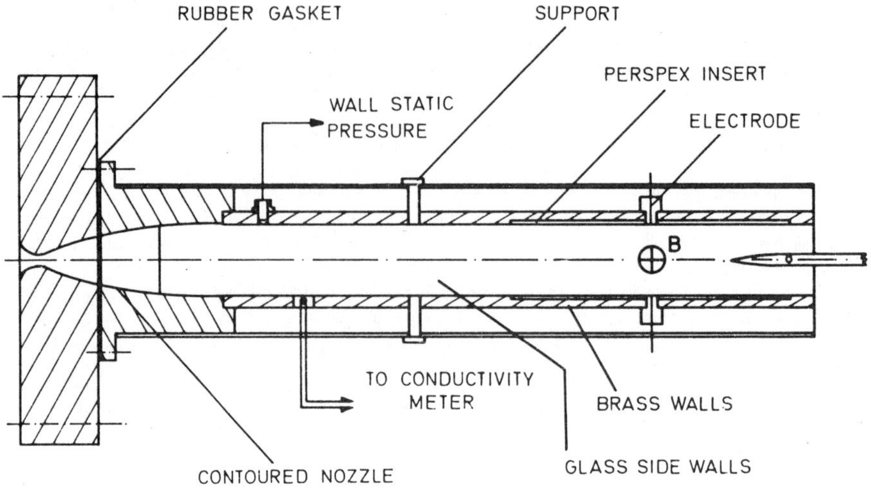

Fig. 1-4-10 Mach 4 test section

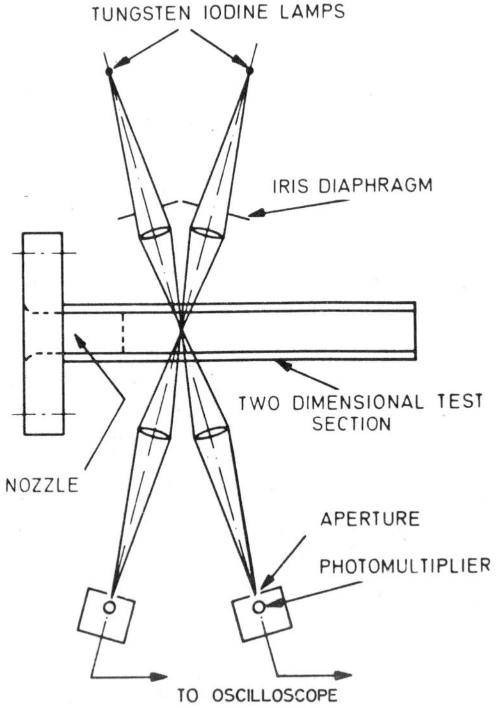

Fig. 1-4-11 Double beam sodium line reversal apparatus

the test area. The fuel jet flow can then be analysed by the usual methods employed for jets in an infinite stream, and compared with the experimental results of flame location, temperature distribution and concentration distribution. Figure 1-4-19 shows such an injector mounted through the throat of a Mach 3 nozzle, together with the nozzle expansion bell.

With near stoichiometric fuel injection, pressure, temperature and velocity changes are produced throughout the flow field, and a much more elaborate analysis is required. In general the fuel injector configuration will be complex, possibly involving multiple vortices to promote rapid mixing, and a semi-empirical theoretical mixing analysis must be used. The change of duct area at the fuel injection station will often be significant, for example Ref. 3 shows that for stoichiometric hydrogen fuel, the fuel jet area may be as much as 30 percent of the duct area. The impulse that results from downstream fuel injection is a significant part of the total impulse at the higher hypersonic speeds, and it is therefore likely that such a geometry would be used in practice. In the shock tunnel test section, it is convenient to study this geometry by wall slot injection along the upper and/or lower walls, thus giving a two dimensional flow field. The instrumentation described below can be used to determine the results from such a test, bearing in mind that boundary layers, friction and area change can have a profound effect on the flow pattern. In general, shock tunnel tests are conducted with the walls at room temperature, whereas the walls in a flight engine will be hot and the heat transfer rate reduced. Since wall heating has a significant effect on boundary layer profiles and transitions, as test techniques become more sophisticated it may be necessary to preheat the test section walls electrically before firing the tunnel.

When fuel is to be introduced into the test section, the timing of injection must be carefully synchronized with firing. Early injection results in depletion of the vacuum, so that the time must be controlled to about 0.5 millisecs. This may be achieved by fast acting solenoid valves which require about 1 millisecond to open; however an alternative system consists of a multiple sheet plastic diaphragm, which is removed by passing a large current through a wire introduced between the sheets. This technique was developed at GASL (21) for supersonic combustion research using a 3" shock tunnel.

The flow rate of the hydrogen is easily controlled by an orifice connected to a small reservoir having a capacity of a few hundred cm^3. This reservoir is precharged to a known pressure, and following opening of the valve will discharge exponentially in a period of about one second. The pressure, and hence the flow rate, remains essentially constant during the tunnel running time, and is readily monitored by means of a pressure transducer. If heated fuel is required, the reservoir can be placed in a furnace, and the ducts to the test section heated.

Standard Instrumentation of the Shock Tunnel

In most shock tunnels used in supersonic combustion research, the test section gasdynamic conditions are evaluated by measuring only two properties downstream of the nozzle. Usually these are temperature and pressure, which, when combined with a knowledge of the stagnation conditions, can be used to calculate all the physical properties of the test gas. The use of a Mollier Chart, or real gas data in some form, is therefore required.

Temperature is monitored by a double-beam reversal technique (22), it is usual to employ the spectral lines of sodium for this although the chromium and cesium lines are equally suitable. Figure 1-4-11 is a diagram of the equipment used to measure the temperature on the Shock Tunnel at Sheffield, based on the sodium doublet. Sodium was introduced into the tunnel in the form of sodium chloride. It is preferable to seed the test gas with a smoke of a sodium compound, but with the above method there did not seem to be any lack of sodium vapour. Quartz iodine lamps were used, as these gave a higher brightness temperature than the tungsten filament lamps previously used (12). The vibrational and translational temperatures were assumed to be in complete equilibrium, so that the temperature measured by this method, the vibrational temperature, was taken as equal to the true gas temperature.

Piezo-electric transducers were used to measure the pressure levels throughout the shock tunnel. In the test section two separate methods of measurement were used. The wall static pressures were found by Kistler 701 transducers set in antivibration mounts, and exposed to the flow via small holes. Alternatively a sting arrangement permitted the use of a static or pitot pressure probe built to a National Physical Laboratory design (11). The signals from the charge amplifiers were usually filtered to reduce the noise level from mechanical vibrations, despite precautions taken to avoid this noise.

Two alternative mechanisms were used to measure the shock of the primary speed, depending on whether the tunnel was driven by combustion heated helium or cold hydrogen. For combustion driving with the primary shock speed $M_s > 10$, ionisation gauges were used. These consisted of a central electrode held at 300 volts and surrounded by an earthed electrode. On passage of the shock, the gap broke down and a pulse was obtainable. At lower shock speeds pressure transducers were used because of the low level of ionisation present.

The Measurement of Gas Velocity by Electromagnetic Induction

The possibility of measuring velocity by electromagnetic induction has been known for a long time and commercial instruments are available to meter liquid flows. The application of the method to metering ionised gas flows has not received so much attention, and this study presents the application of this technique to velocity measurement in the supersonic combustion test section. The test gas after expansion will be ionised and therefore the method should be applicable to measuring its velocity. If this test gas is passed between the poles of a magnet so that the magnetic field is perpendicular to the velocity vector, the magnitude of the induced voltage is given by:

$$v = B\ V_m\ D \text{ volts} \qquad \text{(Eq. 1-4-8)}$$

This equation gives the magnitude of the measured voltage only if all the walls of the test section are non conducting (23); if this is not so, the observed signal will be less than that given by equation 1-4-8. The ratio of the observed signal to the theoretical signal is defined as the sensitivity.

J.M. Shaw of the Royal Aircraft Establishment, Farnborough, has reported privately that early workers in the field (24) experienced difficulty due to the appearance of a voltage when there was no imposed magnetic field. This voltage was of the same order of magnitude as the induced voltage (ca 1 volt) and made interpretation of the results difficult. Holbeche (24) investigated this phenomenon in a 2 inch shock tube, and found that the voltage could be of either polarity. It therefore seems necessary to have the induced voltage as large as possible and also to use a different measuring circuit.

Cason (25) used the technique to measure velocity in a plasma jet. He found results to be in agreement with calculations, although no direct comparison with other forms of velometer was reported. The results using a d.c. excited magnet were found to be affected by thermionic emmission at the electrode surfaces and an a.c. excited magnet was substituted to eliminate such effects. A shock-tube induction flowmeter is described by Croce (26) but once again the results given are only compared with theoretical induced voltage values although the sensitivity of this type of instrument can be over 100 percent (27). B. Edwards has informed us privately that at Rolls-Royce Ltd, Derby, velocities have been measured with an induction system during studies on ethylene-oxygen detonations. By comparison with drum camera measurements they found that it was necessary to apply a calibration factor of between 1.5 and 1.7 to the velocity predicted from the induced voltage.

Conductivity of the Test Gas

Equation 1-4-8 predicts the induced voltage provided that the fluid has a finite value of electrical conductivity. The level of the conductivity in the test section will determine the external impedance necessary to accurately determine the induced voltage. If a 10 percent loss in uncorrected voltage is acceptable, then the equivalent circuit yields $R_L \geqslant 9R_g$ where R_L is the load resistance and R_g the gas resistance. Even if σ, the gas conductivity is known, the determination of R_g presents some difficulty, because only when $A >> D^2$, where A is the area of the electrodes, is it equal to $D/A\sigma$. The technique used by Pain and Smy (28) was therefore adopted.

A small perspex box was constructed, with the same internal dimensions as the test section, and with provision for placing a pair of electrodes in the side walls. The box was filled with salt solution of known conductivity (σ') and the resistance between the electrodes (R_g') found using an a.c. bridge circuit, then $R_g = R_g' \sigma'/\sigma$. Using flush electrodes made of 1.59 mm. diameter silver steel rods, it was found that $R_g = 10^3/\sigma$. All types of electrode can be tested in this way before fitting to the test section.

The expression for R_g is a function of σ, the gas conductivity and some estimate of its value is needed. Viegas and Peng (29) have calculated the conductivity of thermally ionised air above 3000°K. for various pressures. Very little attempt has been made to find conductivities below this level of temperature as the degree of ionisation becomes so small. Coombe (30) gives an expression for the conductivity of a slightly ionised gas as:

$$\sigma = \frac{7.05 \times 10^{-13}\, T^{3/4}}{\alpha\sqrt{P}\, \exp\,(eV_i/2kT)} \qquad \text{(Eq. 1-4-9)}$$

where V_i is the ionisation potential of the gas and α an appropriate average for the electron-atom collision cross-section. Using this equation it has been possible to extend the data given in Ref. (12) down to a temperature of 1500° K. Figure 1-4-12 shows the result of these calculations for three pressures. The graph shows the effect of the exponential term in equation 1-4-9 and it can be seen that the resistance of the gas will be very high over most of the range of static temperature of interest.

The conductivity of the gas is therefore likely to be in the order of 10^{-8} mho/m so that the resistance between the flush electrodes will be $\sim 10^{11}$ ohms. The measuring circuit must therefore have an impedance of at least 10^{12} ohms unless either an estimate of signal loss can be made or the conductivity increased. Using a 100:1 or 1000:1 attenuating circuit, the input impedance of a sensitive oscilloscope can be raised to $10^8 \rightarrow 10^9$ ohms. For higher values of input impedance, an amplifier based on an electrometer valve must be used.

Seeding of the test gas with materials of low ionisation potential such as caesium or potassium can increase the conductivity by several orders of magnitude. Frost (31) has calculated the conductivity of various seeded atmospheres and the results for argon can be compared with the results of Lim, Resler and Kantrowitz (32) despite the pressure differences. Seeding with 0.02 atm K., increases the con-

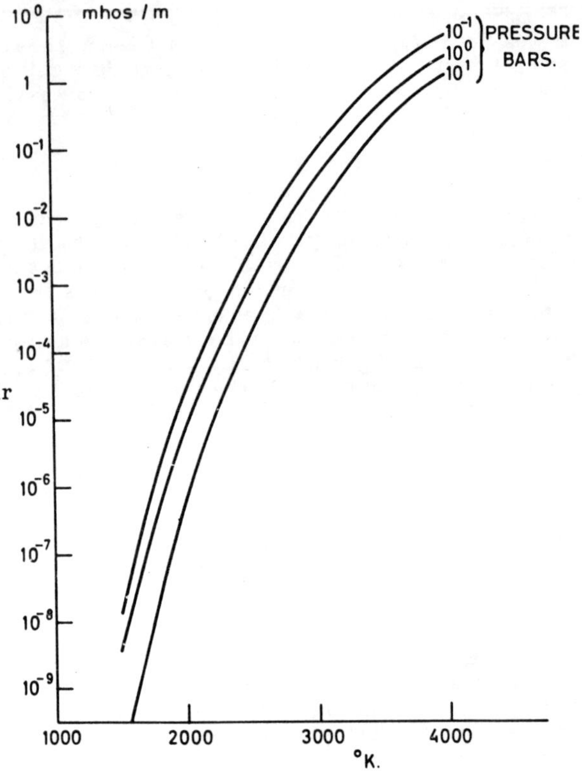

Fig. 1-4-12 Conductivity of air

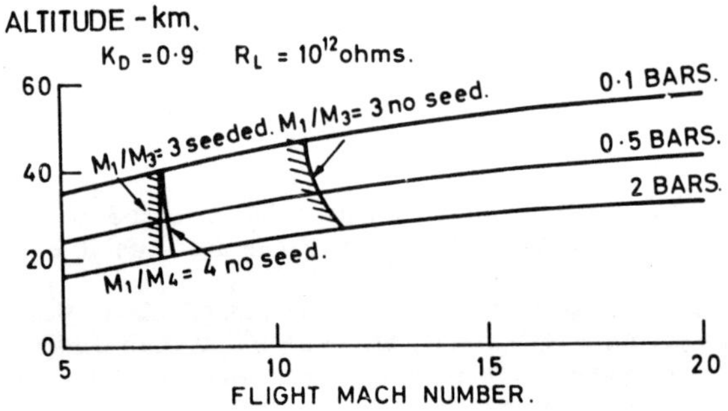

Fig. 1-4-13 Limits of inductio
flowmeter applicability

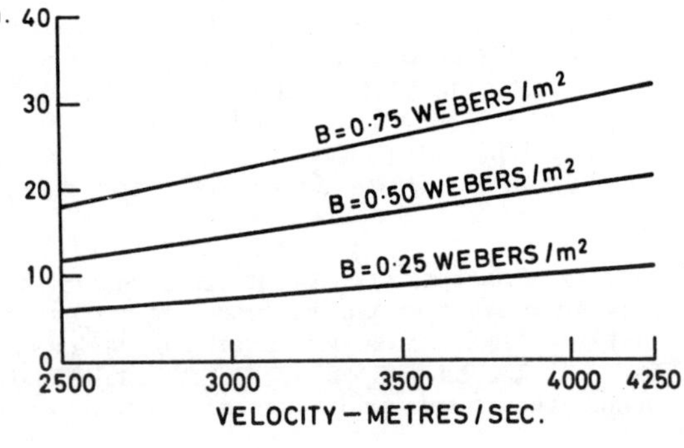

Fig. 1-4-14 Induced voltage
versus velocity

ductivity 10^3 times. Since the same improvement can be expected with air then the limit on the external measuring impedance will be lowered to $\sim 10^9$ ohms. This arises because the air is only very weakly ionised and the majority of the ionisation comes from the seed (31), and is therefore generally independent of the carrier gas. An exception arises when the gas contains OH which strongly absorbs electrons, hence difficulties are anticipated when the electromagnetic velocity meter is used in supersonic hydrogen flames. On the other hand hydrocarbon flames are electrically conducting due to the reactions:

$$CH + O \longrightarrow CHO^+ + e$$

$$CHO^+ + H_2O \longrightarrow H_3O^+ + CO$$

which produce electrons and thus may provide an alternative source of conductivity.

Range of Applicability of the Method

The conditions within the test section must correspond to the simulated conditions within a scramjet combustion chamber. For a specific flight Mach number, the combustor conditions at any altitude can be found from figures 1-4-3 and 1-4-4 and hence the conductivity found from figure 1-4-12. Within the range $M_1 = 5$ to 20 and with $K_D = 0.9$, $M_1/M_3 = 3, 4$ the combustor pressure will lie between 10 and 0.1 bars, hence conductivity will not be greatly affected by pressure within this range.

From the previous discussions it appears that for a measuring circuit of impedance 10^{12} ohms and the resistance between electrodes equal to $10^3/\sigma$, the lower limit of applicability lies between $M_1 \sim 7.0$ and $M_1 \sim 11$ depending on M_1/M_3. This limit is illustrated on figure 1-4-13. The limit may be further reduced by increasing the external impedance or reducing the internal impedance by (a) different electrode structure or (b) seeding of the gas. Seeding is easily accomplished in the shock tunnel but is impracticable in the scramjet. It is likely that seeding can reduce the limit to $M_1 \sim 5$ when $M_1/M_3 = 4$. This would bring the electromagnetic velocity measuring technique within the range of some continuous facilities such as the pebble bed heated hypersonic wind tunnel.

Electrode Geometry

Shercliffe (23) has examined many types of electrode configurations to give some estimate of the sensitivity to be expected from any particular arrangement. Theoretical calibration is practically impossible and all types of meter need empirical calibration. One of the simplest cases is that of a channel with non-conducting walls and internal wetted electrodes. This geometry is also very suitable for supersonic application as no obstacle is presented to the flow. Such a meter can have a sensitivity greater than 1 depending on the aspect ratio (a/b) and on the velocity profile within the duct. In a shock tunnel test section with a contoured nozzle the velocity profile should be reasonably flat, hence reducing the sensitivity to near unity. Growth of the boundary layer in the duct may have adverse effects on the voltage recorded by wall electrodes and it may be necessary to use projecting electrodes. The possibility of using a pair of closely spaced electrodes to explore the duct velocity profile can also be examined.

In order to overcome the electrode effects present, it will be necessary to use a large value of B. The maximum value of B attainable will therefore limit the extent to which D may be reduced.

At the edges of the magnetic field the induced voltage will decrease and hence cause internal currents to flow. This means that the electrodes must be sufficiently re-

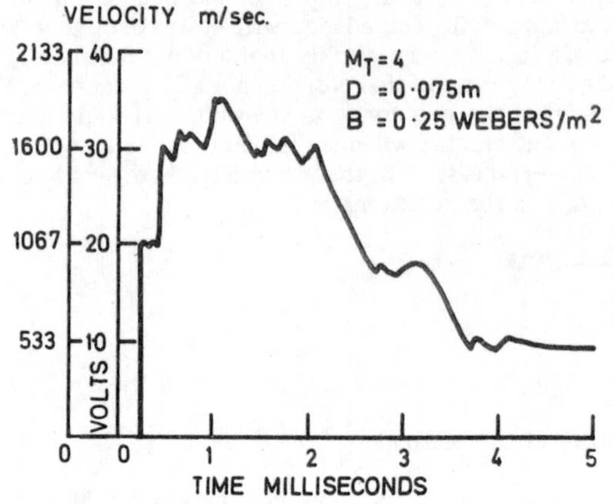

Fig. 1-4-15a) Induced voltage trace

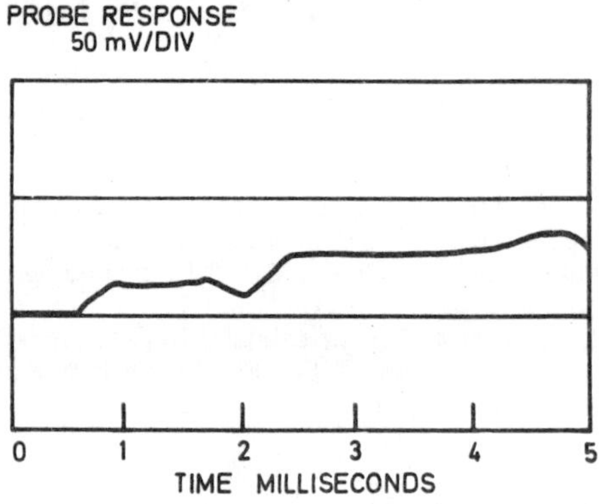

b) Conductivity meter trace

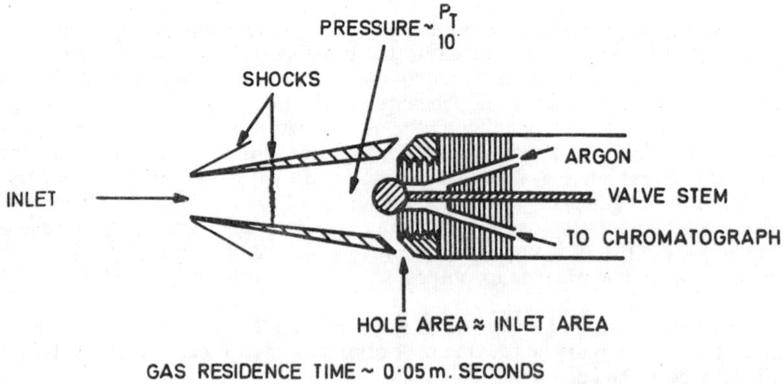

Fig. 1-4-16 Sampling probe head

mote from the edges of the field to eliminate these end effects. Shercliffe (33) has analysed the two-dimensional case with an abrupt field and with a fringed magnetic field and finds that for a 1 percent loss in signal, the ratio of length of steady magnetic field to the interelectrode distance must be at least 3.04. Current leakage through the Hartmann boundary layer is negligible with the ionised gas flows being considered, because the distance taken for such a boundary layer to form is many times greater than the length of magnetic fields likely to be used.

Apparatus used

A test section, designed and built to give Mach 4 flow with two-dimensional expansion only (34), was manufactured from non magnetic materials, and the internal surfaces were electrically non conducting. This rectangular section was chosen so that the whole test section could be placed between the poles of a large electromagnet (Fig. 1-4-19). The interelectrode distance could be varied from zero to 7.5 cms. and B could be altered to a maximum value of 0.75 webers/m^2. The expected range of velocity of the test gas could be computed from a knowledge of the stagnation conditions, assuming isentropic expansion through the nozzle. Alternatively it could be found from the static temperature and pressure by the use of real gas data. In the Mach 4 test section, the velocity range would be from 2300 m/sec. to 4000 m/sec. if scramjet combustor conditions were reproduced in the shock tunnel.

Figure 1-4-14 shows the theoretical values calculated from equation 1-4-8, of the induced voltage per cm., as a function of velocity, for three values of B.

If the maximum interelectrode distance is used (D = 7.5 cms) then from this graph it can be seen that the induced voltage is likely to be approximately 150 volts.

Results

We found that at supersonic combustor conditions, the electrode voltage in the absence of a magnetic field was less than 0.1 volts, which at the maximum field corresponded to a minimum D of 0.25 cms., for 2 percent accuracy. Thus reasonably accurate velocity profiles were obtainable and if some means could have been found to eliminate the spurious voltage, even higher resolution should have been possible. Preliminary results using a magnetic field confirmed that the output voltage was two orders of magnitude greater than the spurious voltage, (Fig. 1-4-15) (a)).

The Measurement of Conductivity

Several methods have been used to monitor conductivity levels in shock tubes, each having been developed to suit a particular piece of apparatus. The simplest form of a conductivity meter consists of two probes held at different potentials. The voltage across them and the current flowing is recorded continuously. The passage of a medium of different conductivity causes a change in current andthe unknown conductivity can therefore be found. Unfortunately the probes may upset the supersonic flow in the test section, and frequently prove noisy.

Lin *et al* (32) (35) measured the conductivity of ionised argon and air by means of an electrodeless technique. A search coil surrounding the shock tube was placed upstream of a field coil. The passage of the conducting medium gave rise to an induced voltage across the search coil due to displacement of the magnetic field. A knowledge of the velocity of the gas is required for calibration, hence this method is unsuitable for use where it is required to find the velocity.

Fig. 1-4-17 General view of shock-tunnel

Fig. 1-4-18 Mach 4 nozzle assembly

A most promising electrodeless technique is that of Olson and Lary (36) in which the dissipation of energy in a 20 Mc/s radio frequency field is related to the conductivity of the surrounding fluid. This technique has been adapted to the test section of our shock tunnel as follows.

To avoid flow disturbance the R.F. coil, 1.27 cm. x 3.18 mm was set in a piece of perspex 3.8 cm. diameter and the surface covered with an insulating material. By using a diameter of perspex that was large relative to the size of the coil size the effects of the brass walls, into which the probe fitted, were minimised. The associated circuit was that used by Harris, the details of which are given in (34). Calibration was accomplished by means of electrolytic solutions of known conductivity and an output of about 2 volt metre/mho was obtained. This design has been used extensively at our laboratory for MHD research and is capable of measuring conductivity down to $\sim 10^{-4}$ mhos/metre at frequencies above 10 kilocycles. Although this sensitivity is not as high as would be desirable, it is sufficient for feasibility studies. Results to date showed that the conductivity was of the order of 10^{-3} mhos/metre (Fig. 1-4-15 (b)) and therefore adequate for the velocity measuring technique. There are still some doubts as to the extent to which this measurement was influenced by high enthalpy boundary layers.

High Speed Gas Sampling

A knowledge of the composition of the gases in the test section is essential for extensive supersonic combustion research. Two particular aspects require composition information:

a) Mixing of the driven test gas (air) with the driver gas.

b) Mixing of the fuel with the air in the combustor.

These two aspects may be investigated independently and although various optical and spectroscopic methods are applicable, gas sampling and analysis provides a valuable technique. Since the running time of the tunnel is only a few milliseconds, the gas sampling process must be particularly rapid. A fast acting solenoid valve developed by B.P. with a sampling time of 0.5 milliseconds was available, and this period was adequately short. The valve was triggered electrically by suitable delay circuits controlling the discharge of a 300 μF condenser charged to 200 v. via a thyratron. The solenoid valve motion was monitored by a capacity transducer mounted on the valve stem. The gas sample of about 0.5 ml. could then be analysed by a gas chromatograph using argon as the carrier gas. Hydrogen, helium, oxygen and nitrogen could be detected with a hot wire catharometer. Water is absorbed in the column. It follows therefore that the concentration of all the gases of interest could be determined.

The design of the sampling head is particularly important and a supercritical 'intake' design is proposed. (Figure 1-4-16). The time constant of the sampling chamber must be small compared to the running time of the tunnel, so that the areas of the inlet and exhaust holes must be carefully calculated. This apparatus has not yet been used in the shock tunnel, however the results will be reported in due course.

High Speed Cine Optical Techniques

The starting process in the test section, and the fuel/air mixing process both require a certain time to become established. These time dependent phenomena can be evaluated very conveniently by high speed cine optical techniques, using framing speeds in excess of 10,000 frames/sec. The Beckman and Whitley Dynafax camera is capable of speeds up to 35,000 frames/sec. and is therefore very suitable for this purpose. The total number of frames is 224, corresponding to a duration of

Fig. 1-4-19 Injector mounted through throat of Mach 3 nozzle and nozzle expansion bell

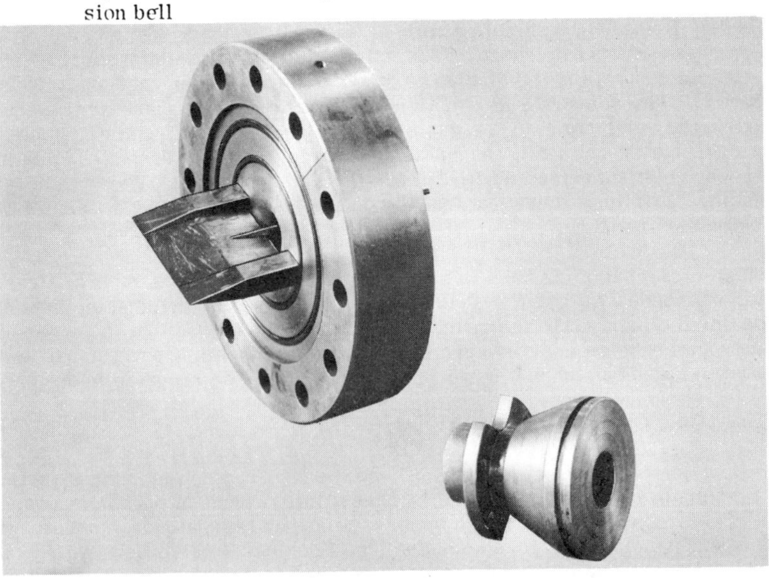

Fig. 1-4-20 Test section between poles of electromagnet

6.4 milliseconds at the maximum framing rate. Synchronization of the camera to the test period presents a problem, however, since there is insufficient light from the combustion to expose the film, a separate light source must be used which can therefore be used for synchronization. A flash source, giving a square wave light output of variable duration (8-22 milliseconds), is available and has been successfully used with the Dynafax camera in a conventional Toepler Schlieren system.

An off-axis, single pass optical system based on 15 cms. parabolic mirrors was set up to photograph events in the test section a short distance downstream of the nozzle. As an alternative to the cine Schlieren, a single exposure was often taken using an argon-spark light source having a duration of less than 1 μsec. (constructed from details givenin (37)). Triggering of the light sources was accomplished from either an ionisation gauge or direct from an oscilloscope gate using appropriate delays. A mechanical shutter was arranged to close after the light source had triggered, to prevent overexposure of the film due to the luminosity of the driver gases.

When the Dynafax camera is placed in the optical system, direct projection of the image on the film must be made to obtain sufficient exposure and retain good quality. During setting up the image must be observed directly on the film because of the nature of the camera optics.

Conclusions

(a) The tailored interface hypersonic shock tunnel is a suitable facility for supersonic combustion research corresponding to scramjet operation between Mach 10 and Mach 20.

(b) Since the combustion chamber operates at about one third of the flight Mach number, connected testing can be carried out in a test section very similar to the conventional supersonic wind tunnel. With this technique reasonable simulation can be carried out with stagnation pressures $\geqslant$ 200 bars and stagnation temperatures $\geqslant$ 5000°K.

(c) These conditions can be realized with hot hydrogen or hot helium driving air at shock Mach numbers in the range 7 to 13. Operation above tailoring will result in curtailment of testing time due to contact surface instability. A driven tube length of about 10 metres is required to obtain sufficient running time for combustion experiments.

(d) The dimensions of the connected test section are determined by (i) mixing and kinetic processes which indicate that 1 metre is a convenient length, and (ii) friction effects which indicate that the maximum length/diameter ratio should be about 10.

(e) The starting phenomena in the test section require

- (i) a starting pressure ratio about an order of magnitude greater than the running pressure ratio for rapid starting, and

- (ii) pre-evacuation of the test section to about 1/30th of the safe operating pressure to prevent damage by the starting shock.

(f) The flow is expected to freeze in the nozzle just before entering the test section resulting in an above equilibrium concentration of oxygen atoms.

(g) A solenoid valve or electrically ruptured diaphragm can be used to introduce fuel into the combustion test section at the required instant. Detailed knowledge of the flow field is required for satisfactory interpretation of performance.

(h) The measurement of gas velocity by electromagnetic induction has been investigated. An 18 cm. diameter magnet (0.75 Wb/m^2) placed around the non-magnetic test section has been used to produce Faraday voltages proportional to velocity. An output in excess of 20 volts/cm. can be generated.

(i) This technique is applicable above (i) equivalent flight Mach numbers of 11 without seeding, (ii) equivalent flight Mach numbers between 5 and 7 with seeding.

(j) Electrical conductivity of the gas has been investigated using an electrodeless radio frequency probe technique. The results confirm that the gas in the test section is sufficiently conductive to permit the use of a load impedance of $> 10^8$ ohms in the velocity measuring instrument.

(k) A high speed gas sampling valve (0.5 milliseconds sampling time) can be used to study the concentration of species in the test section.

(l) High speed (35,000 frames/sec) cine photography can be used for time resolved flow visualization techniques.

References

1. **Franciscus, L.C.,** 'Off design performance of hypersonic supersonic combustion ramjets', NASA TM X-52032, June 1964.

2. **Mordell, D.L., Swithenbank, J.,** 'Hypersonic ramjets', Second Int. Cong. of the Aer. Sci., Pergamon Press, N.Y., 1960, 831-848.

3. **Swithenbank, J.,** 'Hypersonic air-breathing propulsion'. Progress in Aer. Sci., Vol. 8, Ed. D. Kuchemann, Pergamon Press, Oxford and N.Y., 229-294.

4. **Curran, E.T., Swithenbank, J.,** 'Really high speed propulsion by scramjets'. Aircraft Eng., Jan. 1966, 36-41.

5. **Curran, E.T., Bergsten, M.B.,** 'Discussion of inlet efficiency parameters'. Wright-Patterson A.F.B. Rep. No. ASRPR TM 62-68, April 1963.

6. **Ferri, A., Libby, P.A., Zakkay, V.,** 'Theoretical and experimental investigation of supersonic combustion'. Third Int. Cong. of the Aer. Sci. Stockholm, 1962, Spartan Books Inc. 1964.

7. **Scaggs, N.E., Dunn, R.G.,** 'Development of a facility for supersonic combustion simulation'. Proc. twenty-sixth meeting of the Supersonic Tunnel Ass., AIAA Paper 66-743, Sept. 1966.

8. **Dobrowolski, A.,** 'Analysis of nonconstant area combustion and mixing in ramjet and rocket-ramjet hybrid engines'. Lewis Res. Centre, NASA TN-3626, 1966.

9. **Schetz, J.A., Jannone, J.,** 'The ignition of flowing hydrocarbon/air mixtures by a hydrogen pilot flame'. Pyrodynamics, Vol.2, 1965, 1-14.

10. **Townend, L.H., Reid, J.,** 'Some effects of stable combustion in wakes formed in a supersonic stream'. Supersonic Flow Chem. Proc. and Radiative Transfer, Ed. D.B. Olfe and V. Zakkay, Pergamon Press, 1964, 137-56.

11. **Pennelegion, L., Cash, R.F., Bedder, D.F.,** 'Design and operating

features of the N. P. L. 6 in. shock tunnel', Aer. Res. Council R. and M. No. 3449, 1967.

12. **Wood, M.P.,** 'The shock tunnel as a supersonic combustion test facility'. Fuel Soc. J. 1966, Univ. of Sheffield, Also: **Wood, M.P.,** 'Ph.D. Thesis'. Univ. of Sheffield, 1967.

13. **Nagamatsu, H.T., Martin, E.D.,** 'Combustion investigation in the hypersonic shock tunnel driver section'. Gen. Elec. Research Lab., (Schenectady), Reprint No.3263, 1959.

14. **Davies, L.,** 'The interaction of the reflected shock with the boundary layer in a shock tube and its influence on the duration of hot flow in the reflected-shock tunnel, Parts I and II'. A.R.C. Current Papers Nos. 880-881. H.M.S.O. London, 1967.

15. **Bray, K.N.C.,** 'Chemical reactions in supersonic nozzle flows'. 9th Int. Symp., on Combustion, Aug. 1962.

16. **Ackroyd, J.A.D.,** 'A study of the running times in reflected shock tunnels'. A.R.C. Current Paper No. 883, H.M.S.O. London, 1967.

17. **Smith, C.E.,** 'An analytic study of the starting process in a hypersonic nozzle'. Proc. Heat Transfer and Fluid Mech. Inst. Standford Univ., Press 1964. also 'The starting process in a hypersonic nozzle'. Univ. of Oxford, Dept. of Eng. Science Rept. 1000, July 1965.

18. **Chisnell, R.F.,** 'The motion of a shock wave in a channel with application to cylindrical and spherical shock waves'. J. Fluid Mech. Vol. 2, 1957, 286.

19. **Shapiro, A.H.,** 'Compressible Fluid Flow' Ronald Press, New York, 1953.

20. **Billig, F.S.,** 'The design of supersonic combustors based on pressure-area fields'. 11th Int. Symp. on Combustion, Aug. 1966.

21. **Tamagno, J., Lindemann, O.,** 'Experimental results on supersonic combustion'. Gen. Applied Sci. Lab. Tech. Rep. No. 322, Dec. 1962.

22. **Gaydon, A.G., Hurle, I.R.,** 'Measurement of times of vibrational relaxation and dissociation behind shock waves in N_2, O_2, Air, CO, CO_2 and H_2'. 8th Symp. on Combustion. The Williams and Wilkins Co., Baltimore 1962, 309-318.

23. **Shercliff, J.A.,** 'The theory of electromagnetic flow-measurement'. Cambridge Univ., Press, Cambridge 1962.

24. **Holbeche, T.A.,** 'Some experimental studies of the behaviour of electrostatic probes in shock tube flow'. R.A.E. Tech. Note No. Met. Phys. 338, Plasma Physics Symposium at R.A.E. Farnborough, June 1961.

25. **Carson, C.,** 'Gas velocity probe for moving ionised gases'. J. Appl. Phys. 36, No. 2, Feb. 1965, 342-7

26. **Croce, P.A.,** 'Shock tube induction flowmeter'. Rev. Sci. Instruments 36, No. 11, Nov., 1965, 1561-4.

27. Reference 23 31-32.

28. **Pain, H.J., Smy, P.R.,** 'Experiments on power generation from a moving

plasma'. J. Fluid Mechanics Vol. 10, 1961, 51-64.

29. **Viegas, J.R., Peng, T.C.,** 'Electrical conductivity of ionised air in thermodynamic equilibrium'. American Rocket Soc., J., May 1961, 654-7.

30. **Coombe, R.A.,** 'Magnetohydrodynamic generation of electrical power'. London, Chapman and Hall, 1964, 70-1.

31. **Frost, L.S.,** 'Conductivity of seeded atmospheric pressure plasmas'. J. Applied Physics, Vol.32, No.10, Oct. 1961, 2029-36.

32. **Lin, S., Resler, E.L., Kantrowitz, A.,** 'Electrical conductivity of highly ionized argon produced by shock waves'. J. Applied Physics, Vol. 26, No. 1, Jan. 1955, 95-109.

33. Reference 3, pp. 35-41.

34. **Parsons, R.J.,** 'The design and construction of a shock tunnel test section for velocity measurement studies'. H.I.C. 74 part 2. Dep. of Fuel Technology and Chemical Eng. Univ. of Sheffield, April 1966.

35. **Lin, S., Lamb, L.,** 'Electrical conductivity of thermally ionized air produced in a shock tube'. J. Applied Physics Vol. 28, No.7, July 1957, 754-9.

36. **Olson, R.A., Lary, E.C.,** 'Electrodeless plasma conductivity probe apparatus'. Rev. of Sci., Instruments Vol. 33, No.12, Dec. 1962.

37. **North, R.J.,** 'The Argon-Jet Spark Light Source'. NPL/Aero 419/1960.

Acknowledgements

Supersonic combustion research at Sheffield University is sponsored jointly by a Ministry of Technology Contract and United States Air Force Grant No. AF/EOAR 67-5. This support is gratefully acknowledged.

List of Symbols

A	area
a	semi-width of test section (in field direction)
B	magnetic field strength
b	semi-width of test-section (perpendicular to field direction)
C_f	skin friction coefficient
D	interelectrode distance
e	charge of the electron
h	enthalpy
K_D	process efficiency
k	Boltzmann's constant
M	Mach number
P	stagnation pressure
p	static pressure
q	dynamic pressure
R_g	gas resistance
R_L	load resistance
T	stagnation temperature
t	static temperature
V	velocity
V_i	ionisation potential
v	voltage
α	electron-atom collision cross-section
ϵ	exponent in the pressure area relationship defined in eqn. 1-4-7.
γ	specific heat ratio
η	kinetic energy efficiency
σ	gas conductivity

Subscripts

1, 3, 5, 7	stations defined in Fig. 1-4-2.
D	diffuser
m	mean value
N	conditions in nozzle prior to firing
S	shock
T	value in test section
y	immediately behind the nozzle starting shock

Superscripts

*	conditions at nozzle throat
'	values for salt solution

Commentary on Chapter 1-4

G. Winterfeld

I'd like to make two general remarks on experimental investigation of supersonic combustion. First, presently used facilities allow stagnation temperatures that correspond to flight Mach numbers of about 6-7. This is at the lower limit of the range of application of supersonic combustion. Dr. Swithenbank has shown the way to simulate combustor inlet conditions corresponding to flight Mach numbers above 10. Although the technique itself is not new, it provides versatile facilities at relatively low cost. The disadvantage of this method is the extremely short running time, which, as one has to deal with combustion in non-premixed gases, requires extensive development work in the field of testing and measuring techniques. However, for flight Mach numbers above 10 there is no alternative, if one disregards flight tests.

However a considerable number of problems, which have to be solved experimentally, lie in the range of flight Mach numbers between 6 and 10. Examples include the problem of ignition and starting of the reaction processes. In order to study these problems it is sufficient to equip the existing facilities with arc heaters, thus avoiding all the complications in test techniques. Additionally this method offers the advantage of longer running times as well as the use of more conventional measuring techniques.

The second problem arises in connection with the increase in stagnation temperatures and is related directly to our method of experimental investigation. At the inlet of a real scramjet combustor the air is in chemical equilibrium. On the other hand, in a ground test facility, a high enthalpy equilibrium gas is expanded through a de Laval nozzle in a time that is relatively short compared with recombination times. This results in a non-equilibrium gas at the combustor inlet, the state of which is not very well known. As a consequence, all ignition processes are influenced, as has been shown by experiments on shock induced combustion. Although in these cases the special method of introducing the fuel has played a considerable role, a rough calculation shows that in the range of 1800 to 2000°K stagnation temperature the deviation from equilibrium of the air itself can be of importance. If we wish to transfer results on ignition tests in a ground facility to a real scramjet combustor, we have to consider these deviations from equilibrium. A possible way to do this could be to define a higher equivalent temperature that corresponds to the static temperature at the combustor inlet so as to make the ignition delays for both temperatures the same. This requires further investigation of recombination processes in nozzle flow as well as systematic investigations on ignition delays in shock induced combustion.

1-5

Shock Tube Techniques for Fuel Droplet Combustion Studies

F. JAARSMA and W. MERKSEN
N. L. R., Amsterdam, Netherlands

Summary

Several investigators have been interested recently in the interactions of shockwaves with several physical disturbances in the medium such as vortices, pressure waves, heat sources, flames and liquid or solid particles. The passage of a shockwave over such a disturbance usually causes distortions of the wave, of the flow field behind the shock and of the disturbance itself. The present paper treats the interaction between shockwaves and burning or non-burning fuel droplets in a pure oxygen atmosphere. Particular attention is given to the distortion of the initial disturbance due to the passing shockwave. This means that the droplet shattering process is studied in the burning and non-burning mode as well as the distortion of the flame and the wave.

In the open literature much work is done on the droplet shattering process behind shockwaves, but the initial shockwave Mach numbers are only limited to about 2. On the other side it has been shown that a detonation wave can be generated in oxygen containing a fuel mist, so that there exists a gap of experiments on the mechanism of droplet shattering and flame dynamics between M=2 and the detonation Mach number of about 5. Therefore the investigations which have been carried out cover the range of extreme weak shocks (M = 1.04) up to shockwaves of the detonation speed in pure oxygen atmosphere at different initial pressures, containing a single, well defined burning or non-burning fuel (Diethylcyclohexane) droplet.

The first part of the Chapter pays attention to the calibration of the shock tube. The determination of the properties of the flow behind the shockwave becomes difficult when the shockwave Mach number approaches unity since the accuracy of these properties becomes one or two orders of magnitude less than the accuracy at which the Mach number of the shock can be determined. This problem has been solved by proper detector signal amplification and pressure measurements behind the shock.

Furthermore a description is given of the shock tube, droplet injection system, ignition system, Cranz-Scharding optical observation system, triggering sequence and safety precautions. The experimental technique is compared with other techniques on studies of the droplet and liquid jet shattering process.

The experimental results show that at almost all shockwave Mach numbers and initial pressures (from 0.2 to 4.7 ata) the fuel droplets shatter in the sheartype mode. Except at very low Weber numbers the characteristic deformation time and break-up time or burning time depend primarily on the dynamic pressure of the flow behind the shock and also weakly on the pressure step across the shock. These characteristic times can be reasonably well predicted from a simple theoretical model.

The results indicate further that the question of flame blow-off or no flame blow-off from a burning droplet depends strongly on the initial oxygen pressure in the shock tube and only weakly on the shockwave Mach number and initial droplet size. The critical blow-off pressure is about one atmosphere. Below this pressure the flame consistently blows off; at higher pressures the flame stays attached to the droplet while it is shattering.

It has been shown that the flame, even if it stays attached to the droplet, does not alter the shattering process. Also the drag coefficient of the displacement of the front of the droplet is not influenced by the existence of a flame. This drag coefficient stays constant at about 2.5 at Reynolds numbers (based on the initial droplet diameter) ranging from 10^3 to 6×10^5 in subsonic, transonic and supersonic flows.

Series of unique time instant photographs of the interaction process between shocks and flames around fuel droplets are given. The pictures clearly show the development of the torus shaped vortex of the initial flame behind the shock. Also unique pictures of the burning and shattering droplets in subsonic, transonic and supersonic flows are presented.

Sommaire

De nombreux expérimentateurs étudient actuellement l'interaction des ondes de choc avec diverses perturbations physiques dans des milieux tels que des tourbillons, des ondes de pression, des sources de chaleur, des flammes et des particules solides ou liquides. Le passage d'une onde de choc sur une telle perturbation amène généralement une distorsion de l'onde de choc, de l'écoulement derrière le choc et de la perturbation elle-même. Dans le présent exposé on examine l'interaction d'ondes de choc et de gouttelettes de combustible brûlant ou non dans de l'oxygène pur. Une attention particulière est donnée à la distorsion de la perturbation initiale sous l'effet du passage d'une onde de choc. Le processus de fragmentation de la gouttelette est étudié en présence ou non de flamme, ainsi que la distorsion de la flamme et de l'onde de choc.

Il existe de nombreuses publications sur la fragmentation des gouttelettes derrière une onde de choc, mais celle-ci ne dépasse pas un nombre de Mach de 2. On peut par ailleurs montrer qu'on peut produire une onde de détonation dans une suspension de combustible dans de l'oxygène, de telle sorte qu'il n'existe pas de données expérimentales sur la fragmentation des gouttelettes et la dynamique des flammes entre M = 2 et M = 5 (onde de détonation). C'est pourquoi les essais rapportés ici comprennent le domaine de Mach allant des chocs très faibles (M = 1,04) jusqu'à la détonation dans l'oxygène pur, à diverses pressions initiales, mais toujours avec une gouttelette d'un fluide bien défini (diethylecyclohexane) avec ou sans combustion.

La première partie de l'exposé définit l'étalonnage du tube à choc. La détermination des propriétés de l'écoulement en aval de l'onde de choc devient difficile lorsque le nombre de Mach du choc est voisin de l'unité, puisque la précision sur la détermination des diverses grandeurs est d'un ou deux ordres inférieure à la précision de la mesure du nombre de Mach. Ce problème a été résolu par emploi d'un détecteur convenable pour les mesures de pression en aval du choc.

On décrit en détail le tube à choc, l'injection de gouttelette et le système d'allumage, ainsi que la visualisation par le procédé Cranz-Scharding, le cycle de mesure et les précautions prises pour la sécurité des essais. On compare la technique expérimentale utilisée aux autres techniques de fragmentation de gouttelettes ou de jets liquides.

Les résultats expérimentaux montrent qu'à presque tous les nombres de Mach et toutes les pressions initiales (de 0, 2 à 4, 7 bars) les gouttelettes se fragmentent par cisaillement. Sauf pour des nombres de Weber très faibles, le temps caractéristique de déformation et le temps caractéristique de fragmentation ou de combustion dépendent essentiellement de la pression dynamique de l'écoulement en aval du choc et un peu du saut de pression à travers le choc. Ces temps caractéristiques peuvent être prévus avec une bonne précision par la théorie.

Les résultats indiquent aussi que l'extinction ou non-extinction de la flamme dépendent de la pression initiale de l'oxygène dans le tube à choc et très peu du nombre de Mach du choc et du diamètre initial de la goutte. La pression d'extinction est de l'ordre du bar. Pour des pressions inférieures la flamme est systématiquement soufflée; à des pressions supérieures la flamme reste attachée à la gouttelette qui se fragmente.

On a pu montrer que bien que la flamme rests attachée, le processus de fragmentation n'est pas altéré. De même le coefficient de résistance correspondant au déplacement du front de la gouttelette n'est pas modifié pas l'existence de la flamme. Ce coefficient de résistance à une valeur constante de l'ordre de 2,5 pour des nombres de Reynolds (calculés avec le diamètre initial de la goutte) allant de 10^3 jusqu'à 6.10^5 en écoulements subsonique, transsonique et supersonique.

On présente des prises du vues instantanées montrant le processus d'interaction entre le choc et la flamme entourant des gouttes de combustible. Les photographies montrent clairement le développement d'une tourbillon torique dans la flamme initiale en aval du choc. Des vues instantanées de la combustion et de la fragmentation des gouttelettes en écoulements subsonique, transsonique et supersonique sont également présentées.

Introduction

Shock tubes have so far been used mainly to measure physical and chemical phenomena within the shock wave or in the region compressed and heated by shock at high shock wave Mach numbers. This Chapter however deals with one aspect of the interaction of relative weak shocks with physical disturbances in the medium such as vortices (1), heat sources (2), (3), flames (4) and liquid or solid particles (5), (6), namely the interaction of shock waves with burning and non-burning fuel droplets.

A great deal of work has already been done on the fragmentation or shattering of liquid droplets by a passing shock wave. These experiments were motivated by two interests; the disintegration of rain drops in the vicinity of airfoils flying at supersonic speeds; and the disintegration of fuel droplets in liquid rocket engines operating under unstable conditions. The latter was particularly strong because in the development of large liquid rocket engines, combustion instabilities in the screaming mode appeared to be very destructive. This screaming mode was usually associated with shock type pressure fronts and therefore often referred to as a transverse or spinning detonation (7). The sustaining mechanism of these detonative instabilities is not understood but its conduct is very nonlinear. One of the driving mechanisms is very probably the shattering process of propellant jets and droplets at the injection plate (7).

Webber (8) has shown that a burning kerosine mist in an oxygen atmosphere can easily amplify a shock wave if its initial strength is above a critical, though small value. He also showed that non-burning fuel mist in oxygen can sustain a detonative wave. This has been verified by Cramer (9) and Nicholls (10) under more controlled test conditions. From these experiments it was concluded that the burning rate of liquid fuel droplets behind shock waves is much higher than can be predicted from droplets burning under forced convection only. This increase in burning rate must

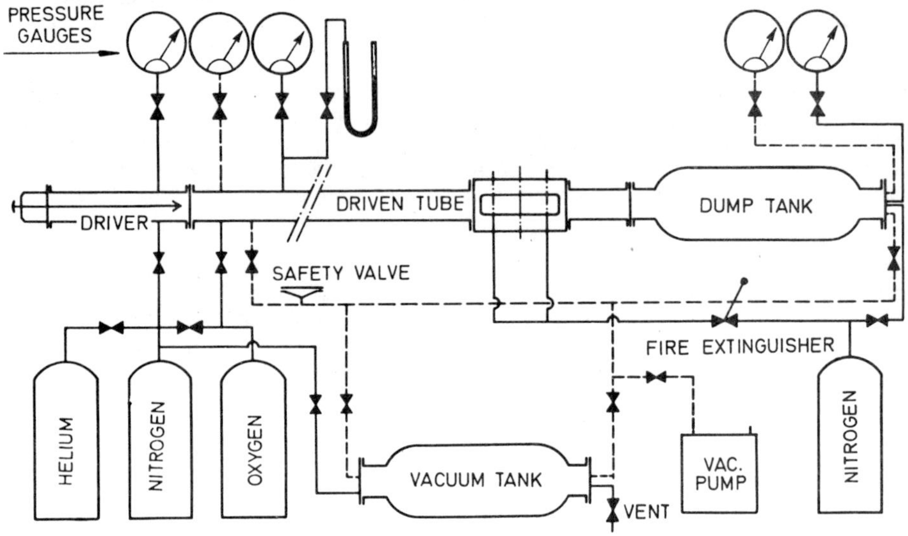

Fig. 1-5-1 Diagram of the shock tube

Fig. 1-5-2 The shock tube

be due to distortion and fragmentation of the individual fuel droplets.

Engel (11), Hanson (12), Dickerson (13), Wolfe (14) and their co-workers have studied the disintegration of liquid drops behind shock waves in shock tubes varying the diameter, surface tension, density and viscosity of the droplets. Rabin *et al* (15) were the first to determine the interaction process between shock waves and burning or non-burning fuel droplets. Gordon (16) proposed a theoretical model of the shattering process and derived an equation for the total breakup time as a function of the dynamic pressure of the flow around the droplet, droplet diameter, liquid surface tension, density and viscosity. The experimental work by Wolfe (14) indicated that the trend of this theoretical model was correct, but that the theoretical breakup times were a factor three times the average experimental values. Morell (16) (17) and Clark (18) studied the similar problem of liquid jet breakup caused by a transverse flow. Their results can be compared to some extent with the results of liquid droplet shattering.

All these experiments have been performed at relatively low primary shock wave Mach numbers, never exceeding M_s=2, or low flow velocities. In case of a detonation in a fuel spray in oxygen, the detonation front Mach number is about 5 and the flow behind the leading shock front is supersonic. Droplet shattering under these extreme conditions has not yet been studied. Also the question of what happens to a burning droplet and the flame around it if it is passed by a shock front of considerable strength, has not been yet answered.

A simple realistic model of the shattering process and the resulting change in burning rate and burning time should be very helpful in theoretical work, examples being that by Busch, Laderman and Oppenheim (19), describing the transient in a two-phase detonation, and by Chinitz and Agosta (20), treating the movement of a shock front through a liquid fueled combustion region of a rocket engine. A series of experiments has therefore been carried out at NLR to study the interaction between burning and non-burning liquid fuel droplets and shock waves ranging from M_s = 1.04 to 4.7 at three initial pressure levels of 0.2, 1.04 and 4.7 kg/cm^{2}abs. The fuel used was diethylcyclohexane (DECH). This is a pure compound fuel with properties similar to ordinary hydrocarbon fuels. It has a low vapor pressure so that no excessive decrease in droplet diameter occurs before firing. Many investigators (9), (10), (15) have used DECH for these reasons. The test gas used was commercial oxygen in nearly all experiements.

Experimental Equipment and Procedure

Shock Tube

The experiments were carried out in a conventional brass shock tube 62 mm square with fillet radii of 5 mm in the corners. The driver had a length of 90 cm and could withstand a pressure of 100 kg/cm^2. The driven section had a length of 324 cm. The centre of a 30 cm long transparent test section was located at a distance of 267.5 cm behind the diaphragm separating the low and high pressure sections. The side walls of this test section were formed from 45 mm high by 25 mm thick schlieren quality glass plates mounted flush with the inside of the tube. This section could withstand a pressure of at least 25 kg/cm^2. The testing time was 2 to 4 milliseconds depending on test conditions, which was sufficient for the interaction studies. Figures 1-5-1 and 1-5-2 and also (21) give more details of the shock tube, including the associated systems.

Upon firing, the diaphragm between the driver and the driven section was punctured by a spring actuated spear on the axis of the driver. For high pressure differentials across the diaphragm, aluminum or soft copper sheets were used: these were grooved across the diagonals by a 60 V-shaped roller under constant pressure. For

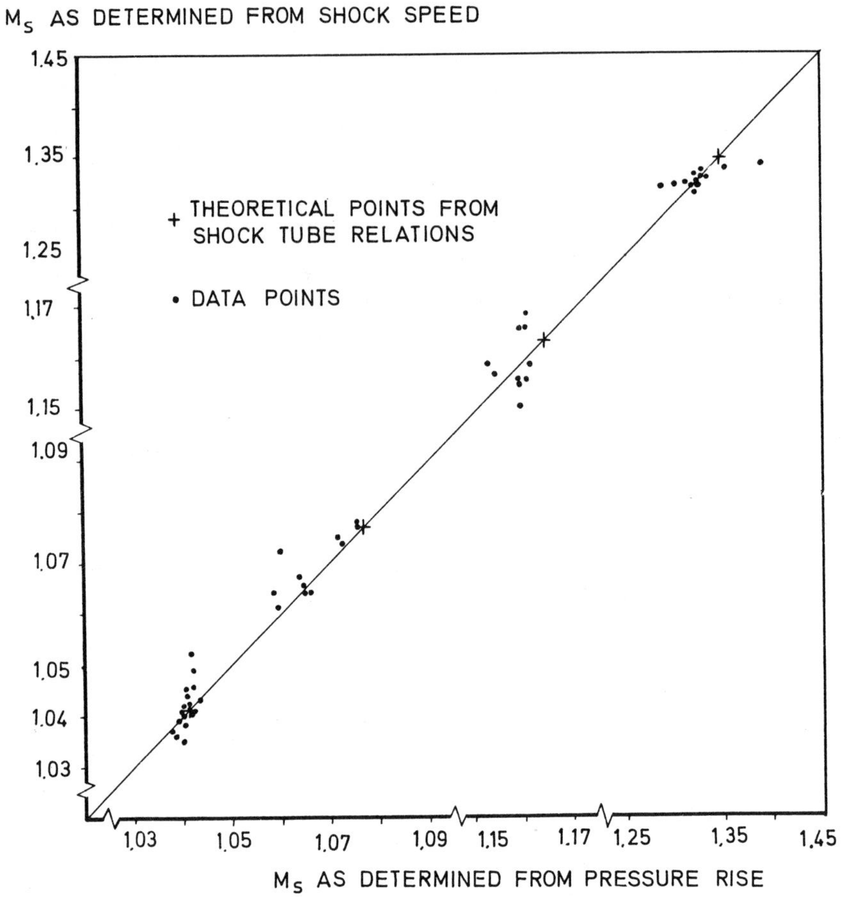

Fig. 1-5-3 Comparison of shock wave Mach numbers (M_S) as determined from different sources.

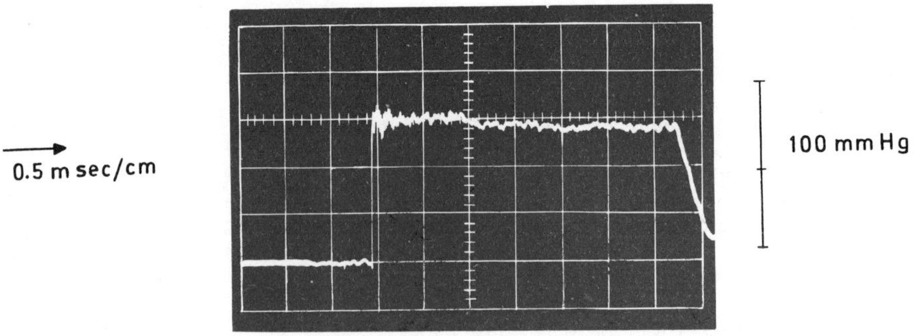

Fig. 1-5-4 Registration of a pressure rise across a M_S = 1,040 shock wave

low pressure differentials, slightly grooved Melinex or cellophane diaphragms had been used. In many cases however, shattering of the diaphragm or non-complete opening occured, causing a substantial reduction in the Mach number of the shock wave. Much better results were obtained with grooved hard brass foils (0.025 mm and 0.05 mm thick) which opened completely, and from which diaphragm material did not become detached. Furthermore,shock wave Mach numbers as measured were closer to the theoretical values.

Safety Precautions

Downstream of the driven section, a stainless steel dump tank with a volume of five times that of the driven section was connected to the tube. The dump tank was separated from the driven section by a weak diaphragm that broke at primary shock wave contact. This prevented the windows of the test section from being over-loaded by the reflected shock. Furthermore the oxygen containing the burning or non-burning fuel droplets would flow into the nitrogen filled dump tank, so that a burning droplet was extinguished by mixing and cooling, as for any fires that might have been caused by the droplet flame. In case one of the two diaphragms did not rupture after droplet ignition, the flame of the droplet could be extinguished by a hand operated nitrogen shower in the test section. It was found that this was necessary when the initial oxygen pressure in the driven section was higher than atmospheric.

When a low initial oxygen pressure in the driven tube was required, the pressure could be rapidly released by dumping the excess pressure into a vacuum, tank since the droplet had to be applied at atmospheric pressure. This prevented the flow of oxygen into the vacuum pump and the evaporation of excess fuel before firing.

The tube operators were protected from possible shattering of the window by thick steel screens.

Determination of Shock Wave Properties

In many shock wave interaction experiments it is desirable to generate very weak shock waves in the shock tube. If the shock wave Mach number approaches unity many properties of the flow behind the shock are proportional to M_s^2-1. This means that by measuring the shock speed only, the accuracy with which these properties can be calculated is one or two orders of magnitude less than that with which the shock wave Mach number (M_s) can be determined. For example, an inaccuracy of 1 % in M_s means a 50 % inaccuracy in the dynamic pressure of the flow behind the shock if $M_s = 1.04$.

Three methods are available for the determination of flow properties: a) the properties may be derived from the one-dimensional shock tube relations starting at the pressure ratio across the diaphragm; b) an attempt may be made to measure the shock speed and speed of sound with extreme accuracy; and c) a property may be measured that is also approximately proportional to M_s^2-1.

With respect to the first method, which was used by Morell (17), appreciable deviations from the theoretical values may occur in practice.

The second method has been followed by Selberg (22), who measured the average shock speed between two points, using as sensors two piezo-electric pressure transducers coupled to two very similar amplifiers. As shock position identification is very important here, it is thought that hot film gauges with a sensiting length of only 1 mm (versus a few millimeters for a pressure transducer) are probably more suitable for this purpose than pressure transducers. The hot film is placed in a Wheatstone Bridge. A small change in electrical resistance excites a highly

Fig. 1-5-5 Test section

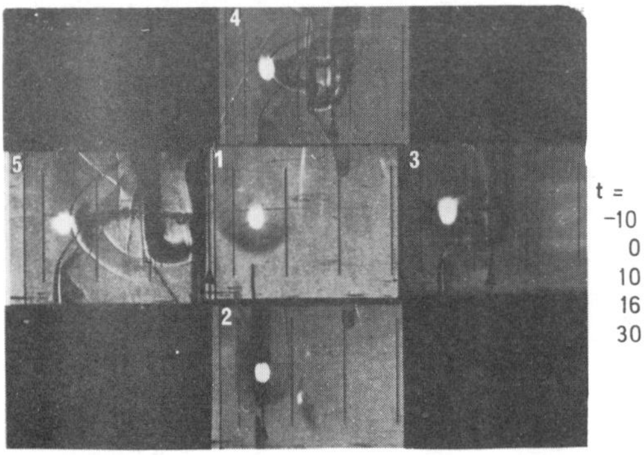

BURNING M_S = 2,57 p_1 = 1,03 D_o = 0,8

▲ = SHOCK WAVE POSITION
t = TIME AFTER SHOCK WAVE PASSAGE (μ sec). PICTURE NUMBERING INDICATES SEQUENCE

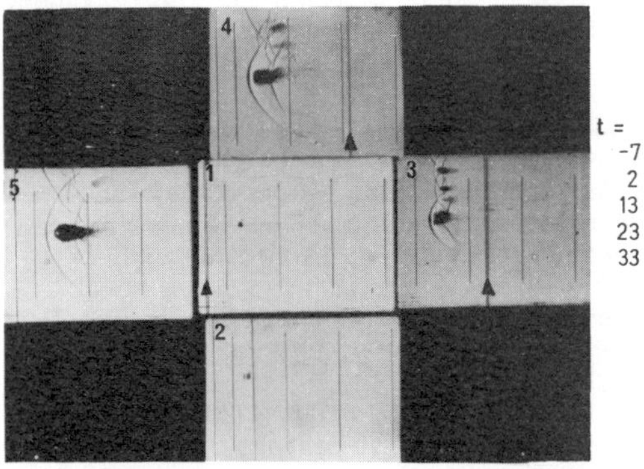

NON - BURNING M_S = 2,59 p_1 = 1,03 D_o = 0,8

Fig. 1-5-6 Example of two film plate showing five schlieren pictures of the fragmentation process of a burning and non-burning droplet.

damped 0.25 MC oscillator, the first pulses of the oscillators stopping or starting a 10 MC counter. Using these hot film detectors, shock waves as slow as $M_s = 1.04$ at NTP, yielding a temperature rise of the film of only 0.04°C, can be detected. Precautions must be taken, however, not to have the oscillator excited by disturbing electromagnetic fields (21). Hot film signal amplifiers based on the use of tunnel diodes are even more sensitive to resistance changes, and less sensitive to external disturbances. Such amplifiers are now under development. Care must be exercised in determining the speed of sound of the test gas, which means that the composition and temperature must be accurately known, the temperature for example, to better than 0.5°C. It is therefore advisable to locate the complete shock tube and gas reservoirs in a temperature controlled room.

The third method of determining the properties of shock waves has turned out to be more accurate at low Mach numbers, as may be seen in figure 1-5-3. In this case, the static pressure behind the shock wave is measured by a piezo-electric pressure transducer (Vibro meter 12 QP 250 c) mounted flush with the wall. These pressure gauges can be accurately calibrated in a quasi-dynamical way and show excellent linearity and reproducibility for pressures amounting to only one percent of their range. Figure 1-5-4 depicts such a pressure recording. The accuracy obtained in this way is better than 5 percent and the uniformity of the pressure level can easily be observed. Dewey approached this shock tube calibration problem by recording the particle paths of smoke traces behind the shock at different stations along the tube (23). The determined flow velocity was again proportional to $M_s^2 - 1$ and he obtained an accuracy of 5 percent.

Droplet Injection and Ignition

In all experiments dealing with the interaction between shock waves and some physical disturbance, it is important to define the disturbance accurately. The characterizing parameters such as position, size and intensity must either be preset or measurable.

For droplet injection into a shock tube many methods have been used. Engel (11) and Wolfe (14) worked with freely falling droplets formed at the tip of a hypodermic needle. In that case the droplet size was not preset, but the approximate size was related to bore diameter.

More uniform freely falling droplets of a predictable size can be generated by vibrating capillars (10). These devices for producing a stream of uniform droplets are based on Rayleigh's analysis of the instability of capillary jets. The production of single droplets is not possible by this method.

For his droplet shattering experiments, Hanson developed an acoustic liquid drop holder (24). The principle of which employed the pressure resulting from acoustic vibration: the vibration field (50 kC) was maintained between a vibrator and a reflector. The droplets stayed at rest at the pressure node and could be injected at these locations by a hypodermic needle.

Other techniques of droplet injection make use of a sudden retraction of a thin wire (15), endothermic needle (13) or web (25) from which the droplet is suspended. The shock wave passes the droplet before it starts to fall. Although the initial volume of the droplet can be accurately predetermined using a syringe, the droplet diameter will be less after retraction because of the wetting of the suspension device and the formation of secondary droplets (satellite droplets). In spite of these disadvantages this technique has been used in the NLR experiments since it is an easy and controllable way of droplet injection, in addition, the droplets can be easily ignited before wire retraction. Burning droplets cannot be supported by an acoustic holder, and free droplets are difficult to ignite. Droplet suspension from

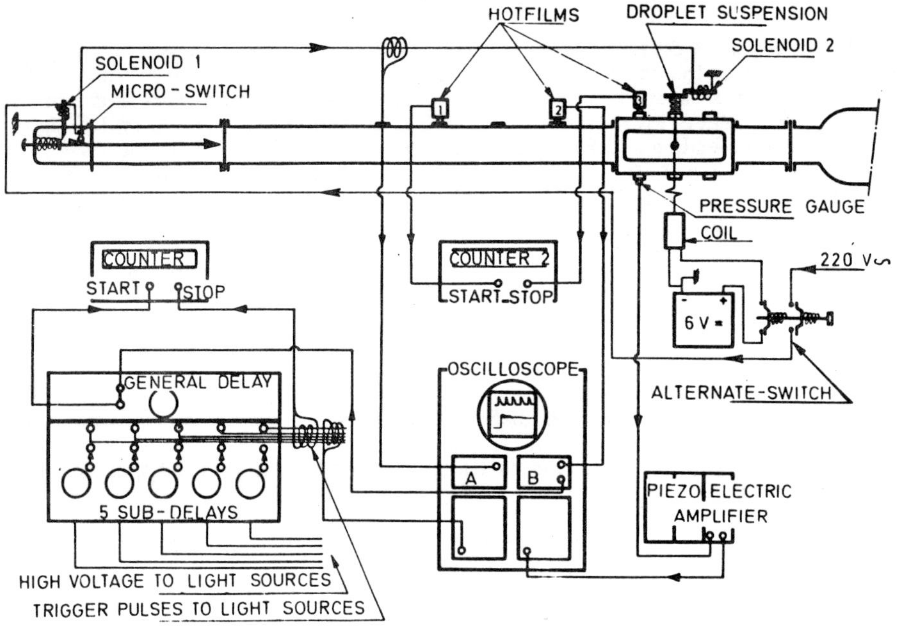

Fig. 1-5-7 Electrical system of the shock tube

syringes is not possible, because after ignition vapour forms in the tube and the droplet immediately falls. The droplets were suspended from a thin (0.5 mm) stainless steel wire which was withdrawn by a strong spring (Figure 1-5-5).

Ignition was accomplished by a short duration spark from an automobile coil. The spark gap was formed by the droplet suspension wire and a stainless steel electrode rising from the bottom of the test section. A few milliseconds before the wire was retracted, the droplet was ignited leaving a free burning droplet. No disturbance of the droplet or flow field behind the shock due to the spark has been detected.

In the initial experiments, an electrically exploding wire beneath the droplet was used as an ignition source. However molten metal particles struck the glass walls pitting the glass surface, as may be seen in some photographs.

Other ignition techniques such as the use of small hydrogen flames (15) seemed to be more troublesome than the method employed.

Optical System

The optical system consisted of a five-fold Cranz-Schardin shadowgraph-schlieren system. This has the advantage over framing and high-speed cameras, which are usually used in the U.S. (13), (14), because the spark sources can be triggered by the event, in this case the shock wave. Furthermore the time intervals between the flashes can be arbitrarily chosen between 1 and 2000 μ sec at each run by adjustable time delays, built by Lunatron, U.K. The effective flash duration of 0.2μsec is short enough for good time resolution. The disadvantage of the Cranz-Schardin system is the change of observation angle of each picture. This angle is however less than one degree with respect to the system axis and can be easily accounted for. The linear magnification of the test section on the plate camera (Polaroid Land) can be continuously changed from 0.1 to 10.

The Cranz-Schardin system is easy and fast to operate. By properly selecting the time interval between the pictures at a series of similar runs a complete insight into the process can be obtained. Figure 1-5-6 yields two picture sequences of a burning and non-burning droplet under equal test conditions. As is shown, the flame of the stationary droplet exposes the polaroid film so that no details in this vicinity can be observed. This film exposure by luminous events could also be considered a disadvantage of the Cranz-Schardin system.

At some runs, streak films were made of the luminescence of the droplet flame by means of a Früngel drum camera running at different speeds.

Operation and Trigger Sequence

Before filling the tube with oxygen and driver gas, the complete tube was evacuated to 1 torr. As soon as the oxygen pressure in the driven section became atmospheric, the fuel droplet was suspended from the spring loaded pin in the test section by putting the syringe through a hole underneath the pin. During this operation a small oxygen out flow from the tube was maintained, preventing air from entering the test section. Immediately after this procedure the oxygen pressure was increased or decreased to its value (p_1) and the driver was loaded. After adjustment of the pressures, the manual shutter of the camera was opened, causing two microswitches to close (Fig. 1-5-7). The first switch caused ignition of the droplets in such a way that the light of the spark did not expose the film. The second switch powered a solenoid which unlocked the spear in the driver. At some adjustable position, the spear, on its way towards the diaphragm, closed a third microswitch which powered the locking solenoid of the droplet suspension pin. Before the shock wave reached the fuel droplet it had passed the two hot films (820 mm and 114 mm from the test

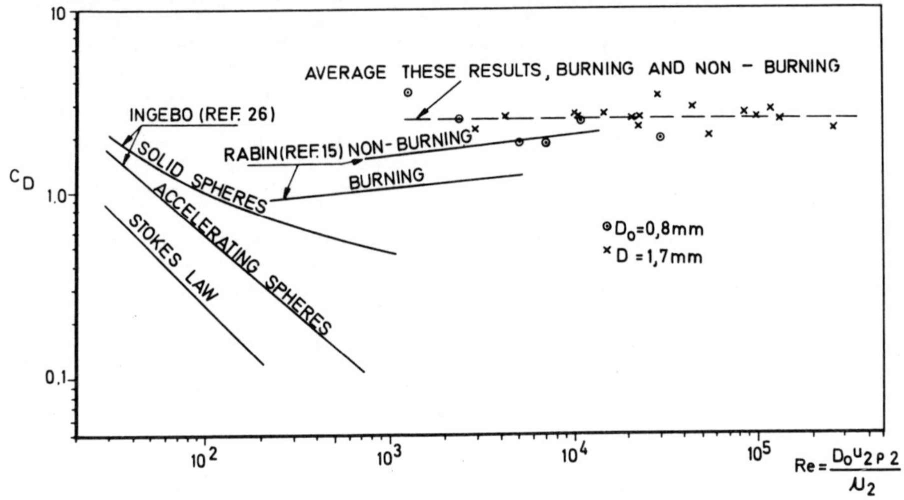

Fig. 1-5-8 Drag coefficients of shattering droplets as determined from the leading side displacements and based on the initial droplet diameter D_o

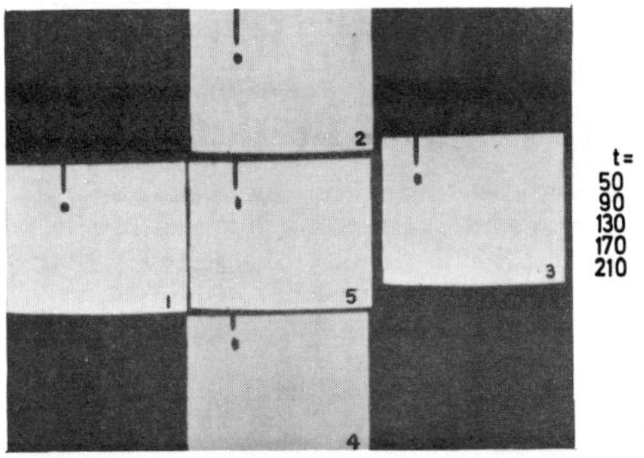

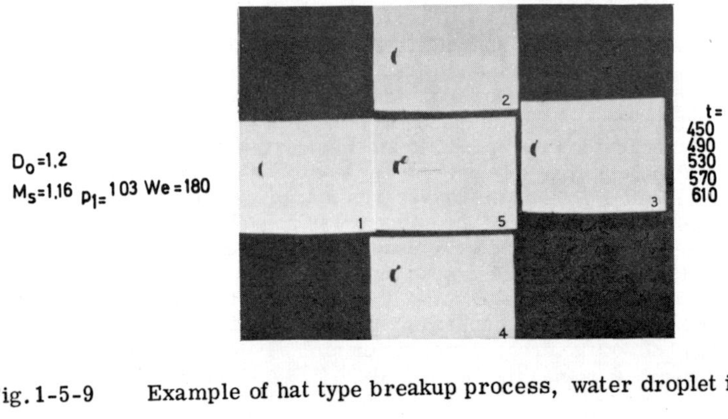

Fig. 1-5-9 Example of hat type breakup process, water droplet in air.

section centre) for shock velocity determination, and a third hot film (520 mm from the centre) for oscilloscope triggering purposes. The Tektronix 555 dual beam oscilloscope recorded the pressure rise across the shock and the times of the spark flashes of the schlieren system. The exact value of the time interval between the trigger pulse and the first spark flash was registrated by a 1 M.C. counter. Determination of the shock position on the first schlieren photograph also yielded information on the shock speed.

Test Results

The features of the interaction process of fuel droplets with shock waves of interest to scientist and engineers working in the field of liquid rocket engines, scramjets and detonation waves in two phase flow systems are the following:

a) The displacement of a characteristic point of a shattering droplet, for example, the windward face.

b) The shattering process that gives the type of shattering (bag, hat or shear type), characteristic deformation time, first microdroplet appearance, droplet widening and complete breakup times.

c) The question whether an already burning droplet will lose its flame by the action of a passing shock wave, the nature of the characteristic process and how it influences the shattering process.

d) The deformation, displacement and widening of the initial hot flame region around a droplet due to the passage of a shock wave and the resulting distortion of the shock wave.

From previous work on droplet shattering (10 to 18) the significant parameters appears to be the droplet diameter, the relative velocity, density and viscosity of the gaseous medium, and the density, surface tension and viscosity of the liquid. The resulting dimensionless parameters turned out to be the Reynolds number ($Re = \rho u D_0/\mu$), Weber number ($We = \frac{\rho u^2 D_0}{\sigma}$) and Mach number of the flow behind the shock.

In the experiments under consideration two droplet diameters were used, $D_0 = 1.7$ mm and $D_0 = 0.8$ mm. The fuel used was diethylcyclohexane, except in the earlier runs where water drops in air were used. Parameters such as liquid surface tension and density were not studied since it was felt that the surface tension had little influence on the fragmentation process if the Weber number was high, as was the case in the NLR experiments, while the density of many practical fuels does not vary much. The variables of gaseous media depend on shock wave Mach number and the initial pressure. The shock wave Mach numbers have been chosen such that the flow velocity behind the shock (u_2) is roughly doubled at each subsequent Mach number. Table 1-5-1 gives the range of test conditions and the characteristic flow properties.

The next paragraphs describe the interesting phenomena and the influence of the important parameters.

Droplet Displacement

Many authors have determined the drag coefficients of droplets and spherical particles. Rabin *et al* (15) and Selberg (22) give comparisons of different data that show a wide scatter. From these comparisons it appears that deviations from the steady state values occur if the particles are accelerating. Burning or evaporating

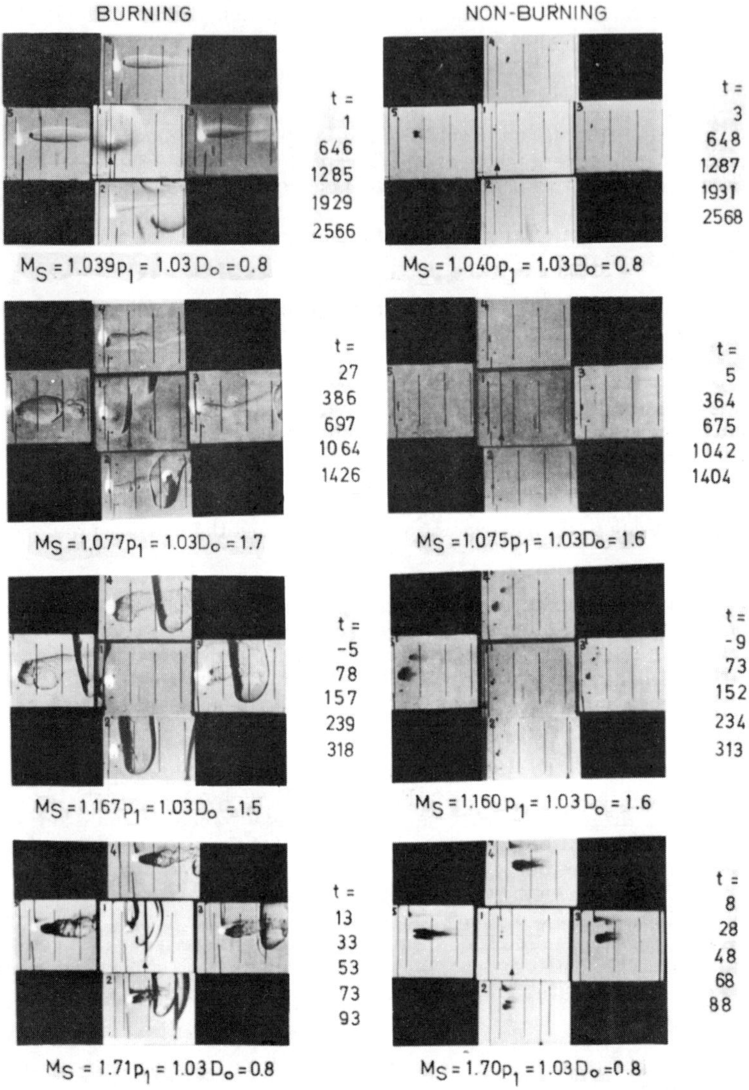

Fig. 1-5-10 Picture sequences showing the shock wave burning and non - burning interaction process under various conditions. See also fig. 1-5-6

droplets show drag coefficients that are somewhat smaller than under steady diameter conditions. Drag coefficients under shattering conditions at high Reynolds numbers have not yet been determined.

From the present experiments, the drag coefficients of the shattering droplets were determined from the displacement of the windward faces, and based on the initial droplet diameter. Measurements on the individual photograph series indicated that these windward faces of the droplets had a more or less constant acceleration during fragmentation. This means that the drag coefficient based on the initial droplet diameter was constant during fragmentation. As far as could be observed, burning and the flow conditions (subsonic, transonic and supersonic) relative to the moving droplet did not appear to have any influence on the drag coefficient. Figure 1-5-8 shows that the drag coefficient was independent of Reynolds number (from 10^3 to 3×10^5) and had an average value of 2.5. Figure 1-5-8 can be considered an extension of Rabins results.

The total droplet displacement before complete breakup is less than 15 initial droplet diameters, so that in many practical cases the shattering droplet can be considered stationary.

Shattering process

If a droplet is subjected to the effects of a sudden increase in the velocity of its surrounding medium, three types of fragmentation of the droplet may occur. At low Weber numbers or low relative velocities, the droplet first deforms into a torus shaped ring with a very thin liquid membrane inside. This membrane expands, taking the shape of a bag and finally breaks down into many microdroplets like a soap bubble. Subsequently the ring, having its axis parallel to the main flow, disintegrates into small droplets which are larger however than those originating from the bag. This bag type breakup process is very slow and occurs only if the aerodynamic forces are just capable of deforming the droplet against surface tension forces (15). This process is adequately described in (12) but was hardly observed at all in these experiments.

If the relative velocity increases, giving an increase of the Weber number, a hat type fragmentation process occurs. In this case the leeward face of the droplet flattens and even becomes concave. The windward face also flattens but stays convex. Around the periphery a small liquid ring is formed which moves streamwise to the droplet back. Here the ring leaves the droplet and fragmentates. In the meantime the droplet is flattened completely, perpendicularly to the flow direction, and loses small droplets at the periphery. Figure 1-5-9 gives a picture sequence of this process of a 1 μ litre water droplet in air. This process occurs at Weber numbers up to 200 approximately.

The most important and most frequently observed shattering process is the shear type process that occurs at all Weber numbers in excess of 200. Microdroplets are sheared off the liquid surface, probably due to aerodynamic shear forces, while the droplet is flattened perpendicularly to the flow direction. On the surface of the droplet, small waves are generated which yield the microdroplets. Figure 1-5-10 gives twelve picture sequences of the fragmentation process in subsonic, transonic and supersonic flows at different pressure levels.

The time interval between the passage of the shock wave and the first appearance of microdroplets is an important factor, since it determines for example the beginning of the reaction zone in a two-phase detonation if it is assumed that it is shattering that limits the reaction rate. These time intervals could not be determined accurately in the present experiments due to insufficient optical resolution and lack of information on the sizes of the first microdroplets. What could be determined was

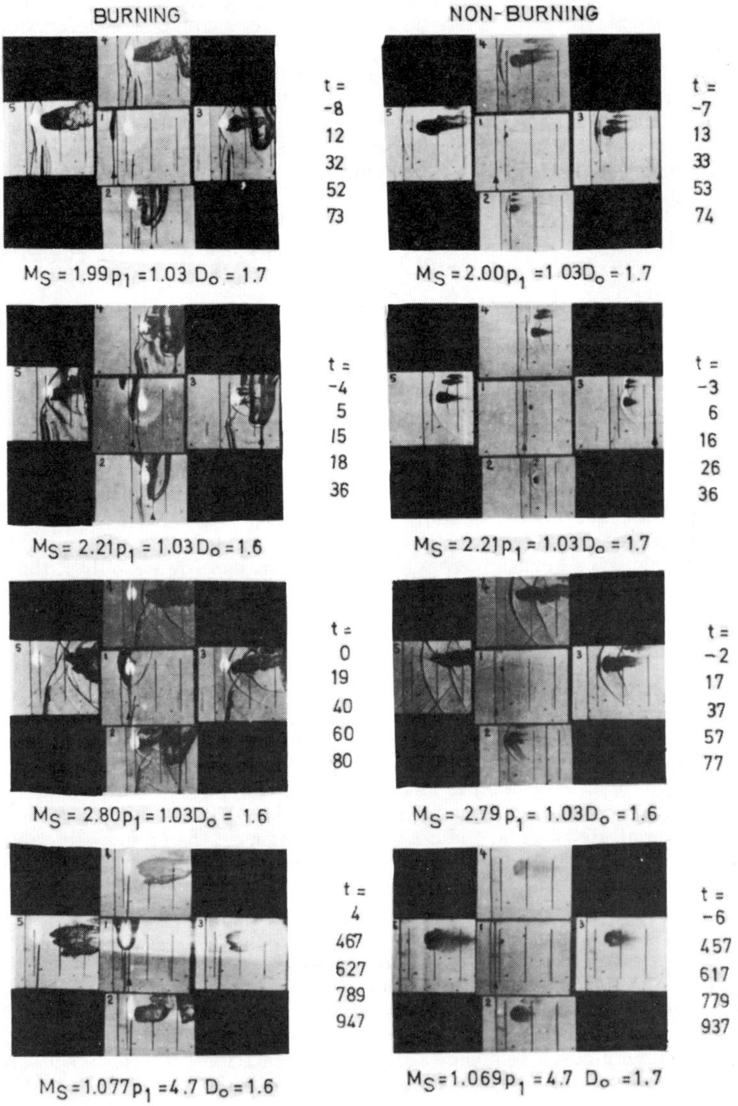

Fig. 1-5-10 Continued

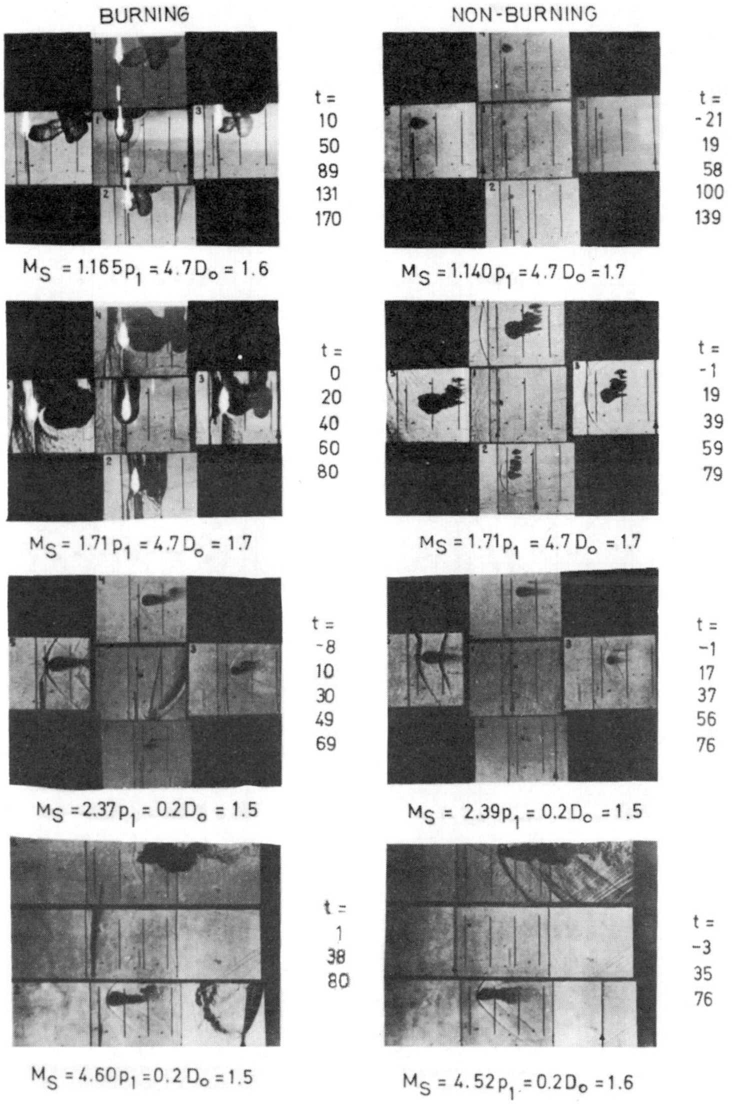

Fig. 1-5-10 Concluded

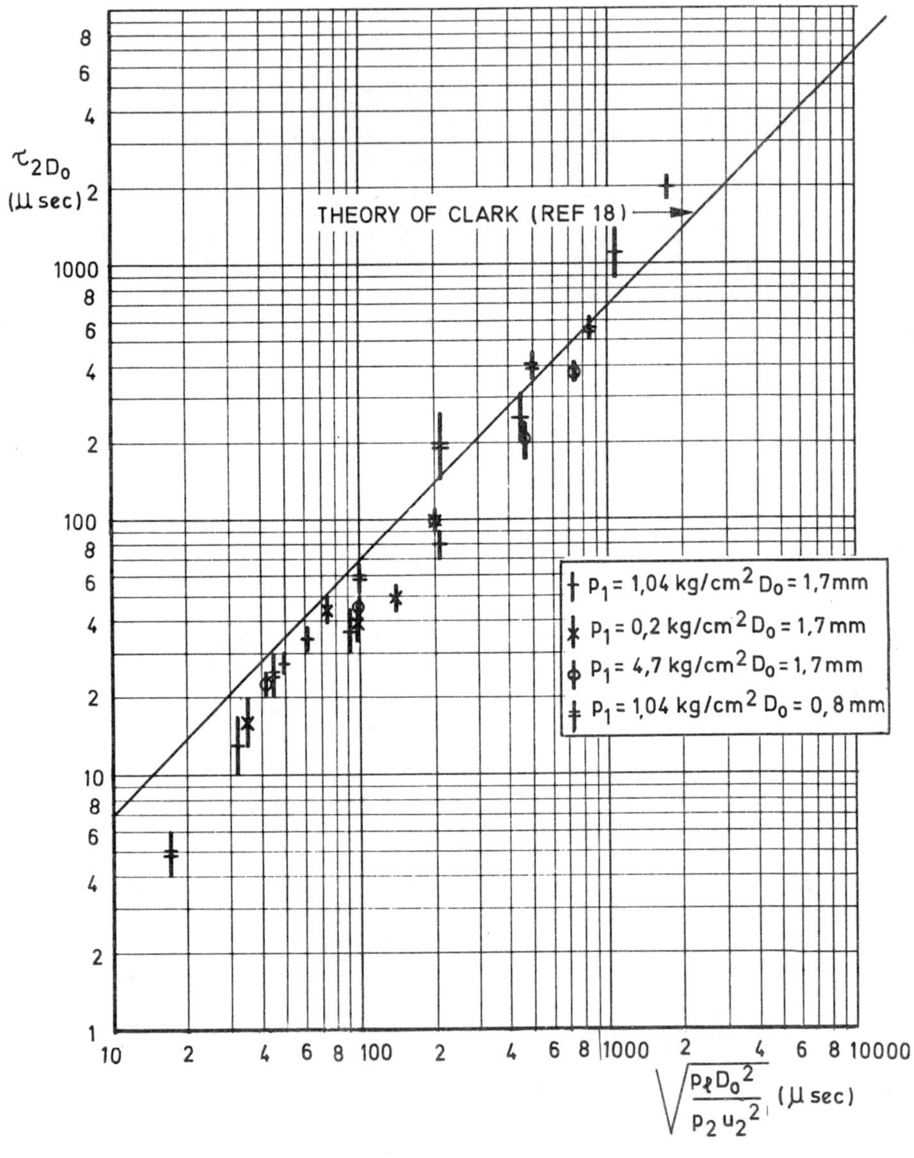

Fig. 1-5-11 Characteristic deformation time

the time interval in which microdroplets could be definitely observed. Table 1-5-2 gives these time intervals for the various test conditions, also including the appearance times of microdroplets released from satellite drops. It appears that the smaller the droplet, the sooner microdroplets were formed, indicating that inertia forces of the liquid were involved and not the surface tension. At strong shock waves microdroplets are formed within only a few microseconds they assume almost instantaneously the flow velocity.

It is felt that the time of first microdroplet appearance is closely related to a characteristic deformation time, which can be determined with a higher accuracy. In this case the characteristic deformation time has been defined as the time interval in which the droplet width has doubled (τ_{2D_0}). If the relative gas velocity (u_2), the gas density (ρ_2), liquid density ($\rho \ell$), droplet diameter (D_0) and the characteristic time (τ) are assumed to be the important parameters in deformation and shattering, the dimensionless number $\rho_2 u_2^2 \tau^2 / \rho \ell D_0^2$ can be formed according to ref. (18). This number can be derived from dimensional analysis but also from considerations of liquid motion perpendicular to the flow direction due to aerodynamic pressure forces. Neglecting viscosity and surface tension, Gordon (16) arrived at the same result. Making use of Gordon's analysis it has turned out that viscosity can be completely neglected here, but at low Weber numbers (< 100) the dimensionless group may be expressed as:

$$\rho_2 u_2^2 \left(1 - \frac{16}{We}\right) \tau^2 / \rho \ell D_o^2$$

Figure 1-5-11 represents the characteristic deformation time τ_{2D0} as a function of $\sqrt{\frac{\rho \ell D_0^2}{\rho_2 u_2^2}}$ showing to some degree a linear relationship. The simple theoretical model of Clark (18) is also included and shows a good order of magnitude fit of the experimental data. However, when the flow behind the shock has low dynamic pressure, the model underestimates the characteristic deformation times, with at high dynamic pressures the model gives an overestimation. If at low Weber numbers surface tension is included, the characteristic deformation time τ_{2D0} can be expressed as:

$$\tau_{2D0} = 0.5 \sqrt{\frac{\rho \ell D_0^2}{\rho_2 u_2^2 \left(1 - \frac{16}{We}\right)}}$$

Under the same experimental conditions it has been shown that flame attachment does not alter the fragmentation process as may be seen in figure 1-5-10. The time of complete burning and breakup was found to be equal, as for the displacement of burning and non-burning droplets. In order to determine more accurately the burning or breakup time, streak films were made of the self luminosity of the flame attached to the droplet: Figure 1-5-12 is a typical example, showing accurately the time of complete breakup. Figure 1-5-13 gives these burning times (τ_B) as a function of $\rho \ell D_0^2 / \rho_2 u_2^2$. Probably at low Weber numbers a correction similar to the above can be applied. Again a linear relationship is obtained for various pressures and droplet diameters yielding:

$$\tau_B = 4.5 \sqrt{\frac{\rho \ell D_0^2}{\rho_2 u_2^2}}$$

These data are compared in figure 1-5-13 with the burning times theoretically due

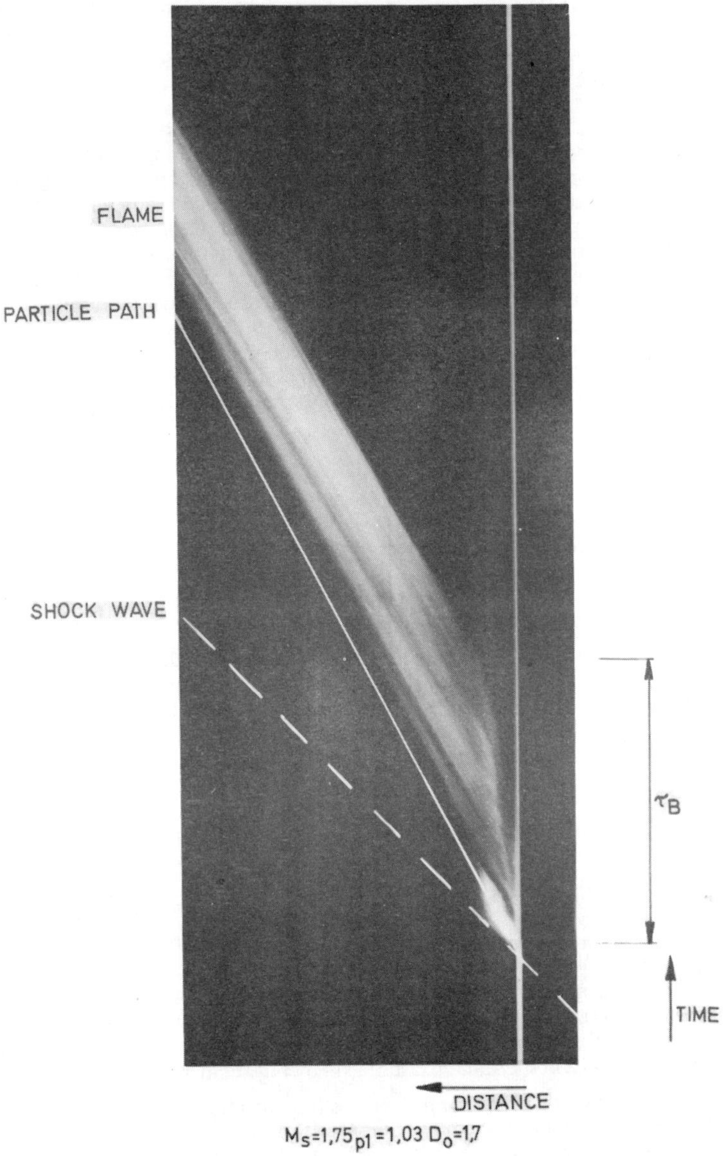

Fig. 1-5-12 Example of streak photograph

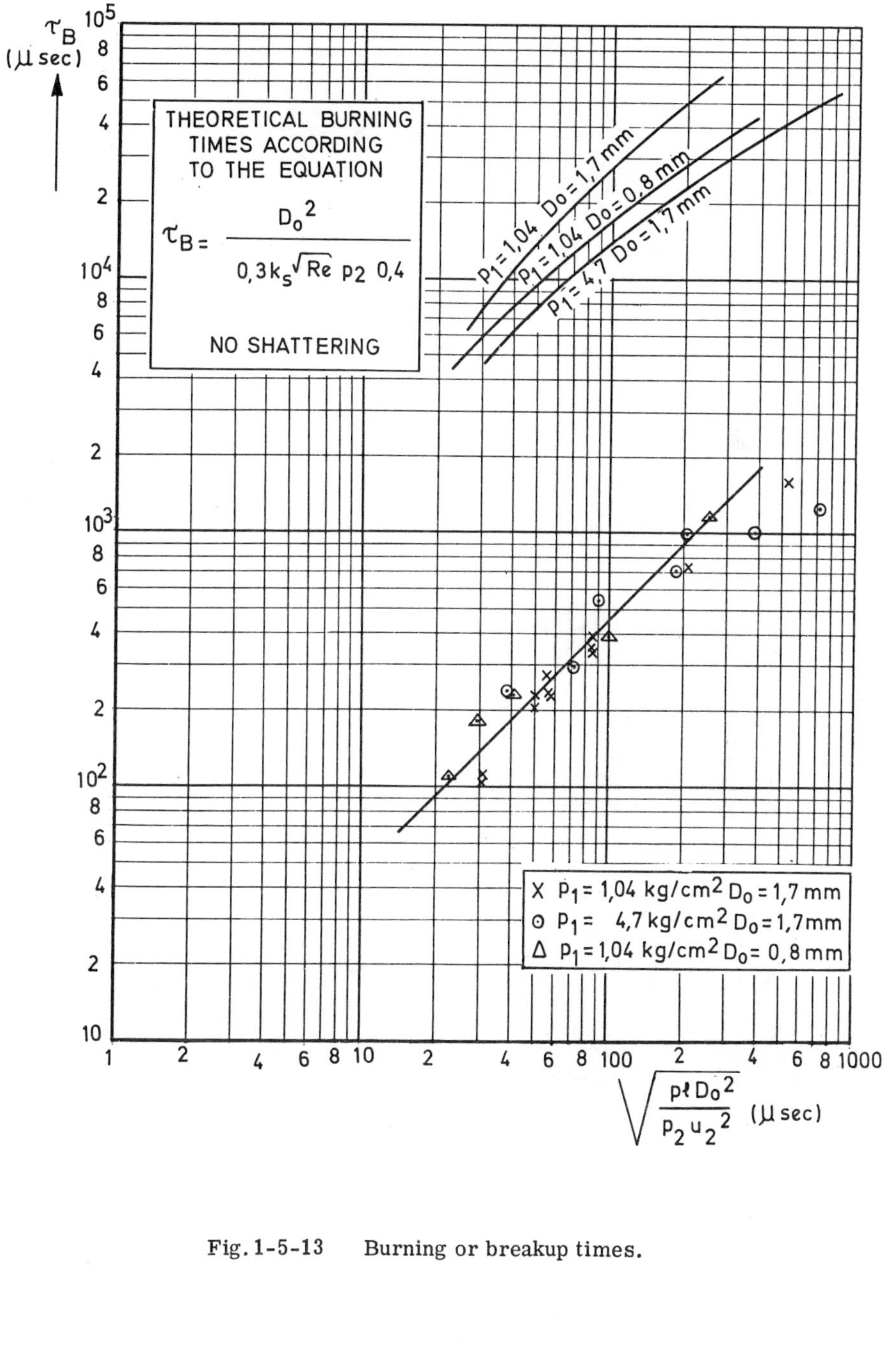

Fig. 1-5-13 Burning or breakup times.

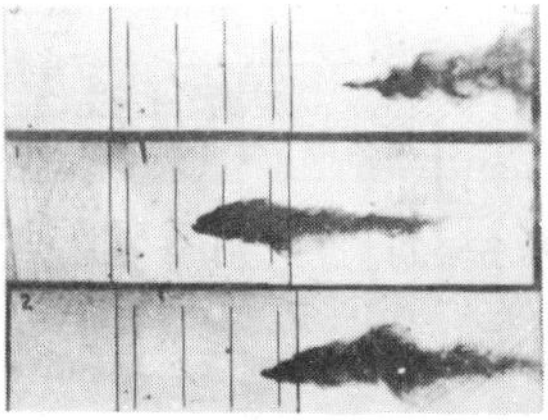

t =
1200
1296
1402

$M_S = 1.67\ p_1 = 1.03\ D_o = 1.6$

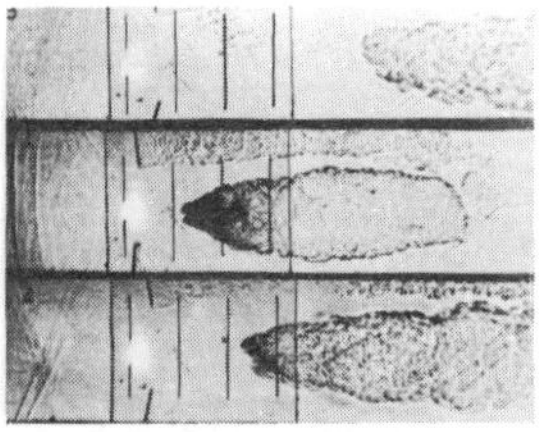

t =
1200
1296
1402

$M_S = 1.67\ p_1 = 1.03\ D_o = 1.7$

Fig. 1-5-14 Complete breakup and burning

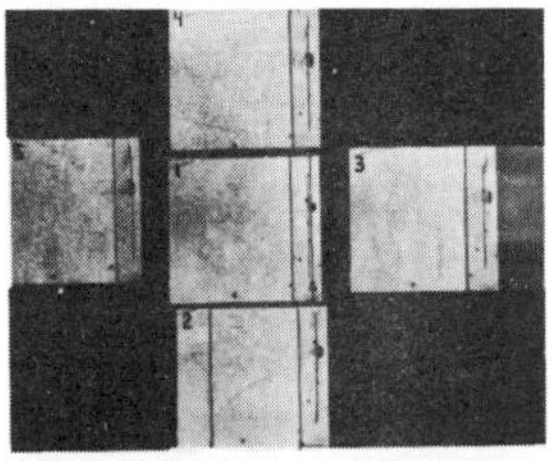

t =
0
80
160
240
320

$M_S = 1.56\ p_1 = 1.03\ D_o = 1.6$

Fig. 1-5-15 Droplet deformation due to two shock wave passages only. Water in air

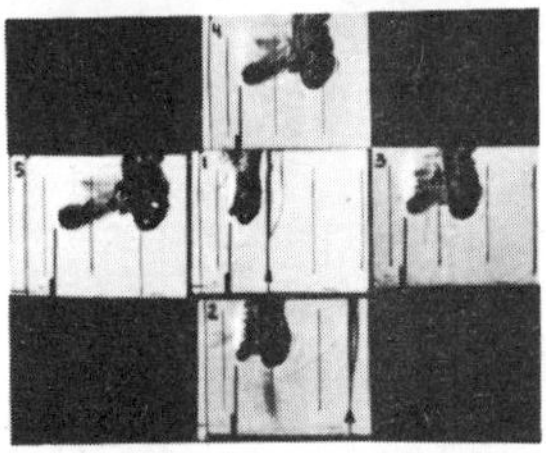

t =
19
60
78
121
160

$M_S = 1.15\ p_1 = 4.7\ D_o = 1.7$

Fig. 1-5-16 Possible flame attachment of DECH fuel in air at elevated pressure

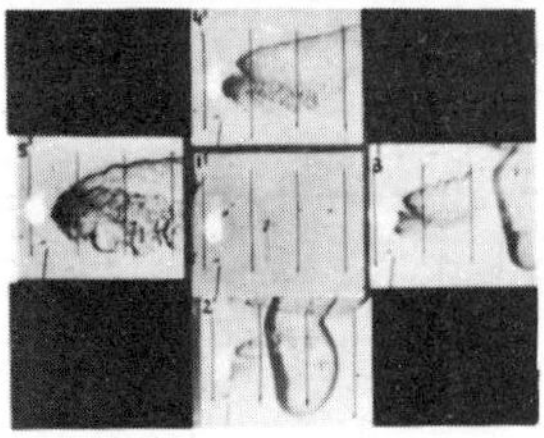

t =
-11
149
309
470
625

$M_S = 1.15\ p_1 = 1.03\ D_o = 1.5$

Fig. 1-5-17 Reignition of the micro-droplet cloud by a satellite droplet flame to forced convection if shattering did not occur. This empirical relationship is obtained from (27) and (28) by taking into account a 0.4 power law of the pressure dependence. The burning rate (k_s) has been taken as 0.084 cm^2/sec as experimentally determined. It can be seen that the burning rate controlled by shattering is 2 orders of magnitude larger than that due to convection only.

It is apparent from the picture series of figure 1-5-10 that the Mach number of the flow behind the primary shock (M_n) did not influence the fragmentation although the flow field altered appreciably. The fragmentation rate could not be observed in these experiments. In general however, the rate of microdroplet formation increased with time after shock passage and reached a maximum at about half the complete breakup time. Subsequently the rate was decreasing due to a decrease in droplet size, until complete breakup, as may be seen in figure 1-5-14. The microdroplet distribution was not uniform in space, clouds usually being formed. The total width of the complete microdroplet cloud was about one order of magnitude larger than the initial droplet diameter.

Although the characteristic deformation and breakup times can be correlated with properties of the flow behind the shock, the influence of shock wave passage can probably not be neglected in every respect. During the time interval of shock wave passage (order of one μ s) the droplet is subjected to a high pressure differential across its faces. This pressure pulse will start to flatten the droplet perpendicular to the flow direction. Order of magnitude estimation indicates that this phenomenon causes a characteristic deformation time that is 3 to 10 times longer than that observed. This has been confirmed by some runs in which a suspended water droplet was located extremely close to the end plate of the shock tube. A shock wave passed the droplet twice (primary and reflected shock) leaving the droplet in a motionless atmosphere except for the time interval between both shock passages, taking only a few microseconds. The droplet was also flattened as may be seen in figure 1-5-15. The deformation rate was about 30 percent of that experienced at a single shock passage.

Flame Blow-off

So far no information was available as to whether a burning droplet would lose its flame if a shock wave were passed over it. If the flame stayed attached to the droplet, shattering would have meant a substantial increase in burning rate as has been shown in the preceding paragraph. However, if the flame blew off and if no other flame or hot region was available to reignite the microdroplets leaving the main droplet, a shock wave passage would have meant a complete quenching.

Under the conditions existing in the shock tube (pure oxygen atmosphere, single droplet and DECH fuel) it was observed that the answer to this question depended primarily on the initial pressure. At all Mach numbers at 0.2 kg/cm^2abs the flame consistently blew off the droplet. At N.T.P. the flame stayed attached to the droplet in about half the runs, independent of Mach number. At 4.7 kg/cm^2abs the flame did not blow off the droplet. This phenomenon was also observed at the much smaller satellite droplets so that it can be concluded that flame blow-off is independent of droplet diameter and shock wave Mach number.

Estimation of order of magnitude indicates that the blow-off did not depend on shattering which can be considered an aerodynamic process. Some runs were performed in air at N.T.P. and at 4.7 kg/cm^2abs. At the lower pressure the flame blew off but at the higher pressure the flame had a tendency to stay attached, as may be seen in figure 1-5-16. This indicates that flame blow-off depends on partial oxygen pressure and that the process is governed by chemical kinetics. Figure 1-5-17 shows the rapid ignition of the microdroplet cloud by the flame which stayed attached to a satellite droplet indicating a high flame propagation speed in such a cloud, and fast chemical reactions.

The last pictures of figure 1-5-10 show the shattering and ignition of a non-burning droplet by a shock wave which approached the detonation speed. The intense luminosity was clearly visible during the run.

Interaction of Shock Waves with the Hot Initial Flame Regions

The phenomenon of the interaction of shock waves with hot regions is well described for similar cases by Hamernik (2), Dosanjh (3) and Markstein (6) although the shock wave Mach numbers in these investigations were small, not exceeding 1.35. Little attention will therefore be paid to it here.

The shock front assumes a curved, convex shape when passing through the flame. Due to the acceleration of this shock wave an expansion wave is emitted upstream from the cold-hot interface, causing the centre of the flame region to move at greater velocity than the average velocity behind the shock. This generates a toroidal vortex of the flame region as it is moving down the tube. This vortex formation is clearly shown in the photographs of figures 1-5-6, 10 and 17. The flow acceleration due to expansion during the first instant of interaction, causes the droplet to experience a higher flow velocity than under non-burning conditions. This increase in flow velocity has no noticeable influence on the shattering process.

Concluding Remarks

It is shown that under liquid rocket engine conditions, the burning rate of the liquid droplets near the injection plate can be increased many orders of magnitude by shock waves. The increase depends on the dynamic pressure of the flow behind the shock and it is due to shattering and flame attachment. From these results the length of the reaction zone behind the shock can be computed if the shock strength and droplet size distribution are known. It is felt that if the length of this reaction zone is less than a characteristic length of the combustion chamber, for example, diameter, a back and forth bouncing or spinning detonation like wave may develop. This occurs when an initial pressure disturbance is strong enough to produce the required short reaction zone length. On the other hand if the characteristic length of the burner is small a strong disturbance will not cause instability.

This trend has been well observed in the experiments of Ref. (7). The effectiveness of baffles in the combustion chamber near the injection plate may also be explained in the same way.

References

1. **Dosanjh, D.S., Weeks, T.M.**, 'Interaction of a Starting Vortex as well as a Vortex Street with a Travelling Shock Wave', AIAA J., Vol. 3, No. 2.

2. **Hamernik, R.P.**, 'Interaction of an Advancing Shock Front with a Concentrated Heat (Entropy) Source', M.Sc. Thesis, Syracuse University, Jan. 1967.

3. **Dosanjh, D.S., Hamernik, R.P.**, 'Interactions of an Advancing Shock Front with a Concentrated Heat (Entropy) Source', 6th Int. Shock Tube Symp., Freiburg, April 1967.

4. **Markstein, G.H.**, 'A Shock Tube Study of Flame Front-Pressure Wave Interactions', Sixth Symp., (Int.) on Combustion, Reinhold, N.Y. 1956.

5. **Marble, F.E.**, 'A Gas Containing Small Solid Particles', Fifth AGARD

Colloquium, Braunschweig 1962, Pergamon Press.

6. **Nayfoh, A.H.,** 'Shock Wave Structure in a Gas Containing Ablating Particles' The Physics of Fluids, Vol.9, No.12, Dec. 1966.

7. **Levine, R.S.,** 'Experimental Status of High Frequency Liquid Rocket Combustion Instability', Tenth Symp. (Int.) on Combustion, Combustion Inst., 1964.

8. **Webber, W.T.,** 'Spray Combustion in the Presence of a Travelling Wave', Eighth Symp., (Int.) on Combustion, Baltimore Williams and Wilkins, 1960.

9. **Cramer, F.B.,** 'The Onset of Detonation in a Droplet Combustion Field', Ninth Symp. (Int.) on Combustion, Academic Press, 1962.

10. **Nicholls, J.A., Dabora, E.K., Ragland, K.W.,** 'A Study of Two Phase Detonation as it Relates to Rocket Motor Combustion Instability', NASA CR-272, Aug. 1965.

11. **Engel, O.G.,** 'Fragmentation of Waterdrops in the Zone Behind an Air Shock' J. Res., of Nat. Bureau of Standards, Vol.60, No.3, March 1958.

12. **Hanson, A.R., Domich, E.G., Adams, H.S.,** 'Shock Tube Investigation of the Breakup of Drops by Air Blast', The Physics of Fluids, Vol.6, No.8, Aug. 1963.

13. **Dickerson, R.A., Schuman, M.N.,** 'Rate of Aerodynamic Atomization of Droplets', J. of Spacecraft and Rockets, AIAA, Vol.2, No.1, Jan.-Febr. 1965.

14. **Wolfe, H.E.,** 'Photographic Study of Breakup of Liquid Drops', J. Soc. of Motion Pixture and Television Eng., Vol.75, 1966.

15. **Rabin, E., Schallenmuller, A.R., Lawhead, R.B.,** 'Displacement and Shattering of Propellant Droplets', AFOSR TR 60-75, 1965.

16. **Gordon, G.D.,** 'Mechanism and Speed of Breakup of Drops', J. Applied Physics, Vol.30, No.11, Nov. 1959.

17. **Morell, G., Povinelli, F.P.,** 'Breakup of Various Liquid Jets by Shock Waves and Applications to Resonant Combustion', NASA TND-2423, Aug. 1964.

18. **Clark, B.J.,** 'Breakup of a Liquid Jet in a Transverse Flow of Gas', NASA TND-2424, Aug. 1964.

19. **Busch, C.W., Laderman, A.J., Oppenheim, A.K.,** 'Parametric Study of the Generation of Pressure Waves by Particle Fueled Combustion', AIAA Second Annual Meeting, San Francisco, July 1965, Paper 65-357.

20. **Chinitz, W., Agosta, V.D.,** 'Shock Wave Propagation in Liquid Propellant Rocket Engines', Pyrodynamics, Vol.1, Oct. 1964.

21. **Derksen, W.,** 'Dynamische metingen aan verschillende typen drukopnemers m.b.v. een schokbuis', NLR TM-C.34, 1966 (in Dutch).

22. **Selberg, B.P.,** 'Shock Tube Determination of the Drag Coefficient of Small Spherical Particles', NASA CR-418, April 1966.

23. **Dewey, J.M., Anson, W.A.,** 'The Calibration of Shock Tubes for Studying Aerodynamic Loading Problems', Sixth Int. Shock Tube Symp., Freiburg, April 1967.

24. **Hanson, A.R., Domich, E.G., Adams, H.S.,** 'Acoustical Liquid Drop Holder', The Review of Scientific Instruments, Vol.35, No.8, Aug. 1964.

25. **Jenkins, D.C., Booker, J.D.,** 'The Time Required for High Speed Air-stream to Disintegrate Water Drops', ARC C.P. No.827, 1965.

26. **Ingebo, R.D.,** 'Drag Coefficients for Droplets and Solid Spheres in Clouds Accelerating in Airstreams', NACA TN 3762, 1956.

27. **Lambaris, S., Combs, L.P., Levine, R.S.,** 'Stable Combustion Process in Liquid Propellant Rocket Engines', Fifth Colloquium of the AGARD Combustion and Propulsion Panel, Braunschweig, 1962.

28. **Priem, R.J., Heidmann, M.F.,** 'Propellant Vaporization as a Design Criteria for Rocket Engine Combustion Chambers', NASA TR R-67, 1960.

List of Mathematical Symbols

C_D	Drag coefficient of a fragmentating droplet as determined from the displacement of the windward face and based on the initial droplet diameter
D_o	Initial droplet diameter (mm)
k_s	Burning rate constant, under stagnant conditions
M_s	Shock wave Mach number
p	Static pressure (kg/cm^{2}abs)
Re	Reynolds number = $\rho\ u\ D_o/\mu$
t	Time after shock wave passage (μsec)
T	Static temperature (°K)
u	Velocity (m/sec)
We	Weber number $\rho\ u^2 D_o/\sigma$
ρ	Density
σ	Surface tension of the liquid
τ_B	Breakup or burning time
τ_{2D_o}	Characteristic deformation time (width of the droplet = 2 D_o)
μ	Gaseous viscosity

Subscripts

1 Initial conditions ahead the shock

2 Conditions behind the shock

ℓ Liquid

Table 1-5-1. Range of Test Conditions and Some Important Properties of the Conditions behind the Shock Wave

p_1	M_s	p_2	T_2	u_2	M_2	D_0=1.7 mm		D_0=0.8 mm	
kg/cm^2 abs		kg/cm^2 abs	°K	m/sec		Re	We	Re	We
0.2	2	0.9	500	410	1	19000	8000		
0.2	2.4	1.3	600	540	1.2	25700	15400		
0.2	2.8	1.8	725	670	1.3	31500	25700		
0.2	4.6	4.9	1500	1200	1.6	50800	119000		
1.03	1.04	1.15	300	23	0.07	2800	41	1330	19
1.03	1.07	1.2	310	36	0.11	4800	140	2250	65
1.03	1.16	1.5	325	80	0.23	10800	630	5000	310
1.03	1.3	1.9	350	150	0.4	24300	2900	11650	1400
1.03	1.7	3.3	430	300	0.8	60000	16500	29000	8000
1.03	2	4.6	500	410	1	93000	37400		
1.03	2.2	5.7	550	480	1.07	117000	59000		
1.03	2.6	8.1	670	610	1.24			73000	52000
1.03	2.8	9.3	725	670	1.3	160000	134000		
4.7	1.04	5.2	300	23	0.07	12900	190		
4.7	1.07	5.5	310	36	0.11	23700	640		
4.7	1.16	6.5	325	80	0.23	51000	2900		
4.7	1.3	9	350	150	0.4	120000	15000		
4.7	1.7	15.5	430	300	0.8	291000	81000		

Table 1-5-2. Time Interval After Which Microdroplets could be Clearly Observed (μ sec).

	p_1 kg/cm^2 abs	M_s	satellite droplets	main droplets
D_0=1.7mm	0.2	2	30	44
	0.2	2.4	17	28
	0.2	4.6		13
	0.2	4.6		7
D_0=1.7mm	1.03	1.04		3375
	1.03	1.07	506	1530
	1.03	1.16	160	315
	1.03	1.3		80
	1.03	1.7		12
	1.03	2		11
	1.03	2.2		6
	1.03	2.8		4
D_0=1.7mm	4.7	1.04		481
	4.7	1.07		163
	4.7	1.16		32
	4.7	1.3		15
	4.7	1.7		9
D_0=0.8mm	1.03	1.04		2570
	1.03	1.07	480	640
	1.03	1.16	161	245
	1.03	1.3	45	66
	1.03	1.7	8	21
	1.03	2.6		4

1-6

Shock-Induced Combustion by High Speed Shots in Explosive Gas Mixtures

H. BEHRENS, H. LEHR, W. STRUTH and F. WECKEN.
Deutsch-Französische Forschungsinstitut St. Louis.

Summary

Shock-induced combustion phenomena were studied by firing spheres or cone-cylinders into mixtures of hydrogen with air or with oxygen and taking shadow photographs of the disturbances they created. The tests were carried out at pressures of 0.24-0.72 atm and velocities up to 2900 m/s. The missiles of 9 mm calibre of the earlier experiments were fired without spin, later cone-cylinder configurations of 15mm calibre were spin-stabilised. Among the topics studied were the relative amount of heat added to the flow, ignition temperatures and boundary layer combustion.

Experiments with cone-boattail configurations show separation of flow. This is explained by the generation of a pressure increase in the flow direction because of the addition of heat by combustion in supersonic flow. The implications for hypersonic propulsion with external burning are considered.

While a qualitative discussion of the experiments demonstrates the potential power of the missile technique to study shock-induced combustion of gas mixtures at high Mach numbers at well-defined conditions difficulties remain in the quantitative theory of conical flow of detonating gases under real gas conditions, taking into consideration ignition and reaction times. An outlook is given on the theory of conical flow with heat addition in the shock front neglecting real gas effects.

Sommaire

Les phénomènes de combustion induits par onde de choc ont été étudiés dans un montage balistique où on tire des projectiles sphériques ou en forme de cônes-cylindres dans des mélanges d'hydrogène et d'air ou d'oxygène à des pressions allant de 0.24 à 0.72 bars et une vitesse inférieure à 2900 m/s, en utilisant la méthode du ombres. Les projectiles de calibre 9 mm des premiers essais ont été tirés sans rotation, des configurations cône-cylindre des essais ultérieurs de calibre 15 mm ont été stabilisées par rotation. Parmi les caractéristiques étudiées, on doit citer l'accroissement d'enthalpie de l'écoulement, la température d'inflammation et la combustion de la couche limite.

Des expériences effectuées avec des cônes munis de dérive montrent une séparation de l'écoulement. Ceci s'explique par l'accroissement de pression dans le sens de l'écoulement, par suite de l'addition d'enthalpie due à la combustion supersonique et on considère le problème de la propulsion hypersonique avec combustion externe.

Bien que l'analyse qualitative des expériences montre l'intérêt de cette technique pour étudier la combustion induite par choc dans des mélanges gazeux aux nombres de Mach élevés, il reste des difficultés pour la théorie quantitative de l'écoulement conique des gaz réels détonants, si on veut faire intervenir le délai d'allumage et le temps de réaction. On presente une analyse sommaire de l'écoulement conique avec rapport de chaleur, où les effets de gaz réels sont négligés dans l'onde de choc.

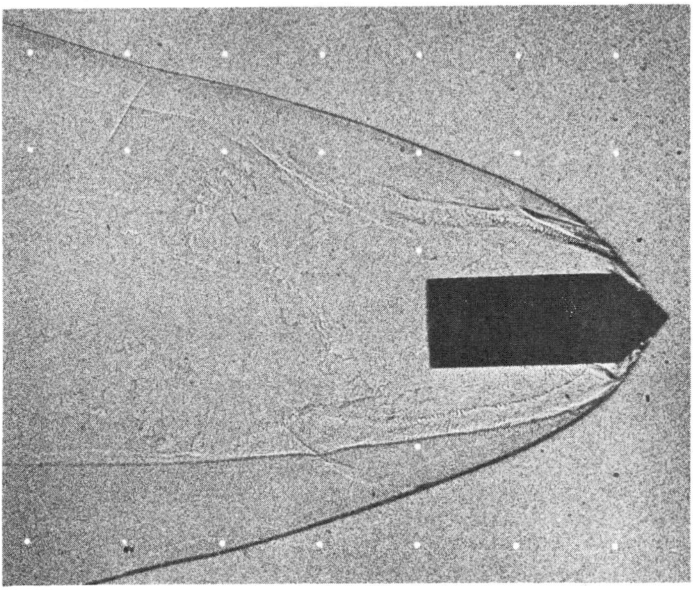

Fig. 1-6-1 Combustion in supersonic flow, 9mm projectile, $u_\infty = 2555$ m/s, $M_\infty = 6.30$ $\delta_c = 40°$, $\sigma = 52°$, $P_\infty = 0.55$ atm

u_∞ = missile velocity or free stream velocity
M_∞ = free stream Mach number, P_∞ = free stream pressure
δ_c = half cone angle, σ = half shock angle

Fig. 1-6-2 Combustion in subsonic flow, 9mm projectile, $u_\infty = 2080$ m/s $M_\infty = 5.14$, $\delta_c = 40°$, $\sigma = 67°$ $P_\infty = 0.55$ atm

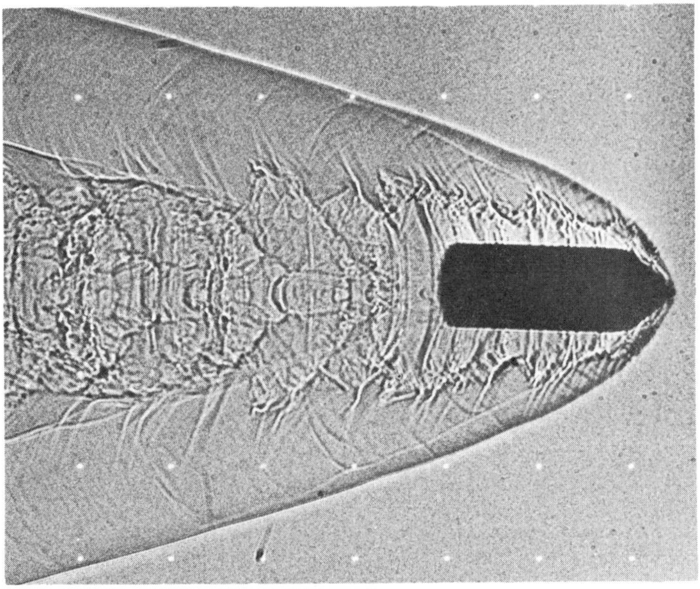

Introduction

While experiments in shock-induced combustion by means of a combustion tunnel require an expensive facility, the missile technique is a rather simple method. As missile velocities up to 3000 m/s are obtainable by solid propellants, corresponding to Mach numbers of 7 - 10 and to shock wave temperatures up to 3000°K, ignition with an adequate short delay of the explosive gas mixture is assured. The method allows perfect mixing of the components and avoids turbulence, or disturbances from the walls. Experiments may be carried out in gases that are difficult to handle, since small quantities only are necessary. In this paper experiments in mixtures of hydrogen with air or oxygen are discussed, the phenomena being studied using high-speed shadow photography.

Experiments with 9 mm Projectiles without Spin

The material of the projectiles was Lexan, a polycarbonate plastic. In the earliest experiments the projectiles were spherical and some of the phenomena observed have been described elsewhere (1) (2). This report describes experiments using projectiles of cone/cylinder configuration. When these projectiles were fired without spin, like the spheres, they normally flew in a more or less oblique attitude. A photograph of a projectile in a nearly horizontal attitude at $M_\infty = 6.3$ in a stoichiometric hydrogen-air mixture of 0.55 atm, is shown in figure 1-6-1. The half cone angle was 40° and the half-shock angle was 52° which compares with the half-shock angle of 47° when heat is not added. On the basis of the theory of detonative conical flow, the relative amount of heat added, as expressed by the Damköhler parameter q, may be calculated (3). Behind the shock the flow was supersonic so that the heat addition occurred in supersonic flow.

Figure 1-6-2 was obtained in an experiment at $M_\infty = 5.14$ resulting in a half-shock angle $\delta = 67°$. There was subsonic flow behind the shock, the shock front had a convex shape, and large disturbances arose at the boundary of the combustion gas in contrast to the rather smooth combustion shown in figure 1-6-1. Combustion in supersonic flow is smooth because supersonic flow will not propagate pressure disturbances upstream, unlike subsonic flow in which instabilities may occur (4).

However, periodic combustion may occur in supersonic flow behind the shock, similar to the phenomena found with spherical projectiles. Figure 1-6-3 shows a projectile in a mixture $1.7\ H_2 + O_2$ with $p_\infty = 0.24$ atm and $M_\infty = 5.30$. An approximate evaluation suggests that the length of the periods may be due to ignition delays. With hydrogen-air mixtures however we have not found this periodic burning.

The photograph in figure 1-6-4 shows a firing in which the angle of attack was as usually found: the behaviour behind the bow wave was evidently asymmetric.

Experiments with Spin-Stabilised Projectiles of 15 mm Calibre

Because of difficulties caused by the oblique attitudes of the projectiles, an attempt was made to stabilise the projectiles by spin, at the same time increasing their calibre to 15 mm. The twist of the rifled barrel used was 2°40'. Because the projectile travelled only 0.40 m in gas at a pressure of about 0.5 atm, this stabilisation proved to be adequate.

Subsequent experiments with projectiles having half-cone angles (δ_c) of 35° or 40°, fired at Mach numbers around $M_\infty = 6$, confirmed the results gained in earlier experiments with 9 mm calibre projectiles. A decrease of the angle (δ_c) to 32.5° revealed the phenomenon shown in figure 1-6-5. Parallel to the cylinder wall, a sharp dark line can be seen extending some distance beyond the end of the projectile.

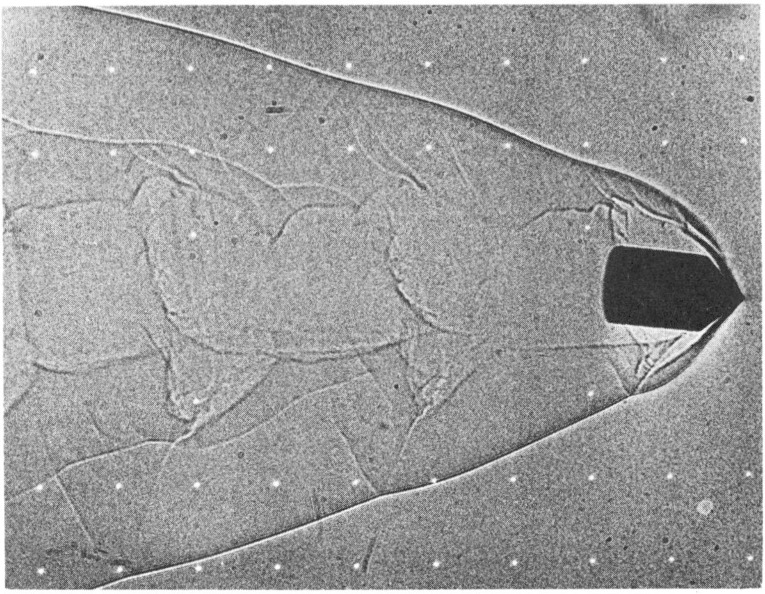

Fig. 1-6-3 Unstable periodic combustion. Mixture $1.7\ H_2 + O_2$, 9 mm projectile. $P_\infty = 0.24$ atm, $u_\infty = 2695$ m/s, $M_\infty = 5.30$, $\delta_c = 40°$, $\sigma = 55°$.

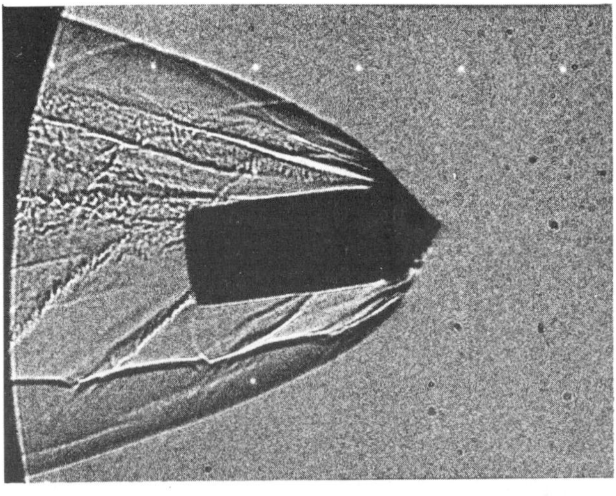

Fig. 1-6-4 Combustion in supersonic flow, separation of flow on lee side. 9 mm projectile.

$u_\infty = 2450$ m/s, $M_\infty = 6.05$, $\delta_c = 40°$, $\sigma = 54°$, $P_\infty = 0.55$ atm.

We interpret this as boundary layer combustion. The shock angle is in accord with the theoretical value for conditions without heat addition, suggesting that there may have been no combustion in the conical flow region.

To check this assumption about the existence of a boundary flame, projectiles were fired at the same velocity into an inert atmosphere having the composition $2H_2$ + $4.78N_2$, i.e. the oxygen in the explosive mixture had been replaced by nitrogen. Since oxygen and nitrogen have approximately equal molecular weights, sonic velocities in the two mixtures are nearly the same, and equal missile velocities correspond to nearly equal Mach numbers and temperatures. Figure 1-6-6, the photograph of this experiment, showed no indication of a similar line. We conclude then that the phenomenon in figure 1-6-5 was not caused by ablation, one further possible cause of such a dark line.

A further experiment in a hydrogen-air mixture, but with a missile velocity about 100 m/s lower than in the preceding experiments, gave results shown in figure 1-6-7. The fact that this also showed no flame line suggests that there may be an ignition temperature that was not attained under the conditions of this experiment.

The projectile shown in figure 1-6-8 had an angle (δ_c) of 32.5° and an angle of attack of about 2.5°. The shock angle was again consistent with the theoretical shock angle without heat addition. On the lee side turbulent burning can be seen starting at the cone/cylinder shoulder while on the windward side there was laminar boundary layer combustion at first, turning to turbulent combustion towards the end of the cylinder. Flow separation occurred first on the lee side and then propagated obliquely round the cylinder to the opposite side. Ahead of the separated combustion layer there was a shock wave evidently caused by the expansion of the burning layer.

Reverting now to figure 1-6-4 a separation of flow can again be seen combined with turbulent combustion on the lee side. In that case combustion was already established in the conical flow region, demonstrating that flow separation was not dependent on the absence of heat addition in the conical flow region. Flow separation with heat addition does seem to be helped by a small angle of attack.

We deduced from these results, however, that flow separation ought to occur also in experiments with projectiles in a horizontal attitude provided they had surfaces inclined in relation to the flow direction. A projectile was therefore constructed with the nose cone followed by a boat-tail as shown in figure 1-6-9. Figure 1-6-10 shows the result of experiments using it: flow separation caused by combustion can be seen, not starting at the shoulder but towards the rear of the projectile.

The cork-like body following the projectile is the sabot needed to fire such a model. However the distance between the projectile and the sabot is rather short. Recent experiments in inert gases have shown that the sabot interferes with the projectile, causing separation of flow, if the distance between projectile and sabot is as short as in this figure. Further experiments with an increased distance are in progress to demonstrate separation of supersonic flow with heat addition.

Figure 1-6-11 shows a projectile like that used in the preceding figures but with the point broken off. Because of the bluntness, ignition occurred behind the detached shock front, where the flow was subsonic. Further around the corner the flow became supersonic. In contrast to the conditions shown in figure 1-6-10 there was no separation of flow from the rear part of the projectile.

Discussion of Separation of Supersonic Flow with Combustion

Flow separation is a well-studied phenomenon caused by an increase in pressure

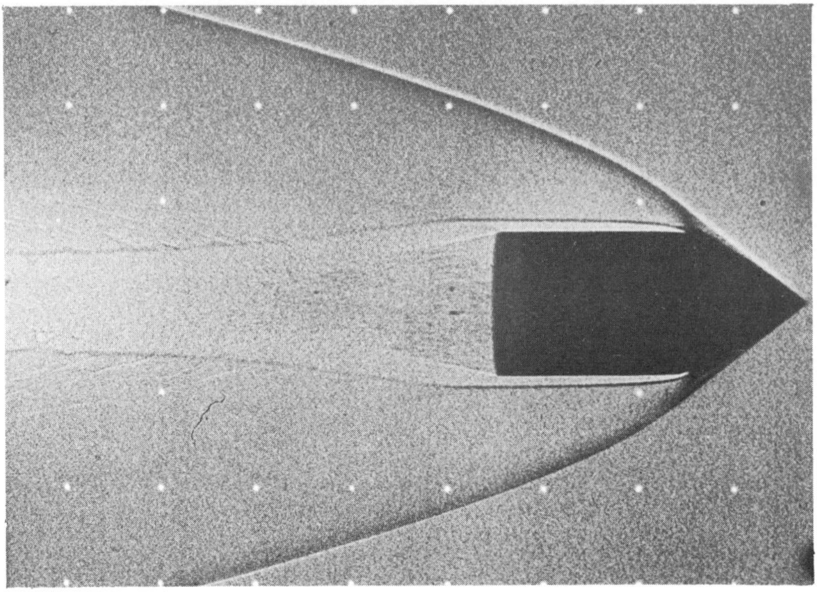

Fig. 1-6-5 Laminar boundary layer combustion, 15mm spin stabilized projectile

$u_\infty = 2410$ m/s, $M_\infty = 5.95$, $\delta_c = 32.5°$, $\sigma = 37.5°$, $P_\infty = 0.55$ atm

Fig. 1-6-6 No combustion or ablation, inert atmosphere. Mixture $2H_2 + 4.78\ N_2$, 15mm spin stabilized projectile
$u_\infty = 2395$ m/s, $M_\infty = 5.91$, $\delta_c = 32.5°$, $\sigma = 37.5°$, $P_\infty = 0.55$ atm

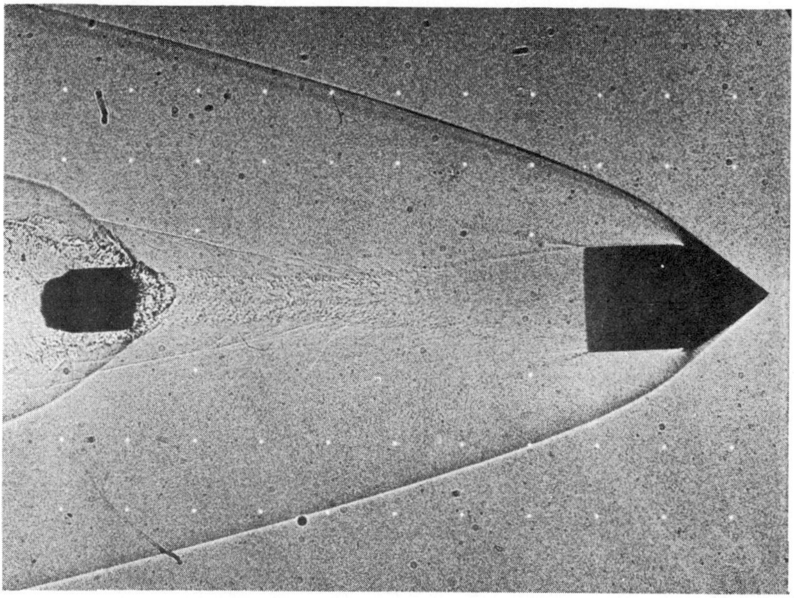

Fig. 1-6-7 No ignition or combustion, because of lower missile velocity (cf. Fig. 1-6-5), 15mm spin stabilized projectile.

$u_\infty = 2295$ m/s, $M_\infty = 5.66$, $\delta_c = 32.5°$, $\sigma = 37.5°$, $P_\infty = 0.55$ atm

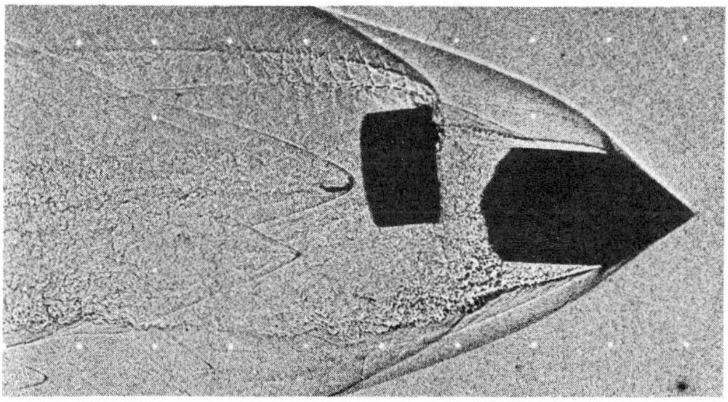

Fig. 1-6-8 Laminar boundary layer combustion on one side, separated flow with turbulent combustion on opposite side, 15mm spin stabilized projectile.

$u_\infty = 2387$ m/s, $M_\infty = 5.90$, $\delta_c = 32.5°$, $\sigma = 37.5°$, $P_\infty = 0.55$ atm

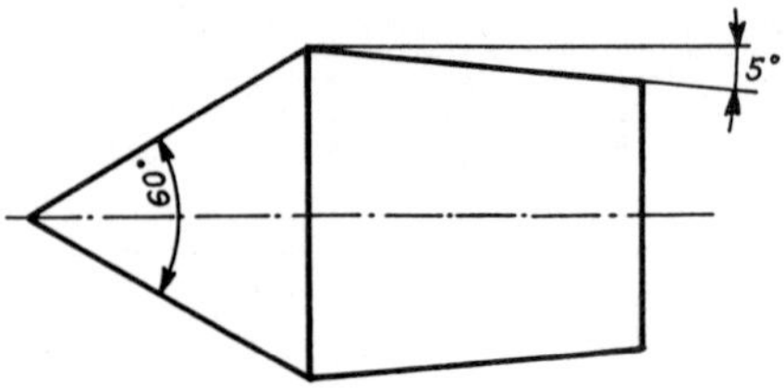

Fig. 1-6-9 Diagram of cone boattail projectile

Fig. 1-6-10 Cone boattail projectile. Separation of flow with turbulent combustion, 15 mm spin stabilized projectile.

$u_\infty = 2425$ m/s, $M_\infty = 6.0$, $\delta_c = 30°$, $\sigma = 35.0°$, $P_\infty = 0.55$ atm

Fig. 1-6-11 Cone boattail projectile with the point broken off. No separation of flow, 15 mm spin stabilized projectile.

$u_\infty = 2290$ m/s, $M_\infty = 5.66$, $\sigma = 35.0°$, $P_\infty = 0.55$ atm.

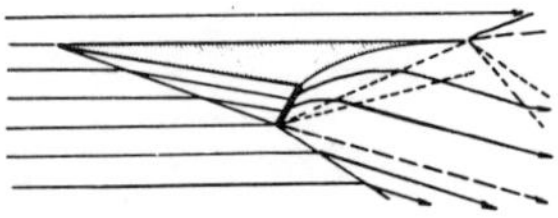

Fig. 1-6-12 A lifting propulsive body with detonation wave. (Küchemann (7))

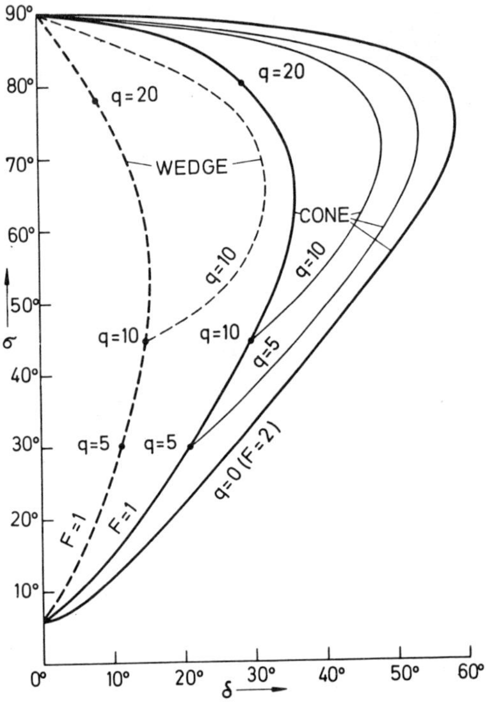

Fig. 1-6-13 Shock angle vs. cone or wedge angle; $M_\infty = 10$, $\gamma = 1.34$, F = 1 and various relative heat additions

$$q = \frac{Q}{c_p \cdot T_\infty}$$

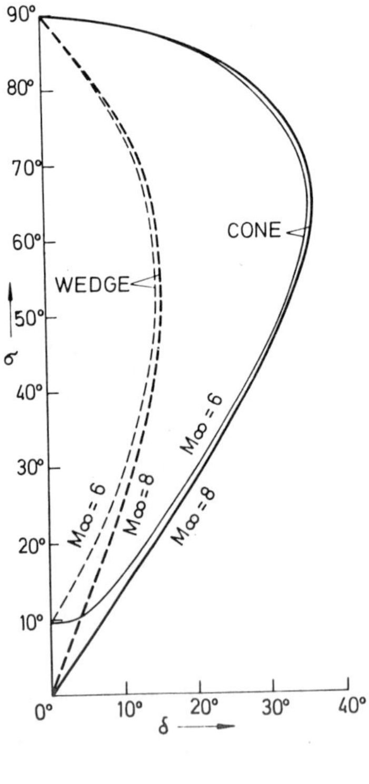

Fig. 1-6-14 Shock angle vs. cone or wedge angle; M_∞ = 6 and ∞, γ = 1.34, F = 1

in the direction of flow (dp/dx O) (A decrease in pressure has a stabilising effect). Consequently flow has a tendency to separate in subsonic flow around the corner, where the expansion induced by the corner effects a pressure increase. In supersonic flow around a corner however there is a pressure decrease; this stabilises the flow and hence there is no separation. Now, heat addition in supersonic flow increases the pressure. Strong heat addition by combustion may more than compensate for the effect of expansion, so that the pressure may increase around the corner. That this is a real case has been shown by quantitative studies of non-ideal detonations with lateral expansion (5).

The pressure increase caused by combustion is still more pronounced when the flow has passed the corner and pressure decrease by expansion is no longer effective. Separation may also occur when the reaction has started in the conical flow region, as in figure 1-6-4, provided that the reaction is still continuing. However, when the reaction is complete there is no further pressure increase and no separation, as can be seen in figure 1-6-11.

Combustion, associated with separation of flow, has been studied previously by Townend (6) (7) who called it 'base-burning' because it was observed at the end of a cylinder where the flow inevitably separated. His studies showed that the pressure in the wake of the body was increased by combustion just to ambient pressure or slightly above, but was clearly less than the pressure that would have been established with combustion in laminar attached flow. His results are interesting in connexion with the development of hypersonic ramjets having external combustion. Well-known sketches of hypersonic lifting propulsive bodies (Fig. 1-6-12) assume detonative combustion to be stabilised at the shoulder of the body, the rear part forming a half-open nozzle (7).

The cone/boat-tail projectile shown in figures 1-6-9 and 1-6-10 may be considered as a simple model of such a propulsive body converted into a symmetrical body of revolution, so that it produced no lift. The experiments using it demonstrated that when combustion occurred there was no attached flow in the rear part: the separated flow might reduce the drag but would give no propulsive force. It is necessary to achieve the combustion without separation of flow, e.g. in a combustion chamber before the gases will pass to the tapered rear part in a manner that contributes to propulsion. In such a case the combustion gases would not separate as illustrated in figure 1-6-12, because there would no longer be heat addition along the tapered rear part.

Some Remarks on the Theory of Hypersonic Flow of Detonating Gases

The experiments described above are particular examples of the general case of hypersonic flow around bodies with either external heat addition or heat release by combustion, which has the same physical effect. When combustion is induced by shock waves, the phenomena may be classified as oblique detonations. When combustion occurs in laminar boundary layers or in turbulent flow separation, the phenomena belong to the field of viscous supersonic flow with heat addition.

It would have been more convenient theoretically had these experiments used wedge-shaped bodies like those used by Gross (8) in his supersonic tunnel experiments. However in photographic studies of high speed projectiles, the use of wedge shapes introduces so much complication that experimenters prefer conical missiles, even though the analysis of conical flow fields does involve more complex mathematics.

We have extended the theoretical work of Bartlett (9) on the flow of detonating gases around conical bodies, covering the whole range of shock angles up to 90° and Mach numbers up to infinity (i.e. $M_{\infty} = \infty$). Like Bartlett we have used the dimensionless parameter F as a measure of heat addition, defined as:

$$F = 1 + \sqrt{\left(1 - \frac{Q}{Q_{max}}\right)}$$

where Q is the amount of heat added per unit mass of gas and Q_{max} its maximum value for a given upstream flow. The value of F for a shock wave without heat addition is 2, and with maximum heat addition, as in a Chapman-Jouguet detonation is 1. For strong or overdriven detonations $1 < F < 2$ is valid. Townend (10) showed the usefulness of this parameter, which had been introduced into the theory of detonation by Adamson and Morrison (11).

Some of our results (12) are shown in figure 1-6-13. To demonstrate their relationship to the second Damköhler parameter curves of constant q have been added defined as $q = \frac{Q}{c_p \cdot T_\infty}$ where c_p = specific heat capacity at constant pressure and T_∞ = absolute static temperature of the upstream flow. Once a computation programme has been worked out, it is very easy to evaluate experiments of conical flow with heat addition, varying M_∞, γ (the ratio of the specific heats) F (or q) and the cone angle δ_c (or shock angle σ).

The usefulness of the parameter F is shown in the case of $M_\infty \to \infty$ for F = const. < 2, where $q \to \infty$. There are no difficulties in the numerical procedure and figure 1-6-14 gives results for Chapman-Jouguet detonations (F = 1) in wedge and conical flows at $M_\infty = 6$ and $M_\infty = \infty$. As in flow without heat addition (F = 2) there is a definite limiting behaviour for $M_\infty \to \infty$. The flow pattern 'freezes' in the hypersonic regime with infinitely increasing Mach numbers. But in contrast to flow with F = 2, the density ratio $\rho_1/\rho_2 = \epsilon$ does not approach zero even when $\gamma \to 1$

$$\lim_{M_\infty \to \infty} \epsilon = \lim \frac{\rho_1}{\rho_2} = \frac{\gamma + 1 - F}{\gamma + 1} = \frac{(\gamma - 1) + (2 - F)}{\gamma + 1}$$

From the expression on the right it is seen that for $\epsilon \to 0$ it is necessary for $\gamma \to 1$ and $F \to 2$. Consequently approximations that are justified for small density ratios, namely for $\epsilon \to 0$, break down for strong or Chapman-Jouguet detonations ($F < 2$).

However, the theory in this form assumes that heat addition proceeds immediately behind the shock front, and does not allow for the ignition delay and reaction time. A preliminary analysis of the shock angles measured in our experiments shows that chemical equilibrium is in fact not established. It is evident therefore that the theory in its simple form does not adequately describe our experimental findings. In principle, theoretical methods already evolved for dealing with the flow of relaxing gases should be applicable to flow with heat addition without having to restrict the latter to a discontinuity. A more advanced theory that could allow for real-gas effects like this should also be appropriate for analysing the flow of detonating gases around blunt bodies like spheres. The careful measurements that Ruegg and Dorsey (13) made of the stand-off distances of bow waves are helpful here. Samozvantsev (14) also, who investigated the role of ignition delay in flow round spheres, showed that this caused a detonation wave to decay into a shock wave and not to be transformed into a Chapman-Jouguet wave expected from the theory. From his work it seems that the diameter of a sphere would need to be at least 400 mm to stabilise a Chapman-Jouguet wave in a hydrogen-air mixture.

Hence, for use in experiments to investigate combustion in supersonic flow, at any rate as far as boundary phenomena are concerned, there appear to be no essential differences between wedge-shaped and cone-shaped projectiles.

In conclusion we suggest that the technique we have described may be comparable in its range of application in supersonic flow with heat addition to the Bunsen burner method in the subsonic field. Both of these methods use premixed systems, so setting aside the problems of mixing during the experiment.

References

1. **Struth, W.,** 'Kurzzeitaufnahmen von Schüssen mit Hyperschallgeschwindigkeit in reagierende Gase.' Sixth Int. Cong. of High-Speed Photography, Willink en Zoon, 1963, 443.

2. **Behrens, H., Struth, W., Wecken, F.,** 'Studies of Hypervelocity Firings into Mixtures of Hydrogen with Air or Oxygen.' Tenth Symp. on Combustion, The Combustion Inst. 1965, 245.

3. **Behrens, H., Struth, W., Wecken, F.,** 'Shock-induced Combustion in the Bow Waves of High-Speed Missiles.' ISL-Ber. 2/66.

4. **Ferri, A.,** 'Review of Problems in Application of Supersonic Combustion.' J. Roy. Aeron.Soc., 68, 1964, 575.

5. **Wecken, F.,** 'Non-ideal Detonation with Constant Lateral Expansion.' The Fourth Symp. on Detonation, N.O.L., Silver Spring, Md. 1965. also ISL - T 19/65.

6. **Townend, L.H.,** 'Some Effects of Stable Combustion in Wakes Formed in a Supersonic Stream'. R.A.E. Techn. Note No. Aero.2872 1963 .

7. **Küchemann, D.,** 'Hypersonic Aircraft and their Aerodynamic Problems.' Progr. Aeron. Sci., Vol.6, Pergamon Press 1965, 271.

8. **Gross, R.A.,** 'A Study of Combustion in Supersonic Flow'. Res. 12, 1959, 381.

9. **Bartlett, R.S.,** 'Tables of Supersonic Symmetrical Flow around Right Circular Cones, with and without the Addition of Heat at the Wave.' R.A.E. Techn. Rep. No.66278, 1966.

10. **Townend, L.H.,** 'An Analysis of Oblique and Normal Detonation Waves.' R.A.E. Techn. Rep. No. 66081, 1966.

11. **Adamson, T.C., Morrison, R.B.,** 'On the Classification of Normal Detonation Waves.' Jet Prop. 25, 1955, 400 & 403.

12. **Wecken, F., Behrens, H., Adam, B.,** 'Numerische Untersuchungen zur detonativen Kegelstromung.' ISL - Techn.Mitt 1967.

13. **Ruegg, F.W., Dorsey, W.W.,** 'A Missile Technique for the Study of Detonation Waves.' J. Res. Nat. Bureau of Standards 66 C 1962, 51.

14. **Samozvantsev, M.P.,** 'The Stabilisation of Detonation Waves by Means of Bluff Bodies.' PMTF No.4, 1964. R.A.E. Library Transl. 1088, 1965, 126.

1-7

Aerophysics Research Based on Free Flight Range Measurements

A. Q. ESCHENROEDER, H. H. KING, and K. S. WEN
AC Electronics Defense Research Laboratories, California

Summary

Recent advances in technique have significantly extended the application of free-flight ranges to the study of high temperature gas-dynamic phenomena. Progress in ballistics permits the launching of complex models at high velocity. Improved sensitivity and resolution in physical measurements have enriched the information gained from laboratory scale flight investigations. These factors combine to place the range in a unique position between most of the laboratory experiments and full scale test programs.

Two approaches are followed in relating sub-scale and full scale observations: (a) direct scaling and (b) evolution of theoretical description. In the first approach, classical results of similitude must be extended to include modeling of nonequilibrium chemistry. Because of the complexity of real gases, this procedure is usually limited to binary collision dominated flows. The second approach requires a detailed mathematical model that contains the essential physical features of interest. The model may be refined using the experimental findings. Within the scope of this theoretical system observations of both small and large scale events are predictable. Evidently, the analytical approach is the more general of the two.

Examples of research based on free flight range measurements can be found in problems of flight vehicles and in the study of fundamental processes. Reentry body wake structure and ionization phenomena have been extensively investigated. Experiments are in progress to study the performance of supersonic combustion ramjets. Future work in some of the basic areas may now be contemplated using free flight range techniques. Flying grids have been employed to generate turbulence fields under hypersonic conditions. Models with base flow injection have been successfully launched. Within feasible rates of injection, the wake is transformed to a jet-like flow. Introduction of reactive gases above the critical rate permits the study of turbulent combustion. Neutral composition, temperature and ionization measurements are of primary interest in these experiments.

Sommaire

Les progrès récents des techniques expérimentales ont permis d'étendre de facon importante l'application des essais en vol libre à l'étude de la dynamique des gaz à température élevée. Les progrès en balistique permettent le lancement de modèles complexes à grande vitesse. La sensibilité accrue et la bonne résolution des mesures physiques ont permis de compléter les informations obtenues à partir de maquettes en vol à l'échelle du laboratoire. Ces facteurs permettent de bien placer les essais en vol entre les essais au laboratoire et les essais en grandeur.

Pour relier les essais obtenus sur maquette et sur le montage réel on peut: (1) tenir compte directement de l'échelle du montage ou (2) approfondir la description théorique du phénomène. Dans le premier cas, les méthodes classiques d'étude de l'effet d'échelle doivent être étendues au cas des réactions chimiques hors-équilibre. Par suite de la complexité des gaz réels, ce procédé est généralement limité aux écoulements où les collisions binaires prédominent. Le second procédé nécessite un schéma mathématique détaillé qui comprend les particularités physiques essentielles du modèle. Le schéma peut être amélioré au fur et à mesure de l'obtention des résultats. Dans la mesure où ce système théorique est valable, les phénomènes à petite ou grande échelle peuvent être prédits. C'est naturellement cette méthode analytique qui est la plus puissante des deux.

Des exemples de recherches basées sur des mesures en vol libre peuvent être trouvés dans l'étude des véhicules aussi bien que dans celle des processus fondamentaux. La structure du sillage des corps de rentrée et les phénomènes d'ionisation ont été étudiés en détail. Des essais sont en cours pour étudier la combustion supersonique dans les statoréacteurs. Des travaux dans des domaines fondamentaux peuvent être envisagés maintenant par emploi de la technique du vol libre. Des "grilles volantes" ont été utilisées pour créer des champs de turbulence en vol hypersonique. Des maquettes avec injection au culot ont été lancées avec succès. Dans un domaine réalisable de taux d'injection, le sillage est transformé en un écoulement en forme de jet. L'introduction de gaz réactifs avec une concentration supérieure à la valeur critique permet l'étude de la combustion turbulente. La composition des neutres, les mesures d'ionisation et de température sont du plus grand intérêt dans ces essais.

Introduction

In the solution of propulsion and energetics problems the complexity of the interaction of fluid dynamics with chemical kinetics constitutes a major obstacle. Theoretical syntheses based on a combination of fundamental information from the fields of atomic and molecular physics, fluid dynamics, and combustion, have met with only limited success. It has become apparent that the gap between basic laboratory data and full scale flight behaviour is often too broad to span satisfactorily with existing theory.

Highly resolved observations of sub-scale flight tests in the laboratory range provide an intermediate step that promises to bridge this gap. The ballistic range provides a synthesis of the physical processes in flight phenomena.

The free flight of an object through a controlled environment is intuitively appealing, but the object is so much smaller than its prototype that the results must be interpreted with care. Bulk fluid dynamic behaviour is reproduced in sub-scale, in the free flight range, but the chemical kinetic phenomena do not completely submit to scaling. However, theoretically based mathematical models can remedy this deficiency when they are developed in conjunction with range results.

Recent advances in techniques have extended the application of free flight ranges to the study of high temperature gas dynamics. Progress in ballistics permits the launching of complex models at high velocity and improvements in the sensitivity and resolution of the physical measurements have enriched the information obtainable from laboratory scale flight investigations.

Operational Roles of the Free Flight Range

Detailed reviews of the problems of scaling are available (1-5) the present treatment therefore summarises briefly just the significant physical aspects, the limits of scaling, and the experimental consequences, all in the context of free flight subscale experimentation.

Compressibility and viscosity influence the motion and energy partitioning in the classical sense of fluid dynamics. Wave phenomena which establish the distribution of pressures in high velocity flows are scaled by the Mach number. Shear stresses and heating arise from viscous effects in the flow. The Reynolds number and various combinations of it with the Mach number provide the scaling parameters for the latter. Duplication of the fluid dynamic parameters is essential in order to represent faithfully the aerodynamic flow that results in lift, drag, or thrust forces. More significant in the present context, however, are the controlling effects exerted by compressibility and viscosity on the pressure and temperature levels in the flow field. These levels ultimately determine the frequency and energy of the molecular collisions that control chemical activity.

Chemical reactions give rise to some effects of primary interest in propulsion and energetics applications and the simulation of kinetic processes, as well as of the necessary fluid dynamical conditions, is therefore needed for scaling. We may treat kinetics and fluid dynamics separately only if conditions of weak coupling prevail i.e., energetic feed-back of the chemistry is negligible. An example of weak coupling is provided by ionisation in rocket exhaust or re-entry wakes. By contrast, strong coupling must occur in chemical propulsive applications because the desired effect is the generation of force by reactive energy deposition. In this case, the flow not only acts as a heat bath, but receives major perturbations from reactions via the temperature and density, primarily through the energy budget. If the flow is supersonic or faster, momentum perturbations introduc ed by reactions are usually secondary in magnitude.

Simultaneous scaling of both flow and chemistry therefore seems to be an indicated requirement. For rate processes, two sets of Damköhler numbers must be preserved (1): one set consists of ratios of residence times to reaction times; and the other, of ratios of chemical heat release to total enthalpy. Each member of each set corresponds to an elementary reaction step. The second set of Damköhler numbers measures the potential of the chemical system for perturbing the sensible energy of the flow, while the first set measures the extent to which the reactions drive to completion during a specified flow event e.g. diffusion through a boundary layer, or convection through a hot stagnation shock layer. In principle then, the Damköhler numbers, Mach number, and Reynolds number must be simultaneously preserved for geometric similitude of flow fields.

Limits of simplification are usually set if some parameter becomes particularly small or large. Examples are boundary layers for large Reynolds numbers, Mach number independence for high Mach numbers, and incompressible flow for chemically frozen gases at low Mach number. Direct scaling of reacting flows is possible only under certain of such limiting conditions and only for certain specific regions of the flow. The most useful special case is that of binary collision modeling. The small parameter in this case is the Damköhler first number for each three-body reaction path. In hypersonic flows with nonequilibrium chemistry, the free stream velocity is the main determinant of temperature levels in decelerated portions of the field. Thus if we are confronted with conflicting requirements in temperature dependence among the first set of Damköhler numbers, we may resolve the difficulty by matching model and prototype velocities. If two-body collisions control the reaction processes the remaining simulation requirements can then be met by an appropriate choice of ambient density in relationship to the model size. To increase binary collision number in proportion to the decreased residence time, the ambient density in the model experiment must be at a high level. Temperature dependent properties such as local speed of sound, transport coefficients and rate coefficients and velocity dependent properties, such as inertial force and residence time are automatically adjusted by duplicating the flight speed. References (3) and (4) contain discussions of the binary scaling limit showing that viscous and compressible interactions scale simultaneously under these constraints.

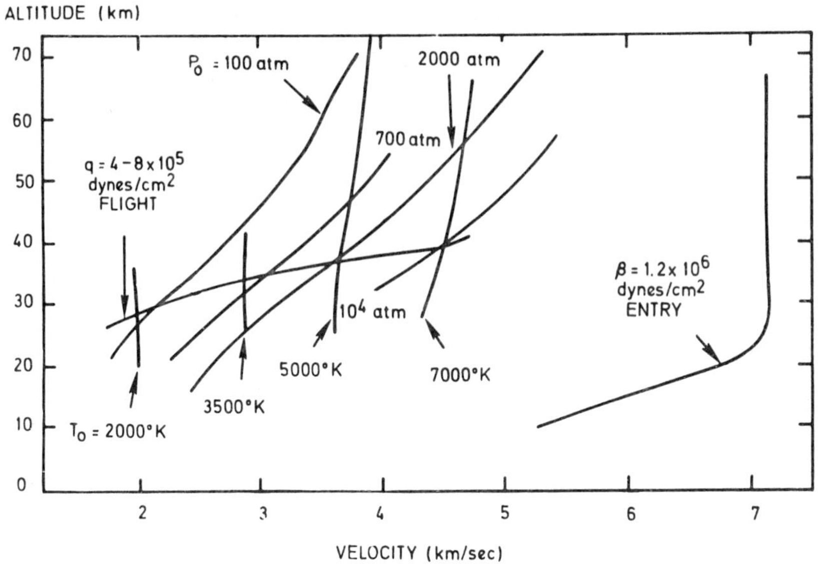

Fig. 1-7-1 Reservoir requirements for wind tunnel flight duplication

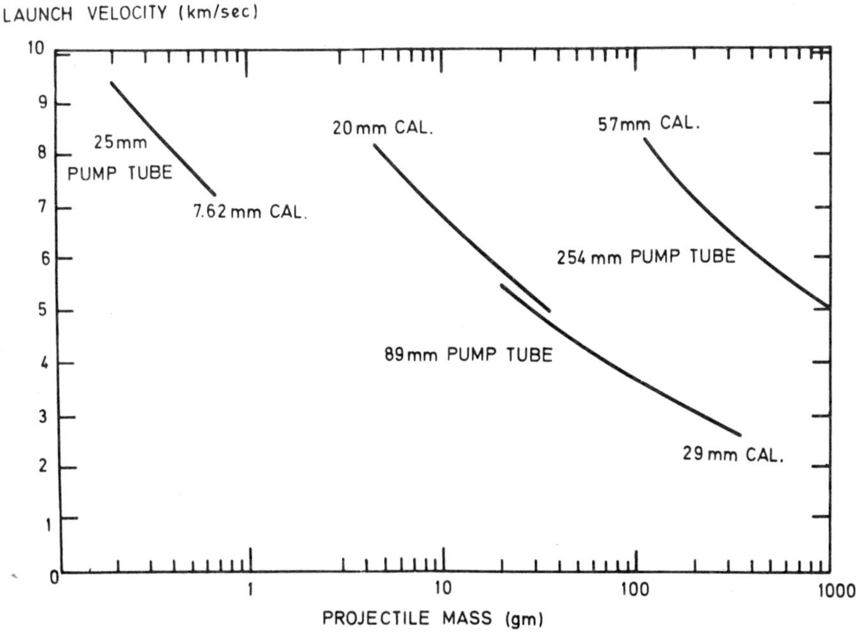

Fig. 1-7-2 Launch capabilities - AC-DRL light gas guns - saboted composite models

Theoretical Models

The laws of conservation and the constitutive equations for reacting gas flows serve as foundation stones on which theoretical models are constructed to enable prototype conditions to be predicted. The procedure lies somewhere between an abstract analysis and a numerical analogue to the experiment. The formal structure of the partial differential equations is complete in the macroscopic sense but the detailed description of certain coefficients is not. The lack of detail in both the kinetic molecular and turbulent transport representations must therefore be supplemented by laboratory data of a more fundamental nature than those obtained in flight tests.

For example, shock tube measurements may provide chemical rate data for each of a large set of elementary reaction paths. Folding these data into a flow calculation of gas composition permits their consistency with other notions of fluid behaviour to be checked when free flight range observations are available. In this sense laboratory flight measurements may be employed to improve fundamental understanding of unit processes in hypersonic flows. The inverse approach of deriving basic data from measurements of the flow field has elements of danger however because independent control of masking effects is difficult.

Turbulent flow is only approximately represented in a mathematical statement of the model. Since eddy scale sizes are not far removed from the macroscopic scale range, turbulence is influenced primarily by the specific characteristics of the boundaries and gradients in any flow configuration. Fluid dynamic scaling is usually accessible so that turbulent mixing relationships derived from range experiments describe similar behaviour in many other systems.

A converse function of the theoretical model is to feed information back to the experiment. One of the objectives of simulation studies is the definition of an optimum set of data. High enthalpy flows are characterised by an overwhelming variety of state variables which are spatially distributed. The choice of measurements must be restricted to something less than the complete list and must deal with finite sets of independent parameter values. The theoretical model assists the identification of the quantities of interest and the estimation of the essential parameters of the instrumentation e.g. dynamic range and sensitivity and is, therefore, an invaluable tool for the design of the experiment.

Choice of Laboratory Range Flight Conditions

Assume, as a first premise, that fluid dynamic similarity is assured for continuum flows by duplicating the flight speed and the density length product of the prototype (5). Then, as a second premise, accept the restriction that only the binary collision chemical kinetics will be similarly scaled, the tertiary-controlled effects requiring the theoretical model. At sufficiently high pressures there is then chemical equilibrium throughout the field and either viscous or chemical simulation must be abondoned because of conflicting density requirements. There may be an exception to this if the dissociation is complete over a wide range of ambient densities (2).

As examples of free flight range simulation consider: a) a supersonic combustion ramjet (scramjet) climbing along an acceleration trajectory restricted by fixed aerodynamic pressure (4.8×10^5 dynes/cm^2): and b) a heavy object decelerating in the terrestrial atmosphere (ballistic coefficient of 1.2×10^6 dynes/cm^2) after entering at sub-meteoric initial velocity. Figure 1-7-1 shows these trajectories on the familiar map of flight altitude versus velocity. Phantom lines show loci of constant reservoir pressure and temperature for expanding flow-type tunnel facilities. One millisecond melting time for a typical nozzle throat lies along a nearly vertical line from the centre (7). Limitations of physical test length for wakes, and of purity

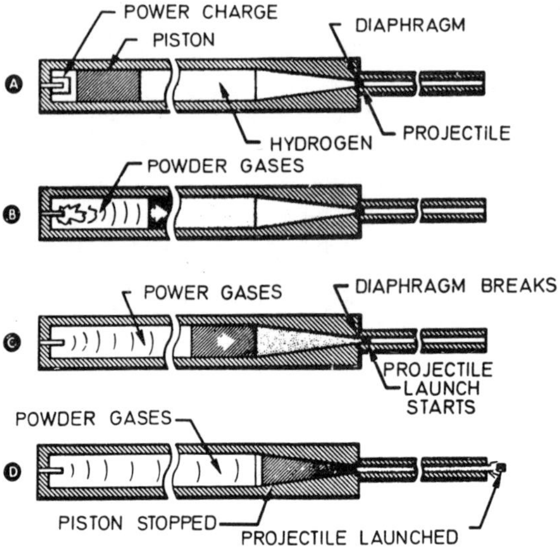

Fig. 1-7-3 Accelerated reservoir light gas gun operating sequence

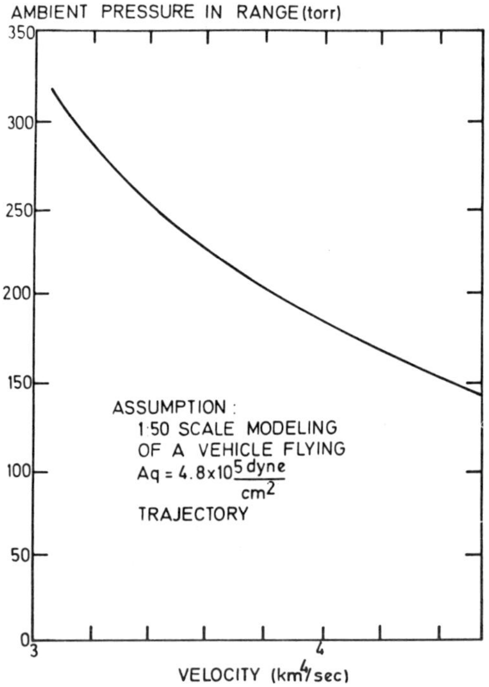

Fig. 1-7-4 Range pressure requirements for binary scaling flight conditions of scramjet

for reaction-sensitive flows apply to the right hand side of the map.

Unlike the wind tunnel, the free-flight range experiment almost completely decouples flow velocity from free-stream pressure. Impurities and nonequilibrium excitation of the test medium are either absent or independently controllable. The velocity capability of light gas guns covers the right hand portion of the map. Easily accessible ambient pressures satisfy total Reynolds number (binary-collision parameter) duplication due to recent developments of large gun facilities (8). The facility described in (8), for example, has launched slender conical models at speeds exceeding 6.5 km/sec. Total launch mass of model plus sabot approximates 300 grams in a 57 mm bore launch tube (9). Figure 1-7-2 summarises launch-mass performance of light gas guns employing the accelerated-reservoir mode of operation (10). Figure 1-7-3 shows the operating sequence of such a gun in a self-explanatory manner.

Let the scramjet scale model to be flight tested in a laboratory range at speeds of 3.2 and 4.8 km/sec, simulating the 4.8 x 10^5 dynes/cm^2 acceleration trajectory, be 3 cm diameter. Then, if the full scale prototype to be simulated is, for example, 150 cm in diameter, it is necessary to employ the range ambient pressures shown in figure 1-7-4 to preserve binary collision similarity. The ambient pressure lies in the range 148 to 320 torr, which is easily attainable using present practice.

Hypersonic engine configurations have mostly external flows which lend themselves well to laboratory-fixed observational instruments. The flight medium is of controlled composition and is free of vitiation, nonequilibrium and erosive contamination, any one of which would throw doubt on the experimental conclusions. Scramjet performance which is extremely sensitive to component efficiencies is influenced by chemical perturbations arising from these test-device artifices. Besides being an adjunct to wind tunnel tests, therefore, free flight range testing can serve usefully as an economical intermediate between laboratory research and full scale flight tests.

Sub-scale studies of observable phenomena in re-entry vehicle wakes, along with measurements of aerodynamic coefficients, are among the earliest applications of free flight ranges. The example chosen has a ballistic coefficient of 1.2 x 10^6 dynes/cm^2 with a 25° entry angle from an initial speed of 6.9 km/sec. Figure 1-7-1 shows that there is essentially no deceleration until some altitude well below 30 km. For a fixed prototype scale length and model scale length, the required range pressure increases with decreasing altitude in direct proportion to the atmospheric pressure. This leads to rather large values, since the ambient atmospheric pressure increases tenfold from 30 to 15 km in altitude. In planning simulation tests, therefore, an additional degree of freedom may be employed: as indicated on figure 1-7-2 lower velocities permit larger models, and larger models permit lower pressures for a fixed value of the binary similitude parameter. Gun performance must be calculated for various model and launch tube sizes during the design of the experimental procedure. The Richtmeyer-von Neuman technique (11) has been adapted and applied to guns of various geometries (12) and from this a curve of mass versus velocity can be obtained for each gun size. Assuming that a family of model-sabot designs for slender conical flight vehicles retains geometrical similarity, launch mass can be related to model size. Consequently, a relationship between model size and launch velocity is obtainable. In the present example, model-size to bore-size ratio is conservatively taken to be 2:3.

Assuming that the diameter of the prototype was 60 cm, and that the gas pump tube was 25.4 cm internal diameter by 32.5 metres long and varying the gun size to take advantage of larger models at lower speeds, range pressures have been held down to a few atmospheres to simulate flight conditions down to 12 km. Figure 1-7-5 gives the pressure requirements under these constraints. Vehicles with lower

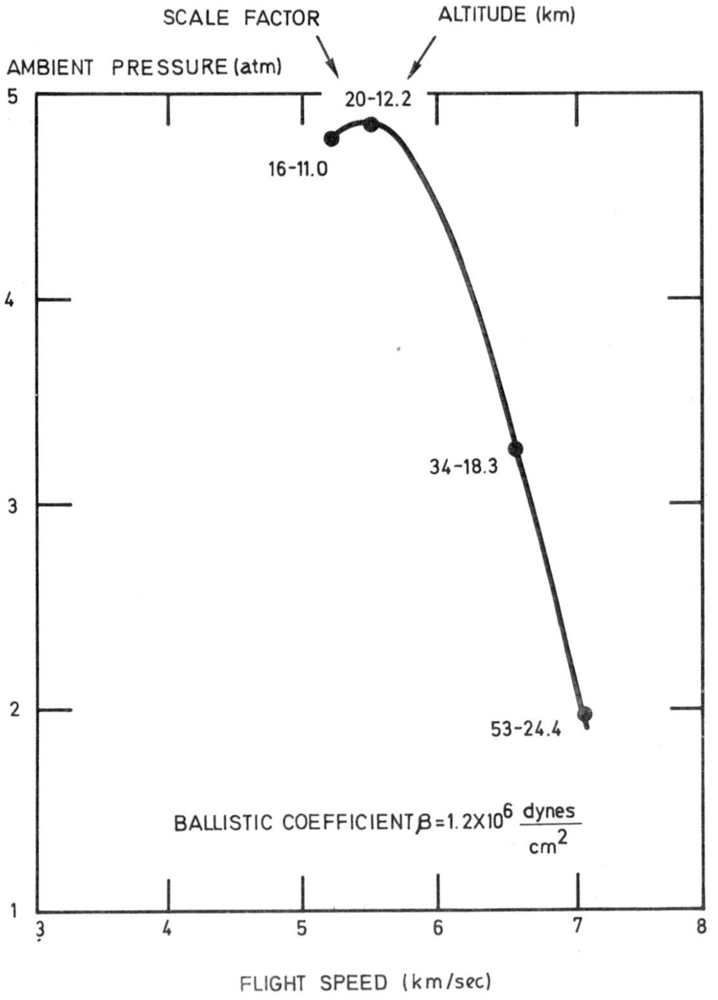

Fig. 1-7-5 Range pressure requirements for binary scaling reentry object with maximum model size for each speed.

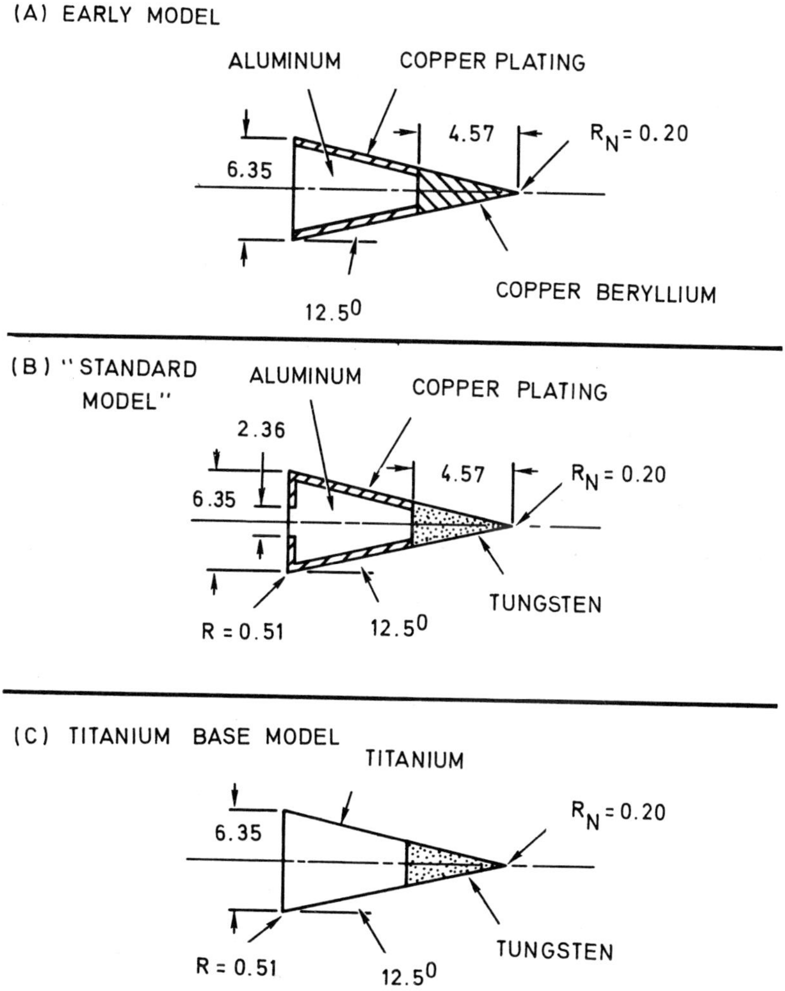

Fig. 1-7-6 Evolution of cone model designs - all dimensions in millimeters

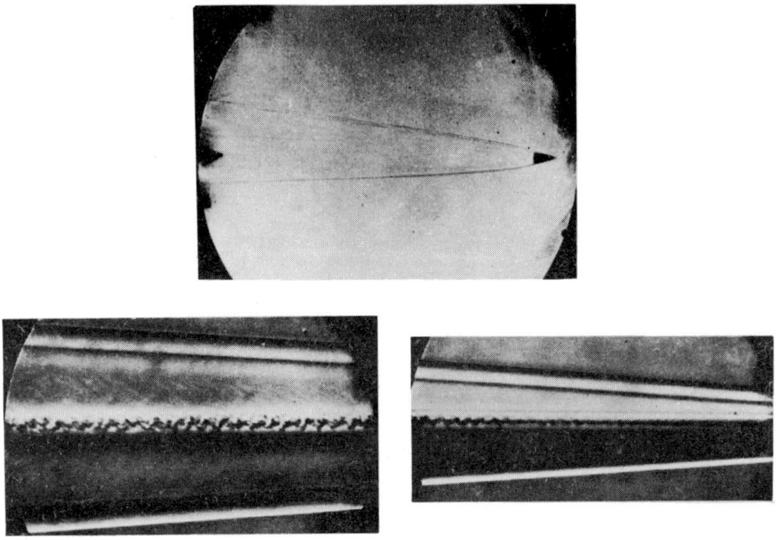

$\theta_C = 12.5^0$, $D_B = 6.35\,mm$, $R_N/R_B = 0.06$, $P_\infty = 75$ torr, $U_\infty = 5.1$ km/sec

Fig. 1-7-7 Hypersonic wake behind clean cone in air

ballistic coefficients decelerate at still higher altitudes thus permitting still lower pressures to be used in the range for the same scale length-velocity restrictions on the model.

Cone Wake Ionization

Objectives of the Study

For many years physicists have used the scattering of radio waves as a diagnostic of meteor trails. The residual ionization provides information about the body and its flight: blunt bodies ionize the air primarily through sudden heating behind the strong detached shock wave wrapped around the forward surfaces while slender bodies generate electrons primarily in the viscous boundary layer through frictional heating; however, even slightly blunt leading surfaces can ionize the air by shock as vigorously as the boundary layer produces electrons.

For altitudes well in excess of 30 km, the behaviour of a slender conical body can be simulated in the free flight range using models of the order of a centimeter in size at sub-atmospheric ambient pressures. Full re-entry velocities (6 to 7 km/sec) are also attainable. The object is to learn how the wake spreads in the laminar and turbulent regimes, and how the free electron distributions decay. As such experiments (13) and their interpretations (14) are described in the literature, a brief summary will suffice.

Experimental Methods

Note that during the launch of a slender conical model the reflection of stress waves from the conical tip leads to high tensile pulses, which can cause failure.

Initial tests were directed toward zero angle-of-attack, non-ablating flights in order to idealize the wake conditions so that analysis was possible. Ballasted composite model designs shown in figure 1-7-6 were evolved to meet these requirements, and the final basic design has served successfully for some years. Problems are that: angles of attack exceeding a moderate fraction of the cone-angle cause unacceptably large turbulence asymmetries in the wake; and that tip or edge ablation of the model deposits metallic impurities leading to spurious initial levels and decay rates of wake ionization; attempts to utilise hollow-based cones to meet model design objectives must be viewed with caution until controlled tests have been carried out to study sensitivity of the wake ionization to base flow geometry.

Two types of instrumentation are appropriate for the wake measurements: flow visualization apparatus and microwave plasma diagnostics. Schlieren photography using a single mirror double-pass system has yielded the information needed on flow structure, as reported by Wilson (15) (16). Both spark and giant-pulse laser light sources have been used. Figure 1-7-7 shows examples of the photographs obtained from the system. The flight conditions were those to which the discussion below is directed. Behind the slender cone, laminar wake flow was followed by persistent periodic instabilities. Low turbulence Reynolds numbers for such a case indicate that inertial randomisation barely outpaced the dissipation by viscous stresses. Finally, some semblance of turbulent flow appeared. The microwave transmission measurements were carried out using millimeter wave focused interferometer probes and resonant focused probes as described by Primich, Hayami, and others (17) (18). Although the transverse line integral of electron density is the instrumental output considered here, radial profiles have also been obtained using instrumentation of this type (19).

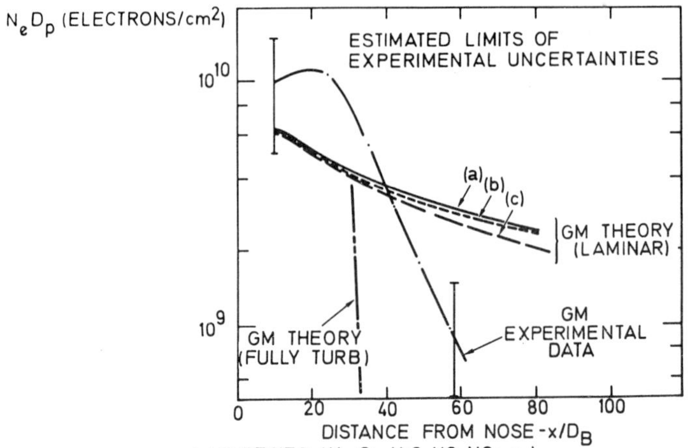

(a) 7 SPECIES ($N_2, O_2, N, O, NO, NO^+, e^-$)

(b) 7 SPECIES PLUS O^-, O_2^-

(c) 7 SPECIES PLUS O^-, O_2^-, NO_2 & NO_2^-

CLEAN CONE IN AIR

$P_\infty = 75$ torr, $\theta_c = 12.5°$, $D_B = 6.35$ mm, $R_N/R_B = 0.06$, $U_\infty = 5.85$ km/sec

Fig. 1-7-8 Cone wake ionization in air

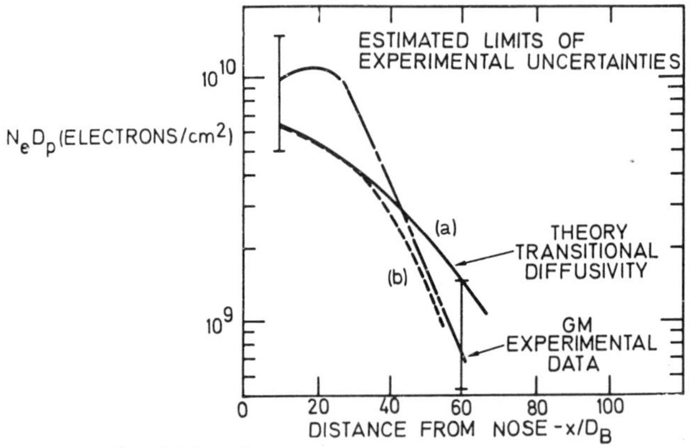

(a) 7 SPECIES (N_2, O_2, NO, NO^+, e^-)

(b) 7 SPECIES PLUS O^- & O_2^-
OR
7 SPECIES PLUS O^-, O_2^-, NO_2 & NO_2^-

FORM OF TRANSITIONAL DIFFUSIVITY IS :

$$\left[\mu_\epsilon = K\delta(\rho_e U_e - \rho_c U_c),\ K = K_i + (K_T - K_i)\frac{(x-x_i)^2}{(x_T - x_i)^3}(3x_T - x_i - 2x)\right]$$

CLEAN CONE IN AIR

$P_\infty = 75$ torr, $\theta_c = 12.5°$, $D_B = 6.35$ mm, $R_N/R_B = 0.06$, $U_\infty = 5.85$ km/sec

Fig. 1-7-9 Kinetics in transitional wake of a cone

Interpretation of Results

Wilson (15) (16) showed that by duplicating flight speed and total Reynolds numbers, full simulation of the wake flow structure was obtained. Ionization, at levels high enough to be governed by binary reactions, also scales under the same rules (20) (21). Behind cones however ionization is low and three-body electron attachment is an important feature, and it is then necessary to construct a theoretical model, which may be tested in the range.

Early attempts to analyse the cone wake results using conventional laminar and turbulent transport approximations were unable to explain the observations. Laminar transport gives too slow a decay and although fully turbulent transport gives a very rapid rate at first, it also breaks down eventually because the velocity deficiency calculated for the flow rapidly becomes too small. Figure 1-7-8 shows data obtained from our experiments plotted within the alternative limits predicted from theories of laminar and fully turbulent flow.

Using an axisymmetric, finite-difference, viscous-wake code (22) (23) and a kinetic-overlay code (24), K.S. Wen has studied individually the effects of initial conditions, mixing model, and kinetic model on the cone wake ionization (14). As suggested by the Schlieren photograph (Fig. 1-7-7) he found that a transitional form of the transport process was needed to describe wake growth and ionization between the incipiently unstable region and the fully turbulent region. Attachment kinetics must also be included in the calculation that is shown in figure 1-7-9.

In this example theoretical analysis was needed because the dissimilarity of certain microscopic mechanisms precluded direct sealing. However experimental evidence was relied on to establish confidence in the theory used.

Supersonic Combustion Ramjet Performance

Objectives of the Study

In contrast with the foregoing example, the use of the free flight range to investigate the performance of supersonic combustion ramjets is still in the preliminary development stage. This discussion deals mainly with future work that appears feasible on the basis of evaluation studies of new experimental techniques. The most novel aspect is the use of models having complex hollow shells that contain fuel storage and delivery systems. The instrumentation required is based on extensions of and improvements in existing methods.

The discovery that fuel can be mixed and burnt within a practical space in a supersonic airflow has suggested vast extensions of airbreathing engine performance (25) (26) (27). The principal processes within a scramjet include compression, mixing, heat addition, and nozzle expansion, but in an actual engine these components are difficult to identify. The compression occurs on inclined forward surfaces. High Reynolds and Mach numbers lead to interactions between a turbulent boundary layer and a shock layer on the large configurations under consideration. Separation, due to the reimpingement at shock from the inner surface of the cowl to the centre-body, can be suppressed by boundary layer turbulence in many cases. Gaseous fuel injected into a turbulent mixing zone burns in the after portions of the combustor and releases heat in the nozzle. At high Mach numbers the thrust margins are small and high component efficiencies are essential.

Experimental Methods

Compared with the model used in the re-entry physics problem, the scramjet model is more delicate because it must have a thin cowl, a fuel cavity and an injection

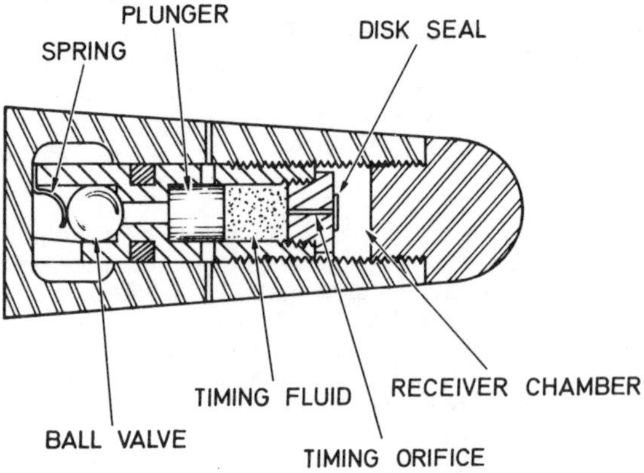

Fig. 1-7-10 Fuel injection model

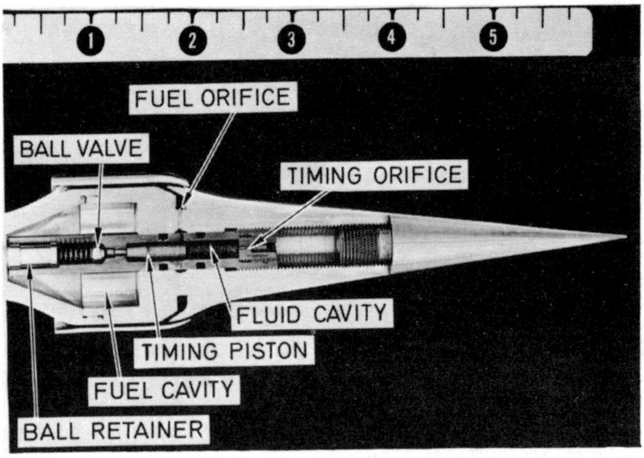

Fig. 1-7-11 Scramjet engine model mockup (scale in inches)

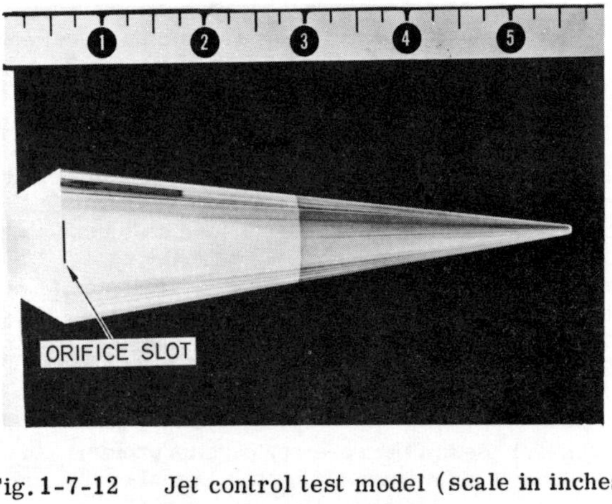

Fig. 1-7-12 Jet control test model (scale in inches)

mechanism. Fortunately, the required flight speeds are only about one half the re-entry velocities, and with the longer launch tube lengths on the newer facilities, (8) the launch accelerations are nearly an order of magnitude less than the 10^6 g-levels encountered in re-entry physics testing.

The model must deliver fuel at an equivalence ratio near to unity during flight through the instrumented portion of the range. Technically the most interesting fuel is hydrogen, which is most conveniently stored in the model as a gas at room temperature. For simulation of a full scale engine at 3 to 5 km/sec with a 3 to 6 cm diameter annular combustor model, the required hydrogen flow rate may be as high as several hundred grams per second.

Since the storage density of gaseous hydrogen is at best of the order of .06 grams/ cm^3, at most a few grams can be stored conveniently in the model. At 3 km/sec under uniform delivery conditions of approximately 100 grams per second, 1 gram of fuel is delivered for every 30 metres of model flight. For accurate measurement of thrust a range length of at least 30 metres is probably required. The conflicting requirements of high fuel delivery rate, small available storage volume and fuel delivery over a reasonable distance therefore dictate a fuel system with a time delay feature. Typically the delay required is about 20 to 35 milliseconds, which allows sufficient of the length of the range to be used to measure the drag of the model without fuel injection.

Figure 1-7-10 illustrates a test vehicle for a fuel delivery system having a time-delay mechanism. When the model is launched, the acceleration in the launch tube drives the ball valve into a locked position by lodging it in the tapered section. The plunger remains in the aft position during acceleration but when the model leaves the launch tube the inertia becomes negligible and the pressure force predominates. At this point the loosely held disk seal breaks free of the timing orifice plug, and the timing fluid flows forward into the receiver chamber. A computer code has been developed to predict the dynamics of the system, delays of 10 to 50 milliseconds being practicable. Appropriate contouring of the aft end of the plunger permits the gas flow rates to be kept constant for several milliseconds.

A proposed design for a scramjet flight test vehicle nominally designed for Mach 12 is illustrated in figure 1-7-11. The extremely narrow cowl passage illustrates the extreme narrowness of the combustor channel that results from axisymmetric compression. Although practical limitations militate strongly against axisymmetric configurations in actual engines (27), this form is chosen to maximise success in launching. Note that the valve piston is contoured to flatten the top of the injection pulse by increasing the flow area as the pressure decays. The tungsten nose mounted in a titanium afterbody provides ballasting.

A mock-up of an external burning manoeuvre control vehicle is shown in figure 1-7-12. The valve design is similar to that in the scramjet but asymmetric injection occurs at the slot. A test model like this could provide data on inert or burning jet control. Pressure builds up on the afterbody, altering the trim of the vehicle so that lift forces deflect it from its original flight path.

Both mock-ups are preliminary designs: flight tests await completion of the development of the launch system.

The range instrumentation and the analysis needed to reduce the data into aerodynamic coefficients is given in detail in the review by Charters (28). Studies are presently in progress to evaluate laser illuminated ranging devices for measuring flight dynamics at high data rates. Spark - and laser - illuminated Schlieren photography has been extensively employed for flow visualisation, giving information on shock angles, wave cancellation effects, shock-boundary layer interactions, and plug nozzle expansion structure. Absorption spectroscopy is useful in determining the concentrations of key chemical species in the exhaust stream. Line reversal techniques (29) have been adapted for wake temperature measurements, and a new technique involving the capture and analysis of wake gases is under study to give an indication of overall combustion efficiency in flight. X-ray techniques used in conjunction with a timing piston for the measurement of combustor static pressure in flight are also being investigated.

Interpretation of Results

Compressibility and viscosity effects in the inlet act to produce laminar and turbulent hypersonic interactions. Duplication of Mach number and total Reynolds number will scale these, as well as the further interactions within the cowl. Modeling is necessary for the latter process to produce the correct separation-reattachment behaviour of the boundary layer under conditions of shock impingement.

Turbulence structure in the fuel mixing zone is also susceptible to Mach-Reynolds scaling; however the ensuing chemical kinetic steps demand that the ratios of residence time to reaction time be preserved for each elementary process. These ratios are called Damköhler numbers and may be defined in terms of either diffusively or convectively determined residence times. When comparing ignition rate with mixing rate the turbulent mixing time is appropriate in the Damköhler numbers. Chemical time for ignition is approximately proportional to the reciprocal of the density, and to a mixed algebraic dependence on temperature, which is usually dominated by an exponential. The complexity of the ignition Damköhler number demands that the velocity, temperature, and the density x length product are individually duplicated. However, as Ferri has indicated (27), if mixing is the rate controlling process for both model and prototype conditions, it is unnecessary to be concerned with ignition delays.

Completion of combustion in the stages of heat release and recombination during expansion is controlled by three-body reactions. Hence, chemical times are proportional to reciprocal density^2 x length products, accompanied by a power law temperature dependence. Residence time is simply the average convection velocity divided by the characteristic length. The Damköhler parameters derived for these processes require the preservation of velocity, the temperature, and density^2 x length for proper modeling. The higher order density dependence for the tertiary reactions poses a conflict with the previous requirements.

As in the re-entry case, the final reactions in the flow field do not obey binary scaling, and theoretical models must be developed to use range measurements for full scale predictions. Since it is probably necessary to operate near equilibrium conditions, a computer code for combustor flows has been developed to solve simutaneously the mass-action laws and the conservation equations. The boundary constraint is a Crocco pressure-area duct family as employed by Billig (30). The basic programme is adapted from existing programmes for equilibrium flow (31). A characteristic code with finite rate processes is used for the nozzle flow but a simpler approximate approach is the calculation of a pressure field assuming frozen characteristics, followed by streamtube (32) calculations of reaction histories. These analytical methods must be used for the portions of the flow where reaction scaling no longer applies.

The provisional methods of predicting pressure distribution that have been developed use the shock-expansion procedure for the forebody, the quasi-one-dimensional code for the combustor, and Prandtl-Meyer relationships for the afterbody surface conditions. These approximate techniques are used during preliminary design, when numerous production computer runs are needed for various shapes and flight conditions. Computer techniques that utilise tabular readouts from design layouts give the mass distributions, so that static stability estimates are also available.

The design of the fuel injection system evolves from a code that couples fluid dynamics with rigid body dynamics for the internal parts accounting for non-ideal gas behaviour at high pressures. A viscous layer code is being developed for inlet flow fields.

Since thrust measurements require the optimisation of the deployment of the range instrumentation, a random variable analysis is in hand to identify the best configuration of spark stations. Radar techniques are being evaluated as an alternative to spark stations because of their potentially higher data rates. They hold out the possibility of using a shorter test section which may be necessary because of the limitation of fuel storage within the models.

Mixing and Reaction Experiments

Objectives of the Studies

Free-flight experimental techniques, while being unsuitable for fundamental kinetics measurements, are unparalleled for examining the chemical process in a turbulent flow under conditions appropriate to high speed flight.

The experiment may be considered at three levels of complexity, the cases being:

a) Homogeneous turbulence in a reacting gas;

b) Supersonic mixing of an inert contaminant having a low Schmidt number;

c) Diffusion combustion of a reactive gas in a supersonic turbulent flow.

Detailed information regarding the concentration field and the velocity field is required in each case. The first case considers the influence of chemical energetics on the turbulent motion (33) (34). Investigators at several laboratories are devising steady flow systems in an attempt to study such effects, but limitations on Damköhler number hamper their efforts. The second and third cases introduce inhomogeneities and anisotropies, which render their interpretation difficult in terms of statistical turbulence theory. They serve as a basis for testing theoretical models proposed to describe the rms fluctuation levels and the profiles of mean quantities in flow fields.

Experimental Methods

One technique suitable for studying homogeneous turbulence utilises a perforated 'flying grid model'. This was first employed in the so-called 'swiss cheese' experiment of Uberoi and Kovasznay carried out by Charters at the US Army Ballistic Research Laboratories (35) to study compressibility effects in a non-reacting gas. Attempts to achieve similar results by using a group of small spheres flying together were quickly ruled out in hypervelocity flight conditions because the flight pattern was not reproducible.

The model we chose was a 22-1/2° truncated cone of 13 mm base diameter with five symmetrically disposed holes bored parallel to the axis. A Schlieren photograph of the model in flight at a speed of approximately 5 km/sec in air at 150 torr pressure is shown in figure 1-7-13. The model was constructed of copper plated aluminum. The holes however were unplated, and the microwave probe records indicated that the bare aluminum surfaces were heavily ablated. Ablation can be suppressed by taking appropriate precautions in model design and by choosing suitable operating parameters.

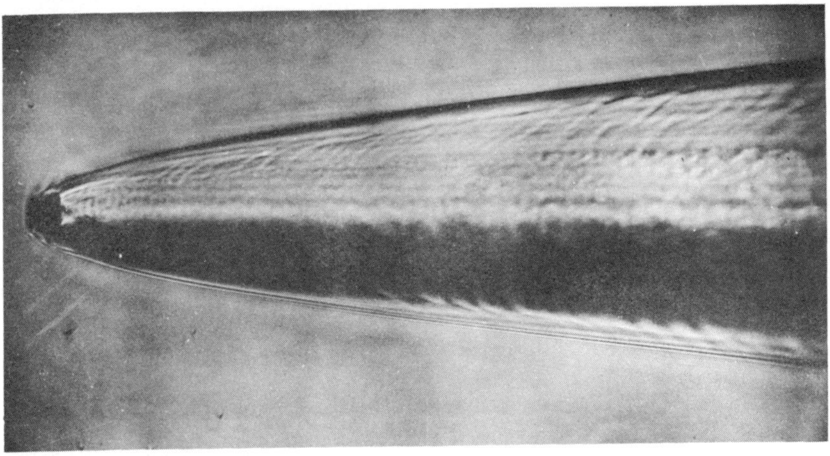

Fig. 1-7-13 Schlieren photograph of flying grid model in flight - velocity 5.05 km/sec, pressure 150 torr

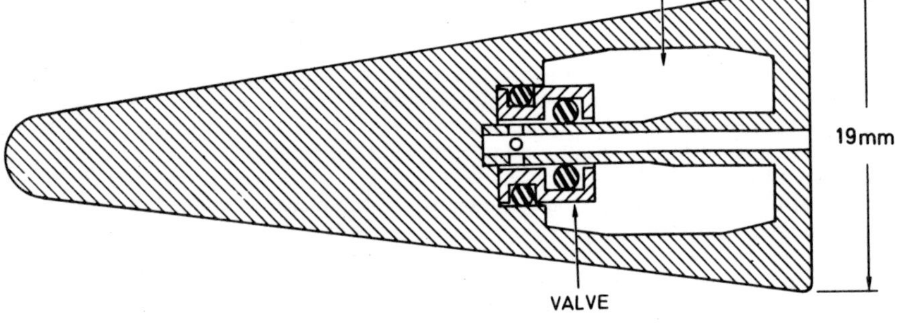

Fig. 1-7-14 Base injection model - schematic

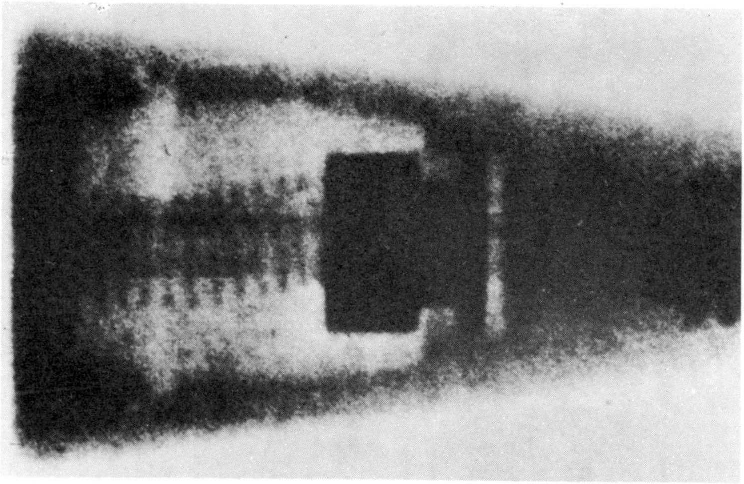

(a) Before launch

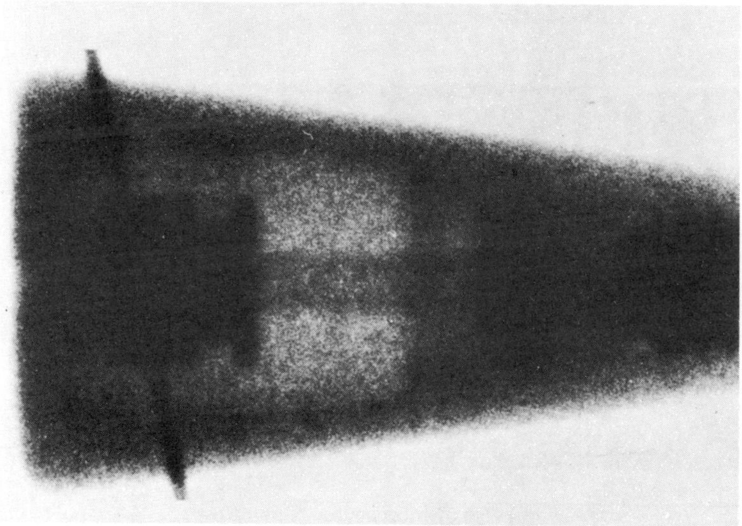

(b) After launch

Fig. 1-7-15 Valve position in X-ray photograph of base injection

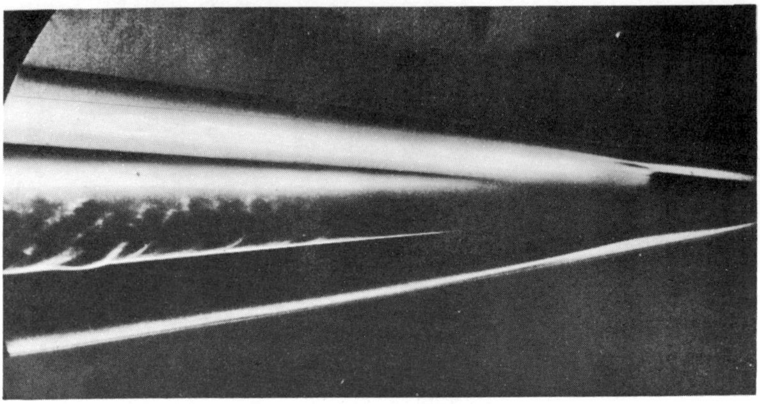

(a) Zero injection

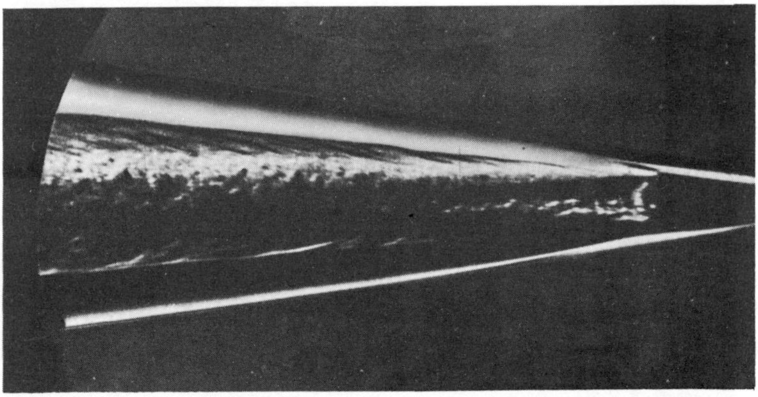

(b) 5% of intercepted mass flow

Fig. 1-7-16 Base injection effects (model shown in Fig. 1-7-14 in flight at 3.2 km/sec through air at 100 torr ambient pressure)

By bleeding gas from the stagnation region into the base flow the nose drag was reduced and the recompression pulse in the near wake was suppressed. Consequently nearly homogeneous turbulent flow at constant pressure was obtained for some distance behind the body. The ratio of reaction time to eddy lifetime can be selected by adjusting the ambient pressure. Different gas compositions yield different values of the ratio of the chemical energy stored in overdissociation to the kinetic energy present in the turbulence. Seeding the central hole with an ablative tracer creates diffusion into a homogeneous turbulence field.

Experiments into the second and third cases rely on injecting a foreign gas into the base flow of a model. Zakkay (36) and others have demonstrated that an injected mass flow equal to a few percent of free stream flow is sufficient to remove the recirculation region. Under these conditions the flow behaves like a jet, the initial profiles of which are known far better than are those of a wake at hypersonic speeds. The wake-jet effect can be obtained in the free-flight range by using the base-injection model shown diagrammatically in figure 1-7-14. Since a porous base on a plenum chamber is more desirable than the single orifice, it will be incorporated in future designs. The internal collar with the 'O' ring seals is set back during launch acceleration, and the exposure of the holes in the central tube allows the stored gas to flow out of the base orifice. The collar remains stationary after it has been set back, because it is driven up an external taper. A. C. Charters and J. Andon first defined the design of this model, and O. P. Prachar developed it into flight hardware.

Flight test results on such a model are illustrated photographically in figures 1-7-15 and 1-7-16. The X-ray pictures in figure 1-7-15 show the valve collar positions previous to firing, and during free flight respectively. The set back of the collar was sucessful and the model maintained its integrity. The flight speed was 3.2 km/sec and the ambient pressure was 100 torr. The 19 mm base diameter model had a blunted tip for ease in launching. The mass flow of gas injected was approximately five percent of the mass flow of air swept out by the frontal area of the model. The Schlieren photographs in figure 1-7-16 permit a comparison of wake structures. The sensitivity of the instrumentation to gradients highlights the turbulence structure. The influence of injection on transition is reminiscent of the conditions in a boundary layer.

Helium is a suitable inert gas to inject into the base flow in the inert gas experiment, but hydrogen could be used if the flight speed is low enough to prevent ignition or if the carrier gas is inert.

Hydrogen is best used in the reactive mixing experiment because of ease of ignition and because the high temperature kinetics of the hydrogen-oxygen system have been studied with a high degree of success in recent years.

In the mixing problem where velocity fluctuation fields are the main interest, doppler measurements of laser light scattered from small particles have been employed for statistical studies in liquid flows (37). Application to high speed flows is now under study. One difficulty is the optical mixing of the scattered light with the reference light. The inclusion of sub-micron particles is accompanied by problems of agglomeration before the test, and the thermochemical survival in the flow field. Statistics on species concentration may be accessible by measuring fluorescence fluctuation along an intense pencil beam of light. The determination of mean properties or their integrals follows closely the pattern of previous free-flight range practice.

Interpretation of Results

The output of these experiments is dependent on the sensitivity and resolution of the measurements. Assuming that both mean flow and statistical information become available, we have theoretical developments available for interpreting the results. For experiment (a) our statistical theories of turbulence spectra in the presence of chemical energy release (34) have yet to be tested. Statistics on low Schmidt number concentration are important in the studies of light gas mixing for combustion and of ionization fluctuations in re-entry wakes. Experiment (b) could yield information to resolve the disagreement among theoreticians (38-41), about the inertial-diffusive sub-range. Also many predictions are available but untested regarding rms concentration fluctuation levels in wakes and experiment (c) covers the statistics of a reacting contaminant, which have been the subject of other recent studies. Experiments (b) and (c) could be pivotal in pointing out the best path for future theoretical development.

References

1. **Penner, S.S.,** 'Combustion Researches and Reviews', London, Butterworths, 1955, 140.

2. **Gibson, W.E., Marrone, P.V.,** 'A Similitude for Nonequilibrium Phenomena in Hypersonic Flight,' The High Temperature Aspects of Hypersonic Flow, Oxford, Pergamon Press, 1964.

3. **Birkhoff, G.,** 'Hydrodynamics,' Princeton Univ. Press, 1961, 109.

4. **Eschenroeder, A.Q.,** 'Nonequilibrium Scaling of High Enthalpy Gas Flows,' General Motors Tech. Rep. 64-02C, June 1964.

5. **Wittliff, C.E., Sundaram, T.R., Rae, W.J., Lordi, J.A.,** 'Study of High-Density Hypervelocity Flows and Similitudes,' Cornell Aeronautical Lab., Rep. No. AF-2081-A1. AEDC TR67-72, April 1967.

6. **Eschenroeder, A.Q.,** 'Reaction Kinetics in Hypersonic Flow,' Advances in Chemical Physics, New York, John Wiley and Son.

7. **Leonard, R.L., Rose, P.H.,** 'Feasibility of a High Performance Aerodynamic Impulse Facility,' Avco-Everett Res. Rep. 265, Feb. 1967.

8. **Charters, A.C.,** '2-1/4" Gun Project at GM Defense Research Laboratories,' Aeroballistic Range Ass. Paper, Also GM DRL CTN 65-04. Oct. 1965.

9. **Barrett, B.J., Rogers, W.K.,** 'Second Interim Report on GM DRL's 2-1/4" Gun Project,' General Motors Tech. Rep. 66-01P. Sept. 1966.

10. **Curtis, J.S.,** 'An Accelerated Reservoir Light-Gas Gun,' Nat. Aer. and Space Admin. Tech. Note D-1144, Feb 1962.

11. **Piacesi, R., Gates, D.F., Siegel, A.E.,** 'Computer Analysis of Two-Stage Hypervelocity Model Launchers,' Naval Ordnance Lab. TR62-87, Aug. 1963.

12. **Collins, D.J., Christman, D.R., Charters, A.C., Sangster,** D.K., Proceedings, Hypervelocity Techniques Symp. 5, Mar 1967.

13. **Primich, R.I., Hayami, R.A., Auston, D.H.,** ***et al*** , 'Free Flight Range Measurements of Ionization Behind Slender Hypersonic-Velocity Cones,' Gen. Motors Tech. Rep. 65-19A. Apr. 1965.

14. **Wen, K.S., Chen, T., Hayami, R.A., Primich, R.I.,** 'An Eddy Diffusion Model for Predicting Rapid Wake Ionization Decay Behind Hypersonic Cones.'

AIAA paper No. 67-21, Jan 1967.

15. **Wilson, L.N.,** AIAA J. 4, Oct 1966, 1741.

16. **Wilson, L.N.,** 'Far Wake Behavior of Hypersonic Bodies,' AIAA Paper No. 67-31. Jan. 1967.

17. **Primich, R.I., Auston, D.H., Hayami, R.A., McLeod, J.D., Zivanovic, S.,** 'Two New Techniques for Determination of Transient Plasma Properties,' Proc. Seventh Int. Conf. on Ionization in Gases, Beograd, 1966.

18. **Hayami, R.A., Kelley, K.J.,** IEEE Transactions on Aerospace and Electronic Systems, AES-3, No.2, Mar 1967, 339.

19. **Auston, D.H.,** 'A Quasi-Optic Imaging Resonator for Plasma Diagnostics,' IEEE Trans. on Aerospace and Electronic Systems, AES-3, No.2, May 1967.

20. **Hayami, R.A., Primich, R.I.,** 'Wake Electron Density Measurements Behind Hypersonic Spheres and Cones,' AGARD Specialist Meeting on Fluid Physics of Hypersonic Wakes. May 1967.

21. **Eschenroeder, A.Q., Hayami, R.A.,** 'Scaling Experiments on Wake Ionization Behind Nonablating Hypersonic Spheres,' General Motors Tech. Rep. 64-02L. Nov. 1964.

22. **Wen, K.S., Chen. T.,** 'Axisymmetric Viscous Wake Analysis for a Hypersonic Reentry Body,' General Motors Tech. Rep. 66-12A, Apr. 1966.

23. **Chen. T., Wen. K.S.,** 'Fortran Computer Code for the Axisymmetric Viscous Wake Analysis of a Hypersonic Reentry Body,' Gen. Motors Tech. Rep. 66-66, Oct. 1966.

24. **Lieu, B.H., Wen, K.S.,** 'A Kinetics Overlay Program for the Hypersonic Wake,' Gen. Motors Tech. Rep. 67-02B, Feb. 1967.

25. **Ferri, A.,** J. Royal Aero. Soc. 68, Sep. 1964, 575.

26. **Avery, W.H., Dugger, G.L.,** 'Astronautics and Aeronautics 2, Jun 1964.

27. **Ferri, A.,** 'Review of Scramjet Propulsion Technology,' AIAA Paper 66-826, Dec. 1966.

28. **Charters, A.C.,** 'High Speed Problems of Aircraft and Experimental Methods,' Princeton, Princeton University Press, 1961. Sec. 0, 905.

29. **Rockman, C.M.,** 'Line Reversal Measurements of Hypersonic Wake Temperatures,' General Motors Tech. Rep. 66-01B, Mar. 1966.

30. **Billig, F.H.,** 'Design of Supersonic Combustors Based on Pressure-Area Fields,' paper at the Eleventh Int. Symp. on Combustion, Berkeley. Aug. 1966.

31. **Farmer, M.D., Eschenroeder, A.Q.,** 'A Unified Matrix Approach for the Computation of Real Gas Flows,' Gen. Motors Tech. Rep. 64-02E, Aug. 1964.

32. **Chen, T., Eschenroeder, A.Q.,** 'A Fortran Computer Code for Inviscid,

Nonequilibrium Streamtube Flow, ' Gen. Motors Tech. Rep. 65-01P, Dec. 1965.

33. Eschenroeder, A.Q., Physics of Fluids, 7, Nov. 1964, 1735.

34. Eschenroeder, A.Q., AIAA J. 3, Oct. 1965, 1839.

35. Uberoi, M.S., Kovasznay, L.S.G., J. Appl. Phys. 26, 1955, 19.

36. Fox, H., Zakkay, V., Sinhi, 'A Review of Some Problems in Turbulent Mixing, ' New York University Rep. AA-66-63, Sep. 1966.

37. Welch, N.E., Tomme, W.J., 'The Analysis of Turbulence from Data Obtained with a Laser Velocimeter, ' AIAA Paper No. 67-179, Jan. 1967.

38. Batchelor, G.K., Howells, I.D., Townsend, A.A., J. Fluid Mech. 5, 1959, 134.

39. Pao, Y.H., AIAA J. 2, Sep. 1964, 1550.

40. Corrsin, S., Physics of Fluids 7, Aug. 1964, 1156.

41. Gibson, C.H., 'A Unified Spectral Theory for Turbulent Scalar Fields of Arbitrary Diffusivity, ' University of California, San Diego Rep. IRPA 65-68. Mar 1966, revised Dec. 1966.

1-8

A Facility for Hypersonic Flow Simulation over Blunt Nosed Bodies

V. ZAKKAY and W. H. MAK
New York University, U.S.A.

Summary

A facility for simulating re-entry flow fields over bluntnosed bodies is presented here. By using H_2O-O_2 mixture, it is shown that simulation may be achieved at lower stagnation temperatures. In this manner, the test gas has not dissociated and therefore the absence of the complicated chemistry in trying to understand the fluid mechanics aspect of the problem. The criteria for simulation of supersonic flow over blunt-nosed bodies using any gaseous test medium have been found to be the dynamic pressure and the density ratio across the normal shock. The effect of γ for different gases chiefly affects the surface Mach number distribution. Combustion gases may be used as an aerodynamic test medium, provided that water vapor condensation is avoided and that equilibrium flow conditions are ensured. In the H_2O-O_2 mixture family, a mixture of equivalence ratio $\gamma = 0.2035$ has been found to yield density ratios of the order of ten across a normal shock in equilibrium flow by suitably adjusting the pressure level. A buffered shock tunnel test facility is proposed to use the H_2O-O_2 ($\gamma = 0.2035$) mixture as the test medium. Flow Mach numbers of the order of ten can be attained at a static temperature of 300°K. Operating conditions have been determined for the heated air driver, the helium buffer and the H_2O-O_2 mixture, such that tailored condition of the reflected shock is achieved for maximum testing time.

Sommaire

On présente dans cet exposé un montage expérimental pour la simulation de l'écoulement autour de corps à bord d'attaque émoussé dans les conditions de la rentrée atmosphérique. L'emploi du mélange H_2O-O_2 permet la simulation à basse température d'arrêt et le gaz actif n'est pas dissocié, ce qui evite les équations chimiques compliquées et facilite l'analyse de l'aspect aérodynamique du phénomène. Les critères de simulation de l'écoulement supersonique autour de corps émoussés sont quels que soient les gaz utilisés, la pression dynamique et le rapport de masse volumique à la traversée d'un choc droit. La variation de γ ne joue de rôle que sur la répartition du nombre de Mach sur la surface de l'obstacle. On peut utiliser des gaz de combustion comme fluide actif, pourvu que l'on évite la condensation de la vapeur d'eau. Dans la famille des mélanges H_2O-O_2, un mélange de rapport d'équivalence $\lambda = 0,2035$ fournit un rapport de densités de 10 à travers une onde droite lorsque le niveau de pression a une valeur correcte. Un tunnel à choc avec réservoir tampon et le mélange H_2O-O_2 ($\lambda = 0,2035$) sont proposés pour ces essais. On atteint un nombre de Mach de l'ordre de 10 à la température statique de 300°K. Les conditions de fonctionnement ont été déterminées pour de l'air chauffé comme gaz moteur, un tampon d'hélium et le mélange H_2O-O_2 de façon telle que le choc réfléchi ne perturbe pas l'écoulement durant la majeure partie du temps d'essai.

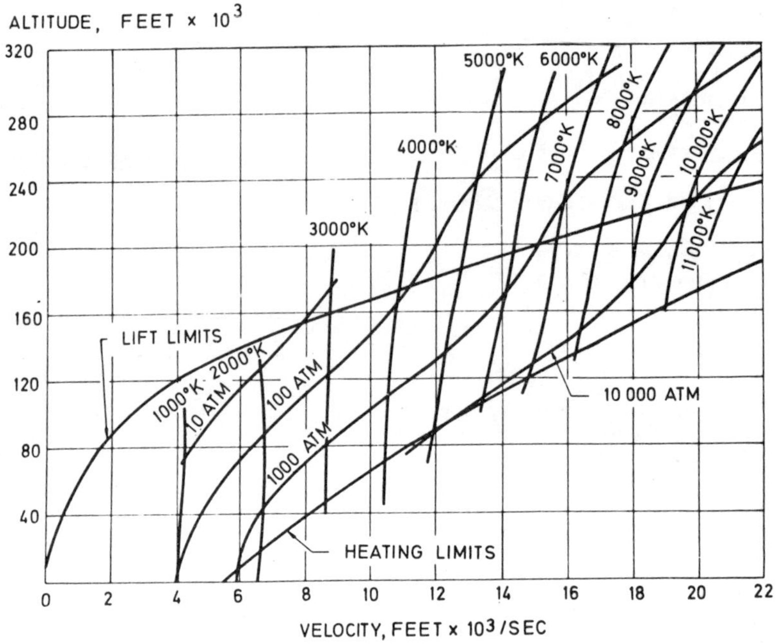

Fig. 1-8-1 Tunnel reservoir pressures and temperatures required for flight duplication in equilibrium air

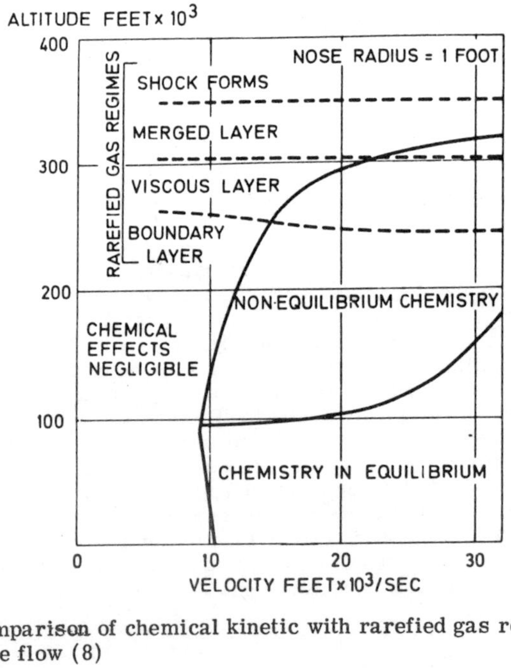

Fig. 1-8-2 Comparison of chemical kinetic with rarefied gas regimes for blunt-nose flow (8)

Introduction

The development of hypersonic vehicles such as intercontinental ballistic missiles, satellites and space probes has given rise to numerous environmental problems, especially in the areas of fluid dynamics and physics of high temperature gases. When the gas temperature is above about 2000°K, dissociation as well as caloric 'imperfections' are present. At satellite speeds all of the oxygen and a considerable fraction of the nitrogen molecules are dissociated. In addition, thermal ionization and chemical reaction between some of the constituents need to be taken into consideration. These phenomena are referred to as real gas effects.

Current methods of hypersonic flow simulation usually consist of expanding high temperature and high pressure air to high velocity. Various facilities have been developed. Bray (1) has evaluated the hypersonic gun tunnel which works on the principle of a free piston compressor. The piston strength requirements limit the stagnation temperatures to low values. Recently, Bird *et al* (2) discussed the performance limit of the reflected shock tunnel and concluded that reservoir conditions will be limited to about 2000 atmospheres pressure at 9000°K. The problem associated with this type of facility is the nonequilibrium phenomenon in the nozzle expansion and the short testing time. Direct heating of the air by passing it through a high energy electrical arc can produce high enthalpy flows for a period of time up to two minutes. However Cann and Buhler (3) showed that the pressure will not exceed about 100 atmospheres. The development of magnetohydrodynamic accelerators has shown that velocities of 30,000 to 40,000 ft/sec can be attained at densities corresponding to altitudes greater than 90 km. The estimated performance of such an accelerator has been reported by Ring (4). Both the arc heater and MHD accelerator have serious disadvantages. Electrode contamination of the flow; local variation in the high energy air properties due to nonuniform heating; effect of high energy electric and magnetic field on measuring instruments; and the problem of accurately calculating theoretical conditions are some of the problems yet to be overcome.

In recent years considerable interest has been shown in the utilization of other gases as a test medium for the simulation of aerodynamic data. For low temperature conditions, the use of helium has been quite successful in simulating air under the hypersonic approximation, and the theoretical work and experimental results have been reported by Love *et al* (5) and Ladson (6 and 7), respectively. At higher temperatures and pressures where real gas effects become important, attention has been focused on using combustion gas mixtures for simulating combined loading and heating such as that obtained in hypersonic flight. There has been considerable knowledge gained on the thermodynamics and flow properties of reacting gas mixtures expanding through a nozzle, e.g. the expansion of combustion gases through a nozzle to obtain thrust as in the case of turbojet and ramjet engines. In this case however the combustion generally takes place under very high pressure in order to achieve the desired flow for simulation. The purpose of this study is to seek a suitable gas mixture which meets closely the criteria for simulation of re-entry flowfields, and the production of which involves few engineering difficulties.

Flow Simulation

Simulation Requirements in Air

In wind tunnel testing, complete duplication in both scale and ambient flow conditions over the wide portion of the flight corridor indicated in figure 1-8-1 is impossible to achieve. Simulation is the technique of duplicating only those dimensionless parameters or flow conditions most intimately associated with the phenomenon being studied, e.g. Mach number M_∞, Reynolds number Re_∞, hypersonic similarity parameter $M_\infty \alpha$ (where α is the thickness ratio or angle of attack), viscous inter-

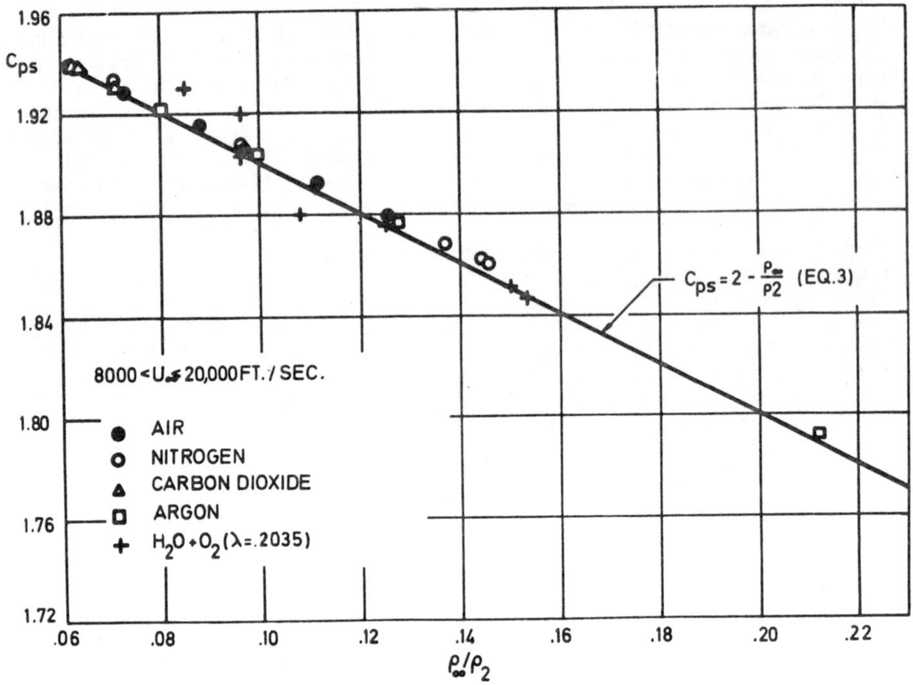

Fig. 1-8-3 Variation of Cp_s vs. $\frac{\rho_\infty}{\rho_2}$ across a normal shock wave in equilibrium gas mixtures

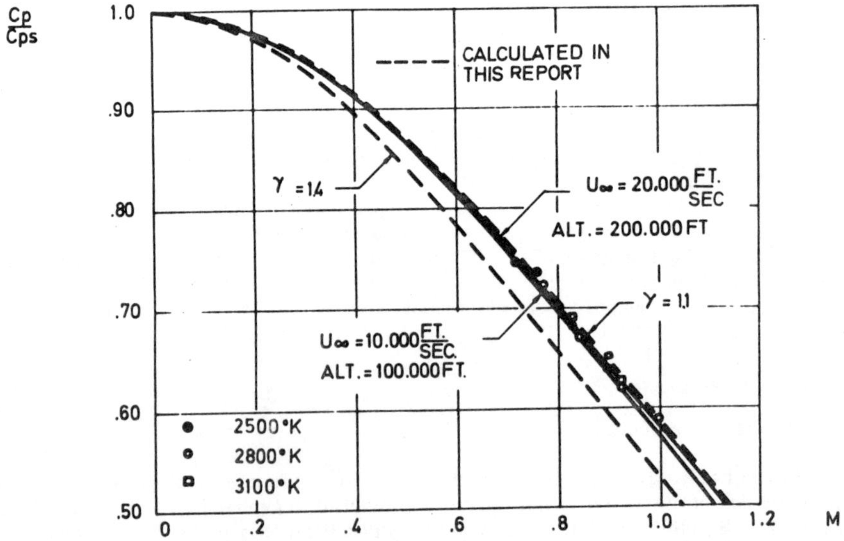

Fig. 1-8-4 Variation of $\frac{Cp}{Cp_s}$ with surface Mach number on a blunt-nosed body (hemisphere)

action parameter M^3_∞/Re_∞, flow velocity U_∞, and ambient density ρ_∞. Consequently each experiment is set up in order to determine the important parameters or conditions requiring duplication, and an analysis of the flow phenomenon is necessary.

For aerodynamic testing in the regime where real gas effects are important, Harvey (8) has shown that, for flight velocity above 13,000 ft/sec and in the altitude range of 100,000 ft to 30,000 ft, the flow chemistry is neither frozen nor in equilibrium. Thus, in addition to the parameters and ambient conditions cited, 'chemical kinetic simulation' must be provided. It is not well understood how this can be accomplished. One of the most serious problems in hypersonic wind tunnels is chemical nonequilibrium in the test section freestream. In these wind tunnels, the test gas in the reservoir reaches chemical equilibrium due to the high temperature and pressure and the long residence time. On expansion, chemical equilibrium is maintained at local temperature and pressure for some small distance through the nozzle. Farther downstream, however, the temperature and pressure decrease to a point where the chemical reactions are no longer fast enough to maintain equilibrium. As a result, the composition departs from the equilibrium values and finally 'freezes' out in some nonequilibrium state. Much effort has been applied to solve the nonequilibrium flow chemistry in a nozzle. Therefore in the regime where 'chemical kinetic simulation' is desired, the freestream dissociation level should be sufficiently low so that the energy in dissociation is small compared to the total freestream energy, and also the freestream dissociation should be small compared to the maximum dissociation fraction in the flow over the body. The former condition ensures that the freestream velocity corresponds with that of the flight condition, and the latter ensures that the flow behind the shock wave approaches equilibrium in nearly the same manner as that encountered in the flight; thus the aerodynamic properties of the flow are not changed.

The next problem in simulating the flowfield is deciding whether model scaling is permissible. Gibson (9) has shown that when chemical recombination can be neglected for nonequilibrium flows about blunt bodies at high altitudes, binary scaling keeping $\rho_\infty L$ = constant, where L is the characteristic body length, is possible for a given flow speed. Since the model is usually scaled down one to three orders of magnitude in wind tunnel testing, the corresponding increase in flow density is difficult to obtain. Moreover, three-body recombination will become important, so that the binary scaling is not valid where one would like to use it.

Figure 1-8-2 is the velocity-altitude chart of Harvey (8) showing the viscous and chemical kinetic regimes of hypersonic, high altitude flight. By superposing figure 1-8-2 on figure 1-8-1, one may notice that the flight corridor is divided into a portion above 100,000 ft altitude and 13,000 ft/sec flow speed in which nonequilibrium and frozen chemistry dominates. Thus to simulate the hypersonic flight conditions, one must provide:

1. Freestream velocity and density on a large scale model,

2. A low freestream dissociation level, and

3. Sufficient time for testing to obtain the desired data.

Velocity duplication, however, requires high stagnation temperatures. To produce velocities of 16,000 ft/sec and 20,000 ft/sec requires temperatures of about 7,500°K and 10,000°K, respectively, as illustrated in figure 1-8-1. To duplicate the ambient density simultaneously along the flight corridor, stagnation pressures in excess of the 2,000 atm limit are necessary. This difficulty may be overcome by expanding from a total enthalpy slightly higher than flight total enthalpy to a Mach number low than flight Mach number, and an ambient temperature greater than flight temperature in order to achieve the desired flow velocity.

Flow Simulation in Other Gaseous Test Media

The phenomenon in high enthalpy, hypersonic flow generally involves temperatures above 1800°K at the subsonic region behind the detached shock of a blunt-nosed body. One of the differences between air and other test gases is the difference in γ, the effective ratio of specific heats or the isentropic exponent, before and behind the normal shock. Another difference is in the density rise across a normal shock at the same freestream conditions of ρ_∞ and U_∞. These differences may cause significant changes in shock detachment distance, shock shape, pressure distribution, etc. in the inviscid flow. The properties that will influence the viscous flowfields will be the differences in the viscosity and thermal conductivity of the test gases. Considering the pressure recovery behind the shock wave and the stagnation point we can easily show the stagnation pressure coefficient to be:

$$Cp_s = 2 \frac{\rho_\infty}{\rho_2} \qquad \text{(Eq. 1-8-1)}$$

which depends on γ_2 only on the second order term. The density ratio ρ_∞/ρ_2 is of the order of 0.12, so that the pressure changes by about six percent across the shock layer. The values of Cp_{max} vs. ρ_∞/ρ_2 are plotted in figure 1-8-3 for equilibrium air, nitrogen, carbon dioxide, argon, and an oxygen-steam (λ = 0.2035) mixture. It is seen that equation 1-8-1 is a good fit to the computed points and that the stagnation pressure coefficient depends only on the density ratio across the shock.

In flight simulation, the problem is to find out the variation of the aerodynamic data with the type of test gas used. Table 1-8-1 shows the values of Cp_s and ρ_∞/ρ_2 for five different test gases at the same freestream velocity and at two different freestream densities. It is evident that Cp_s and ρ_∞/ρ_2 differ quite markedly from the value for air.

The shock detachment distance Δ has been used frequently for comparison of various flow theories and for comparing theories with experiments. For simulation in the regime of equilibrium flow chemistry, Δ must be large compared to the dissociative relaxation lengths, so that there is a lower limit on the model radius in wind tunnel testing, below which nonequilibrium effects cannot be neglected.

In order to show the simulation of pressure and Mach number distribution around the body Inouye (10) gives a curve fitted to the numerical results for equilibrium air, nitrogen, argon and carbon dioxide, independent of :

$$Cp = Cp_s\,(1.0 - 1.25 \sin^2 \Phi + 0.284 \sin^4 \Phi) \qquad \text{(Eq. 1-8-2)}$$

The variation of Cp/Cp_s with the flow Mach number along the surface is shown in figure 1-8-4. The points are calculated for the H_2O-O_2 (λ = 0.2035) mixture at stagnation conditions of T_s = 2500, 2800, 3100°K and p_s = 0.1, 0.3, 1.0 and 3.0 atmospheres by assuming isentropic expansion. They are seen to agree well with the results for air, so that both the pressure and Mach number distribution are expected to be unaffected by using a combustion gas mixture.

The value of the stagnation point velocity gradient is important in heat transfer problems. The modified Newtonian Theory predicts it to be:

$$\frac{(du)}{(dx)}_{x=0} = \frac{U_\infty}{R_s} Cp_s^{\frac{1}{2}} \left(\frac{\rho_\infty}{\rho_s}\right)^{\frac{1}{2}} \quad \text{(Eq. 1-8-3)}$$

where R is the body radius of curvature at the stagnation point. The values resulting from the solutions of (10) are 9 percent to 15 percent higher than the values obtained from equation 1-8-3. For a given freestream condition, differences in the velocity gradient among the test gases (see Table 1-8-1) may be as high as 38 percent.

For the case of stagnation point heat transfer, the most widely used result on stagnation point heating was developed by Fay and Riddell (11). For axisymmetric bodies, this can be used in the form:

$$\dot{q}_s = \frac{0.759}{(Pr)^{0.6}} \left(\frac{\rho_\infty \mu}{\rho_s \mu_s}\right)^{0.1} (\mu Cp_s)^{0.5} \frac{\rho_\infty^{1.1} U_\infty^{3}}{2R_s} \left[1 + (Le^{0.52}-1)\frac{h_{de}-h_{dw}}{h_e-h_w}\right] \quad \text{(Eq. 1-8-4)}$$

in which equation 1-8-3 has been substituted for $\left(\frac{du}{dx}\right)_{x=0}$. It can be seen that $\dot{q}_s$ is weakly dependent on the thermodynamic and transport properties except for the bracket term. This term will make a large contribution when the dissociation energy becomes an appreciable fraction of the total energy. For example, at 3,000°K and 1 atm for H_2O+O_2 ($\lambda = 0.2035$) Le = 1.728, $\frac{h_{de}-h_{dw}}{h_e-h_w} = 0.30$ and the bracket term is 1.099. For air, Le = 1.327 and $\frac{h_{de}-h_{dw}}{h_e-h_w} = 0.08$, so that the bracket term is 1.013. The ratio $\dot{q}_{H_2O+O_2}/\dot{q}_{air}$ turns out to be 1.18, i.e. the Fay-Riddell theory predicts a significantly higher heat transfer rate in the combustion mixture. Therefore, in choosing a particular mixture as a test medium, it is desirable to choose one with the least dissociation, while meeting with the other criteria of equilibrium flow and large density ratio across the normal shock.

Flow Simulation in H_2O-O_2 Mixtures

The use of steam-oxygen mixtures has several advantages. The upper temperature limit can be extended in view of the relatively simple and well understood H_2O-O_2 reaction. In order to suppress nonequilibrium effects in the nozzle expansion at temperatures above 2200°K, high pressures are required. When short running time is not a restriction, then pressures up to 1000 atm can be readily achieved in the shock tunnel type of operation. In the following section an experimental facility will be described, based on the above principle.

Further remarks need to be made on the use of H_2O-O_2 mixtures as a test medium at high temperatures. First, the oxygen atoms are chemically very reactive at these temperatures, therefore the test duration should be kept short in order to reduce oxidation and corrosion effects. Secondly, from the consideration of the chief quantities to be simulated we may conclude that the density ratio is the most important parameter that affects the data. The dependence on γ alone shows up in the pressure and Mach number distribution. However, in the range of temperature and pressure considered, both the combustion gas mixtures and air have comparable values of γ (see Table 1-8-1). Therefore, in using another gaseous test medium in flow simulation, the density ratio must be simulated in addition to Gibson's criteria of the freestream density and velocity.

Thirdly, the chemical relaxation distance must be small compared with the shock detachment distance in order to ensure equilibrium flow. Figure 1-8-5 shows the

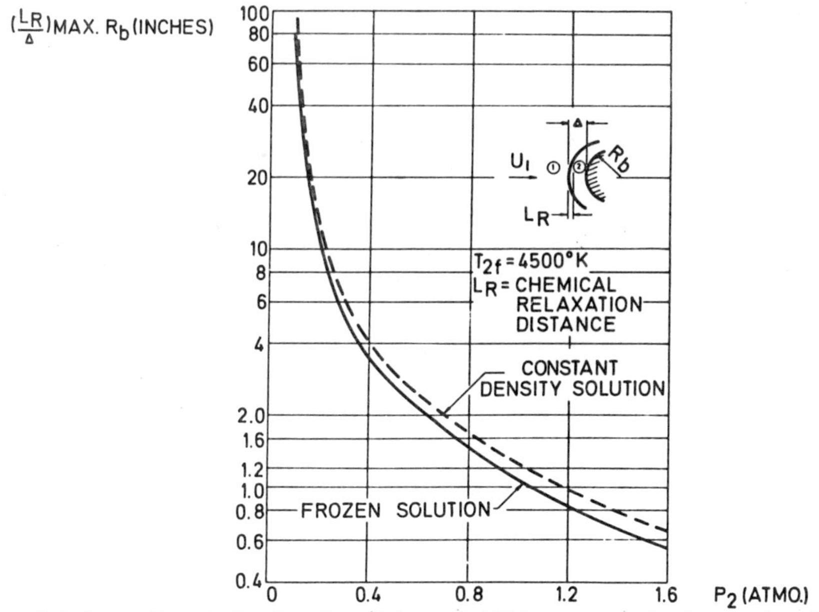

Fig. 1-8-5 Chemical relaxation distance behind a normal shock wave in a H_2O-O_2 ($\lambda = 0.2035$) mixture

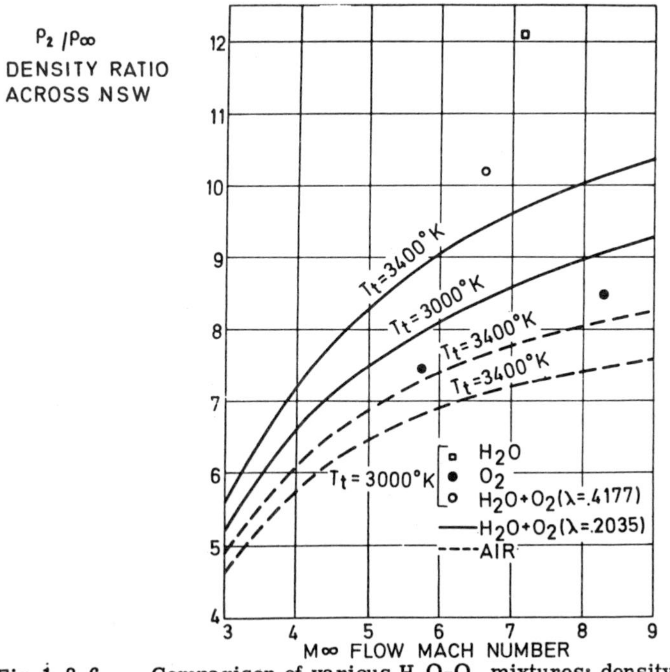

Fig. 1-8-6 Comparison of various H_2O-O_2 mixtures: density ratio across a normal shock P_t = 300 Atm

variation of the relaxation distance with pressure behind the normal shock for a typical temperature of $T_t = 4500°K$ in a H_2O-O_2 ($\lambda = 0.2035$) mixture. It can be seen that simulation of the equilibrium flow can be achieved for stagnation pressures behind the normal shock of the order of an atmosphere.

Finally it is necessary to decide on the specific composition of the mixture. The four mixtures chosen for comparison are pure O_2, pure H_2O (steam), H_2O-O_2 ($\lambda = 0.2035$) and H_2O-O_2 ($\lambda = 0.4177$). Their thermodynamic and transport properties are conveniently tabulated by Svehla (12). All the four gases are assumed to have the same reservoir temperature and pressure (3400°K, 300 atm). They are allowed to expand isentropically and the shock characteristics are determined for at least two different Mach numbers. The results are tabulated in Table 1-8-2 and the density ratio is plotted in figure 1-8-6. Dissociation is less in oxygen and high in H_2O. However there is a considerable increase in the density ratio by comparing H_2O-O_2 ($\lambda = 0.2035$) with oxygen. The difference between H_2O-O_2 ($\lambda = 0.2035$) and H_2O-O_2 ($\lambda = 0.4177$) is not much, but the latter shows a faster decrease in the pressure. This means that nonequilibrium flow is more likely to occur. Thus it seems that H_2O-O_2($\lambda = 0.2035$) is comparatively better than the others from the point of view of high density ratio and the requirement of equilibrium flow.

A Proposed Buffered Shock Tunnel Facility

The proposed experimental facility will utilize a H_2O-O_2 mixture of equivalence ratio $\lambda = 0.2035$ as the test medium. It will be capable of generating a short duration, high enthalpy, equilibrium gas flow simulating conditions that correspond to a density ratio of ten or more across a normal shock.

In order to generate the combustion gas mixture in the range of temperatures greater than 3000°K and pressures in excess of 100 atm , the shock tube is the most promising. While the testing time that can be achieved is brief, in the order of milliseconds, the flexibility and convenience of this device has made it an attractive research tool. The basic theory and techniques of operation of shock tubes have been extensively reported in the literature (13) (14) (15) and therefore will not be discussed here. In this section, we shall determine the performance characteristics of such a shock tube with a H_2O-O_2 mixture as the driven gas.

The proposed experimental facility is shown diagramatically in figure 1-8-7. It consists of a driver section, a buffer section and a driven tube. The following performance calculation has been carried out using air as the driver, helium as the buffer and a H_2O-O_2 mixture of mass ratio one to thirty-nine as the driven gas. A double diaphragm system separates the driver from the buffer gas from the combustible mixture in the driven tube. If the conditions in the helium buffer are such that p_2 and u_2 are equal to or greater than the Chapman-Jouget values, the detonation in the H_2O-O_2 mixture will be followed by a region of quasi-steady flow. The wave diagram is shown in figure 1-8-8.

In order to increase the testing time, the conditions of the helium buffer and the burnt mixture at the interface are matched ('tailored') to avoid additional waves from the interaction of the reflected shock and the interface. The tailoring conditions for the helium buffer gas are presented in Table 1-8-3 as well as the conditions behind the reflected shock. From these results it is seen that the pressure level is particularly sensitive to the detonation velocity.

The basic arrangement of the 'unsteady-expansion' type of buffered shock tube is shown in figure 1-8-9. In this case, the stagnation sound speed and pressure in

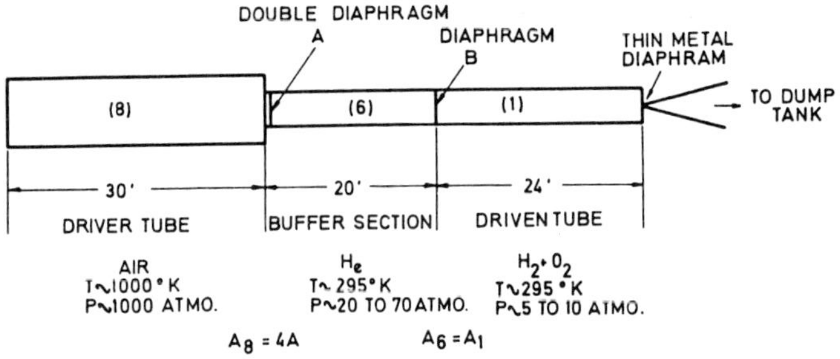

Fig. 1-8-7 Schematic arrangement of the proposed buffered shock tunnel facility

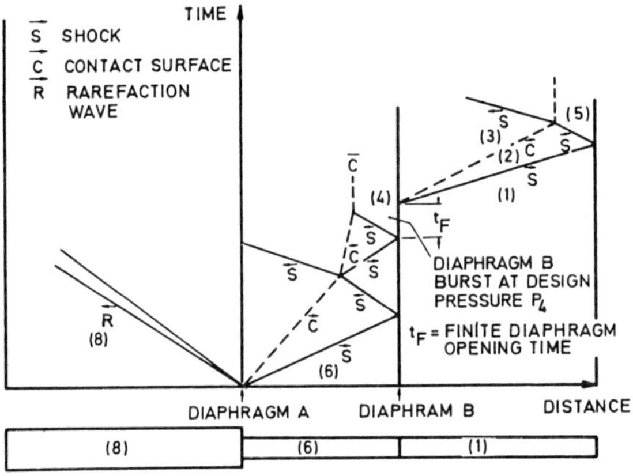

Fig. 1-8-8 Idealized initial flow in the buffered shock tube

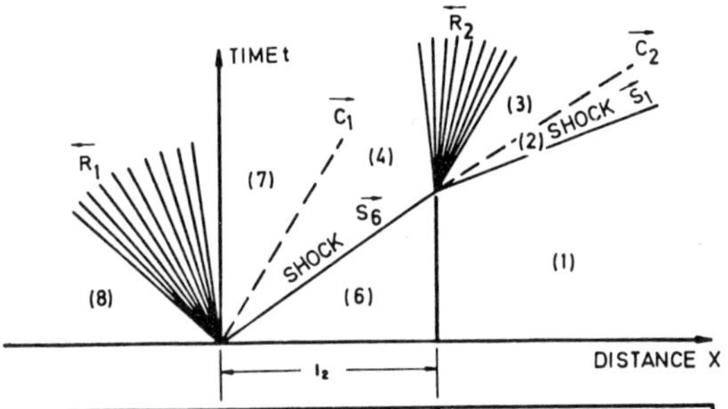

Fig. 1-8-9 Initial flow in constant area buffered shock tube, unsteady expansion method

region (7) have to match those required for tailoring. A second possibility is the reflected shock type of buffered shock tube, which is a shock tube with an area contraction at diaphragm B. The shock reflection occurs because of this area contraction, and the wave diagram is shown in figure 1-8-10. In general the analysis of the second configuration is a more difficult problem. A detailed treatment of the optimum performance of buffered shock tubes is given by Fetz (16). The first configuration is not preferred, even though there are fewer disturbances arising from the interaction at the diaphragm B station. This is due to the difficulty in matching u_7, a_7 and p_7 to u_3, a_3 and p_3. for various tailoring requirements. The second configuration affords the flexibility for various tailoring requirements because the helium is brought to rest behind the reflected shock.

Table 1-8-4 gives the required conditions for the operation of the buffered shock tunnel. The method of analysis is reported in detail (16) and will not be outlined here. A temperature of 1000°K has been assumed for the driven air, which gives a sound speed of about 2370 ft/sec, taking into account the bulk compressibility at high pressures (~ 600 atm). This temperature is currently attainable in the N.Y.U. hypersonic tunnel facility, which has an operating pressure of 20,000 lb/in^2. An area reduction of four at the double diaphragm station A was also assumed in the analysis. The wave diagram for the flow at an area reduction is shown in figure 1-8-11. From the results in Table 1-8-3, it is seen that less pressure ($P_8 < P_5$) is required in the driver to attain the desired pressure P_5 in the H_2O-O_2 mixture behind the reflected shock.

The above calculations ignore the phenomena of non-ideal diaphragm opening boundary layer growth and interfacial mixing of the gases. These will reduce somewhat the testing time, which is of the order of six milliseconds for most shock tunnels in operation. However, present day instrumentation, capable of giving time resolution in the microsecond range, is adequate for the type of operation anticipated here.

Equilibrium Nozzle Flow Calculations

In order to design appropriate hypersonic nozzles and to define the test conditions, various flow properties must be determined. The thermodynamic and transport properties of this H_2O-O_2 system have been tabulated by Svehla (12). The purpose of this subsection is to present for specific reservoir pressures and enthalpies the following properties as a function of the flow Mach number: temperature, pressure, density, velocity, area ratio, Reynolds number and isentropic exponent. An isentropic expansion of the gas from the reservoir to downstream stations in the nozzle was assumed. An iterative procedure was used to obtain the equilibrium pressure ratio, density ratio and temperature ratio across a normal shock. For brevity, the results of the nozzle flow properties calculation are given in figures 10 to 19 of (17) for T_t = 3400°K and 3800°K at reservoir pressures of 300 and 1000 atm , respectively. The results for the variation of density ratio, and γ_2 with flow Mach number across a normal shock, are given in figures 1-8-12 and 1-8-13. It is clearly seen that a density ratio of the order of 12 may be obtained with this test gas without dissociation.

Freezing Criterion

In an actual nozzle, it is necessary to determine whether the flow is in complete equilibrium. Two effects have to be considered, namely molecular dissociation and molecular vibration, and the problem is reduced to the calculation of their apparent freezing points. Recently Fine (18) studied the kinetics of hydrogen oxidation downstream of lean, flat hydrogen-air flames and the result of gas sampling has yielded a rate equation for an overall reaction:

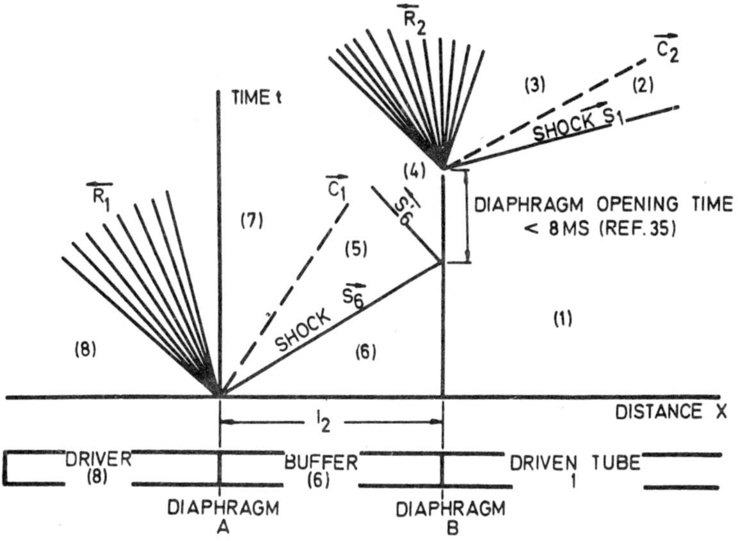

Fig. 1-8-10 Initial flow in constant area buffered shock tube reflected shock method

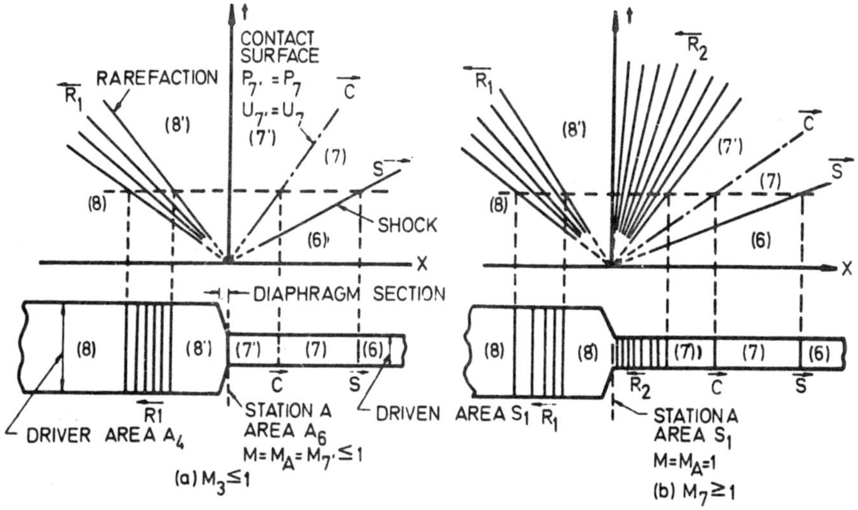

Fig. 1-8-11 Idealized flow with monotonic area reduction at diaphragm A

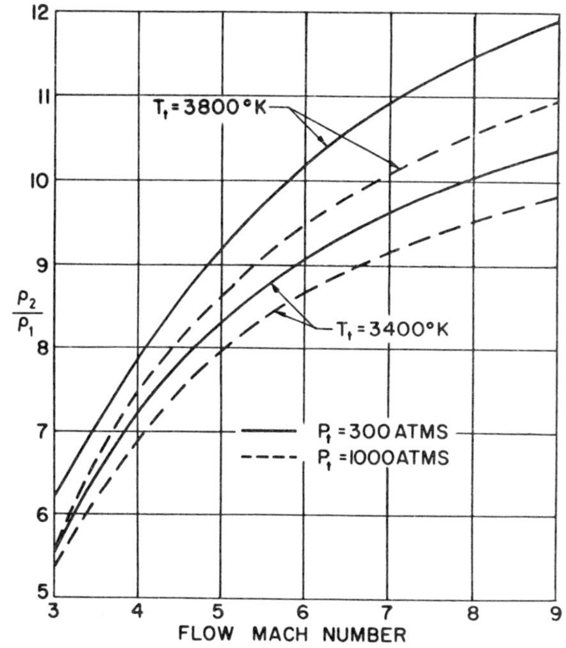

Fig. 1-8-12
Variation of density ratio with flow Mach number across a normal shock

Fig. 1-8-13
Variation of isentropic exponent behind a normal shock (γ_2) with flow Mach number

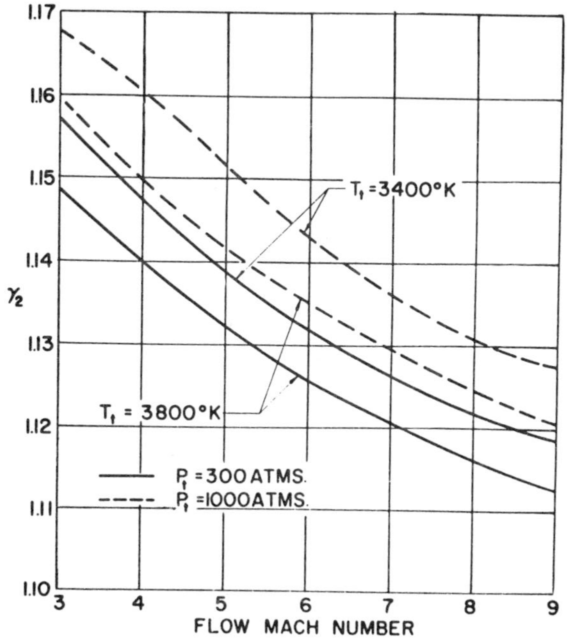

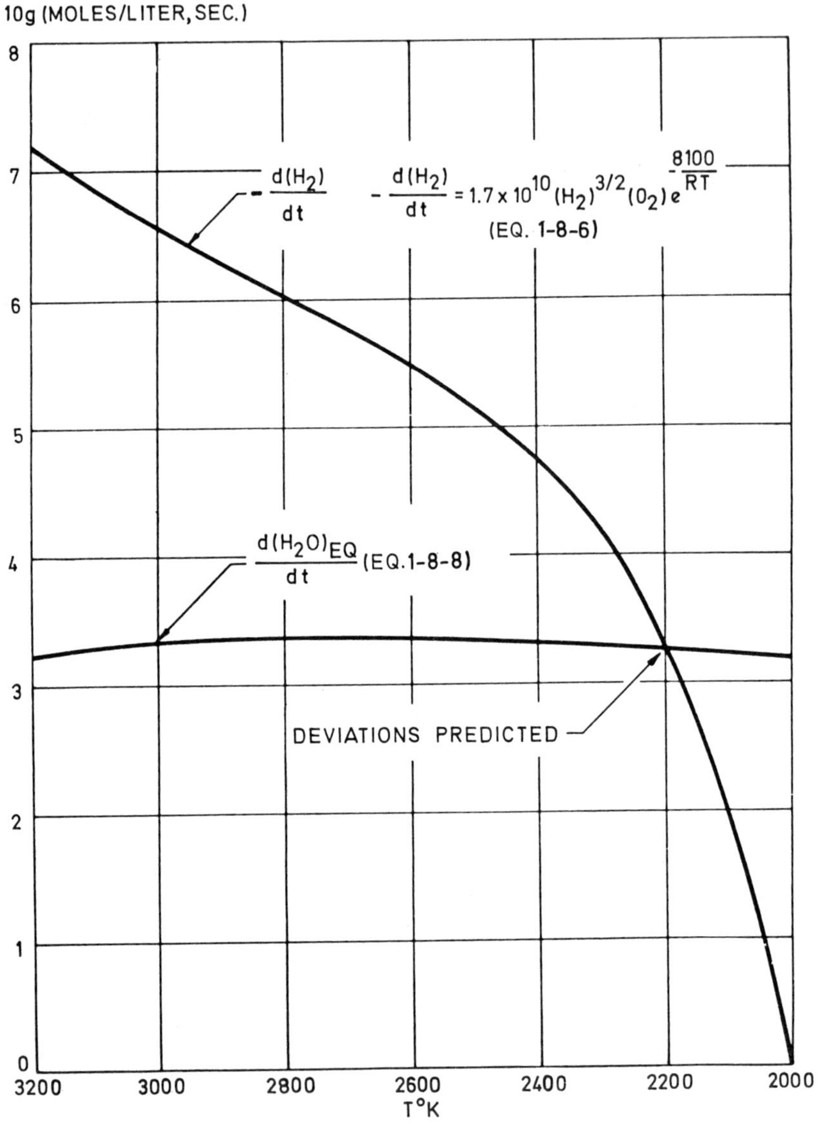

Fig. 1-8-14 Determination of freezing point

$$H_2 + \tfrac{1}{2} O_2 \underset{k_D}{\overset{k_R}{\rightleftharpoons}} H_2O \qquad \text{(Eq. 1-8-5)}$$

A rate equation for the decay of hydrogen downstream of a flat flame was found to be best described by

$$-\frac{d[H_2]}{dt} = 1.7 \times 10^{10} [H_2]^{3/2} [O_2]_e - \frac{8100}{RT} \qquad \text{(Eq. 1-8-6)}$$

The effect of diffusion has been neglected, so that the above rate probably represents a lower limit. Since the concentrations of hydrogen downstream of these flames far exceed their equilibrium values, the rate equation is essentially a recombination rate. The Bray criterion (19) for the freezing point in nozzle expansion is defined as:

$$\frac{-d\alpha}{dt} \sim r_D \approx K r_R \qquad \text{(Eq. 1-8-7)}$$

where K is of the order of unity and the left side represents the net rate of change of the dissociating gas. For the overall reaction given by equation 1-8-5, the net rate of change of $[H_2O]$ equals the total rate minus the change in the partial molar density due to the expansion:

$$\frac{d[H_2O]}{dt} = U \frac{d}{dx} \left[[H_2O]_e - [H_2O]_t g(\rho) \right] \qquad \text{(Eq. 1-8-8)}$$

where the concentrations are given by:

$$[\ ]_i = \frac{c_i \rho}{m} \qquad \text{(Eq. 1-8-9)}$$

$$[\ ]_t g(\rho) = \frac{c_t \rho}{m_t} \qquad \text{(Eq. 1-8-10)}$$

A graphic solution of equations 1-8-6 and 1-8-8 using the density, velocity and concentrations from the equilibrium flow calculations will give the freezing point. The case chosen to check this dissociation nonequilibrium is the most extreme case studied corresponding to reservoir conditions of p_t = 300 atm and T_t = 3800°K). A conical nozzle of 20° included angle with a throat radius of 0.20 inches was assumed. Equations 1-8-6 and 1-8-8 are plotted in Fig. 1-8-14 against the static temperature. A freezing point is indicated at a temperature of 2200°K (p = 5.4 atm). At this temperature dissociation in the gas mixture is less than ½ percent so that dissociation equilibrium prevails.

It is noted that the rate equation for the decay of hydrogen (Eq. 1-8-6) which Fine found experimentally may not be applicable at temperatures of the order of 3000°K, when the OH concentration is appreciable, such that a more complex reaction of H_2O-O_2 will take place. For example, eight reactions were written to describe the chemical reaction for the hydrogen and oxygen system in (20). It is, therefore, advisable to further check the conclusion reached above by the simple rate equation, by considering the different rates of reaction for this more complicated process.

However the freezing point at ~ 2200°K is close to the temperature (~1600°K) at which the experiments were performed, and the conclusion at least gives us a qualitative, if not quantitative picture of dissociation equilibrium in the nozzle flow.

As for vibrational nonequilibrium, a similar freezing criterion can be defined. Since the vibrational relaxation for the mixture is not known, one can only infer the actual situation by considering H_2O-H_2O and O_2-O_2 relaxations. It is known that small traces of water vapor in oxygen reduce the relaxation time by orders of magnitude, so that the relaxation time of H_2O-H_2O will be used as a guide. Huber and Kantrowitz (21) have investigated water vapor with the jet method at temperatures up to 800°K. The relaxation time, corrected to 1 atm is of the order of:

$$\tau_p \sim 3 \times 10^{-8} \text{ atm sec}$$

The relaxation distance for a particle velocity of 10^4 ft/sec is ~ 3×10^{-4} atm ft. For $P_t = 300$ atm, $T_t = 3800°K$ in H_2O-O_2 ($\lambda = 0.2035$), $P = 0.1$ atm at $T = 1000°K$, i.e. the relaxation distance is ~ 3×10^{-3} ft which is still small compared with any characteristic length in the flowfield. Therefore it may be assumed that the flow maintains vibrational equilibrium as well.

Conclusion

The results of this investigation indicate that the criteria for simulation of hypersonic flow over blunt-nosed bodies using other gaseous test media are the dynamic pressure and the density ratio across the normal shock. The effect of γ shows chiefly on the pressure - Mach number distribution. Since the pressure distribution is independent of γ, the effect of γ then alters the Mach number distribution on the surface.

Combustion gas mixtures may be used as a test medium, provided that condensation of the water vapor formed is avoided and that equilibrium flow conditions are ensured. In the H_2O-O_2 mixture family, a mixture of equivalence ratio $\lambda = 0.2035$ has been found to yield a density ratio of about ten in equilibrium flow by suitably adjusting the pressure level.

A buffered shock tunnel facility is proposed using the H_2O-O_2 ($\lambda = 0.2035$) mixture as the test medium. Operating conditions have been determined for the heated driver air, the helium buffer and the H_2+O_2 mixture, such that tailored conditions of the reflected shock are achieved. This will yield the maximum testing time of the order of six milliseconds for a normal size shock tunnel. This test time will be reduced somewhat when such flow phenomena as boundary-layer growth, interfacial mixing, non-ideal diaphragm opening occurs in the actual shock tube flow. Nozzle flow properties and normal shock characteristics in the test section have also been computed. The computation was terminated arbitrarily at a static temperature of 300°K, even though the temperature of saturation has not been reached. With the increased reservoir temperature which can be produced by the reflected shock method, flow Mach numbers of the order of ten can be achieved at this temperature.

Acknowledgement

The authors wish to acknowledge the support of this work by the Aerospace Research Laboratories, United States Air Force, Wright-Patterson Air Force Base, Ohio, under Contract AF33(615)3236, Project 7065.

The authors thank Professor Antonio Ferri, Director Aerospace Laboratory and Astor Professor of Aerospace Sciences, for suggesting this problem and for his valuable discussions on the work reported here.

References

1. **Bray, K.N.C.,** 'Evaluation of the Hypersonic Gun Tunnel'. Progress in Astronautics and Rocketry Series by ARS, Vol. 7, Aug. 1961, pp 547-579.

2. **Bird, K.D., Martin, J.F., Bell, T.J.,** 'Recent Developments in the Use of the Hypersonic Shock Tunnel as a Research and Development Facility'. Pres. Third Hypervelocity Techniques Symp., Denver, Colorado, March 1964.

3. **Cann, G.L., Buhler, R.D.,** 'A Survey and Prediction of the Performance Capability of Coaxial Arc Heaters'. Pres. AGARD Specialist's Meeting on Arc Heaters and MHD Accelerators for Aerodynamic Purposes, Rhode-Saint-Genese, Belgium, Sept., 1964.

4. **Ring, L.E.,** 'General Consideration of MHD Acceleration for Aerodynamic Testing'. Pres. AGARD Specialist's Meeting on Arc Heaters and MHD Accelerators for Aerodynamic Purposes, Rhode-Saint-Genese, Belgium, Sept., 1964.

5. **Love, E.S., Henderson, A., Bertram, M.H.,** 'Some Aspects of Air Helium Simulation and Hypersonic Approximation'. NASA Rep. TND49, 1959.

6. **Ladson, C.L.,** 'A Comparison of Aerodynamic Data Obtained in Air and Helium in the Langley II Hypersonic Tunnel'. NASA Rep. TM-X666, 1962.

7. **Ladson, C.L., Blackstock, T.A.,** 'Air-Helium Simulation of the Aerodynamic Force Coefficients of Cones at Hypersonic Speeds'. NASA Rep. TND1473, 1962.

8. **Harvey, D.J.,** 'Chemical Kinetic Regimes of Hypersonic Flight Simulation'. AEDC Rep. TDR63-3, Jan. 1963.

9. **Gibson, W.E.,** 'Dissociation Scaling for Nonequilibrium Blunt-Nose Flows'. AEDC Rep. TDR62-25, Feb., 1962.

10. **Inouye, M.,** 'Blunt Body Solution for Spheres and Ellipsoids in Equilibrium Gas Mixtures'. NASA Rep. IND2780, 1965.

11. **Fay, J.A., Riddell, F.R.,** 'Theory of Stagnation Point Heat Transfer in Dissociated Air'. J. Aero. Sci. Vol. 25, No. 73, 1958.

12. **Svehla, R.A.,** 'Thermodynamic and Transport Properties for the Hydrogen-Oxygen System'. NASA Rep., SP3011, 1964.

13. **Lukasiewicz, J.,** 'Shock Tube Theory and Applications'. Nat. Aer. Est. Canada Rep. No.15, 1952.

14. **Glass, J.I., Patterson, G.N.,** 'A Theoretical and Experimental Study of Shock Tube Flow'. J. Aero. Sci. Vol.22, 1955, pp 73-100.

15. **Hertzberg, A., Smith, W.E., Glick, H.S., Squire, W.,** 'Modifications of the Shock Tube for the Generation of Hypersonic Flow'. CAL Rep. No. AD-789-A-2, AEDC Rep TN-55-15, March 1955.

16. **Fetz, B.H.,** 'Analysis of the Optimum Performance of Buffered Shock Tubes'. ARL Rep. 64-41, March 1964.

17. **Mak, W.H., Zakkay, V.,** 'Hypersonic Flow Simulation over Blunt-Nosed Bodies'. NYU Rep. NYU-AA-66-68, Dec., 1966.

18. **Fine, B.,** 'Kinetics of Hydrogen Oxidation Downstream of Lean Propane and Hydrogen Flames'. J. Phys. Chem. Vol. 65, No. 414, 1961.

19. **Bray, K.N.C.,** 'Atomic Recombination in a Hypersonic Wind Tunnel Nozzle'. J. Fluid Mec. Vol. 6, No. 1, 1959.

20. **Libby, P.A., Pergament, H., Bloom, M.H.,** 'A Theoretical Investigation of Hydrogen-Air Reactions'. Pt.I of 'Behaviour with Elaborate Chemistry'. GASL Rep. No. 250, 1962.

21. **Huber, P.W., Kantrowitz, A.,** 'Heat Capacity Lag Measurements in Various Gases'. J. Chem. Phys. Vol. 15, No. 275, 1947.

List of Mathematical Symbols

A	area
a	sound speed
C	mole fraction of the ith species
Cp	pressure coefficient $= \dfrac{p - p_\infty}{\frac{1}{2}\rho_\infty u_\infty{}^2}$
e	internal energy per unit mass
ΔE_c	heat of chemical reaction per unit mass
$g(\rho)$	function of density in equation 1-8-10
h	absolute enthalpy per unit mass
h_D	dissociation energy per unit mass
k	constant
k_R , k_D	recombination, dissociation rate coefficients
L	characteristic length
L_R	chemical relaxation length
Le	Lewis number
M	Mach number
m	molecular weight
p	pressure
Pr	Prandtl number
$\dot{q}$	heat transfer rate in equation 1-8-5
r	radius
R	universal gas constant; also used for body radius
Re	Reynolds number
r_R , r_D	recombination, dissociation rate respectively
S	entropy
T	temperature

t	time
u, U	speed in laboratory coordinates
u_D	detonation speed
v	speed in stationary shock coordinates
x	distance from stagnation point along surface
Z	compressibility factor
α	thickness ratio, angle of attack; also used as dissociation fraction
γ	isentropic exponent, specific heat ratio
Δ	shock detachment distances
λ	equivalence ratio
μ	viscosity
ρ	density
τ	relaxation time
Φ	angle between free stream direction and normal to body surface

Subscripts

1, 2, 5, etc.	denotes regions of quasi-steady flow in shock tubes; 1, 2 are also used as conditions upstream and downstream of a normal shock
b	body
D	detonation
e	external flow at edge of boundary layer, also equilibrium
f	frozen
i	number of the species
p	piston
r	recovery condition
s	stagnation condition
t	reservoir condition also time
max	maximum
*	condition at minimum area, sonic
w	wall condition
o	standard condition at 273.16°k, 1 atm
∞	freestream conditions

Table 1-8-1 - Shock Conditions at Flight Velocity of 10,000 ft/sec in Equilibrium Gas Mixtures

ρ_∞ (slug/ft^3)	Gas	M_∞	$\frac{\rho_\infty}{\rho_2}$	C_{P_s}	γ_s	$\frac{R_s}{U_\infty}\left(\frac{du}{dx}\right)_{x=0}$
2.377×10^{-4}	air	8.95	0.1253	1.879	1.191	.485
	N_2	8.94	0.1458	1.860	1.280	.521
	CO_2	9.51	0.0711	1.931	1.111	.370
	A	8.22	0.2569	1.758	1.511	.672
	$H_2O + O_2$ (λ=0.2035)	9.05	0.0965	1.920	1.118	.430
2.377×10^{-6}	air	8.95	0.1113	1.892	1.147	.459
	N_2	8.94	0.1438	1.862	1.241	.517
	CO_2	9.51	0.0634	1.938	1.090	.350
	A	8.22	0.2402	1.775	1.317	.653
	$H_2O + O_2$ (λ =0.2035)	9.05	0.0850	1.032	1.094	.405

Table 1-8-2 - Comparison of Various H_2O-O_2 Mixtures across a Normal Shock Wave

T_{t1} = 3400°K, P_{t1} - 300 atm

Mole Fraction H_2O	0		.33818		.58918		1.00000	
O_2	1.00000		.66182		.41072		0	
Equivalence Ratio	0		.2035		.4177		1.0000	
T_1 °K	300	600	300	600	300	600	300	600
M_1	8.30	5.73	9.08	6.31	9.32	6.61	10.28	7.14
ρ_1/ρ_2	.118	.134	.096	.108	.091	.098	.0725	.0825
P_2/P_1	85.7	39.3	104.2	48.3	107.9	52.9	131.3	61.6
P_2 atm	8.23×10^{-1}	4.84	2.79×10^{-1}	1.83	1.45×10^{-1}	1.09	4.78×10^{-2}	4.00×10^{-1}
T_2 °K	2915	3070	2770	2945	2700	2895	2545	2735
Z_2	1.046	1.033	1.082	1.063	1.096	1.076	1.134	1.111
γ_2	1.146	1.165	1.118	1.130	1.109	1.119	1.100	1.111

Table 1-8-3 - Requirements of Tailoring Condition for an Overdriven Detonation in a Shock Tube

T_1 = 298.15°K γ_4 = 5/3

M_1	P_1 atm	P_2 atm	P_5 atm	T_5 °K	h_5 cal/gm	u_3 ft/sec	a_3 ft/sec	a_4 ft/sec	$\frac{P_4}{P_5}$
4.9	1.0	19.76	59.1	3410	650.7	3465	4669	5080	.5095
	5.0	93.26	262.5	3511	619.9	3261	4692	5058	.5142
	10.1	182.17	501.6	3545	607.7	3180	4684	5031	.5190
5.0	1.0	21.69	69.6	3490	710.4	3744	4700	5175	.5037
	5.0	104.24	218.8	3617	689.0	3592	4733	5168	.5074
	10.0	204.99	616.1	3663	680.2	3529	4734	5155	.5094
5.1	1.0	23.43	79.6	3557	764.1	3980	4730	5260	.5003
	5.0	113.53	369.9	3703	747.0	3851	4731	5223	.5028
	10.0	224.14	719.2	3759	740.0	3800	4769	5251	.5038

Table 1-8-4 - Operating Conditions in the Driver and Buffer Section

Driver: T_8 = 1000°K A_8/A_4 = 4.0 Buffer: T_6 = 295°K
a_8 = 2370 ft/sec a_6 = 3315 ft/sec
γ_8 = 1.4 γ_6 = 5/3

M_1	P_1 atm	P_5 atm	$\frac{P_4}{P_5}$	a_4	$\frac{T_4}{T_6}$	$\frac{P_7}{P_6}$	M_6	$\frac{u_7}{u_6}$	$\frac{P_8}{P_5}$
4.9	1.0	59.1	.5095	5080	2.348	2.990	1.610	.7416	.592
	5.0	262.5	.5142	5058	2.327	2.957	1.602	.7330	.606
	10.0	501.6	.5190	5031	2.303	2.918	1.592	.7229	.603
5.0	1.0	69.6	.5037	5175	2.436	3.130	1.644	.7774	.637
	5.0	318.8	.5074	5168	2.430	3.121	1.642	.7750	.637
	10.0	616.1	.5090	5155	2.418	3.102	1.638	.7700	.632
5.1	1.0	79.6	.5003	5260	2.517	3.260	1.676	.8091	.659
	5.0	369.9	.5028	5223	2.482	3.203	1.662	.7953	.650
	10.0	719.2	.5038	5251	2.508	3.245	1.672	.8056	.663

The Use of a Gun Tunnel for Hypersonic Intake Calibration

R. HAWKINS and E. CHARLTON
Bristol Engine Division, Rolls-Royce Ltd., Bristol, England.

Summary

The use of a gun tunnel as a low temperature test facility for hypersonic intake research has necessitated the development of many new experimental techniques. Techniques suitable for calibrating intakes for both supersonic and subsonic combustion ramjet engines have been studied, and developed to the extent that an acceptable standard of data has been achieved.

One of the major advantages of the gun tunnel which emerged from this study was the ability to establish supersonic throat flow within the intake, without recourse to variable geometry. This, coupled with the fact that the flow duration is short (typically 30 to 100 milliseconds) has led to great simplification and economy in intake model construction.

Although standard strain gauged diaphragm transducers are suitable for pressure measurement in a gun tunnel, specialised techniques are necessary to ensure an accurate mean pressure is obtained at the combustor entry.

In the case of an intake for a supersonic combustion engine where the blockage of multi-tube rakes would be unacceptable, a high speed pitot traverse technique has been successfully developed.

For the more difficult case of the subsonic combustion intake, it is necessary to cover a range of back pressures up to the critical value. This has been achieved by fitting a specified volume to the intake, in which the rate of pressure rise is slow enough to permit a transient calibration procedure to be adopted.

Tests have been made in a conventional wind tunnel facility to check the validity of the results obtained in this way.

Sommaire

L'emploi des souffleries-canon comme montage expérimental à basse témpérature pour les recherches d'entrée d'air hypersonique a nécessité la mise au point d'un grand nombre de techniques expérimentales nouvelles. Des techniques pouvant s'appliquer à la calibration d'entrées d'air de statoréacteurs à combustion subsonique ou supersonique ont été étudiées et développées jusqu'à un stade où des données convenables ont pu être obtenues.

L'un des principaux avantages des souffleries-canon qui est apparu de ces études est la possibilité d'amorcer un écoulement supersonique au col de l'entrée d'air, sans avoir recours à la géométrie variable. Ceci, couplé avec le fait que la

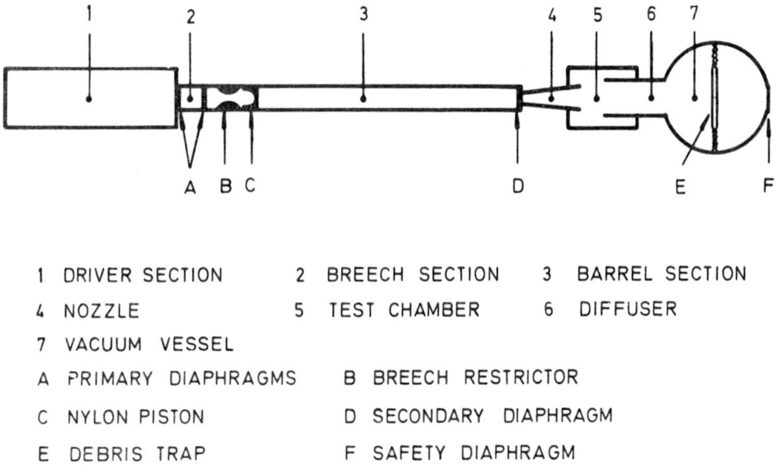

Fig. 1-9-1   B. S. E. L. gun tunnel arrangement

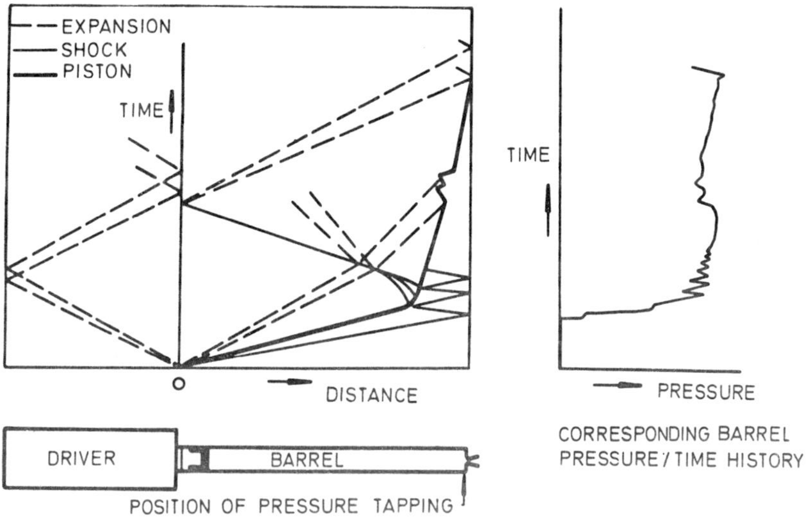

Fig. 1-9-2   Schematic wave diagram for gun tunnel

durée de l'écoulement est brève (30 à 100 millisecondes) a permis une grande simplification et des économies notables dans la réalisation de la maquette de prise d'air.

Bien que des capteurs standards à diaphragmes munis de jauges extensométriques conviennent pour les mesures dans les souffleries-canon, des techniques mieux appropriées sont nécessaires pour s'assurer que la pression moyenne à l'entrée du foyer a la valeur correcte.

Dans le cas d'une prise d'air pour moteur à combustion supersonique et par suite du blocage, des peignes ne peuvent être utilisés, une technique utilisant la traversée rapide d'un pitot a été mise au point.

Dans le cas plus difficile d'une prise d'air pour combustion subsonique, il est nécessaire de pouvoir couvrir un domaine de pression allant jusqu'à la pression critique. On a pu obtenir cela en connectant la prise d'air à une enceinte de volume suffisamment élevé pour que la montée en pression soit lente et on a adopté une méthode de calibration en régime transitoire.

Des essais ont été effectués dans une soufflerie classique pour vérifier la validité des résultats.

Introduction

The assessment of airbreathing propulsion systems for hypersonic flight has led to the need for more detailed performance data on hypersonic intakes.

At low hypersonic speeds where real gas effects in the intake are confined to caloric imperfections, there is a strong case for attempting to separate the gas dynamics from aerodynamic phenomena. If this can be done with realism, conventional hypersonic wind tunnels that provide a 'cold' airstream will be suitable for basic intake research.

One form of 'cold' hypersonic facility is the gun tunnel. Such tunnels have been developed extensively in Europe and the UK during the last decade, and exploratory studies (1) have indicated their suitability for hypersonic intake research. The present paper examines critically the flow conditions in a particular gun tunnel, and describes techniques that have been developed to permit reliable intake calibration studies to be made in it.

Description of the Gun Tunnel

A large gun tunnel, capable of simulating a Mach number/Reynolds number corridor thought typical of that to be followed by airbreathing hypersonic vehicles, has been installed at the Bristol Engine Division of Rolls-Royce Limited.

A diagram of this facility is shown in figure 1-9-1. Principal dimensions of the several components are as follows:

Driver Section:	Length	18 feet;	549 cm
	Internal Diameter	14.2 in;	36.1 cm
	Maximum working pressure	4000 lb/in^2	270 atm
Barrel:	Length	30 feet;	914 cm
	Internal Diameter	3.79 in;	9.63 cm

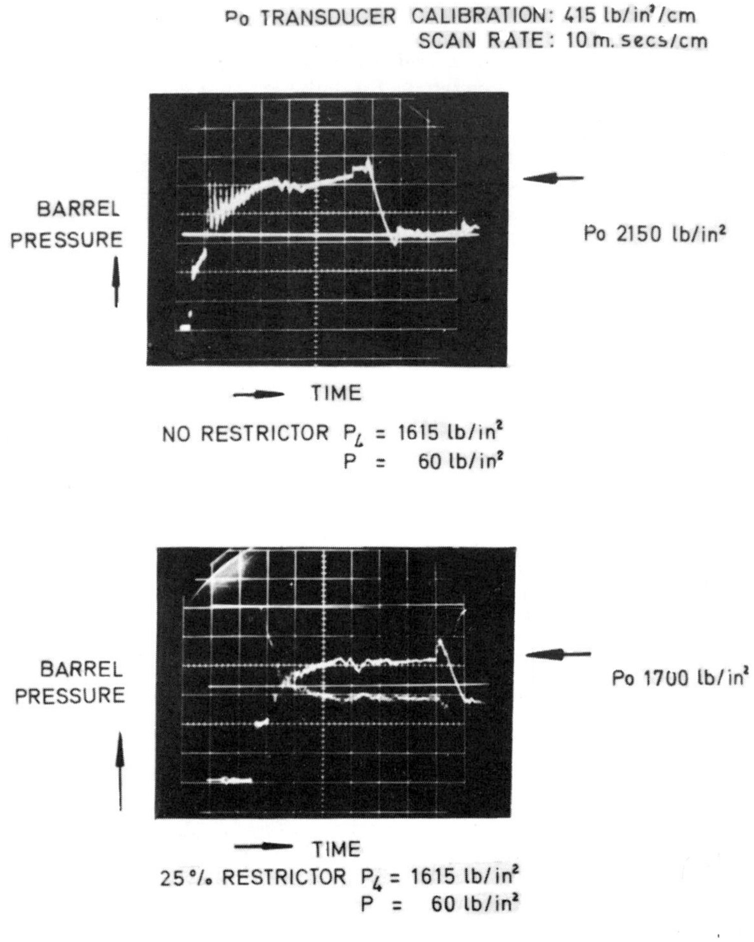

Fig. 1-9-3 Effect of breech restrictor on barrel stagnation pressure

Test Chamber:	Length	3 feet;	91.4 cm
	Width	2 feet;	61.0 cm
	Height	2 feet;	61.0 cm
Diffuser:	Length	9 feet;	274 cm
	Minimum Internal Diameter	12 in;	30.5 cm
Vacuum Vessel:	Volume	1000 ft^3;	28.3×10^6 cm^3
	Minimum working pressure	1mm.Hg	

Nozzles: The tunnel is equipped with interchangeable axisymmetric profiled nozzles having the following characteristics:

Mach Number	Nozzle Throat Diameter		Nozzle Exit Diameter		Maximum Unit Reynolds Number	
	inches	cm	inches	cm	(per foot)	(per cm)
6.0	0.75	1.91	5.86	14.9	5×10^7	5×10^8
7.0	0.75	1.91	8.30	21.1	4×10^7	12×10^8
8.0	0.75	1.91	11.10	28.2	3×10^7	9×10^8

The Mach 7 nozzle was used for the majority of the tests discussed herein. The desired unit Reynolds Number (10^7 per foot) and stagnation temperature (750°K) were obtained with an initial driver pressure of 1600 lb/in^2. For these tunnel operating conditions the following diaphragms and piston were used:

Primary Diaphragms: 0.125" Aluminum sheet L.17 fully annealed and scored to a depth of 0.015".

Secondary Diaphragm: Two sheets of 0.001" thickness Scotch Brand Polyester adhesive tape Type 56.

Piston: Nylon to specification Nilonic Engineering Type 31 (Grade 6) stress relieved during machining. Mass 0.26 lbs.

Uniformity of Stagnation Pressure

When the facility was designed, emphasis was given to all features that could contribute to uniformity of the flow in the test section both in time and space. However, the gun tunnel - along with the shock tunnel - is a facility that depends on transient compression processes to provide the stagnation conditions. Consequently the usable test duration is short (about 50 milliseconds), and the steadiness of the stagnation conditions throughout this interval must be assessed accordingly.

A schematic wave diagram for the RR/Bristol gun tunnel is shown in figure 1-9-2, together with the corresponding barrel stagnation pressure as measured at a wall orifice near the end of the barrel. The interval covered is the time between the bursting of the primary diaphragm and the arrival of the head of the expansion wave reflected from the closed end of the driver. For the first third of this interval, the pressure transducer clearly shows the passage of discrete shockwaves between the piston and the barrel end, and these collectively build up the stagnation pressure to its 'steady' value (P_o).

Half way through the interval there is an expansion resulting from a wave interaction at the change in cross sectional area between the driver and the barrel. This influences the trajectory of the piston, and leads to a significant perturbation of

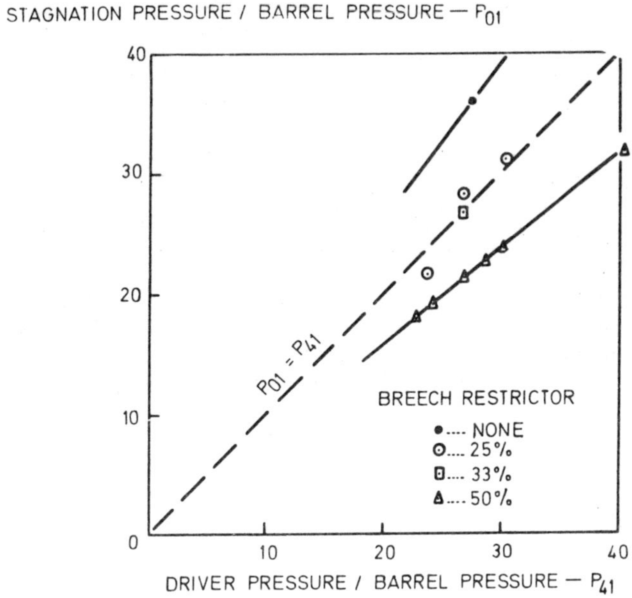

Fig. 1-9-4 Stagnation pressure/initial barrel pressure v. initial driver pressure/initial barrel pressure

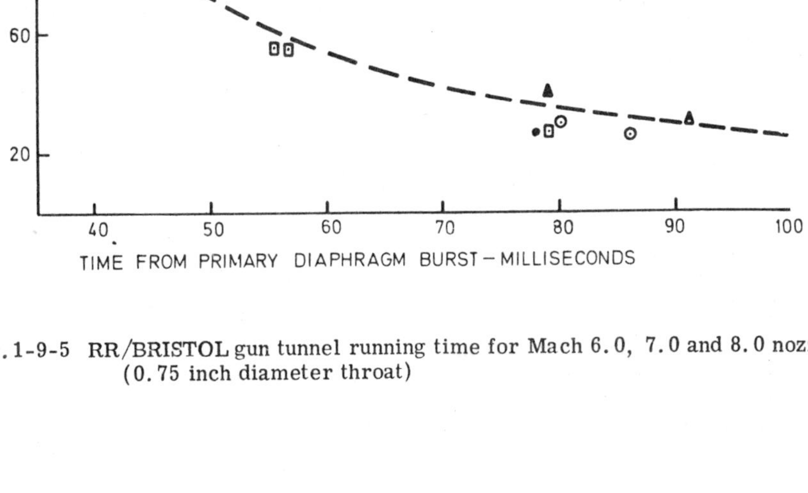

Fig. 1-9-5 RR/BRISTOL gun tunnel running time for Mach 6.0, 7.0 and 8.0 nozzles (0.75 inch diameter throat)

about 10 percent of the stagnation pressure. Following this perturbation, the stagnation pressure remains relatively steady until the end of the run. The run is terminated by the arrival of the piston at the end of the barrel, which occurs typically about 80 milliseconds after the primary diaphragm bursts. This event is not shown in figure 1-9-2, but coincides approximately with the arrival of the reflected head of the expansion wave from the closed end of the driver.

The perturbation of the stagnation pressure that occurs about half way through the run has proved particularly troublesome when using the facility for calibrating intakes. If a mixed compression intake is operating near its maximum pressure recovery, a rapid change in free stream conditions can lead to the intake unstarting. It would be impossible to make reliable intake performance measurements near the critical point under such conditions. Consequently, an exacting maximum tolerance on the fluctuations in the stagnation conditions must be met if the gun tunnel is to be used successfully for the calibration of high performance intakes. In specifying such a tolerance it is more important to consider the maximum acceptable rate of change in stagnation pressure than the overall variation in pressure.

As the pressure history shown in figure 1-9-2 proved to be unacceptable, attempts were made to improve the uniformity of the barrel stagnation pressure by using a restriction in the region of the breech. The function of this restriction was to attenuate (ideally to neutralize) the waves crossing this station. A smoothly profiled convergent-divergent nozzle was used immediately downstream of the second primary diaphragm to provide the local restriction, and some success was achieved in reducing the overall fluctuation of the stagnation pressure over the last 40 milliseconds of the run. Figure 1-9-3b shows a typical oscillogram of the barrel stagnation pressure for a run with a 0.25 breech restrictor fitted, whilst figure 1-9-3a shows the corresponding pressure fluctuations without the restrictor. It will be noted, however, that the relatively sharp-edged perturbation midway through the run was still present, and that the change in pressure associated with it was not significantly reduced by the presence of the breech restrictor.

Profiled convergent-divergent nozzles having restrictor ratios of 0.33 and 0.50 were tried also, but neither of these nozzles influenced significantly the shape or magnitude of the mid-run perturbation. (More recent gun tunnel development studies carried out at the D.V.L. Porz Wahn - Germany, suggest that an orifice plate used instead of a profiled nozzle would be more effective in attenuating waves at the breech section.) Since the 0.33 restrictor appeared to give the most uniform stagnation pressure trace, all further intake calibration tests made were with that restrictor fitted. Using this breech restrictor and an initial primary diaphragm pressure ratio of 27, the overall fluctuation of the stagnation pressure, including the midpoint perturbation, did not exceed ± 5 percent of the mean level; while for the last 30 milliseconds of the run the stagnation pressure was steady to within about ± 3 percent. This is a very uniform stagnation pressure by shock tunnel and gun tunnel standards, and has been shown to be acceptable for intake calibration studies. If the midpoint perturbation could be eliminated, the time for which the stagnation pressure remains sufficiently steady would be increased by 10 milliseconds.

With a breech restrictor fitted, the mean level of stagnation pressure achieved from given initial conditions was reduced. Figure 1-9-4 shows the degree by which the compression ratio was lowered by the use of a restrictor, and figure 1-9-5 the corresponding total running time. It will be noted that the running time was increased by the use of a restrictor.

Duration and Uniformity of Test Section Flow

The running times presented in figure 1-9-5 are not the usable test times but the intervals between the primary diaphragm bursting and the cessation of the tunnel

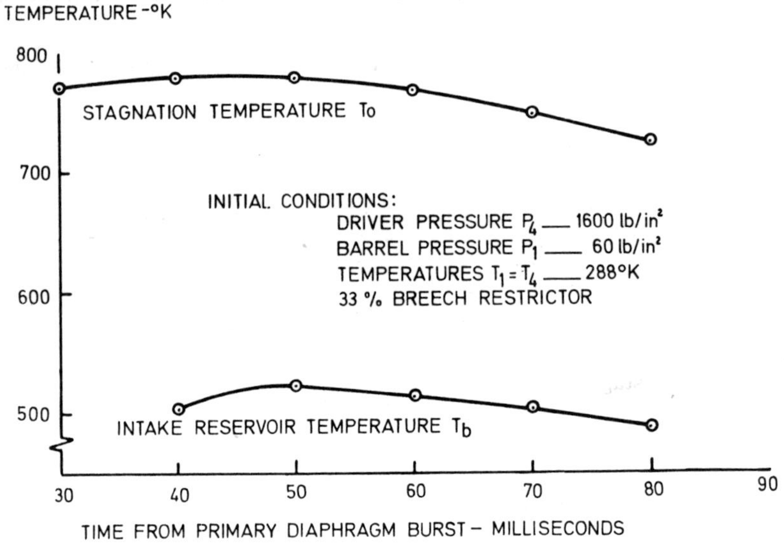

Fig. 1-9-6 Gun tunnel temperature history $M_o = 7.0$

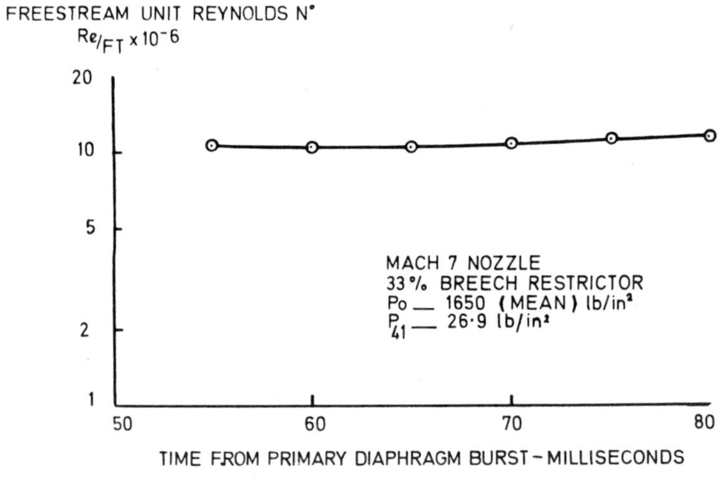

Fig. 1-9-7 Gun tunnel freestream unit Reynolds number history

flow when the piston arrived at the end of the barrel. The first 15 milliseconds or so of the total running time was taken up by the primary shock wave traversing the barrel ahead of the piston. Flow through the test section was fully established by about 20 milliseconds, and became relatively steady 30 to 40 milliseconds after the primary diaphragm burst. The 'mid run' perturbation referred to in the previous paragraph occurred at about 45 milliseconds. Hence, with an effective breech restrictor the usable test duration for which the stagnation pressure remained relatively constant was from 35 milliseconds to the time when the piston arrived at the end of the barrel. This interval was a function of primary diaphragm pressure ratio, and varied from about 5 milliseconds with $P_{41} = 100$, to 50 milliseconds with $P_{41} = 25$.

The uniformity of the test section flow from about 30 milliseconds after the primary diaphragm bursts have been studied in some detail. Pitot pressure surveys have been used to determine the Mach number distribution through the test section, and a thermocouple probe to measure the variation of stagnation temperature with time. Fluctuations in the angularity of the flow in the test section have been studied by taking a sequence of Schlieren photographs, each of one microsecond exposure at a framing rate of 2800 per second. The exposure time for each frame was sufficiently short to freeze the position of any local disturbances that might have existed in the flow. The angularity of the flow through the test section was assessed by examining the linearity and steadiness of the shock wave propagated from a large 9° wedge.

The contoured Mach 7 nozzle gave a test flow distribution of high quality across 85 percent of the nozzle diameter between axial stations at the nozzle exit and at one diameter downstream. The Mach number within this region was $M = 7.0 \pm 2$ percent and was thus comparable with the best uniformity in test section flow achieved in conventional hypersonic wind tunnels.

The variation of the stagnation temperature throughout the usable test duration is shown by the upper curve in figure 1-9-6. The slight rise in temperature between 30 and 45 milliseconds corresponds to the continuing slow rise in stagnation pressure immediately prior to the arrival of the 'mid-run' perturbation (Fig. 1-9-3b). From 60 milliseconds the stagnation temperature fell at a rate of about 2.5°C per millisecond. The free stream unit Reynolds number variation with time for a typical run is shown in figure 1-9-7.

These results show that an acceptable test flow existed for intake calibration studies: at a Mach number of 7, for which the minimum P_{41} required to avoid liquefaction in the test section was 25 : 1, the usable test duration was about 50 milliseconds.

Intake Calibration Technique

Class of Intake Selected for Study

Hypersonic intakes may be divided broadly into two classes; those delivering a supersonic air flow to the engine or combustor, and those in which the delivery air flow is subsonic. In the former class, the pressure recovery in the normal operating regime is independent of the conditions existing downstream of the intake delivery plane. The intake performance is a function solely of its geometry and flight environment. This is not the case with intakes having subsonic diffusers. In this class of intake the pressure recovery depends upon the blockage imposed at the diffuser exit. When such an intake is operating at full capture mass flow, it may give any pressure recovery up to a certain maximum value - the critical pressure recovery. When operating at pressure recoveries below the critical value, an intake is said to be supercritical, whilst attempts to increase the press-

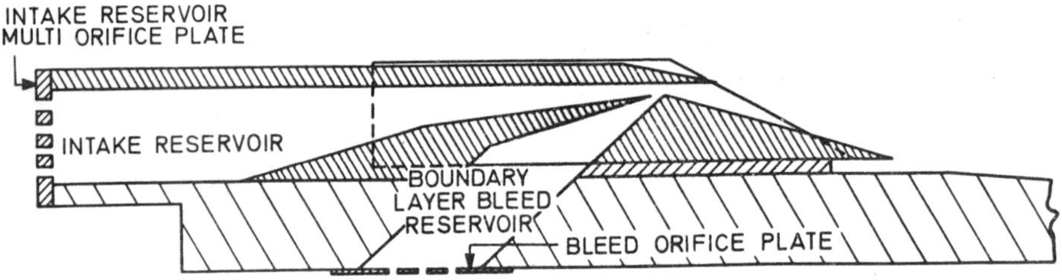

Fig. 1-9-8 Photograph and diagram of intake model and reservoirs

ure recovery beyond the critical value leads to subcritical operation. When operating in the latter regime, the internal flow through the intake may be unstable giving rise to violent fluctuations in capture mass flow and pressure recovery. This phenomenon is known as 'buzz'. In high performance hypersonic intakes the critical pressure recovery usually coincides with the maximum attainable pressure recovery, there being no usable subcritical regime.

Basic calibration of an intake having a subsonic diffuser involves:

(a) Establishing the internal flow with the intake operating supercritically, (establishing internal supersonic flow normally requires the use of variable geometry components in the intake),

(b) Throttling the diffuser exit until the pressure recovery is raised to its critical value, and then

(c) Further throttling the flow until the subcritical stability limit is reached.

Measurements must be taken from which the capture mass flow and pressure recovery characteristics can be computed. More detailed calibration may include pressure surveys and drag measurement.

Attempts have been made to calibrate both classes of intake in gun tunnels. However, more effort has been devoted to perfecting a technique that would permit the calibration of intakes with subsonic diffusers, and the present chapter is mainly concerned with that.

Proposed Calibration Technique

Conventionally, some form of mechanical throttle is used to vary the back pressure when calibrating intakes with subsonic diffusers. Although the duration of a test in a gun tunnel is sufficient to use movable components it was considered desirable if possible to avoid a moving throttle because of the complication in programming its motion. The method proposed required a reservoir attached to the exit of the intake, and a calibrated orifice through which the air in the reservoir could exhaust. The arrangement is shown diagrammatically in figure 1-9-8., fitted to an intake in which the same method is used to throttle the flow through a boundary layer bleed duct.

Prior to the tunnel run the intake and its reservoir were evacuated, along with the test section and dump tank. A few milliseconds after the flow through the test section started, the flow over the intake was fully established and air was swallowed by the intake at a constant rate. Since the reservoir pressure was initially very low, the intake started with a low back pressure and operated very supercritically. The rate at which mass flowed into the reservoir was dictated by the intake and its environment, while the rate at which mass left through the orifices was dependent primarily on the reservoir pressure. Consequently, the reservoir pressure rose rapidly at the start of the run, when the outflow was small compared with the intake flow, but the rate of rise became less as the reservoir filled. If the area of the calibrated orifices had been chosen carefully, the mass flow from the reservoir equalled the intake capture mass flow when the reservoir pressure had risen to the maximum intake pressure recovery, at which point the back pressure remained steady. By selecting a suitable reservoir volume, the time taken to reach this point could be chosen to avoid the fluctuations in tunnel stagnation pressure that occurred in the first 35 to 40 milliseconds of the run. In this way the intake was operating near its maximum pressure recovery only when the tunnel flow conditions were steady. The rate of change of back pressure was then relatively slow throughout the usable test duration, and gave adequate time for sampling intake performance data.

RESERVOIR AND STAGNATION PRESSURE HISTORIES

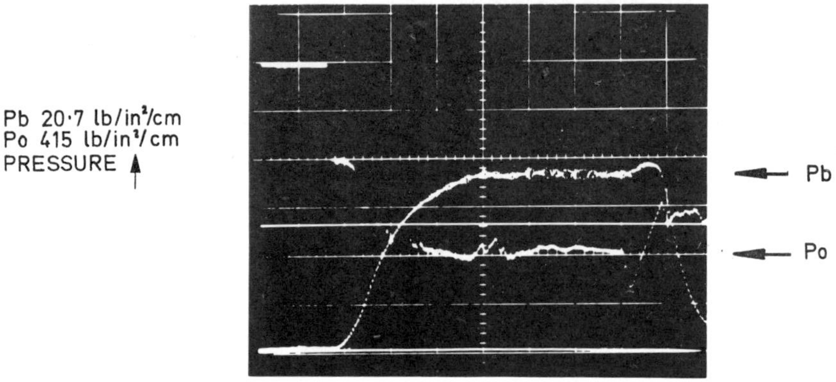

RESERVOIR TEMPERATURE AND PRESSURE HISTORIES

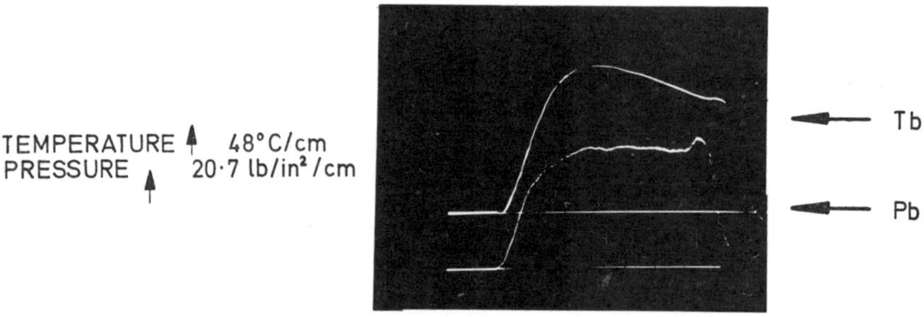

Fig. 1-9-9 C. 32 intake calibration (long reservoir)

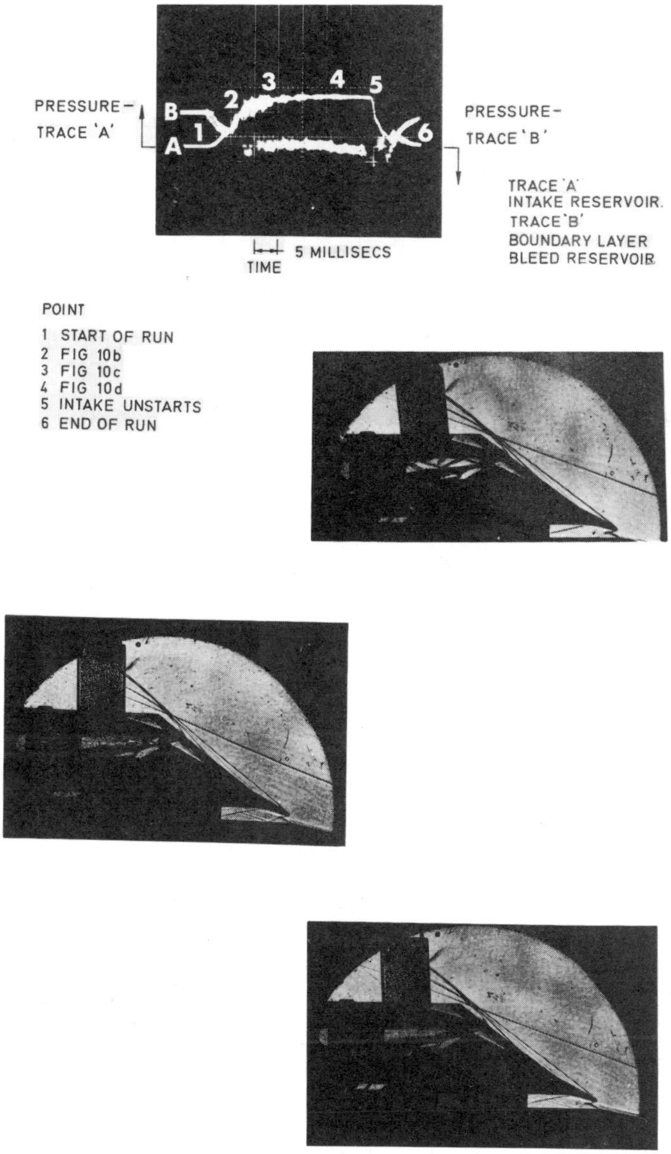

Fig. 1-9-10 (a) Reservoir pressure rise characteristics (b), (c) & (d) Schlieren photographs of intake flow

Oscillograms which illustrate this technique of intake throttling are shown in figure 1-9-9a. Considerable development was needed before a successful calibration test such as this could be made repeatedly.

Assessment of Calibration Technique

There are a number of fundamental limitations to this technique, perhaps the most important of which is that the reservoir filling process is initially of a wave-type nature. Since the wave transit time for a typical intake model is only an order of magnitude less than the time available for filling the reservoir, steps must be taken to ensure that the waves are attenuated rapidly.

During preliminary tests with the model shown in figure 1-9-8, a single orifice was used at the rear end of the reservoir. Although the intake flow was established satisfactorily when the tunnel started, the shock which preceded the internal flow was reflected without significant attenuation from the rear end of the reservoir and travelled back upstream through the intake. On reaching the entry of the intake the capture flow was interrupted and was not re-established during the remainder of the test run. This problem was overcome by replacing the single orifice with a number of small orifices uniformly distributed across the end plate of the reservoir. However, the first 30 milliseconds of the reservoir filling time was still dominated by the passage of discrete waves, and only in the final stages of the run did the rise in reservoir pressure approximate to a continuous process.

Empirical studies of wave attenuation by multi-orifice plates (2) suggested that an optimum ratio for the open to closed area of the plate should be about 1/3. Consequently the cross sectional area of the intake reservoir was modified until, with the orifice area required for mass flow compatibility when the intake was at the critical point, the ratio of open to closed area at the rear end of the reservoir was equal to 0.4. This gave much increased attenuation of the internal waves, and the filling process more rapidly approached a continuous process. Typical results for the model shown in figure 1-9-8 are given in figure 1-9-10. The pressure traces of figure 1-9-10a were given by transducers mounted in the main reservoir downstream of the intake diffuser (trace A) and in the bleed duct reservoir (trace B). The traces show that the intake unstarted when the critical point was reached (point 5), and that approximately 20 milliseconds were available for intake pressure surveys.

Having established the basic feasibility of this method for calibrating intakes with subsonic diffusers, three more elaborate models were designed with which to refine the technique and to measure intake performance:

a) The C32; a two-dimensional mixed compression intake with the internal compression achieved by multiple reflections of an oblique shockwave giving 9° deflections of the flow.

b) The SFA 104; an axisymmetrical mixed compression intake with the internal compression achieved by multiple reflections of a weak shockwave between conical surfaces.

c) The C40; a high performance two-dimensional mixed compression intake with part-isentropic compression.

No provisions for boundary layer bleed were made in any of these intakes.

When designing the intake reservoirs a theoretical analysis of the filling process was made in order to select the optimum reservoir volume and its proportions, i.e. length and cross-section. The analysis, which is given in Appendix I, assumes a continuous filling process to a final steady pressure p_{max} when the mass flow through the reservoir orifices is equal to the intake capture mass flow. The derived relationship between reservoir pressure and time is:

$$\frac{p}{p_{max}} = 1 - e^{-k(t-t_o)}$$

where $k = \dfrac{RQ^*A^*\sqrt{T^*}}{V}$

and t_o is the time from which the reservoir pressure begins to rise. Volumes were chosen so that the pressure rise time would either be equal to the available run time, in which case the critical pressure recovery would be approached slowly, and the intake would operate near to its critical point only when the tunnel stagnation conditions were steady, or so that the pressure rise time was one half of the available run time, in which case the approach to critical would be more rapid but, provided the intake was not unstarted by the mid-run perturbation, there would be sufficient run time still available for making internal pressure surveys and capture mass flow measurements at a constant back pressure.

Four different volume configurations have been studied, two giving long rise times, and two giving shorter times. Their geometries are listed in the table below, and the criteria governing their selection are as follows:

i) The effective exit orifice area (A^*) is specified by the intake capture area and estimated critical pressure recovery,

ii) The internal volume (V) is specified by the required pressure rise time,

iii) The internal length is limited so that the total transit time for a wave should not exceed one millisecond, and

iv) The cross-sectional area of the reservoir at the orifice plate is specified by the requirement that the ratio of open to closed area of the plate should be 0.4.

To some extent these criteria are mutually incompatible, particularly in so far as (i) and (iv) are concerned. Since it was necessary to cover a range of A^* so that different levels of pressure recovery could be obtained, a wire mesh screen having an open/closed area ratio of 0.4 was fitted in the reservoir. This ensured effective attenuation of internal waves and relaxed the restriction on the orifice plate area ratio.

INTAKE TYPE	MAXIMUM CAPTURE AREA (M_o = 7)	INTERNAL VOLUME	INTERNAL LENGTH
C 32 (Short Reservoir)	4.35 in^2 28.1 cm^2	26 in^3 426 cm^3	9.75 in 24.8 cm
C32 (Long Reservoir)	4.35 in^2 28.1 cm^2	52 in^3 852 cm^3	16.5 in 41.9 cm
SFA 104	6.28 in^2 40.5 cm^2	24 in^3 393 cm^3	10.5 in 26.7 cm
C40	10.9 in^2 70.3 cm^2	280 in^3 4590 cm^3	22.0 in 55.9 cm

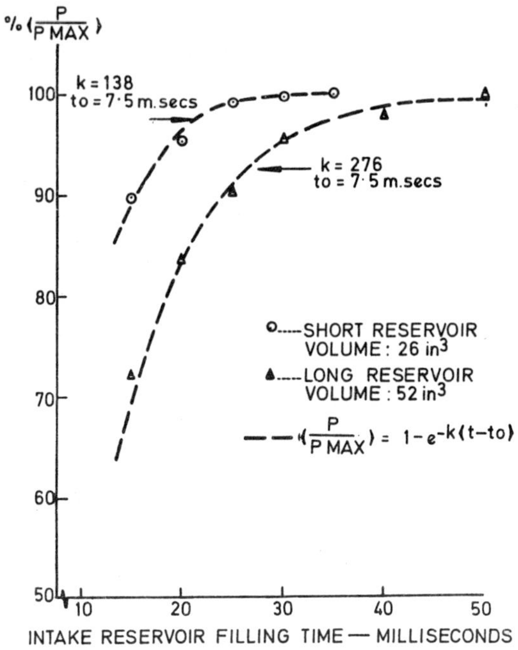

Fig. 1-9-11 Intake reservoir pressure rise characteristics C. 32 intake

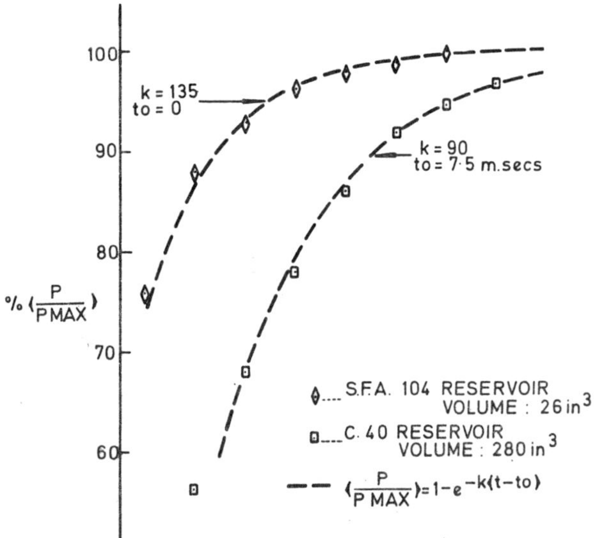

Fig. 1-9-12 Intake reservoir pressure rise characteristics C. 40 and S. F. A. 104 intakes

Typical results for these intake reservoir combinations are shown in figures 1-9-11 and 1-9-12. Also shown in these figures are the corresponding theoretical rise times, the quoted values of k and t_o being empirical. Reasonably good agreement between theory and experiment is obtained when the internal waves have been damped, so justifying the assumption that the latter stages of the reservoir filling process may be considered continuous. This conclusion is supported also by high speed cine sequences which show a steady advance of the terminal shock as the critical point is approached.

Intake Instrumentation

Determination of basic intake performance requires accurate measurement of intake pressures and capture mass flow. Assuming a knowledge of the effective area of the reservoir orifices, the latter parameter depends upon the measurement of pressure and temperature in the reservoir when the inlet and outlet flows have equilibrated.

As the reservoir pressure level for the models studied was generally between 3 and 10 atmospheres, no difficulty was experienced in achieving adequate response from conventional pressure transducers mounted on or close to the reservoir. Solartron NT 4-313 unbonded strain gauge transducers have been used exclusively for the measurement of reservoir pressure. The transducer diaphragm was mounted over a cavity of 0.003 cu.ins. (0.05ccs) with a short communicating drilling to the inside wall of the intake reservoir. Typical rise times to a pressure of 5 atmospheres for this transducer installation were 1 to 2 milliseconds, and the accuracy achieved when used in conjunction with a Tektronix Type 502A oscilloscope was about ± 2 percent.

Stagnation temperature in the reservoir was measured using an unshielded thermocouple probe having a rise time of about 20 milliseconds. The probe was of a type developed by East and Perry (3) and consisted of a Chromel-Alumel thermocouple junction supported by a stainless steel hypodermic tube of 3mm outside diameter. The thermocouple bead was approximately 0.003 ins. (0.076 mm) in diameter and was formed by welding 0.001 ins. (0.025 mm) diameter wires in an inert atmosphere.

A similar thermocouple probe but shielded with a platinum element furnace was used for free stream stagnation temperature measurement. When using this probe it is necessary to pre-heat the shield to a temperature close to that which is to be measured.

Typical records from tests with the C32 intake where temperature measurements were made, are shown in figures 1-9-6 and 1-9-9b. It will be noted that the stagnation temperature of the air in the reservoir was significantly lower than that of the free stream. Heat transfer to the walls of the intake is inevitable since the desired ratio of tunnel stagnation temperature to model wall temperature is 2.7 and remains sensibly constant at this value throughout the test. In calculating the mass flow rate from the reservoir it is important to use the instantaneous local temperature and pressure. Provided these measurements were made when the reservoir flow had equilibrated, repeatability of the derived mass flow rate for a given intake geometry was within ± 3 percent.

In order to record as much pressure data as possible during the time for which the intake back pressure remained constant, the maximum switching rate of a standard Scannivalve was determined. The pressure switch used was the General Design Scannivalve Type 48J. At that time, the normal switching rate for this device was 48 ports per second. For high speed switching the Ledex sequencing solenoid on the Scannivalve was removed and the rotor driven by a Servomex variable speed

Fig. 1-9-13 C. 40 model intake and reservoir

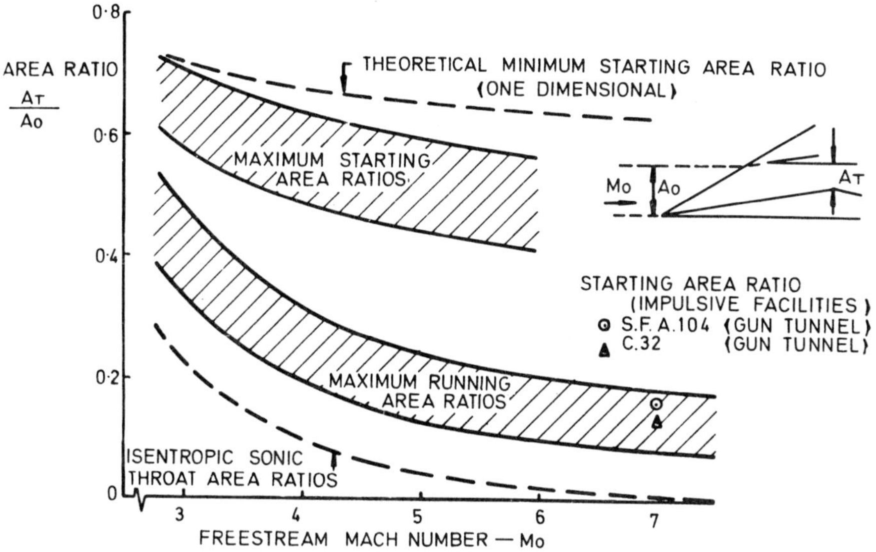

Fig. 1-9-14 Empirical intake starting and minimum running contraction ratios

motor. With this drive, continuous switching speeds up to 500 ports per second have been obtained with no signs of damage to the rotor and stator faces.

The Scannivalve was fitted with a Solartron pressure transducer Type 4-312. In order to obtain the maximum response, the size of the cavity over the transducer diaphragm was optimised using experimental data for a transducer of similar design. The data, which were obtained from a CEC Type 4-327 transducer, showed (4) that with a cavity of 0.008 in^3 (0.13 cm^3) a response time of 2 milliseconds - including a 1 millisecond pulse application time - could be achieved if the final pressure was above 0.5 atmospheres absolute. This is equivalent to a switching rate of 250 ports per second, and would allow reliable measurement of up to 10 internal pressure points during a run.

Test Results and Discussion

The intake calibration technique was used to measure the performance of a number of external/internal supersonic compression intakes, one of which - the S.F.A.104 - had previously been calibrated in a conventional hypersonic wind tunnel. The intakes are described in paragraph 5.3, and a photograph of the C.40 appears in figure 1-9-13. The C.32 intake is similar to the C.40 except in the compression surface design.

The conditions under which these intakes were calibrated in the gun tunnel were:

Free stream Mach number		7.0
Free stream Unit Reynolds number		10^7 per foot
Reynolds number based on height of free stream capture tube	C.32	1.0×10^6
	C.40	2.2×10^6
	S.F.A. 104	2.4×10^6
Ratio of intake surface temperature to free stream stagnation temperature		0.38

Maximum Internal Contraction Ratio for Impulsive Starting

When calibrating external/internal compression intakes in a conventional facility, variable geometry is needed to establish supersonic flow at the internal throat. The degree of variability required expressed in terms of the total contraction ratio, is shown in figure 1-9-14. A typical intake starting sequence would involve first increasing the throat area with low back pressure until supersonic flow is established through the internal contraction. At a free stream Mach number of 7 the throat to capture area ratio required for starting is about 0.6. The throat area is then reduced towards the maximum running contraction ratio and the intake performance measured at various contraction ratios until the limiting ratio is reached.

Preliminary intake tests in the gun tunnel suggested that internal contractions could be started at area ratios very near the running limit. These observations led to a systematic study with the above mentioned intake models at a free stream Mach number of 7.

Performance results for the C.32 intake are shown in figure 1-9-15 where the critical total pressure recovery is presented at various overall contraction ratios. The intake was operated as a fixed geometry intake, the contraction ratio being varied between runs by moving the compression ramp relative to the cowl so

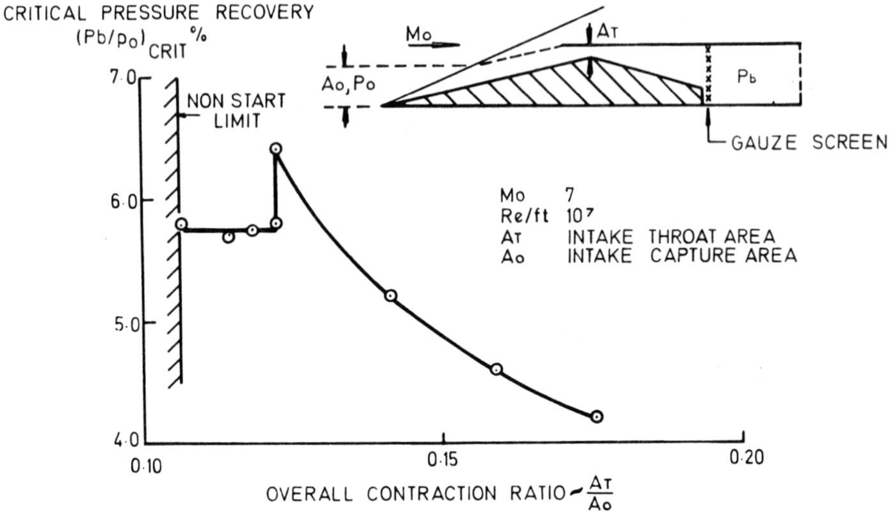

Fig. 1-9-15 Critical pressure recovery characteristic C. 32 intake - original fixed geometry

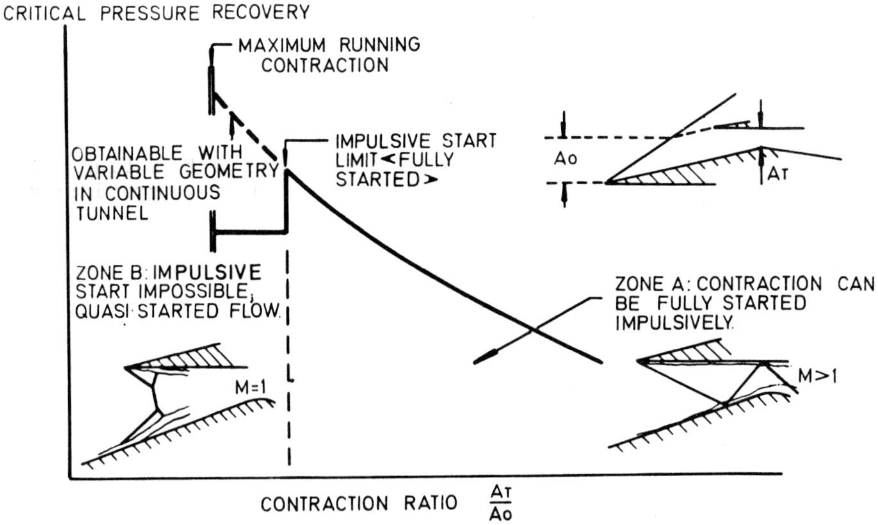

Fig. 1-9-16 Critical pressure recovery characteristic of intake model in a gun tunnel facility

as to increase or decrease the throat height.

The results show that the intake could be started simultaneously with the test section flow at all contraction ratios down to 0.108. The criterion for successful starting is that the intake lip shock be attached, thus ensuring full capture mass flow. With small throat areas, the internal contraction could not be started (i.e. supersonic flow could not be established through the contraction) and the intake operated at reduced mass flow throughout the run. Reference to figure 1-9-14 indicates that the maximum contraction ratio at which the C.32 could be started in the gun tunnel was the same as the expected maximum running area ratio for that intake.

The maximum starting area ratio for the axisymmetric S.F.A.104 intake was determined also in the gun tunnel at $M_o = 7$. This was found to occur at $A_T/A_0 = 0.148$, which agreed with the maximum running area ratio for the same intake when calibrated in a conventional continuous running facility (Fig.1-9-19). These calibration results seemed to confirm the impression that internal contractions could be started impulsively in a gun tunnel provided the contraction did not exceed the maximum running area ratio. However, the critical pressure recoveries obtained from the intakes in the gun tunnel exhibited an anomalous discontinuity close to the contraction limit. Such discontinuities had not been observed in continuous facilities, so a close examination of this phenomenon was made in the gun tunnel using the C.32 intake.

It was found that a stable operating regime existed near the non-start limit in which the intake was technically started, i.e. the lip shock was attached, but the flow at the internal throat was choked. This resulted in the flow through the contraction being transonic rather than supersonic, and on being throttled the intake gave a lower critical pressure recovery than that for the fully started contraction. The different operating regimes are illustrated in figure 1-9-16. The boundary to the quasi-started flow regime was well defined by the discontinuity in critical pressure recovery, and was reproducible. In this regime the critical pressure recovery appeared to be independent of contraction ratio up to the non-start limit and significantly lower than the recoveries obtained by conventional calibration techniques in a continuous facility. Up to this quasi-start limit, impulsive starting of the flow through the contraction led to the same internal flow field and critical pressure recovery as were obtained in the continuous facility.

Attempts were made to avoid the quasi-started flow regime by using a variable geometry contraction. The lip of the C.32 intake was modified as shown in figure 1-9-17, to allow the internal contraction to be increased during the test run. By setting the lip in a lowered position prior to the run, the internal contraction was reduced to a value which could be started impulsively. The aerodynamic force acting on the lip and external flap when the tunnel was operating caused rotation about the pivot so that the lip moved towards its design position. This position was reached about 35 milliseconds after the primary diaphragm burst and the spring loaded locking pin prevented further movement of the lip during the run.

Analysis of high speed cine and internal pressure records showed that, although the actuation of the moving lip and locking device was satisfactory, the flow through the contraction with the lip lowered was not established fully at the commencement of the run. This was traced to the disturbance at the trailing edge of the moving lip which caused local choking of the internal flow, and resulted in a quasi-started condition at all lip positions. Results for the C.32 intake before and after the modification to the lip are shown in figure 1-9-18. It can be seen that the performance of the two geometries was essentially the same, and that the moving lip

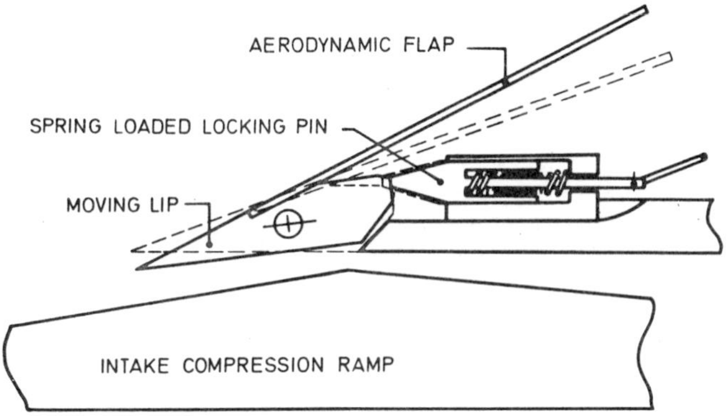

Fig. 1-9-17 C. 32 intake model - longitudinal section through moving lip.

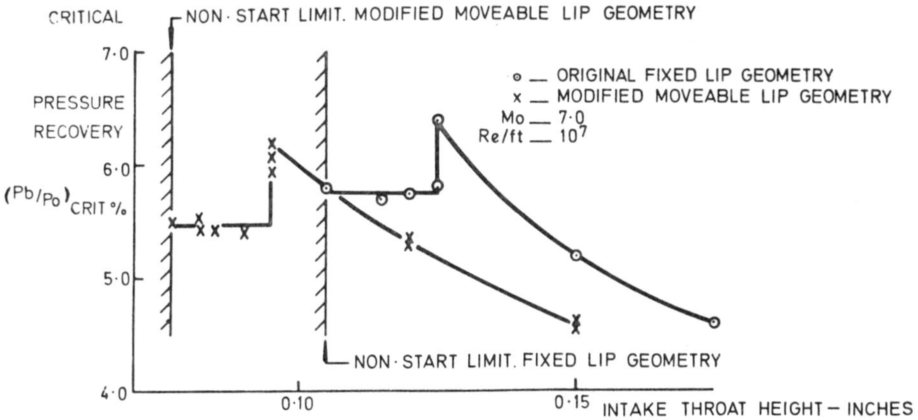

Fig. 1-9-18 Critical pressure recovery characteristics C. 32 intake original and modified geometries

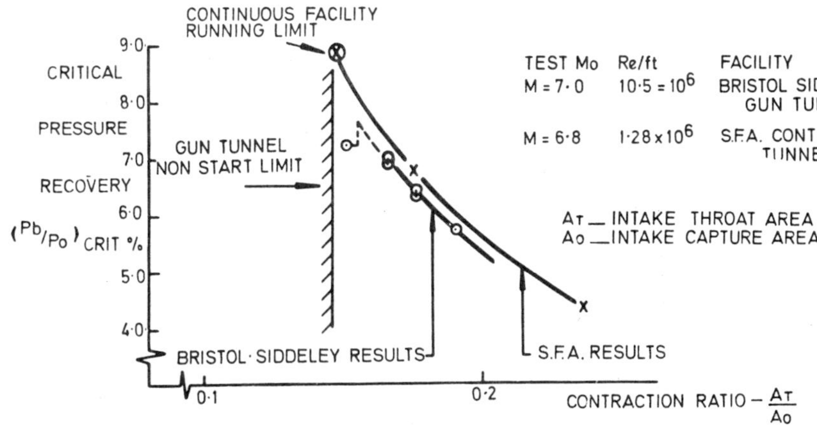

Fig. 1-9-19 Comparison of intake results from Bristol Siddeley gun tunnel and Svenska Flygmotor continuous facility intake model S. F. A. 104

did not permit an extension of the calibration range up to the minimum running area ratio. It is considered that an alternative design of moving lip for which the internal profile remained continuous at all lip positions would have overcome this problem. However until this has been demonstrated, the range of contraction ratios over which impulsive starting will lead to representative internal flows cannot be said to extend to the maximum running limit.

Critical Pressure Recovery

The reproducibility and validity of performance results obtained in the gun tunnel facility have been assessed in some detail. The scatter in capture mass flow and internal pressure measurements was discussed previously and stated to be ± 3 percent and ± 2 percent respectively. This scatter is greater than is usually experienced when calibrating intakes in continuous facilities, but is probably acceptable for many intake studies.

With regard to measurement of critical pressure recovery, provided care was taken to ensure that the rate of change of reservoir pressure was low as the critical point was approached, and that the intake was not unstarted as a result of fluctuations in free stream conditions, the results were very reproducible (Figs. 1-9-18 and 1-9-19.) Critical pressure recoveries for the S. F. A. 104 axisymmetric intake in the gun tunnel are compared with those obtained in a conventional facility in figure 1-9-19. The latter results were obtained by the Aerodynamics Department of Svenska Flygmotor AB., Trollhattan, Sweden, to whom the S. F. A. 104 intake belonged.

The two sets of results compared favourably over the range of contraction ratios where the internal flow could be established fully in the gun tunnel. When calibrated in the gun tunnel the intake was operated with fixed geometry and the contraction started impulsively, whereas axial translation of the compression spike was required to establish the internal flow when tested in the continuous facility. The consistent difference between the results is largely due to the difference in free stream Mach number for the two facilities - $M_o = 6.8$ in the continuous facility and $M_o = 7.0$ in the gun tunnel. Although there was a factor of 10 between the unit Reynolds numbers for the facilities, a transition fixing device was used on the intake when in the continuous facility. Observations of boundary layer and shock/boundary layer interactions made during the tests showed that the device was effective in simulating the higher test Reynolds number of the gun tunnel calibration.

The critical pressure recovery for the quasi-started regime in the gun tunnel is clearly a non-representative result. The boundary to this regime is shown in figure 1-9-14 together with that for the C.32 intake. It will be seen that the majority of the intake contraction ratio range could be covered reliably in the gun tunnel without recourse to variable geometry.

Although it has been shown that large internal contractions may be started impulsively in a gun tunnel this process was possible only if it occurred simultaneously with the commencement of the test section flow. If for any reason the intake unstarted during a test, the internal flow could not be re-established. The subcritical characteristics of the S. F. A. 104 intake when calibrated in the gun tunnel were in qualitative agreement with those from the continuous facility. With contraction area ratios between about 0.18 and the maximum running contraction, unstarting the internal flow led to stable subcritical operation. Buzz occurred with contraction ratios beyond 0.18, the frequency being about 260 cycles per second.

Comparison of Gun Tunnel with Wind Tunnel

The main disadvantages which the gun tunnel appears to have compared with a conventional hypersonic wind tunnel for intake studies, are a shorter running time and possibly a poorer quality test flow. The advantages of the gun tunnel include:

a) The ability to start internal contractions impulsively almost up to the maximum running limit without the need for variable geometry,

b) The ability to simulate the ratio of intake compression surface temperature to free stream stagnation temperature without cooling or heating the model,

c) The ability to maintain the model and its associated instrumentation at ambient temperature.

These features combine to yield a considerable simplification in intake model design, with consequent reduction in cost.

Acknowledgements

The authors acknowledge the assistance given by Svenska Flygmotor A.B., Trollhattan, Sweden, who kindly loaned the S.F.A.104 intake model for calibration studies in the gun tunnel, and in particular Mr. G. Rosander, of that Company, who calibrated the S.F.A.104 intake model in the F.F.A. hypersonic wind tunnel and provided the results for comparison with those obtained from the gun tunnel.

They also acknowledge the assistance of Mr. J.H. Perry of Southampton University in obtaining gun tunnel temperature measurements.

References

1. **Fry, W.A.S.,** 'The Gun Tunnel as an Intake Test Facility and Preliminary Assessment of Ramjet Intake Performance at M = 10'. Rolls-Royce Ltd., Bristol Division Rep. AP 5146, Oct. 1965.

2. **Henshall, B.D.,** *et al*, 'Development of Very High Enthalpy Shock Tunnels with Extended Steady-State Test Times'. A.V.C.O. Corporation RAD Report No. TR 62-16. April 1962.

3. **Perry, J.H., East, R.A.,** 'A Short Time Response Stagnation Temperature Probe'. Southampton University A.A.S.U., Rep. No. 264, March 1966.

4. **Harper, L.R.,** 'Response of a Strain Gauge Pressure Transducer and Mounting to a Fast Rising Pressure Pulse'. Rolls-Royce Ltd. Bristol Division Rep. AP 5394, Oct., 1966.

List of Mathematical Symbols

A area

k $\equiv \frac{RA^*Q^*\sqrt{T^*}}{V}$

M Mach number

p reservoir pressure during filling process

P stagnation pressure

Q	$\equiv \frac{\dot{W}\sqrt{T}}{AP}$
R	gas constant
Re	Reynold's number
t	time
T	stagnation temperature
v	velocity
V	intake and reservoir volume
W	air mass
$\dot{W}$	air mass flow rate

Subscripts

o	free stream conditions
1	initial barrel conditions
4	initial driver conditions
b	reservoir conditions
T	intake throat conditions

Superscript

*	choked reservoir orifice conditions

Definitions

$$P_{41} \equiv \frac{P_4}{P_1} \; ; \qquad P_{o1} \equiv \frac{P_o}{P_1}$$

Appendix

Analysis of Reservoir Filling Time

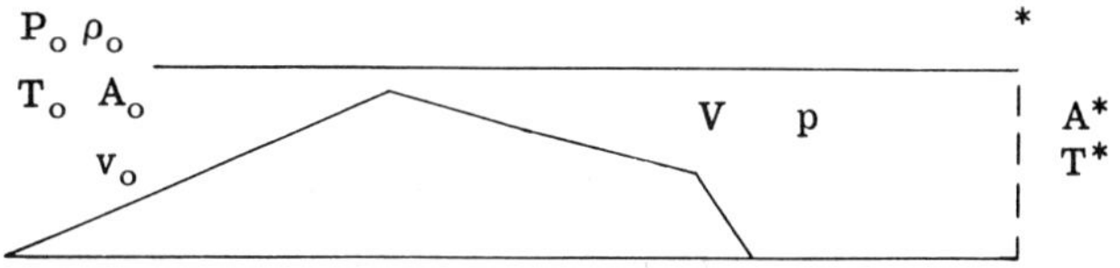

Capture mass flow rate = $\dot{W}_o$ lbs/sec = constant where:

$$\dot{W}_o = \frac{\rho_o A_o v_o}{g}$$

$$= \frac{Q_o A_o P_o}{\sqrt{T_o}} \quad \text{= rate at which mass enters the intake and reservoir.}$$

Let the final steady pressure (at $t = t_\infty$) be p_{max} when the rate of mass flow through the choked reservoir orifices will be equal to $\dot{W}_o$ lbs/sec. Then maximum mass flow rate leaving reservoir:

$$= \frac{Q^*A^*p_{max}}{\sqrt{T^*}} = \dot{W}_o \qquad \text{(Eq. 1-9-i)}$$

At any general time (t) and reservoir pressure (p) the mass of air in the intake and reservoir whose combined volume is (V) will be:

$W = \frac{Vp}{RT}$ lbs and assuming this air is at near-stagnation conditions and that its temperature is steady with time:

$$\frac{dW}{dt} = \frac{V}{RT}\frac{dp}{dt} \qquad \text{(Eq. 1-9-ii)}$$

The instantaneous rate at which mass leaves the reservoir will be:

$$\dot{W} = \frac{Q^*A^*p}{\sqrt{T^*}}$$

so that:

$$\frac{dW}{dt} = \dot{W}_o - \dot{W} = \frac{Q^*A^*p}{\sqrt{T^*}}(p_{max} - p) \qquad \text{(Eq. 1-9-iii)}$$

Equating 1-9-ii and 1-9-iii:

$$\frac{V}{RT} \cdot \frac{dp}{dt} = \frac{Q^*A^*}{\sqrt{T^*}}(p_{max} - p)$$

Assuming no loss in stagnation temperature within the intake and reservoir, then $T = T^*$ and :

$$\frac{dp}{(p_{max}-p)} = \frac{RQ^*A^*\sqrt{T^*}}{V} \cdot dt \qquad \text{(Eq. 1-9-iv)}$$

Integrating, putting $p = o$ when $t = t_o$

$$-\log_e (p_{max} - p) = k(t - t_o)$$

or:

$$\frac{(p_{max} - p)}{p_{max}} = e^{-k(t - t_o)}$$

$$\frac{p}{p_{max}} = 1 - e^{-k(t - t_o)} \qquad \text{(Eq. 1-9-v)}$$

where $k = \frac{RQ^*A^*\sqrt{T^*}}{V}$

Commentary on Chapter 1-9

I.M. Auriol

L'utilisation de moyens de simulation aérodynamique fonctionnant pendant des temps courts est très tentante, car elle conduit à des investissements modérés pour des performances d'installation qui peuvent être élevées. Les possibilités de telles installations ainsi que leurs défauts ont été mis clairement en évidence par les auteurs.

En France, dans le domaine des entrées d'air, des essais analogues ont été faits à l'O.N.E.R.A. et je crois qu'il est intéressant de comparer les méthodes et les résultats. L'installation et les performances des souffleries de l'O.N.E.R.A. utilisées pour ces essais sont indiquées sur la figure A1-9-1.

On remarquera que l'installation R'_o a des performances et des dimensions tout à fait analogues à celles de la soufflerie qu'ont décrite les auteurs. A noter, toutefois, une différence: Cette soufflerie est à piston lent. La compression obtenue dans le tube est une compression sensiblement isentropique. Cette solution a été choisie pour éviter les irrégularités mentionnées par les auteurs et dues à la formation d'ondes de choc dans le tube de compression.

Mais ce n'est pas mon but de discuter des avantages comparés des souffleries à piston lent ou à piston rapide. Dans ce domaine, comme dans celui des canons à gaz légers, on peut discuter longuement des avantages et inconvénients des deux systèmes. Je noterai seulement que l'on a peu de renseignements sur le taux de turbulence dans l'un et l'autre genre d'installation et que - comme le soulignent les auteurs - le degré de turbulence pourrait avoir une influence sur les résultats. A mon avis, la turbulence devrait être plus faible dans les souffleries à piston lent.

Les essais faits à l'O.N.E.R.A. concernaient essentiellement les conditions d'amorçage et ils n'ont pas comporté de mesure de débit.

La figure A 1-9-2 donne des renseignements sur ces essais. L'entrée d'air du type pitot de révolution était placée dans la veine et une observation strioscopique permettait de voir si l'entrée était amorcée ou non. On a noté l'importance considérable sur l'amorçage de l'entrée d'air de la pression existant dans le caisson avant l'amorçage de la soufflerie. Pour une maquette donnée, on a fait varier le rapport de cette pression caisson à la pression de veine en cours de fonctionnement. On a déterminé quel était le rapport limite d'amorçage.

Les auteurs mentionnent la possibilité d'un faux amorçage avec décollement dans le convergent. Nous n'avons jamais observé un tel phénomène et la figure A 1-9-3 montre qu'une strioscopie soignée permet de surveiller efficacement l'écoulement et aurait dû permettre de détecter de tels désamorçages.

La figure A1-9-4 indique les résultats obtenus. En abscisse est porté le rapport de la pression caisson à la pression de veine en cours de fonctionnement $\frac{Pc}{Po}$ ainsi que le rapport de cette pression caisson à la pression reservoir $\frac{Pc}{P_{io}}$, en ordonnée le rapport de la section de sortie à la section d'entrée.

Une variation sur l'axe des ordonnées correspond donc pour un Ψ donné à une variation de longueur de l'entrée d'air. Les courbes correspondent aux conditions limites d'amorçage. Elles montrent l'intérêt d'un abaissement de la pression caisson. La courbe supérieure est relative à une entrée conique de révolution de

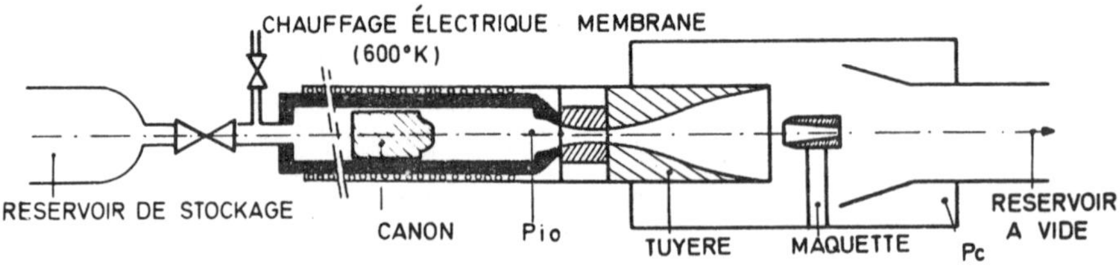

	R'o	R4
MACH	7	14
Φ SORTIE TUYERE	135 mm	325 mm
PRESSION VIDE Pc	Pc ≫ 3mb	Pc ≫ 0.1mb
PRESSION GÉNÉRATRICE	Pio = 145 bars	Pio ≪ 200 bars
T° GÉNÉRATRICE	Tio = 800°K	Tio = 1650°K
REYNOLDS/cm	400·000	≪ 30·000
DURÉE DE RAFALE	50 ms	200 ms

Fig. A1-9-1 O.N.E.R.A. gun tunnel

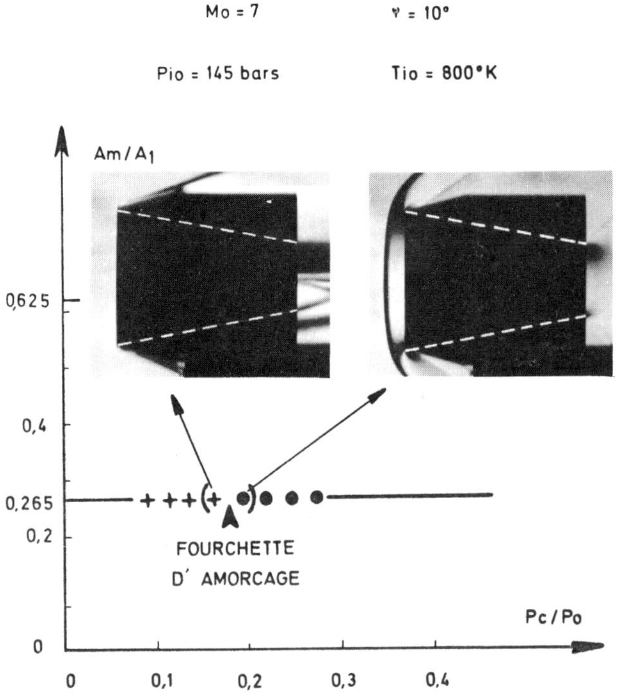

Fig. A1-9-2 Amorçage

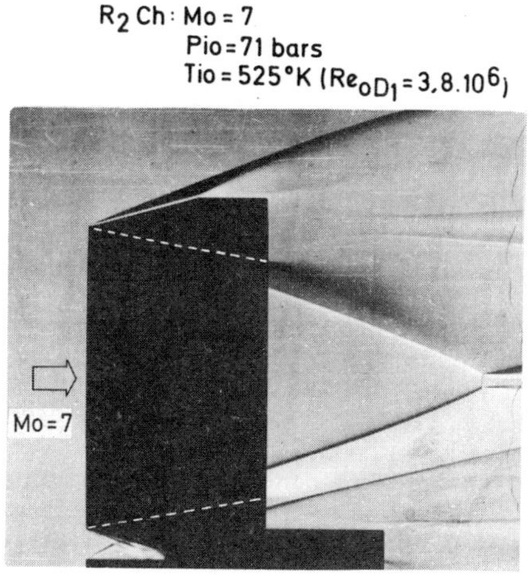

Fig. A1-9-3 Prise d'air de revolution a C.S.I. (= 10)

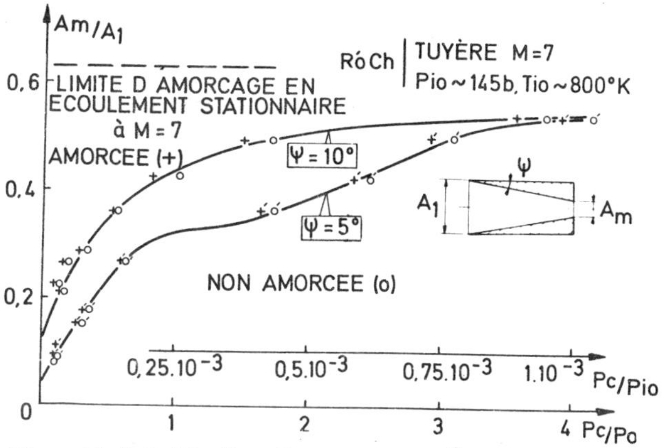

Fig. A1-9-4 Limites d'amorçage en écoulement instationaire

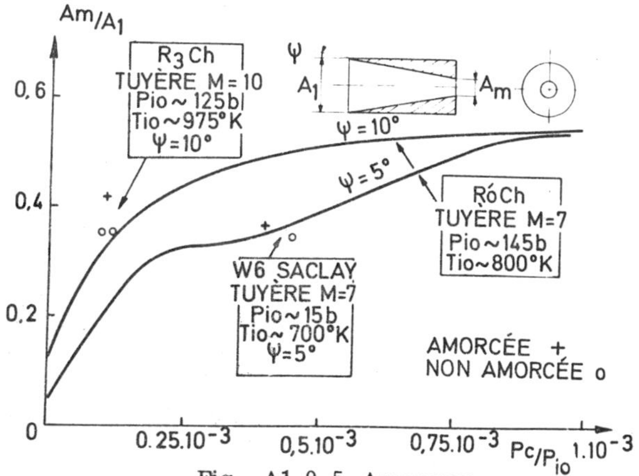

Fig. A1-9-5 Amorçage

demi-angle au sommet $\Psi = 10°$, la courbe inférieure à $\Psi = 5°$.

La figure A1-9-5 compare les résultats obtenus dans plusieurs souffleries de caractéristiques différentes et montre que les conditions d'amorçage sont tout a fait comparables pour ces souffleries malgré des géométries assez différentes.

Les auteurs ont indiqué qu'ils avaient observé de faux amorçages avec onde de choc avalée, mais décollement dans le convergent. Ce régime n'a jamais été observé dans les essais de l'O.N.E.R.A. Les entrées d'air ne sont pas du même type, mais la cause de cette différence peut être due à l'existence d'un reservoir derrière l'entrée d'air. Il pourrait y avoir une sorte de résonance acoustique conduisant dans certaines conditions à un désamorçage partiel : les auteurs ont noté des désamorçages dus à la réflexion de l'onde de choc au fond du réservoir et y ont rémédie en introduisant un grillage dans le réservoir. Mais ce grillage est peut-être inefficace à certains régimes ou pour des ondes acoustiques plus faibles. Il est possible que le niveau de turbulence de la soufflerie joue aussi un rôle. Il serait intéressant de savoir si les conditions du faux amorçage ne sont pas influencées par la capacité et la forme du réservoir, ainsi que par la présence de grillages amortisseurs.

De toute maniére, qu'il soit dû à l'installation, ce qui me paraît probable, ou qu'il soit possible en vol, le régime mentionné par les auteurs est très important. Les expérimentateurs devront, à l'avenir, tenir compte de cette possibilite.

En resume, les résultats obtenus par les auteurs confirment, comme ceux obtenus à l'O.N.E.R.A., la possibilité d'utiliser des moyens d'essai à temps de rafale court pour les études d'entrée d'air. Les moyens de mesure actuellement disponibles permettent d'obtenir des résultats quantitatifs.

Enfin, notons que les aérodynamiciens doivent rester vigilants pour détecter les phénomènes fondamentaux ou parasites que ce genre d'essais peut mettre en évidence.

Part 2

Optical Techniques

p

2-1

Interferometric Measurements in Optical and Microwave Frequency Ranges on Electromagnetically Accelerated Shock Waves

H. MUNTENBRUCH.
Institut für Plasmaphysik, München, Germany.

Summary

Fast shock waves can be produced in light gases electromagnetically. These are then often of a blast wave character. With such shock waves a distinction has to be made between two regions - the shock wave heated gas immediately behind the shock front and the plasma cloud heated in the electric discharge. Interferometric measurements on shock waves in hydrogen show that these regions are only separated from one another up to about M = 20. The blast wave character of the waves can be clearly observed both in the time dependence of the shock front velocity and in the density distribution behind the front. These results are confirmed by microwave measurements, which, however, also show different behaviour of the electron density owing to relaxation effects.

Sommaire

Des ondes de choc rapides peuvent être produites par intéraction électromagnétique dans des gaz légers. Avec de tels chocs, une distinction doit être faite entre deux régions, le domaine du gaz chauffé par choc immédiatement en aval du front de l'onde et le nuage de plasma chauffé par la décharge électrique. Des mesures interférométriques sur des ondes de choc dans l'hydrogène montrent que ces régions ne sont séparées que jusqu'à des nombres de Mach de l'ordre de 20. Le caractère d'onde balistique peut être clairement mis en évidence à la fois par la loi d'évolution en fonction du temps de la vitesse du front d'onde et par la distribution de densité derrière le choc. Ces résultats sont confirmés par des mesures effectuées au moyen d'ondes ultra-courtes, qui mettent également en évidence le comportement particulier de la densité d'électrons, dû aux phénomènes de relaxation.

Introduction

Since 1900, when Vieille (1) invented the diaphragm shock tube, it has been possible to produce shock waves under well-defined conditions. To a good approximation, such shock waves are stationary i.e. the velocity remains constant during the passage of the wave, and all variables of state at points a fixed distance from the shock front remain constant with time. However in light gases the attainable shock Mach number, which is a function of the initial pressures in the high and low pressure sections, and of the types of gas used, is not very high.

Fowler and co-workers (2) therefore attracted a great deal of attention in 1951 with their first experiments using fast electric discharges to generate shock waves. It soon became clear that in this way velocities of a few million cm/sec, i.e. Mach numbers of, say, M = 50, could be attained in hydrogen, deuterium and helium.

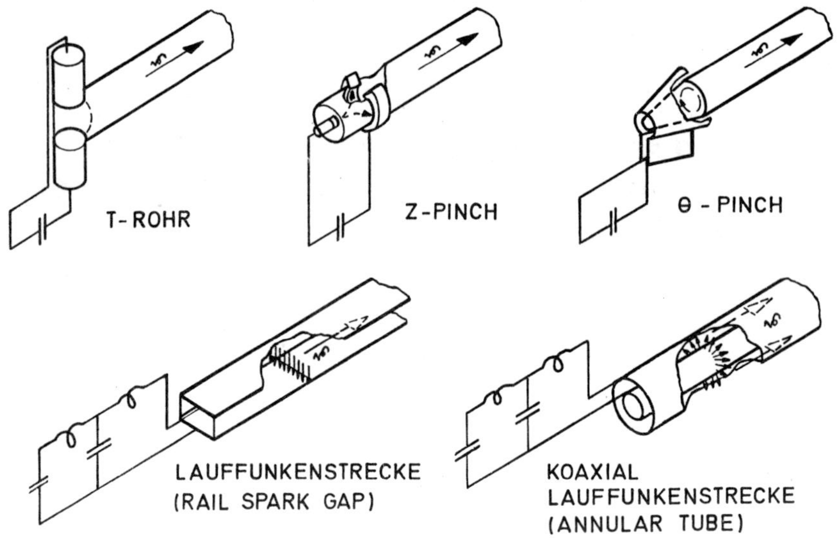

Fig. 2-1-1 Various driver systems for electro-magnetic acceleration of shock waves

Fig. 2-1-2 Streak photograph of a shock wave in hydrogen

Further improvements were then made to the driver systems used in the electromagnetic production of shock waves, Kolb (3), for instance, demonstrating in 1957 that, in addition to the input of energy by ohmic heating, an input of momentum obtained from the magnetic field of the backstrap in T-tubes also could be used to generate shock waves. Josephson (4) used the conical Z-pinch as a driver system, and a whole number of other drivers were developed as well.

Driver systems can be divided into two groups. The first (Fig. 2-1-1, top) includes those used to supply a gas or plasma with energy or momentum in a very short time, e.g. in a few μsec. In this way the shock wave is given a push, so to speak, and continues more or less on its own after the driver has been switched off.

The second group comprises those systems in which the accelerated plasma is continuously supplied with energy and momentum so that a steady state can be achieved with the right kind of energy storage. This group of electromagnetic shock tubes has been used for a large number of experiments with a pronounced bearing on plasma physics and magnetogasdynamics.

We are concerned here only with shock waves produced with T-tubes, i.e. with a driver system from the first group. Attempts to describe such shock waves in terms of theories applicable to stationary shock waves have presented serious contradictions between theory and experiment. However, the free-running shock waves produced by instantaneous energy or momentum input can now be successfully described, i.e. experimental results and theoretical predictions are in good agreement. The experiments discussed here contributed to this success.

Shock Fronts and Luminous Fronts

Electromagnetic shock waves were not completely understood in the past because luminous fronts were invariably identified as shock fronts. It was the luminous fronts, an example of which is given in the streak photograph (Fig. 2-1-2) that most experimentalists investigated. Velocities of a few million cm/sec were measured, or up to a few tens of millions cm/sec near the starting point. At such high velocities the gas in the shock front should of course have been sufficiently excited for luminescence.

Streak photographs are simply x-t diagrams and the velocity distribution derived experimentally from them yielded the relation:

$$x \propto t^{\alpha}$$

as one would expect for instantaneously accelerated shock waves, owing to the conservation of either momentum or energy as shown by Harris (5), Sedov (6) and others.

Quantitatively however there were many discrepancies, both in the determination of the exponent α and primarily between the values of the measured state variables behind the front in the luminous plasma and the values calculated on the basis of the Rankine-Hugoniot relations and the equations of state.

Like Blackman and Niblett (7) and other experimentalists, we concluded that the true shock front must actually precede the luminous front. To make the shock front itself visible, therefore, we decided to use an interferometric method and chose a Mach-Zehnder interferometer, the beam path of which is reproduced in the experimental set-up shown in figure 2-1-3.

The light beam of the lamp, which is pulsed to increase its intensity, is split and superimposed again, producing an interference fringe system in the image plane.

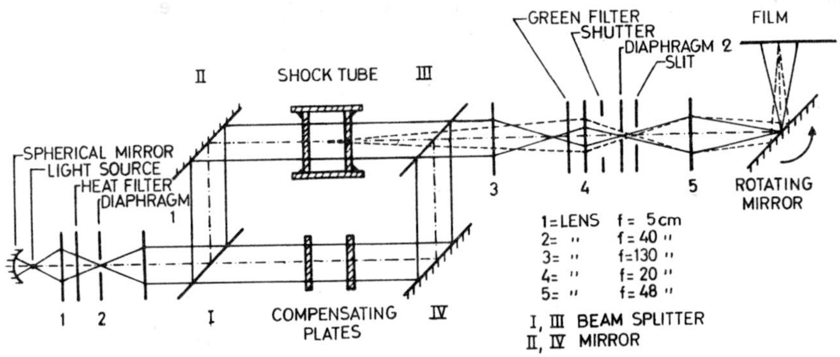

Fig. 2-1-3 Beam path in the interferometer/streak camera set-up

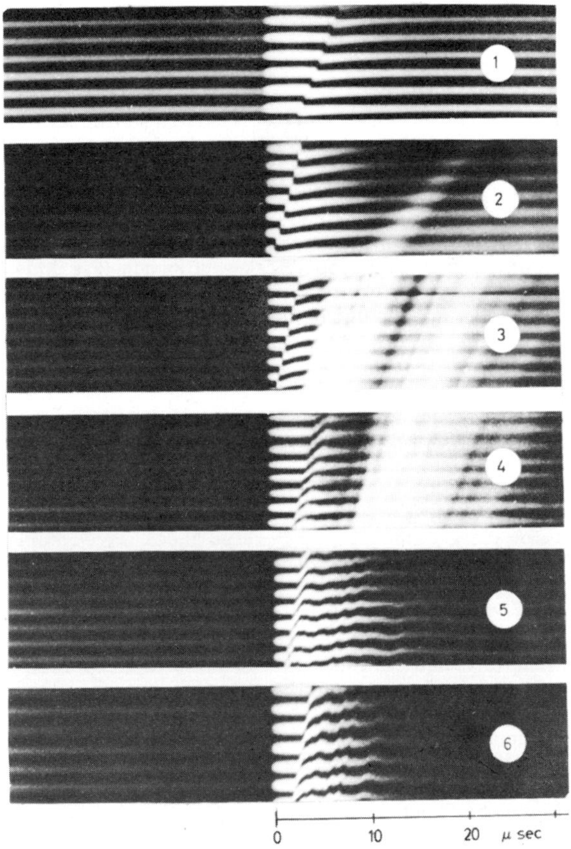

Fig. 2-1-4 6 interferograms of shock waves in hydrogen (p_o = 5 torr) with various velocities

Change in the optical path of the measuring beam leads to a displacement of the fringes. Only a narrow slit,to which the fringes are perpendicular, is imaged. This is done with a streak camera so that the fringe shift can be resolved in time. An interference filter allows or prevents simultaneous imaging of the plasma luminescence, depending on its position.

Using an interferometer/streak camera set-up like this is practically the same as photographing x-t diagrams. A few of the pictures taken by Brinkschulte (8) are shown in figure 2-1-4. All of them show shock waves in hydrogen at an initial pressure of 5 torr with Mach numbers M = 6.5, 10.4, 11.8, 12.2, 17.1, 18.5. In the first four the luminescence was photographed as well. The bright region in the fringes is due to the pulsing of the lamp and constitutes a time mark. The sudden deflection of the interference fringes indicates a jump in the density, i.e. the shock front. At M = 6.5 the luminous front lags so far behind that it is not seen in the figure. At M = 10.4 the luminous front can be recognized and there is a clear difference in velocity. At M = 12 the discharge plasma defined by the luminous front approaches the shock front, parts of it sometimes breaking away and advancing even closer to the shock front (4th set of fringes), and for Mach numbers up to M = 20 the luminous and shock fronts remain merged, this being why the luminescence was suppressed here, otherwise the fringes could not have been recognized.

Velocity of the Shock Front

Like the luminous front, the actual shock front, which is often made visible only by an interferogram, has a velocity distribution of the form:

$$x \propto t^{\alpha}$$

This can be seen particularly well in figure 2-1-5. The photographs are seen to have been arranged alongside one another to form a continuous shock front, but in fact each section was placed in the position previously calculated for it with respect to time. Extremely good reproducibility of the shock waves was needed. For a travel time of about 70 μsec Brinkschulte was able to obtain differences of only ± 1/2 μsec in the arrival time at the end of the tube.

From these and similar photographs it was possible to determine the exponent of the similarity solutions that describe these non-stationary shock waves and as already mentioned provide information on the velocity distribution. Brinkschulte found that one particular solution, namely the 'standard' solution among the 'homology' solutions that were obtained by von Weizsacker and co-workers, best described the shock waves:

$$x = \text{const} \cdot t^{1-K_o}$$

or:

$$v = \text{const} \cdot t^{-K_o}$$

with:

$$K_o = K_o(\gamma) \text{ and } 0.37 < K_o < 0.5 \text{ for } 3 > \gamma > 1$$

The value of the exponent $K_o = 0.45$, which was measured by Brinkschulte for hydrogen agrees very well with the theoretical value.

Density Profile Behind the Shock Front

The similarity solutions also provide information on the behaviour of the variables of state behind the shock front. They show, for instance, that pressure and density should drop again with increasing distance from the shock front, and that the temperature should rise. The solutions for $\gamma = 1.4$ are shown in figure 2-1-6.

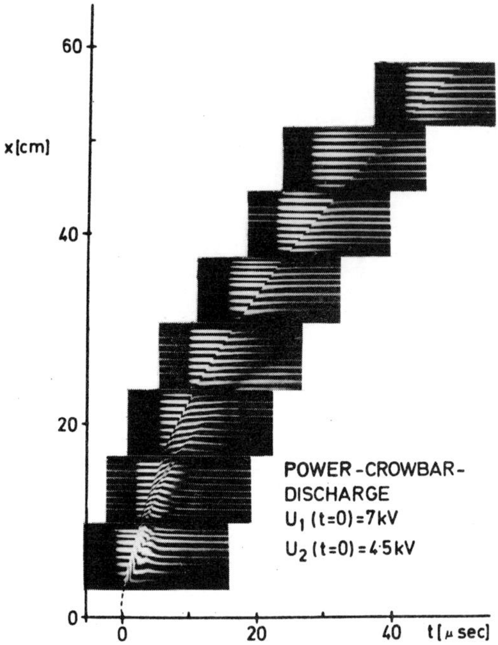

Fig. 2-1-5 Interferogram in 8 sections showing the complete progress of the shock wave (hydrogen, 5 torr, Power Crowbar Discharge U_1 (t=0) = 7 kV, U_2 (t=0) = 4, 5 kV

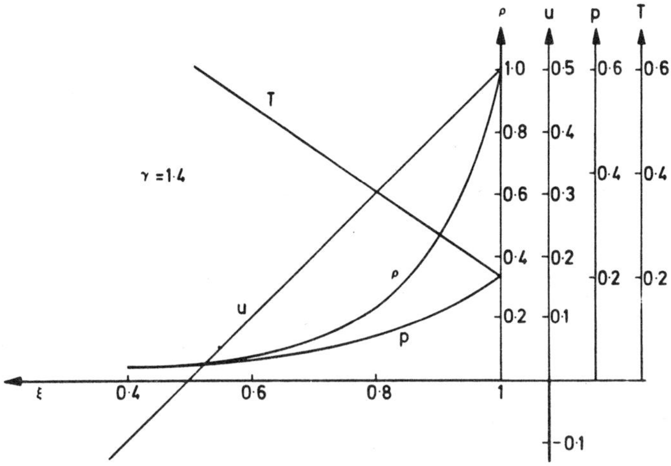

Fig. 2-1-6 Homology solutions for γ = 1.4: Behaviour of various parameters behind the shock front

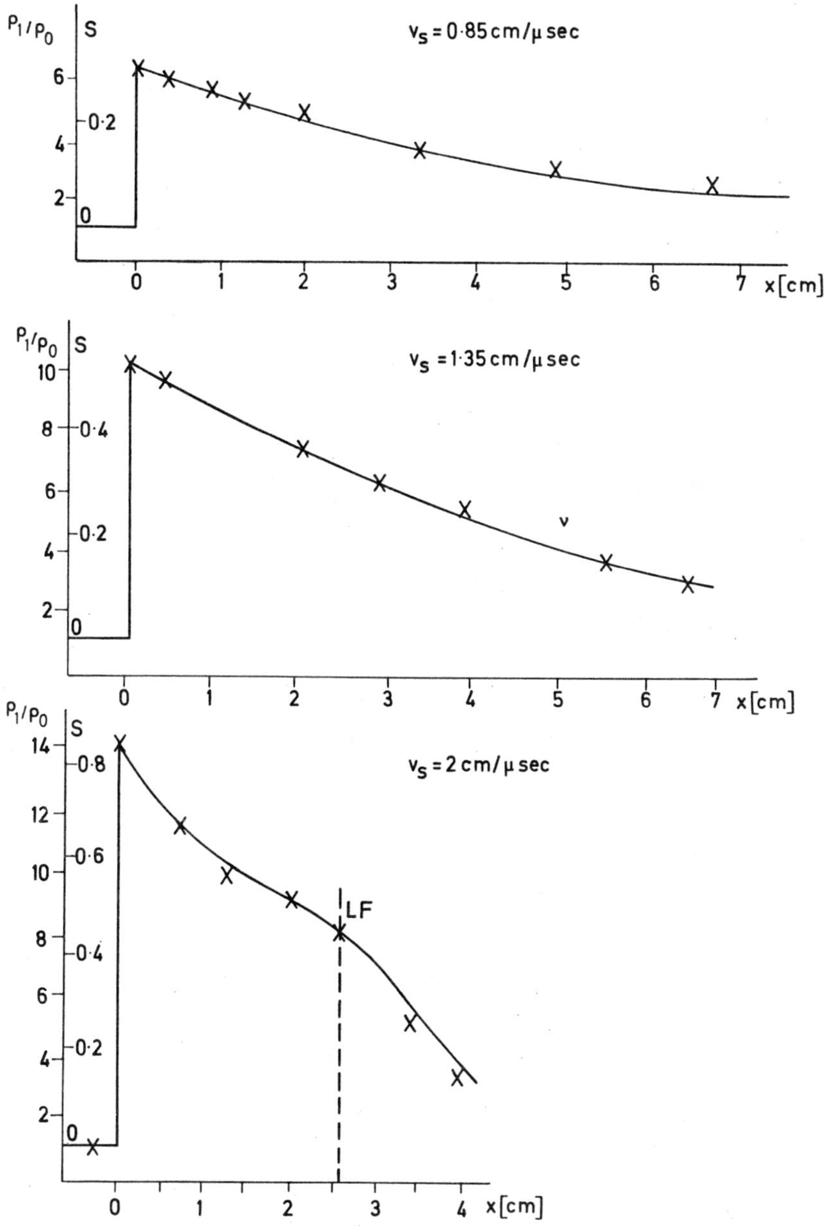

Fig. 2-1-7 Evaluation of the fringe shift in 3 cases

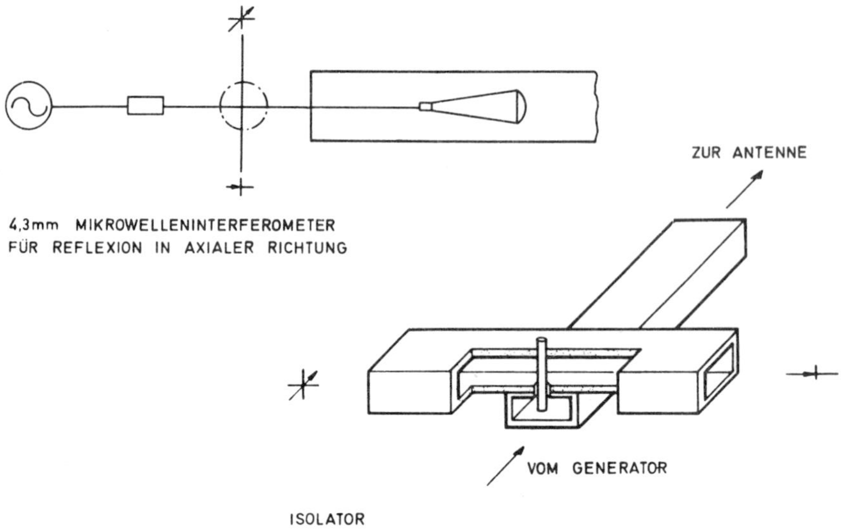

Fig. 2-1-8 Reflectometer with variable coupler for Doppler measurements of the velocity

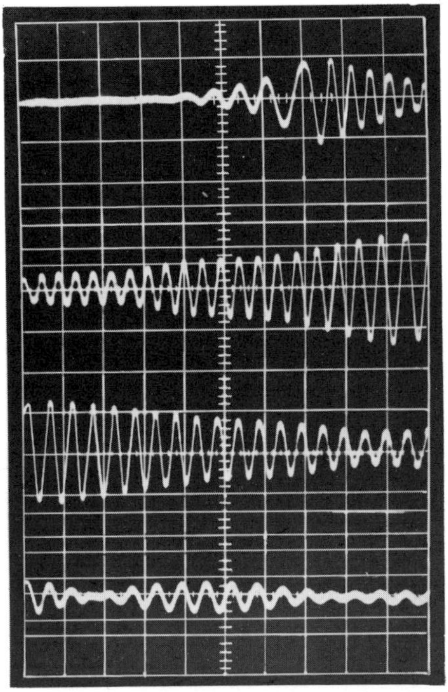

Fig. 2-1-9 Doppler signals of the velocity distribution of the reflecting fronts

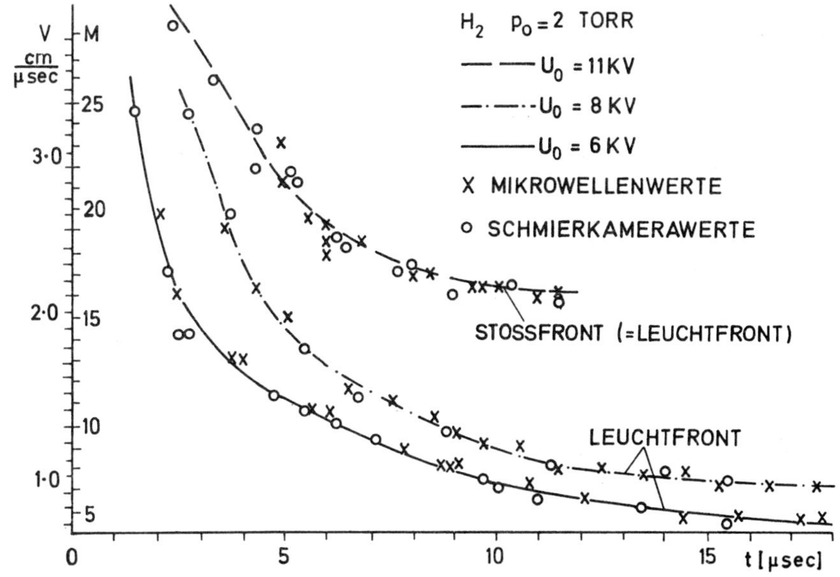

Fig. 2-1-10 Doppler velocities of luminous fronts and fast shock fronts

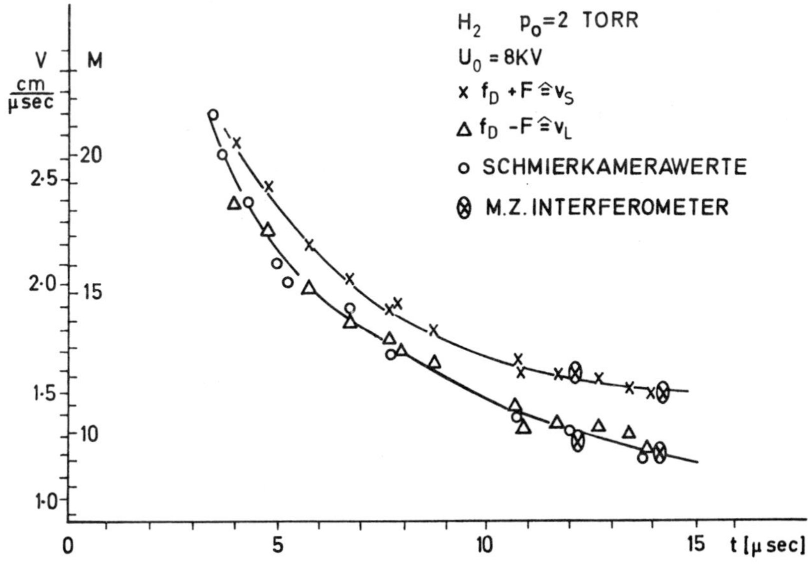

Fig. 2-1-11 Doppler velocities of the luminous front and shock front in the "intermediate" region

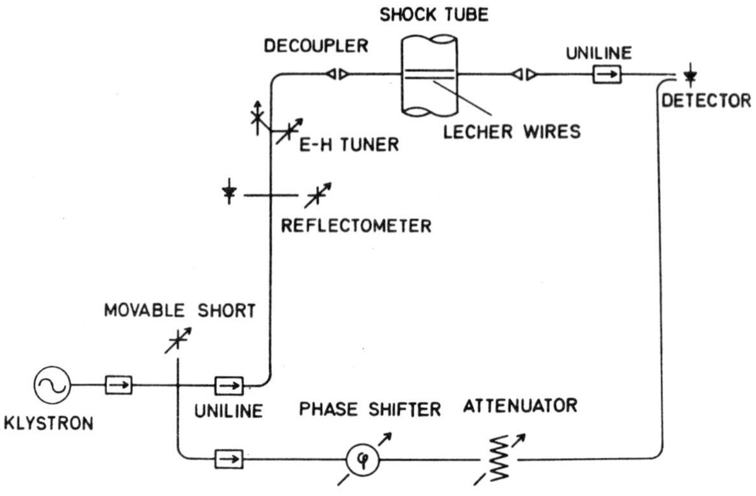

Fig. 2-1-12 Transmission and reflection interferometer with high spatial resolution (with Lecher wires) for n_e and ν_e measurements

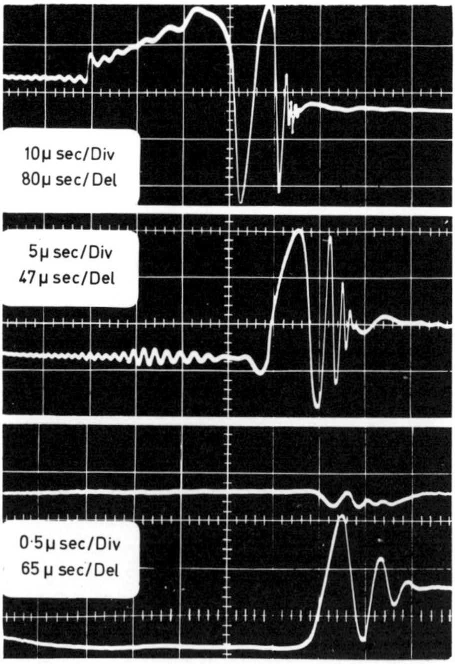

Fig. 2-1-13 Transmission and reflection signals of shock waves in argon

ξ is essentially a spatial coordinate. Discussion is limited here to the drop in density. This drop can be clearly seen in the interferograms of figure 2-1-4 where the return of the fringes to their original position, which starts immediately behind the shock front, shows the drop in density in exactly the form corresponding to the theory. Quantitative evaluations, a few examples of which are given in figure 2-1-7, provide even more details. The density jumps to values conforming exactly to the jump conditions, taking into account the law of mass action for the dissociation, i.e. the dissociation equilibrium sets in immediately without relaxation. Since the degree of dissociation rises with increasing velocity, with the result that γ should become smaller and K_o larger, the theory shows that the density should drop and this also is clearly seen.

The interferometric investigations thus confirm so many details of von Weizäcker's homology solutions that we believe that these solutions correctly describe other variables as well.

Velocity Measurements with Microwaves

Like the theory of stationary shock waves, the homology theory does not take relaxation phenomena into account, and so it does not have anything to say, for example, about the electron density distribution behind the front. It was therefore necessary to study the behaviour of the electrons separately. Since in the frequency range of visible light electron densities below 10^{16} cm^{-3} did not cause any fringe shifts that could be measured in practice, we were obliged to measure at other frequencies. For his investigation Makios (9) used various versions of the 4 mm interferometer.

The first question to answer was whether in the shock front the jump in the mass density was accompanied by a jump in the electron density. Reflection measurements with an interferometer of the type sketched in figure 2-1-8 showed that at low velocities no such jump existed (Doppler signals observed by Makios (Fig. 2-1-9) gave velocities corresponding to those of the luminous front, i.e. of the front of the plasma cloud produced in the discharge gap, and not to the velocity of the non-luminous shock front travelling ahead of it.)

At high velocities, on the other hand, the velocities measured with microwaves corresponded to those obtained interferometrically by Brinkschulte for the shock front itself. This means that there was an electron density jump as well. Such velocity distributions are shown in figure 2-1-10.

In an intermediate region, beats in the Doppler signals allowed the velocities of both the shock front and luminous front to be recognized, i.e. the electron density jump in the shock front was sufficient to cause reflection, but part of the microwave radiation crossed this threshold and was not reflected till it reached the luminous front. The two velocities obtained with beats from a Doppler signal are shown in figure 2-1-11.

Distribution of n_e and ν_e behind the shock front

Once the early microwave measurements had shown that at higher velocities free electrons were present immediately behind the shock front, transmission measurements were made to find the electron density distribution behind the shock front. The interferometer used for this purpose is shown in figure 2-1-12.

In the case of argon, which exhibits pronounced relaxation behaviour, signals of the form contained in figure 2-1-13 were evaluated and compared with calculations of the transmission coefficient as a function of the electron density for various collision frequencies ν_e (Fig. 2-1-14) in order to determine the electron density distri-

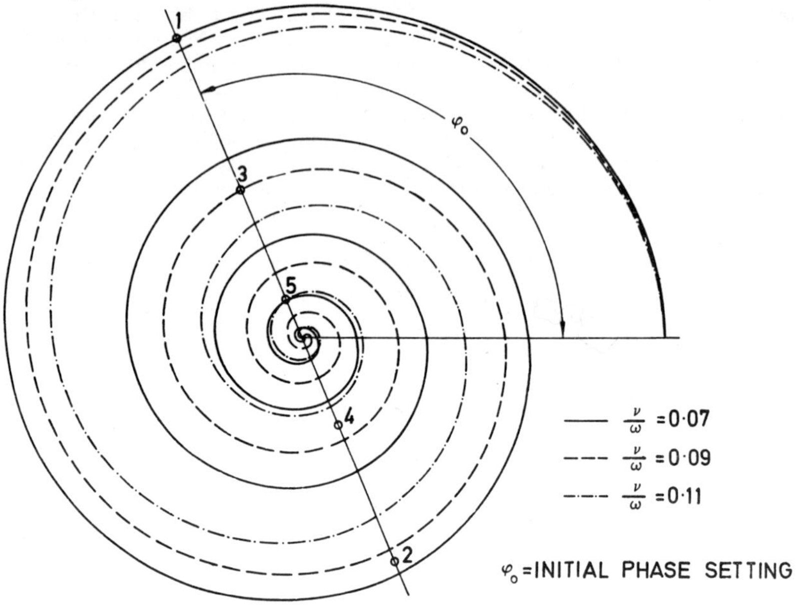

Fig. 2-1-14 Transmission coefficient as a function of the electron density for various collision frequencies ν_e

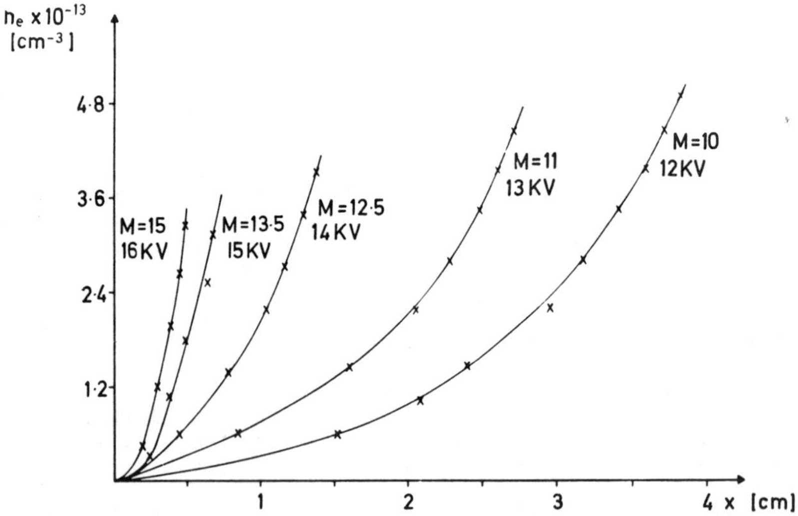

Fig. 2-1-15 Electron density distribution behind shock fronts of various velocities in argon

bution behind the shock front (Fig. 2-1-15). As already stated, the distribution displayed the familiar relaxation behaviour.

As the bottom signal in figure 2-1-13 clearly shows, electrons left over from the precursor ionization are also compressed in the shock front. This explains why the microwaves are reflected at the shock front in spite of the relaxation effects. It also becomes clear, however, that these electrons are not released in the shock front. The phase difference that can be seen at the jump in the signal shows a jump in the electron density that again agrees exactly with the density jump calculated from the jump conditions of the shock front.

In this respect, too, the picture of the shock waves that is obtained by microwave interferometry thus agrees with that obtained by Mach-Zehnder interferometry.

Acknowledgement

This work has been undertaken as part of the joint research program of Institut für Plasmaphysik and EURATOM.

References

1. **Vieille, P.,** C.R.129, 1228. 1899.

2. **Fowler, R.G., Goldstein, J.S., Clotfelder, B.E.,** Phys. Rev. 82, 879. 1951.

3. **Kolb, A.C.,** Phys. Rev. 107, 345. 1957.

4. **Josephson, V.,** J. Appl. Phys. 29, 30. 1958.

5. **Harris, E.G,** NRL-Rep. 4858. 1956.

6. **Sedov, L.I.,** 'Similarity and Dimensional Methods in Mechanics'. Infosearch London. 1961.

7. **Niblett, B., Blackman, V.H.,** J. Fluid Mech. 4, 191. 1958.

8. **Brinkschulte, H.,** Proc. 7th Int. Conf. on Ioniz. Phen. in Gases, Belgrade 1965.

9. **Makios, W.,** Z. Naturforschg. 21a, 2040. 1966.

2-2

Visualisation à faible masse volumique

CLAUDE VERET
Office National d'Etudes et de Recherches Aerospatiales Chatillon sous Bagneux, France

Summary

The schlieren method is frequently used for the visualization of supersonic flow, but its sensitivity is much reduced in hypersonic flow experiments because of the low gas density that normally has to be used for these. In such cases, the phase contrast method offers certain advantages.

Both methods are described and their sensitivities are compared. Examples of wind-tunnel hypersonic flow visualizations are given.

The conditions for the application of these methods to the visualization of flows that are themselves luminous (plasmas and flames) are indicated.

Sommaire

La strioscopie est une méthode de visualisation des écoulements aérodynamiques très utilisée en supersonique. En hypersonique et dans tous les cas où la masse volumique du milieu est faible, la sensibilité est très réduite. Il est alors avantageux d'utiliser la méthode du contraste de phase.

Ces deux méthodes sont décrites et leurs sensibilités sont comparées. Des exemples d'application à la visualisation d'écoulements hypersoniques en soufflerie sont présentés. On indique les conditions dans lesquelles ces méthodes sont applicables à la visualisation d'écoulements lumineux par eux-mêmes : plasmas ou flammes.

Introduction

La strioscopie est une méthode de visualisation des écoulements aérodynamiques utilisée dans le domaine supersonique. Compte tenu de la limitation pratique des pressions génératrices, plus les vitesses d'écoulement sont élevées, plus l'air dans lequel on les obtient est raréfié. La sensibilité de la strioscopie, qui est proportionnelle à la masse volumique du milieu, décroît donc progressivement lorsque la vitesse de l'écoulement augmente.

En hypersonique, la sensibilité devient très faible, mais il est possible de l'améliorer en faisant appel à une méthode plus sensible, le contraste de phase.

La détermination de la limite de sensibilité de la strioscopie et la comparaison avec le contraste de phase font l'objet d'études à la Division Optique de l'O.N.E.R.A. depuis plusieurs années (1).

Cet exposé présentera les méthodes utilisées pour visualiser un écoulement hypersonique et montrera les résultats obtenus. On indiquera les possibilités d'applica-

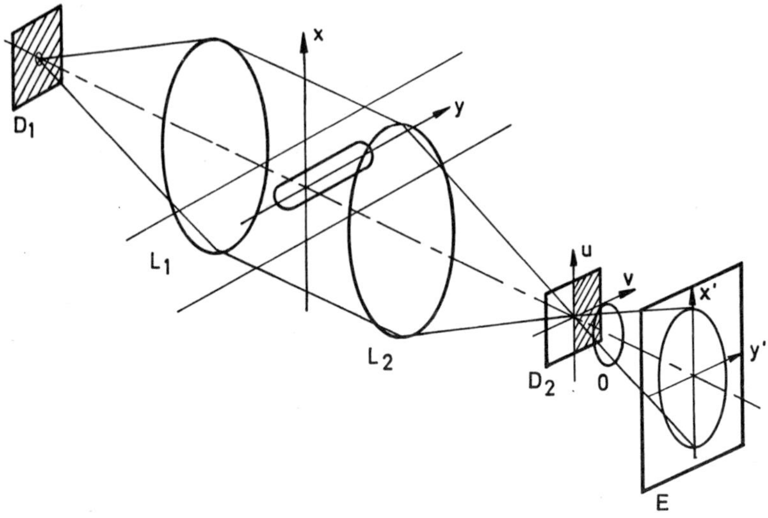

Fig. 2-2-1 Schéma optique d'une strioscopie

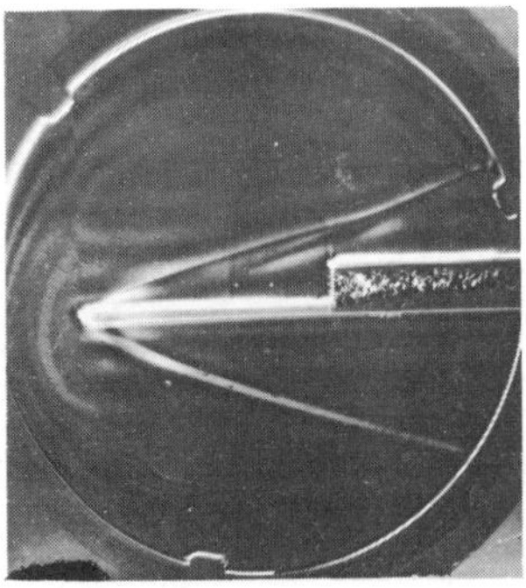

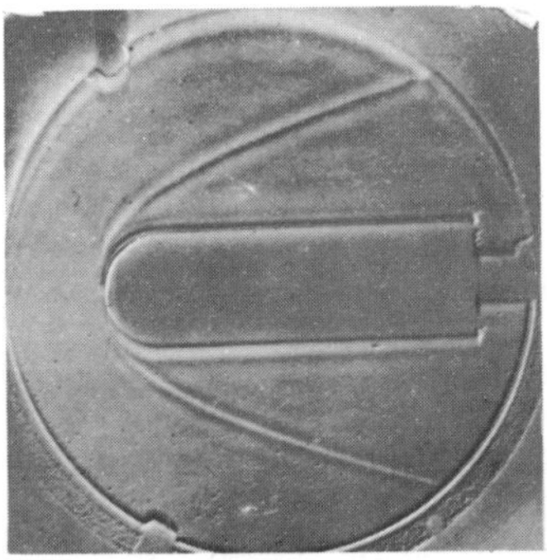

$M = 8{,}2 - \rho_0 = 8{,}7 . 10^{-7} g . cm^{-3}$

Fig. 2-2-2 Ecoulement bidimensionnel

Fig. 2-2-3 Ecoulement de révolution

tion de ces méthodes à la visualisation d'écoulements dans des milieux lumineux par eux-mêmes : flammes ou plasmas.

Sensibilite limite des méthodes optiques

Strioscopie

Considérons le schéma optique d'un strioscope de soufflerie (Fig. 2-2-1). Un diaphragme d'entrée D_1 trou, source, fente ou couteau éclairé par une source lumineuse intense est placé dans le plan focal d'une lentille L_1. La lumière issue de L_1 traverse la chambre d'expérience aérodynamique perpendiculairement à la direction de l'écoulement et tombe sur une seconde lentille L_2 L'image du diaphragme D_1 se forme dans le plan focal de L_2 où est placé un diaphragme de sortie D_2 masque ou couteau. Un objectif O conjugue ensuite la chambre d'expérience et le plan d'un écran d'observation E.

L'écoulement aérodynamique autour d'une maquette entraîne une certaine répartition de masse volumique à l'intérieur de la chambre d'expérience. Comme, d'après la loi de Gladstone, l'indice de réfraction est une fonction de la masse volumique, l'onde lumineuse plane incidente sur l'écoulement est accélérée ou retardée localement selon la valeur de l'indice en chaque point. L'onde lumineuse émergente est ainsi déformée, cette déformation pouvant être caractérisée par les différences de chemins optiques Δ sur chaque rayon lumineux à la traversée du milieu :

$$\Delta = \Sigma\, n dl \qquad \text{(Eq. 2-2-1)}$$

dl : longueur élémentaire prise sur un rayon lumineux au voisinage d'un point de ce rayon où l'indice est n.

Cette différence de marche Δ peut varier d'un rayon à l'autre de sorte qu'elle est fonction des coordonnées (x, y) des points d'un plan perpendiculaire à l'axe optique de l'appareil.

A une difference de marche Δ correspond un déphasage φ de l'onde lumineuse transmise donnée par :

$$\varphi\,(x,\ y) = \frac{2\pi}{\lambda}\ \Delta\,(x,\ y) \qquad \text{(Eq. 2-2-2)}$$

λ : longueur d'onde de la lumière.

Cela signifie que l'onde lumineuse qui, en l'absence d'écoulement serait représentée par la fonction :

$$s = a\ e^{i2\pi\nu t} \qquad \text{(Eq. 2-2-3)}$$

devient, en présence de l'écoulement :

$$s\,(x,\ y) = a\,e^{i[2\pi\nu t+\varphi\,(x,\,y)]} = ae^{i\varphi\,(x,\,y)}\,e^{i2\pi\nu t} \qquad \text{(Eq. 2-2-4)}$$

Cela signifie que l'écoulement aérodynamique constitue un objet optique dit 'de phase' car il n'agit que sur la phase de l'onde sans modifier la partie réelle a de l'amplitude complexe $ae^{i\varphi}$ comme ce serait le cas si le milieu traversé était absorbant.

Au voisinage de la limite de sensibilité, les déphasages sont petits de sorte qu'on peut simplifier l'expression de l'amplitude complexe b :

$$b(x, y) = ae^{i\varphi(x, y)} = a(\cos\varphi + i\sin\varphi)$$

$$\simeq a[1 + i\varphi(x, y)] \qquad \text{(Eq. 2-2-5)}$$

C'est la somme de deux termes dont l'un représente l'amplitude de l'onde non déformée et l'autre, en quadrature avec le premier, est proportionnel au déphasage.

Cette onde lumineuse se propageant dans le système optique atteint le plan focal de la lentille L_2. Dans ce plan, l'expression de l'amplitude complexe est la transformée de Fourier (TF) de celle obtenue dans le plan x, y (2).

D'après equation 2-2-5, on a :

$$TF\ b(x, y) = a[\delta(u, v) + i\phi(u, v)] \qquad \text{(Eq. 2-2-6)}$$

u et v coordonnées des points du plan focal de L_2.

$\delta(u, v)$: fonction de Dirac, égale à 1 pour u = 0, v = 0 et à 0 en tout autre point, qui représente l'image du diaphragme d'entrée D_1 supposé trou source ponctuel.

$\phi(u, v)$: TF de la fonction objet de phase $\varphi(x, y)$ qui correspond à son spectre de fréquences spatiales.

Le diaphragme de sortie D_2, placé dans ce plan (u, v) a pour effet d'atténuer l' amplitude de l'image de la source sans modifier le répartition $\phi(u, v)$.

L'amplitude complexe de l'onde après traversée du diaphragme D_2 devient ainsi :

$$B(u, v) = a[\theta\delta(u,v) + i\phi(u, v)] \qquad \text{(Eq. 2-2-7)}$$

θ : facteur de transmission en amplitude du diaphragme D_2 qui est la racine carrée du facteur de transmission en intensité.

L'objectif 0 conjuguant le plan objet (x, y) et le plan d'écran (x', y'), l'amplitude dans ce dernier plan est, de nouveau, la transformée de Fourier de celle du plan (u, v) :

$$b'(x', y') = TF[B(u,v)]$$

$$= a[\theta + \varphi(x', y')] \qquad \text{(Eq. 2-2-8)}$$

La répartition des intensités dans le plan image est ainsi :

$$I(x',y',) = b'.b'^{*} = a^2[\theta^2 + \varphi^2(x', y')] \qquad \text{(Eq. 2-2-9)}$$

b'^{*} étant l'imaginaire conjuguée de b'.

Le sensibilité de la méthode est caractérisée par le contraste de l'image dans le plan d'écran E :

$$C_1 = \frac{I - I_F}{I_F} \qquad \text{(Eq. 2-2-10)}$$

I_F : intensité du fond obtenue pour $\varphi = 0$

d'ou : $C_1 = \frac{\varphi^2}{\theta^2}$ (Eq. 2-2-11)

En pratique, l'intensité du fond contient une certaine quantité de lumière parasite due aux diffusions sur les surfaces optiques ou à la diffraction au bord des objets matériels placés dans le champ. L'intensité parasite est proportionnelle à l' intensité d'éclairage A^2, de sorte que si P est un facteur de lumière parasite, le contraste image devient:

$$C_1 = \frac{\varphi^2}{\theta^2 + P} \qquad \text{(Eq. 2-2-12)}$$

La limite pratique s'obtient pour un contraste de l'ordre de 0,1 et un facteur de lumière parasite de l'ordre de 0,05. Le déphasage minimal est ainsi de l'ordre de 7.10^{-2} ce qui correspond d'après l'equation 2-2-2 à une différence de marche voisine de $\lambda/100$.

Contraste de phase

Le contraste de phase s'obtient avec le même appareillage que celui décrit pour la strioscopie. Seul, le diaphragme de sortie D_2 est changé. Il est constitué par un support transparent sur lequel est déposée une couche mince, en forme de masque ou de couteau, déphasante de $\pi/2$ et absorbante.

Si est facteur de transmission en amplitude du dépôt, l'expression de l'amplitude complexe après traversée du diaphragme devient, au lieu de l'equation 2-2-7:

$$B_2 (u,v) = a[\tau e^{i\pi/2} \delta(u,v) + i\phi (u,v)]$$

Comme: $e^{i\pi/2} = i$ (Eq. 2-2-13)

on a : $B_2 (u,v) = ai[\tau \delta(u,v) + \phi (u,v)]$

Dont la TF représentant l'amplitude dans le plan d'écran est:

$$b'_2 (x',y') = ai[\tau + \varphi (x', y')] \qquad \text{(Eq. 2-2-14)}$$

L'intensité dans l'image est ainsi:

$$I_2 (x',y') = b'_2 . b'^*_2 = a^2 [\tau + \varphi (x',y')]^2 \qquad \text{(Eq. 2-2-15)}$$

Compte tenu d'un facteur de lumière parasite P comme précédemment, le contraste par rapport au fond est donné par:

$$C_2 = \frac{(\tau + \varphi)^2 - {}^2}{\tau^2 + P} = \frac{\varphi^2 + 2\tau\varphi}{\tau^2 + P} \qquad \text{(Eq. 2-2-16)}$$

Dans les conditions limites pour la strioscopie, soit:

$\varphi = 7.10^{-2}$

$\tau^2 + P = 5.10^{-2}$

et en prenant : $\tau = 10^{-1}$

on obtient : $C_2 = 0,38$

Le contraste de l'image est environ 4 fois plus fort en contraste de phase qu'en strioscopie.

Le déphasage minimal pour obtenir un contraste limite de 0,1 est de l'ordre de 2.10^{-2} auquel correspond une différence de marche limite de $\lambda/300$ soit 3 fois plus faible qu'en strioscopie.

Amélioration de la sensibilité

D'après les relations (Eq. 2-2-12) et (Eq. 2-2-16), on voit que le facteur de lumière parasite P joue un rôle prépondérant dans la limitation du contraste. En effet, si P était nul, le contraste serait augmenté en réduisant les facteurs de transmission θ ou τ.

Examinons donc les moyens de réduire ce facteur de lumière parasite.

Diffusion par les surfaces optiques - L'une des causes de lumière parasite est la diffusion sur chacune des surfaces optiques rencontrées par la lumière utile entre le diaphragme d'entrée et le diaphragme de sortie. Cette diffusion provient soit des défauts résiduels du poli de ces surfaces, soit de dépôts gras ou de poussières sur elles. Il est donc très important de réduire au minimum le nombre de ces surfaces, de les nettoyer au mieux et de les préserver de toute salissure.

C'est pourquoi, pour les études en hypersonique, il est nécessaire de placer l'appareillage optique à l'intérieur de la chambre d'expérience, des hublots étant placés, pour l'entrée de la lumière, avant le premier diaphragme D_1 et pour la sortie de l'image, en arrière du second diaphragme D_2. Les seules surfaces optiques sont ainsi celles des lentilles L_1 et L_2, ou des miroirs qui peuvent les remplacer, et ces surfaces étant à l'intérieur de la chambre en dépression, se salissent moins.

Diffraction par les objets matériels - Dans l'étude de l'écoulement aérodynamique autour d'une maquette, la présence de cette maquette dans le champ éclairé a pour effet de diffracter la lumière. Cette lumière diffractée se présente dans l'image sous l'aspect d'une ligne brillante le long du contour extérieur de l'objet et d'un certain nombre de franges parallèles à cette ligne. La ligne et les franges paraissent d'autant plus intense par rapport au fond que le coefficient de transmission θ ou τ du diaphragme de sortie est plus faible. La sensibilité de la visualisation est ainsi réduite dans ces régions qui sont cependant très intéressantes du point de vue aérodynamique puisqu'elles sont proches de la surface de la maquette.

L'effet nuisible de ces franges de diffraction peut être fortement atténué par l'emploi d'une méthode de superposition de clichés photographiques (3).

Un premier cliché de référence est pris avant mise en marche de la soufflerie sur lequel sont enregistrées les variations d'éclairement dues à tous les défauts résiduels de l'appareillage (poussières, taches ...) et aux franges de diffraction produites par le contour des objets.

Les réglages étant rigoureusement conservés, un second cliché est ensuite pris en présence de l'écoulement.

Après développement des deux clichés dans les mêmes conditions, une copie positive par contact est effectuée sur le cliché de référence. Cette copie est ensuite mise au contact du second cliché avec écoulement et l'ensemble est copié pour obtenir un positif. De cette manière, toutes les variations d'éclairement qui ont impressionné les deux clichés sont éliminées et il subsiste uniquement les variations produites par l'écoulement. Cette méthode permet de réduire considérablement l'effet de la lumière parasite. Appliquée en contraste de phase, elle permet de détecter des différences de phase de l'ordre de 6.10^{-4} et des différences de marche voisine de $10^{-4}\lambda$, soit environ 100 fois plus faibles qu'en strioscopie courante.

Exemple de Visualisation d'un Ecoulement Hypersonique

Conditions limites de visualisation d'une onde de choc

L'application de la méthode du contraste de phase sera illustrée par la visualisation d'une onde de choc en hypersonique.

Si ρ_o est la masse volumique de l'air en amont de l'onde de choc et ρ la masse volumique en aval, le rapport $(\rho - \rho_o)/\rho_o$ tend vers une limite constante et égale à 5 lorsque la vitesse de l'écoulement augmente indéfiniment (1). Cette limite est très approximativement atteinte à partir d'un nombre de Mach de 10.

En faisant intervenir la relation de Gladstone:

$$n - 1 = k\rho \qquad \text{(Eq. 2-2-17)}$$

n : indice de réfraction

k : constante égale à $0,227 \ g^{-1} \ cm^3$ pour l'air

La différence de marche entre l'amont et l'aval de l'onde de choc est:

$$\Delta - \Delta_o = 5 \ ek\rho_o \qquad \text{(Eq. 2-2-18)}$$

e : longueur traversée par la lumière dans l'onde de choc approximativement égale à la longueur d'un profil bidimensionnel perpendiculairement à la direction de l'écoulement.

La différence de phase est donc :

$$\varphi - \varphi_o = \frac{10\pi ek\rho_o}{\lambda} \qquad \text{(Eq. 2-2-19)}$$

En contraste de phase, une valeur approchée du contraste donné par (16), obtenue en admettant P petit devant τ^2 est :

$$C_2 = \frac{2\varphi}{\tau} = \frac{20\pi ek\rho_o}{\tau\lambda} \qquad \text{(Eq. 2-2-20)}$$

Ainsi, la valeur limite du produit $e\rho_o$, obtenue en prenant:

$C_2 \ \tau = 0,1$

$\tau = 3.10^{-2}$

$\lambda \ \tau = 5.10^{-5}$ cm

$k \ \tau = 0,227 \ g^{-1} \ cm^3$

est: $e\rho_o \ \tau = 10^{-8} \ g \ cm^{-2}$

Exemple de visualisation d'un ecoulement hypersonique

Des expériences sont en cours à la soufflerie hypersonique du CNRS à Bellevue (4). Le nombre de Mach de l'écoulement est de 8, 2 et la masse volumique en avant de l'onde de choc est de 8, 7. 10^{-7} g/cm^3.

Le cliché 2 montre l'écoulement autour d'un profil bidimensionnel de 6 cm de longueur et le cliché 3 celui autour d'une maquette cylindrique de 20 mm de diamètre.

Ces clichés ont été obtenus par la méthode de superposition.

Conclusion

Les méthodes optiques sont utilisables même pour la visualisation des écoulements hypersoniques en milieu raréfié.

Pour les très faibles masses volumiques, inférieures à 10^{-6} g/cm^3, le contraste de phase apporte un gain notable de sensibilité et sa limite de sensibilité est environ 100 fois plus basse que celle de la strioscopie.

Cependant, comme pour toute méthode très sensible, il importe d'éliminer au mieux les effets parasites. C'est pour cela que le montage de l'appareillage à l' intérieur de la chambre d'expérience est préconisé et que la méthode de superposition de deux clichés, l'un pris à l'arrêt, l'autre en marche,a été mise au point.

Ces méthodes de visualisation sont susceptibles d'être appliquées à l'étude de milieux lumineux par eux-mêmes, flammes ou plasmas, en utilisant une source lumineuse d'éclairage intense et monochromatique, laser à gaz ou à impulsion. Il est alors possible de placer devant l'appareil de prise de vues un filtre sélectif centré sur la longueur d'onde d'émission de la source qui a pour effet d'atténuer fortement la lumière émise par le milieu étudié.

Reférénces

1. **Philbert, M.**, 'Visualisation des écoulements à basse pression'. La Recherche Aérospatiale no. 99. Mars-Avril 1964. p. 39-48.

2. **Maréchal, A., Francon, M.**, 'Diffraction Structure des images'. Edition de la Revue d'Optique Théorique et Instrumentale, Paris, 1960.

3. La méthode de superposition a été proposée par **M. Philbert** et mise en application par **R. Beaupoil** et **G. Cadinot** de la Division Optique.

4. Les expériences aérodynamiques sont effectuées par **B. Monnerie** de la Direction Aérodynamique et l'appareillage optique a été mis en oeuvre par **M. Philbert, J. Surget, R. Beaupoil** et **G. Cadinot** de la Division Optique.

Commentary on Chapter 2-2

H. Oertel.

In order to stimulate discussion, I should like to have a few words in view of Mr. Veret's very interesting paper and its particular importance in the field of experimental hypersonic research.

Hypersonic flight always is high altitude flight. The higher the altitude is, the less we know about the flow. The air density is very low there, hence the mean free path of the air particles is long. Consequently, the boundary layers are very thick and the relaxation times of the molecular oscillations as well as of the chemical and electronic reactions are so long that we are no longer in a position to rely upon thermal equilibria. Thus, we focus our attention on studies at gas densities which are as low as possible. These studies can only be done in the wind tunnel at very low densities. This is due to a number of reasons; the most important of them is that we are unable to increase the reservoir pressure arbitrarily. It is seen from figure A-2-2-1 that the reservoir pressure p_o must be much higher than the stagnation pressure p_s. For example, at a $p_o = 1000$ atm, a flow of $U = 5000\ ms^{-1}$ can only be generated with an air density equal to that existing at an altitude of $H > 67$ km. If we want to generate this flow with an air density like it exists at an altitude of $H = 37$ km, we have to compress the air to $p_o = 10\,000$ atm.

Experiments conducted in the field of hypersonic flows, however, will become problematic if there is no possibility of visualizing at least the bow wave. Unfortunately, within short blowing times and with model dimensions of some centimeters, one is able to do so only in the case of air densities existing at altitudes below 50 km. We, therefore, are very interested in all studies on visualisation of low densities.

There are quite a series of theoretical and experimental investigations of the limits of visualisation with respect to shadow, schlieren and interference methods. Figure A-2-2-2 lists some results (1) (2) (3) that have been obtained recently. All studies conducted until now suggest nearly the same conclusion: these techniques leave us in the lurch when we have to deal with air densities at altitudes of more than 50 km. Directly comparable information on phase-contrast methods were not available even though the formulas to do the calculations can be found in many books. In this connection, I should like to mention a hand-book article of Wolter (4).

Now Mr. Veret has made a calculation in order to demonstrate that in the case of an optimum set-up, the phase-contrast method would allow operation at altitudes up to 67 km. When superposing two phase-contrast images, it would even be possible to reach an altitude of about 84 km.

This would be sensational if one really happened to do so. At an altitude of 84 km, the mean free path $\bar{s}$ is about 9 mm. The thickness of the shock front has the same order of magnitude. It would then even be possible to take photographs of the density variation in a shock front. Nevertheless, I should like to discuss a few doubts:

1. The shock front is many light wave lengths thick. The question arises whether a calculation taking into account only the light phase ahead of and behind the shock front is corresponding to reality.

2. The light source was supposed to be very narrow. In reality it must always have a certain width. Isn't it a matter of fact that as this width increases, the contrast of the image sharply decreases?

3. We need a good reproduction not only of the shock, i.e. of a phase object, but also of the edge of the model, i.e. of an amplitude object. We need it, for instance, if we wish to measure the detachment distance of the bow wave. However, the phase-contrast methods suffer from the disadvantage of reproducing the amplitude objects with a large halo. The proposed superposition of two pictures compensates the influence of the halo on the reproduction of the shock. But where is the edge of the model? For the sake of comparison with the phase contrast images shown even now, I should like to project a differential interferogram (Fig. A-2-2-3). It shows a hypersonic flow of the same Mach number and of the same density around a 30 mm diameter cylinder of 10 cm in length. The edge of the body is reproduced in such a good way that it is possible to observe even the stagnation boundary layer. Is it to be hoped that a reproduction of this quality can be achieved by means of the phase-contrast method?

References

1. **Philbert, M., Dubois, G.,** 'La visualisation des ecoulements aerodynamiques a faible masse specifique'. La rech. aeronautique 81. 1961.

2. **North, R.J., Stuart, C.M.,** 'Flow visualisation and high-speed photography in hypersonic aerodynamics'. NPL-Aero Rep 1029, ARC 23, 954 Hyp. 263. 1962.

3. **Oertel, H.,** 'Ein Differentialinterferometer fur Messungen im Hyperschallstoßrohr'. ISL-Techn. Mitteilung T 17, 1961.

4. **Wolter, H.,** 'Schlieren, Phasenkontrast- und Lichtschnittverfahren'. Handbuch der Physik, Bd. XXIV. 1956.

5. **Oertel, H.,** 'Storsrohre, Shock tubes, Tubes à choc', Springer Verlag, Wien, New York, 1966.

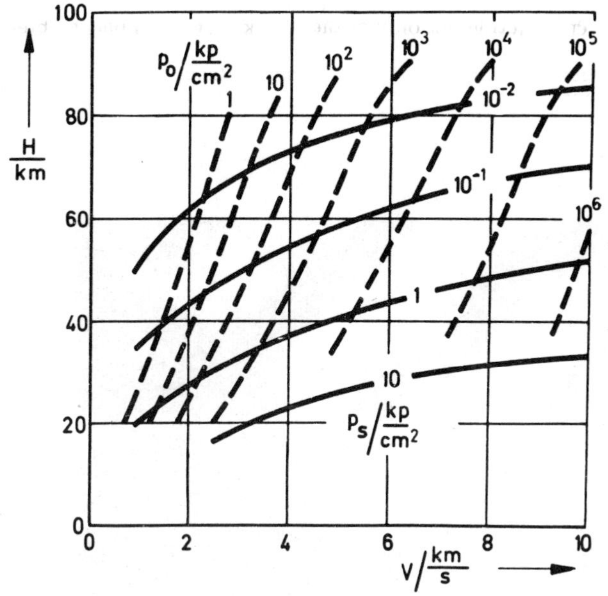

Fig. A2-2-1 Reservoir pressures p_o which are needed to produce hypersonic flows with stagnation pressures p_s, velocities V and densities existing at altitudes H.

Method	$\Delta\varphi/2\pi$	ρ/ρ_o	h km	$\bar{s}/\lambda$	Author	Basis
Shadow	$3\cdot10^{-1}$	10^{-2}	33	10	North	E 62
Schlieren	$2\cdot10^{-2}$	$8\cdot10^{-4}$	52	200	North	E 62
	10^{-2}	$4\cdot10^{-4}$	58	400	Véret	T 67
Interference	$6\cdot10^{-2}$	$2\cdot10^{-3}$	44	80	North	E 62
Diff. Interference	10^{-1}	$4\cdot10^{-3}$	39	40	Oertel	E+T 61
					North	E 62
Diff. Interference	$6\cdot10^{-2}$	$2\cdot10^{-3}$	44	80	North	E 62
Colour	$2\cdot10^{-2}$	$8\cdot10^{-4}$	52	200	Philbert	E+T 61
Phasecontrast	$3\cdot10^{-3}$	10^{-4}	67	1000	Véret	T 67
Phasec. + Comp.	$2\cdot10^{-4}$	$8\cdot10^{-6}$	84	20 000	Véret	T 67

$\Delta\varphi$ = observable light phase difference $\bar{s}$ = particle mean free path

ρ = density, ρ_o = sea level density λ = light wave length

h = corresponding altitude

Fig. A2-2-2 Results of some theoretical (T) and experimental (E) investigations on the limits of visualisation of strong shock waves observed with an optical path of 1 cm.

Fig. A2-2-3 Differential interferogram of Mach number 8 flow about a circular cylinder. Diameter 30 mm. Length 100 mm. Density 6.10^{-4} sea level density. Stagnation temperature 3000°K (5)

Réponse de M. Véret

1. Le saut de masse volumique de part et d'autre d'une onde de choc a été choisi comme élément de comparaison car c'est lui qui caractérise le mieux la sensibilité d'une méthode de visualisation d'un écoulement aérodynamique. Il est certain que, si à faible masse volumique, l'onde de choc est épaissie, les conditions de propagation de la lumière ne sont pas les mêmes que pour une onde de choc très mince. Cependant, le saut de masse volumique de part et d'autre du choc reste un élément de comparaison valable entre différentes méthodes de visualisation.

2. La source lumineuse utilisée en contraste de phase est effectivement très étroite; c'est une fente fine dont la largeur est de l'ordre de grandeur de la frange centrale de diffraction dans son image sur la lame de phase. Il est vrai que, si cette largeur augmente, le contraste dans l'image visualisée diminue.

3. Le but de la méthode de superposition des clichés est d'éliminer l'effet des phénomenès de diffraction au bord des objets d'amplitude (maquette). La qualité de la restitution du bord de ces objets dépend à la fois des conditions photographiques pour l'obtention du contretype du cliché sans écoulement et de la précision de superposition de ce contretype et du cliché avec écoulement. Il est théoriquement possible d'obtenir une aussi bonne définition du bord des objets que par les autres méthodes. L'obtention du résultat pratique optimal est le but d'une étude en cours.

2-3

Diagnostics Laser, situation actuelle des méthodes utilisant la diffusion Thomson de la lumière

F. ROSTAS
Centre de Recherches de la C.G.E. Marcoussis, Essonne, France

Summary

The availability of lasers has brought about a rapid evolution in optical methods of plasma diagnostics. We have endeavoured here to summarize the results already obtained by the different workers making use of Thomson diffusion. This method has reached a degree of maturity which allows its use in many practical applications. It is particularly interesting because of the wealth of information obtainable : depending on conditions, one can deduce the electron density and temperature and the ion temperature. Time and spatial resolutions are both excellent (a few nanoseconds and a few mm^3); measured densities are of the order of 10^{15} to 10^{17} cm^{-3}. The light sources used are in general Q-spoiled solid state lasers with powers ranging from 10 to 100 MW and pulse durations of 10 to 30 nsec. All of the theoretically predicted characteristics of the diffused spectrum have been observed, and this justifies the confidence granted to this diagnostic method.

Sommaire

L'apparition des lasers a provoqué une évolution rapide des méthodes optiques de diagnostic des plasmas. On s'efforcera de faire le point sur les résultats obtenus à ce jour par les différents expérimentateurs utilisant la diffusion Thomson. Cette méthode arrive actuellement à un degré de maturité qui rend possible son emploi dans de nombreux cas pratiques. Elle présente un intérêt tout particulier du fait de la richesse des informations qu'elle fournit : suivant les conditions on peut obtenir la densité et la température électroniques ainsi que la température ionique. Les résolutions temporelle et spatiale sont excellentes (quelques nanosecondes et quelques mm^3) et les densités mesurées sont de l'ordre de 10^{15} à 10^{17} cm^{-3}. Les sources employées sont en général des lasers solides déclenchés dont la puissance peut aller de 10 à 100 MW, la durée d'impulsion étant en général de 10 à 30 nsec. Toutes les caractéristiques de la lumière diffusée prévues théoriquement ont maintenant pu être mises en évidence, ce qui justifie la confiance que l'on peut avoir dans ce genre de mesures.

Introduction

Depuis l'apparition des sources intenses de lumière cohérente que constituent les lasers, les diagnostics optiques des plasmas ont suscité un grand intérêt et fait de grands progrès. On peut distinguer trois grandes méthodes mettant en oeuvre des lasers:

L'interférométrie optique, qui permet d'atteindre la densité électronique par la mesure de l'indice de réfraction.

La mesure de la dispersion rotatoire des électrons dans un champ magnétique : on mesure ici la rotation du plan de polarisation de la lumière et l'on en déduit le produit (N_e B) du champ magnétique et de la densité électronique.

La diffusion de la lumière par les électrons, enfin. La méthode repose ici sur le fait que la diffusion dite de Thomson, c'est-à-dire provoquée par les électrons mis en mouvement par le champ électrique de l'onde, est bien plus importante que la diffusion de Rayleigh provenant de la polarisation des particules neutres. La lumière diffusée dépend donc uniquement des électrons libres, et son étude permet d'atteindre leurs caractéristiques.

Nous nous efforcerons, dans la présente contribution, de définir l'état actuel des études concernant la diffusion Thomson envisagée comme diagnostic de plasma.

Cette méthode avait d'abord été proposée et essayée aux fréquences radioélectriques pour faire des mesures dans l'ionosphère. La théorie complète du phénomène avait été faite par différents auteurs (1), (2) et assez bien vérifiée par l'expérience (3), (4), (5).

Il n'était guère concevable cependant de l'étendre aux plasmas de laboratoire, vu leurs dimensions forcément restreintes, sans la transposer à des longueurs d'onde beaucoup plus faibles. L'apparition des lasers à cristaux qui constituent une source de lumière monochromatique extraordinairement intense permettait d'envisager de telles expériences et plusieurs groupes se mirent au travail dès que la suggestion fut avancée par T.P. Hughes en 1962 (6).

De très grands progrés ont été réalisés depuis ces débuts difficiles et toutes les caractéristiques du spectre de diffusion prévu ont pu être observées. La diffusion Thomson constitue donc maintenant un diagnostic sûr et extrêmement riche puisqu'elle permet d'atteindre, simultanément, la densité et la température électroniques et aussi la température ionique. Des études sont en cours actuellement pour chercher à en tirer également des informations sur la distribution des vitesses électroniques.

Caractéristiques de la lumière diffusée par les electrons d'un plasma

Diffusion par un électron seul

Suivant le résultat classique concernant la diffusion dite 'de Thomson' un électron soumis à l'action d'une onde électromagnétique représentée par son champ électrique (Fig. 2-3-1).

$$\vec{E} = \vec{E}_o \exp j\,(\omega_o t - \vec{K}_o . \vec{r})$$

rayonne dans une direction θ et à une distance R, une onde :

$$\vec{E}' = \vec{E}_\theta \exp j\,(\omega_o t - \vec{K}_1 . \vec{r})$$

avec:

$$\left|\vec{E}_\theta\right| = \frac{r_o}{R} \sin\theta \left|\vec{E}_o\right|$$

$$\left|\vec{K}_1\right| = \left|\vec{K}_o\right|$$

$$r_o = \frac{e^2}{mc^2} = 2,82.10^{-13}\text{ cm}$$

r_o est le rayon classique de l'électron, m est sa masse et e sa charge.

L'intensité diffusée est donc liée à l'intensité incidente par:

$$I\ (\theta) = I_o \frac{r_o^2 \ \sin^2 \theta}{R^2}$$

et la puissance rayonnée dans un angle solide $d\Omega = \frac{dS}{R^2}$ est :

$$dP\ (\theta) = I\ (\theta)\ dS = r_o^2\ \sin^2 \theta\ I_o\ d\Omega$$

On définit la section efficace différentielle de diffusion par:

$$dP\ (\theta) = \sigma_o\ (\theta)\ I_o\ d\Omega$$

et la section efficace totale par :

$$P = \int_{4\pi} \sigma_o\ (\theta)\ I_o\ d\Omega = \sigma_o I_o$$

d'où :

$$\sigma_o\ (\theta) = r_o^2\ \sin^2 \theta = 0,795.10^{-25} \sin^2 \theta\ (cm^2)$$

$$\sigma_o = \frac{8\pi}{3} r_o^2 = 6,65.10^{-25}\ (cm^2)$$

Diffusion par un ensemble d'électrons indépendants soumis à l'agitation thermique

Chacun des électrons, du fait de sa vitesse d'agitation, voit une fréquence déplacée par effet Döppler par rapport à la fréquence de l'onde incidente; en outre, l'onde qu'il réémet est de nouveau déplacée en fréquence. Si l'on peut admettre que les électrons sont distribués au hasard sur des distances de l'ordre de la longueur d'onde, les intensités diffusées s'ajoutent.

On montre alors aisément que, si les vitesses de ces électrons sont réparties suivant une distribution maxwellienne, l'intensité diffusée dans une bande de fréquence $d\omega$ suivant une direction définie par les angles θ et Φ (Fig. 2-3-1) est donnée par:

$$dI\ (\omega, \theta, \Phi) = \frac{\sigma_o\ (\theta)}{R^2}\ I_o\ N_e \left(\frac{m}{2\pi k T_e}\right)^{1/2} \frac{1}{2K_o \sin \frac{\Phi}{2}} \times$$

$$\exp \left\{ - \frac{m}{2kT_e} \left[\frac{\omega - \omega_o}{2K_o \sin \frac{\Phi}{2}} \right]^2 \right\} d\omega$$

où T_e est la température des électrons et N_e leur densité.

L'intensité diffusée dans toute la bande de fréquence ne dépend que de θ et non plus de Φ, elle est égale à:

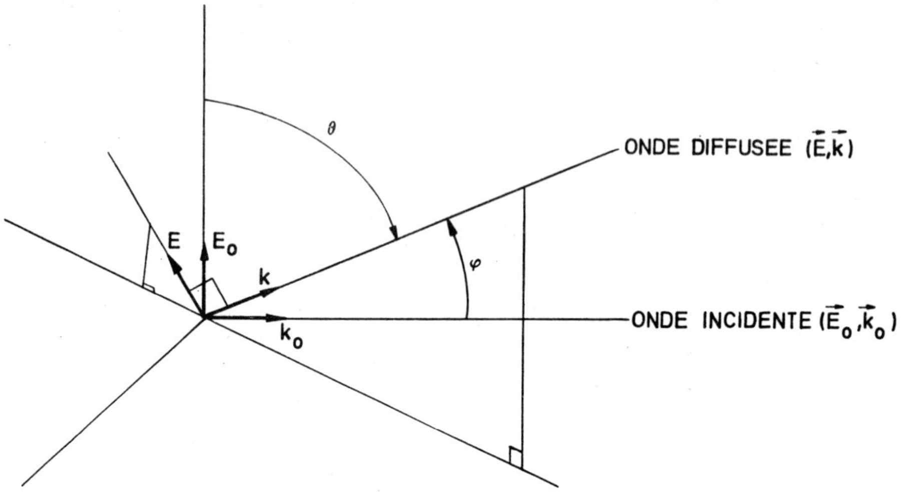

Fig. 2-3-1 Géométrie de la diffusion

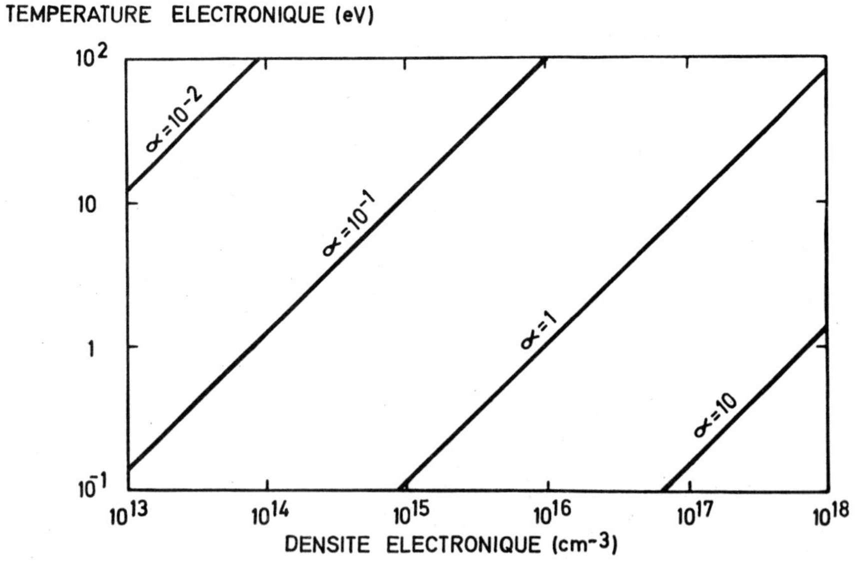

Fig. 2-3-2 Droites iso - α dans le plan N_e - T_e

$$I(\theta) = \int_{\omega} dI(\theta, \Phi, \omega) = \frac{N_e \sigma_o(\theta)}{R^2} I_o$$

La largeur de la raie diffusée dépend, elle, de Φ mais non de θ

$$\Delta\omega = 4 K_o \sin \frac{\Phi}{2} \left[\frac{2 k T_e}{m} \operatorname{Log}_e 2 \right]^{1/2}$$

(largeur totale à mi-hauteur).

Influence des effets collectifs

On ne peut pas toujours considérer que les électrons sont répartis au hasard dans le plasma. Les phénomènes d'interactions collectives provoquent, en effet, des fluctuations régulières de densité sur des distances qui sont de l'ordre de la longueur de Debye:

$$\lambda_D = \left[\frac{k T_e}{4\pi N_e e^2} \right]^{1/2}$$

Tant que la longueur d'onde de la lumière incidente est très inférieure à λ_D, tout se passe comme si les électrons étaient indépendants. Si λ devient égal ou supérieur à λ_D on doit considérer que la diffusion se fait, non pas sur les électrons individuels, mais sur les "quasi-particules" que représentent les fluctuations de densité.

Salpeter (3) fait intervenir un paramètre fondamental qui définit la nature de l' interaction envisagée:

$$\alpha = \frac{\lambda}{4\pi \sin \frac{\Phi}{2} \lambda_D} = \frac{\lambda}{\sin \frac{\Phi}{2}} \left[\frac{N_e e^2}{4\pi k T_e} \right]^{1/2}$$

et il montre que la section efficace de diffusion est donnée par:

$$\sigma(\theta) = \sigma_o(\theta) \frac{1 + Z\alpha^2}{1 + (Z+1)\alpha^2}$$

où Ze est la charge de l'ion.

On vérifie que, pour $\alpha = 0$, on retrouve la section efficace σ_o et on constate que $\sigma(\theta)$ tend vers $\frac{1}{2} \sigma_o(\theta)$, pour α très grand (si $Z = 1$).

Si la valeur globale de la puissance diffusée dépend peu du coefficient α, il n'en va pas de même de sa répartition spectrale. Celle-ci est constituée de deux parties distinctes dont l'importance relative varie suivant la nature et l'état du plasma.

La première partie, que nous qualifierons d'électronique, prédomine aux faibles valeurs de α, elle provient des fluctuations de la densité électronique sans corrélation avec les fluctuations ioniques. La largeur de son spectre de fréquence est relativement grande et elle correspond approximativement à l'élargissement Döppler dû aux vitesses thermiques des électrons.

Si l'on rapporte l'intensité diffusée à une variable réduite:

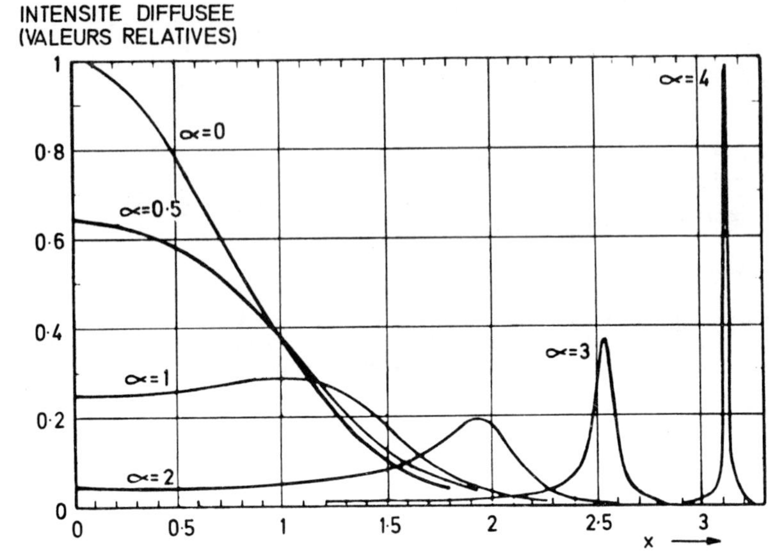

Fig. 2-3-3 Répartition spectrale de la lumiére diffusée (3).

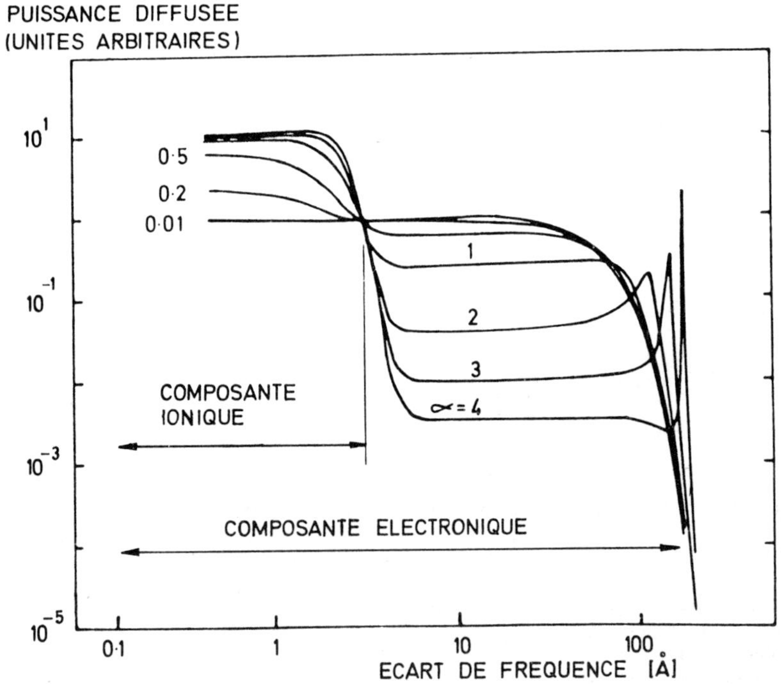

Fig. 2-3-4 Répartition spectrale de la lumière diffusée (5).
Cas d'un plasma d'hydrogène - $\theta = \gamma = \pi/2$, $T_e = T_i = 10$ eV $\lambda = 0.69\ \mu$. Les intensités relatives, correspondant aux diverses valeurs de αne sont pas respectées.

$$x = \frac{\omega - \omega_o}{\omega_e}$$

où :

$$\omega_e = 2 K_o \sin \frac{\Phi}{2} \left[\frac{2 k T_e}{m_e} \right]^{1/2}$$

son spectre a la forme représentée sur la figure 2-3-3 et sa valeur totale est proportionnelle à la section efficace électronique.

$$\sigma_o (\theta) = \sigma_o (\theta) \frac{1}{1 + \alpha^2}$$

Le restant de l'intensité diffusée est dû aux électrons dont les fluctuations de densité sont en phase avec les fluctuations des ions. Sa largeur spectrale est donc liée a l'élargissement Döppler correspondant aux vitesses thermiques des ions, et son intensité globale est proportionnelle à une section efficace que nous qualifierons d'ionique.

$$\sigma_i (\theta) = \sigma_o (\theta) \frac{Z \alpha^4}{(1 + \alpha^2) [1 + (Z + 1) \alpha^2]}$$

Le spectre de la composante ionique a la même forme que celui de la composante électronique, mais la variable réduite α doit être remplacée par :

$$y = \frac{\omega - \omega_o}{\omega_i}$$

où:

$$\omega_i = 2 K_o \sin \frac{\Phi}{2} \left[\frac{2 k T_i}{M_i} \right]^{1/2}$$

et le paramètre α par :

$$\beta = \left[Z \frac{T_e}{T_i} \frac{\alpha^2}{1 + \alpha^2} \right]^{1/2}$$

Pour en revenir à la composante électronique, on remarque que, lorsque α croît, son énergie se concentre dans les raies satellites qui sont écartées de la fréquence centrale d'une quantité:

$$\Delta \omega = \left[\omega_p^2 + \frac{3 k T_e}{m_e} 4 K_o^2 \sin^2 \frac{\Phi}{2} \right]^{1/2} \sim \omega_p$$

où :

$$\omega_p = \left[\frac{4 \pi N_e e^2}{m} \right]^{1/2}$$

est la pulsation de plasma.

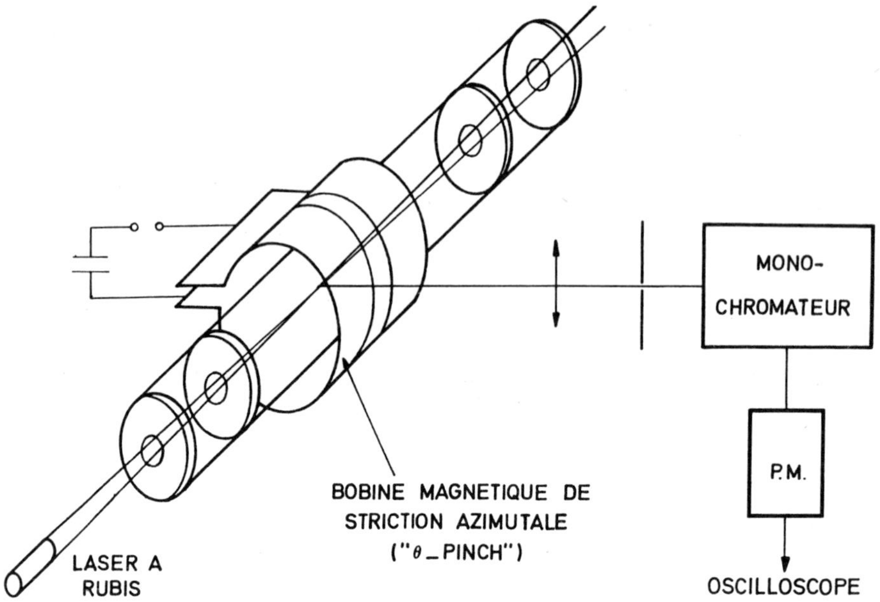

Fig. 2-3-5 Diffusion à angle droit

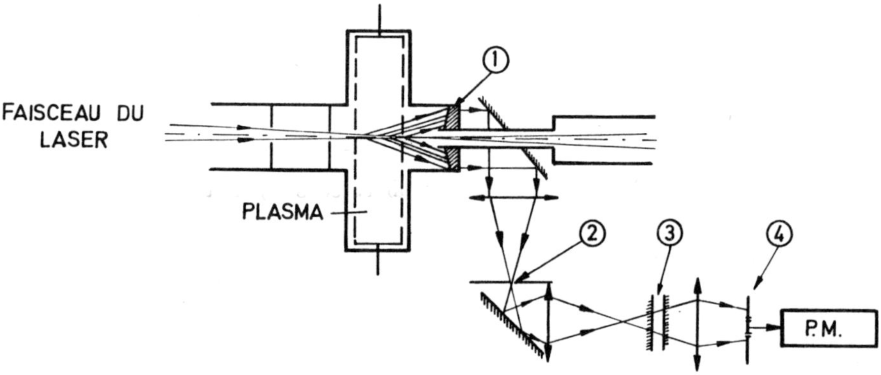

1 - Lentille conique
2 - Diaphragme de sélection angulaire
3 - Interféromètre de FABRY-PEROT
4 - Diaphragme de sélection de fréquence

Fig. 2-3-6 Diffusion aux petits angles (montage typique)

On retrouve ici l'équation de dispersion des oscillations de plasma. On peut donc interpréter la présence des raies satellites comme provenant de l'interaction de l'onde lumineuse incidente et des oscillations de plasma ayant le vecteur d'onde ($2K_o \sin \Phi/2$). Par contre, la distribution obtenue pour $\alpha = 0$ est bien, comme on s'y attendait, de forme gaussienne avec une largeur totale à mi-hauteur :

$$\Delta \omega = 4 K_o \sin \frac{\Phi}{2} \left[\frac{2 k T_e}{m_e} \text{Log}_e 2 \right]^{1/2}$$

comme pour une assemblée d'électrons sans corrélation.

Le rapport de la composante électronique à la composante ionique est, pour Z = 1 :

$$\frac{\sigma_e (\theta)}{\sigma_i (\theta)} = \frac{1 + 2 \alpha^2}{\alpha^4}$$

La composante ionique prédomine donc dès que α s'approche de 2.

On a porté sur la figure 2-3-4 la distribution spectrale de la lumiére diffusée, compte tenu des deux composantes pour Z = 1, Te = Ti = 10eV et une masse ionique correspondant à celle de l'hydrogène.

On voit que la composante ionique est extrêmement étroite par rapport à la composante électronique. Elle le serait encore plus pour des ions plus lourds. Il est donc à prévoir qu'elle est très difficile à observer car elle risque d'être perdue dans la lumière parasite provenant directement du laser.

Résultats Expérimentaux

Conditions expérimentales

La distribution spectrale de la lumière diffusée dépend entièrement, nous l'avons vu, de la valeur du paramètre α. Pour un plasma de densité et de température données, α peut être modifié en changeant la direction d'observation par rapport à la direction de la lumière incidente. Dans les premières expériences, la direction d'observation était toujours perpendiculaire au faisceau incident ($\Phi = \pi/2$). On obtenait ainsi, en général, des valeurs de α inférieures à l'unité.

Le dispositif employé par Fünfer et ses collaborateurs (7) était de ce type (Fig. 2-3-5).

En réduisant suffisamment la valeur de l'angle Φ, c'est-à-dire en observant la lumière diffusée aux petits angles, il est, en principe, toujours possible d'augmenter suffisamment α pour voir apparaitre toutes les structures prévues du spectre diffusé. Ascoli-Bartoli et ses collaborateurs (8) ont été les premiers à utiliser de tels montages (Fig. 2-3-6) et à mettre en évidence des raies satellites nettement séparées. Depuis, la diffusion aux petits angles a pris une grande importance et les résultats les plus intéressants ont été obtenus sur ce genre de montage.

Le section efficace de diffusion des électrons est très faible et ceci entraîne de grandes difficultés expérimentales. Pour que la lumière diffusée puisse être détectée et analysée, il faut qu'un certain nombre de conditions soient remplies :

La puissance diffusée et reçue sur le photomultiplicateur doit être supérieure à la puissance équivalente de bruit du récepteur. Pour des densités de l'ordre de 10^{15} cm^{-3} et des longueurs de plasma de l'ordre du centimètre, il faut une puissance incidente minimum de 10^5 watts environ.

La luminosité propre du plasma ne doit pas couvrir la lumière diffusée. Ici on peut profiter du fait que la lumière diffusée est polarisée alors que celle émise par le plasma ne l'est pas. Dans les premières expériences, des montages différentiels étaient utilisés à cet effet. Ils ont pu être abandonnés ultérieurement lorsque la puissance des lasers a pu être augmentée. Avec un laser déclanché de 10 MW et un plasma dont la densité électronique est de 10^{15} cm^{-3} environ et la température électronique de quelques électrons volts, l'intensité diffusée est supérieure à l' intensité émise par le plasma dans le fond continu Bremsstrahlung.

La lumière parasite provenant du laser doit être soigneusement éliminée par un choix judicieux du montage expérimental. En effet, la puissance diffusée pour une densité de 10^{15} cm^{-3} et une longueur de plasma de 1 cm est pratiquement de 10^{10} à 10^{11} fois plus faible que la puissance incidente. Les premiers montages ne permettaient pas d'observer le centre du spectre diffusé car la lumière parasite y était souvent 10^2 à 10^3 fois plus intense que la lumière diffusée. Fort heureusement le spectre de la lumière incidente est toujours beaucoup plus étroit que celui de la lumière diffusée, ce qui a permis de les distinguer, même dans les premières expériences. Depuis, la géométrie des enceintes a été améliorée, de même que le taux de réjection des monochromateurs, de telle sorte que l'on observe maintenant le spectre 'ionique' diffusé qui a une largeur de l'ordre de 1 Å.

Exploitation des résultats

Les renseignements qui peuvent être obtenus à partir du spectre de la lumière diffusée dépendent de la valeur du coefficient α. Nous pouvons distinguer les cas suivants:

A. Observation de la composante 'électronique'

A-a) $\alpha \ll 1$ - La répartition de l'intensité dans le spectre est gaussienne. On obtient T_e à partir de la largeur du pic et N_e à partir de l'intensité totale diffusée. Cette intensité est en général comparée à celle qui est diffusée par un gaz neutre dont on connaît la section efficace de Rayleigh, ce qui permet de calibrer d'un seul coup tout systéme optique.

A-b) $\alpha \leqslant 1$ - Le spectre présente des épaulements de part et d'autre de la raie centrale (apparition de raies satellites). On obtient simultanément N_e et T_e en cherchant la courbe théorique qui s'adapte le mieux aux mesures.

A-c) > 1 - Le spectre électronique est réduit aux raies satellites. Leur écartement permet de déterminer N_e. T_e est obtenu en observant le spectre ionique ou en effectuant simultanément des observations suivant deux directions. Le plus souvent, la condition $\alpha > 1$ est obtenue en observant la lumière diffusée vers l'avant ($\Phi \geqslant 10°$). En général l'observation à angle droit permet d'obtenir $\alpha < 1$ et l'on peut alors déterminer T_e.

B. Observation de la composante 'ionique'

Cette composante s'observe principalement lorsque $\alpha \geqslant 1$ car son intensité est alors grande devant celle de la composante électronique. Son interprétation est la plus simple lorsque α est nettement supérieur à l'unité. Dans ce cas :

$$\beta = [T_e/T_i]^{1/2}$$

Si α n'est pas grand, il faut connaître sa valeur pour interpréter le spectre ionique et en déduire T_e et T_i.

B-a) $\beta << 1$ - La largeur du pic ionique (~ 1 Å) permet de déterminer directment T_i - (T_e et N_e doivent être déterminés par ailleurs).

B-b) $\beta \sim 1$ - L'étude de la forme du spectre comparée aux spectres théoriques permet de déterminer T_i et β et donc T_e si α est connu ou si l'on sait que $\alpha >> 1$.

B-c) $\beta >> 1$ - L'écart en fréquence des deux pics composant le spectre ionique permet de determiner la densité ionique car il est égal à la fréquence de plasma ionique.

Expériences mettant en évidence la diffusion incohérente sans effets collectifs

Ce type de diffusion se caractérise, comme nous l'avons vu, par une distribution gaussienne de l'intensité dans le spectre; il apparaît pour des valeurs de α inférieures à l'unité. On obtient ces spectres, en général, lorsque la direction d'observation est perpendiculaire à la direction de propagation de la lumière ($\Phi = 90°$). Lorsque la densité du plasma est forte, et sa température relativement basse, on peut obtenir, même à 90°, des spectres mettant en évidence des phénomènes collectifs ($\alpha \sim 1$). Les premières expériences faites utilisaient toutes la diffusion à 90° et la plupart permirent d'observer la diffusion incohérente.

Thomson et Fiocco (9) essayèrent tout d'abord de mettre en évidence la diffusion par un faisceau d'électrons, mais la densité et la puissance du laser étaient trop faibles pour que l'expérience soit concluante. Les mêmes auteurs ont ensuite (10) effectué des mesures sur un arc à cathode creuse (Ne ~ 10^{13} cm^{-3}, Te ~ 5 à 10 eV) mais, là encore, les résultats furent décevants. Funfer et coll. firent leurs premiers essais sur un plasma de striction azimutale où la température électronique était relativement faible; α était de l'ordre de 2 mais les raies satellites ne furent pas observées. La puissance du laser était de 100 kW. Dans une deuxième série d'expériences (11) la température électronique était bien supérieure (~10eV) la puissance lumineuse avait été portée à 10, puis 100 MW, en utilisant un laser déclanché (30 ns), la valeur de α était alors de 0, 5 à 1 et ils purent observer les deux spectres de la figure 2-3-7 où l'on voit nettement l'apparition des effets collectifs cas (b).

Ces spectres ont été relevés avec un polychromateur à 6 canaux, ce qui permet d'en faire l'analyse au cours d'une seule décharge, d'où il résulte une bien meilleure précision.

Davies et Ramsden (12) ont fait également une mesure dans des conditions semblables sur une striction azimutale. Le laser avait une puissance de 10 MW et une durée d'impulsion de 30 ns. La puissance diffusée était 10^{13} fois plus faible que la puissance incidente et la lumière diffusée était 10 fois plus intense que la luminosité propre du plasma. La lumière parasite, par contre, était 10^3 fois plus intense au centre de la raie (sur 20 Å) que la lumière diffusée. Le spectre était relevé en effectuant un certain nombre de décharges et en décalant le monochromateur à chaque fois. La température mesurée était de 3, 3 eV et la densité électronique évaluee à 5.10^{15} cm^{-3} en comparant l'intensité diffusée par les électrons à celle diffusée par un gaz neutre.

Certains auteurs se sont attachés à faire des mesures dans des plasmas moins denses, tels que ceux que l'on rencontre dans les post-décharges ou dans des arcs à basse pression.

S.E. Schwartz (13) a étudié la post-luminescence de décharges dans l'hélium à des densités de l'ordre de 10^{13} à 10^{14} cm^{-3}. Dans des expériences de ce type, la lumière émise par le plasma ne constitue pas la limitation la plus draconienne. En effet, elle varie comme le carré de la densité électronique alors que la lumière

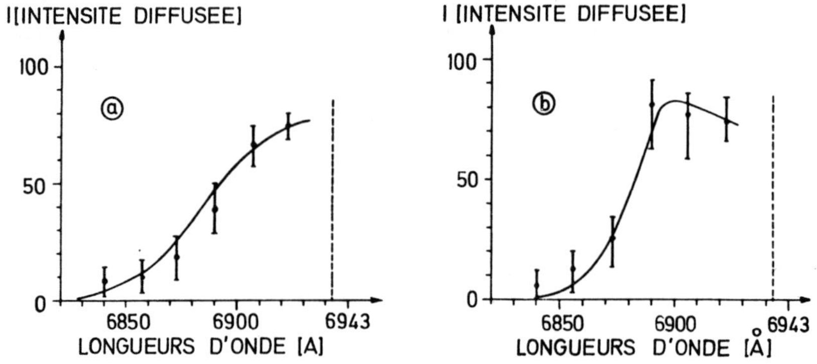

Fig. 2-3-7 Résultats obtenus par FUNFER et coll. (11). Les courbes tracées correspondent aux répartitions théoriques qui rendent le mieux compte des points expérimentaux, on en déduit les paramétres suivants:

a) $T_e = 90\ 000°K$, $N_e, \sim 2.10^{16}\ cm^{-3}$, $\alpha \sim 0.53$

b) $T_e = 55\ 000°L \pm 20\%$, $N_e = 4.10^{16} \pm 30\%\ cm^{-3}$, $\alpha = 0.97$

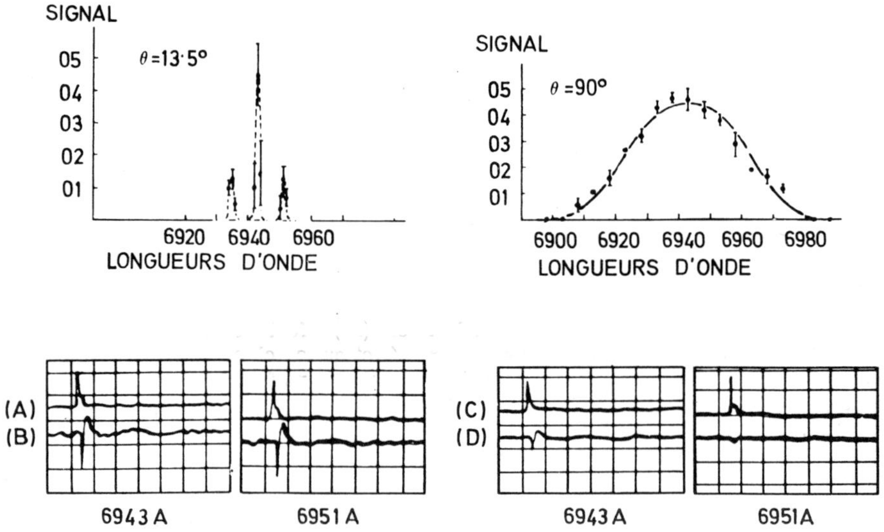

Fig. 2-3-8 Mise en évidence des raies satellites. Résultats de RAMSDEN (20).

Oscillogrammes A et C - Signal de référence (Intensité du laser)
Oscillogramme B - Lumière diffusée
Oscillogramme D - Lumière parasite.

Echelle 0.5 μs/cm - A 6943 Å, un filtre gris de densité unité a été interposé devant le récepteur diffusée varie linéairement avec N_e. Par contre, la lumière parasite doit être éliminée très soigneusement, ce qui n'est pas possible dans toutes les configurations. L'auteur a mesuré ainsi les densités électroniques en comparant la puissance totale diffusée à celle fournie par un gaz neutre.

Gerry et Rose (14) ont fait des expériences sur un arc à basse pression à cathode creuse entretenu dans de l'argon. L'arc fonctionne de façon continue et le spectre diffusé est analysé en faisant tourner progressivement un filtre interférentiel. La densité électronique est de 10^{13} à 10^{14} cm^{-3} et la température de 4 eV. La lumière diffusée est plus intense que la lumière parasite (10 à 100 fois), dès que l'on s'écarte de la raie laser de plus de 3 à 4 Å. Par contre la luminosité propre du plasma est toujours plus intense (2 à 10 fois) que la lumière diffusée, à cause de la proximité d'une forte raie d'émission de l'argon.

D'autres auteurs encore ont utilisé la diffusion incohérente à 90° : Consoli et coll. (15) sur une striction azimutale dans l'hydrogène Patrick (16) sur une onde de choc magnétohydrodynamique dans l'argon ($p \sim 0,1$ Torr, $M \sim 2$, $N_e \sim 10^{16}$ cm^{-3} T_e = 60 eV), et Malyshev (17) sur un arc dans l'hélium à 0, 2 Torr.

Izawa et coll. (18) sur un tube à choc électromagnétique, ($10 < M < 20$, $10^{16} < N_e < 10^{18}$, T_e 23 000°K), ont mis en évidence un léger épaulement sur le spectre diffusé, correspondant a $\alpha = 0,85$.

On peut dire, d'une facon générale, que pour les valeurs de α inférieures à ou proches de l'unité, les mesures faites par diffusion Thomson sont en bon accord avec les autres mesures et que cette technique, si elle est employée avec un soin suffisant, permet de mesurer des densités électroniques à partir de 10^{13} cm^{-3} et des températures à partir de 1 à 2 eV.

Mise en évidence des effets collectifs

Ces effets apparaissent lorsque le coefficient α est supérieur à l'unité. La théorie prévoit alors l'apparition de raies satellites espacées de la fréquence incidente de $\pm \omega_p$, ω_p étant la fréquence de plasma. Nous avons vu que la diffusion à 90° avait fait apparaître dès les premières expériences des raies satellites peu marquées correspondant à des valeurs de α proches de l'unité. Pour obtenir des valeurs plus élevées, Ascoli-Bartoli (8) et collaborateurs mirent au point un montage permettant de recueillir la lumière diffusée vers l'avant ($\Phi \sim 3°$). Ce montage qui utilisait des lentilles prismatiques 'Axicon' a servi de modèle aux autres expérimentateurs qui ont pu l'améliorer suffisamment pour obtenir des résultats très intéressants. Notons toutefois l'expérience de Chan et Nodwell (19) effectuée sur un chalumeau à plasma ($N_e \sim 10^{16}$, $T_e \sim 1$ eV, $\Phi = 45°$ $\alpha \sim 4,4$) et qui se classe tout à fait à part, mettant en jeu un plasma de faible énergie mais ayant une forte densité. L'expérience fait apparaître deux fortes raies satellites mais la précision des mesures est insuffisante pour en déduire de façon certaine la densité électronique. Les raies semblent élargies du fait de l'inhomogénéité du plasma.

Ascoli-Bartoli et coll. (8) purent mettre en évidence les raies satellites, mais leur montage présentait des imperfections dûes à un manque de reproductibilité du plasma et des impulsions lasers. Ils purent aussi mettre en évidence la composante ionique sur le même montage. Un dispositif semblable a été utilisé par Ramsden et Davies (20) leur permettant d'observer le même plasma à 13°5 ± 0, 3° et 90° (Fig. 2-3-8).

Ils ont alors pu mettre en évidence, sur une striction azimutale, des pics satellites extrêmement nets et, de leur séparation, en déduire la densité électronique. La raie ionique est visible et se distingue bien de la lumière parasite mais elle n'est pas résolue dans cette expérience. Les observations à 90° donnent un spectre de

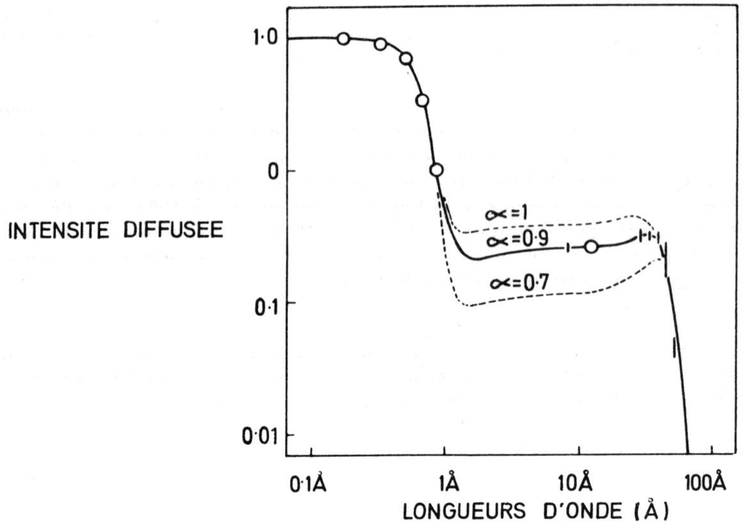

Fig. 2-3-9 Mise en évidence simultanée du pic ionique et de la composante électronique. (24).
Les courbes tracées sont des courbes théoriques correspondant à trois valeurs d'essai du paramétre α

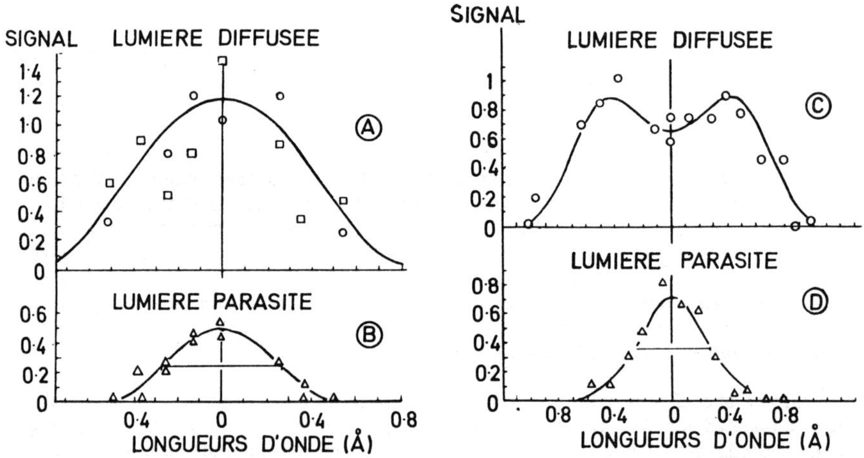

Fig. 2-3-10 Analyse du pic ionique (26). Les courbes tracées sont des courbes théoriques:

a) $\beta = 1$, $T_e = T_i = 200$ eV

b) $\beta = 2$, $T_e = 800$ eV, $T_i = 200$ eV

diffusion incohérente dont la largeur permet de determiner T_e. L'obtention simultanée de ces deux spectres permet des recoupements intéressants: le spectre à 13°5 permet de déterminer la densité électronique d'après l'écartement des raies satellites. Dans l'exemple donné, les raies sont distantes du centre de 8 Å ce qui correspond à une densité électronique de 2, 4. 10^{15} cm^{-3}. Le rapport de l'intensité des raies satellites à celui de la raie ionique permet de déterminer α . On trouve ici α = 3, 0 ± 0, 25, ce qui, compte tenu de la densité trouvée plus haut, permet de déterminer la température électronique, soit 1, 1 eV. Le spectre relevé à 90° donne une température de 1, 0 eV, ce qui, pour une densité de 2, 4.10^{15} cm^{-3} correspond à α = 0, 5, en accord avec la valeur prévue d'après le rapport des sinus des angles. Enfin, l'intensité totale diffusée, étalonnée d'après la diffusion de l'azote, conduit à une densité N_e = (2, 4 ± 0, 5) 10^{15} cm^{-3}. L'excellente cohérence de tous ces résultats montre la valeur des prévisions théoriques mais elle n'est possible que grâce au soin apporté au système optique qui permet de réduire la lumière parasite et d'obtenir des résultats reproductibles.

Evans et coll. (21) ont également observé la lumière diffusée vers l'avant sur une striction azimutale. L'angle d'observation est de 5°. En effectuant les mesures à divers instants au cours de la compression, les auteurs ont pu relever des spectres pour des valeurs de α égales à 1, 2, 1, 3 et 1, 5. Pour ces valeurs proches de l'unité, on peut déterminer simultanément N_e et T_e. Pour l'un de ces spectres, les auteurs trouvent les résultats suivants : α = 1, 3 ± 0, 05 écart entre une raie satellite et la raie centrale : $\Delta\lambda$ = 16 Å ± 1 Å , $N_e = (6,0^{+1}_{-0,6})$ 10^{15} cm^{-3}, T_e = 103^{+29}_{-17} eV.

Ici, comme précédemment, la précision des mesures atteint 25% environ, ce qui est excellent.

Mise en évidence du pic ionique

Dès que α est de l'ordre de 2 unités, la puissance diffusée dans le pic 'ionique' est du même ordre de grandeur que celle qui est répartie dans le spectre électronique. La difficulté de sa mesure tient surtout à sa faible largeur (1 Å environ). Il est donc indispensable, pour l'atteindre, d'éliminer presque totalement la lumière parasite diffusée. Il faut également disposer d'un laser ayant une grande finesse de raie ($\leq$ 0, 5 Å). On sait que la structure du pic ionique ressemble à celle du spectre électronique, elle permet donc d'atteindre la température ionique, ce qui est capital dans les expériences de fusion thermonucléaire. Plusieurs expérimentateurs ont pu mettre ce pic en évidence et certains ont même pu étudier en détail sa structure.

Ascoli-Bartoli et coll. (22) ont pu le montrer, grâce à un système original de polychromateur, sur le montage expérimental qu'ils avaient utilisé précédemment pour la diffusion à 3°et en stabilisant convenablement leur laser vis à vis des dérives thermiques de longueurs d'onde.

Auparavant, De Silva et coll. (23) avaient détecté sous un angle de 10° un pic très étroit (< 2 Å) qu'ils avaient attribué à l'effet ionique, mais la résolution de leur système dispersif était insuffisante pour permettre une analyse détaillée.

Anderson (24) a pu, sur une striction linéaire applatie 'Triax', tracer un spectre complet de la lumière diffusée, y compris la composante ionique pour une valeur de α proche de l'unité (Fig. 2-3-9).

Kronast et coll. (25), d'une part, et Ransden et coll. (26), d'autre part, ont pu analyser de facon plus détaillee ce pic et en tirer la valeur de la densité électronique et des températures ioniques et électroniques.

Considérons, à titre d'exemple, les résultats de Ramsden (Fig. 2-3-10). Le laser à rubis employé avait une puissance de 50 MW, une largeur de raie de 0, 3 Å et une durée d'impulsion de 30 ns. La lumière diffusée était recueillie entre 4, 9° et 5, 1°, elle était 10^{11} fois moins intense que la lumière incidente. Le spectre était analysé à l'aide d'un interféromètre de Fabry-Pérot.

Les résultats présentés sur la figure 2-3-10 se rapportent à deux instants distincts (4 et 5 μs) d'une décharge de striction azimutale (100 kJ). Les courbes sont tracées en répétant l'expérience pour divers réglages de l'interféromètre, alors que Kronast et coll. utilisent un polychromateur. Pour la mesure faite a 4 μs la courbe théorique qui s'ajuste le mieux aux résultats donne $\beta = 1$ donc $T_e = T_i$. La largeur de la raie donne T_i = 200 eV, son intensité totale permet de calculer $N_e = 2, 0 \pm 1, 4 . 10^{17}$ cm^{-3}, la valeur de α correspondante est de 5, 5.

A 5 μs on constate un changement important de l'allure du spectre. On observe deux pics, ce qui correspond à une valeur β supérieure à l'unité. La courbe théorique qui s'ajuste le mieux donne $\beta = 2$, T_i = 200 eV, donc $T_e = 4\ T_i$ = 800 eV. L'intensité totale diffusée donne $N_e = 2 . 10^{17}$ cm^{-3} donc $\alpha = 3$. On vérifie *a posteriori* que l'on est bien dans les conditions où les effets collectifs dominent, ce qui justifie l'analyse faite.

On voit donc, avec ce dernier groupe de résultats, que toutes les caractéristiques prévues du rayonnement diffusé ont pu être mises en évidence et exploitées de façon à donner des renseignements précis sur le plasma.

Tout récemment Evans et coll. (27) ont pu observer un pic ionique nettement séparé en deux parties d'intensités inégales. Cette observation avait déjà été faite par Kronast et coll. (25). Ils ont pu en déduire des renseignements semi-quantitatifs sur le phénomène supposé être la cause de cette asymétrie. Celle-ci est attribuée à une instabilité ionique provoquée par un déplacement relatif des électrons et des ions. Appliquant une théorie de Rosenbluth et Rostoker (4), ils ont pu calculer la vitesse relative de ce déplacement. On voit donc apparaître la possibilité d'obtenir par ces méthodes de diagnostic des renseignements touchant certains phénomènes profonds apparaissant dans le plasma.

Nouvelles expériences envisagées

A notre connaissance, un certain nombre d'expériences sont actuellement en cours de réalisation ou ont été récemment proposées.

A.A. Offenberger (28) a entrepris de déterminer la distribution des vitesses électroniques dans un arc à cathode creuse en étudiant en détail la lumière diffusée à 10, 6 μ. Il ne s'agit plus ici d'expériences pulsées mais de mesures en continu. La possibilité d'utiliser des méthodes de détection synchrone compense le fait que la puissance du laser est beaucoup plus faible que dans les expériences déjà réalisées (100 W au lieu de 10^7 à 10^8 Watts).

W.B. Johnson (29) a également entrepris des expériences utilisant un laser continu. Il s'agit ici d'un laser à argon. Le but recherché est également une détermination plus fine de la distribution des vitesses électroniques grâce à une étude plus détaillée du spectre diffusé (30).

Kroll et coll. (31) et Schkarofsky (32) ont proposé et étudié les modalités d'une détermination de la distribution des vitesses utilisant la diffusion d'un faisceau laser sur des oscillations de plasma provoquées par un autre faisceau.

Conclusion

Nous nous sommes efforcés de résumer ici les principaux travaux publiés jusqu'à la fin de 1966 sur la diffusion de la lumière par les électrons d'un plasma.

Il semble que cette technique soit maintenant bien au point, tant sur le plan expérimental que sur celui de l'interprétation des résultats. Toutes les caractéristiques prévues pour le spectre de la lumière diffusée ont pu être qualitativement et souvent quantitativement vérifiées. Les développements en cours permettent d'attendre de nouveaux perfectionnements et des renseignements plus détaillés encore sur les caractéristiques du plasma. Les derniers résultats semblent montrer la possibilité de mettre en évidence des phénomènes d'instabilité et d'oscillation prévus théoriquement.

Après avoir été étudiée pour elle-même, cette méthode de mesure a maintenant atteint un stade où elle peut servir dans la pratique et fournir des renseignements utiles sur des plasmas rencontrés dans des expériences réelles.

Références

1. **Gordon, W.E.,** P.I.R.E. 46,1824, 1958.
2. **Bowles, K.L.,** Phys. Rev. Lett. 1,454, 1958.
3. **Salpeter, E.E.,** Phys. Rev. 120,1528, 1960.
4. **Rosenbluth, N.N., Rostoker, N.,** Phys. Fluids 5,776, 1962.
5. **Gerry, E.T., Patrick, R.M.,** Phys. Fluids 8,208, 1965.
6. **Hughes, T.P.,** Nature 194,268, 1962.
7. **Funfer, E., Kronast, B., Kunze, H.J.,** Phys. Lett. 5,125, 1963.
   Funfer, E., Kegel, W.H., Kronast, B., Kunze, H.J., - VI° Conf. Int. Phen. Ionis. Gaz, Paris 1963, Vol IV. p. 119
   Kunze, H.J., Naturf, Z.F., 20 A,801, 1965.
8. **Ascoli- Bartoli, U., Katzenstein, J., Lovisetto, L.,** - Nature 204,672, 1964.
9. **Fiocco, G., Thompson, E.,** - Phys. Rev. Lett. 10,89, 1963.
10. **Thomson, E., Fiocco, G.,** - VI° Conf. Int. Phen. Ionis. Gaz, Paris 1963, Vol IV. p. 111.
11. **Kunze, H.J., Funfer, E., Kronast, B., Kegel, W.H.,** Phys. Lett. 11,42, 1964.
12. **Davies, W.E.R., Ramsden, S.A.,** Phys. Lett. 8,179, 1964.
13. **Schwartz, S.E.,** P.I.E.E. 51 1362, 1963. J.A.P. 36,1836, 1965.
14. **Gerry, E.T., Rose, D.J.,** J.A.P. 37,2715, 1966.
15. **Consoli, T., Gormezzano, C., Slama, L.,** Phys. Lett. 20,267, 1966.
16. **Patrick, R.M.,** Phys. Fluids 8,1985, 1965.

17. Malyshev, G.M., Ostrovskaya, G.V., Razdobarin, G.T., Sokolova, L.V., Sov. Phys. Doklady 11, 441, 1966.

18. Izawa, Y., Yokoyama, M., Yamanaky, C., J. Phys. Soc. Japan 21, 1610, 1966.

19. Chan, P.W., Nodwell, R.A., Phys. Rev. Letters 16,122, 1966.

20. Ramsden, S.A., Davies, W.E.R., Phys. Rev. Letters 16, 303, 1966.

21. Evans, D.E., Forrest, M.J., Katzenstein, J., Nature 211, 23, 1966.

22. Ascoli-Bartoli, U., Katzenstein, J., Lovisetto, L., Nature 207, 63, 1965.

23. de Silva, A.W., Evans, D.E., Forrest, M.J., Nature 203, 132, 1964.

24. Anderson, O.A., Phys. Rev. Letters 16, 978, 1966.

25. Kronast, B., Rohr, H., Glock, E., Zwicker, H., Funfer, E., Phys. Rev. Letters 16,1082, 1966.

26. Ramsden, S.A., Benesch, R., Davies, W.E.R.,John, P.K., I.E.E.E. J. Quant. Electr. Q.E.2. 267, 1968.

27. Evans, D.E., Forrest, M.J., Katzenstein, J., Nature 212, 21, 1966.

28. Offenberger, A.A., M.I.T. Quarterly Progress Reports No. 81, April 1966.

29. Johnson, W.B., IEEE Trans. Antennas and Propag. AP 15,152, 1967.

30. Brown, T.S., Rose, D.J., J.A.P. 37,2709, 1966.

31. Kroll, N.M., Ron, A., Rostoker, N., Phys. Rev. Letters 13, 83, 1964.

32. Schkarofsky, I.P., Bull. Am. Phys. Soc. 11, 375, 1966.

Commentary A on Chapter 2-3

W. H. Kegal

Light scattering has become a very powerful tool in plasma diagnostics due to the high spatial and time resolutions achievable and due to the fact that the interpretation of the measured data is not complicated in most cases. Dr. Rostas gave a survey on recent experiments. He also discussed the most important results of the theory and demonstrated the excellent agreement between theory and experiment.

I would like to add a few more remarks on the theory essentially pointing out the underlying assumptions and by that the limitations of the method.

The light scattered by an ensemble of particles may be considered as the interference pattern which one obtains by summing up the waves scattered by the different single particles. In doing this summation one derives the well known (1) scattering formula:

$$dI_s (\omega_{s_1} \vec{k}_s) = \sigma (\theta) I_o (\omega_{o1} \vec{k}_o) < |n(\omega_1 \vec{k})|^2 > do \qquad \text{(Eq. A2-3-1)}$$

with:

$$\vec{k}_s = \vec{k}_o \pm \vec{k} \quad ; \quad \omega_s = \omega_o \pm \omega \qquad \text{(Eq. A2-3-2)}$$

where dI_s is the intensity of the light with frequency ω_s scattered into the solid angle do, I_o is the primary intensity, $\sigma_e(\theta)$ is the differential scattering cross-section for a single particle and:

$$n (\omega_1 \vec{k}) = \int_V n (\vec{r}, \omega) e^{i \vec{k} \vec{r}} d^3 r \qquad \text{(Eq. A2-3-3)}$$

is the Fourier transform of the density fluctuations, where ω and $\vec{k}$ are determined by equation A2-3-2.

Equation A2-3-1 is a linear approximation corresponding to the first Born approximation. Nonlinear effects and multiple scattering are neglected and the linear dimensions of the particles are assumed to be small compared to the wavelength. Otherwise the formula is quite general. It shows that from a light scattering experiment one obtains direct information about the density fluctuations of the scattering particles. In order to connect these fluctuations with quantities as density and temperature further restrictive assumptions have to be made.

If one considers free streaming noninteracting particles the fluctuation spectrum gives directly the velocity distribution in $\vec{k}$-direction.

In the case of Thomson scattering in a plasma one has to take into account the Coulomb forces between the particles. In this case one may derive the fluctuations from the linearized Vlasov equation, assuming a stable and homogeneous plasma. In this theory one further assumes that there are many particles in the Debye volume:

$$n_e D_e^3 >> 1 \quad ; \quad n_i D_i^3 >> 1 \qquad \text{(Eq. A2-3-4)}$$

and that the average spacing between particles is small compared to the wavelength of the fluctuation considered:

$$n_e k^{-3} >> 1 \qquad \text{(Eq. A2-3-5)}$$

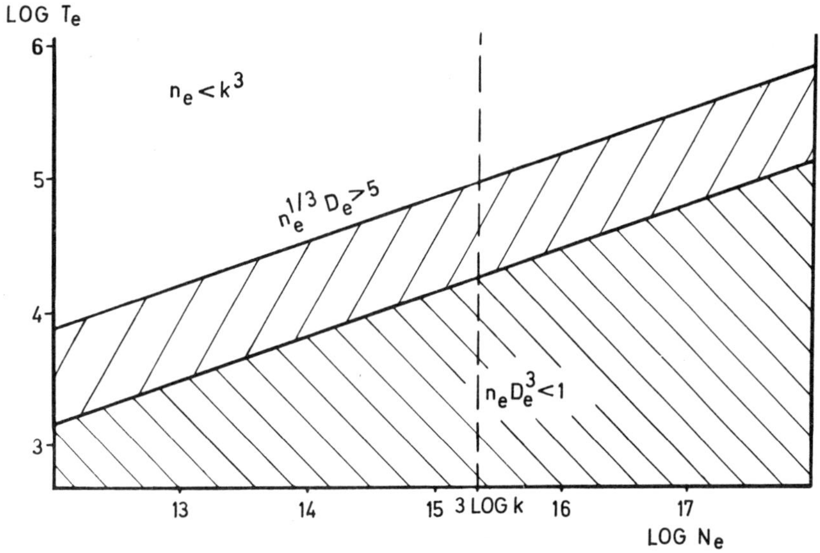

Fig. A 2-3-1 Range of validity of the light scattering theory based on the linearized Vlasov equation

A result of the theory is, that if $\alpha \equiv (k D_e)^{-1} << 1$, (i.e. in the case that the wave-length of the considered fluctuation is small compared to the Debye length) the electrostatic interaction of the particles plays no role (i.e. the spectrum is that of free streaming particles). As in the case of noninteracting particles the condition of equation A2-3-5 is not necessary, one is led to conclude that the same is true in the case of a plasma as long as $\alpha << 1$. In a plasma in which the Debye length D_e is so large, that for $k = n^{1/3}$ one obtains $\alpha << 1$, let say $\alpha \leq 0.2$ which is equivalent to the condition:

$$n_e^{1/3} D_e > 5 \qquad \text{(Eq. A2-3-6)}$$

there is no limitation to k as either the condition of equation A2-3-5 is fulfilled or $\alpha << 1$. Figure A2-3-1 gives the range of validity of the theory in terms of density and temperature. In the hatched part to the lower right of the diagram the theory is not applicable as there is less than one particle in the Debye volume. In the part to the upper left, the condition of equation A2-3-6 is fulfilled. In the intermediate range there is (because of the condition in equation A2-3-5) a lower limit to the density for a given k. For ruby laser light (λ = 6943 Å) and a scattering angle of 90° this limit is $2.05 \cdot 10^{15}$ cm^{-3}.

With these assumptions the theory gives general expressions for the fluctuation spectra, allowing for arbitrary stable distribution functions. In actual numerical calculations (1) (2) one usually assumes a Maxwell distribution for the electrons and ions but allows for different temperatures and a drift velocity between the two species. The results one obtains in this case, with the additional assumption that the mean velocity of the electrons is large compared to that of the ions, have been discussed by Dr. Rostas.

It should further be noted that if one is interested in narrow features in the spectrum ($\alpha >> 1$) the influence of collisions have to be considered (3).

Finally I would like to remark, that in applying light scattering as a diagnostic method one assumes that the plasma is not seriously disturbed by the laser light. On the other hand one actually takes the most intense light source one knows, so that in some cases the above assumption may become a serious limitation to the method. For example if one takes a 10 MW laser with a pulse duration of 10^{-8} sec and focusses its beam in a hydrogen plasma with $n_e = 10^{17}$ cm^{-3} and T = 60 000°K to an area of 1 mm^2, the absorbed energy is about 27 percent of the thermal energy in the scattering volume.

Summarizing this discussion it may be said that if one applies the method of light scattering to cool and dense plasmas, the interpretation of the spectra becomes questionable.

1. **Rosenbluth, M.N., Rostoker, N.,** 'Scattering of Electromagnetic Waves by a Nonequilibrium Plasma.' Phys. Fluids 5, 776, 1962.
Kegel, W.H., Lichtstreuung und Mischung in einem Plasma. Inst. f. Plasmaphysik, Garching, Rep. IPP 6/21, 1964.

2. **Kegel, W.H.,** 'Kurven zur Bestimmung von Plasmaparametern durch Lichtstreuexperimente' Inst. f. Plasmaphysik, Garchung, Rep. IPP 6/34, 1965.

3. **Dubois, D.F., Gilinski, V.,** 'Incoherent Scattering of radiation by Plasmas.' II. Effect of Coulomb Collisions on Classical Plasmas. Phys. Rev. 133, A 1317, 1964.

4. **Taylor, E.C., Comisar, G.G.,** 'Frequency Spectrum of Thermal Fluctuations in Plasmas.' Phys. Rev. 132, 2379 , 1963.
Farley, D.T., 'The Effect of Coulomb Collisions on Incoherent Scattering of Radio Waves by a Plasma.' J. Geoph. Res. 69, 197, 1964.

Commentary B on Chapter 2-3

B. Fontaine

La méthode de diagnostic que nous rapportons ici est analogue dans son principe à celle bien connue proposée par Ashby et Jephcott (1) (2). Elle présente l'intéret d'une bande passante nettement plus grande (> 5 MHz) tout en gardant les avantages de l'utilisation de l'infrarouge ($\lambda = 3,39\,\mu$), c'est-a-dire une meilleure sensibilité pour la mesure de la densité électronique et l'élimination de l'effet des atomes neutres sur l'indice de réfraction.

Le schéma de l'interféromètre est présenté dans la figure B2-3-1. Le montage utilise la longueur d'onde 3, 39μ d'un laser à gaz hélium-néon. Le faisceau du laser, après avoir traversé le plasma d'épaisseur d, est réfléchi par un miroir plan M_3. Ce miroir M_3 forme avec le miroir M_2 du laser une cavité résonante M_2 - M_3 couplee à la cavité du laser M_1 - M_2. Lorsque l'indice de réfraction de la cavité annexe est modifié par la présence du plasma, son mode de résonance varie également, de telle sorte que l'intensité du faisceau laser décrit un cycle complet de modulation lorsque ce mode change d'une unité.

Le faisceau modulé sortant à l'autre extrémité du laser est dirigé sur la partie sensible d'un détecteur infra-rouge à antimoniure d'indium, à effet photomagnéto-électrique, fonctionnant à température ambiante. Le temps de réponse de l'interféromètre est de l'ordre de 0, 2 μs (1) (2). Par contre, le montage d'Ashby et Jephcott, utilisant le couplage entre les longueurs d'onde 0, 6328 μ et 3, 39 μ se trouve limité à une bande passante de 500 KHz, la détection s'effectuant sur la radiation visible.

L'indice de réfraction μ est lié aux caractéristiques du plasma par la relation:

$$\mu - 1 \sum k_i N_i$$

où N_i est la densité du ième composant et k_i sa réfractivité spécifique

Dans le cas présent, pour l'argon:

$$(\mu - 1)_{\text{plasma}} = (\mu - 1)_A + (\mu - 1)_{A^t} + (\mu - 1)_{\text{électrons}}$$

La contribution des atomes neutres peut s'écrire, lorsqu'on est éloigné d'une raie de résonance de l'argon, ce qui est le cas ici, et en utilisant les coefficients numériques donnés par Alpher et White (3).

$$(\mu - 1)_A = (1{,}03\ 10^{-29} + \frac{0{,}58\ 10^{-43}}{\lambda^2})\ N_a \qquad \text{(MKSA Units)}$$

La contribution des électrons est exprimée par la relation classique:

$$\mu^2 = 1 - \frac{\omega^2{}_p}{\omega^2}$$

ou ω_p est la "fréquence plasma" : $\omega_p^2 = \dfrac{N_e\, e^2}{m_e\, \mathcal{E}_o}$ et $\omega = \dfrac{2\pi c}{\lambda}$

de plus, $\omega >> \omega_p$ d'où :

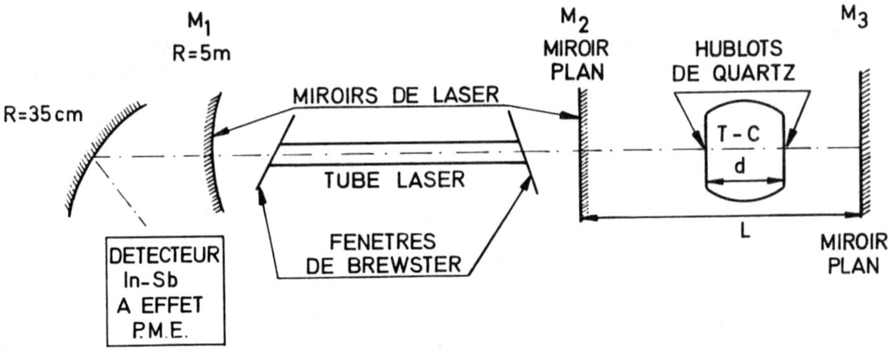

Fig. B-2-3-1 Montage interférométrique

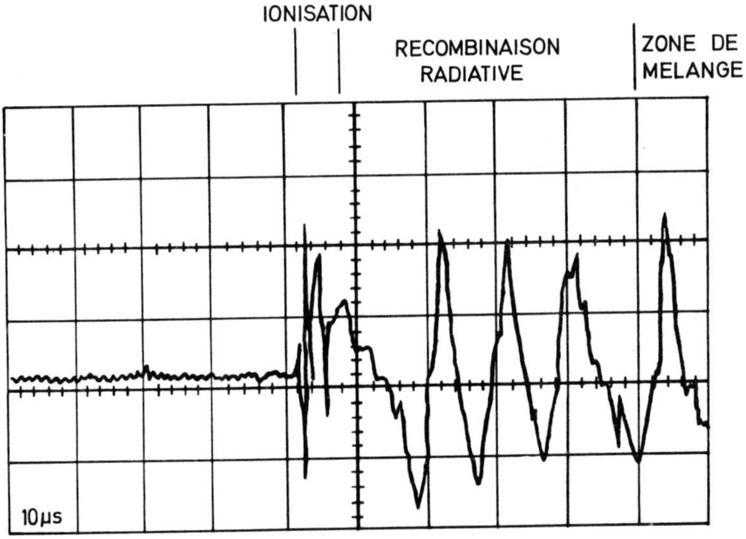

T C 5

M_{OC} = 14·5
P_i = 2·5 mbars
N_e MESURE = 4·10 cm^{-3}

Fig. B-2-3-2 Enregistrement laser dans l'infrarouge

$$(\mu - 1)_{N_e} = - \frac{e^2 \lambda^2}{8 \pi^2 m_e c^2 \mathcal{E}_o} N_e = 4,46\ 10^{-16} \lambda\ 2\ N_e \qquad \text{(MKSA Units)}$$

La contribution des ions, $(\mu - 1)_{A^+}$, est négligeable car leur réfractivité spécifique est du même ordre que celle des atomes neutres et leur nombre est égal à celui des électrons.

Les effets relatifs des électrons et des atomes neutres sur l'indice de réfraction peuvent être exprimés par le rapport:

$$\frac{(\mu - 1)_{N_e}}{(\mu - 1)_{N_a}} = - 4,33\ 10^{13} \lambda^2 \alpha$$

où $\alpha = \frac{N_e}{N_a}$ est très approximativement le degré d'ionisation. Pour $\lambda = 3,39\mu$ et $\alpha = 0,1$ alors $\frac{(\mu - 1)_{N_e}}{(\mu - 1)_{N_a}} = - 50.$

L'effet des atomes neutres est donc négligeable par rapport à celui des électrons pour la longueur d'onde choisie, ce qui est très important pour l'utilisation d'un interféromètre à une seule longueur d'onde. Il faut remarquer que l'effet des neutres ne serait plus du tout négligeable pour $\lambda = 6328$ Å.

Le montage précédent a été utilisé pour la mesure de la densité électronique du plasma d'argon produit par une onde de choc ionisante en tube à choc.

Le tube à choc est du type à combustion; il comporte une section de mesure de section droite carrée, 45 x 45 mm^2, de longueur 500 mm. Des hublots en quartz ont été placés sur les parois latérales. Une 'frange' correspond donc à une variation de la densité électronique:

$$\Delta N_e = \frac{1,12\ 10^{15}}{\lambda\ d} = 7,33\ 10^{15}\ cm^{-3}$$

Dans la plupart des expériences effectúées, une colonne de gaz ionisé d'environ 30 cm de long, se propageait devant la section d'essai, a une vitesse de 4000 m/s. Les conditions des essais étaient alors les suivantes:

gaz d'essai : argon

M_{oc} = 15

P_1 = 2,5 mbars

P_2 = 0,86 bars

T_2 = 11800 °K

α = 0,12

$N_{e_{théorique}}$ = 5,6 10^{16} cm^{-3}

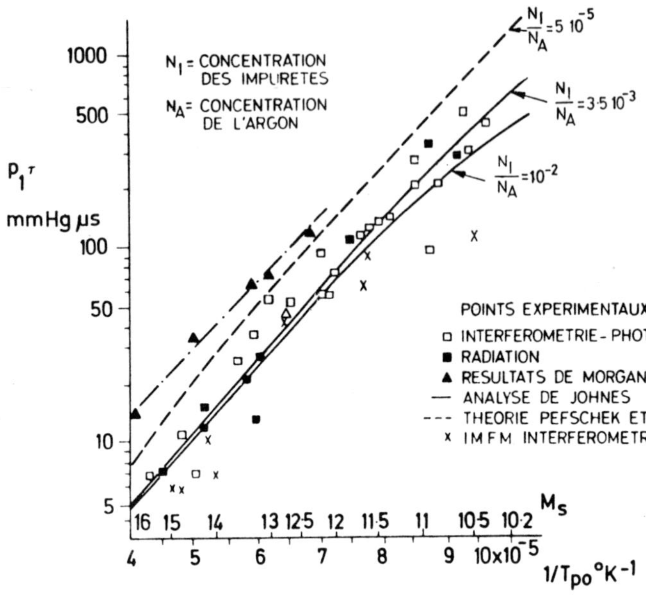

Fig. B-2-3-3 Relaxation d'ionisation dans l'argon

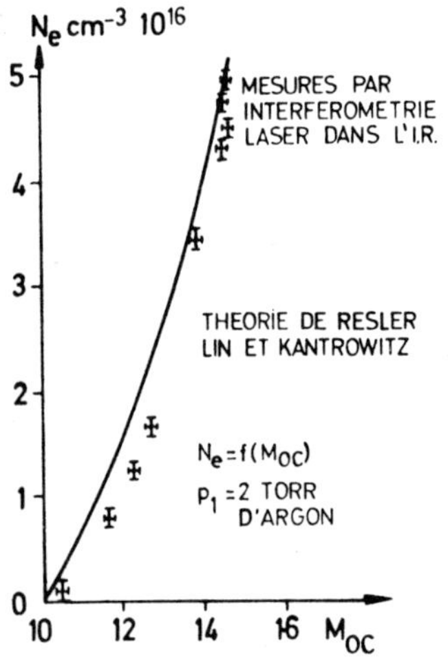

Fig. B-2-3-4
Densité électronique
à l'équilibre

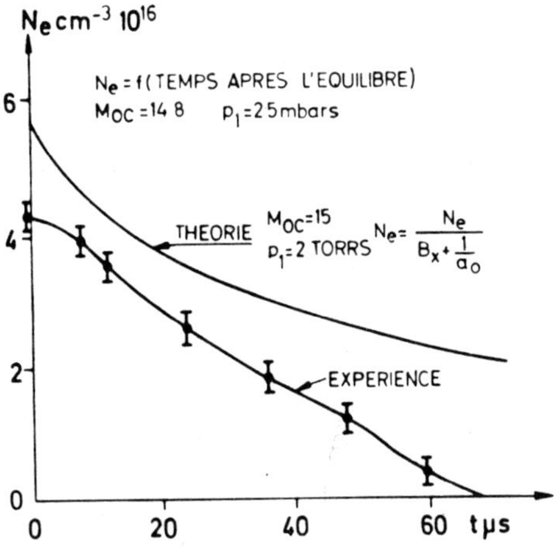

Fig. B-2-3-5
Recombinaison radiative

Quelques expériences ont, toutefois, été effectueés à des nombres de Mach plus faibles, toujours avec P_1 = 2, 5 mbars.

La figure B2-3-2 représente un enregistrement-type obtenu à l'aide de l'intérferomètre. On distingue bien le saut de densité volumique derrière le choc, la zone de relaxation et la recombinaison jusqu'à la surface de contact.

Les mesures ont fourni simultanément 3 séries de résultats:

En premier lieu, une estimation du temps de relaxation d'ionisation derrière l' onde de choc qui concorde assez bien avec les résultats de Johnes (4) pour des taux d'impuretés élevés (Fig. B2-3-3).

Ensuite, une mesure de la densité électronique à l'équilbre qui montre une bonne concordance (Fig. B 2-3-4) entre les valeurs obtenues à partir des relations de Rankine-Hugoniot (5) et l'éxpérience. L'incertitude dans le dépouillement est d'un quart de frange.

Enfin, la mesure du profil de désionisation à partir de l'équilibre. La figure B2-3-5 sur laquelle a été porté un résultat expérimental, met en évidence un certain désaccord avec la courbe théorique portée également sur la figure. Il y a lieu pourtant de remarquer que la courbe théorique a été calculée en prenant seulement en considération la recombinaison radiative par émission d'un spectre continu (6) alors qu'il y a évidemment émission d'un spectre de raie. D'après Horn (7) cette dernière émission pourrait contribuer pour 20 à 30 percent aux pertes radiatives totales se produisant dans les conditions de la présente étude.

L'interférométrie laser dans l'infrarouge a permis, grâce à son faible temps de résponse et à sa sensibilité, d'observer des phénomènes très rapides, de l'ordre de la fraction de microseconde tels que la relaxation d'ionisation ou d'autres phénomènes transitoires se produisant dans les plasmas en écoulement en tube à choc pour des valeurs de la densité électronique se prêtant peu, jusqu'ici, à des mesures précises.

L'extension du présent travail à l'étude de l'évolution de la densité électronique du plasma d'argon produit en tube à choc et soumis à l'influence de champs électriques et magnétiques, conversion directe ou accélération, est actuellement en cours.

References

1. **Ashby, D.E.T.F., Jephcott, D.F.,** Appl. Phys. Letters, 3, 13, 1963.

2. **Ashby, D.E.T.F., Jephcott, D.F., Malein, A., Raynor, F.A.,** J. Appl. Phys. 36, 29, 1965.

3. **Alpher, R.A., White, D.T.,** Phys. Fluids, 2, 162, 1959.

4. **Jones, N.R.,** Ph. D., Liverpool, 1964.

5. **Resler, E.L., Lin, S.C., Kantrowitz, A.,** J. Appl. Phys. 23, 1390, 1952.

6. **Rutowsky, R.W., Bershader, D.,** Phys. Fluids, 7, 5,68, 1964.

7. **Horn, K.P.,** Res. Rpt. S.U.D.A.A.R. No. 268, Stanford University, Calif., May 1966.

Part 3

Enthalpy Probes

3-1

Measurements with Aerodynamic Probes in Plasma Jets produced by Electrothermal and Hall Current Accelerators

S. KRAUSE,
Deutsche Versuchsanstalt für Luft- und Raumfahrt e.V.
Stuttgart - Vaihingen, Germany.

Summary

The measurement of macroscopic aerodynamic quantities in plasma jets may contribute largely in obtaining a better understanding of the phenomena in electric plasma accelerators. A survey is given of early and recent experiments performed at the Institut fur Plasmadynamik, D.V.L., Stuttgart, with common and new types of aerodynamic probes. Radial distributions of Pitot pressure, stagnation enthalpy, mass and energy fluxes and other quantities in plasma jets emerging from electrothermal and Hall current accelerators are presented. A graphical iteration method is described for the determination of any thermodynamic values and the plasma velocity from the measured quantities. An interesting example shows the calculation of an unknown quantity by simultaneous use of aerodynamic and non-aerodynamic experimental results: the axial velocity in a Hall current accelerator is derived from aerodynamic yaw probe determination of the swirl angle and spectroscopic Doppler effect measurement of the azimuthal velocity.

Sommaire

La mesure de grandeurs aérodynamiques macroscopiques dans des jets de plasma peut aider considérablement la compréhension des processus élémentaires dans les accélérateurs de plasma. On décrit les expériences effectuées à l'Institut für Plasmadynamik, D.V.L., Stuttgart, avec des sondes aérodynamiques classiques et des sondes de conception nouvelle. On présente les distributions radiales de pression Pitot, d'enthalpie d'arrêt, de débit-masse, de flux d'énergie et d'autres grandeurs dans un jet de plasma issu d'un accélérateur électrothermique et d'un accélérateur de Hall. Une méthode graphique d'approximations successives est délicate pour la détermination des grandeurs thermodynamiques et de la vitesse à partir des mesures effectuées. Un exemple intéressant montre le calcul d'une grandeur inconnue par emploi simultané de résultats expérimentaux aérodynamiques et non-aérodynamiques: la vitesse axiale dans un accélérateur de Hall est déduite de l'angle de la ligne de courant déterminé par une sonde à incidence et de la vitesse azimutale mesurée spectroscopiquement par l'effet Döppler.

Introduction

Plasma producing and accelerating machines are suitable for wind tunnel purposes and for propulsion in space. To obtain a better understanding, so as to permit optimization of these machines, plasma diagnostics must be used.

In addition to spectroscopic and microwave techniques which essentially do not disturb flowing plasmas, mechanical probes working on the basis of electromagnetic or aerodynamic principles may be used. If such probes are sufficiently small, the disturbances resulting from their insertion into a plasma jet may be negligible.

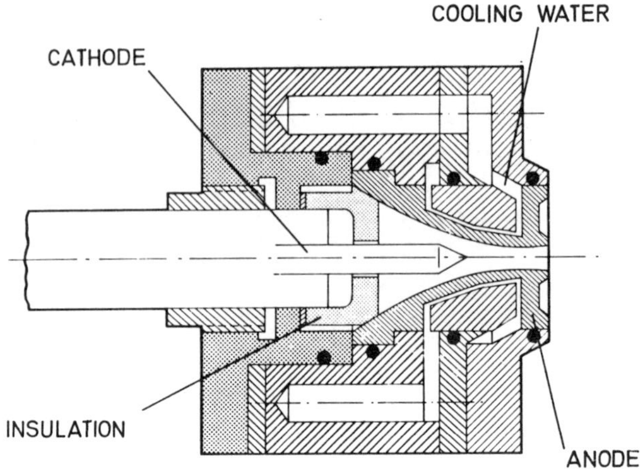

Fig. 3-1-1 Device for atmospheric plasma jet

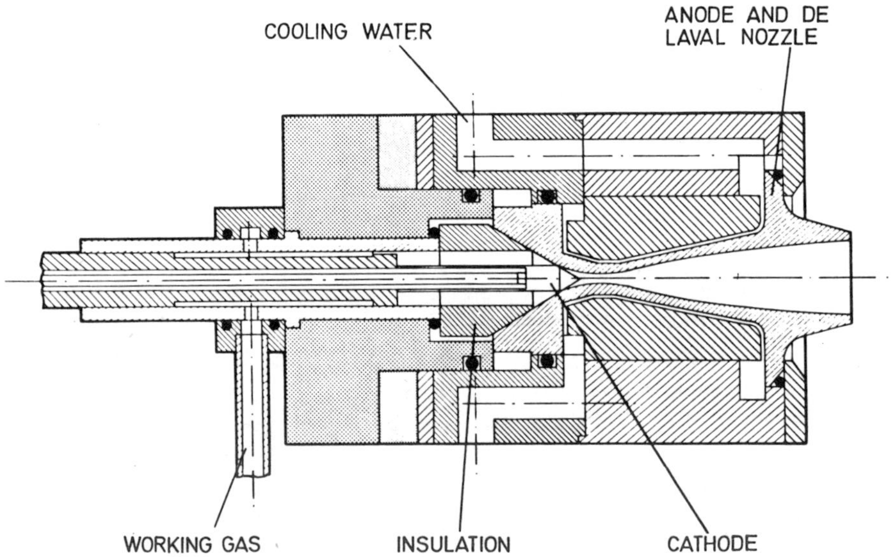

Fig. 3-1-2 Electrothermal plasma accelerator. Nozzle designed for uniform parallel jet

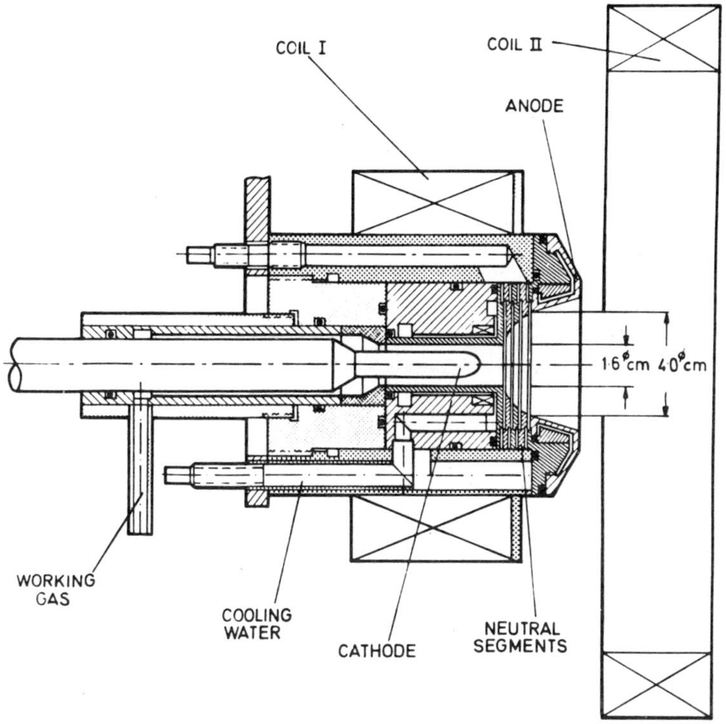

Fig. 3-1-3 Hall current (MPD) plasma accelerator

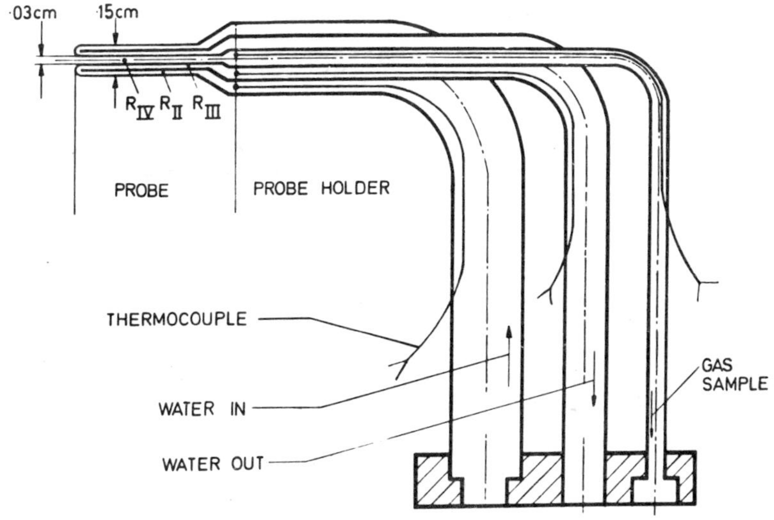

Fig. 3-1-4 Coaxial calorimetric probe. Scheme

Being particularly or even exclusively appropriate to the measurement of certain quantities, none of the diagnostic methods mentioned above may be called superfluous. For example, macroscopic quantities like Pitot pressure, stagnation enthalpy, and mass flux are directly accessible only by aerodynamic probes.

A development of aerodynamic probes as well as of other diagnostic methods was therefore started parallel to the development of electric propulsion engines at Institut für Plasmadynamik, D.V.L., Stuttgart.

The size of the probes was determined by the diameter of the plasma jet to be investigated. With probes that were too small, the theoretically difficult phenomena of slip and molecule flow would be encountered whereas with probes that were too large, local measurements would become impossible. An acceptable compromise was found in all cases.

A survey will be given of the most important types of probes and the measurements performed in plasma jets from electric propulsion devices which have undergone rapid development during the last few years. In all cases the working gas was argon. In figure 3-1-1 through 3-1-3, three types of electric arc engine are presented, all examined by aerodynamic probes. Engines that can be operated for periods of days produce stationary plasma jets. Measurement techniques are therefore somewhat simplified though difficult cooling problems are encountered.

Plasma Jet at Atmospheric Pressure

Apparatus and Purpose

The pressure in the jet of an electric propulsion engine for space is of course many orders of magnitude less than one atmosphere. However, the pressure encountered at the throat of a plasma wind tunnel supersonic nozzle or in some industrial applications of plasma jets is frequently about one atmosphere, so that an ability to measure quantities in atmospheric plasma jets is necessary. Because energy supplies are often limited, atmospheric plasma jets tend to be restricted in diameter, and the probe diameters need to be at least one order of magnitude smaller still. Consequently, the technical problem is that of making a probe having a diameter as small as possible.

The arc engine shown produces a 0.6 cm diameter subsonic plasma jet (2) discharging into the atmosphere (Fig. 3-1-1). To avoid the destruction of the probe placed at 0.08 cm from the nozzle exit, the nozzle exit was surrounded by a large open plexiglass cylinder into which a small amount of nitrogen was constantly injected parallel to the jet. The direct current arc within the engine (Fig. 3-1-1) was kept at a power of 4.29 kW during all experiments, with the mass flow rate of $1.59 gs^{-1}$

Probes

Three different types of probes were used to determine the stagnation enthalpy by calorimetric means, i.e. the hot plasma drawn into the water-cooled probe transferred its energy to the cooling water, the temperature rise of which was measured.

Coaxial Calorimetric Probe - The coaxial calorimetric probe was first developed by Grey *et al* (3). W. Fischer-Schlemm succeeded in reducing its overall diameter from about 0.3 cm to 0.15 cm.

As may be seen in figure 3-1-4, it consists of three concentric thin walled tubes forming three ducts. The central duct is used for Pitot pressure measurements or as a suction channel for measurements of stagnation enthalpy. The two annular ducts lead the cooling-water into and out of the probe. The in-going and out-going water temperatures are measured by two thermocouples. A third thermocouple is used to determine, from its temperature, the small amount of energy remaining in the gas. The mass of gas that passes through the probe in a certain time interval is determined by measurements of temperature and pressure in a glass burette.

To determine the energy of the gas sucked into the probe, the rise in temperature of the cooling water when no gas is sucked in, ΔT_{WI}, is deducted from the temperature rise when it is sucked in, ΔT_{WII}, taking into account the energy remaining in the gas, $C_{pG} \cdot T_{oG}$. This assumes that the flow about the exterior of the probe does not differ significantly and consequently whether gas enters the probe or not. The stagnation enthalpy is then given by:

$$h_o = \frac{C_W \dot{m}_W (\Delta T_{WII} - \Delta T_{WI})}{m_G / \Delta t} + C_{pG} \cdot T_{oG} \qquad \text{(Eq. 3-1-1)}$$

where:

C_W = specific heat of water

C_{pG} = specific heat at constant pressure of gas at normal temperatures

$\dot{m}_W$ = mass flow of water in unit time

m_G = mass of gas sucked in during time Δt

$T_{oG} \approx$ stagnation temperature of gas

Perpendicular Calorimetric Probe - In the co-axial probe, discussed above, the cooling water is led to the probe tip and then returned. In the new perpendicular probe it passes the suction hole in one direction only (Fig. 3-1-5) and therefore needs only two concentric tubes instead of three; this allows the diameter of the probe to be made considerably smaller (0.1 cm in the present case). However, because the main axis of the probe is now placed perpendicular to the jet, the latter is perhaps disturbed more than by the co-axial probe, but the difference in the flow around the probe, with and without suction is now much smaller. The probe may be used for Pitot pressure measurements also, provided the suction hole is properly machined.

Injector Type Probe - The next step in the simplification of the enthalpy probe is to suck the gas sample directly into the cooling water. This is done by placing a baffle in the cooling water stream just ahead of the suction orifice in order to produce the necessary subpressure (Fig. 3-1-6).

Two thermocouples measure by difference the temperature rise of the cooling water. This temperature rise results from heat transfer through the copper-walls of the probe as well as from the energy of the swallowed plasma. The temperature rise due to the latter by itself can be determined by deducting the temperature rise due to heat transfer through the copper walls. This can be obtained by measuring the temperature rise in the water cooling of a second probe geometrically similar to the main probe, but without a suction hole, and isolated into the jet at the same position and tested under the same conditions.

Part of the gas swallowed by the cooling water is separated from the cooling water in a burette, while the remainder is dissolved in the water. A theoretical correction must be made for the latter.

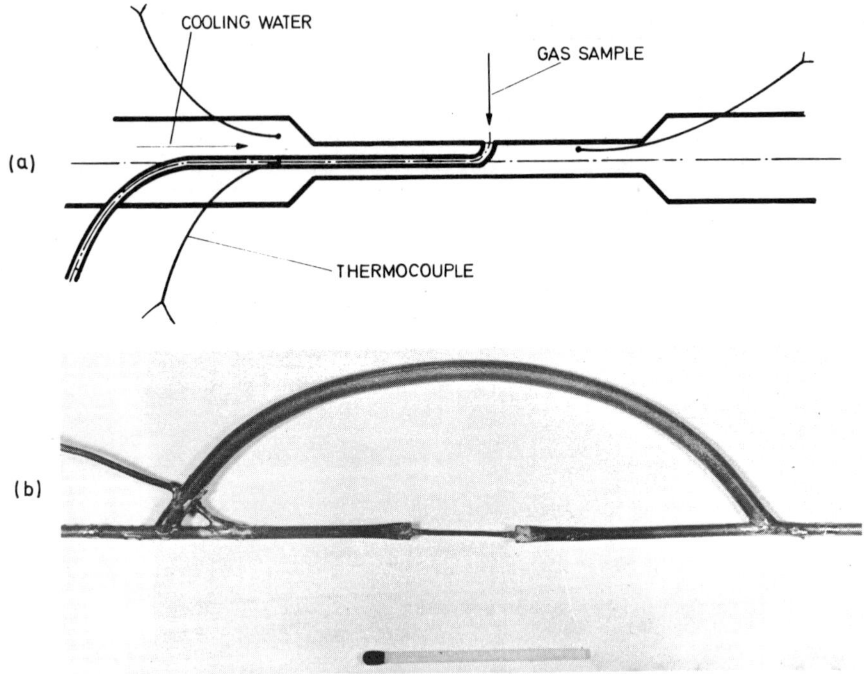

Fig. 3-1-5. Perpendicular calorimetric probe. Diagram and photograph

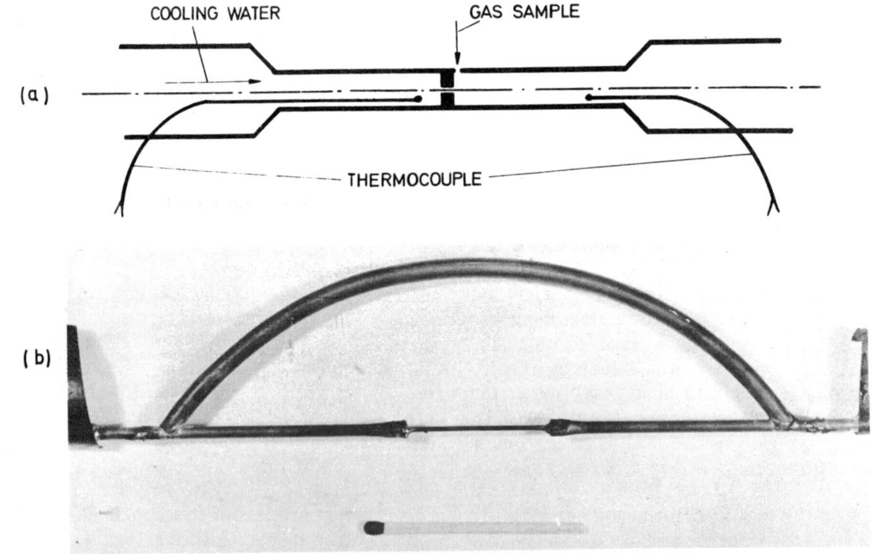

Fig. 3-1-6. Injector type calorimeter. Diagram and photograph

The stagnation enthalpy is now given by:

$$h_o = \frac{C_W \dot{m}_W (\Delta T_{WII} - \Delta T_{WI}) + C_{pG} \dot{m}_{GL} \cdot T_{WII2}}{\frac{m_{GB}}{\Delta t} + \dot{m}_{GL}} \qquad \text{(Eq. 3-1-2)}$$

using the additional expressions:

$\left.\begin{matrix}\Delta T_{WII} \\ \Delta T_{WI}\end{matrix}\right\}$ = Temperature rise of cooling water of the injector type probe and of the probe without suction hole, respectively

T_{WII2} = Temperature of cooling water during suction, taken at the position where no more heat transfer occurs

m_{GB} = Gas accumulated in burette during time Δt

$\dot{m}_{GL}$ = Mass of gas dissolved in and removed per unit time by cooling water.

The mass of dissolved gas, $\dot{m}_{GL}$, may be calculated from:

$$\dot{m}_{GL} = \rho_o \, p_B \, \dot{m}_W \, \lambda \qquad \text{(Eq. 3-1-3)}$$

where:

ρ_o = Standard atmospheric density of dissolved gas

p_B = Pressure in burette

λ = Technical solubility coefficient at temperature in burette

The injector type calorimeter is unsuitable for Pitot pressure measurements.

Test Results and Derived Quantities

Under the conditions of section 2.1 the radial distribution of the stagnation enthalpy, 0.08 cm from the nozzle exit (Fig. 3-1-1), was determined by all three types of enthalpy probes. Fig. 3-1-7 shows that the difference between the measurements obtained by the different probes never exceeded 30 percent. Since the arc discharge was not accurately reproducible, this may be considered a good result.

At atmospheric pressure and with electric and magnetic fields absent the argon plasma under consideration should be in thermodynamic equilibrium. An enthalpy-entropy-diagram including lines of constant thermodynamic quantities may therefore be used in the subsonic case to derive graphically any thermodynamic quantity from the stagnation enthalpy, Pitot pressure, and static pressure measured at each probe position. The static pressure is assumed to be constant throughout the jet and equal to the surrounding pressure. From the static and stagnation enthalpies, the plasma velocity is given by the energy relation:

$$\frac{w^2}{2} = h_o - h \qquad \text{(Eq. 3-1-4)}$$

where w = plasma velocity and h = static enthalpy.

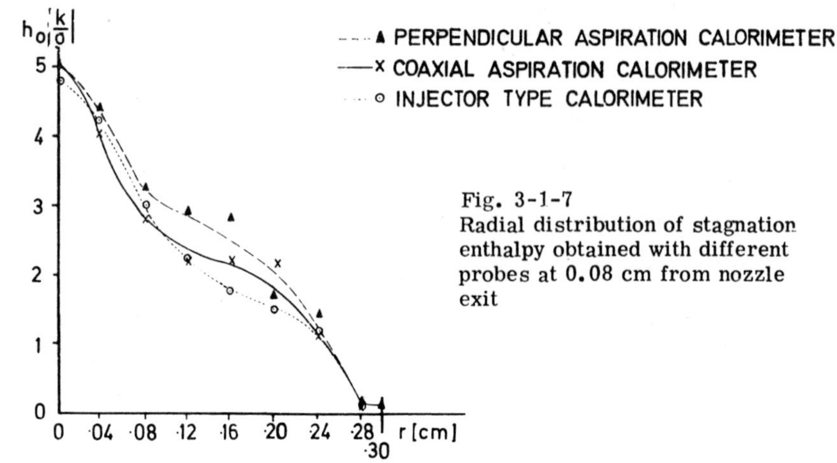

Fig. 3-1-7
Radial distribution of stagnation enthalpy obtained with different probes at 0.08 cm from nozzle exit

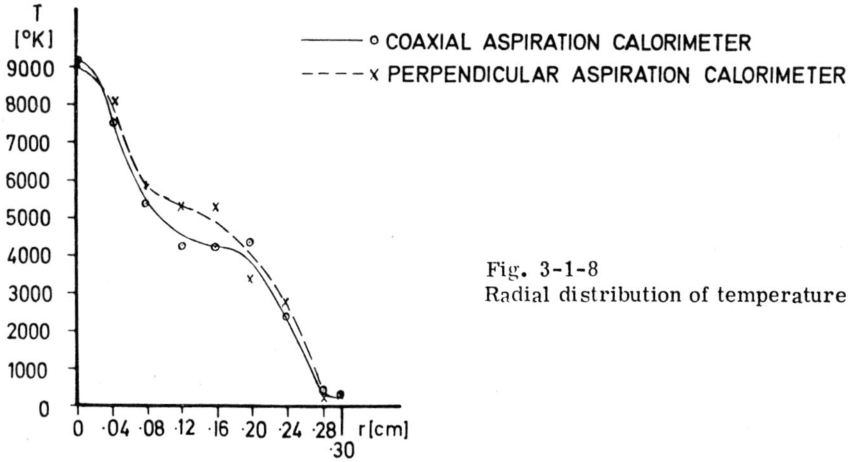

Fig. 3-1-8
Radial distribution of temperature

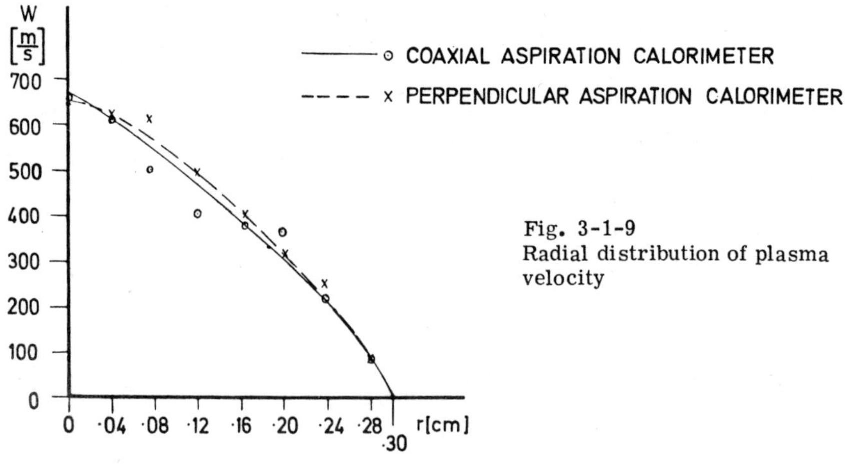

Fig. 3-1-9
Radial distribution of plasma velocity

As examples of derived quantities the radial distributions of static temperature and plasma velocity are given in figures 3-1-8 and 3-1-9 respectively.

Conclusions

The stagnation enthalpy and temperature distributions show that the atmospheric plasma jet may be regarded as a hot core coated by a cooler layer. To obtain the uniform plasma jet needed for wind tunnel purposes, a settling chamber with a second nozzle should be added to this plasma producing mechanism.

Integration of Mass Flux and Energy Flux

Denoting the plasma density by ρ , the mass and energy fluxes are ρ w and ρ wh$_o$, respectively. The total mass flow, $\dot{m}$, is obtained by integrating ρ w over the cross section of the plasma jet,

$$\dot{m} = \int_A \rho\ w\ dA = 2\pi \int_r \rho\, wr dr \qquad \text{(Eq. 3-1-5)}$$

Similarly, the total power in the jet is given by:

$$N_j = \int_A h_o \rho w\ dA = 2\pi \int_r h_o \rho\, wr dr \qquad \text{(Eq. 3-1-6)}$$

With h$_o$, ρ, w all known as shown previously, and by graphical integration of equations 3-1-5 and 3-1-6, the result may be compared with the corresponding directly measurable values.

In the present case, the mass flow as calculated by equation 3-1-5 was 8.3 percent and 9.5 percent larger than the directly measured value of 1.59 g/s for the coaxial and the perpendicular calorimetric probes respectively. Integration of the energy flux resulted in a total jet energy 0.5 percent and 7.5 percent larger than the directly measured value for the coaxial and perpendicular probes respectively. The measured total jet power was taken as being equal to the electric arc power supplied to the engine minus the heat lost by water-cooling the cathode and the nozzle, and the heat losses were determined from the water flow as measured using flow meters and the water temperature rise as measured using thermistors.

Supersonic Plasma Jet Emerging from an Electrothermal Plasma Accelerator

Apparatus

For a long time the electrothermal arc jet engine was the only stationary arc device available for accelerating a gas to high velocities. Within that engine, the gas is heated by an electric arc at high pressure with subsequent, preferably isentropic expansion to supersonic velocities by a de Laval nozzle.

Figure 3-1-2 shows such an engine. It is essentially the same engine as that shown in figure 3-1-1 except for the additional supersonic part of the nozzle. The supersonic nozzle was designed using the method of Foelsch (5) to produce a uniform parallel jet with constant static pressure within the whole jet. The nozzle throat and nozzle exit diameters are 0.3 and 1.8 cm respectively. In figure 3-1-10, the jet which was examined by aerodynamic probes in this study (4), is shown. The axial positions (0.25 cm and 5.5 cm from nozzle exit) where measurements of the radial distributions of some quantities were made, are indicated on the photo. The following operating conditions obtained throughout:

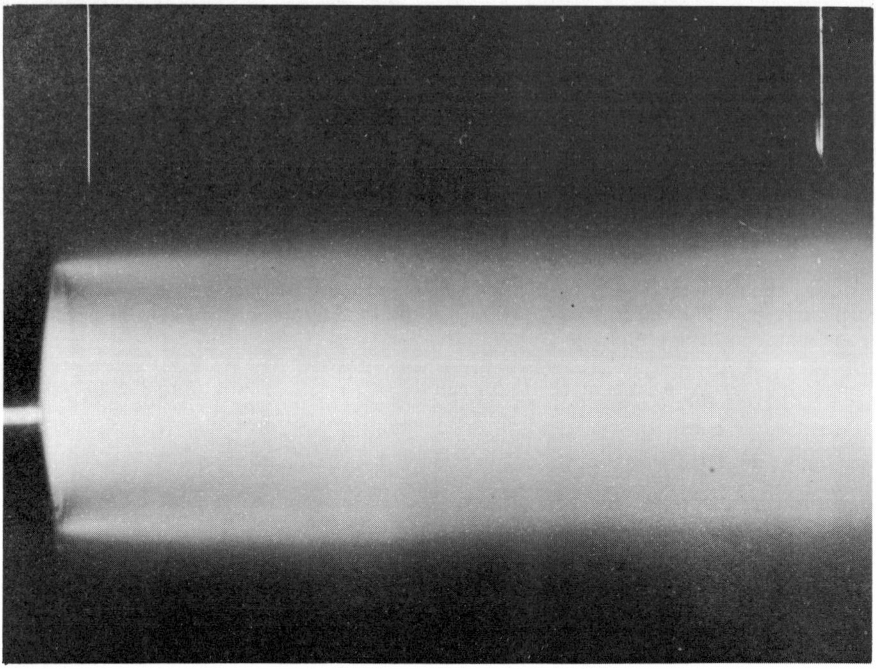

Fig. 3-1-10 Supersonic plasma jet with parallel streamlines

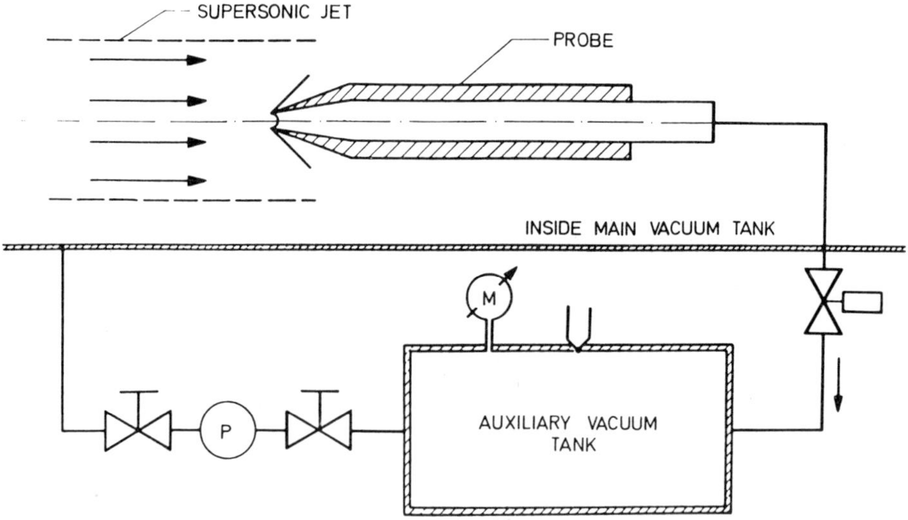

Fig. 3-1-11 Scheme of mass flux measurement

direct current	140 A
arc voltage	32 V
mass flow, argon	0.987 g/s
static pressure within jet = pressure in surrounding vacuum tank	3 Torr

Probes

Pitot Pressure Probe - The impact tube used for Pitot pressure measurements in gas dynamics may also be used in plasma dynamics. For stationary flow, water cooling is necessary.

If small pressures are met during the experiments, outgassing of long pressure leads, and thermal transpiration at high temperatures may cause errors in Pitot pressure measurement. In the present case it was estimated that those errors were negligibly small.

At probe Reynolds numbers below about 100 considerable viscosity effects are to be expected. In this region, with decreasing Reynolds number the measured Pitot pressure first decreases to a value some percent smaller than the theoretical Pitot pressure, and then, after a minimum, it rises again to a value considerably above the theoretical. The errors depend largely on the nose shape of the probe. This effect is described in detail by Potter and Bailey (6).

The exact experimental determination of the Reynolds number in plasma flow is difficult. In the present case it was estimated to be eventually smaller than 100. Thus it was necessary to verify that the probe Reynolds number was not so small that errors were introduced due to viscosity effects. For that purpose the Pitot pressure distribution at one axial station of the jet was measured by two probes with nose shapes that were geometrically extremely different: essentially similar results were obtained, which means that viscosity errors were not involved.

Mass Flux Probe - The mass flux may be measured directly by sucking gas into a small auxiliary vacuum tank for a certain period of time (Fig. 3-1-11). To make this possible, thin-lipped, externally chamfered probe tips are used to meet the following requirements:

(a) The gas is contained in the stream tube, that has the same diameter as the probe mouth and no more and no less must enter the probe.

(b) The flow ahead of the probe tip must be undisturbed.

In other words, the axisymmetric bow shock forming ahead of the probe should be swallowed by the probe during suction, leaving the oblique part of the shock attached to the probe lips (Fig. 3-1-11).

The mass of gas sucked into the probe during the time interval Δt is given by:

$$m_G = \rho \, w \, A_p \, \Delta t \qquad \text{(Eq. 3-1-7)}$$

where A_p is the probe tip area. m_G may be determined from the pressure and temperature rises in the auxiliary tank:

$$m_G = \frac{V}{R}\left(\frac{p_n}{T_n} - \frac{p_v}{T_v}\right) \approx \frac{V}{RT_R}\left(p_n - p_v\right) \qquad \text{(Eq. 3-1-8)}$$

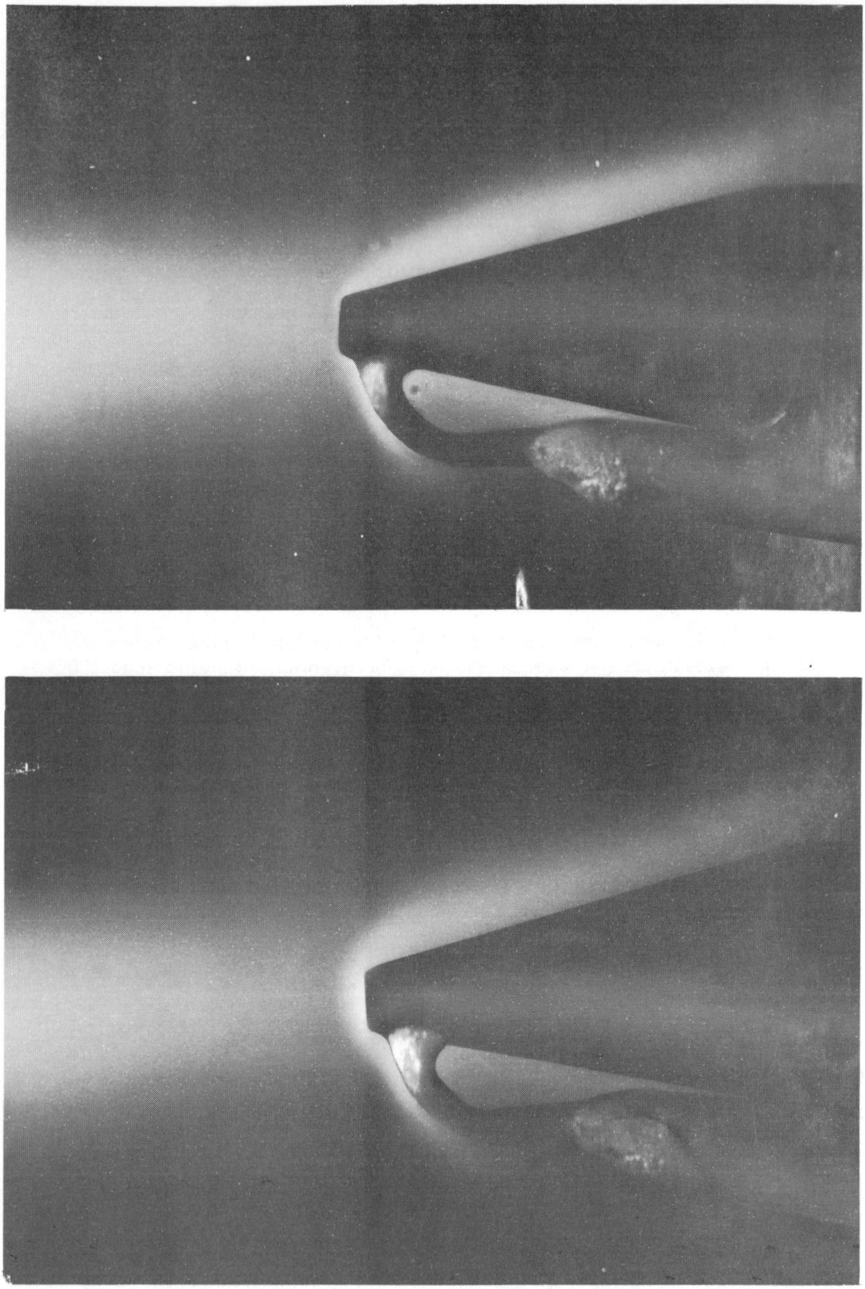

Fig. 3-1-12 Mass flux probe during suction (above) and non-suction

where:

V = volume of auxiliary tank

R = specific gas constant of cold gas

$p_{v,n}$, $T_{v,n}$ = pressure and temperature, respectively, of gas in auxiliary tank before and after suction

T_R = tank wall temperature $\approx$ room temperature

Equations 3-1-7 and 3-1-8 give:

$$\rho w = \frac{V}{R T_R \; A p} \cdot \frac{p_n - p_v}{\Delta t} \sim \frac{\Delta p}{\Delta t} \qquad \text{(Eq. 3-1-9)}$$

Stationary Calorimetric Probe - The stationary calorimetric probe uses the same principle as the coaxial calorimetric probe designed for subsonic atmospheric plasma flows, as shown previously. It is shown in figure 3-1-13.

A supersonic plasma jet is generally considerably larger than the jet described previously, and the probe diameter may therefore be much larger also. Heat insulation can thus be incorporated between two now separated coolant flows, the outer one protecting the probe, the inner one being used for the measurement. The comparative measurement without suction (Eq. 3-1-1) drops out and the stagnation enthalpy becomes:

$$h_o = \frac{c_w \; \dot{m}_w \cdot \Delta T_w}{m_G / \Delta t} + C_{pG} \cdot T_{oG} \qquad \text{(Eq. 3-1-10)}$$

With the bow shock swallowed by the probe, an interesting quantity, i.e. the energy flux, may also be directly obtained by the measurement. It is an energy flow related to an area and is given by:

$$S = \frac{c_w \cdot \dot{m}_w \, \Delta T_w + C_{pG} \, T_{oG} \cdot m_G/\Delta t}{A_p} \qquad \text{(Eq. 3-1-11)}$$

so that from equations 3-1-7, 3-1-10, 3-1-11:

$$h_o = \frac{S}{\rho w} \qquad \text{(Eq. 3-1-12)}$$

Equation 3-1-12 is also valid for the subsonic case as shown previously. Thus one single probe meets all requirements for the measurement of four partially interdependent quantities, namely Pitot pressure, mass flux, energy flux, and stagnation enthalpy.

Nonstationary Calórimetric probe - In some cases this stationary calorimetric probe may suffer the disadvantage of the temperature rise of the inner coolant stream during suction, and this may lead to a wide scattering of the measured results.

A considerably larger temperature rise may be obtained by replacing the inner coolant double tubing by one solid copper tube having mass m_{Cu}. The energy will then be stored up in the mass of copper instead of being carried away by the cooling

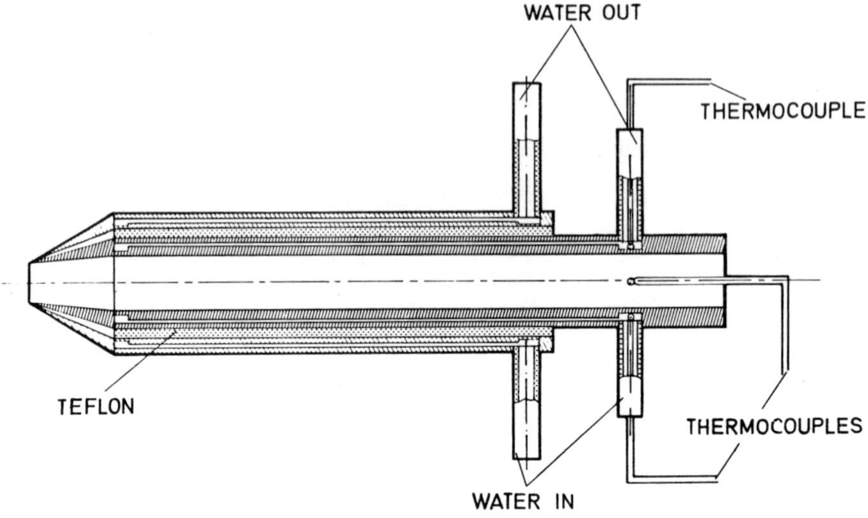

Fig. 3-1-13   Stationary calorimetric probe. Scheme

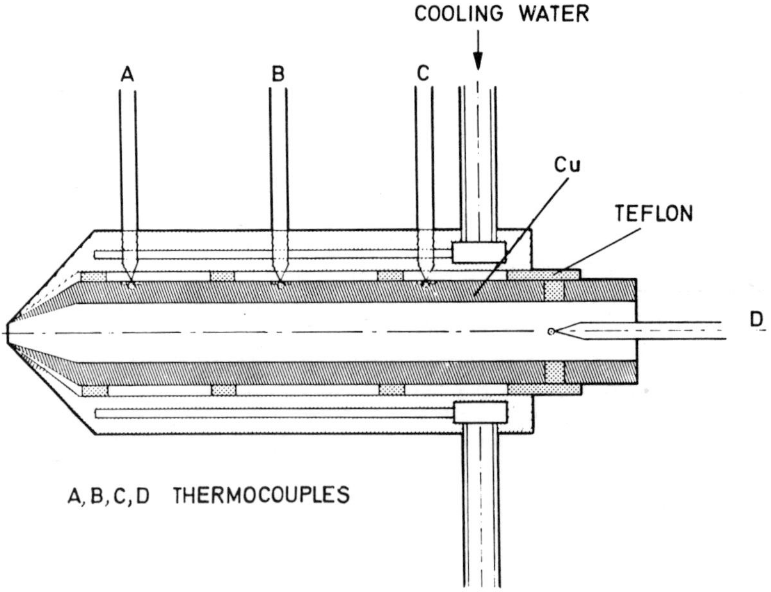

Fig. 3-1-14   Non-stationary calorimetric probe. Scheme

water. As the temperature of the copper now increases with time during suction, the flow through the probe may be called nonstationary.

The increase in temperature of the copper, ΔT_{Cu}, can be determined by thermocouple measurement, but as the temperature of the copper has a gradient with higher values at the probe tip, three thermocouples were imbedded into the copper tube. Stagnation enthalpy and energy flux (the latter only for the supersonic case and with bow shock swallowed) are then given by:

$$h_o = \frac{C_{Cu} \int_{m_{Cu}} \Delta T_{Cu}\, dm_{Cu}}{m_G} + C_{pG} \cdot \overline{T}_{oG} \qquad \text{(Eq. 3-1-13)}$$

and :

$$S = \frac{\frac{1}{\Delta t}\left(C_{Cu} \int_{m_{Cu}} \Delta T_{Cu}\, dm_{Cu} + C_{pG}\, m_G\, \overline{T}_{oG} \right)}{A_p} \qquad \text{(Eq. 3-1-14)}$$

with C_{Cu} written for the specific heat of copper.

If the bow shock remains swallowed by the probe during the whole suction time, the probe may be used also for mass flux measurement.

Experimental Results

Mach Number - Figure 3-1-15 shows Mach number profiles that were obtained from Pitot pressure measurements assuming that the static pressure was constant over the jet. In the subsonic part of the jet, the Mach number was obtained from the isentropic relation between Mach number, Pitot pressure, and static pressure. In the supersonic part of the profile, the Mach number was derived from the Rayleigh equation. Since the isentropic exponent was assumed to be constant throughout the jet, the Mach number distribution is only an approximation.

Figure 3-1-16 shows the measurement described previously proving that the Reynolds number in the probe is large enough to avoid viscosity errors in the interpretation of the Pitot pressure measurements.

Mass Flux - The radial distributions of mass flux are plotted in figure 3-1-17. Close to the nozzle exit, distinct maxima occur at the jet boundary; these may be explained by the water cooling, which creates a cooler layer around the plasma flow within the nozzle. Because the pressure may be regarded as constant over a cross-section within the nozzle, the density is larger in that layer than in the core of the flow. At a distance of 5.5 cm from the nozzle exit, the maxima mentioned above have completely disappeared (Fig. 3-1-17).

The influence of suction time on the ρw-measurement is demonstrated by figure 3-1-18 in which the ratio $\Delta p/\Delta t$ ($\sim \rho w$, see Eq. 3-1-9) is plotted against the suction time interval. A certain mass of gas is contained in the probe and in the tube leading from the probe to the electromagnetically operated valve that regulates the suction time. This gas is sucked into the auxiliary tank with the sample gas and causes positive relative errors that are inversely proportional to the suction time interval. From figure 3-1-18 it may be concluded that the suction time interval in the present case must exceed 25 seconds.

Energy Flux and Stagnation Enthalpy - Radial distribution of energy flux, obtained from measurements with the stationary calorimetric probe by way of equation 3-1-11

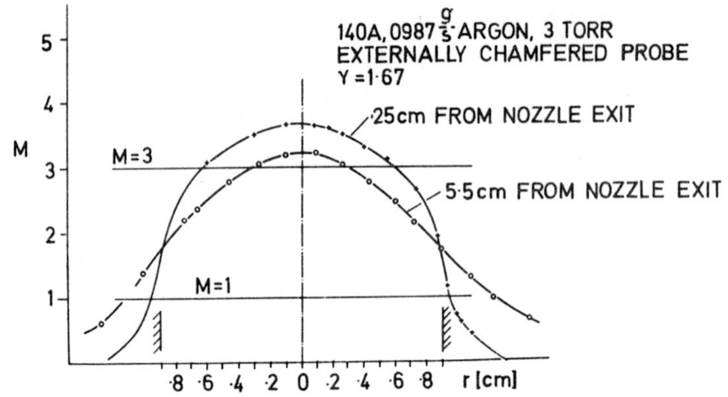

Fig. 3-1-15 Radial Mach number distributions

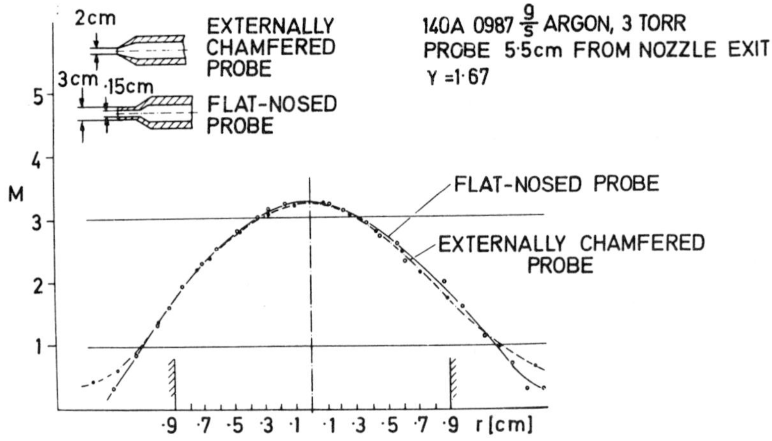

Fig. 3-1-16 Mach number distributions obtained with different probes

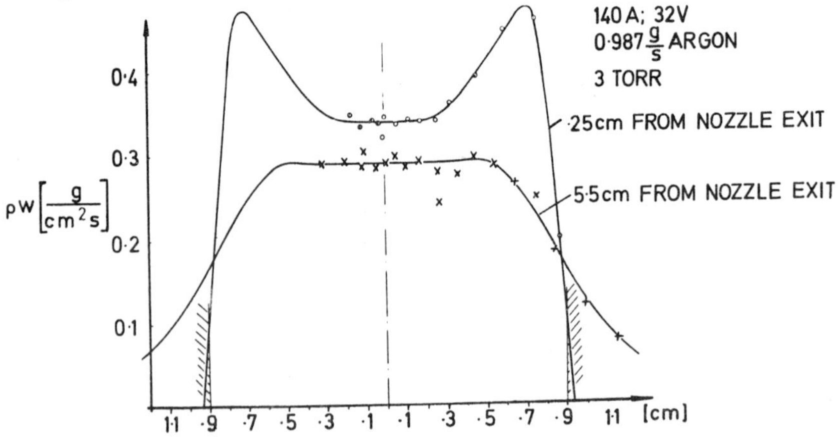

Fig. 3-1-17 Radial distribution of mass flux

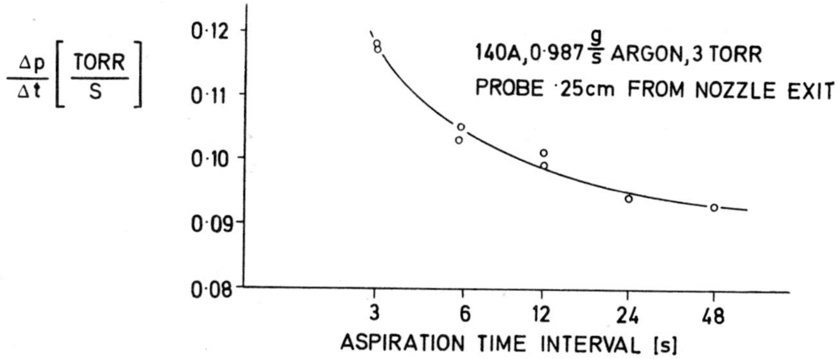

Fig. 3-1-18 Influence of suction time

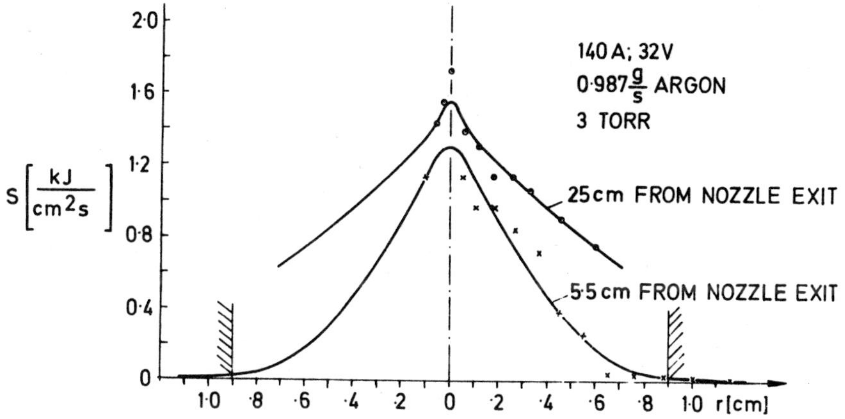

Fig. 3-1-19 Radial distribution of energy flux

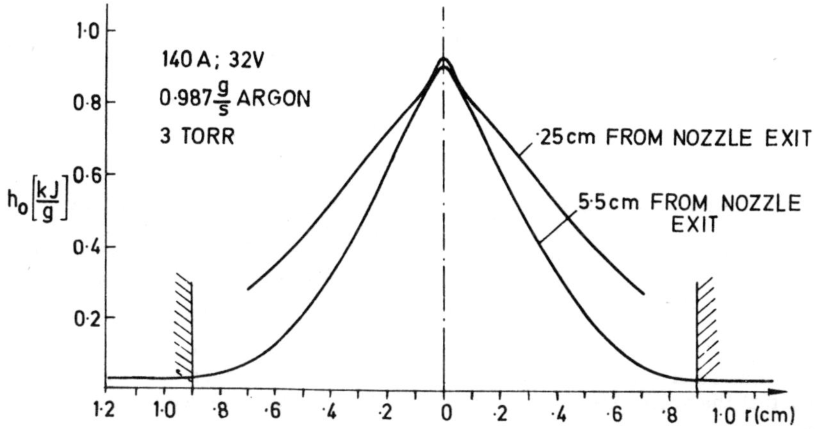

Fig. 3-1-20 Radial distribution of stagnation enthalpy

are shown in figure 3-1-19. Figure 3-1-20 shows the corresponding radial distributions of the stagnation enthalpy as calculated from the mean curves of figures 3-1-17 and 3-1-19 (mass and energy fluxes respectively) by the aid of equation 3-1-12.

With increasing distance from the nozzle exit the stagnation enthalpy in the centre of the jet remains essentially unchanged within the range of the present experiments. However, considerable enthalpy losses occur at the jet boundary within the same range.

Up to now, the non-stationary calorimetric probe has been used only for rather crude experiments, for instance the temperature rise of the thermocouple in the central portion of the copper heat capacitor was taken as representative of the temperature rise in the whole of the copper tube. These experiments yielded enthalpy distributions about 40 percent below those obtained with the stationary probe. As expected, the scattering of the plotted points was much less when the non-stationary probe was used instead of the stationary probe.

Derived Quantities

If thermodynamic equilibrium is applicable to the present supersonic case, the enthalpy-entropy-diagram mentioned previously can be used for the unambiguous determination of the remaining thermodynamic quantities. For the supersonic case, however, and with knowledge of the aerodynamically measurable quantities, p'_o; h_o; p, ρw, the graphical method becomes iterative. It is described in detail in (4) and consists essentially of the following steps:

Estimate the total pressure, p_o by using p, p'_o and the Rayleigh formula: this gives to a first approximation, the static conditions on the line of constant static pressure in the enthalpy-entropy diagram, whence first approximations of enthalpy, h, and density, ρ , may be read.

Calculate by equation 3-1-4 the first approximation of the plasma velocity, w.

Compare the first approximation of $\rho . w$ with the measured value (ρw).

Choose better approximations on the line of constant static pressure in the neighbourhood of the first approximation until both become equal.

For the present case, in the jet centre, the above method yielded a static temperature of 1586°K, a stagnation temperature of 8200°K, a total temperature of 8400°K, and a Mach number of 3.724. The Mach number agreed well with the value derived from the Rayleigh formula for conditions in the centre of the jet shown in figure 3-1-15 to be 3.67 ($\gamma = 1.67$).

Strictly, at least within the de Laval part of the nozzle, relaxation effects should be taken into account. The distribution of energy among the different degrees of freedom, for example, cannot equilibrate within the short time available to the plasma as it passes at high velocity through the nozzle, where the pressure undergoes rapid decrease. An enthalpy-entropy-diagram cannot be designed in these circumstances, nor should the equilibrium diagram be used.

Fortunately, in this case, the temperatures were relatively low and the static pressure was relatively high. The good agreement of the Mach numbers tends to confirm that the values obtained using the equilibrium diagram were a good approximation.

Integration of Mass Flux and Energy Flux

After graphical integration of mass flux and energy flux following equations 3-1-5 and 3-1-6, the comparisons described previously may be made:

The integration yielded:

(a) Mass flows differing from the integrally measured values by percentages between -13.4 percent and + 10 percent.

(b) Ratios of the jet power to the power supplied by the electric arc that agree well with some integrally measured ratios obtained earlier with similar engines at similar conditions.

Plasma Jet Emerging from a Hall Current Accelerator

Apparatus

At present, the Hall current accelerator of which a recent variation is shown in figure 3-1-3, is the most promising and most intensively investigated electric arc thrustor for space applications. It may be shown from theory that extremely high plasma velocities, e.g. more than 10^5 m/s (Hydrogen) may be achieved at extremely low flow rates.

Apart from some aerodynamic contribution, the main acceleration of the plasma with the Hall current accelerator is performed by electromagnetic body forces given by the vector product $\vec{j} \times \vec{B}$, where $\vec{j}$ and $\vec{B}$ are the electric current density and the magnetic field strength respectively. The latter may be composed of the self-magnetic and a superimposed external magnetic field. A detailed description of the principle of this type of plasma acceleration may be found in (7).

In the jets of the atmospheric plasma engine and the thermoelectric engine, electric currents and magnetic fields do not exist. In the free jet of the Hall current accelerator, however, particularly close to the nozzle exit, considerable electric currents are present, and these interact with the two types of magnetic fields. Figure 3-1-21 shows schematically the geometric form of the electric stream lines ranging far out into the free jet.

It follows that, when aerodynamic probes are inserted into a jet containing electric currents, measurements made may depend on whether or not some current flows through the probe body. Furthermore, the behaviour of the electric current depends largely on the size and material of the probe.

The measurements to be described were performed at the following conditions of the Hall current accelerator:

electric current	800 A
arc voltage with magnetic field present	46 V
mass flow rate	.22 g/s
pressure in the jet surroundings	. 5 Torr
magnetic field strength produced by 2 coils in a typical point	1350 Gauss.

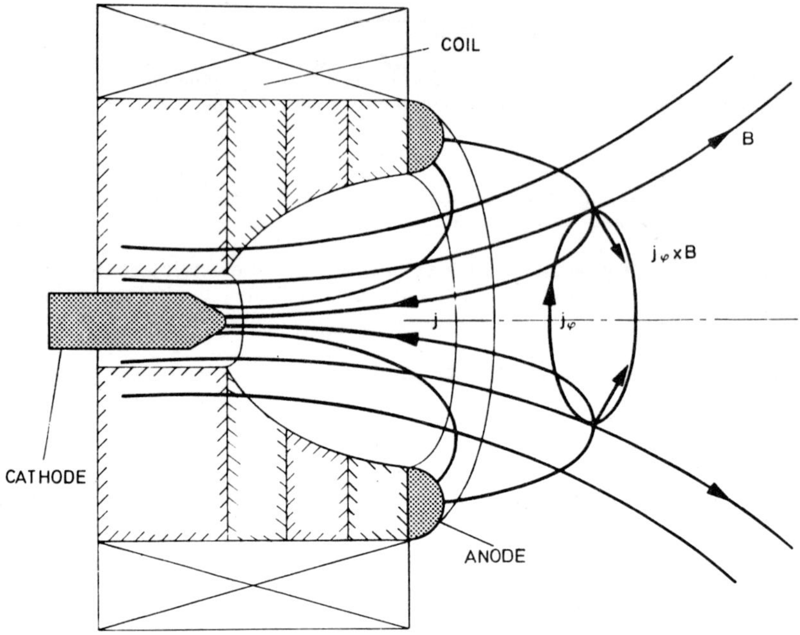

Fig. 3-1-21   Electric streamlines and magnetic lines in the jet from Hall current accelerator. Scheme

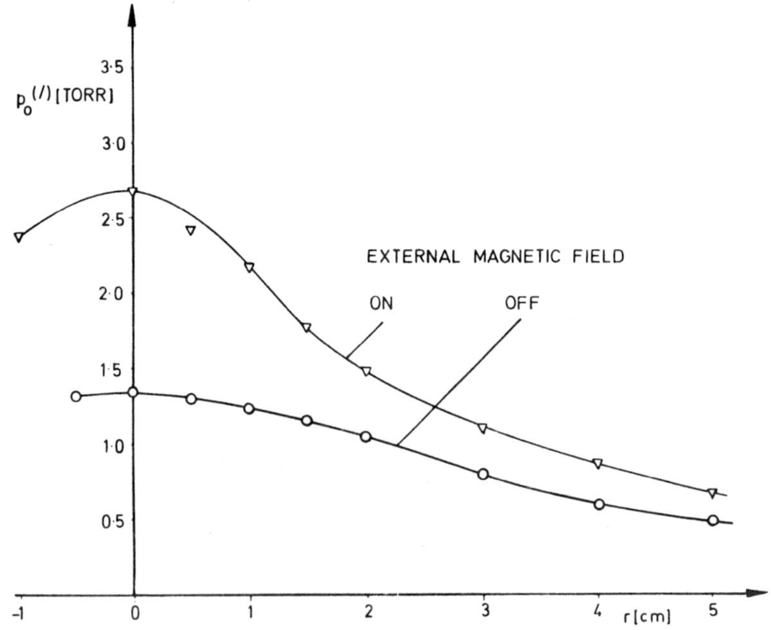

Fig. 3-1-22   Radial distribution of Pitot pressure, 17 cm from nozzle exit

Pitot Pressure, Mass Flux, Stagnation Enthalpy

The diameter of the jet was about 8 cm. Probes could therefore have the large diameters (1 cm at the mouth), that were needed at the low Reynolds numbers resulting from the low densities.

The radial distribution of Pitot pressure at a position 17 cm from the nozzle exit is presented in figure 3-1-22 to show the influence of the external magnetic field.

Because of technical difficulties, no quantitative results could be obtained from mass flux and stagnation enthalpy measurements. Qualitative measurements, however, showed that the mass flux profile at this particular axial position should have been very similar to the Pitot pressure profile of figure 3-1-22, with external magnetic field present.

Swirl

Production of Swirl - Consider the generalized Ohms law:

$$\vec{j} \times \vec{B} = \frac{\sigma}{1 + (\omega \tau)^2} \left[\vec{E}^* \times \vec{B} + \frac{\omega \tau}{B} \left((\vec{B} \times \vec{E}^*) \times \vec{B}\right)\right]$$

$$\vec{E}^* = \vec{E} + w \times \vec{B} \qquad \text{(Eq. 3-1-15)}$$

The above equation gives the direction of the force $j \times B$. The acceleration and the plasma velocity should have the same direction.

Now, the second expression in the brackets of equation 3-1-15 is essentially a vector directed axially and radially, while the first term is a vector with azimuthal direction. Consequently it may be expected that some swirl is produced within the range of action of the electromagnetic forces. Patrick and Powers (8) confirmed the existence of swirl in the jet of their annular arc, which was similar in principle though different geometrically from the Hall current accelerator.

In order to verify experimentally the existence of swirl in the exhaust jet of the present engine, a suitable swirl probe was constructed. As in (8), it was used to determine the local swirl angle between the axial and azimuthal components of the macroscopic plasma velocity. Then the swirl angle α is obtained from

$$\tan \alpha = \frac{w_\varphi}{w_z} \qquad \text{(Eq. 3-1-16)}$$

The radial component of the velocity is not of interest in this connection, but it can be assumed that it is small at large distances from the nozzle exit.

Swirl Probe - The swirl probe is a variation of the well-known aerodynamic yaw probe. Essentially it consists of an intensively water-cooled copper cylinder 1.6 cm diameter and 5 cm long in which two orifices of 0.2 cm diameter are located 1 cm from one end, their center lines forming an angle of 90° where they intersect at the axis of the cylinder.

Slip flow or transition flow deviations need not be considered since absolute pressures are not measured. The swirl angle is measured by rotating the cylinder until the pressure differential between the two orifices is zero.

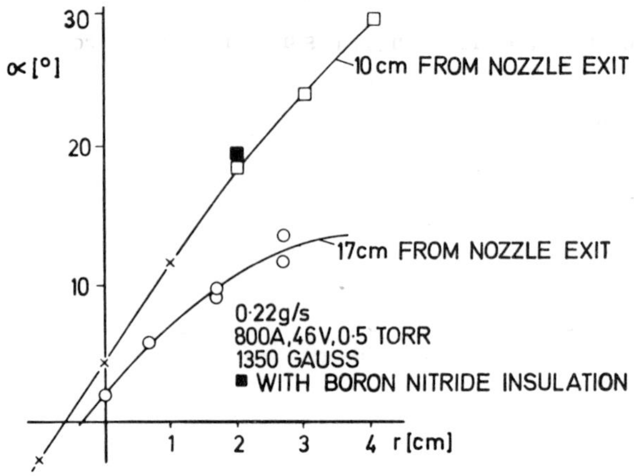

Fig. 3-1-23 Radial distributions of swirl angle

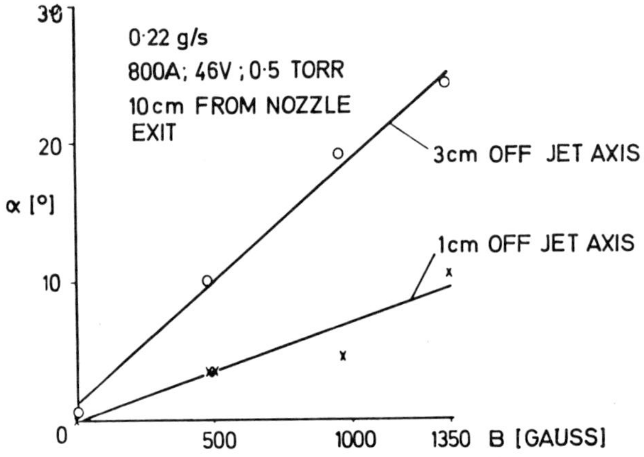

Fig. 3-1-24 Swirl angle vs. external magnetic field of the Hall current accelerator

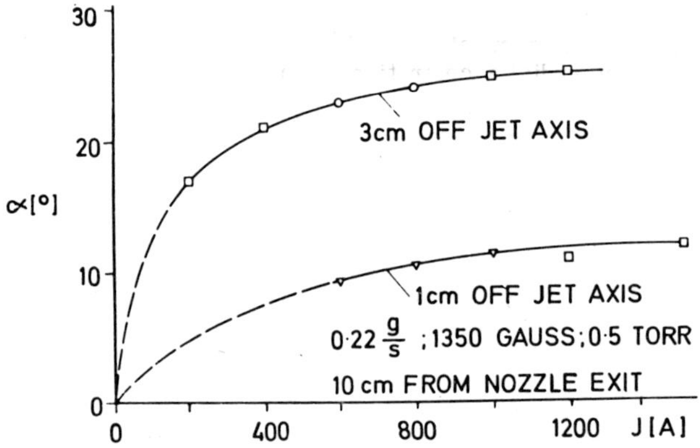

Fig. 3-1-25 Swirl angle vs. arc current of the Hall current accelerator

There was no evidence of any electric current passing through the probe. No difference was observed in the non-insulated case when a great part of the probe was insulated with boron nitride (Fig. 3-1-23). Also there was no measurable change in arc voltage when the probe was inserted into the jet.

Measurements - In figures 3-1-23 through 3-1-26 different symbols are used for different sets of experiments to demonstrate the repeatability of the results.

Figure 3-1-23 shows the radial distribution of the swirl angle, indicating that when w_z is assumed constant the jet is rotating like a solid body.

Figure 3-1-24 demonstrates that the swirl angle is proportional to the external magnetic field, B.

Figure 3-1-25 shows the swirl angle growing very slowly with arc current in the interesting region of high currents.

Figure 3-1-26 shows the swirl angle becoming smaller as the distance from the nozzle increases. Thus the swirl seems to be produced essentially within or close to the nozzle.

Radial Distribution of Axial Velocity as Derived from Swirl Angle and Spectroscopic Doppler Effect Measurements of the Azimuthal Velocity

While measurement of the radial distribution of the axial plasma velocity of a jet still presents difficulties using the spectroscopic Doppler method. Beth. *et al* (9) have succeeded in using this method to determine the radial distribution of the azimuthal velocity. Knowing the swirl angle and the azimuthal velocity from (9), the axial velocity can be obtained readily from equation 3-1-16. Figure 3-1-27 shows the results of such an evaluation, the radial distribution of the axial velocity being given at two axial stations. The operating conditions were the same as those mentioned previously, except that the magnetic field was 1200 Gauss and one of the axial stations was 8 cm from the nozzle exit instead of 10 cm. To account for these small differences, interpolation corrections were made by the aid of figures 3-1-24 and 3-1-26.

Conclusions

In all cases the probe measurements gave a swirl direction consistent with the swirl direction that is predicted by the first term in the brackets of equation 3-1-15. Reversing the direction of the external magnetic field should reverse the swirl direction with the axial and radial acceleration remaining unchanged.

Figure 3-1-27 shows that there is:

(a) A considerable increase of w_z between stations 8 cm and 17 cm from the nozzle exit indicating that a significant part of the axial velocity is developed at a great distance from the nozzle;

(b) A velocity minimum in the centre of the jet,

(c) A relatively uniform distribution of velocity at 17 cm from the nozzle,

(d) Excellent agreement between the measured axial velocity and the mean axial velocity obtained from $\overline{w}_z = \frac{F}{\dot{m}}$ using values F = thrust and $\dot{m}$ = mass flow rate, as measured by Ungerer (10); from this it is assumed that the maximum velocity has been attained at a station 17 cm from the nozzle.

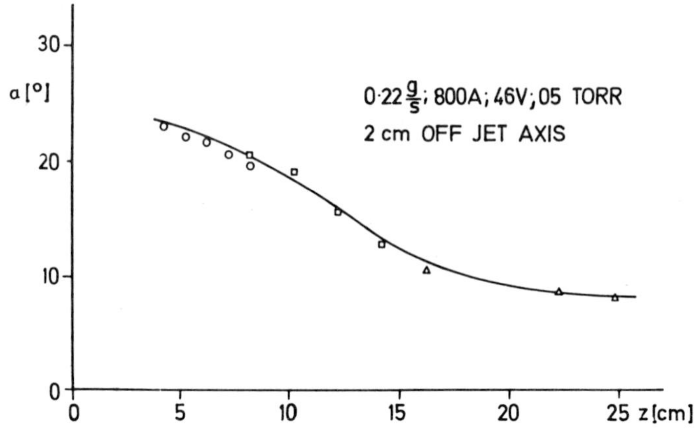

Fig. 3-1-26 Swirl angle vs. axial distance from the nozzle exit

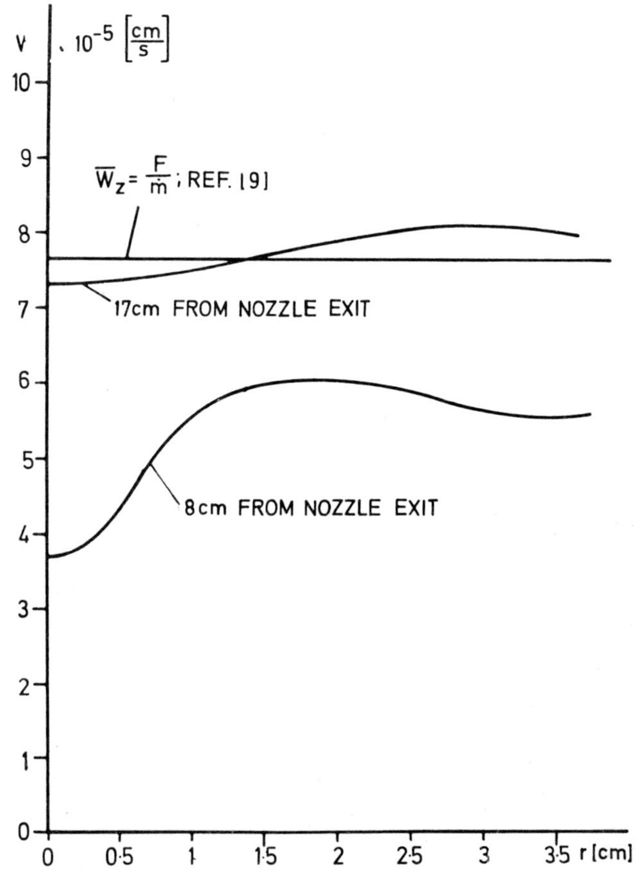

Fig. 3-1-27 Radial distribution of axial plasma velocity

Because of the good agreement it may be concluded that no significant entrainment of ambient gas exists in this special case, otherwise $\dot{m}$ would be larger than Ungerer measured.

Assuming that at the station of 17 cm downstream of the nozzle exit electromagnetic forces are no longer important, in other words that an essentially constant distribution of static pressure has developed over the jet radius, an interesting conclusion regarding the radial temperature distribution may be drawn from two independent measurements.

i A maximum Pitot pressure measured in the centre of the jet (Fig. 3-1-22) at constant static pressure signifies a peak in the Mach number distribution. Because $M = \frac{w}{a}$ with M and a written for the Mach number and the sonic speed respectively, and because $w \approx$ const. (Fig. 3-1-27), it follows that both a and the temperature are at a minimum in the centre of the jet.

ii With the first measurements of ρ w indicating a maximum in the centre of the jet (see section 4.2), and w being approximately constant (Fig. 3-1-27), ρ also must be a maximum. With p assumed constant and ρ exhibiting a maximum, the temperature in the centre of the jet must be a minimum.

The deduction that there is a minimum in the temperature distribution at 17 cm from the nozzle has been further supported by a photograph showing that the jet has a relatively dark cone when compared with the brighter outer regions.

Acknowledgements

The readiness of Dipl.-Ing. W. Fischer-Schlemm to make available the results of his experiments concerning the atmospheric plasma jet is appreciated.

Special mention is due Mr. E. Martin for valuable assistance in performing all other experiments.

References

1. **Fischer-Schlemm, W.,** 'Sonden zur Messung von Enthalpie und Druck hocherhitzter Gase'. DLR-FB 67-17, March 1967.

2. **Peters, T.,** 'Erzeugung von Plasmastrahlen hoher Temperaturen und Geschwindigkeiten', Astronautica Acta VII, 1961, 150.

3. **Grey, J., Jacobs, P., Sherman, M.,** 'Calorimetric Probe for the Measurement of Extremely High Temperatures', Rev. Sci. Inst., 33, 1962.

4. **Krause, S.,** 'Messungen mit gasdynamischen Sonden an einem Uberschall-plasmastrahl', DLR-FB 66-49, July 1966.

5. **Foelsch, K.,** 'The Analytical Design of an Axially Symmetric Laval Nozzle for a Parallel and Uniform Jet', J.A.S., March 1949.

6. **Potter, J.L., Bailey, A.B.,** 'Pressures in the Stagnation Regions of Blunt Bodies in the Viscous Layer to Merged-Layer Regimes of Rarefied Flow,' AEDC-TDR 63-168, ARO Inc., Sept. 1963.

7. **Krülle, G., Ungerer, E.,** 'Untersuchungen an kontinuierlich arbeitenden, achsensymmetrischen Plasmatriebwerken mit elektromagnetischer Beschleunigung,' Raumfahrtforschung 11, 1, 1967.

8. **Patrick, R.M., Powers, W.E.,** 'Plasma Flow in a Magnetic Arc Nozzle', 3rd Symp. on Advanced Propulsion Concepts, Cincinnati, Ohio, USA, Oct. 1962.

9. **Beth, M.U., Kling, M.G., Nedder, G., Bohn, W.L.,** 'Angular Velocity Profiles of a Rotating Argon Plasma Jet', 8th Int. Conf. on Phenomena in Ionized Gases, Vienna, Sept. 1967.

10. **Krülle, G., Ungerer, E.,** DVL-Institut f. Plasmadynamik, Stuttgart, unpublished experiments, 1966.

List of Mathematical Symbols

A	area
a	speed of sound
B	magnetic field
C	specific heat
C_p	specific heat at constant pressure
E	electric field
F	thrust of Plasma accelerator
h	enthalpy
j	current density
M	Mach number
m	mass
N	power
p	pressure
$p_o^{(\prime)}$	Pitot pressure, meaning p_o for subsonic and p_o' for supersonic speeds, respectively
R	specific gas constant
r	jet radius
S	energy flux
T	temperature
t	time
V	volume of auxiliary vacuum tank
w	velocity
z	downstream coordinate, with origin at nozzle exit
α	swirl angle
γ	isentropic exponent
λ	technical solubility coefficient
ω	cyclotron frequency
ρ	density
σ	electrical conductivity
τ	collision time of electrons

Subscripts and Superscripts

Cu	copper
G	gas
J	jet
L	dissolved
P	probe
R	room
w	water
n, v	conditions in auxiliary vacuum tank after and before suction, respectively
o	total
' o	stagnation
r, p, z	cylinder coordinates

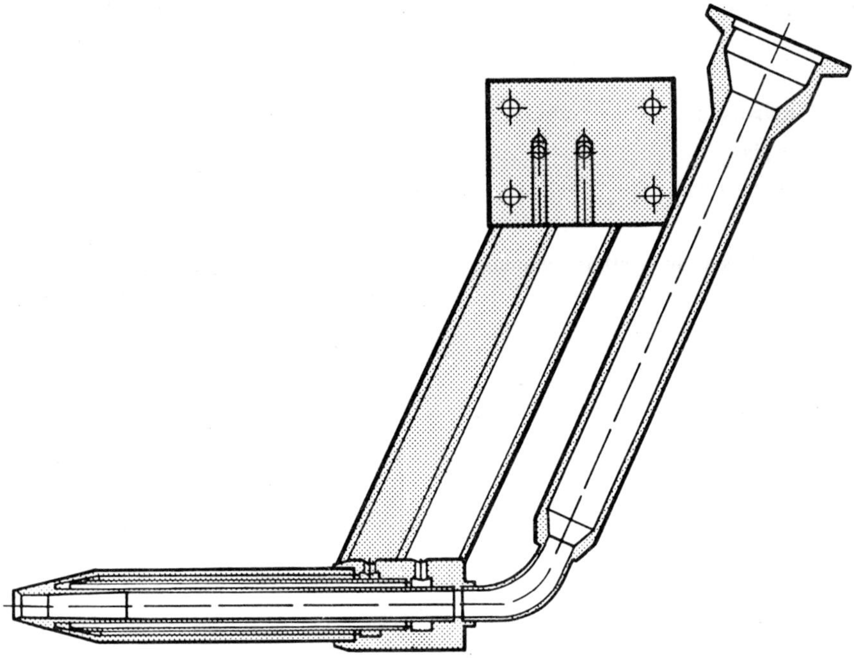

Fig. A3-1-1 Mass flow probe. Scheme

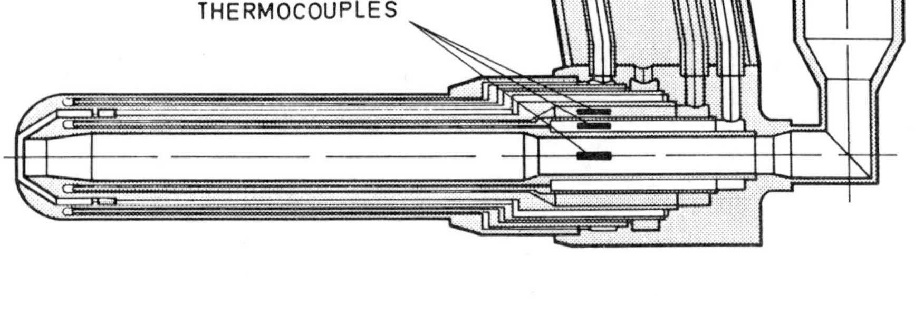

Fig. A3-1-2 Total enthalpy probe. Scheme

Commentary A on Chapter 3-1

Aerodynamic Probes for the DVL Low-Density Arc Tunnels

M. Fiebig, K. Kindler.

Introduction

The DVL arc tunnels have been designed for the study of low density flows of high-temperature and high-Mach number. Typical operating conditions for the tunnels are (1):

stagnation temperature (°K)	$3000° < T_o < 10\,000$ °K
Mach number	$5 < M < 15$
static pressure (Torr)	$10^{-3} < p < 10$

Aerodynamic probes have been designed for profile measurements of mass flow total enthalpy and pitot-pressure in the free jet test section. In the following the main design features and measurement criteria for the probes will be discussed.

Mass flow Probe

A cross-section of the mass flow probe is shown in figure A3-1-1. The outer probe diameter is 20 mm, the entrance diameter 9 mm. The high arc jet temperatures make a cooled probe necessary. The inlet design is of special interest. To fix the probe's capture area to its lip in a low-density, high Mach number flow it is necessary to design an entrance lip which possesses cylindrical symmetry and which is also symmetrical about the lip itself, (2). Therefore a 15° half-angle blunted double wedge has been chosen for the lip, the radius of curvature of the blunted lip being 0.2 mm. Even when the front shock is swallowed by the probe an unsymmetrical wedge for the lip geometry is not sufficient to force the stagnation streamline to the tip of the lip, as an attached lip shock cannot be established at low densities, high Mach numbers and high temperatures. In a high Mach number low density flow the shock stand-off distance cannot be neglected, as the high stagnation temperature will erode any sharp lip to a blunted one. If an unsymmetrical lip design is chosen the finite shock stand-off distance will lead to a flow field where the probe's capture area is not identical with the nominal entrance area of the probe, and the exact capture area cannot be determined. For small probes an uncertainty of the capture diameter of a few tenths of a millimeter will result in an uncertainty of the mass flow on the order of 10 percent.

The low static pressure in the free jet test section necessitates large cross-sectional area tubing and a vacuum pump which operates down to 10^{-4} Torr at a capacity of 200 m^3/h. The mass flow can be measured continuously by metering the mass flow pumped from the vacuum tank with turbometers and monitoring the pressure and the pressure and temperature history of the vacuum tank. The mass flow through the probe is determined either from the conservation mass flow through the system or from the energy flux through the vacuum tank.

Total Enthalpy Probe

The principal difference between the total enthalpy and the mass flow probe lies in their cooling system, (Fig. A3-1-2). The geometry of the gas-sampling channel is identical for both probes.

Extreme care has been taken to insulate the outer and inner cooling systems. The

Fig. A3-1-3 Pitot pressure probes

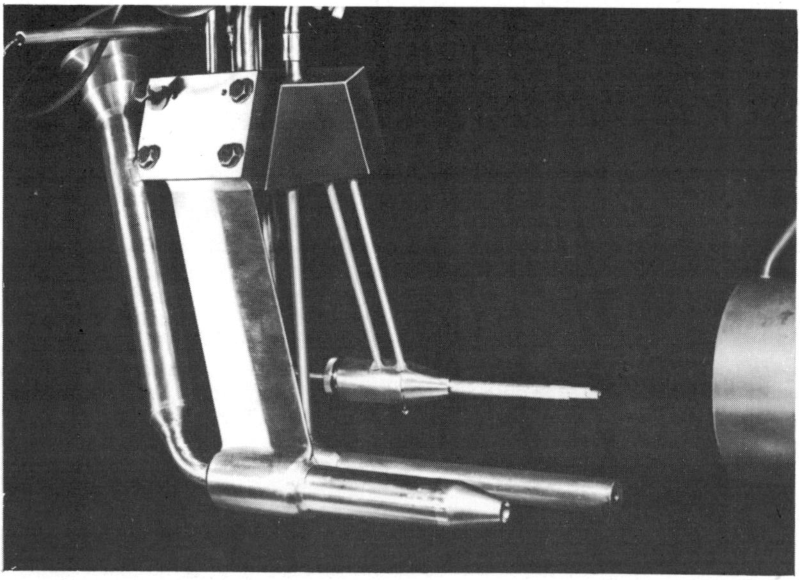

Fig. A3-1-4 Triple probe selector

outer cooling system has the sole purpose of cooling the outer probe structure. The inner cooling system should absorb all the heat energy from the gas that enters the probe. To achieve this purpose while allowing for continuous measurements of the total enthalpy, the inner and outer cooling systems are insulated by stagnant gas. The stagnant gas in the probe's head stems from the hot free stream while the stagnant gas in the main body of the probe stems from room-temperature ambient air. The stagnant ring of hot gas is designed so that the free stream stagnation streamline terminates within the stagnated gas region. This facilitates separation of the heat flow from the outer and inner gas flow to the outer and inner cooling system. The capture area itself does not enter the total enthalpy calculations. A ceramic or synthetic insulator serves as heat resistance between the hot and cool stagnant gases.

To establish the heat balance for the sampled gas flow, which determines the free stream total enthalpy, water inlet and outlet temperatures and sampled gas temperature are measured at the same axial position at the rear of the probe. In addition the water flow rate and gas mass flow have to be monitored. To avoid any heat flow between the two cooling systems and to achieve the water temperature rise for exact determination of the heat balance, the water flow rates can be adjusted. Because the inlet areas of the mass flow and total enthalpy probe are identical, comparison of the measured mass flows gives information on the location of the stagnation ring for the total enthalpy probe.

Pitot Pressure Probes

Both the mass flow and total enthalpy probe can be operated as impact pressure probes. In addition a series of impact pressure probes of different geometry have been designed to determine low Reynolds number effects on impact pressure measurements. The viscous effects on the probe measurements change with the external geometry of the pitot probe. Therefore flat-nosed and hemispherical probes were investigated. The diameters of the three water cooled probes were 15 (4) mm, 10 (3) mm, 5 (2) mm. The values in brackets are orifice diameters. The condition for reasonable response time limits the probe size. The time lag for the smallest probe of the pressure system (Fig. A3-1-3) was below five minutes in the range of 1 to 10 mm Hz.

Further one must study the effect of orifice-to-probe diameter ratio. As shown in (3), in low density flow, this has a remarkable influence on the measured impact pressure. For changing the orifice-to-probe diameter ratio, short tubes with different bore holes were inserted into the probe orifice.

The probes were mounted on a triple probe selector shown in figure A3-1-4. By this probe holder three probes could be tested in a given flow. The two lateral probes were interchangeable, e.g. for mounting mass flow, total enthalpy or pitot probes of other shapes.

Pitot Pressure Measurements

Impact pressure measurements at low Reynolds numbers require to take into account viscous effects for data reduction. To determine these viscous effects probes of different diameters are used to establish the Reynolds number effect.

Considerations of (3) concerning the correlation parameter have shown that in hypersonic flow $Re_2 \sqrt{\frac{\rho 2}{\rho 1}}$ is a reasonable parameter to avoid Mach number dependance of the data. If the pressure ratio of the measured and inviscid pitot pressure $\frac{p_{t2m}}{p_{t2i}}$ is plotted as a function of the free stream Reynolds number Re_1

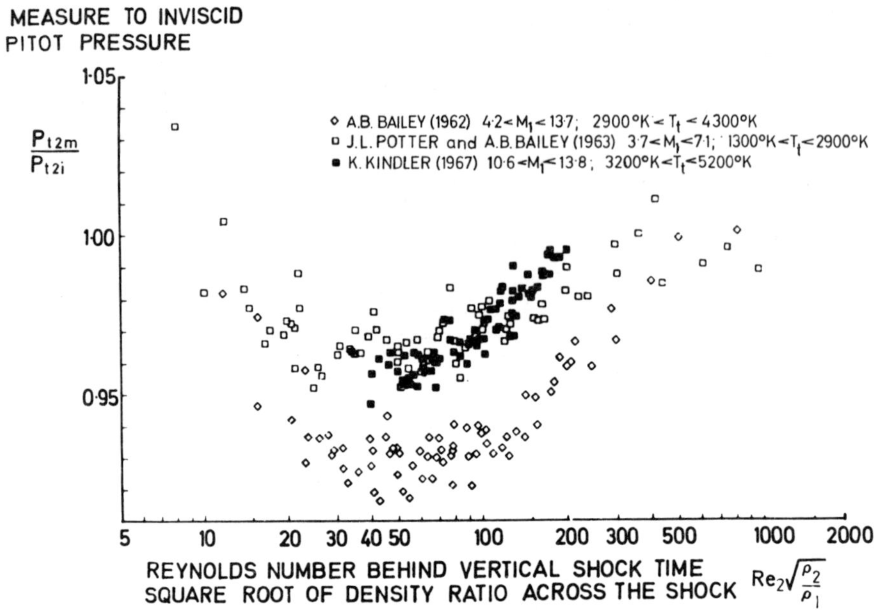

Fig. A3-1-5 Relationship of pressure ratio and Reynolds Number

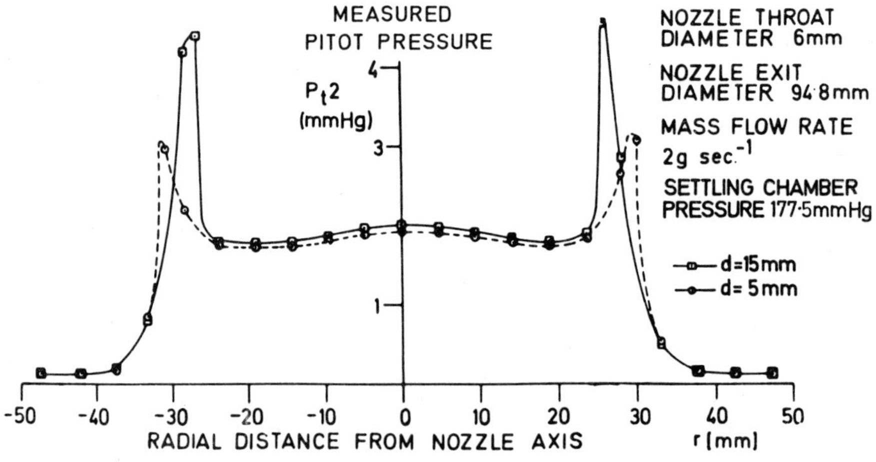

Fig. A3-1-6 Pitot pressure profiles of one arc jet

this ratio increases with increasing Mach number.

In a flow with an axial Mach number gradient the influence of varying shock stand-off distance has to be eliminated before the measured data can be interpreted. If the Mach number is increasing axially a larger pitot pressure will be measured with probes having a larger shock stand off distance. The correction which results by taking into account the variation in shock stand off distance can be seen from figure A3-1-5. The axial pressure gradient was neglected in (4), but was considered in (3), (5).

Figure A3-1-5 shows a decreasing pressure ratio with decreasing Reynolds numbers to about $Re_2\sqrt{\frac{\rho 2}{\rho 1}} \approx 50$. This behaviour is caused by two effects: viscous and rarefaction effects. The decrease of the pressure ratio accounts for the viscous effects in the shock layer predicted by theoretical studies (6), (7), (8). The increasing pressure ratio is caused by additional rarefaction effects on the shock. The shock thickness and stand-off distance increase and we get more and more the pressure ratio behaviour of low Reynolds number subsonic flow.

Pitot Pressure Profiles

Two pitot pressure profiles of the same arc jet are shown in figure A3-1-6. The measurements were taken with different diameter probes by rotating the triple probe holder. The explanation of the differences in the two profiles may be as follows:

If probes of different size are inserted into the arc jet it can be expected that the flow rate pumped from the test chamber is changed. This effect should be pronounced when the distance between probe and diffusor inlet is relatively short. Any obstacle in the test chamber should increase the static pressure of the chamber. The test conditions for the jet figure A3-1-6 were that static pressure, calculated from isentropic relations using settling chamber pressure and pitot pressure, is a factor of two lower than the pressure sensed with the probes at the jet boundary. This latter pressure corresponds to the test chamber pressure. Hence, the static pressure in the outer part of the jet will be pressed upon the jet by the test chamber while the test chamber pressure will not influence the hypersonic jet core. The jet core pitot profile is, therefore, insensitive to probe size. The edge of the jet will have a decreased Mach number due to compression from the test chamber; this in turn gives rise to the peaks in the pitot pressure profiles. The test chamber pressure will be slightly higher for the larger probe than for the smaller probes; hence the peaks in the pitot pressure profiles will be more pronounced for the larger probe because a slight variation of test chamber pressure can cause a large pitot pressure rise at the edge of the jet core.

References

1. **Becker, M., Heyser, A.,** 'Der DVL-Plasmawindkanal PK1, eine Anlage fur hypersonische Geschwindigkeiten, geringe Dichten und hohe Temperaturen'. DVL-Bericht Nr. 669. DLR-FB-67, June 1967.

2. **Huber, F.J.A.,** 'Probes for measuring mass flux, stagnation point heating, and total enthalpy of high temperature hypersonic gas flows.' AIAA Aerodynamic Testing Conf. Los Angeles, Calif., Sept. 1966.

3. **Bailey, A.B., Boylan, D.E.,** 'Some experiments on impact-pressure probes in a low-density, hypervelocity flow.' AEDC-TN-61-161, Dec. 1961.

4. **Potter, J.L., Bailey, A.B.,** 'Pressures in the stagnation regions of blunt

bodies in the viscous-layer to merged-layer regimes of rarefied flow.' AEDC-TDR-63-168, Sept. 1963.

5. **Bailey, A.B.,** 'Further experiments on impact-pressure probes in a low-density, hypervelocity flow.' AEDC-TDR-62-208, No.1962.

6. **Kindler, K.,** 'Pitotsondenmessungen in Hypergeschwindigkeitsströmungen geringer Gasdichten und hoher Gesamtenthalpien.' DLR-Mitt.-67.

7. **Probstein, R.F., Kemp, N.,** 'Viscous aerodynamic characteristics in hypersonic rarefied gas flow.' J. Aerospace Sci. vol. 27, 1960.

8. **Levinsky, E.S., Yoshihara, H.,** 'Rarefied hypersonic flow over a sphere' 'ARS Progress in Astronautics and Rocketry: Hypersonic Flow Research,' ed. by F.R. Riddell. Academic Press Inc., New York, vol.7, 1962.

9. **Cheng, H.K.,** 'The blunt-body problem in hypersonic flow at low Reynolds number.' IAS Preprint 63-92, Jan. 1963.

Commentary B on Chapter 3-1

V. Sprengel

I am myself working with the type of co-axial calorimetric probe developed by Grey, and I wish to question whether the teflon cylinder used by the author as heat insulation in his stationary and non-stationary probes is adequate in view of the small size of the gas samples sucked in, which make very small changes to the measured temperature of the coolant water flows.

As shown in figure 3-1-4 there is in the Grey probe only one single coolant circuit for cooling both the outer surface and the inner, calorimetric one. The difference between this probe and the stationary probe is that two measurements must be made, one with and one without suction of the sample. With suction the heat transfer to the coolant flow is increased just by the enthalpy of the gas sample. Provided that the conditions of flow at the probe tip remain the same for both of these measurements, the stagnation enthalpy can be obtained in this way, with the advantage that a possible transfer of heat to the inner flow of coolant is compensated for.

If the same principle is adopted for the author's stationary and non-stationary probes, the difficulty of insulating the inner coolant from the outer coolant can be evaded. A second measurement without suction would indicate just what that additional heat flux is. Alternatively, would it be possible to replace the teflon cylinder by a vacuum in a simple manner, so as to obtain more effective heat insulation?

When blunt or chamfered probe tips are used for measuring stagnation enthalpy a further problem is to know where the stagnation point is, when the resulting shock at supersonic flow is sucked into the probe.

With the Grey probe, it is assumed that both the shock and the stagnation point are located in front of the probe tip. The suction must therefore be not too strong, and the probe must be sensitive enough to measure limited temperature changes.

The exact location of the stagnation point is of great interest, precisely because it is the stagnation enthalpy that it is necessary to measure.

3-2

Etude expérimentale d'une couche limite laminaire figée avec réaction catalytique à la paroi

G. LASSAU et E.A. BRUN
Laboratoire d'Aérothermique, C.N.R.S. Meudon, France

Summary

During the re-entry of a vehicle into the atmosphere, the components of the air undergo a transformation across the shock wave because of the marked increase in temperature. One may assume that there is an equilibrium flow in the shock layer, but in the vicinity of the wall the temperature is lower; chemical equilibrium no longer exists, the reaction time being too large in comparison with the residence time in the boundary layer. Along the wall the gas is in a metastable state. Generally the wall has a catalytic effect on the reaction rate that is sufficiently big to re-establish equilibrium state.

Experimental evidence of the catalytic effect of the flat wall parallel to the flow on the concentration in the boundary layer is given. The theory shows that from the concentration profiles in the layer, the wall concentration can be obtained.

A gas flow in a metastable state is obtained from a low pressure arc in nitrogen. A flat plate is placed in the plasma jet. The luminous phenomena show the catalytic effect of the plate.

Concentration curves are obtained by three different methods:

1. By means of an enthalpy probe giving the activated nitrogen concentration in the boundary layer;
2. By injecting a carbide in the activated nitrogen and studying the luminosity;
3. By mass spectrometry of cyanhydric acid.

The three methods give similar results.

Sommaire

Lors de la rentrée des engins dans l'atmosphère, les composants de l'air subissent une transformation à la traversée de l'onde de choc, par suite de la forte élévation de température. On peut admettre qu'il existe un écoulement en équilibre dans la couche de choc mais, au voisinage de la paroi, la température s'abaisse; l'équilibre chimique n'existe plus, la constante de temps de la réaction étant très grande devant le temps de séjour des gaz dans la couche limite. Nous aurons donc, le long de la paroi, un gaz dans un état métastable. Généralement, la paroi a un effet catalytique qui rend la vitesse de réaction suffisamment grande pour rétablir l'état d'équilibre.

Un écoulement de gaz dans un état métastable est obtenu par un arc à basse pression dans de l'azote. Une plaque plane est placée dans le jet de plasma. La lum-

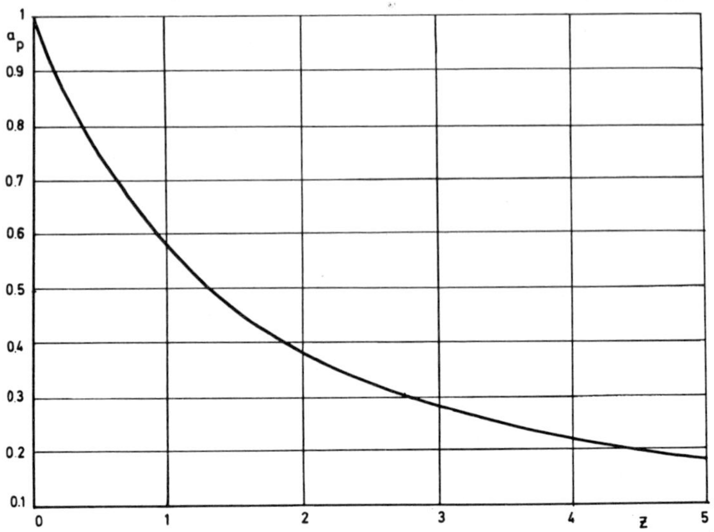

Fig. 3-2-1 La concentration pariétale donnée par la relation universelle

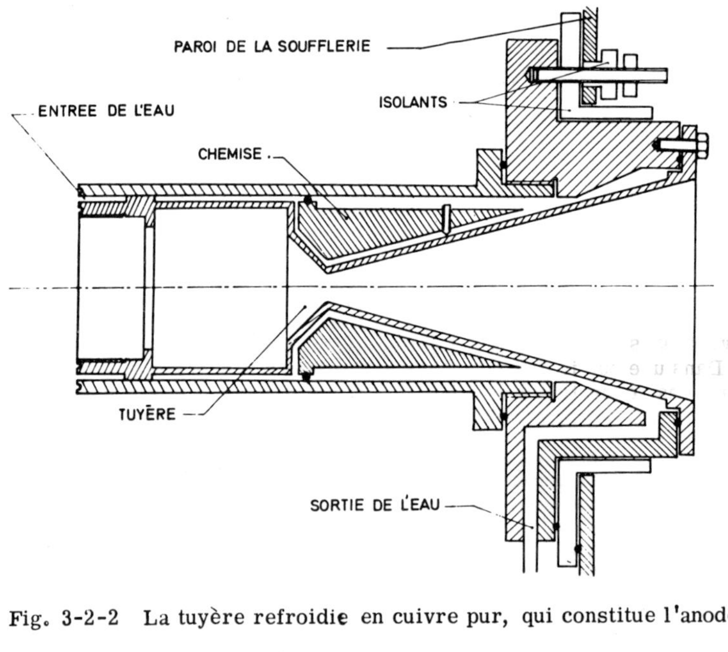

Fig. 3-2-2 La tuyère refroidie en cuivre pur, qui constitue l'anode

inosité du phénomène montre l'effet catalytique de la paroi.

Les courbes de concentration peuvent être obtenues de trois façons différentes:

1. par une sonde d'enthalpie donnant la concentration en azote activé dans la couche limite;

2. par injection d'un carbure dans l'azote activé et en étudiant la luminosité obtenue;

3. par spectrométrie de masse de l'acide cyanhydrique.

Les trois méthodes donnent des résultats voisins.

Introduction

L'étude des réactions catalytiques dans une couche limite laminaire est justifiée par diverses applications. De telles réactions interviennent notamment dans l'échauffement des capsules spatiales au cours de leur rentrée dans l'atmosphère.

Lors de cette rentrée (1), les composants de l'air subissent, par suite de la forte élévation de température, certaines transformations à la traversée de l'onde de choc (dissociation, formation de NO, ionisation). L'équilibre chimique dépend de la vitesse et de l'altitude.

Au voisinage de la paroi, la température subit au contraire une diminution rapide, mais le gaz ne prend pas en chaque point l'état thermodynamique correspondant à l'équilibre chimique, car le temps des réactions est grand devant le temps de séjour des gaz dans la couche limite. La couche limite est alors formée de gaz en état métastable.

Cependant, en général, la paroi a un effet catalytique qui rend la vitesse de réaction suffisamment grande, au contact même de la paroi; il y aura alors dans la couche limite, non seulement un transfert de chaleur, mais encore un transfert de masse, puisque la concentration n'est pas la même aux deux frontières de la couche limite.

Nous nous attacherons ici au problème des relations entre la diffusion d'un constituant et sa production ou sa destruction sur une paroi placée dans un écoulement.

On trouvera dans les références (2) à (20) les travaux théoriques effectués à ce sujet. Dans une publication précédente (20), il a été montré que l'on peut connaître les profils des concentrations par une méthode d'intégration générale des équations de la diffusion et par une résolution analogique de ces équations. En outre, la distribution des concentrations pariétales réduites a_p ne dépend que d'une variable Z définie par:

$$Z = 2,98\, \mathcal{D}\, S^{2/3} \frac{U_\infty x}{\nu} = 2,98\, \mathcal{D}\, S^{2/3} \mathcal{R}_x{}^{1/2} \qquad \text{(Eq. 3-2-1)}$$

Dans cette formule:

S représente le nombre de Schmidt, rapport de la viscosité cinématique ν à la diffusivité massique D:

$$S = \frac{\nu}{D} ,$$

$\mathcal{D}$ est le nombre de Damköhler catalytique, défini par l'expression:

$$\mathcal{D} = \frac{k_d + k_i}{U_\infty}$$

où k_d et k_i sont les vitesses de la réaction chimique directe et inverse.

Ainsi il existe une relation universelle qui donne la concentration pariétale relative en fonction de Z (Fig. 3-2-1) ce qui entraîne une relation entre x et Z.

Bien entendu, pour trouver l'expression numérique de ces profils et de ces distributions, il faut introduire dans les équations, les valeurs des grandeurs qui caractérisent la réaction. Le but de ce travail est de déterminer expérimentalement les coefficients cinétiques de la réaction catalytique de l'azote atomique sur une plaque, soit:

$$N + N + \text{Paroi} \qquad N_2 + \text{Paroi}.$$

Nous sommes partis d'un écoulement figé d'azote atomique, obtenu à basse pression. La basse pression a pour effet d'augmenter considérablement les dimensions de la couche limite (5 cm d'épaisseur à la distance x = 7 cm du bord d'attaque) et, par suite, d'y faciliter les mesures de gradient de concentration. Par contre, les mesures de température et de pression sont très délicates du fait de la raréfaction du gaz.

Nous allons décrire d'abord l'appareillage utilisé pour obtenir un écoulement d'azote à basse pression, partiellement dissocié, ensuite les dispositifs de mesure des concentrations en azote atomique. Nous verrons enfin que les résultats obtenus par trois méthodes différentes concordent entre eux et vérifient la théorie signalée ci-dessus.

Réalisation d'un Jet de Plasma à basse Pression

Soufflerie

Les mesures ont été effectuées dans la soufflerie SR 2 du laboratoire d'Aérothermique du C.N.R.S. Nous rappelons brièvement ses caractéristiques (21).

Le groupe de pompage est à deux étages. Il fournit un débit volumique de 3 m^3s^{-1} pour une pression d'aspiration comprise entre 50 et 500 microns de mercure. Le caisson d'expérience a un débit de fuite de l'ordre de 1 torr . ls^{-1}. Le vide limite qui peut être atteint est de l'ordre de 5 microns de mercure.

Nous avons diminué le gradient de vitesse longitudinale dans le jet en placant le point d'aspiration des pompes à un mètre environ de la sortie de la tuyère. Le jet garde ainsi une section sensiblement constante. Ce long tube introduit une perte de charge non négligeable qui divise le débit d'aspiration au niveau du caisson par un facteur 3.

Obtention de l'azote atomique

Un écoulement d'azote est chauffé par un arc électrique.

La tuyère, en cuivre pur, est refroidie par une circulation d'eau; elle constitue l'anode; sa section au col est de 11 mm et son diamètre de sortie est de 80 mm (Fig. 3-2-2). La cathode est un barreau de tungstène thorié emmanché sur un barreau de cuivre pur refroidi par une circulation d'eau (Fig. 3-2-3). L'ensemble est isolé de la masse suivant le montage de la figure 3-2-2.

L'arc est alimenté par une batterie pouvant donner jusqu'à 2 000 ampéres sous 120 volts. Un générateur haute fréquence, placé en série, permet d'amorcer l'arc. Une résistance refroidie est incorporée dans le circuit pour stabiliser l'intensité du courant (22). La difficulté principale est le refroidissement au col. Une double chemise permet d'obtenir un meilleur échange thermique dans cette région. D'autre part, un champ magnétique d'un dixième de tesla est produit par une bobine centrée sur l'axe du jet (22); il a pour rôles de faire tourner l'arc et de restreindre les dimensions de l'écoulement au col (22), (23); ainsi, nous avons pu fonctionner avec la tuyère précédemment décrite sans aucune usure du col durant plusieurs semaines.

Des instabilités d'arc peuvent apparaître après un certain temps de fonctionnement; elles disparaissent lorsque l'on augmente ou diminue la distance entre électrodes. Ce phénomène peut s'interpréter par une augmentation de la résistance de surface de l'anode, où se forment des nitrures de cuivre. Cependant, ces variations de la distance entre électrodes ont peu d'influence sur la puissance dissipée dans l'arc (23).

Le jet de plasma

Le jet de plasma ainsi obtenu est cylindrique. Le diamètre du jet augmente légèrement jusqu'à une distance de l'ordre de 10 cm en aval de la tuyère, puis il devient constant et égal à 14 cm.

La réaction catalytique est étudiée sur une plaque plane d'acier, refroidie par une circulation d'eau et placée dans la partie cylindrique du jet, entre la tuyère et le conduit d'aspiration des pompes; le convergent du conduit est refroidi par une circulation d'eau.

Le débit massique total d'azote injecté est de 4, 5 10^{-4} $kg.s^{-1}$. La pression dans le caisson est alors de 200 microns de mercure. L'intensité de l'arc est de 230 A sous une tension de 60 V; celle de la bobine, de 600 A sous une tension de 20 V; on obtient ainsi une température dans le jet voisine de 2 000°C.

Mesure des Concentrations

Mesure par le spectrographe de masse

(a) Le mélange gazeux que nous avons à analyser a une masse volumique très faible puisqu'à basse pression et à haute témperature.

Le spectrographe de masse est le seul moyen d'analyse qui soit adapté à ces faibles densités; cet appareil ne nécessite en effet, qu'une pression totale des gaz de l'ordre de 10^{-5} mm de mercure. Il reste la difficulté de transporter les gaz dans le spectrographe de masse sans modifier leur état. Or, l'azote atomique n'est pas en équilibre chimique dans l'écoulement: sa concentration est supérieure à la concentration à l'équilibre et, même si un tel équilibre existait dans l'écoulement, le refroidissement au cours du trajet dans la sonde conduirait à une concentration nulle au niveau du spectrographe de masse. Il faut donc transformer l'azote atomique instable en un corps stable avant tout transport à l'endroit même où l'on veut que l'analyse soit faite (24) (25).

(b) On sait que les carbures d'hydrogène donnent, avec l'azote actif, des composés à base d'acide cyanhydrique et des traces de radicaux C N et de gaz cyanogène C_2N_2. La réaction (26) est suffisamment rapide pour être considérée comme instantanée. L'idée est de faire réagir l'azote atomique avec l'hydrocarbure, puis

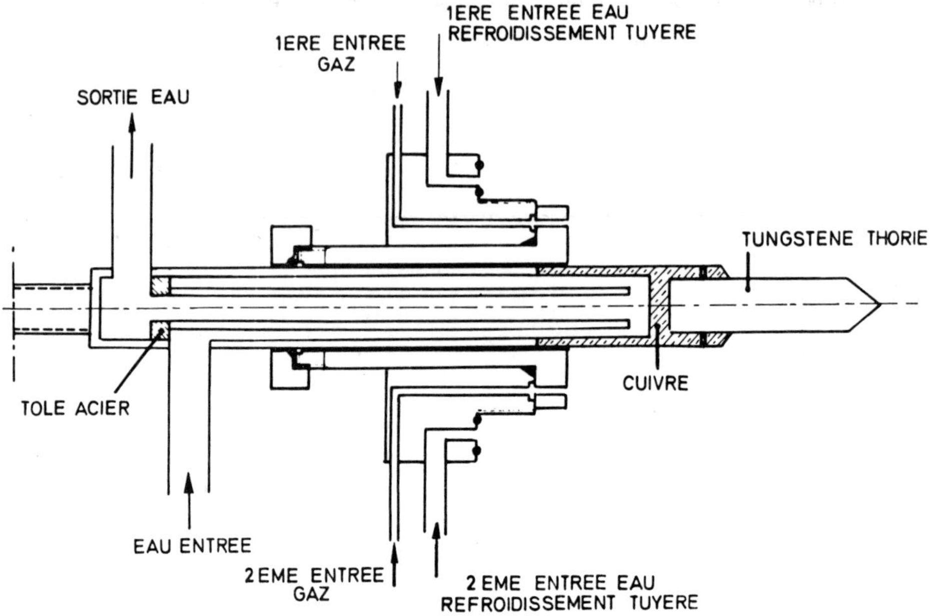

Fig. 3-2-3 La cathode refroidie qui est isolée de la tuyère

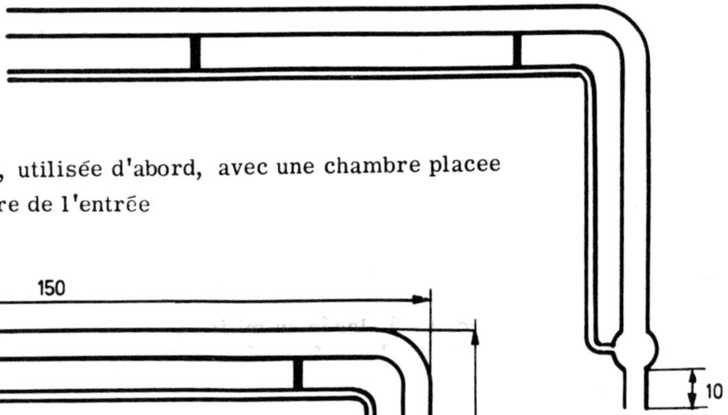

Fig. 3-2-4 La sonde, utilisée d'abord, avec une chambre placee en arrière de l'entrée

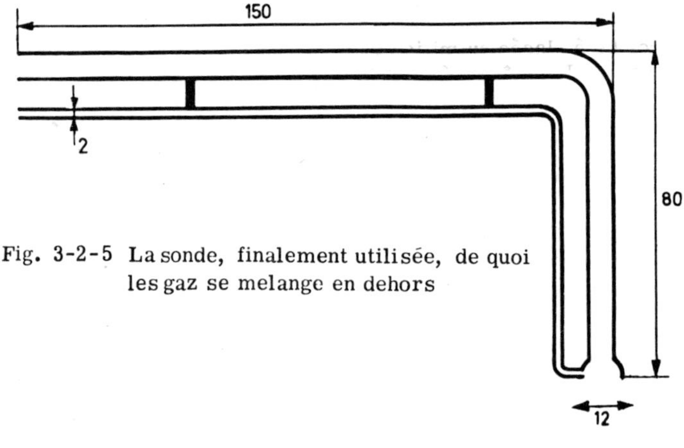

Fig. 3-2-5 La sonde, finalement utilisée, de quoi les gaz se melange en dehors

de transporter les produits stables dans le spectrographe de masse.

La sonde utilisée, où s'effectue la réaction, est en quartz: ce matériau, qui a une constante de réaction catalytique très faible et un coefficient de rayonnement notable, ne fond pas dans l'écoulement, alors que la surface d'un barreau d'acier non refroidi bout en moins d'une minute.

Dans un premier dispositif, l'hydrocarbure était introduit (Fig. 3-2-4) dans une chambre placée en arrière de l'entrée de la sonde. Avec cette géométrie, nous n'avons pu observer aucune formation d'acide cyanhydrique dans le spectrographe de masse. Ce résultat était dû au fait que, les gaz raréfiés ayant une viscosité cinématique très grande, la vitesse dans le tube d'entrée de la sonde est faible, et l'azote atomique a le temps de s'y recombiner avant d'entrer en contact avec le carbure.

La sonde que nous avons finalement employée est représentée sur la figure 3-2-5 : les gaz se mélangent en dehors de la sonde, et l'on prélève, derrière cette zone, une fraction de gaz stable que l'on analyse dans le spectrographe de masse. Pour diminuer le temps de réponse, le gaz à analyser est aspiré par une pompe à vide; le spectrographe est branché entre cette pompe et la sonde.

(c) Dans le spectrographe, les molécules à détecter sont ionisées. Cette ionisation s'accompagne d'une cassure d'un certain nombre de molécules, de sorte que des pics parasites apparaissent dans le spectre obtenu. L'essentiel est de ne pas avoir de composés autres que l'acide cyanhydrique de masse moléculaire 27. La plupart des carbures donnent l'ion $C_2 H_3^+$ qui a précisément la masse moléculaire 27. Même l'acétylène de masse 26, donne un certain pourcentage de molécules de masse 27, par suite de la recombinaison de molécules d'acétylène avec des atomes d'hydrogène provenant du craquage. L'expérience nous a montré que seul le méthane ne donne pas le pic vingt-sept parasite. Le spectre obtenu à l'aide de l'azote actif et du méthane présente en outre, le pic 26 négatif, que l'on attribue à des ions CN^- provenant de la décomposition de H C N. (Fig. 3-2-6).

Dans ces conditions, la concentration molaire en azote atomique (nombre de moles d'azote atomique/nombre de moles total) sera obtenue en faisant le rapport de la somme des hauteurs des pics 26 (C N) et 27 (H C N) à la somme des hauteurs des pics 26 (C N), 27 (H C N), 28 (N_2) et de la moitié de la hauteur du pic 14 dû à la dissociation de N_2. Dans les conditions expérimentales employées, la concentration des atomes d'azote provenant de la dissociation de N_2 dans le spectrographe de masse est égale à 0,08 fois la concentration de N_2.

(d) La sonde a d'abord été placée au milieu du jet, c'est-à-dire dans la zone de concentration maximale. Pour être sûr que tout l'azote atomique aspiré par la sonde a réagi avec le méthane, nous avons étudié comment variaient les concentrations d'azote atomique mesurées, en fonction de la quantité de méthane injectée dans la sonde, lorsque toutes les autres conditions restent égales. On a d'abord fait la mesure dans les conditions de concentration maximale (au centre du jet, en l'absence de plaque). On voit, sur la figure 3-2-7, que, dans ces conditions, pour une concentration de méthane aspiré supérieure à 40%, le pourcentage d'acide cyanhydrique mesuré ne varie plus.

Le spectrographe travaillant à pression constante, les mesures seront d'autant plus précises que la proportion d'azote et d'acide cyanhydrique dans le mélange sera plus élevée. On a donc intérêt à travailler avec le débit minimal de CH_4 nécessaire à la transformation complète de N en H C N (40% du melange). Ceci correspond à une injection de méthane de 40 $mm^3 s^{-1}$ rapporté aux conditions normales.

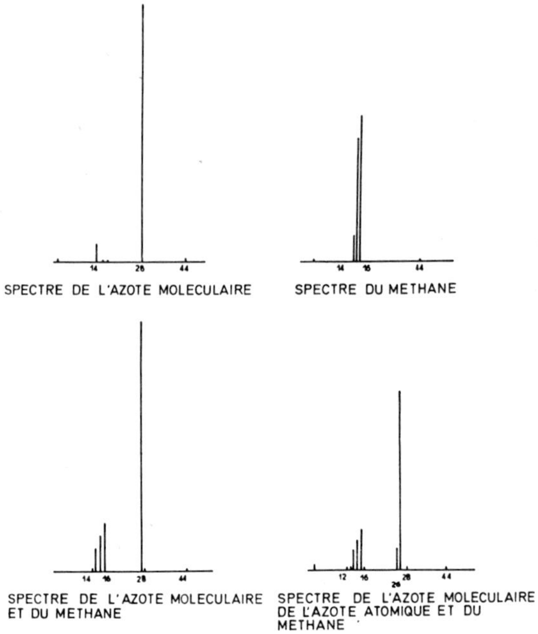

Fig. 3-2-6 Les spectres des composés obtenu à l'aide de l'azote et du methané

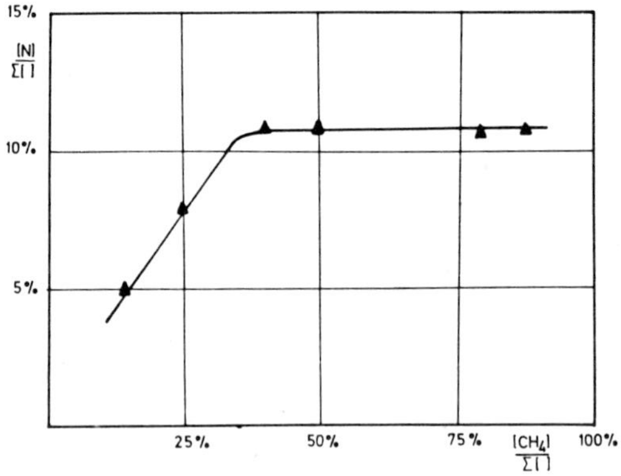

Fig. 3-2-7 La concentration d'acide cyanhydrique mesuré en fonction de la concentraion méthane aspiré

Cette méthode a l'avantage de fournir une valeur absolue de la concentration massique, ou mieux du titre massique, en azote atomique dans le mélange. Sa valeur maximale est 0.15.

Méthode optique

La réaction dont nous avons déjà parlé:

$$N + CH_4 \longrightarrow HCN + \frac{3}{2} H_2$$

est lumineuse et peut être utilisée pour doser l'azote atomique par une méthode optique (27). Si la concentration en CH_4 est suffisante, on peut considérer que l'énergie rayonnée est proportionnelle à la concentration en azote atomique (28).

La sonde employée est un tube de quartz d'axe perpendiculaire à la plaque. On distingue parfaitement à l'oeil nu une trés forte coloration bleue au point d'injection du méthane. On observe la même coloration bleue avec tous les hydrocarbures, par contre, une injection d'ammoniac donne une coloration orangée. La coloration bleue est donc associée à l'apparition d'acide cyanhydrique. L'analyse spectroscopique prouve que ce sont les bandes CN qui donnent cette coloration (28).

Pour mesurer l'intensité lumineuse de la réaction, on forme, sur un photomultiplicateur, l'image de la région voisine de la sortie de la sonde.

Il faut cependant noter que l'on intègre ainsi la lumière émise par tout le reste du plasma sur l'axe optique du système. La mesure sera d'autant plus précise que, dans le rayonnement reçu par le photomultiplicateur, la part due à la réaction chimique est plus grande. Il faut donc que l'intensité I_o, obtenue sans injection de méthane, soit petite devant l'intensité I obtenue avec l'injection. En choisissant convenablement la zone spectrale, nous avons pu obtenir un rapport I/I_o de l'ordre de 6. Si l'on admet que l'émission lumineuse est proportionnelle à la concentration en azote atomique, la concentration relative *a* est donnée par l'expression:

$$a = \frac{I - I_o}{I_\infty - I_o}$$

où I_∞ est l'intensité lumineuse de référence pour l'étude de la couche limite (intensité lumineuse prise en dehors de la couche limite).

Cette méthode a l'avantage de ne nécessiter que l'emploi d'un tube de quartz, pour injecter le gaz et d'un appareillage optique simple.

Méthode calorimétrique

(a) L'énergie de dissociation de l'azote est libérée lors de la recombinaison des atomes. En mesurant la quantité de chaleur due à cette recombinaison pour une masse donnée d'azote, on peut remonter, si l'on connaît la température et la vitesse du gaz, à la concentration en azote atomique (2). Il faut donc prélever une partie du gaz aussi ponctuellement que possible et mesurer la quantité de chaleur qu'il peut transmettre à une paroi froide.

En principe, les sondes employées (29) (30) (31) sont constituées de tubes concentriques. On aspire un certain débit massique par le tube intérieur, et le gaz transmet son énergie calorifique à de l'eau qui circule autour du tube. Des thermocouples permettent de connaître l'élévation de température de l'eau.

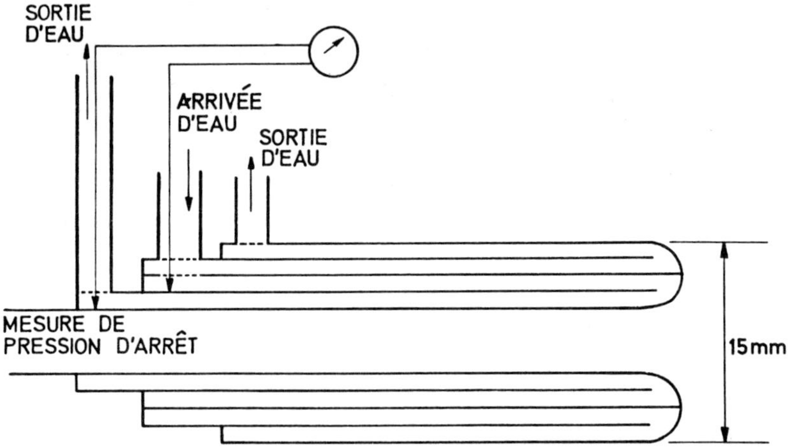

Fig. 3-2-8 La sonde calorimétrique refroidie par deux circulations d'eau

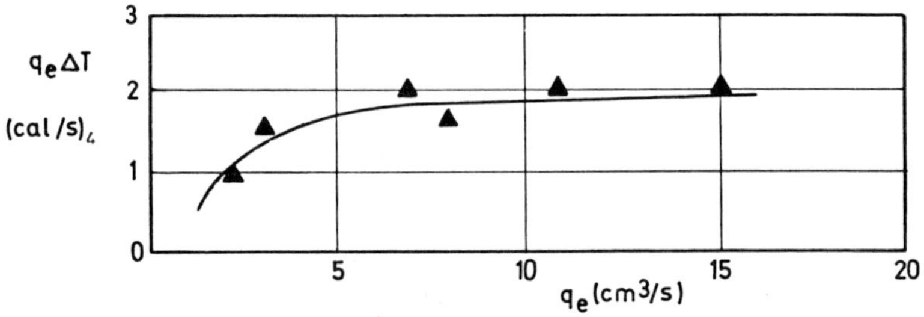

Fig. 3-2-9 L'énergie reçue par l'eau du circuit intérieur de la sonde calorimétrique, en fonction du débit d'eau dans ce circuit

En mesurant les débits d'eau et de gaz aspiré, ainsi que les témpératures de l'eau, avec ou sans pompage dans la sonde, on peut, par différence, en déduire le flux de chaleur reçu par la face interne de la sonde.

(b) La sonde, ici employée, est refroidie par deux circulations d'eau (Fig. 3-2-8). L'existence de cette double circulation épaissit la sonde et rend l'analyse moins ponctuelle; le diamètre de la sonde reste encore acceptable, compte tenu de l'épaisseur de la couche limite dans cet écoulement de gaz raréfié.

Dans le circuit extérieur, le débit d'eau est très grand; ainsi l'élévation de température de l'eau est très faible. Dans le circuit intérieur, le débit d'eau est faible, ce qui permet, même pour un très faible apport de chaleur en bout de sonde, d'avoir une différence appréciable de température aux extrémités de ce circuit d'eau. L'arrivée de l'eau est commune aux deux circuits; de ce fait, la température de l'eau reste pratiquement identique dans les deux circuits jusqu'à l'extrémité de la sonde et la perte thermique est fortement réduite.

Pour estimer les échanges thermiques entre les circuits d'eau, $q_e \Delta T$, nous avons tracé, pour un flux de chaleur donné, l'énergie reçue par l'eau du circuit intérieur, en fonction du débit d'eau q_e dans ce circuit. On voit, sur la figure 3-2-9, que, au-dessus de $q_e = 5 \text{cm}^3 \text{s}^{-1}$, le produit $q_e \Delta T$ est constant: les pertes calorifiques sont donc négligeables au-delà de ce débit d'eau.

(c) Pour éviter les erreurs dues aux différences de potentiel parasites qui peuvent apparaître dans le circuit de mesure, on opère de la manière suivante. Avant chaque mesure, on augmente le débit dans le circuit intérieur de telle manière que les thermocouples d'entrée et de sortie soient à la même température, et l'on règle alors le voltmètre de mesure à zéro.

(d) Par des calculs approchés (32) (33), nous avons montré que le flux de chaleur transmis est proportionnel à la concentration en azote atomique sans qu'il y ait lieu de tenir compte du flux de chaleur transmis par d'autres mécanismes (conduction, rayonnement, etc.). Si Φ est le flux de chaleur mesuré par le courant d'eau en un point de la couche limite et Φ_∞ ce même flux hors de la couche limite, la concentration relative sera:

$$a_i = \frac{\Phi}{\Phi_\infty}$$

(e) Cette méthode a l'avantage de s'accompagner d'une mesure de pression d'arrêt faite dans le tube intérieur.

Resultats

Résultats concernant le jet sans plaque

Nous avons mesuré les variations des flux de chaleur relatifs et de la pression d'arrêt le long du jet de plasma à l'aide de la sonde thermique.

Nous avons porté, sur la figure 3-2-10, les variations de la pression cinétique en fonction de la distance à l'axe et cela pour diverses distances à la sortie de la tuyère. L'erreur commise en mesurant la pression d'arrêt d'un écoulement à faible nombre de Reynolds, à un nombre de Mach voisin de l'unité, et à une température de 2 000°C, a été estimée inférieure à 10 % dans les conditions de l'expérience (34). La variation longitudinale de la pression d'arrêt est faible; ce fait est confirmé par la forme cylindrique du jet.

On vient de voir précédemment que le flux de chaleur capté par la sonde est

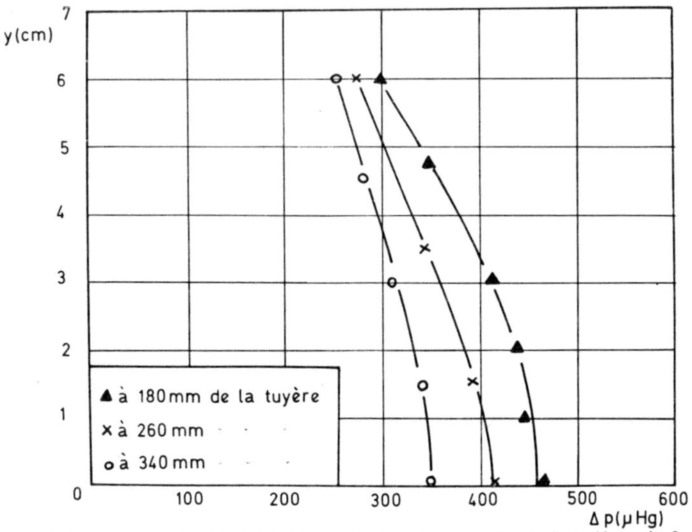

Fig. 3-2-10 Les variations de la pression kinetique en fonction de la distance à l'axe, et celà pour diverses distances à la sortie de la tuyère

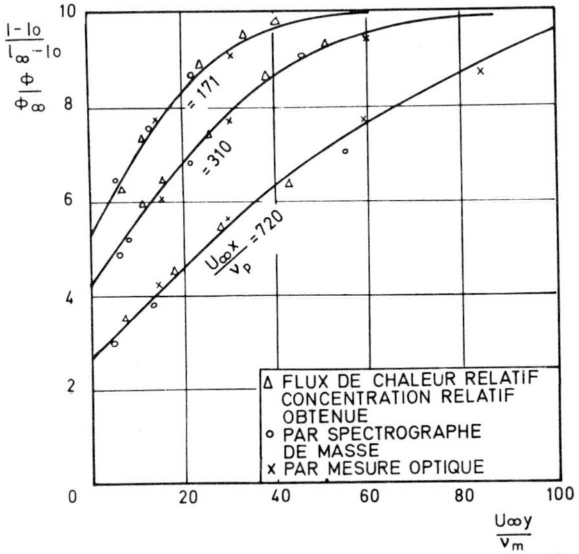

Fig. 3-2-11 La comparaison des résultats donnés par les trois methodes différentes

Fig. 3-2-12 Le flux de masse à la paroi en fonction de la concentration pariétale relative

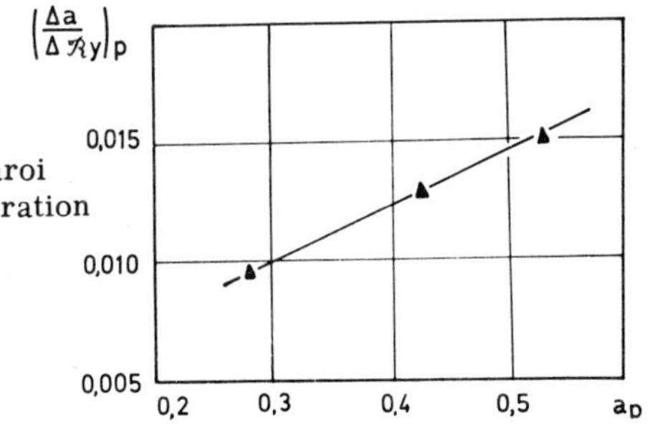

proportionnel à la concentration en azote atomique. La variation longitudinale de la concentration est faible, ce qui correspond à l' hypothese d'un écoulement figé.

Le gradient longitudinal de concentration qui existe est dû à la diffusion de l'azote atomique. En effet, on peut faire le bilan du flux de masse qui sort de la surface fermée définie, d'une part, par les sections circulaires situées à 180 mm et à 340 mm de la tuyère et, d'autre part, par la surface, supposée cylindrique et de rayon R_e, limitant le jet. On a, en supposant une distribution de vitesse parabolique dans le jet,

$$2 \pi R_e D \int_{x_1}^{x_2} \left(\frac{\partial a}{\partial R}\right)_x dx = \int_0^{R_e} 2 \pi a_1 (R).u (R).R\, d R \qquad \text{(Eq. 3-2-2a)}$$

$$- \int_0^{R_e} 2 \pi a_2 (R). u (R). R\, d R$$

$$2 \pi U_{max} \int_0^{R_e} [a_1 (R) - a_2 (R)] [1 - (\frac{R}{R_e})^2] R\, d R \qquad \text{(Eq. 3-2-2b)}$$

$(\partial a/\partial R)_x$ et $[a_1 (R) - a_2 (R)]$ sont calculés à partir des courbes expérimentales de la figure 3-2-10.

On vérifie à 10 % près que les deux membres de l'equation 3-2-2 son identiques.

Ce résultat montre que le plasma reste à l'état figé, ce que justifie d'ailleurs la comparaison du temps de recombinaison de l'azote atomique en phase homogène, de l'ordre de la seconde et du temps de parcours, de l'ordre du millième de seconde.

Résultats d'exploration de la couche limite

On a mesuré par les trois méthodes, les profils de concentration relative. Aux erreurs d'expérience près, ces différentes méthodes donnent le même résultat, ce qui justifie les hypothèses faites pour les mesures d'intensité lumineuse et pour la sonde thermique.

Les résultats sont portés sur la figure 3-2-11. Les viscosités cinématiques dans les nombres de Reynolds qui figurent en abscisses sont calculées à partir de la température moyenne dans la couche limite (1200°C), tandis que celles qui définissent les paramètres des courbes sont calculées à partir de la température de la plaque (500°C).

On détermine une valeur de la concentration pariétale en extrapolant les profils de la couche limite jusqu'à la paroi. Ceci est justifié par la quasi-linéarité de ces profils au voisinage de la paroi, linéarité prévue par la théorie.

On constate alors que le flux de masse à la paroi (da/dy)p est proportionnel à la concentration pariétale relative a_p. (Fig. 3-2-12). Cela prouve que la réaction est du premier ordre.

Confrontation avec la théorie

L'expression de Z en fonction de la concentration pariétale a_p permet donc d'établir la relation entre $\mathcal{R}_x$ et Z. On trouve ainsi que Z est proportionnel à $\sqrt{\mathcal{R}_x}$, et cela, comme le montre le tableau suivant, à une précision de 6 %.

$\mathcal{R}_x$	$\left(\frac{\Delta a}{\Delta \mathcal{R}y}\right)_p$	a_p	Z	$\sqrt{\mathcal{R}_x}$	$\frac{Z}{\sqrt{\mathcal{R}_x}}$
720	0, 0095	0, 280	3	26, 9	0, 11
310	0, 013	0, 425	1, 7	17, 6	0, 095
171	0, 015	0, 525	1, 2	13, 1	0, 092

La valeur du rapport $Z/\sqrt{\mathcal{R}_x}$ conduit alors, d'après l'équation 3-2-1, à la connaissance du nombre de Damköhler, si l'on se donne le nombre de Schmidt.

Exemple d'application

Si l'on veut obtenir la valeur de la constante k_d, car k_i est nul, à partir du nombre de Damköhler, il faut connaître la vitesse U_∞ et la température du plasma.

Pour déterminer ces grandeurs dans un jet, nous avons à notre disposition certains résultats: le débit massique total m, la distribution des pressions cinétiques et des concentrations.

Ayant opéré avec une plaque en acier et une température moyenne dans le jet de plasma d'environ 2 000°C, on trouve, dans le cas de S = 0, 5,

$$k_d = 45 \text{ m.s}^{-1}$$

et, dans le cas de S = 0, 7,

$$k_d = 35 \text{ m.s}^{-1}$$

C'est l'ordre de grandeur de la vitesse de réaction qui correspond à un tel cas, obtenu par Goulard (2) à l'aide d'une sonde thermique placée dans un écoulement hypersonique.

Références

1. **Brun, E.A.,** 'Problèmes de rentrée'. Cours de l'Ecole Supérieure de l'Aéronautique, 1963.

2. **Goulard, R.,** 'On catalytic recombination rates in hypersonic stagnation heat transfer'. Jet Propulsion, vol.28, No.11, Nov. 1958.

3. **Chung, P.M., Anderson, P.M.,** 'Heat transfer to surfaces of finite catalytic activity in frozen dissociated hypersonic flow'. N.A.S.A. T.N.D. 350, Jan. 1961.

4. **Chambre, P.L., Acrivos ,** 'On chemical surface reaction in laminar boundary layer.' J. of App. Phys. vol.27, No.11, Nov. 1956.

5. **Chambre, P.L., Acrivos ,** 'Laminar boundary layer flows with surface reactions.' Industrial and Eng. Chemistry, vol.49, No.6, June 1957.

6. **Lighthill, M.J.,** 'Contribution to the theory of heat transfer through a laminar boundary layer'. Proc. Royal Soc. London A 202, 1950.

7. **Chung, P.M., Liu, S., Mirels, H.,** 'Effects of discontinuity of surfaces catalycity on boundary layer flow of dissociated gas'. Internat. J. of Heat and Mass Transfer, vol.6, No.3, March 1963.

8. **Libby, P.A., Fox, H.,** 'Some perturbation solution in laminar boundary layers theory.' J. of Fluid Mechanics, vol.17, part 3, Nov. 1963.

9. **Fox, H., Libby, P.A.,** 'Dissociated laminar layers with heterogeneous recombination.' Physics of Fluids, vol.9, No.1, Jan. 1966.

10. **Libby, P.A., Fox, H.,** 'A moment method for compressible laminar boundary layers and some applications.' Pibal 615, Dep. of Aerospace Eng. and App. Mechanics Polytechnic Inst. Brooklyn, New York, 1964.

11. **Ingers, G.R.,** 'Dissociated laminar boundary layer flows over surfaces with arbitrary continuous distributions of catalycity'. Int. J. of Heat and Mass Transfer, vol.6, No.9, Sept. 1963.

12. **Ingers, G.R.,** 'Highly nonequilibrium boundary layer flows of a multicomponent dissociated gas mixture'. Int. J. of Heat and Mass Transfer, vol. 7, No.11, Nov. 1964.

13. **Simpkins, P.G.,** 'Diffusion of species over surfaces with discontinuous catalycity'. Physics of Fluids, vol.9, No.6, June 1966.

14. **Simpkins, P.G.,** 'Some experiments in partially dissociated boundary layer.' J. of Fluid Mechanics, vol.26, part 1, Sept. 1966.

15. **Rosner, D.E.,** 'The apparent kinetics of surface reactions in external flow systems; diffusional falsification of activation energy and reaction order'. A.I.CH.E. Journal, vol.9, No.3, May 1963.

16. **Rosner, D.E.,** 'Convective as an instrude in kinetics studies of surface catalysed reactions'. A.I.A.A. no.4, April 1964.

17. **Li, T.Y., Kirk,** 'Asymptotic method of solution of the frozen dissociated laminar boundary layer flow over a flat plate surface with arbitrarely distributed catalycity'. Int. J. of Heat and Mass Transfer, vol.10, No.3, March 1967.

18. **Freeman, N.,** 'The diffusion of species in frozen hypersonic boundary layer fundamental phenomena in hypersonic flow.' Gordon Hall, 1966.

19. **Lassau, G.,** 'Expression approchée des profils de vitesse et de concentration dans une couche limite laminaire'. C.R. Acad. Sc. Paris, 261, 4617-4620, Nov. 1965.

20. **Lassau, G., Le Fur, B.,** 'Profils de concentrations dans une couche limite laminaire figée avec une répartition continue et discontinue de catalyseur sur la paroi'. Int. J. of Heat and Mass Transfer, vol. 10, no.7, July 1967.

21. **Seurin, P.,** 'Effet de la température de paroi sur l'écoulement dans une tuyère supersonique de soufflerie à basse densité.' Thèse soutenue à la Faculté des Sciences de Paris, June 1966.

22. **Jacquelin, M.H.,** 'Réalisation et étude d'un écoulement continu de plasma dans une tuyère en présence d'un champ magnétique axial.' Thèse soutenue

à la Faculté des Sciences de Paris, No. 726, 1962.

23. **Alliot, J.M.,** 'Fluctuations de température dans un jet de plasma'. Thése soutenue à la Faculté des Sciences de Paris, May 1967.

24. **Herron, Franklin, Bradt.,** 'Etude par spectrométrie de masse de la réaction de quelques hydrocarbures avec l'azote actif.' Canadian Journal Chemical, vol. 37, 1959.

25. **Fontijn, Rosner, Kurzius.,** 'Chemical scavenger probe determinations of atom and excited molecule concentrations in nonequilibrium supersonic stream of active nitrogen'. Aerochem. Res. Lab, August 1962.

26. **Blades, Winkler, C.A.,** 'Mesure des vitesses de la réaction de destruction du méthane par l'azote atomique'. Canadian Journal Chemical, vol. 29, 1951.

27. **Young, Sharpless,** 'Chemiluminescent reactions involving atomic oxygen and nitrogen.' J. of Chemical Physics, vol. 39, No. 4, August 1963.

28. **Van Tiggelen, A., Feugier, A.,** 'Chimiluminescence et chimiionisation dans les flammes, Revue de l'Institut Français du Pétrole, vol. XX, No. 7, July 1965.

29. **Au, G., Sprengel, U.,** 'Kalorimetrische Messungen von örtlichen temperaturen und geschwindigkeiten in einem stickstoff-plasmastrahl.' Z. Flugwiss, 14, Heft 4, 1966.

30. **Haenig.,** 'Use of a catalytic probe for detection of dissociated non-equilibrium states in hypersonic flow.' A.R.S. Journal, vol. 29, No. 5, May 1959.

31. **Rosner, D.E.,** 'Catalytic probes.' A.R.S. Journal, vol. 32, 1962. A.I.A.A. Journal, vol. 2, No. 4, 1964.

32. **Acrivos, A., Goddard.,** 'Asymptotic expansions for laminar forced convection heat and mass transfer low speed flow.' J. of Fluid Mechanics, vol. 33, No. 2, Oct. 1965.

33. **Hartunian, Liu.,** 'Slow flow of a dissociated gas about a catalytic probe'. Physics of Fluids, vol. 6, No. 3, March 1963.

34. **Enkenhus, K.R.,** 'Pressures probes at very low density.' University of Toronto, U.T.I.A., Rep. 43, Jan. 1957.

List of Mathematical Symbols

a	concentration massique, rapport de la masse volumique de l'azote atomique à la masse volumique de l'azote atomique à l'extérieur de la couche limite,
D	coefficient de diffusion,
$\mathcal{D}$	nombre de Damköhler catalytique, $\mathcal{D} = \frac{k_i + k_d}{U_\infty}$
I	intensité lumineuse,
k_d	constante de la réaction catalytique directe,
k_i	constante de la réaction catalytique inverse,
q_e	débit d'eau dans la sonde thermique,
R	rayon du jet,
R_x	nombre de Reynolds xU_∞/ν,
S	nombre de Schmidt : rapport du coefficient de viscosité cinématique au coefficient de diffusion,
T	température,
U_∞	vitesse,
x	abscisse longitudinale,
y	abscisse verticale,
Z	variable mathématique définie par $Z = 2,98\,\mathcal{D}S^{2/3}\sqrt{R_x}$,
ν	coefficient de viscosité cinématique,
Φ	flux de chaleur recueilli par la sonde thermique.
1, 2	section droite du jet d'abscisse x_1 et x_2,
p	se rapporte à la paroi,
∞	se rapporte à l'extérieur de la couche limite,
max	se rapporte au centre du jet,
e	extérieur du jet.

Commentary on Chapter 3-2

F. von Burger

Pour les problèmes de la rentrée des satellites dans l'atmosphère il faut bien connaître l'écoulement au voisinage de l'engin et surtout dans la couche limite. Un des phénomènes importants est la recombinaison des gaz dissociés dans la couche limite et l'effet catalytique de la paroi.

L'étude théorique de ces problèmes a été faite par M. Lassau et d'autres auteurs cités dans ce papier. Ici, M. Lassau et M. Brun vérifient expérimentalement les résultats de la théorie par l'étude de la recombinaison de l'azote atomique sur une paroi dans un écoulement à basse pression.

On peut faire quelques remarques sur les problèmes particuliers à la couche limite et sur les questions théoriques: par exemple, cet écoulement a-t-il déjà la nature d'un écoulement de glissement? Quant à moi, je voudrais seulement dire quelques mots concernant les méthodes de mesure.

Pour ceux d'entre nous qui ne sont pas familiarisés avec ces problèmes, il paraît étonnant qu'on puisse atteindre un écoulement figé, à 2 000°C suffisamment stable dans la veine d'essais, à une pression aussi basse.

A cette basse pression et avec un aussi faible nombre de Reynolds, les expérimentateurs atteignent une épaisseur de couche limite presque égale à la longueur de la paroi jusqu'au point de mesure; ces conditions permettent la mesure des profils dans la couche limite avec des sondes de quartz; elles ne fondent pas dans cet écoulement à 2 000°C car leur coefficient de rayonnement est à peu près égal à un et le coefficient de convection dans cet écoulement est très faible.

Je voudrais demander à M. Lassau de nous dire:

d'une part, quelques mots sur le dispositif de la figure 3-2-4: comment la vitesse de l'écoulement peut-elle diminuer ainsi dans ce tube de 10 mm de longeur, malgré l'aspiration, pour que le temps de transfert soit de l'ordre de la seconde?

d'autre part, avec la méthode chimique, vraiment ingénieuse, on mesure un rapport de la masse N à la masse totale du mélange de 15 % Ce mélange, est-ce la masse totale d'azote ou est-ce le mélange d'azote avec le méthane? Ceci pour connaître le rendement de l'arc électrique.

La sonde d'enthalpie de diamètre de 15 mm n'est-elle pas trop grande par rapport à l'épaisseur de la couche limite, bien que ce diamètre soit nécessaire pour augmenter la sensibilité par la double circulation d'eau déjà proposée par Grey?

Enfin, peut-on éviter les perturbations provoquées dans l'écoulement par la sonde en réglant l'aspiration à l'intérieur de la sonde?

En conclusion, les auteurs ont obtenu des résultats presque identiques avec trois méthodes de mesure différentes, ils ont vérifié la formule théorique et montré que le nombre Z est proportionnel à $\sqrt{R_x}$.

M. Lassau

Le libre parcours moyen calculé avec les données physiques de l'écoulement est de l'ordre de un millimètre, c'est-à-dire que la zone de glissement est très faible devant les dimensions de la plaque.

Il est certain que l'aspiration est un moyen de diminuer les perturbations créées par une sonde, mais ce procédé ne peut être utilisé dans nos essais. En effet, le débit d'un gaz raréfié à travers une canalisation de faible diamètre, et dont la longueur est celle de la soufflerie, est fonction de la conductance du tube, soit, dans le cas présent, de l'ordre de un litre par seconde. On ne peut donc aspirer un débit suffisamment grand dans la sonde pour atteindre, avant l'injection du méthane, une vitesse de l'ordre du centimètre par seconde.

La dimension des sondes peut paraître importante, mais elle est de l'ordre de 1/5 de l'épaisseur de la couche limite. On sait que l'on trouve les profils de Blasius avec une sonde de diamètre extérieur de 2/10 de mm dans une couche limite de 1 mm d'épaisseur.

L'écoulement, bien que subsonique, est figé car nous sommes à un nombre de Mach voisin de 0,8 et à basse pression, 200 microns de mercure. Dans ces conditions, le rapport du temps de recombinaison chimique au temps de parcours de l'azote atomique est de l'ordre de 10^3, ce qui implique le figeage. D'autre part, les variations de la concentration dans le jet libre montrent qu'il n'y a qu'un phénomène de diffusion et non de recombinaison en phase homogène. Même dans la couche limite, les profils expérimentaux sont des profils dus à la diffusion et non des profils tenant compte de la recombinaison en phase homogène; ces profils ont un point d'inflexion.

Enfin, la concentration de 0.15 est le rapport de la masse de l'azote atomique à la masse totale (azote atomique + azote moléculaire) contenue dans un élément de volume de l'écoulement. Ceci conduit à un rendement de l'ordre de 50 %.
Sur la figure 3-2-7, puisqu'il s'agit de contrôler l'avancement de la réaction, on a porté le rapport de la masse de l'azote atomique à la masse totale dans la zone de réaction (azote atomique + azote moléculaire + méthane).

3-3

Sonde de température pour écoulement à haute enthalpie

D. GAUTROT
Office National d'Etudes et de Recherches Aérospatiales. Chatillon Sous Bagneux - France

Summary

The stagnation temperature of combustion gases is determined by means of a chromel-alumel thermocouple which can be cooled by an airflow. The rate of temperature increment is recorded during the brief duration of non-cooling. From this record, the equilibrium temperature is determined with an analogue computer. The time of measurement is about 1 second.

Sommaire

La détermination de la température d'arrêt dans les gaz issus de foyers de combustion est faite au moyen d'un thermocouple chromel-alumel pouvant être refroidi par un courant d'air. On enregistre la montée de température pendant le temps très court où le refroidissement est interrompu. Un calculateur analogique détermine, à partir de cet enregistrement, la température d'équilibre. Le temps nécessaire à la mesure est d'environ 1 seconde.

Introduction

Parmi les paramètres fondamentaux qui régissent les lois de l'aérothermodynamique, la température des gaz est une grandeur essentielle à laquelle il est toujours difficile d'avoir accès. La haute enthalpie et la vitesse de ces gaz dans une installation de combustion interdisent toute introduction de sondes thermoélectriques classiques.

D'autre part, si la mesure de la température moyenne est fondamentale, il apparaît que la distribution du champ thermique est tout aussi importante. Aussi est-il de première importance de mesurer des températures locales afin de définir un gradient de température dans une direction donnée.

Principe

Considérons un thermocouple chromel-alumel dont une des soudures (soudure froide) est thermostatée à une température connue (0°C par exemple). Supposons qu'au temps $t = 0$ la soudure chaude soit plongée dans un milieu à la température T_o. L'expérience montre que, sous certaines conditions, la réponse électrique du thermocouple suit une loi exponentielle, ainsi que l'indique la théorie.

En effet, soit m la masse de la jonction, c sa chaleur massique. Au temps t la température de la jonction est T et le bilan thermique permet d'écrire:

$$m\, c\, dT = h\, s\, (T_o - T)\, dt$$

ou h est le coefficient de convection et S la surface de la jonction soumise au flux

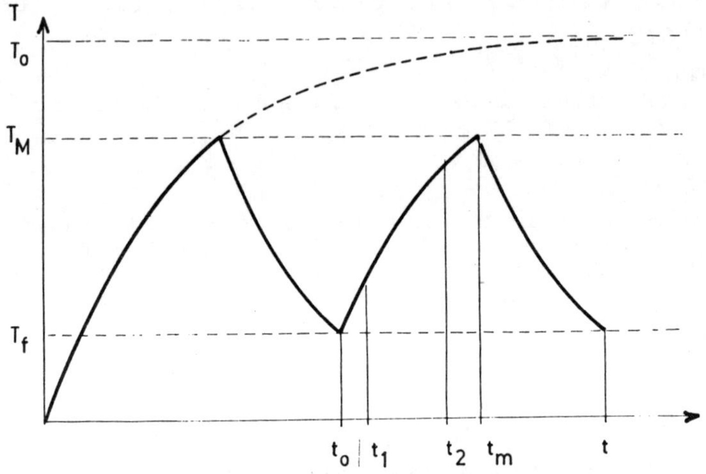

Fig. 3-3-1 Allure du cycle du température

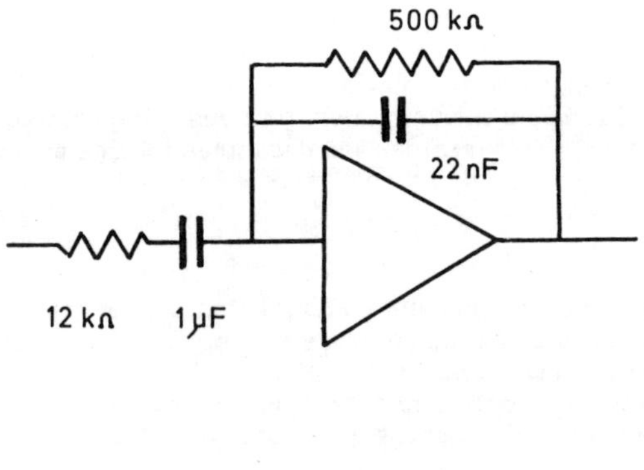

Fig. 3-3-2 Système dériviteur

thermique. On a aussi:

$$\frac{dT}{T_o - T} = \frac{hs}{mc} \, dt$$

En posant $\frac{mc}{hs} = \theta$ et en intégrant, il vient (condition initiale) : pour $t = o$ $T = O$

$$T = T_o \left(1 - e^{-\frac{t}{\theta}}\right)$$

Il apparaît immédiatement que deux effets thermiques ont été négligés: pertes par rayonnement de la soudure et pertes par conduction le long des fils. En fait, les conditions expérimentales permettent de réduire ces pertes à une valeur négligeable par l'emploi d'une sonde à col sonique: le col de la sonde joue le rôle d'écran vis-à-vis des parois; d'autre part, l'emploi de fils fins soumis sur une grande longueur au flux thermique réduit les pertes par conduction.

Définition du cycle

Les approximations faites étant justifiées, considérons à nouveau le thermocouple plongé brusquement au temps $t = 0$ dans un milieu où la température T_o peut être supérieure au point de fusion des matériaux constituants. La loi exponentielle montre que le thermocouple n'atteindra pas immédiatement cette température T_o. Supposons en outre que nous soyons capables de le refroidir à partir du moment où il aura atteint une température T_M, température suffisamment élevée pour pouvoir enregistrer un arc d'exponentielle exploitable de $T = T_f$ à $T = T_M$ et suffisamment basse pour éviter la destruction du thermocouple.

De $T = T_o \left(1 - e^{-\frac{t}{\theta}}\right)$ on déduit:

$$T_o = \theta \frac{dT}{dt} + T$$

Cette équation différentielle montre que la température du gaz T_o est la somme de la température instantanée et du produit de la constante de temps de la fonction par la derivée de la loi de montée.

Il faut remarquer que la constante de temps n'est pas une constante en général: elle est liée au coefficient de convection qui dépend des conditions génératrices amont, donc de l'installation d'essai.

L'expérience montre qu'il est licite de la considérer comme un invariant au cours de la montée en température de $T = T_f$ à $T = T_M$ mais qu'elle doit être calculée à chaque expérience.

L'arc d'exponentielle fournissant toutes les données nécessaires à la détermination de T_o le calcul analogique en permet l'extrapolation. Pour que la méthode soit exploitable il faut en outre un organe de commande qui permet de cycler la sonde comme montré figure 3-3-1.

Réalisation pratique du cycle

Ceci est pratiquement réalisé de la façon suivante: la sonde est précédée d'une électrovanne qui commande le refroidissement de la jonction par jet d'air dès que celle-ci atteint la température T_M, choisie à 1000°C; le refroidissement cesse lorsque la jonction atteint une température T_f choisie à 200°C. Ces valeurs ont été choisies pour respecter les impératifs suivants:

bonne tenue du thermocouple,

refroidissement suffisant du couple dans l'intervalle de temps $t_c - t_m$ qui doit être le plus court possible, puisque aucune mesure n'est possible au cours de cette période, pour un débit d'air donné,

intervalle de temps $t_m - t_o$ suffisamment long afin de pouvoir extrapoler l'arc d'exponentielle.

Cela est d'autant plus critique qu'un phénomène transitoire apparaît pendant environ 100 ms. Son origine résulte des éléments suivants:

temps de réponse pneumatique de la sonde: gaz chauds et air de refroidissement s'écoulent alternativement et axialement dans la sonde: le temps qui s'écoule entre les deux amorçages du col constitue ce temps de réponse.

légère différence de masse thermique entre le chromel et l'alumel (le rapport $\frac{(\rho c)_c}{(\rho c)_A} = 0,868$) entrainant un échauffement plus rapide du chromel que de l'alumel.
Il faut donc exclure du calcul analogique cette période transitoire.

Commande du Cycle de la Sonde. Etalonnage en Tension

Le rôle de l'organe de commande est double: il doit être capable de commander l'électrovanne suivant un cycle, il doit en outre donner l'ordre au calculateur d'effectuer les opérations sur l'équation differéntielle afin d'extrapoler la valeur T_o.

Cet ensemble comprend:

un amplificateur continu de gain 100 qui élève le signal issu du thermocouple à une valeur exploitable. C'est l'amplificateur d'entrée, l'ensemble du système électronique travaillant à partir de la tension de sortie de cet amplificateur.

un trigger de Schmitt à changement automatique de seuil qui déclenche le relais de commande de l'électrovanne d'air lorsque le thermocouple atteint 1000°C, soit 4, 1 V à la sortie de l'amplificateur d'entrée, et qui l'enclenche lorsque le thermocouple atteint 200°C, soit 0, 81 V à la sortie de l'amplificateur d'entrée. Ces deux seuils sont en outre réglables manuellement.

Les mises au point, réglages et étalonnages en tension peuvent être effectués sans thermocouple grâce à un simulateur de sonde. Il se compose essentiellement d'une source basse tension, alimentant un double circuit RC commuté par le relais du trigger.

Couplé sur l'amplificateur d'entrée, ce circuit émet un signal exponentiel analogue à celui de la sonde, signal dont on peut régler la valeur asymptotique par le niveau de la tension d'entrée (analogie avec la température T_o) et la constante de temps RC par réglage de R (analogie avec θ).

Ainsi commuté, l'organe de commande fonctionne automatiquement, comme un système bouclé dont l'amplificateur d'entrée constitue la chaîne de contre-réaction.

Traitement Analogique du Signal

Dérivation

La dérivation du signal constitue un élément fondamental du calcul puisque, en plus de son rôle dans l'équation différentielle, elle permet le calcul automatique de la constante de temps.

La résolution du problème des parasites dans un circuit de dérivation constitue la difficulté principale. En effet, la dérivée traduisant la pente du signal, celui-ci doit être électriquement très pur et toute modulation parasite doit être évitée. Le montage de la figure 3-3-2 a donné entière satisfaction.

Un tel système est dérivateur dans la bande de fréquence 0 - 10 Hz ce qui est suffisant pour le signal de sonde; ce montage permet en outre de réduire l'induction à 50 Hz.

Détermination de la constante de temps

La constante de temps d'une exponentielle se définit par:

$$|\theta| = \frac{\frac{dT}{dt}}{\frac{d^2T}{dt^2}}$$

Pour l'analogie électrique, la difficulté apparaît immédiatement: le calcul de la dérivée seconde est très difficile, voire impossible pour des phénomènes à variations temporelles rapides.

On peut néanmoins tourner cette difficulté en ramenant le calcul de θ à celui d'un temps.

La figure 3-3-3 montre la courbe: $(C_1)\ \frac{dT}{dt} = f\,(t)$.

Soit (C_2) une courbe affine de (C_1) définie par un rapport k constant, $k < 1$. A partir de l'instant t_1 (ce qui exclut du calcul le phénomène transitoire) considérons le niveau de tension atteint représenté par l'horizontale coupant (C_1) au temps t_2.

On peut écrire:

$$k\,\frac{T_o}{\theta}\quad e^{t_1/\theta} = \frac{T_o}{\theta}\quad e^{t_1/\theta}$$

En simplifiant et en passant aux logarithmes: $t_2 - t_1 = \theta \text{ Log} \frac{1}{k}$ d'où $\theta = \lambda\ (t_2 - t_1)$ avec $\lambda = [\text{Log} \frac{1}{k}]^{-1}$ λ est le coefficient analogique de la constante de temps.

Il est donc possible à partir de la mesure de l'intervalle de temps $t_2 - t_1$ de déduire θ.

Ce calcul est effectué électroniquement de la façon suivante:

la rapport k est simulé par un potentiomètre. Au temps t_1 le niveau de tension atteint par (C_2) est mis en mémoire dans une capacité. Simultanément un chronomètre fournissant une caractéristique linéaire tension-temps est enclenché. Un

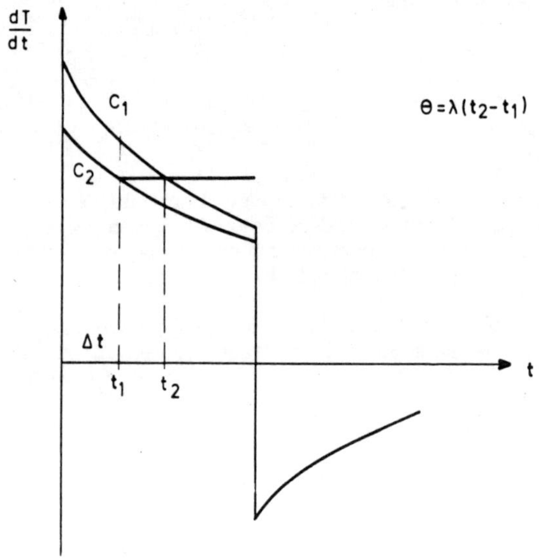

Fig. 3-3-3 Calcul de la constante de temps

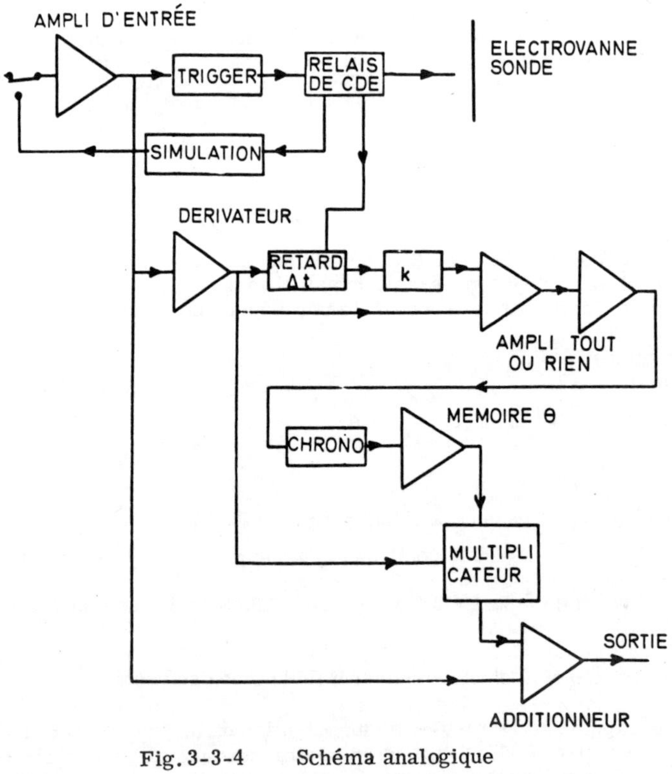

Fig. 3-3-4 Schéma analogique

amplificateur opérationnel effectue alors la différence de tension entre (C_1) et l'horizontale. Cette différence est transformée en un créneau rectangulaire dans un amplificateur travaillant en tout ou rien qui, par l'intermédiaire d'un relais, arrête le chronomètre au temps t_2 dont la tension de sortie proportionnelle à t_2-t_1 donc à θ est mis en mémoire.

La nécessité d'un signal dérivé non parasite apparaît clairement: le chronomètre étant arrêté au moment où les deux courbes se coupent, toute modulation parasite transforme le retour au zéro de l'amplificateur tout ou rien en une variable aléatoire et affecte fortement la précision du système.

Détermination de l'asymptote

Il s'agit tout d'abord du calcul du produit $\theta \frac{dT}{dt}$ puis de la somme $\theta \frac{dT}{dt} + T$.

Le produit $\theta \frac{dT}{dt}$ s'effectue dans un multiplicateur à créneaux dont le principe est le suivant: les signaux rectangulaires fournis par un multivibrateur sont modulés en hauteur par une des variables et en largeur par l'autre. L'intégration du signal résultant est proportionnelle au produit des deux variables.

Ce multivibrateur oscille à 5 kHz ce qui permet une bande passante de 200 environ.

Ce produit est additionné dans un amplificateur opérationnel avec le signal instantané fourni par la sonde. On obtient à la sortie de cet amplificateur un niveau de tension proportionnel à T_o.

On profite de la période de refroidissement de la sonde pour effacer les signaux mis en mémoire.

La schéma analogique de l'ensemble est représenté figure 3-3-4.

Il résulte de ce qui précède que l'acquisition de T_o se manifeste dans l'intervalle de temps t_m -t_2. Pratiquement, le temps de montée entre T_f et T_M doit être au moins de 400 ms.

Le temps de montée étant évidemment fonction de θ et de T_o il est nécessaire que la jonction du thermocouple soit massive. Les essais qui ont été effectués avec des jonctions de 3 mm de diamètre permettent des temps de montée supérieurs à 400 ms. dans les cas les plus critiques, faible θ et haute température.

Technologie de la Sonde. Résultats Obtenus

La sonde à col sonique utilisée mesure la température d'arrêt de l'écoulement. Technologiquement, la sonde se présente de la façon suivante: le thermocouple chromel-alumel est isolé à l'aide d'une perle d'alumine et placé dans un tube central lequel sert de conduit à l'air de refroidissement de la jonction (Fig. 3-3-5). L'air est pulsé sous une pression surpassant de 3 bars environ la pression d'arrêt au droit de la sonde. L'ensemble est entouré d'une gaine en cuivre à l'intérieur de laquelle un débit d'eau sous une pression de l'ordre de 11 bars assure le refroidissement et, grâce à la conductibilité thermique du cuivre, une bonne tenue thermique. Afin de bénéficier d'une constante de temps de la jonction suffisamment grande, même lorsqu'elle est soumise à des flux thermiques importants, le diamètre de la soudure, approximativement sphérique, est de 3 mm. Cette grandeur est un compromis qui permet l'obtention de temps de montée compatibles avec la vitesse de calcul sans exiger de grandes dimensions pour la sonde. La jonction est réalisée par soudure de chromel et d'alumel à laquelle on rapporte une perle d'alumel car l'alumel possède une chaleur massique supérieure à celle du chromel. Les fils du thermocouple ont un diamètre de 0, 62 mm. On évite ainsi les pertes par conduction.

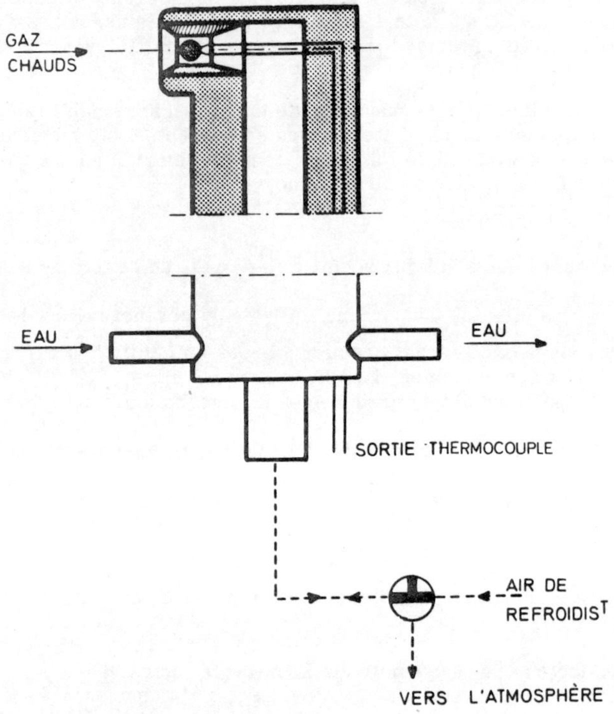

Fig. 3-3-5 Sonde et son circuit - Schéma

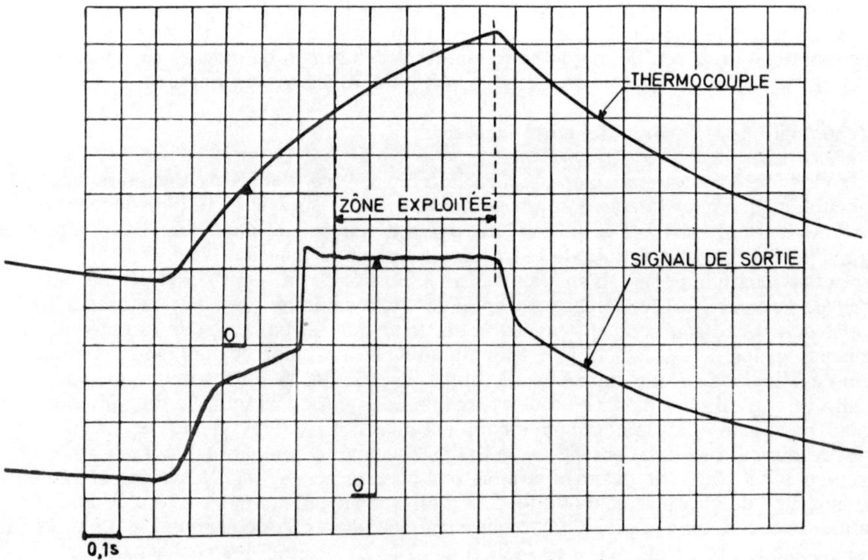

Fig. 3-3-6 Enregistrements des signaux

Cette sonde a d'abord été utilisée dans un écoulement supersonique à Mach 2. La pression d'arrêt était de 3 bars et la température suivant les essais comprise entre 1100°C et 1400°C. Ce niveau de température, modeste compte tenu des possibilités théoriques de la méthode, a néanmoins permis de définir une sonde, de déterminer sa tenue mécanique et thermique et de tester l'électronique.

Dans les conditions précitées, le temps de montée en température est de l'ordre d'une seconde, ce qui conduit à un point de mesure toutes les 2 secondes environ.

D'autres essais sont actuellement effectués à 2000°C et au delà, essais qui permettent de montrer la bonne tenue de la sonde. Ces essais doivent d'autre part, définir les limites d'application de la méthode.

Les enregistrements se présentent sous la forme donnée par la figure 3-3-6. Durant l'intervalle de temps t_2-t_0 on obtient une image de la loi de montée en température. Ceci résulte du fait que la constante de temps n'est pas déterminée et qu'elle se présente sous forme d'une tension nulle. Au temps t_2 la tension de sortie de l'amplificateur sommateur passe brusquement à une valeur dont la grandeur est proportionnelle à l'amplitude T_o. L'horizontalité du palier ainsi obtenue permet en outre la vérification du bon fonctionnement de l'ensemble: toute erreur sur la loi de montée (variation de T_o ou de θ pendant la mesure) entraine une distorsion du signal de sortie.

Actuellement, les erreurs estimées sont de l'ordre de 3% au niveau du calculateur électronique. Les comparaisons que l'on a pu effectuer entre la temperature calculée et celle déterminée par le bilan thermique de l'installation entraînent des erreurs de l'ordre de 5%.

Ces essais sont conjugués avec l'étude de la détermination de la dérivée seconde. Si cette méthode s'avère fructueuse le champ d'application se trouvera élargi. Il est en effet possible, dans ce cas, d'extrapoler plus vite, tout système séquentiel dans l'électronique étant supprimé.

Corrélativement l'étude du déplacement de la sonde dans l'écoulement, déplacement lié au cycle, est entreprise.

Si ces études aboutissent on peut espérer la mesure de températures locales dans un écoulement à 3000°C, mesures sinon absolues, du moins comparatives.

Références

1. **Von Burger-Scheidlin, F., Stottman, P.,** 'Eine dynamische Temperaturen methode mit Thermoelementen.' Zeitschrift für Flugwissenschaften, April 1966.

2. **Wormser, A.F., Pfuntner, R.A.,** 'Pulse technique extends range of Chromel-Alumel to 7000°F.'Soc. Auto. Eng., Rep. 524 A, 1962.

Commentary on Chapter 3-3

F. Von. Burger

Le dispositif présenté par M. Gautrot est intéressant et constitue un progrès notable sur un montage analogue que nous avons réalisé à la D. F. L.

Nous avons estimé seulement la constante de temps à l'aide d'un oscilloscope et fait l'intégration du signal à l'aide d'un circuit résistance capacité, cette intégration étant nécessaire en raison des fluctuations de la température dans la chambre de combustion.

Je voudrais demander à M. Gautrot s'il est possible avec cet appareil de mesurer une valeur quasi-moyenne de la température toujours variable dans une chambre de combustion.

M. Gautrot

1. Il y a en général variations rapides et simultanées de la température et de la constante de temps d'une jonction de thermocouple placée dans une chambre de combustion: en ce cas il n'est pas possible d'extrapoler l'arc de montée fourni par la sonde: la loi de montée est aléatoire.

2. Pour tourner cette difficulté, l'emploi d'une soudure massive permet d'intégrer ces variations instantanées. La température indiquée est donc la moyenne des températures dans un intervalle de temps égal au temps de montée. La bande passante maximum résultante est actuellement d'environ 2 Hz, soit 2 points de mesure de température moyenne par seconde. Si on cherche à dépasser cette limite il semble qu'il y a dégradation de la pureté de l'exponentielle, donc du signal de sortie.

Part 4

Mass Spectrometry

4-1

Investigation of Reacting Gases with the Mass Spectrometer

K. H. HOMANN
Institut für Physikalische Chemie, Universität Göttingen, Germany.

Summary

The application of mass spectrometric analysis together with suitable methods of sampling from reacting gases is reviewed. The question as to how well a sample is representative for the stage of reaction is discussed in the cases of microprobe sampling of stable products and molecular beam sampling for free radicals and active atoms. A supersonic nozzle as source for a molecular beam has been used for sampling from flames and isothermal flow systems at pressures ranging from, a few mm Hg to more than one atmosphere. For effective sampling of free radicals the conditions of flow at the entrance orifice must be such that the sample is not taken in from a boundary layer. Calibration measurements are necessary for possible mass separation effects in the beam. The influence of gas temperature on the mass spectra and the identification and analysis of free radicals and atoms are illustrated by some examples.

Sommaire

L'emploi de l'analyse par spectrométrie de masse et l'échantillonnage convenable des gaz réactifs à analyser sont discutés dans cet exposé. La question de savoir jusqu'à quel point un échantillon est représentatif de l'avancement de la réaction est discutée dans le cas du prélèvement par microsondes de produits stables et l'analyse des atomes actifs et des radicaux libres par faisceaux moléculaires. On a utilisé un faisceau moléculaire issu d'une tuyère supersonique pour l'étude de flammes et d'écoulements isothermes à des pressions allant de quelques torrs à plus d'une atmosphère. Pour l'étude correcte des radicaux libres, les conditions à l'entrée de l'orifice doivant être telles qu'on ne fasse pas de prélèvement dans la couche limite. Des étalonnages sont nécessaires pour étudier la séparation de masse dans le faisceau moléculaire. L'influence de la température du gaz sur la spectrométrie de masse, l'identification et l'analyse des radicaux libres et d'atomes sont illustrées par quelques exemples.

Introduction

A characteristic of the mass spectrometer that makes it particularly useful for the study of chemical reactions in the gas phase is its ability to identify an unknown substance or components that are not stable under normal conditions, such as active atoms, free radicals and some sorts of ions.

Mass spectrometers of the time-of-flight type give(per second)about 10^5 mass spectra of the gas components in their ion source. This time resolution can be used for the study of rapid reactions. If a slow scanning instrument is to be used, the time resolution must be achieved by using a swiftly flowing reaction system. Unlike optical measurements of reacting gases, which can be performed *in situ*, mass spectrometric analyses usually require sampling, because the ion sources usually operate at a pressure lower than that in the reaction system. At very low pressures (~ 10^{-3} torr) however ion-molecule reactions can take place inside the ion source (1) (2). Mass spectrometers can also be used to determine the distribution of reactants and products in crossed molecular beams experiments.

Inlet Systems and Problems of Sampling

Mass spectrometric analysis of reacting gases requires the withdrawal and quenching of representative samples from the system during the progress of reaction, in a time that is only a small fraction of the whole time of reaction. The mass spectrum must then be obtained from the parts of these samples that are still representative of the conditions being studied.

The inlet system used in commercial mass spectrometers is often not suitable for this purpose. Usually it consists of a reservoir, made of metal or glass, into which the sample of the gas is expanded before being admitted into the ion source through a molecular leak in a gold foil. This equipment can be used only if the sample has already been taken from the reaction system by some other means, and kept in a container. Thus only stable reaction products, and gases that do not react at the walls of the inlet system or adhere to them, can be determined quantitatively.

For the analysis of substances such as water, formaldehyde, peroxides, ammonia, acids, and others that are strongly adsorbed at walls or react with metal surfaces, it is necessary to provide a continuous flow of the sample from the reaction system to the inlet orifice of the ion source. This requires a tube with inert walls and without an intermediate reservoir. The sample is analysed after all adsorption equilibria have been established. Fristom and his collaborators (3) used a 'Teflon' (PTFE) tube and a leak in a sapphire disk to avoid reactions on metal surfaces in the inlet system.

Similar inlet systems combined with microprobes have frequently been used for the measurement of concentration profiles in stationary flames (4, 5). The manufacture and use of these microprobes, mostly made of quartz, has been described by Fristom (6). A flowing isothermal reaction system develops concentration gradients, and in flames there are temperature gradients also. When sampling from systems like these, the pressure drop across the probe orifice accelerates the gas to sonic velocity. This influences the gas flow in front of the probe, and in fact the concentration in the sample no longer corresponds exactly to what it would have been at the position of the probe orifice in the undisturbed system. It is in fact averaged over a region extending about four orifice diameters in front of it (7).

To minimise the influence of the probe on the flow of the reacting gases, 'isokinetic' sampling must be used, whereby the sample flows into the probe with the velocity of the undisturbed system. Using this type of probe, the sample must be quenched from the cooled inner walls; this is inadequate for rapid reactions.

A temperature gradient at the probe orifice may be disturbed, or caused, partly by conduction of heat to the probe, which is usually cooler than its surroundings, and to a smaller extent by adiabatic cooling if the gas is accelerated. The reactions may therefore be already partly quenched before the sample enters the orifice. Those reactions with larger activation energies are quenched more strongly than those - such as free radical reactions - with smaller activation energies. Thus there may be uncertainties in the relative concentrations of the different species measured. Further care must be taken in the interpretation of mass spectra of species sampled at different temperatures, since the pattern of ionic fragments might be strongly influenced by the vibrational temperature of the molecules. However, this applies only if vibrationally hot molecules reach the ion source (see below).

In constructing a suitable probe, therefore, a compromise has to be made between minimum influence on the flame and maximum quenching of the reactions. The microprobes mentioned above are slim and sharply tapered at the tip, so that the profiles of concentration and temperature are little disturbed in front of them. On the other hand their shape and the drop of pressure attainable behind their sampling orifices are inadequate for quenching radical reactions. Thus species such as H, OH, O and hydrocarbon radicals appear as H_2, O_2, H_2O, CO, hydrocarbons etc.

Conditions inside the probe must be prevented from changing from viscous flow to free molecular flow. If all species enter the probe by continuous flow with the same mean velocity, but each is pumped off according to its molecular velocity, the measured density of the heavier species will have increased (8).

Molecular Beam Inlet Systems

These same requirements must be met for free radical sampling, but in addition wall effects must be eliminated as far as possible. The outer surface and shape of the probe may have an influence on the radical concentration of the unsampled gas, while the inner surface of the probe can affect the radical concentrations in the sample. Practically, the latter can be eliminated only by admitting a collimated beam of molecules into the ion source. When the density in the molecular beam is such that the mean free path is much larger than the distance to the ion source, there is no excessive loss of radicals or other unstable particles. Their concentration in the beam can be determined, provided that the beam signal can satisfactorily be discriminated from its background.

Two principal sources have been used to generate molecular beams; the molecular leak and the supersonic nozzle.

Molecular leak sampling

Molecular leaks are used as orifices for sampling from both static and flowing reaction systems. Since a molecular leak should not be much larger in diameter than the mean free path of the molecules in front of the orifice, its use is limited to low pressures of some 10 torr or less. A gas having a mean molecular mass of 30 amu (atomic mass unit) at room temperature has a mean free path of $10\mu m$ at 5 torr. At higher pressures smaller holes are needed, but it is impracticable to make holes that are not channels of a size smaller than $5\ \mu m$ in diameter in a material like quartz. Moreover, if the ratio of mean free path to diameter is kept constant, the mass flow through the leak decreases linearly with diameter.

A 180° opening behind the entrance hole, as indicated in figure 4-1-1, is the best arrangement for obtaining a minimum flow resistance behind the orifice. Since the density in the axis of the beam decreases with distance according to an inverse square law, the ion source must be situated close to the orifice. In a typical example, the pressure in the axis of the beam falls to 10^{-6} torr after a path of one centimetre from a beam source of 5 torr having an orifice diameter of about $10\ \mu m$. To obtain sufficient discrimination between beam and background particles, the pressure around the ion source must therefore be considerably less than 10^{-6} torr, otherwise the beam signal must be modulated by a chopper followed by a phase sensitive amplification. An arrangement like this has been used by Foner and Hudson for the detection of the HO_2 radical and the study of its ionisation potential (9).

A static reaction system separated by a molecular leak from a rapid-scanning mass spectrometer was described by Marsden (10). In his work the fast display of the mass spectrum on an oscilloscope was achieved by a rapid variation of the voltage accelerating the ion. Marsden used this system to study the oxidation of hydrogen sulphide near the third explosion limit. By using $D_2S/^{16}O_2$ and $H_2S/^{18}O^{16}O$ mix-

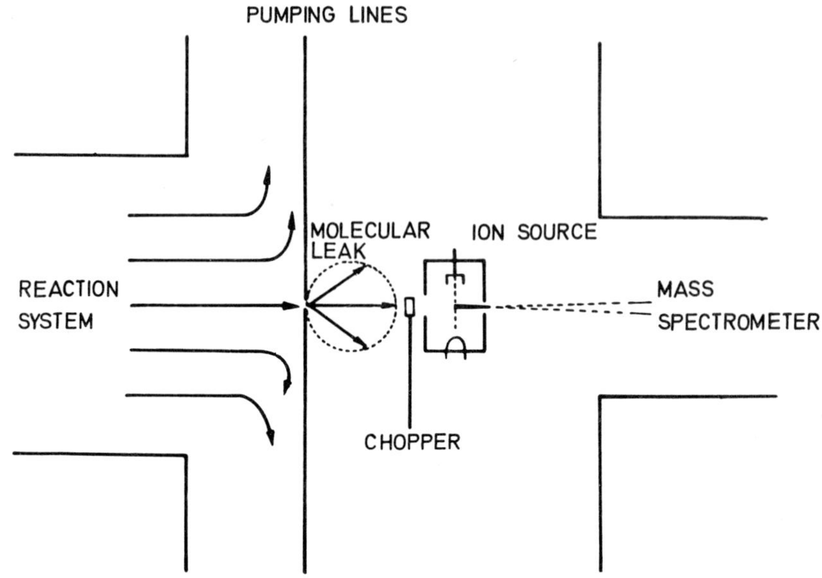

Fig. 4-1-1 Mass spectrometer with molecular leak sampling system

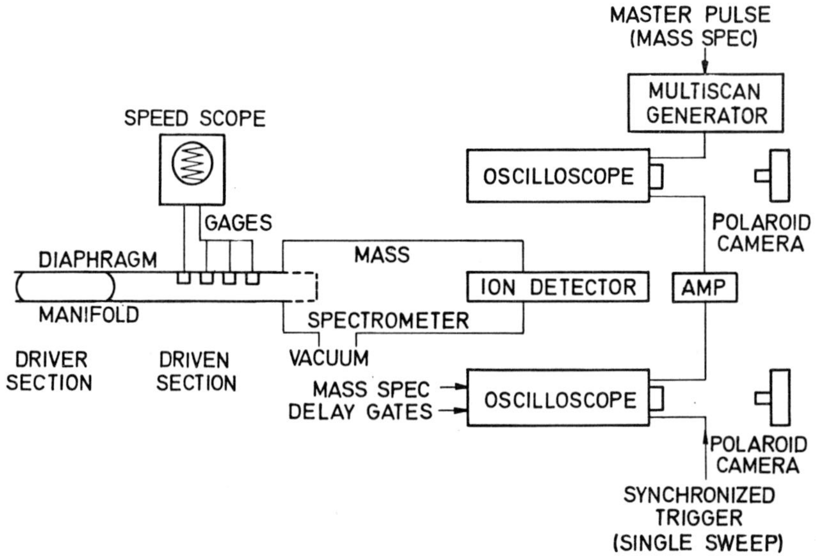

Fig. 4-1-2 Arrangement for mass spectrometric analysis in shock tube experiments

tures, the species H_2, S_2O, H_2SO, S_2, and H_2S_2 could be identified during an induction time of 8-12 sec. Only when the explosion took place was there evidence for the presence of the SO radical.

Reactions behind reflected shock waves have been studied by means of a time-of-flight mass spectrometer. By sampling through a nozzle in the plane end plate of the shock tube, Kistiakowsky and Bradley studied the decomposition of nitrous oxide, following the concentration of N_2O, N_2, O_2, and O-atoms (11). The concentration of oxygen atoms was determined in arbitrary units. More recently Diesen, using a similar set-up, has studied the influence of boundary layer effects at the orifice on the process of sampling (12). He observed an increase in the density of the sampled gas, due to a temperature decrease extending to a distance of about one orifice diameter, but he could not find a boundary layer effect in terms of chemical kinetics.

A comparison of studies of the kinetics of the thermal decomposition of nitrous oxide behind shock waves, using optical absorption methods (13) and mass spectrometric analysis, shows that up to a temperature of 2500°K the results are the same (14). However at temperatures around 3000°K, measurements made using sampling methods give reaction rates that are too low, indicating the increasing influence of the walls. The mass spectrometric method of studying reactions behind shock waves has also been used by Modica (15) and Dove (16).

When using a nozzle instead of a molecular leak for sampling a static reaction system, such as the gas behind a reflected shock wave, the location of sampling must be considered. During the formation of the stationary flow into the sampling nozzle, after the shock wave has been reflected, the place of sampling at first is close to the orifice and then extends out for some diameters in front of it. This change of the place of sampling may have an influence on the concentration measured at the beginning of the reaction. Although the overall kinetics might be measured correctly by mass spectrometric analysis, the results of free radical and atom measurements should be treated with reserve. The influence of the boundary layer and the wall might be much larger on their concentration than on the overall reaction rate.

A schematic diagram of the experimental set-up for mass spectrometric analysis in shock tube experiments is given in figure 4-1-2.

Nozzle-beam Sampling

For sampling free radicals, a molecular leak orifice in a more or less planar wall has the disadvantage of taking in the gas from a region too near to the wall and extending over a radius of about one diameter from the orifice. When such a leak is introduced into a gas flow system, the sample will diffuse into the leak from a stagnation point of the flow in front of the sampling probe. Tapering the probe tip to avoid damming up the flow would increase the resistance to the free molecular flow inside the probe.

On these accounts the use of a nozzle in the tip of a tapered probe is more favourable, especially when sampling from rapidly flowing gases. The gas mixture then enters the nozzle from a region extending several diameters in front of the nozzle (7). This means that the percentage of sampled gas that has come into contact with the wall is greatly diminished. Due to the continuous flow through the nozzle, the mass flow is centred on the axis more than it is with molecular effusion, so that the probe tip may be made more tapered without inordinately increasing the flow resistance behind the nozzle. That major part of the gas, which is not taken in directly through the orifice, flows around the tip and does not diffuse back into the nozzle after having come into contact with the outer probe wall. This has been confirmed

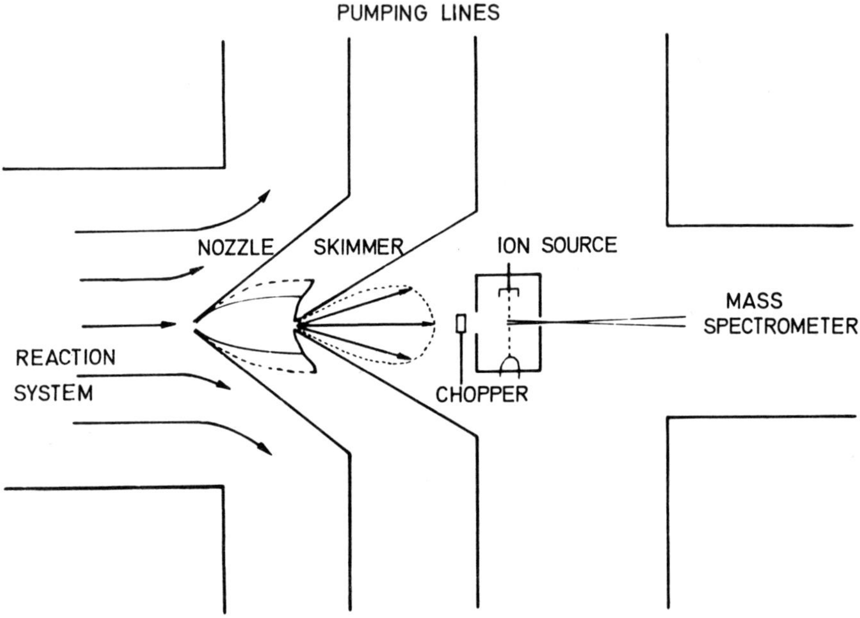

Fig. 4-1-3 Mass spectrometric analysis of flowing reaction system using a nozzle beam sampling system

by sampling a flowing gas mixture containing oxygen atoms through a nozzle probe, the outside of which was coated with a thin layer of carbon. Very little of the CO and CO_2 that was formed at the surface was taken in through the nozzle (17). As Becker has shown, it is not necessary to use a de Laval nozzle for generation of a supersonic jet. A hole with smoothed edges serves the same purpose (18).

Theoretical and experimental work on free expanding jets and nozzles as sources for molecular beams has been much intensified (19, 20, 21), starting with the experiments of Becker (18, 22) and with the theoretical paper of Kantrowitz and Grey (23). The molecular beam is not generated in the conventional manner by effusion through an orifice, but by skimming a sample from a supersonic jet. The sample is then expanded into high vacuum. The skimmer is a cone with a sharp-edged hole at the tip. It is situated between the nozzle and the first Mach disk. The flow between nozzle and skimmer must be such that no shock is detached from the skimmer. The gain in intensity on the axis of the final beam relative to that of a conventional beam is a function of the Mach number of the flow at the skimmer entrance. It increases with M^2 for $M > 3$ (24). The experimental arrangement is schematically shown in figure 4-1-3.

Nozzle beam sampling systems have been used for the study of rapid reactions at both low and high pressure. The mass flow of gas into the nozzle chamber and also into the high vacuum chamber behind the skimmer is much larger than for a conventional molecular beam, especially when sampling from higher pressure. A considerably larger pumping capacity is therefore needed to maintain the necessary vacua. We have used a nozzle beam system to sample a number of flames at pressures up to 60 torr, where the reaction zone extends to about one centimeter (25) (26) (27). The experimental set up is similar to that shown in figure 4-1-3. The object of investigation was the formation of solid carbon and the intermediates in this process. We analysed various rich hydrocarbon-oxygen flames for such hydrocarbon radicals as CH_3, C_2H, C_4H_3, acetylenic polymers C_4H_2 to $C_{12}H_2$, and for very reactive cyclic hydrocarbon species in the 100-600 amu range. The profiles of some highly reactive hydrocarbon intermediates could be quantitatively related to the increase in solid carbon concentration. In hydrogen-oxygen flames, concentration profiles of O, H, OH, and H_2O_2 could be measured. Some evidence for the presence of the HO_2 radical in rich H_2-O_2 flames was obtained. In the same manner, hydrazine decomposition flames were investigated for NH_2 and N_2H_2 intermediates (28).

Greene and Milne have used a nozzle beam for sampling from flames at atmospheric pressure (29) (30). For the expansion of the gas from a pressure of one atmosphere a further pumping chamber had to be inserted between the skimmer and the ion source. A number of flames (H_2-O_2-N_2, CH_4-O_2-Ar, CO-O_2) were analysed both for stable reaction products and the reactive species. Although spatial resolution of the main reaction zone was not too good (the sampling orifice diameter was 0.24 mm), the authors could measure quantitatively the profiles of all species, including H, O, and OH in the recombination zone of the flames. Experimental values for chemical equilibria between stable species and radicals agreed within the limits of experimental error with those calculated from the adiabatic flame temperature. A quantitative temperature-pressure history of the expansion of the sample can be obtained from subsonic and supersonic calculations. For a monatomic gas at 2000°K expanding through a 0.125 mm orifice, the translational temperature drops to 30°K and the pressure to $\sim 7.6 \times 10^{-3}$ torr within about 1.5 μsec from the first appreciable change. Gases with larger heat capacities expand more slowly.

The temperature drop does not quench reactions with zero or negative activation energies, such as condensations when sampling high pressure systems. Condensation of heavier species such as argon and carbon dioxide has been observed (31) (32). Generally, only a fraction of the order of 1 percent of the substance will condense, and the degree of condensation becomes even less when systems at lower pressures are sampled.

A nozzle beam has also been used when sampling reacting gases in an isothermal flow reactor (33). A fast flow around the probe tip favours the removal of the gas that has hit the wall around the orifice at the probe tip, and diminishes the thermal influence of the probe on the reacting gases, which can never be totally excluded when sampling the reaction zone of flames. In this way, the mechanism of the oxidation of CS_2 by O_2, strongly diluted with argon, has been investigated at temperatures up to 1400°K. CS_2 was injected into the hot stream of argon already containing the oxygen (33). The concentration profiles of all stable species and radicals could be measured by changing the distance between the mixing point and the sampling probe in the stationary reaction system. SO_2 and CO were the main reaction products, with little CO_2 being formed after the CS_2 had been consumed completely. Besides the comparatively stable intermediate COS, the concentration of radicals such as SO, S, O, CS, and S_2O could be measured. The main mass peak of S_2O is superimposed on that of SO_3 at 80 amu but by mass spectrometric analysis of the isotope peaks that are different for both compounds, the mass 80 peak could be accounted for mainly by S_2O, with only a very small contribution from SO_3.

More recently the rate of single steps in this complex reaction have been measured in a low pressure fast flowing system (34). Sampling was again done using nozzle beam system combined with a Bendix time-of-flight mass spectrometer. The elementary step O + COS — CO + SO, by which the COS is oxidised, could be distinguished completely from consecutive reactions of SO to SO_2 by mass spectrometric analysis. The reaction O + CS_2 — CS + SO could also be followed separately. With the same experimental set-up the reactions rates O + H_2, O + C_2H_2, O + NH_3 and others have been measured (35). By following quantitatively the concentrations of each participant in the reactions, primary reaction steps could be distinguished from consecutive reactions and parallel steps. This is a preliminary condition for the determination of rate constants for elementary reactions.

The rapid quenching of reactions by expansion through a nozzle could also be used for sampling the exhaust gases of a rocket motor if sufficient cooling of the nozzle itself could be provided. Similarly it would be possible to study the chemical reactions in the subsonic flow regime that lies behind stationary shock waves when heat is added during supersonic flow.

Mass Spectrometric Analysis of Samples

Mass Separation

The transition from supersonic flow into free molecular flow in the final beam is of considerable consequence to what the mass spectrometer sees, if there are large differences in the molecular masses of the components of the sample.

Several communications concerning this effect have come from Becker (22) and Stern (36). Milne and Greene also have observed mass separation while studying the composition of gas mixtures by means of a nozzle beam sampling system (37). By sampling from a freely expanding jet, an enrichment of the heavier components in the axis of the jet has been observed, or not observed, according to the conditions of sampling (38) (39). There is still no full agreement among the various investigators as to the causes of this mass separation. The different radial diffusion velocities of the species, and any deceleration and deflection of the jet by the skimmer all play roles. Fenn and Andersen have recently called attention to the migration of background molecules into the jet, opposing the preferential effusion of light species away from the beam axis (40). This mass separation depends on several parameters of the system and can hardly be predicted exactly, so that calibration measurements with standard gas mixtures are necessary.

Interpretation of Mass Spectra of Radicals

The possibility of using a mass spectrometer for analysing atoms and free radicals has been under discussion since Eltenton first analysed flames for reactive species, and identified these by mass analysis (41). Atoms and free radicals are ionised just like any other substance, if the bombarding electrons have sufficient energy; and the mass spectra of radicals do show fragment ions, depending on the electron energy applied. Frequently the mass lines of atoms and radicals are superimposed on those of ion fragments of stable molecules. This will be the case unless the electron energy lies in the range between the ionisation potential of the radical, and the appearance potential of the same ion from a stable molecule. This energy range is usually between 10 and 15 eV. At such low electron energies, however, the yield of ions is very small, so that an electron multiplier with a high sensitivity must be used for amplification. It is more convenient to use an electron energy of between 20 and 25 eV, so that the fragmentation of many stable molecules - especially di-and triatomic molecules - is not too large, while the production of ions is sufficient for good analytical accuracy. There is no general rule about the use of any particular electron energy: variation can often give additional information about the origin of a mass peak. If the peaks superimposed on the radical mass peaks are due only to fragments of stable components, the parts due to the radical can usually be determined by subtracting the measured distribution of fragment ions obtained from mass spectra of the pure stable compounds. When sampling an isothermal reaction system, the contribution of any radical to a fragment mass peak can often be detected by comparing the shape of the profiles of the parent peak of the stable substance with that of its fragment peak across the reaction zone. A different shape would indicate a contribution by a second substance, which might be a radical or another stable substance that gave an ion fragment of the same mass. If fragments from many known species were superimposed on a radical mass peak - often the case for hydrocarbon radicals and hydrogen atoms in the oxidation of hydrocarbons - the accuracy of radical analysis would decrease because experimental errors in measuring individual fragments are additive. Moreover the sensitivity of the mass spectrometer for hydrogen atoms is low because of their small ionisation cross section, as well as the mass separation effects.

In some cases the labelling of a substance by a stable isotope helps to avoid a superposition of too many mass peaks. For example, the radicals C_2H_5 and HCO which have nearly the same mass (= 29 amu) can be distinguished by using ^{18}O instead of ^{16}O. The use of a highly resolving mass spectrometer to separate peaks that differ by only a small fraction of an atomic mass unit is possible, but it has not yet been applied to the analysis of free radicals.

When sampling gases at different temperatures the variation of their fragment patterns with temperature must be taken into account for the determination of radical contributions. The fragment pattern is dependent on the vibrational energy of the molecule (42). The relative abundance of light fragments of polyatomic molecules increases more rapidly with temperature than those of oligo-atomic molecules (43). The measurements of Greene and Milne on flames showed that the fragmentation of CO_2 to CO varied at least from 12 percent at room temperature to 17 percent at 2000°K, as determined in their instrument at 70 eV electron energy (30). On the other hand we could not find any change in the mass spectrum of acetylene sampled through a nozzle from burned gases at 1700°K, compared with the spectrum obtained from a sample at room temperature (27). In a supersonic jet the sample is cooled very strongly, but the vibrational relaxation in many cases is comparatively slow, so that vibrationally hot molecules may be ionised and give rise to different mass spectra. This effect can be eliminated only by calibration.

Calibration for Radical Analysis

An exact mass spectrometric determination of radical concentrations requires the mass spectrometer to be calibrated specifically for the reactive species in question. It is not yet possible to derive more than a rough value of the ionisation cross section for radicals from quantum mechanical calculations. To generate a defined concentration of radicals or atoms the chemical equilibrium between suitable reactive species at high temperature has sometimes been employed (44). It is more convenient, however, to use a method in which the radical is formed rapidly and quantitatively by the reaction of a stable molecule with another reactive species. The concentration of the product radical is then calculated using the stoichiometric relation. Well known examples of this are the so called 'titration reactions' for generating oxygen atoms and hydroxyl radicals:

$$N^{\bullet} + NO \longrightarrow N_2 + O$$

and

$$H^{\bullet} + NO_2 \longrightarrow NO + OH$$

A prerequisite for the use of an elementary reaction of this kind is that secondary reactions of the products with the reactants be slow compared with the primary reaction. This can often be arranged by using an excess concentration of the stable reactant.

Another technique is to generate an unknown concentration of the radical or atom - by a microwave discharge for example: the radical then reacts quantitatively with a known amount of a stable substance in a reliable stoichiometric relation. The calibration for SO, for example, is performed by:

$$SO_2 \xrightarrow{\text{microwave discharge}} SO + O$$

$$SO + NO_2 \longrightarrow SO_2 + NO$$

The mass spectrometric sensitivity for SO relative to that for SO_2 can then be measured directly (34).

There is no general rule for applying reactions of this kind to the production of any given atom or radical. Many of them have to be found empirically. For the determination of the ionisation potential of a free radical - which is usually accomplished by using an ion source, after Fox (45) - it is not necessary to know the concentration of the radical. A greater variety of methods, such as photo-dissociation, thermal decomposition, dissociation by electrodeless discharges and others, can therefore be applied (46). It might be supposed that in the energy range 20 to 70 eV the mass spectrometric sensitivity for a radical and that for a similarly structured stable molecule (e.g. $CH^{\bullet}_3$ - CH_4, $CN^{\bullet}$ - HCN) would seldom differ by more than a factor of two. Thus radical concentrations can frequently be estimated with an accuracy sufficient for many kinetic calculations.

Conclusion

This short survey of the potentiality of mass spectrometric analysis for studying gas reactions shows that it has great versatility for the solution of many different problems. Detailed application is, however, governed very strongly by the subject to be studied, and the evaluation of the results is in many cases not straightforward. The conditions of sampling and mass spectrometric analysis must be well adapted to the problem posed.

Acknowledgments

This work was supported in part by the Deutsche Forschungsgemeinschaft and by the Fraunhofer Gesellschaft. The author expresses his gratitude to Prof. W. Jost and Prof. H. Ag. Wagner for their advice and encouragement.

References

1. **Stevenson, D.P.,** 'Ion-Molecule-Reactions in Mass Spectrometry', Ch.A. McDowell ed. McGraw-Hill, New York 1963.

2. **Melton, C.E.,** 'Ion-Molecule Reactions in Mass Spectrometry of Organic Ions', F.W. McLafferty ed. Academic Press, New York 1963.

3. **Fristrom, R.M., Grunfelder, C., Favin, S.,** J. Phys. Chem. 64, 1386, 1960.

4. **Fristrom, R.M., Grunfelder, C., Favin, S.,** J. Phys. Chem. 65, 587, 1961.

5. **Fenimore, C.P., Jones, G.W.,** J. Phys. Chem. 65, 993, 1961.

6. **Fristrom, R.M., Westenberg, A.A.,** 'Flame Structure', McGraw Hill, New York, 1965.

7. **Westenberg, A.A., Raeser, A.A., Fristrom, R.M.,** 'Combustion and Flame', 1, 467, 1957.

8. **Tine, G.,** 'Gas Sampling and Chemical Analysis in Combustion Processes'. AGARDograph 47, Pergamon Press, New York 1961.

9. **Foner, S.N., Hudson, R.L.,** J. Chem. Phys. 21, 1374, 1953. ibid. 36, 2681, 1962.

10. **Marsden, D.G.H.,** Rev. Sci. Instr. 33, 288, 1962. Can. J. Chem. 41, 2607, 1963.

11. **Bradley, J.N., Kistiakowsky, G.B.,** J. Chem. Phys. 35, 256, 1961.

12. **Diesen, R.W., Felmlee, W.J.,** J. Chem. Phys. 39, 2115, 2121, 1963.

13. **Olschewski, H.A., Troe, J., Wagner, H.G.,** Ber. Bunsengesellschaft physik. Chemie 70, 450, 1966.

14. **Gutman, D., Belfort, R.L., Hay, A.J., Pancirov, R.,** J. Phys. Chem. 70, 1786, 1793, 1966.

15. **Modica, A.P.,** J. Phys. Chem. 69, 2111. 1965. J. Chem. Phys. 44, 1585, 1966.

16. **Dove, J.E., Moulton, D. McL.,** Proc. Roy. Soc. (London), Ser. A 283, 216, 1965.

17. **Homann, K.H., Mochizuki, M., Wagner, H.G.,** Z. physik. Chem, N.F. 37, 299, 1963.

18. **Becker, E.W., Beyrich, W., Bier, K., Burghoff, H., Zigan, F.,** Z. Naturforsch. 12 a, 609, 1957.

19. **Anderson, J.B., Fenn, J.B.,** Phys. Fluids, 8, 780, 1965.

20. **Bier, K., Hagena, O.,** 'Influence of Shock Waves on the Generation of High-Intensity Molecular Beams by Nozzles.' Rarefied Gas Dynamics J.A. Laurman, ed. Vol.1, Academic Press, New York, p478.

21. **Campargue, R.,** 'High Intensity Supersonic Molecular Beam Apparatus'. Rarefied Gas Dynamics. J.H. de Leeuw, ed. Vol.2, Academic Press, New York, p279.

22. **Becker, E.W., Bier, K.,** Z. Naturforsch. 9 a, 975, 1954.
Becker, E.W., Bier, K., Burghoff, H., Z. Naturforsch. 10 a, 565, 1955.

23. **Kantrowitz, A., Grey, J.,** Rev. Sci. Instr. 22, 328, 1951.

24. **Parker, H.M., Kuhltau, A.R., Zapata, R.N., Scott, J.E.,** 'The Application of Supersonic Beam Sources to Low-Density High-Velocity Experimentation'. In 'Rarefied Gas Dynamics' F.M. Devienne, ed. Pergamon, New York, 1960, p69.

25. **Bonne, U., Homann, K.H., Wagner, H.G.,** 'Carbon Formation in Premixed Flames.' Tenth Int. Symp. on Combustion, The Combustion Institute, 1965, p503.

26. **Homann, K.H., Wagner, H.G.,** Ber. Bunsengesellschaft physik. Chem. 69, 20, 1965.

27. **Homann, K.H., Wagner, H.G.,** 'Some New Aspects of the Mechanism of Carbon Formation in Premixed Flames' Eleventh Int. Symp. on Combustium, Berkeley 1966.

28. **Homann, K.H., MacLean, D.I., Wagner, H.G.,** Naturwissenschaften, 52, 12. 1965.

29. **Greene, F.T., Brewer, J., Milne, T.A.,** 'Mass Spectrometric Sampling of 1-Atm Flame'. Tenth Int. Symp. on Combustion, The Combustion Inst., 1965, p153.

30. **Milne, T.A., Greene, F.T.,** J. Chem. Phys. 44, 2444, 1966.

31. **Becker, E.W., Bier, K., Henkes, W.,** Z. Physik 146, 333, 1956.

32. **Greene, F.T., Milne, T.A.,** J. Chem. Phys. 39, 3150, 1963.

33. **Krome, G.,** 'Massenspektrometrische Untersuchungen des Reaktionsablaufes der Schwefelkohlenstoff-Oxidation bei hohen Temperaturen'. Diplomarbeit, Universität Göttingen, April 1965.

34. **Wagner, H.G., Wolfrum, J.,** 'Bestimmung der Geschwindigkeit der Reaktion O + COS — CO+SO'. Ber. Bunsengesellschaft phys. Chem. 1967.

35. **Homann, K.H., Hoyermann, K.H., Wolfrum, J.,** 'Versuche zur vollständigen Bestimmung der Geschwindigkeiten bimolekularer Reaktionen', Kurznachrichten Nr. 2, Akademie der Wissenschaften, II. Math.-Phys. Klasse, Göttingden. 1967.

36. **Stern, S.A., Waterman, P.C., Sinclair, T.F.,** J. Chem. Phys. 33, 805. 1960.

37. **Milne, T.A., Greene, F.T.,** J. Chem. Phys. 40, 1488, 1964.

38. **Waterman, P.C., Stern, S.A.,** J. Chem. Phys. 31, 405, 1959.

39. **Reis, V.H., Fenn, J.B.,** J. Chem. Phys. 39, 3240, 1963.

40. **Fenn, J.B., Anderson, J.B.,** 'Background and Sampling Effects in Free Jet Studies by Molecular Beam Measurements' Rarefied Gas Dynamics. J.H. de Leeuw ed. Vol.2, Academic Press 1966, p311.

41. **Eltenton, G.C.,** J. Chem. Phys. 15, 455, 1947.

42. **Fox, R.E.,Hipple, J.A.,** J. Chem. Phys. 15, 208,1947.

43. **Osberghaus O., Taubert, R.,** Z. physik. Chem. N.F. 4, 264, 1955.

44. **Fite, W.L., Brackman, R.T.,** Phys. Rev. 112, 1141, 1958.

45. **Fox, R.E., Hickam, W.M., Grove, D.J., Kjeldaas, T.,** Rev. Sci. Instr. 26, 1101, 1955.

46. **Foner, S.N., Hudson, R.L.,** J. Chem. Phys. 45, 49, 1966.

Commentary by E. M. A. Willhoft

Dr. Homann has presented a valuable review of methods of sampling of gases for mass-spectrometric analysis. He has emphasised the necessity of some degree of sophistication in sampling in order to overcome the problems of non-representative withdrawal of products from a reaction zone.

Progress in mass spectrometry, as applied to the study of the formation of transient species during chemical reactions, has in the past been retarded by the difficulty of transferring a sufficient number of such species into the ion-source of the spectrometer. Dr. Homann has attempted to provide us with information relating to the various techniques of sampling. The drawback of the simple pin-hole technique has been pointed out as well as the advantages of the nozzle-beam sampling principle. Dr. Homann refers to the difficulty of producing pin-holes of diameter less than 5μm. My colleague, Dr. A. J. B. Robertson, has developed a simple procedure for producing holes in glass down to about 1μm diameter.

The technique is as follows. About 20 microscope cover glasses, approximately 0.1mm thick, are waxed, and then etched simultaneously with hydrofluoric acid, so that the acid acts over an area of about 1mm^2. When the acid had penetrated two or three of the slides the rest are swamped in water. Several of the slides will now have the etched part very thin (about 10μm) and suitable ones are punctured with a single spark from an induction coil. The length of the spark determines the diameter of the hole. In favourable cases circular holes having smooth edges are obtained.

The need to use a rapid-scanning mass spectrometer coupled to a molecular-beam inlet system was clearly shown to be necessary if it is desired to obtain an indication of transient species present in the neutral beam prior to ionisation. The elegant experiments of Foner and Hudson who used a chopper for the incoming beam and a phase-sensitive amplifier, thereby increasing the sensitivity for detection of incoming particles by a few orders of magnitude, were appropriately mentioned. This kind of system seems well suited for studying catalysis. The coupling of a shock tube to a time-of-flight spectrometer provides a new important technique for studying homogeneous gas phase reactions at high temperatures.

In conclusion, it would be fair to say that since sampling techniques are now more fully understood and utilised, we can expect to see important strides being made in the near future in physical measurements based on the mass-spectrometric study of unstable particles.

4-2

The Study of Chemical Reactions by Quadrupole Mass Spectrometry

J. N. BRADLEY, W. D. CAPEY, J. R. GILBERT, A. J. PARK
University of Essex, Colchester, England.

Summary

The principle of operation of a quadrupole mass spectrometer is described briefly and the advantages of the instrument for the direct study of chemical reactions are summarized. Some modifications which have been made to a commercial instrument to enlarge its usefulness for kinetic studies are also listed.

The use of the apparatus in three applications is described:

i. The study of flash photolytic reactions. The photolysis of NO_2 leads to an initial rapid formation of nitric oxide and oxygen, followed by a much slower reaction to give similar products. The results are interpreted in terms of a number of reactive intermediates of varying lifetimes.

ii. The study of heterogeneous reactions. The decomposition of ammonia on a tungsten surface has been studied at pressures of 10^{-5} to 10^{-4} mm and temperatures of 1400 - 1800°K. The rate of reaction appears to be unaffected by the presence of products at these pressures. A plot of reaction rate versus reciprocal temperature shows a distinct break, providing evidence for a change in the nature of the rate-controlling step with increasing temperature.

iii. The detection of unstable intermediates. NH and NH_2 radicals (as well as ionic species) have been detected among the products of the reaction above. They have been separated from the fragmentation peaks of stable species by appearance potential measurements. Approximate values for their ionization potentials have also been obtained.

Sommaire

On rappelle rapidement le principe de fonctionnement d'un spectromètre de masse quadripolaire et les avantages de cet instrument pour l'étude directe des réactions chimiques. On décrit également quelques modifications apportées à un instrument du commerce pour en accroître l'utilité pour les études de cinétique chimique. On décrit trois applications du montage:

i. Etude de réactions de photolyse par flash - La photolyse de NO_2 donne rapidement un peu d'oxyde nitrique et d'oxygene et par une réaction plus lente les mêmes produits. On interprête des résultats en supposant l'existence de produits intermédiaires de durées de vie variables.

ii. Etude de reactions hetérogènes. - La décomposition de l'ammoniac sur du tungtène a été étudiée à des pressions de 10^{-5} à 10^{-4} torr et des températures de 1400 à 1800°K. La vitesse de réaction n'est pas affectée par la présence de produits de décomposition à ces pressions. Le tracé de la vitesse de réaction en fonction de l'inverse de la température montre une nette cassure mettant en évidence un changement du processus fondamental de la réaction lorsque la température croît.

(iii.) Détection de produits intermédiaires instables - Les radicaux NH et H_2 (ainsi que les ions correspondants) ont été detectés parmi les produits des réactions précédentes. On les a distingués parmi les espèces fragmentées stables au moyen de leurs potentiels d'apparition. Une détermination approchée de leurs potentiels d'ionisation a également été obtenue.

Description of Technique

Mass spectrometers have now been used for some years for the direct study of chemical processes, rather than for providing endproduct analyses. Conventional magnetic deflection instruments suffer from two major disadvantages in this application: slow scanning speeds and interference from magnetic fields. These two difficulties can be overcome in time-of-flight instruments but additional drawbacks are introduced, notably those of bulk and expense. The advent of commercial quadrupole mass spectrometers has provided new opportunities for the use of mass spectrometry in following chemical processes. This paper is concerned with the adaptation of such an instrument for chemical kinetic studies.

Advantages of quadrupole mass spectrometers

The quadrupole mass spectrometer used in the present study is a Centronic EAL Quadrupole Residual Gas Analyzer Type 200, the particular advantages of which are:

1. High speed: the instrument will scan at rates up to 500 μsec per mass peak and will give approximately microsecond time resolution when monitoring a single mass peak.

2. High temperature operation: the instrument can be operated at temperatures up to 400°C.

3. Absence of magnetic fields: this permits high temperature operation and enables charged species to be analyzed.

4. Separate, small detector heads: this allows the detector to be incorporated into the experimental system and also enables more than one unit to be used with a single electronic console.

5. Adequate resolution: a resolution of one mass unit at mass 500 means that all significant gaseous molecules can be monitored.

6. Adequate sensitivity: 10^{-14} torr of N_2.

7. Absence of organic materials: in addition to permitting high temperature operation this means that, within certain limitations, quite corrosive materials may be investigated.

8. Low cost: the cost is comparable with that of a conventional instrument providing similar resolution and sensitivity. The additional benefits described above are therefore attained very economically.

9. Linear mass scan: assists in identification of unknown mass peaks.

10. Open ion sources: a variety of ion sources are available which permit extremely free access to the electron beam. This means that reactive inter-

mediates are easily introduced into, and removed from, the ion source.

Principles of operation

The major components of the quadrupole mass spectrometer are shown schematically in figure 4-2-1. The ions are generated by the electron beam in the ion source in much the same way as in a conventional mass spectrometer. The ions pass through a series of apertures at appropriate voltages to provide electrostatic focussing and then enter the quadrupole unit. A D.C. and an R.F. potential are applied to the quadrupole as shown. The ions travel through the quadrupole unit at a constant axial velocity but execute transverse oscillations. Unless the ions have a mass/charge ratio within a particular narrow range, these oscillations are unstable and their amplitude increases until the ions strike the quadrupoles and are removed. The ions with a mass/charge ratio appropriate to the D.C. and R.F. potentials selected pass through the unit unaffected and are detected by the electron multiplier. In order to scan over a range of masses, the R.F. and D.C. potentials are varied simultaneously, the resolution of the instrument being controlled by the ratio of the two potentials.

Modifications and adaptions to standard equipment

In order to make the instrument more suitable for kinetic studies, a number of modifications have been incorporated:

1. In the original instrument there was no means of restarting a mass sweep once it had commenced and this proved very inconvenient when using slow scan rates. A restart switch was incorporated and a similar switch is now fitted as standard on new instruments.

2. In the original instrument it was not possible to vary the electron current and electron energy independently. A modification was provided by the manufacturers to permit this operation and is now incorporated in new instruments.

3. The meters for measuring the voltages in the ion source proved too insensitive for reproducing a standard set of conditions and jacks have been fitted to enable these voltages to be measured on external meters.

4. Output recording. Although the output from the mass spectrometer can be displayed on an oscilloscope directly, it proved more convenient to obtain permanent records by using an ultra-violet galvanometer recorder (Southern Instruments M.1300 with Type J galvanometers). This has a paper speed of up to 250 cm/sec. and the overall frequency response is limited to 3 Kc/s by the galvanometer. Because of the low input impedance of the recorder, a Fenlow differential amplifier type AD/103/S (x 10) was inserted between the console output and the recorder input to match the impedance.

5. Increased sensitivity. For some purposes it was found necessary to increase the sensitivity of the instrument. This has been achieved by constructing an electrometer amplifier which provides a signal at the recorder of 2×10^{-6} volt/cm. The bandwidth has a maximum range of D.C. to 300 c/s and can be restricted to a few c/s to improve the signal/noise ratio. At maximum sensitivity, a change in peak height of less than 0.1 percent can be detected.

6. Peak selection. The alternatives of monitoring a single mass peak and carrying out repetitive scans do not meet the requirements of some investigations. A circuit has been constructed (1) which replaces the sawtooth waveform of the D.C. generator by three step voltages. This means that three separate peak heights are repetitively recorded and the changes in concentrations of three components can be

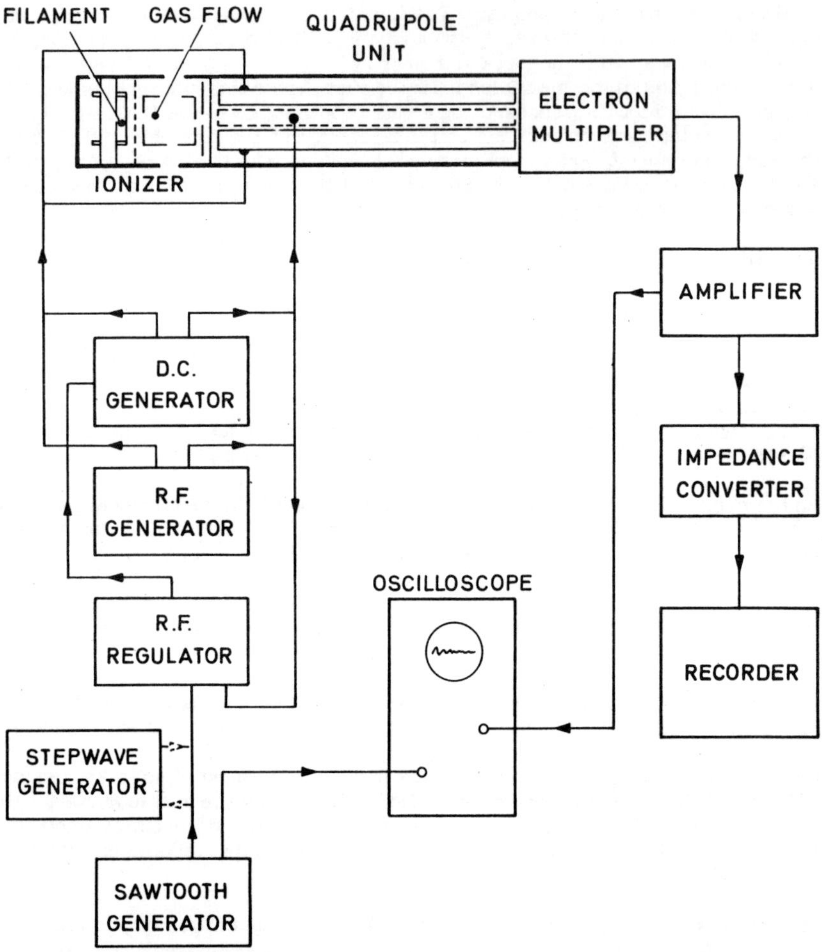

Fig. 4-2-1 Schematic representation of the quadrupole mass spectrometer and its associated circuitry.

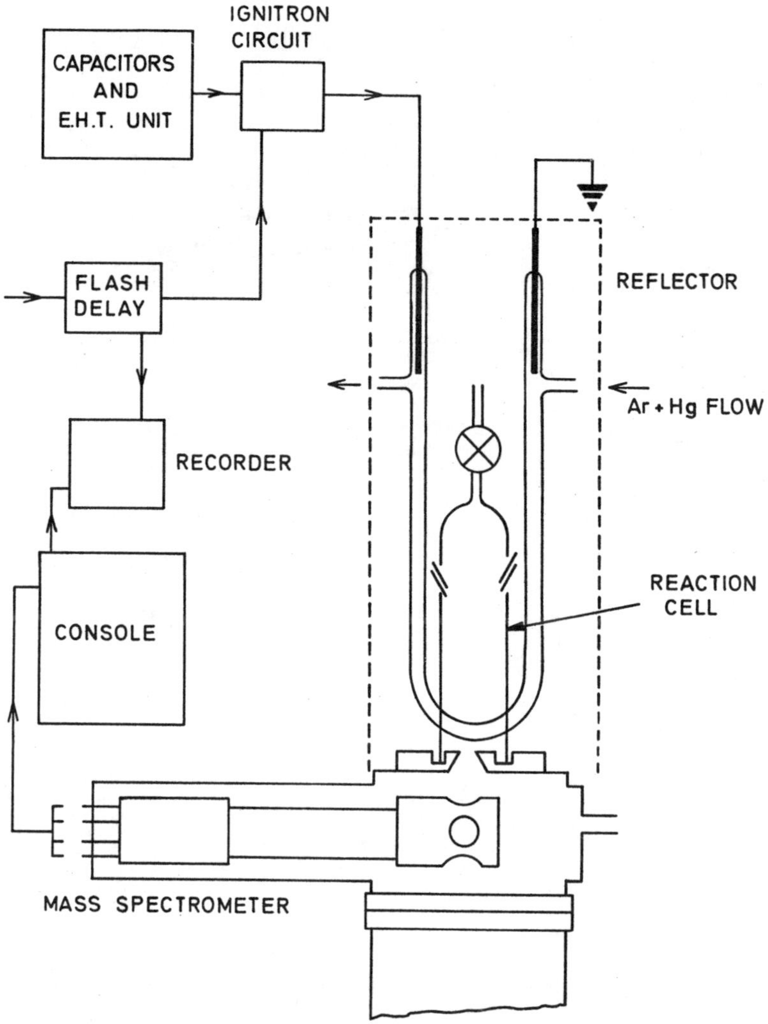

Fig. 4-2-2 Schematic diagram of the apparatus used in the study of flash photolysis reactions.

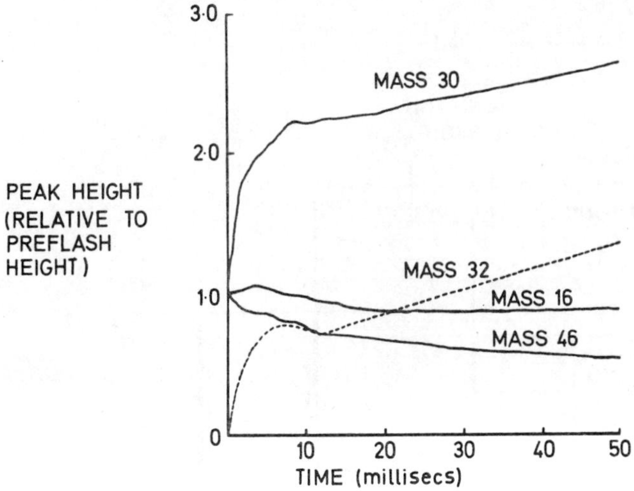

Fig. 4-2-3 Variation of mass peaks with time in the flash photolysis of nitrogen dioxide. The peak height of mass 32 is related to the pre-flash peak for mass 46

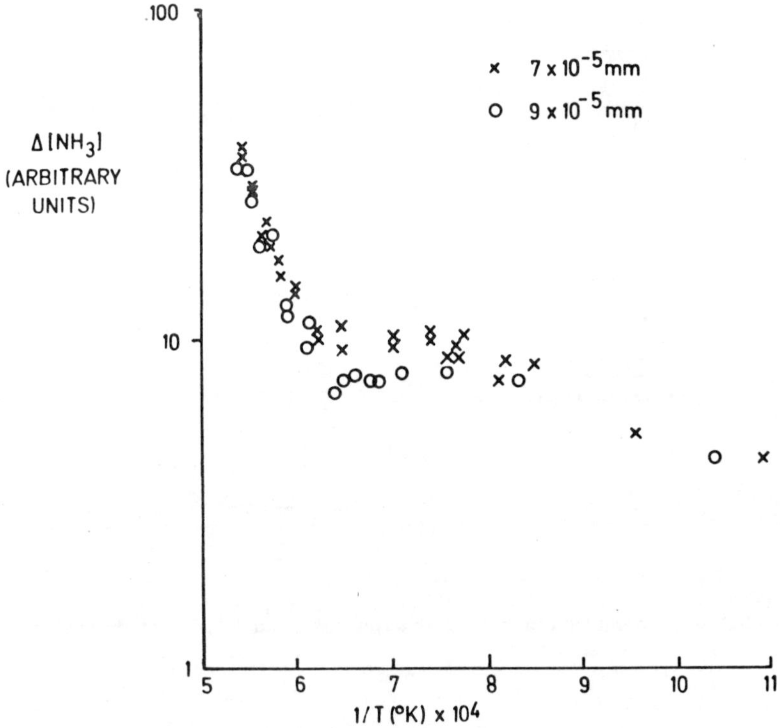

Fig. 4-2-4 Typical plot to rate of decomposition of ammonia versus reciprocal temperature.

measured during the course of a reaction.

Application to Kinetic Studies

The equipment has so far been employed in three different applications and will be introduced in a fourth in the near future. Brief descriptions of the first three and a summary of the results obtained are given to illustrate the use which may be made of the equipment.

The study of flash photolytic reactions.

The flash photolysis technique permits large quantities of energy to be introduced into a chemically-reactive system within very short periods of time. The active species generated will then attain, for comparably short periods, concentrations many orders of magnitude greater than in conventional photochemical systems thus enabling their nature and reactions to be investigated. Traditionally, these active species have been examined by kinetic spectroscopy but the advent of rapid-scanning mass spectrometers has provided an alternative, and very versatile, technique for their analysis.

The nitrogen dioxide system was selected for an initial investigation because the nature and reactions of the likely intermediates are known and because it has been studied previously by this type of technique although with rather inconclusive results (2) (3) (4).

The reaction chamber consisted of a quartz cylinder, 2.5 cm diam. 10 cm long, connected at one end to a conventional gas-handling system and closed at the other by an aluminium flange carrying a molybdenum orifice. The orifice, supplied by Aeon Laboratories Ltd., consists of a molybdenum plate, 0.5 mm thick, containing a conical opening with a .0025 cm diameter hole at the apex. The standard mass spectrometer head, with a cross beam ionizer, is supported in a brass vacuum jacket directly beneath the orifice. A stainless steel cold trap and oil diffusion pump are fitted beneath the ionizer opposite to the orifice to give maximum pumping speed through the ion source.

The photolysis lamp consists of a quartz U-tube, 12 mm diam., 24 cm long, fitted with neon sign electrodes. Argon, saturated with mercury vapour, flows through the lamp at 2 cm pressure. The lamp is driven by a 64 μF capacitor, which may be charged to 15 kV, although a voltage of 6.5 kV is normally used giving a flash energy of 1350 J. The half-width duration of the flash is ca. 40 μs. The complete assembly is illustrated in figure 4-2-2.

When an experiment is conducted, the u.v. recorder is first set in operation and, after a delay of 50 ms to enable the chart paper to reach its maximum speed, a signal is applied to the control circuit for the ignitron (English Electric B.K. 428). The ignitron then discharges the condenser through the flash-lamp, a phototransistor monitor simultaneously producing a time marker on the record.

Initial experiments have been carried out on various mixtures of nitrogen dioxide in argon at total pressures of up to 3 mm. A typical set of records is illustrated in figure 4-2-3, the peaks being corrected for the change in temperature by simultaneously monitoring the argon peak. The major products of reaction are nitric oxide and oxygen, the conversion occurring in two stages: (i) a rapid change, occurring within about 4 ms after the flash during which a significant transient may be detected on the 16 peak as well as small transients on the other peaks. (ii) A slower change occurring over longer periods (up to 150 ms). Careful study of the reaction by both single peak monitoring and repeated scanning has failed to reveal the presence of any other species. These results indicate that two transient species are

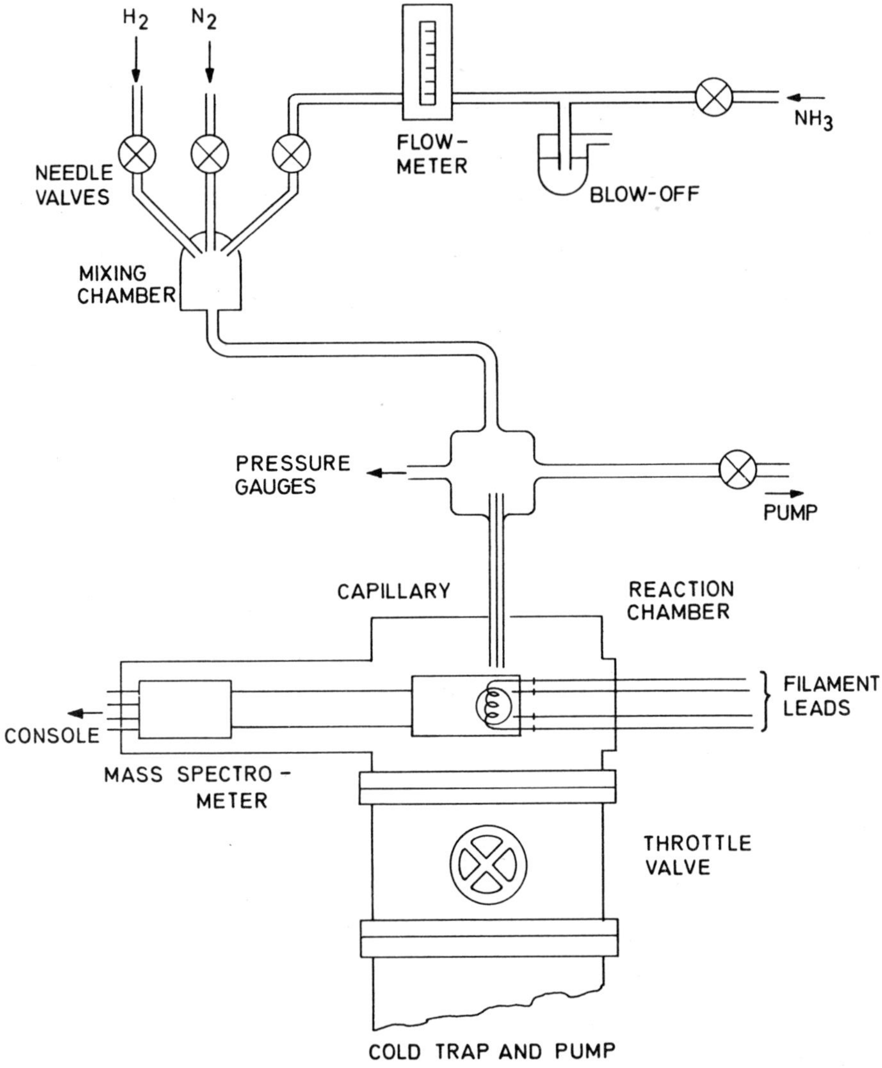

Fig. 4-2-5 Schematic diagram of the apparatus used in the study of heterogeneous reactions.

present, one of which is presumably responsible for the slow reaction and gives significant peaks only at the same mass numbers as the stable species, and a short-lived transient which also gives similar peaks to the stable species but with an enhanced contribution to peak 16.

The cracking pattern of nitrogen pentoxide, N_2O_5, has been measured and it is found that this molecule also gives significant peaks only at the same mass numbers as the stable products, though with a different distribution. If it is assumed that the slow change arises from the decomposition of N_2O_5, then the amount of N_2O_5 present after the initial rapid change can be estimated independently from three separate mass peaks: 46, 32 and 30. The amounts calculated in this way are identical if it is assumed that the overall decomposition may be represented stoichiometrically by:

$$N_2O_5 \longrightarrow NO_2 + NO + O_2$$

In practice the decomposition is likely to be a two-step process

$$N_2O_5 \longrightarrow NO_2 + NO_3$$

$$NO_3 + NO_2 \longrightarrow NO_2 + NO + O_2$$

The evidence for the formation of N_2O_5 in the system thus appears strong, the subsequent decomposition occurring because of the rise in temperature due to flash heating.

The following reactions may be expected to occur during the initial phase:

$$NO_2 + h\nu \longrightarrow NO + O$$

$$O + NO_2 \longrightarrow NO + O_2$$

$$O + NO_2 \ (+M) \longrightarrow NO_3 \ (+ M)$$

N_2O_5 may then be formed by the reaction:

$$NO_2 + NO_3 \longrightarrow N_2O_5$$

It is unlikely that O atoms would be detected with the time resolution available. Because of the reactivity of NO_3, a cracking pattern is not available but it seems reasonable to predict significant peaks at masses 46, 30 and 16 and probably at mass 32. By analogy with N_2O_5, the parent peak at 62 is likely to be unimportant. If the transient behaviour observed on peaks 46 and 30 were attributed to stable species, it would be inadequate to explain the transient on mass 16. It therefore seems probable that the transient peaks are due to the presence of NO_3.

The study of heterogeneous catalysis

The mode of operation of a mass spectrometer enables it to be used to study the kinetics of heterogeneous reactions in flow systems at low pressures. In addition to providing a rapid, complete analysis of the gas leaving the surface, it may also be used to detect unstable intermediates that would decay rapidly at higher pressures or at greater distances from the surface. The advantages of the quadrupole instrument lie mainly in the ease with which it can be incorporated into such a system.

The first experiments have been conducted on the decomposition of ammonia on a tungsten surface. The principle of the technique is to flow a mixture of ammonia, nitrogen and hydrogen over a heated tungsten filament and measure the resulting composition as a function of temperature, pressure and concentration. The composition in the absence of reaction is found simply by switching off the heating current to the filament. Provided that the pumping speed remains constant with respect to the ammonia, and therefore that the mass flux is proportional to the pressure of ammonia, the difference between the recorded ammonia peaks with the filament hot and cold yields the rate of the surface reaction. Since nitrogen and hydrogen are products of the decomposition, it was necessary to add all three gases independently to determine whether the presence of products affects the rate of the reaction.

The design of the apparatus is shown in figure 4-2-5. The gases are premixed in the mixing chamber, the flow of ammonia being controlled by a flowmeter and constant pressure head. The flow of the other gases is less critical and is metered by needle valves alone. The pre-mixed gases are then bled into the reaction chamber via a capillary tube 1 mm internal diameter, 30 cm long. The flow through the capillary is proportional to the pressure in the pre-mixed gas stream provided the viscosity remains essentially constant.

The tungsten filament, 0.025 cm in diameter, 18 cm. long, is situated, as a five-turn coil, 1 cm away from the ion source opening of the mass spectrometer. The exit from the capillary lies 1 cm from the filament and is arranged so that the gas flow is perpendicular to the entrance to the ion source. The temperature of the filament is obtained from its resistivity, by simultaneous measurement of the applied potential and the current. During an experiment the current is recorded to ensure that the temperature remains constant. The reaction chamber is evacuated by a 6 in mercury diffusion pump, via a stainless steel liquid nitrogen trap and throttle valve, giving a maximum pumping speed of ca. 280 litres/sec.

A number of checks were required to ensure that the changes in mass peak heights could be correlated with rates of reaction. Independent experiments have shown that the mass 17 peak is directly proportional to the pressure of ammonia in the system and is independent of any nitrogen and hydrogen present. It was also demonstrated that the pressure at the entry to the capillary was directly propotional to the flow rate into the system. The mass 17 peak was also found to be proportional to the flow rate of ammonia so that the necessary criteria are satisfied. Addition of nitrogen, and of hydrogen plus nitrogen, did not affect the observation although the addition of hydrogen alone did have a slight effect, probably due to a viscosity change in the gas entering the capillary leak.

The rate of decomposition of ammonia was then studied for reaction chamber pressures of 5×10^{-6} to 10^{-4} mm and filament temperatures in the range 900 - 1850°K. The rate was independent of added nitrogen and hydrogen and the 'order' of reaction was unity above 1600°K with respect to the ammonia concentration measured on the mass spectrometer and of complex behaviour below this temperature. Plots of reaction rate versus temperature showed a distinct transition between a reaction with an activation energy of about 45 kcal/mole above 1600°K and a reaction with low temperature coefficient below this value. This indicates either that two competing reactions occur or, alternatively, that the nature of the rate-controlling step changes with temperature.

The detection of unstable intermediates.

The investigation of the surface-catalyzed decomposition of ammonia, described above, provides a valuable example of the way in which the mass spectrometer may be used to detect the presence of radical species.

Ammonia gives contributions to mass peaks 16, 15 and 14 due to the NH_2^+, NH^+

and N^+ ions. NH_2 and NH radicals, if present, will also give contributions to these peaks. In principle, it should be possible to separate the various contributions because the cracking pattern of ammonia is known but, in practice, the concentrations of radicals will be too low to permit a reliable separation. The identification can however be carried out by measuring the appearance potential of the species involved.

The 15 and 16 peak heights have been measured as a function of electron energy under standard conditions using the filament both hot and cold. With the filament heated the ions appear at much lower potentials although the two plots become parallel when the fragmentation of ammonia at higher electron energies dominates the formation of ions, the separation between the two being due to reaction taking place. In addition to permitting an identification of the species present i.e. NH_2 for the mass 16 peak and NH for the mass 15 peak), the appearance potential determination can also provide a rough measure of the ionisation potential of the intermediate. This technique is of value for obtaining thermodynamic properties of reactive species.

References

1. **Arthur, J.R., Lepore, J.J.,** Rev. Sci. Instr., 37, 794, 1966.

2. **Kistiakowsky, G.B., Kydd, P.H.,** J. Amer. Chem. Soc., 79, 4825, 1957.

3. **Rice, F.O., Wunderlich, F.J.,** J. Phys. Chem., 69, 2137, 1965.

4. **Meyer, R.T.,** J. Chem. Phys., 46, 967, 1967.

Commentary on Chapter 4-2

E.M.A. Willhoft.

Some experiments were carried out about a year ago, at the Gas Council Basic Research Group, on ammonia decomposition on tungsten under ultra-high-vacuum conditions. (Base pressures in absence of ammonia $< 2 \times 10^{-9}$ torr). The apparatus used was in fact the one described in the next chapter on hydrazine decomposition. The ammonia pressure range was $4 - 8 \times 10^{-5}$ torr and the temperature of the catalyst was varied from 850 - 2500°K. The catalyst was in the form of a spiralled wire 30 cm long and 75 μm in diameter. Particular attention was given to removing dissolved carbon and other impurities in the wire by exposure of the filament at 2500°K to an ultra-high-vacuum as well as by reduction of the filament at red heat in hydrogen and ammonia.

At temperatures in excess of 2000°K, reaction probabilities equal to or close to unity were obtained. At temperatures of less than 1450°K, a linear Arrhenius plot was obtained with an apparent activation energy of about 25 kcals/mole. At higher temperatures, a deviation from linearity occurred so that the Arrhenius curve approached the $\overline{P}$ axis, at $\overline{P} = 1$, with a continuously decreasing temperature coefficient.

It is noted that Professor Bradley and his co-workers used a filament of 250 μm diameter. With this thickness of wire, it would be extremely difficult to remove all dissolved carbon from the bulk. The difference in the shape of the curves is therefore very likely due to dissolved carbon. The effect of carbon on ammonia decomposition on platinum has been reported elsewhere. The order with respect to NH_3 in our experiments using tungsten was unity, based on a 65 percent change in reactant pressure. In NH_3 decomposition on a carburised platinum wire, the order with respect to ammonia was dependent on the pressure of ammonia and the extent of carburisation of the wire.

The break in our Arrhenius plot is explained in terms of decomposition and evaporation of adsorbed NH_3 molecules becoming comparable with one another in magnitude with increasing temperature. At temperatures of less than 1450°K the rate of evaporation of adsorbed molecules is very much greater than the reaction rate, hence accounts for the linear Arrhenius plot.

4-3

Mass-Spectrometric and other Techniques for Investigation of the Catalytic Decomposition of Hydrazine on Platinum at Low Pressure

E. M. A. WILLHOFT,
The Gas Council, London, S. W. 6.
A. J. B. ROBERTSON,
King's College, London, W. C. 2.

Summary

The investigation by mass spectrometry of atoms, free radicals and transient intermediates evaporating from a heated catalytic surface in the presence of a reactant gas at low pressure is discussed.

Advantages of the quadrupole spectrometer for this work are considered. The decomposition of hydrazine at a pressure of 2×10^{-5} torr on heated platinum gives di-imide, which was observed with a quadrupole mass spectrometer. Other features of the reaction under various conditions are described, particularly the amounts of stable products formed, and the probability of reaction of a hydrazine molecule at a single collision with the platinum surface.

Sommaire

On discute la recherche par spectrométrie de masse des atomes, radicaux libres et espèces intermédiaires issues d'une surface catalytique chauffée en présence d'un gaz réactif à basse pression.

On examine les avantages d'un spectromètre quadrupôle.

La décomposition de l'hydrazine à la pression de 2.10^{-5} torr sur du platine chauffé donne de la di-imide, observée au spectromètre quadrupôle. On décrit d'autres traits de la réaction sous différentes conditions, particulièrement les quantités de produits stables formés, et la probabilité de la réaction de la molécule hydrazine par simple collision avec la surface du platine.

Introduction

Free atoms and radicals are frequently produced when diatomic and polyatomic molecules are decomposed on heated metal surfaces which catalyse the decomposition. The atoms and radicals evaporate from the hot metal surface, and can be detected by mass spectrometry. Free atoms and radicals are often sensitive to surface collisions, and when this is so their detection by mass spectrometry requires that the number of collisions by the particles before detection should be a minimum.

An apparatus for studying such systems therefore requires a line-of-sight path into the ion source of the spectrometer from the place of origin of the radicals. Similar considerations hold for any transient intermediates formed, which may not be free radicals, but which are sensitive to surface collisions. When a heated metal catalyst is used as a source of such transient particles, the reaction can occur at pressures as low as that used in the mass spectrometer ion source, so there is then no problem of sampling from a high-pressure to a low-pressure region. The concentration of transient particles in the ion source of the spectrometer is in-

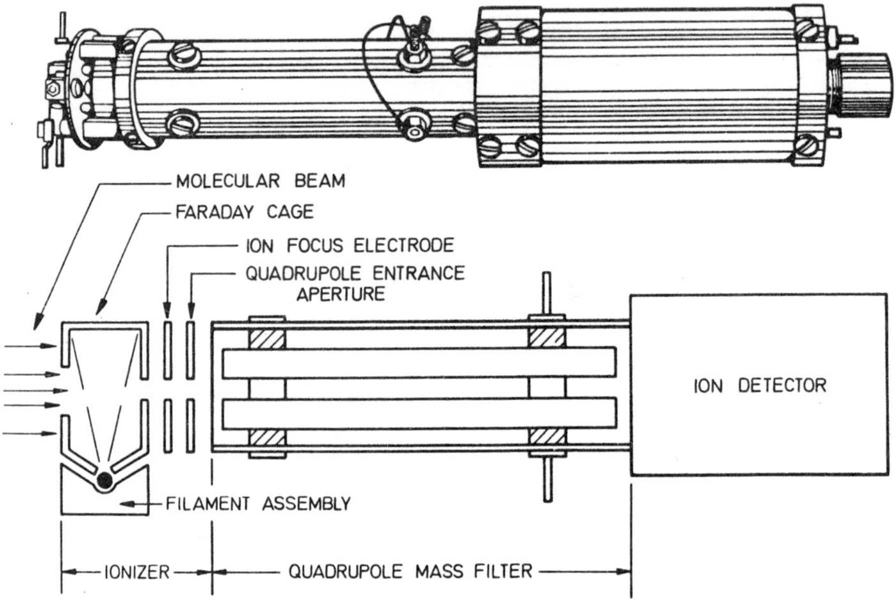

Fig. 4-3-1 Quadrupole assemblies

creased by having the catalytic surface as close to the electron beam of the spectrometer tube as possible, and by pumping the ion source separately. This pumping not only reduces complications due to reaction on the thermionic electron-emitting filament, and on other surfaces in the ion source, but also reduces ion-molecule reactions in the gas phase by effectively allowing a beam of particles to enter the mass spectrometer without undergoing collisions. This method of sampling has been used and discussed in detail by Robertson (1) (2) (3) and by Pentenero and Le Goff (4).

Magnetic deflection mass spectrometers are limited in sensitivity by the usual slit system necessary to provide an acceptable resolution of the incoming ion beam. This disadvantage does not exist with the quadrupole spectrometer. The ion source can be made completely open and allows particles to enter the ionising region through an orifice which can be about 3 mm diameter. In addition, the quadrupole is capable of accepting ions several degrees off the quadrupole axis, a figure quoted by one manufacturer being 30°. In work with transient intermediates with magnetic deflection instruments, contamination of the mass spectrometer slits often leads to troublesome lack of reproducibility. The elimination of fine slits in the quadrupole may then be highly beneficial. A diagrammatic representation of a quadrupole mass spectrometer head is shown in figure 4-3-1. The four quadrupole elements are precisely located in a rectangular array, and when energized by DC and RF voltages act as a mass filter. The figure shows a quadrupole assembly with electron beam ionizer and electron multiplier detector.

With the high-speed mass scan and the associated recording apparatus it is possible to obtain semi-quantitative results for the detection of transient particles provided the mass spectra of these are known. Quantitative analysis, however, may be restricted to the use of energies of the ionising electron beam of about 70 eV, since the ionisation efficiencies of closely related particles are very similar at such energies. However, qualitative identification is possible at low electron energies, just exceeding the ionisation energy, so that a complete analysis requires a low electron-energy source as well as a high-energy source. Further confirmation of the identity of a particle may be obtained from ionisation and appearance potential data which can be obtained with a source providing electrons of low and variable energy.

An investigation was made of the detailed kinetics and mechanism of the heterogeneous decomposition of hydrazine at pressures of about 10^{-5} torr on a heated platinum surface. The low pressure ensured that the flow was molecular, and also enabled the reaction vessel to be kept at room temperature.

In the low electron-energy study, the concentration of transient particles which evaporated directly from the platinum catalyst was increased in the ion source by using the platinum as a thin foil, 8 μm thick and about 2×2 cm^2, in a plane situated normal to the axis of the quadrupole. In the high-energy study, the platinum was used as a coiled filament 30 cm long and 75 μm diameter. Detailed results on the formation of di-imide (N_2H_2) from hydrazine have been reported elsewhere (5). It was possible to determine part of the mass spectrum of the unstable di-imide intermediate. Generally, the determination of free radical mass spectra may be easier when the radicals are produced on a catalytic surface at low pressure since the problems associated with sampling from a high pressure do not arise.

Ionisation efficiency curves for the formation of $N_2H_2^+$ from N_2H_2 and for N_2H^+ formation were constructed and gave an approximate value of 9.9 eV for the ionisation potential of di-imide and an appearance potential of 11.1 eV for the formation of N_2H^+. These values were obtained by assuming a value of 9.00 eV for the ionisation potential of hydrazine (6). Foner and Hudson (7) obtained 9.85 eV for the ionisation potential of di-imide.

In a different investigation of the platinum-catalysed decomposition of hydrazine, a flow system was used similar to the conventional high vacuum system described in the study of ammonia decomposition on platinum (8). This apparatus did not include a mass spectrometer. Hydrazine vapour was passed continuously at pressures of about 10^{-4} torr over a heated platinum wire catalyst. The number ($\underline{N}_s$) of molecules of hydrazine decomposed per second was estimated from the difference between the number collected after each run (lasting up to 1 hour) and the known number ($\underline{N}_e$) of molecules which passed into the reaction vessel per second. The probability $\underline{P}$ of a hydrazine molecule decomposing on colliding with the catalyst wire of surface area $\underline{A}$ cm^2 is (8):

$$\underline{P} = 1.61 \times 10^{-2} \, \underline{N}_s\underline{S}/(\underline{N}_e - \underline{N}_s)\,\underline{A}$$

where $\underline{S}$ is the pumping speed in litres per second at the reaction vessel for gas of unit molecular weight. The value $\underline{P} = 1$ means that every molecule of reactant colliding with the wire catalyst is decomposed. With a hydrazine pressure, in the absence of reaction, of 1×10^{-4} torr, values of $\underline{P}$ close to 3 were obtained with the platinum wire at 1100°K. It seems that an intermediate was formed which caused further decomposition of hydrazine at the walls of the reaction vessel despite the fact that the vessel was at room temperature.

Hydrogen to nitrogen ratios in the non-condensible products of the reaction were measured by pumping these products into a backing volume, and catalytically oxidising the hydrogen on an electrically-heated platinum wire situated opposite a liquid-air trap. Initial ratios of 2.5 were obtained at about 1250°K; they decayed to a steady value of 1.9 after 2 minutes of reaction. Although the hydrogen/nitrogen ratio depended on the catalyst temperature in the range 700 - 1700°K, it always decayed with time of reaction to give a steady value. The time taken to reach this value was about 2 minutes and was independent of temperature.

The ratio of the number of molecules of hydrazine decomposed to the number of molecules of ammonia formed varied from unity at 800°K to approximately 7 at 1660°K. To explain the high ratio of hydrogen to nitrogen (exceeding 2) together with the formation of ammonia, an intermediate must be proposed in which the nitrogen:hydrogen ratio is greater than that in hydrazine. From the mass spectrometric results this intermediate appears to be di-imide which was reported by Foner and Hudson (9) to be formed by the decomposition of hydrazine in the gas phase, and by applying an electrodeless discharge to hydrazine. Although di-imide appears to be an initial product evaporating from the catalyst surface, the reactions at the walls of the vessel which cause P to exceed unity may not necessarily involve di-imide, but possibly free radicals formed by its further decomposition. Ammonia may possibly also be involved in reactions at the vessel walls.

Assuming equal cross-sections for ionisation of N_2H_4 to $N_2H_4^+$ and N_2H_2 to $N_2H_2^+$, the partial pressure of N_2H_2 in our mass spectrometric work was calculated to be 7×10^{-7} torr. This corresponds to a reaction probability of about 0.2 for di-imide formation as a result of a hydrazine molecule colliding with the filament. Since N_2H_2 may react further at the walls of the vessel and its calculated partial pressure corresponds to a steady state condition, this value of reaction probability represents a lower limit. The true value of P for di-imide formation at a single collision of a hydrazine molecule with the filament may therefore be much closer to unity.

Further work is necessary to establish unequivocally the source of the ions N_2H^+, NH_2^+ and NH^+. This might require decreasing the distance between the ion source of the spectrometer and the catalyst to a few millimetres, approximately 250 mm distance was used in the present investigation. Useful information could also be derived by measuring the dependencies of the various ion currents in the mass spectra on the reactant pressure.

Acknowledgements

We thank the Science Research Council for financial support (to E.M.A.W.) during part of this work. E.M.A.W. thanks the Gas Council for permission to publish this chapter.

References

1. **Robertson, A.J.B.,** 'Advances in Mass Spectrometry' ed. J.D. **Waldron.** Pergamon, London, 1959, 559.

2. **Robertson, A.J.B.,** 'Mass spectrometry' Inst. of Petroleum, London, 1952, 47.

3. **Fabian, D.J., Robertson, A.J.B.,** Proc. Roy. Soc. A 237, 1, 1956; Trans. Faraday Soc. 53, 363, 1957.

4. **Pentenero, A., Le Goff, P.,** 'Mass Spectrometry' ed. R.I. Reed, Academic Press, London, 1965, 305.

5. **Willhoft, E.M.A., Robertson, A.J.B.,** Chemical Comm., 1967, 385.

6. **Dibeler, V.H., Franklin, J.L., Reese, R.M.,** 'Advances in Mass Spectrometry' ed. J.D. Waldron, Pergamon, London, 1959.

7. **Foner, S.N., Hudson, R.L.,** J. Chem. Physics. 29, 442, 1958, 443.

8. **Robertson, A.J.B., Willhoft, E.M.A.,** Trans. Faraday Soc. 63, 476, 1967.

9. **Foner, S.N., Hudson, R.L.,** J. Chem. Physics, 28, 719, 1958.

Commentary on Chapter 4-3

K.W. Michel

More specific data about di-imide became available almost 10 years ago. Since the mass-spectrometric detection by Foner and Hudson it has been postulated as an essential intermediate in the homogeneous chain decomposition of N_2H_4 and was determined quantitatively by mass-spectrometry (1).

It is interesting to find that N_2H_2 may also play a role in the heterogeneous decomposition.

Little was known about surface catalyzed decomposition apart from some classical investigations on the stoichiometry of the reaction products and the evidence that the heterogeneous decomposition proceeds in some cases (2) without the participation of free radicals i.e. non-chain mechanism.

Even in view of the results presented by Dr. Willhoft we are far from a conclusive description of the surface reaction. The surface reaction is coupled with adsorption-desorption processes; even if one assumes that the adsorption is fast, the desorption rate of the observed species will influence the overall reaction rate. Investigation of the surface state by methods such as LEED on single crystals might provide additional information in this case.

Despite the care taken by the authors to exclude oxygen and carbon from the reaction vessel, deuterium substitution of hydrazine might help to strengthen belief in the identification of masses 29 and 30. In addition, the distribution of D among the reaction products might indicate whether the chain mechanism suggested here remains valid.

In the case of surface reactions one might consider that the bond-structure of N_2H_2 has some effect. When both H-atoms are linked to one N-atom an unstable species results, which will also explain the large number of species with mass 29.

References

1. Homann, K., MacLean, D.I., Wagner, H.G., Naturwissenschaften 52, 12. 1965.

2. Birse, E.A.B., Melville, H.W., Proc. Roy. Soc. A 175, 164. 1940.

Part 5

Molecular Beams

5-1

New Methods for Producing High Energy Molecular Beams

J.B. ANDERSON AND J.B. FENN
Princeton University, U.S.A.

Summary

Nozzle-source molecular beams are useful in the study of collision processes relating to propulsion and energetics because they can provide the energy and intensity that is necessary for many experiments. This chapter attempts to set forth the art and science of nozzle-beam operation and experiment design as they have emerged from the experience that has been accumulated in many laboratories over the past ten years. In particular, some of the differences between theory and practice will be discussed and a number of useful relationships among the operating variables will be summarized.

Sommaire

Des faisceaux moléculaires issus de tuyères sont utiles pour l'étude des processus de collision relatifs à la propulsion et à l'énergétique car ils peuvent fournir l'énergie et l'intensité nécessaires pour de nombreuses expériences. Dans cet exposé l'art et la science des jets moléculaires issus de tuyéres seront présentés, ainsi qu'une description des montages expérimentaux tels qu'ils apparaissent dans les travaux effectués dans de nombreux laboratoires au cours des dix dernières années. On discutera notamment quelques différences entre théorie et expérience, et on présentera un certain nombre de variables caractéristiques du fonctionnement.

Introduction

Physicists have found it relatively easy to bring about nuclear reactions under precisely controlled conditions in beam scattering experiments with high energy accelerators. They have found it very difficult to effect the same reactions under steady-state conditions except at the catastrophic rates occuring in bombs. Consequently, for elementary collision processes characterized by energies of millions or billions of electron volts we have a great deal of precise, microscopically-detailed information; but the engineering art of using these processes to satisfy human needs must be regarded as in a still primitive state.

At the other end of the energy spectrum, chemists and engineers concerned with chemical reactions and other collision processes between atoms and molecules, characterized by energies below ten or twenty electron volts, have been very successful in carrying out these processes in large populations under steady-state conditions. They have found it much more difficult to bring about individual collisions under the precisely defined conditions which are, in principle, possible with beam scattering experiments. Thus, the engineering art of applying chemical reactions and transport processes has been highly developed but a detailed description of the elementary collisional events in this energy range remains elusive. Thus, we can build and use propulsion systems whose operation is vitally dependent on collision processes about which we know little or nothing.

One of the main reasons we do not have the detailed knowledge of these collision processes that beam scattering experiments can provide is that particle beams with useful intensities have not been available in the energy range between 0.5 and 20 eV. Collisions in this energy range are responsible for many phenomena of interest and importance to propulsion and energetics. Beams from classical effusive or oven sources have a useful upper limit to energy of about 0.5 eV because the source temperature cannot exceed about 3000°K. There are no suitable materials of construction having higher melting points, and the gas temperature cannot exceed the wall temperature at the low source gas densities that must be used. At energies below 100 eV, the intensities of beams formed by charge exchange neutralization of parent ion beams are severely limited by space charge effects. Nevertheless, it has been possible in some cases, in which the detection problem has been solved, to obtain results with charge exchange beams at energies as low as 10 eV (1) (2) (3). In general, however, ion beams at energies below 20 eV, whether used directly or used to generate neutral beams of the same energy by charge exchange, present difficult experimental problems.

Until very recently, except for a few results with very low intensity ion beams and their charge exchange neutral derivatives, the only beam scattering collision studies at energies <10 eV and > 1 eV have been based on the observation of small angle scattering of high energy beams. In this technique, which has been pioneered by Amdur and his collaborators, a beam particle having an energy as high as several keV undergoes a 'grazing' collision with a target particle such that the effective collision energy is only a small fraction of the beam energy in laboratory coordinates (4) (5). By study of such low-angle scattering it has been possible to accumulate a substantial amount of information on the potential energy as a function of distance between various pairs of atoms and molecules in the repulsive region above 1 eV. The method lends itself mainly to elastic scattering, and the results are particularly useful in describing and predicting transport at high temperatures and the equilibrium thermodynamic properties which depend on intermolecular potentials. Inelastic and reactive collisions are difficult to interpret and, in fact, occur only rarely because the collision time is so short. The use of high energy beams is not easily adapted to studying particle-surface interactions in the 1 to 10 eV range because of the difficulty of observing very low angle scattering from a surface.

In the past several years there have emerged several new methods of generating and using molecular beams for the study of collision processes at energies in the range immediately above 1 eV. The importance of this energy range and the power of the beam scattering technique have inspired many 'inventions' and a number of new approaches to the problem of generating particle beams have been evolved and are being pursued. Discussion here is limited to three methods that have recently reached a useful stage of development and with which some worthwhile results have already been obtained. These are (a) the use of cathode sputtering to provide beam atoms, (b) the merging beams technique by which the energy difference between two collinear high energy beams is utilized, and (c) the use of supersonic jets as proposed originally by Kantrowitz and Grey.

Sputtering Sources

The phenomenon of cathode sputtering, by which atoms of an electrode metal are ejected when a high energy ion strikes its surface, has been studied intensively for many years. The yield or number of sputtered atoms per incident ion may be as high as 5 or 10. The energy of the sputtered atoms ranges up to 100 eV or so, depending upon the incident ion energy and the composition of the electrode. The flux of sputtered atoms is generally highest in the direction normal to the surface and varies approximately in proportion to the square of the cosine of the angle relative to the surface normal. These factors led Wehner to suggest in 1954 that sputtered atoms might be collimated to form a beam having a useful intensity in a most interesting energy range (6). The group at the FOM Institute for Atomic and Molecular Physics in Amsterdam has recently been successful in reducing this suggestion to practice (7) (8).

The important features of a sputtering source are indicated schematically in figure 5-1-1. The composition, energy, intensity and direction of incidence of the primary ion beam, and the composition of the target, all represent independent variables whose effects upon neutral beam characteristics have not been thoroughly explored. Politiek and his collaborators have thus far reported results using potassium targets with incident primary beams of argon ions ranging in energy from about 4 to 10 kilovolts. The total intensity of sputtered atoms has been as high as 10^{16} atoms/steradian-sec distributed over an energy range from zero to 50 eV. A rotating-disk velocity filter, with a transmission of thirty-six percent and a selectivity of about six percent, was incorporated in the apparatus. By velocity analysis it was found that a maximum intensity of about three percent of the total flux occurred in the six percent velocity interval at about 2.5 eV, with a drop to about half of this value at 0.5 eV and 25 eV. The extent to which intensities can be increased by increasing the current density of the incident beam has not been established, but there is little doubt that intensities higher than already reported can be obtained. The energy range of the sputtered atoms is determined primarily by the focussing energy of the target lattice and, therefore, cannot be raised much above about 50 eV.

In this apparatus, with the sputter source for high energies and an effusive oven source for low energies, total scattering cross sections of potassium on helium and potassium on argon have been measured over the energy range from 0.01 eV to 17.5 eV (9). The ranges of the two sources overlapped from 0.40 to 0.45 eV. The experimental results are shown in figure 5-1-2. With an improved velocity selector it is planned to extend the energy range to almost 40 eV.

The group at the FOM Institute is vigorously pursuing the development and use of this new beam source (10). They have available a new ion source which provides a ten-fold increase in incident ion current. They have also built a second apparatus to be used for the study of particle-surface interactions including surface ionization of alkali atoms. A third apparatus is under construction. It is designed for the study of chemical reactions which may yield charged products, e.g.:

$$K + Br_2 \longrightarrow K^+ + Br_2^-$$

$$Br_2^- \longrightarrow Br + Br^-$$

The threshold energy for this reaction is estimated to be 3 eV.

No great difficulty is forseen in extending the sputter source technique to target electrodes of other materials including practically all metals. There is some hope that beams of non-metallic atoms and even of compound molecules might be generated. The question of the internal state of the beam particles is not completely settled, but the evidence so far indicates that the sputtered atoms are in the ground state. Sputter source beams apparently require little pumping speed as a consequence of the beam generation process. Beams of relatively refractory materials can be produced which would be extremely difficult in any energy range by any other technique. Because they are produced by electrical means they are convenient to control. Thus, time-of-flight techniques, and experiments in which frequency and phase become important variables, should be practicable.

It is of interest to add here that the sputter source principle is being considered as a possible basis for a space propulsion system of high impulse (11).

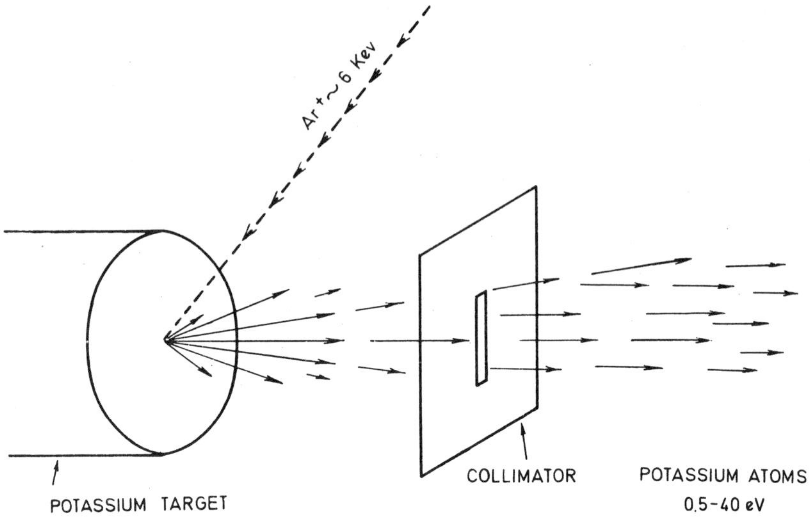

Fig. 5-1-1 Pictorial representation of the sputter source beam technique developed at FOM-Institute for Atomic and Molecular Physics, Amsterdam.

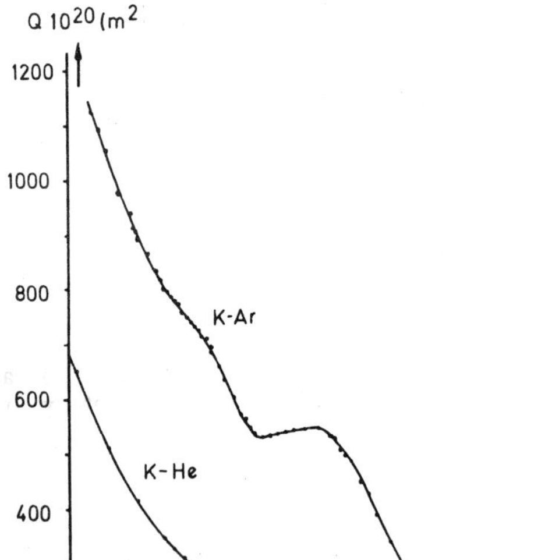

Fig. 5-1-2 Total scattering cross-sections for potassium-helium and potassium-argon determined in experiments using sputter source potassium beams by Politiek, Los, Schipper and Baede. From (9).

The Merging Beams Technique

In this approach to the measurement of collision cross-sections, two beams of high but slightly differing energies are brought into confluence. The resulting interactions are characterized by the velocity difference between the two beams which, in principle, may be made as small as desired. A detector assembly, which is moved axially, measures the flux of collision products as a function of distance along the interaction region of the merged beams. If the intensity and velocity of each beam are known, the cross-section for the interaction to form the product may be calculated.

Figure 5-1-3 shows a schematic arrangement of apparatus with which Trujillo, Neynaber and Rothe measured the cross-section for resonant charge exchange between argon atoms and argon ions (12). They merged a 3465-eV beam of argon atoms with a 2000-eV beam of argon ions. The resulting interaction energy was 100 eV. The parent 2000-eV ions were screened from the detector by a retarding grid to allow detection of the 3465-eV product ions.

Although merged beams of electrons and alpha particles were used many years ago, it was only relatively recently that Cook and Ablow suggested their use in the study of collisions between heavy particles (13). The possibility of a very low interaction energy with very high energy resolution was pointed out by Trujillo *et al* (14). The technique was tested and reduced to practice by Trujillo, Neynaber and Rothe. In addition, Belyaev, Brezhnev and Erastof have carried out a 'single source' experiment for determining the charge exchange cross-section between H^+ and H (15). In this variation a beam of protons is partially neutralized in a charge-exchange chamber. The resulting mixed beam of atoms and ions is passed into an electrically insulated chamber. A potential on this chamber retards the proton beam and thus provides the necessary energy difference between atoms and ions in the interaction region. More recently, Trujillo, Neynaber and Rothe used this single-source technique to extend their argon charge transfer cross-sections to the energy range 0.1 to 20 eV (16).

The collision or interaction energy between two particles is determined by their masses and the velocity with which they approach each other. If the particles in two merged beams have parallel trajectories, the interaction energy W is:

$$W = \tfrac{1}{2}\mu\,(v_2 - v_1)^2 \qquad \text{(Eq. 5-1-1)}$$

where μ is the reduced mass and v_2 and v_1 are the velocities of each beam in the laboratory frame of reference. For equal particle masses equation 5-1-1 can be rewritten in terms of the energies of each beam in laboratory coordinates, E_1 and E_2:

$$W = \tfrac{1}{2}\,(E_2^{\frac{1}{2}} - E_1^{\frac{1}{2}})^2 \cong \Delta E^2/8E \qquad \text{(Eq. 5-1-2)}$$

The latter approximation is good if ΔE, the energy difference between the two beams if small relative to E, the average energy of the two beams. It is clear that the ratio of interaction energy, W, to the energy difference between the two beams, ΔE, is very small as long as the average energy of the two beams, E, is large. This same 'deamplification' applies to any variations in W that may be consequent on fluctuations in ΔE. The net result is that scattering experiments based on the merger of two high energy beams can retain the intensity, energy resolution and detactability advantages of high energy beams, even though the scattering interactions are at very low energies. Figure 5-1-4 is an attempt to represent graphically the relation between interaction energy W and beam energy difference, ΔE, at various energy levels. It is clear that these energy relations imply a great flexibility in the design of experiments.

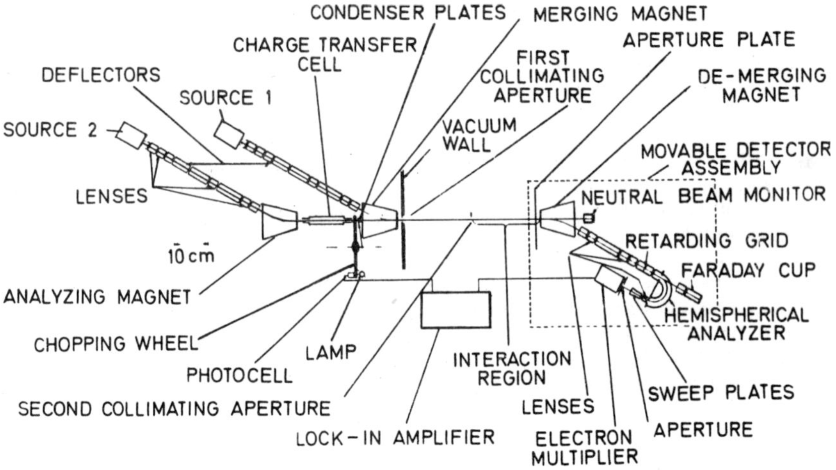

Fig. 5-1-3 Schematic diagram of merging beams apparatus and detection system for measurements of charge exchange cross-sections by Trujillo, Neynaber and Rothe. From (14).

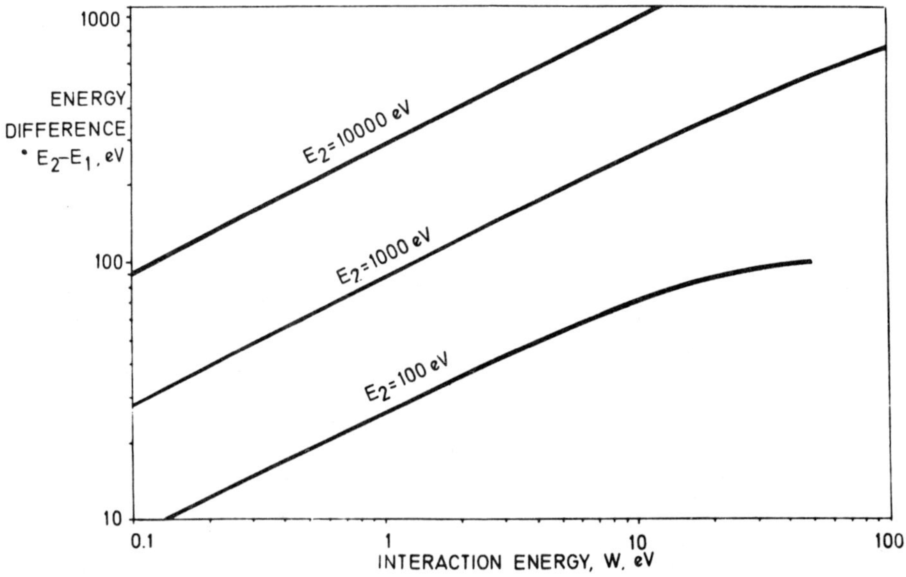

Fig. 5-1-4 Energy relationships for the merging beams technique. Calculated from Eq. 5-1-2.

Unfortunately, the merging beam technique does not provide a solution to the problem of many types of scattering experiments. It lends itself admirably to the measurement of total cross-sections for reactive scattering in which there is a change in structure and, therefore, an intrinsic identifiability of one of the interaction particles. In the argon charge exchange experiment cited, the product ions were identifiable by virtue of their higher energy, i.e. they were formed from neutrals that entered the interaction region at higher energy than the reactant ions. A retarding potential prevented the latter from entering the detector. Because the center of mass is at high velocity relative to laboratory coordinates the solid angle of the reaction product trajectories is very small. This feature is desirable in the study of total reaction cross-sections but would be obviously fatal in any attempt to determine differential cross-sections or total cross-sections for elastic scattering. Such experiments depend upon the ability to resolve scattering angles which should, therefore, be large rather than small in laboratory coordinates. It is also apparent that the method lends itself more readily to cases in which one of the reactants is an ion than to cases in which both reactants are neutral.

Supersonic Jet Sources

The two previous techniques comprised means by which high energy ion beams could be used to bring about collisions at much lower energies; consequently, they allowed the large background of experience with high energy ion beams to be applied to the solution of a new set of problems. We now consider a means of obtaining beams with energies of several electron volts that does not depend upon ion acceleration. Kantrowitz and Grey proposed in 1951 that the quiescent low density gas in the classical effusive oven source be replaced by a low density supersonic jet. The anticipated advantages they set forth included increases in intensity of several orders of ten and a substantial improvement in monochromaticity of the final beam due to narrowing of the molecular velocity distrubution during expansion of the source of gas to high Mach number. The first conclusive experimental demonstration of the Kantrowitz-Grey idea was reported in 1954 by Becker and Bier (17). Since then there has been an ever increasing interest in the development and use of supersonic nozzle beams. (18) (19) (20)

In addition to its ability to provide high useful intensities the Kantrowitz-Grey approach has provided a means by which beam energies could be increased above the limit of about 0.5 eV encountered with effusive sources. First, because the source gas is at high density prior to expansion it can be heated to a temperature much higher than the melting point of the nozzle walls. This possibility has led to the use of shock tubes and high-current arcs as means of increasing the energy of the source gas and, therefore, the energy in the final beam. Second, again because of the high gas density in the source and during the early part of the expansion, heavy molecules in admixture with a light carrier gas can be accelerated aerodynamically in the 'seeded beam' technique.

Nozzle Beam Performance and Design Characteristics

In their initial proposal Kantrowitz and Grey furnished a design for a converging-diverging supersonic nozzle and a skimmer to extract the core of a jet to form a molecular beam (21). A schematic diagram of a typical nozzle beam system is shown in figure 5-1-5. In practice it has been shown more convenient to use as sources free jets formed by expansion of gases from simple converging nozzles or orifices. This has the advantage of avoiding the problems of boundary layer formation in the diverging section of a nozzle and simplifies, to some extent, the gas dynamics of jet formation. Nevertheless, the formation of a free jet involves some of the more difficult problems of gas dynamics: subsonic and transonic flow with viscous effects, supersonic flow in transition from continuum to free-molecular flow and, in some cases, rotational and vibrational relaxation and condensation

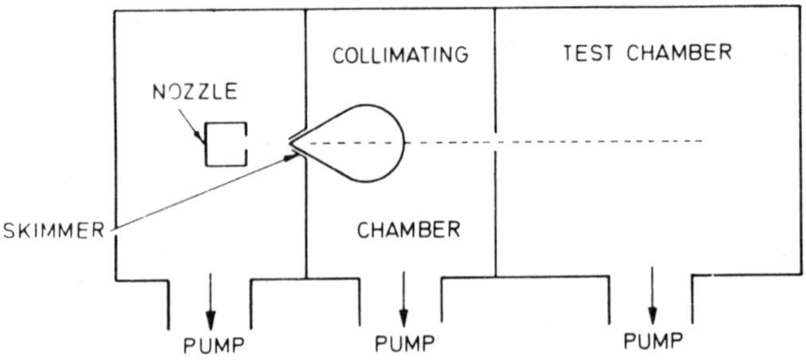

Fig. 5-1-5 Schematic diagram of typical nozzle beam apparatus.

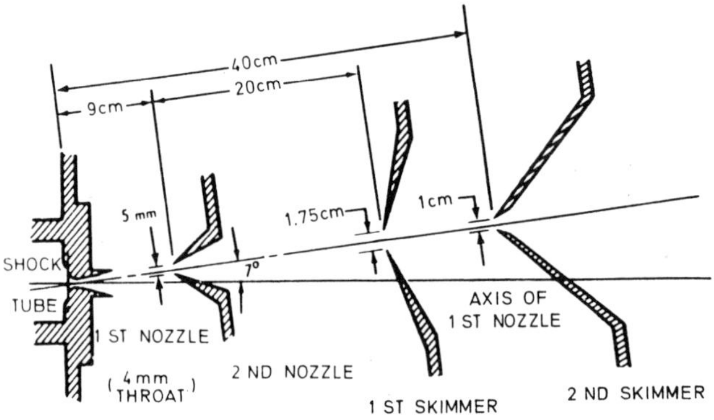

Fig. 5-1-6 Arrangement of nozzle, skimmers and detector used in shock-tube source for molecular beams by Skinner and co-workers. From (67).

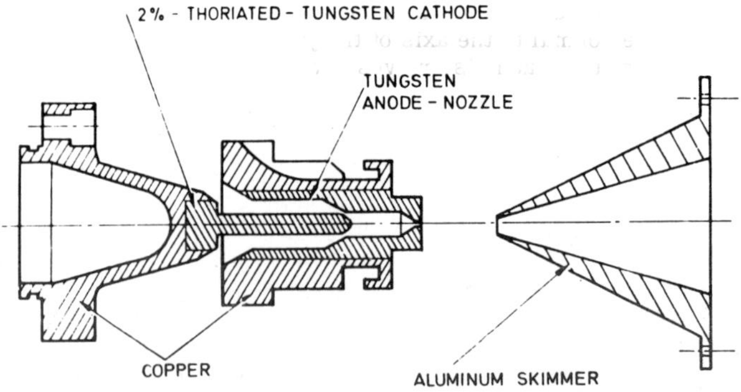

Fig. 5-1-7 Diagram of source electrodes, sonic nozzle and skimmer of arc-heated molecular beam source used by Knuth and Kuluva. From (52).

Measurements of the properties of molecular beams extracted from free jets have contributed to the understanding of these general problems of gas dynamics and stimulated theoretical work in these areas. As a result, many of the properties of free jets are well understood and can be accurately predicted.

The interaction of a jet with a skimmer is not well understood although many of the features of this interaction are clear in a qualitative sense. The Kantrowitz-Grey model for the beam formation process is based on parallel, isentropic gas flow at the skimmer entrance with free-molecular flow downstream of the skimmer entrance. The equations of this idealized model, with refinements made by Parker et al. (22), have been modified to take into account the 'non-ideal' effects of transition to free-molecular flow upstream of the skimmer and divergence of the flow at the skimmer entrance (23) (24) (25), in the absence of interference with the flow by non-ideal skimmer effects. It should be added that equation (1) of (24) has been misinterpreted in (25). Both equation (1) of (24) and equation (6) of (25) include divergence effects. For isentropic flow to the skimmer the equations differ in degree of approximation. At sufficiently low gas densities skimmer interference has been shown to be negligible. Under this condition the molecular beam properties may be calculated directly if the free jet properties are known.

The continuum, isentropic, free jet expansion may be treated by the method of characteristics for inviscid supersonic flow. Solutions of this problem were furnished by Owen and Thornhill and by Sherman (26) (27). Since it is advantageous to operate the nozzle at a high Reynolds number, the continuum treatment is usually satisfactory for the early part of the expansion. As gas density and temperature decrease in the expansion process, the molecular collision rate decreases and continuum flow no longer obtains. A transition to free-molecular flow occurs. The thermal velocity distribution superimposed on the gross flow velocity becomes non-isotropic, and the axial velocity distribution approaches a terminal distribution or 'freezes'. The radial velocity distribution continues to change even in free-molecular flow, a geometric effect. In the transition region the radial velocity distribution becomes increasingly non-Maxwellian but may be approximated by a Maxwellian velocity distribution with excessive 'tails'. The process of translational temperature freezing in free jets has been examined in detail experimentally and theoretically and is fairly well understood (23) (28 through 33). However, quantitative agreement between theory and experiment is not entirely complete.

If the background pressure in the nozzle exhaust chamber is not sufficiently low, a reduction in beam intensity may result from the scattering of jet molecules by background molecules upstream of the skimmer. At high background pressures a Mach disc - a shock wave normal to the axis of the jet - is formed. With decreasing background pressure the Mach disc moves downstream, broadens and, at zero background pressure, disappears. In effect, the jet pumps background gas away from its axis. Our experiments indicated that, as an approximation, the distance cleared by the jet was proportional to the nozzle flow rate, rather than the square root of flow rate; the reason for this has not been explained. Beyond that distance simple scattering by background gas occured (34). The experiments covered a restricted range of conditions which Brown and Heald have recently extended (35). The necessity for a low background pressure in the nozzle chamber can be avoided to some extent by reducing the dimensions of the nozzle-skimmer system, as in Campargue's apparatus (36).

Since skimmer interference imposes an important limitation on nozzle beam performance, it has been investigated widely and a number of explanations of the effect have been proposed. Fenn and Deckers correlated beam intensity results for a fairly wide range of conditions, and proposed the probable existence of a shock in front of the skimmer to explain the loss of beam intensity by skimmer interference

(37). Bier and Hagena, on the basis of schlieren studies of jet-skimmer interaction at higher densities, proposed that the effect be explained by shock waves or 'collisional distortion' downstream of the skimmer entrance (38) (25). Oman proposed an attenuation due to the scattering effects of molecules reflected from the external skimmer surface (39). French and McMichael (40), using an electron beam probe, observed only a slight increase in density in front of the skimmer under conditions corresponding to severe skimmer interference (skimmer Knudsen number $Kn_s < 1.2$, defined as the Lagrangian mean free path of the jet gas molecules at the skimmer entrance divided by the skimmer entrance diameter). Not one of the proposals cited here explains all the effects observed. There is strong evidence for each of them at different flow densities or values of Kn_s.

At very low densities ($Kn_s > 50$) the flow is essentially free-molecular, and skimmer interference is negligible. As density is increased molecules reflected from the external surface begin to scatter the beam molecules as in the Oman model. With a further increase in density the reflected molecules and the scattered molecules contribute to a density build-up upstream of the skimmer. This high density region appears to have the features of the diffuse shock-like structure proposed by Fenn and Deckers. At still higher densities ($Kn_s \sim 1$) there is a sufficient flux of jet molecules to drive molecules reflected from the external skimmer surfaces downstream of the skimmer entrance, and prevent the formation of a detached shock. Under this condition there is very definitely 'collisional distortion' inside the skimmer as proposed by Bier and Hagena. This distortion is probably the result of a diffuse oblique shock extending from the leading edge of the skimmer to the axis downstream.

The range of conditions of direct interest in producing molecular beams is that corresponding to a very slight skimmer interference — a beam attenuation of up to one-third ($Kn_s \gtrsim 10$). In this range, skimmer interference results from scattering in front of the skimmer entrance due to molecules reflected from the external skimmer surface as proposed by Oman and by Fenn and Deckers.

The possibility of self-scattering of beam molecules in the absence of skimmer interference effects has been treated by Valleau and Deckers (41). While the fundamentals of this treatment are certainly correct, the experiments by Anderson *et al* (24), using a cryogen-cooled skimmer to eliminate interference, suggest the beam attenuation predicted by Valleau and Deckers is higher than actually occurs.

The low static temperatures reached in the expansion of a jet may result in the gas becoming supersaturated, and give rise to condensation. The onset of condensation is marked by dimer formation. As noted by Deckers (42) and by Rosner and Fontijn (43) it is possible that dimers are extremely reactive chemically: this may lead to severe problems in studies of chemical reactions with nozzle beams. Our views on dimer formation as well as a simple approximate theory, shown to be reasonably successful, have been given previously (24). The views of Hagena and his co-workers, summarized in (25), probably apply to higher polymers. However, we object to the conclusion that condensation especially dimer formation can be a function of the nozzle pressure diameter product alone (with other variables constant). Because the formation of dimers from monomers requires three-body collisions, both the pressure diameter product and the total pressure must be involved. Recent measurements by Greene and Milne may resolve the question (44).

The optimum design of a nozzle beam system differs markedly, depending on the experiments for which it is to be used. We will not attempt to specify design details here.

Shock Tube Sources

In the isentropic expansion of pure gases in a jet, the average kinetic energy E of the molecules in laboratory coordinates is given by:

$$E = \tfrac{1}{2} MV^2 = \int_{T}^{T_o} c_p \, dT \qquad \text{(Eq. 5-1-3)}$$

where V is the flow velocity, M is the molecular mass, c_p is the molecular specific heat at constant pressure, T_o is the stagnation or source temperature, and T is the static temperature in the jet. For monatomic gases expanded to high Mach numbers from a stagnation temperature of 3000°K, the limit for the material of construction of a nozzle, the energy of the molecular beam formed is about 0.6 eV. In a shock tube, much higher gas temperatures can be obtained for short periods of time, the material limitations can be bypassed, and the beam energy may be increased.

The development of shock tube sources is being carried out in several laboratories notably by Oman (45), Skinner (46), Jones (47), Peng and Liquornik (48) and Dolder (chapter 5-2).

A schematic diagram of the apparatus used by Skinner and co-workers is shown in figure 5-1-6. A converging-diverging nozzle is placed in the end wall of a shock tube. The source gas is thus located in the region behind the reflected shock after passage of the shock. Various arrangements, including diaphragms and quick-opening valves, have been used to prevent the gas flowing through the nozzle before it has been heated. A starting time of 10-50 μ sec is required to establish the nozzle flow. The duration of a run (i.e. the time between the end of the starting period and the arrival at the end wall of the returning wave reflected from the contact front) ranges up to several milliseconds. Skinner and Moyzis found it advantageous to use a double skimmer arrangement to improve intensity, and to prevent effusion of cold gas from the nozzle exhaust section into the beam detection section. Of the various beam detection devices used to investigate beam properties, heat transfer gauges are the most common.

Source tempteratures up to 10,000°K have been obtained. For pure monatomic gases this corresponds to an energy of 2.1 eV. Peng and Liquornik combined the shock tube source with the 'seeded beam' technique to obtain a beam containing 3 eV oxygen atoms. Beam energy is apparently limited only by the maximum temperature that a shock tube can produce.

At source temperatures high enough to produce 1 eV beams of pure gases there will be dissociation, electronic excitation and ionization in the source. Study of these processes and the subsequent relaxation processes in the source gas will be required before the state of the final beam gas can be accurately predicted. Of course, these processes may be studied by investigation of final beam composition also. The experimental problems associated with shock tube sources are severe and the problems of instrumentation and detection are difficult, but the technique is promising.

Arc-Heated Sources

Knuth and his associates at UCLA have pioneered the arc-heated source technique. Argon beam energies of 0.5 to 3 eV have been obtained in their apparatus and investigations of atom-atom collisions and atom-surface collisions have been initiated (10) (49) (50). The performance of the beam system has been evaluated in detail in recent reports (51) (52).

Figure 5-1-7 shows the arrangement of source electrodes, nozzle and skimmer used at UCLA. A commercial plasma discharge device provides the energy input to the arc. Power input is as high as 300 amperes at 14 volts. The stagnation pressure range is 100-1300 torr. Nozzle exit diameters of 1 to 2 mm have been used. The electrodes and nozzle are water-cooled.

Beam intensities, velocities and velocity distributions have been measured with through-flow ionization gauges, rotating velocity selectors and a time-of-flight system. Performance characteristics are similar to those for lower temperature sources in many respects. The correlation of velocity distributions with source conditions found at lower temperatures by Anderson and Fenn (23) has been confirmed for the arc source system (52). As in the shock tube source there is a need to know the states of beam particles. The excitation and ionization in the arc and the relaxation during the expansion process need to be known. Winicur and Knuth have measured excitation and ionization in a 1.52 eV argon beam. Their data 'indicate that the ions are molecular during at least a portion of the free-jet expansion, that nearly all of the ions are neutralized during this expansion, and that most of the excited particles formed as a consequence of this neutralization are long-lived' (51).

The problems of electrode erosion and arc stability encountered in early experiments have been reduced to some extent by an improved electrode configuration. However, contamination of the source gas by electrode material may be a severe problem in many experiments, especially in gas-surface interaction studies. Progress in developing the arc-heated source has been rapid.

Seeded-Beam Techniques

In the isentropic expansion of a gas mixture in a free jet at a density sufficiently high to insure temperature and velocity equilibration between species, equation 5-1-3 gives the average energy of the molecules of both species combined. If the species masses differ, the heavier species will have a greater average energy than the light. In a binary mixture of light and heavy molecules the energy E_h of the heavy species is given by:

$$E_h = \tfrac{1}{2} M_h V^2 = \frac{M_h}{M_m} \int_T^{T_o} c_{pm} \, dT \qquad \text{(Eq. 5-1-4)}$$

where the subscript m indicates the mean value for the mixture and h indicates the heavy species. If differences in specific heats are neglected, it is clear that for the case of a 'solution' of heavy species in a light carrier gas, the translational kinetic energy of the heavy species will be higher than that of a pure gas by the factor M_h/M_m, which in a very dilute solution approaches the ratio of the molecular masses of the heavy and light species.

Measurements of the velocities of molecular beams extracted from jets of gas mixtures were reported first by Becker and Henkes (53) for argon-hydrogen mixtures. Klingelhofer and Lohse (54) and Belyaev and Leonas (55) also reported results for mixtures with a hydrogen carrier gas. These measurements substantiated the predictions of increased velocity for the heavier component in a gas mixture.

We have recently reported velocity measurements for a wide variety of species at varying source temperatures (56). These measurements are summarized in figures 5-1-8 and 5-1-9. In figure 5-1-8, results in terms of heavy species energy are shown for the heavy component in a mixture with 99 percent helium. Figure 5-1-9 shows results for similar mixtures with hydrogen. The solid lines

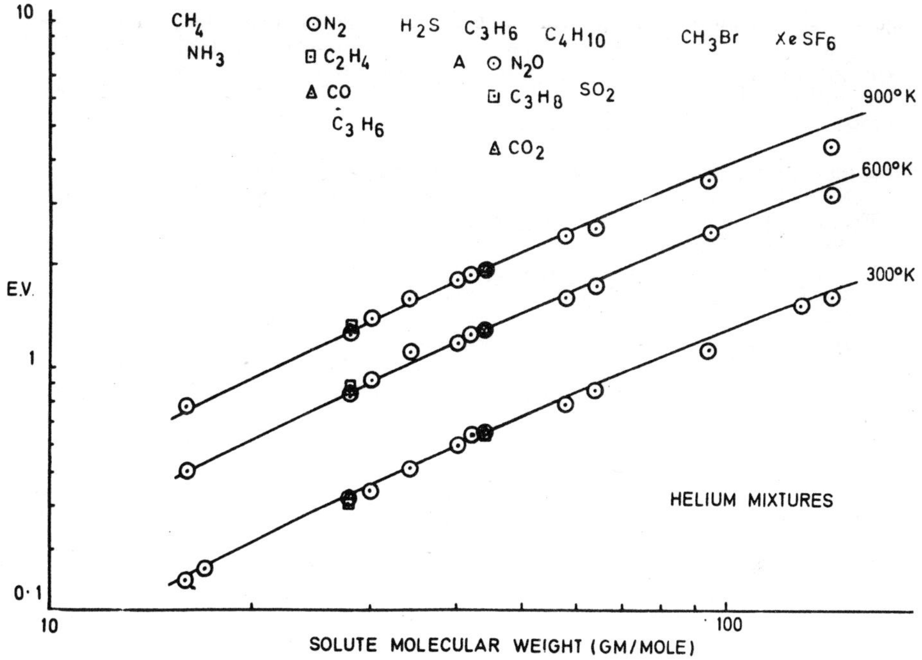

Fig. 5-1-8 Energies of the heavy species in molecular beams produced by the seeding technique using one percent mixtures of the heavy species with helium at varying source temperatures. From (68).

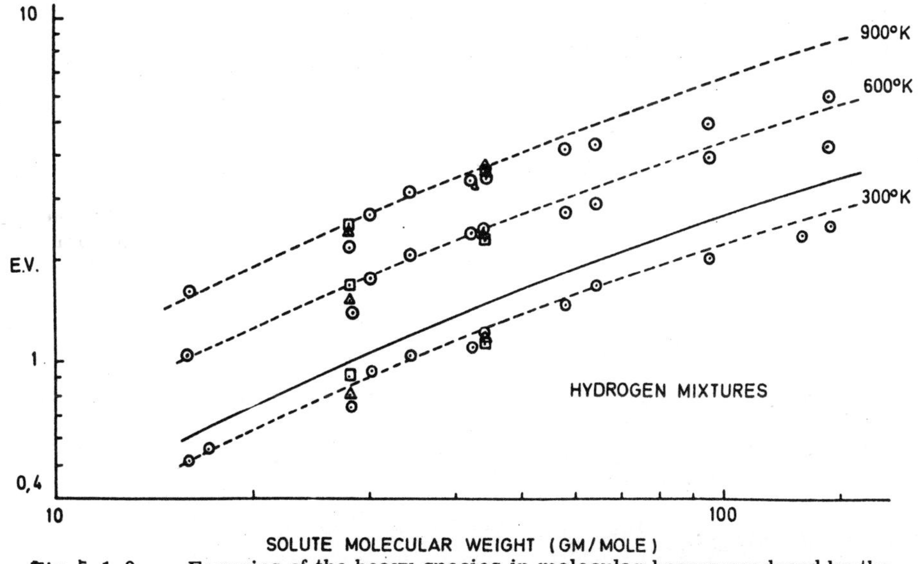

Fig. 5-1-9 Energies of the heavy species in molecular beams produced by the seeding technique using one percent mixtures of the heavy species with hydrogen at varying source temperatures. From (68).

in these figures indicate the "ideal" energies calculated for isentropic, continuum expansion of the jet to high Mach number, assuming no contribution to specific heat by the solute except for translational energy. It is evident that the seeded beam technique can produce beams of 5 to 10 eV energy for species of mass 30 and higher without exceeding a source temperature of 3000°K.

The departures from ideal energies exhibited by the heavier masses in figures 5-1-8 and 5-1-9 are the result of differing species velocities or slip in the expansion process. Incomplete rotational relaxation of the hydrogen carrier may also be responsible for the departures in figure 5-1-9. Velocity slip is primarily the result of pressure diffusion in the expanding jet, and can be overcome by increasing nozzle Reynolds numbers (31)(57).

The dashed lines in figure 5-1-9 indicate energies calculated using a simple approximation to account for the incomplete rotational relaxation of hydrogen. In this approximation the specific heat ratio of hydrogen is taken as 7/5 in the subsonic portion of the expansion and 5/3 in the supersonic portion.

Several theoretical treatments of the expansion of gas mixtures in free jets are available. These have satisfactorily predicted species velocities, as noted above, and diffusive separation effects (58 through 62). An adequate comparison of theory (63)(64) and experiment (31)(57) with respect to the velocity distributions of individual species has not yet been made, and further experimental work is required before definite conclusions can be reached.

The presence of the carrier gas will interfere in some experiments with seeded beams. Fortunately, the heavy species tends to be concentrated on the axis of the beam, so that the final beam is enriched in the heavy species. Mass-discriminating detectors may overcome this problem for many experiments.

The seeded-beam technique has been successfully used in our laboratory in measurements of momentum transfer at surfaces with argon beams at energies up to 4.11 eV (65) and in determining argon-argon total scattering cross-sections at collision energies up to 1.6 eV (66).

Acknowledgements

Preparation of this presentation was made possible by generous support in part from the National Science Foundation under Grant GK 655, in part from the Office of Naval Research through Project SQUID (Contract Nonr 3623(00), NR-098-038) and in part from the Air Force Cambridge Research Laboratories under Contract AF 19(628) - 3887.

References

1. Utterback, N.G., Phys. Rev. 129, 219, 1963.

2. Hollstein, M., Pauly, H., Z. Physik 196, 196, 1966.

3. Utterback, N.G., 'Proc. 7th AGARD/PEP Colloquium on Recent Advances in Aerothermochemistry.' Oslo, 1966.

4. Amdur, I., Jordan, J.E., 'Advances in Chemical Physics,' Vol. X, J. Ross, Ed., Interscience, 1966.

5. Mason, E.A., Vanderslice, J.T., 'Atomic and Molecular Processes,' D.R. Bates, Ed., Academic Press, 1962.

6. **Wehner, G.K.,** J. Appl. Phys. 30, 1762, 1959.

7. **Politiek, J., Rol, P.K., Los, J., Onderdelinden, D., Schipper, J.J.M.,** 'Rarefied Gas Dynamics,' Vol. II, C.L. Brundin, Ed., Fifth Symp., Oxford, 1966, Academic Press, 1967.

8. **Politiek, J., Los, J., Schipper, J.J.M.,** 'Proc. 7th AGARD/PEP Colloquium on Recent Advances in Aerothermochemistry,' Oslo, 1966.

9. **Politiek, J., Los, J., Schipper, J.J.M., Baede, A.P.M.,** 'Proc. 1st Int. Symp. on Molecular Beams of High and Intermediate Energy.' Cannes, 1967.

10. **Los, J.,** FOM-Inst. for Atomic and Molecular Physics, Amsterdam, private communication.

11. **Cheng, S.I.,** Astronautica Acta 12, 272, 1966.

12. **Trujillo, S.M., Neynaber, R.H., Rothe, E.W.,** Rev. Sci. Instr. 37, 1655, 1966.

13. **Cook, C.J., Ablow, C.M.,** 137th Natl. Meeting Am. Chem. Soc., Cleveland, Ohio 1960; cf. AFCRC-TN-59-472. 1959.

14. **Trujillo, S.M., Neynaber, R.H., Marino, L.L., Rothe, E.W.,** Proc. IVth Int. Conf. Phys. Electronic Atomic Collisions, Science Bookcrafters, 1965.

15. **Belyaev, V.A., Brezhnev, B.G., Erastov, E.M.,** Zh. Eksperim i Teor. Fiz. - Pis'ma Redakt. 3, 321, 1966. English transl.: JETP Letters 3, 207, 1966.

16. **Neynaber, R.H., Trujillo, S.M., Rothe, E.N.,** Phys. Rev.

17. **Becker, E.W., Bier, K.,** Z. Naturforsch, 92, 975. 1954.

18. **Anderson, J.B., Andres, R.P., Fenn, J.B.,** 'Advances in Chemical Physics,' Vol. X. J. Ross, Ed., Interscience, 1966.

19. **Knuth, E.L.,** Appl. Mech. Rev. 17, 751. 1964.

20. **French, J.B.,** Amer. Inst. Aeron. Astronaut. J. 3, 993, 1965.

21. **Kantrowitz, A., Grey, J.,** Rev. Sci. Instr. 22, 328, 1951.

22. **Parker, H.M., Kuhlthau, A.R., Zapata, R., Scott, J.E.,** 'Rarefied Gas Dynamics'F.M. Devienne, Ed. Pergamon, 1960.

23. **Anderson, J.B., Fenn, J.B.,** Phys. Fluids 8, 780, 1965.

24. **Anderson, J.B., Andres, R.P., Fenn, J.B., Maise, G.,** 'Rarefied Gas Dynamics,'Vol. II. J.H. de Leeuw, Ed., Academic Press, 1966.

25. **Hagena, O.F., Morton, H.S.,** 'Rarefied Gas Dynamics'Vol. II C.L. Brundin, Ed., Academic Press, 1967.

26. **Owen, P.L., Thornhill, C.K.,** Aer. Res. Council U.K., R and M No. 2616. 1948.

27. **Ashkenas, H., Sherman, F.S.,** Rarefied Gas Dynamics, Vol.II J.H. de Leeuw, Ed., Academic Press, 1966.

28. **Hamel, B.B., Willis, D.R.,** Phys. Fluids 9, 829, 1966.

29. **Cheng, H.K., Edwards, R.H.,** Amer. Inst. Aeron. Astronaut. J. 4, 558. 1966.

30. **Edwards, R.H., Cheng, H.K.,** 'Rarefied Gas Dynamics,' Vol. I, C.L. Brundin, Ed., Academic Press, 1967.

31. **Abuaf, N., Anderson, J.B., Andres, R.P., Fenn. J.B.. Miller, D.R.,** ibid., Vol. II.

32. **Muntz, E.P.,** ibid, Vol. II.

33. **Scott, J.E., Phipps, J.A.,** ibid, Vol. II.

34. **Fenn, J.B., Anderson, J.B.,** 'Rarefied Gas Dynamics,' Vol. II J. H. de Leeuw, Ed., Academic Press, 1966.

35. **Brown, R.F., Heald, J.H.,** 'Rarefied Gas Dynamics,'Vol. II C.L. Brundin, Ed., Academic Press, 1967.

36. **Campargue, R.,** Rev. Sci. Instr. 35, 111, 1964.

37. **Fenn, J.B., Deckers, J.,** 'Rarefied Gas Dynamics,' Vol. I, J.A. Laurmann, Ed., Academic Press, 1963.

38. **Bier, K., Hagena, O.F.,** ibid. 'Rarefied Gas Dynamics,' Vol. II, J.H. de Leeuw, Ed., Academic Press, 1966.

39. **Oman, R.A.,** Phys. Fluids 6, 1030, 1963.

40. **French, J.B., McMichael, G.E.,** 'Rarefied Gas Dynamics' C.L. Brundin, Ed., Academic Press, 1967. Phys. Fluids 9, 1419, 1966.

41. **Valleau, J.P., Deckers, J.,** Can. J. Chem. 42, 225, 1964.

42. **Deckers, J.,** Univ. of Toronto, Private Communication. 1966.

43. **Fontijn, A., Rosner, D.E.,** J. Chem. Phys. 46, 3275, 1967.

44. **Greene, F.T., Milne, T.A.,** J. Chem. Phys. 39, 3150, 1963. Midwest Res. Inst. Reps., 1963 - 67.

45. **Oman, R.A., Bogan, A., Weiser, C.H., Li, C.H., Calia, V.S.,** Grumman Res. Dept. Rept. Re-166, 1963.

46. **Skinner, G.T., Moyzis, J.,** Phys. Fluids 8, 452, 1965.

47. **Jones, T.V.,** 'Rarefied Gas Dynamics,' Vol. II. C.L. Brundin, Ed., Academic Press, 1967.

48. **Peng, T.C., Liquornik, D.L.,** Rev. Sci. Instr. 38, 989, 1967.

49. **Knuth, E.L., Kuluva, N.M.,** 'Proc. 7th AGARD/PEP Colloquium on Recent Advances in Thermochemistry.' Oslo, 1966.

50. **Alcalay, J.A., Knuth, E.L.,** 'Rarefied Gas Dynamics,' Vol. I. C.L. Brundin, Ed., Academic Press, 1967.

51. **Winicur, D.H., Knuth, E.L.,** J. Chem. Phys. 46, 4318, 1967.

52. **Kuluva, N.M.,** UCLA Dept. of Eng. Rept. 67-11, 1967.

53. **Becker, E.W., Henkes, W.,** Z. Physik 146, 320, 1956.

54. **Klingelhofer, R., Lohse, P.,** Phys. Fluids 7, 379, 1964.

55. **Belyaev, Y.N., Leonas, V.B.,** Vestn. Mosk. Univ. Ser. III, Fiz., Astron. 18, 34, 1963.

56. **Abuaf, N., Anderson, J.B., Andres, R.P., Fenn, J.B., Marsden, D.G.,** Science 155, 997, 1967.

57. **Anderson, J.B.,** 'Proc. 1st Int. Symp. on Molecular Beams of High and Intermediate Energy,' Cannes. 1967.

58. **Zigan, F.,** Z. Naturforsch. 17a, 772, 1962.

59. **Mikami, H., Takashima, Y.,** Bull. Tokyo Inst. Technol. 61, 67, 1964.

60. **Sherman, F.S.,** Phys. Fluids 8, 773, 1965.

61. **Rothe, D.E.,** Phys. Fluids 9, 1643, 1966.

62. **Anderson, J.B.,** Amer. Inst. Chem. Enggs. J.

63. **Willis, D.R., Hamel, B.B.,** 'Rarefied Gas Dynamics,' Vol. I. C.L. Brundin, Ed. Academic Press, 1967.

64. **Cooper, A.L., Bienkowski, G.K.,** ibid.

65. **Abuaf, N., Marsden, D.G.,** ibid.

66. **Baratz, B.,** Thesis, Princeton University, 1967.

67. **Skinner, G.T., Fetz, B.H.,** 'Rarefied Gas Dynamics,' Vol. II. J.H. de Leeuw, Ed., Academic Press, 1966.

68. **Abuaf, N.,** Thesis, Princeton University, 1966.

Commentary on Chapter 5-1

O.F. Hagena

More and more laboratories are using molecular beam methods to study problems associated with elastic and reactive scattering of systems gas-gas and gas-solid. The progress in the field of high energy (1-10 eV) molecular beams is evident from the present chapter comparing it with earlier review articles, e.g. [XVIII]. Roman numerals in brackets refer to the previous list of references in (1), while Arabic numerals in brackets refer to the bibliography at the end of these comments. Sputtering sources, merging beams and arc-heated sources have been added to the list of methods being used for a variety of beam experiments. Also, the discussion of some basic areas in nozzle beam performance that - to quote [XVIII] - 'smacks of qualitative opinion and speculation' has now been replaced by more 'quantitative facts and certain knowledge'. This progress in interpretation is especially evident in the discussion of (a) jet-skimmer interaction (see the old discussion of figure 7 in [XVIII], (b) diffusive separation effects, and (c) ideal energies of mixed beams with hydrogen as light component. In view of these improvements it might have been better to avoid reference to [XVIII] which can cause some confusion for a newcomer to the field. The other review papers [XIX] and [XX] show considerable signs of obsoleteness, too.

Some of the many topics discussed may deserve a few additional comments.

Sputtering sources: An interesting modification of the technique used in Amsterdam [VII through X] is reported by Hulpke and Kempter from the Universitat Freiburg (2). They simply used the cathode of a Penning discharge as sputtering target, thus avoiding a separate ion beam system. At energies of the sputtered atoms around 1 eV, this method yielded an intensity increase by more than an order of magnitude.

Skimmer interference: 'Collisional distortion' of the flow field not only downstream, but also upstream of the skimmer is discussed in [XXXVIII]. That the type of skimmer interference depends strongly on the density level at the skimmer is also emphasised in [XXV] and [XXXVIII]. Similar discussions including the effect of background scattering, investigated in detail in [XXIV] can be found in (3) and (4). The skimmer Knudsen number Kn_s for 'conditions of direct interest in molecular beam production' is not a very well defined quantity. For optimum centerline intensity Kn_s was found to vary between slightly less than one (5) and three, from the data of [XXXVIII]. Greater nozzle-skimmer distances and thus higher values of Kn_s are favoured if not only the intensity should be high, but the mass flux through the skimmer should be low. This usually demands a higher pumping speed in the nozzle discharge chamber.

Condensation: For those not familiar with the original literature referenced in (1) the section on condensation of the beam gas is somewhat misleading. In view of the increased interest in using as well as in avoiding condensation in nozzle flows [XLIII] (6-9) a more detailed discussion is justified. The approximate dimer formation theory of Anderson et al. [XXIV] is applicable only to monatomic gases as shown in (10). For simplicity it cannot take into account metastable dimers and orbiting pairs, which are quite frequent at relative kinetic energies that are small compared to the dimer binding energy. The quoted 'reasonable success' (1) refers to dimer-concentration measurements with a mass-spectrometer far downstream of the skimmer, for which the assumed mass separation is doubtful because the larger gas kinetic collision cross section of a dimer compared to a monomer has not been taken into account. More recent measurements by the same group (11) give values differing by more than a factor of two from those used in the comparison of [XXIV]. This is typical of the difficulties of quantitative mass spectrometric sampling. The conclusion reached in [XXIV] 'that significant condensation effects are not to be expected in argon nozzle beams when the source is at room temperature or above, even though the pressure is high enough to result in effective Mach numbers of 17 or above' is not very precise. It makes no prediction how high a Mach number can be obtained at what temperature, and whether this depends on the nozzle diameter. Our own work on condensation was concerned with two different aspects of condensation: (a) For highly condensed beams the mean polymer size was measured as a function of source pressure and nozzle diameter, p_o and d. It was found that the mean size did not scale with the product $p_o \cdot d$ (which is the proper scaling parameter for bimolecular reactions, e.g. freezing of translational and rotational temperature) but is higher for smaller nozzles (10). (b) The conditions under which the onset of condensation influences the Mach number M and the intensity I were investigated by a systematic variation of pressure, nozzle size and skimmer distance. It was found that prior to marked condensation, M no longer increases with p_o but reaches a relative maximum. The position of this maximum was arbitrarily used to define the point of 'onset of condensation'. For argon and CO_2 this point was characterized by a constant value of $p_o \cdot d$ for $0.05 < d < 0.5$ mm (4) [XXXVIII]. The $p_o \cdot d$ criterion for argon was confirmed in a different beam system at the University of Virginia (12). This criterion for the onset of condensation was strictly an empirical result setting a limit for the range of conditions useful for molecular beam experiments. In particular it did not predict a dimer concentration less than a certain value, also it was limited if d was so small that p_o became comparable with the respective vapour pressure. The criterion can be applied to the problem of NO condensation discussed in [XLIII]. Comparing the vapour pressure data it can be estimated that the $p_o \cdot d$ value of NO should fall within the values obtained for CO_2 and N_2, 60 and 500 torr mm, respectively. For the experimental conditions considered in [XLIII] $p_o \cdot d = 830$ torr mm. Therefore, the presence of polymers has to be expected. While the $p_o \cdot d$ criterion seems to be a necessary condition to avoid condensation it is not always sufficient; the tolerable level of polymers will differ from experiment to experiment. To ensure the absence of condensation one has either to measure polymer concentration under the exact experimental conditions, or to prove that changes in source pressure, which will affect the polymer concentration, do not measurably change the experimental results.

References

1. This chapter.
2. **Hulpke, E., Kempter, V.,** Z. Physik 197, 41, 1966.
3. **Hagena, O.F., Angew, Z.,** Phys. 16, 183, 1963.
4. **Hagena, O.F., Schüller, P.G., Angew, Z.,** Phys. 17, 542, 1964.
5. **Scott, J.E., Drewry, J.E.,** 'Rarefied Gas Dynamics,' Vol. I. J.A. Laurmann, Ed., Third Symp. Paris, 1962. Academic Press, 1963.
6. **Greer, H., Griep, D.J.,** J. Spacecraft and Rockets 4, 983, 1967.
7. **Burghoff, H., Gspann, J.,** Z. Naturforschg. 22a, 684, 1967.
8. **Henkes, W.,** Phys. Letters 12, 322, 1964.
9. **Becker, E.W., Klingelhöfer, R.,** Plasma Physics J. Nucl. Energy, Part C, 8, 413, 1966.
10. **Bauchert, J., Hagena, O.F.,** Z. Naturforschg. 20a, 1135, 1965.

10. **Milne, T.A., Greene, F.T.,** 'Mass spectrometric observation of argon clusters in nozzle beams', Sum. Tech. Rep., Midwest Res. Inst. Kansas City, Aug. 1966.

11. **Hagena, O.F., Scott, J.E., Varma, A.K.,** 'Design and Performance of an Aerodynamic Molecular Beam and Beam Detection System', Rep. AST 4038-103-67U, June 1967, Res. Lab. for the Eng. Sciences, Univ. of Virginia.

5-2

Measurements of Velocity and Dissociation in Shock-Heated Molecular Beams

D. SPENCE AND K. DOLDER
University of Newcastle upon Tyne, UK.

Summary

Shock-heated molecular beams of oxygen have been produced in an apparatus similar in principle to that described by Skinner and Moyzis (5). A mass spectrometer has been used to measure the dissociation of oxygen in these beams and it is found that the measured dissociation increases smoothly and consistently with increasing shock strength. The measured dissociation is, however, somewhat higher than that predicted by theory. This discrepancy between theory and experiment can, however, be entirely explained if the cross section for dissociative ionization of oxygen by electron impact increases by a factor two when oxygen is heated to 3, 500°K.

Dissociation has also been measured in beams formed from shock-heated mixtures of argon with either oxygen or hydrogen. The molecular gases are observed to be completely dissociated when they constitute less than about 15 percent of the shock-heated mixture.

Measurements of intensity and particle velocity have been made for a variety of beams. The results agree quite well with theoretical predictions.

Sommaire

Des faisceaux moléculaires d'oxygène chauffés par choc ont été produits dans un montage analogue à celui décrit par Skinner et Moyzis (5). Un spectromètre de masse a été utilisé pour déterminer la dissociation de l'oxygène dans ces faisceaux et on a trouvé que la dissociation mesurée croît lentement mais régulièrement avec l'intensité du choc. La valeur de la dissociation mesurée est cependant plus élevée que celle prévue par la théorie. Cet écart entre théorie et expérience s' explique cependant si l'on double la section efficace d'ionisation dissociative de l'oxygène sous l'effet d'impact d'electrons lorsque l'oxygène est chauffé à 3, 500° K.

Introduction

The production of intense molecular beams by allowing gas to expand isentropically from a reservoir through nozzles was first suggested by Kantrowitz and Grey (1). Nozzle sources are attractive because they can produce beams with particle densities far beyond those attainable from conventional oven sources of the type pioneered by Stern (2). Substantial developments to the original Kantrowitz and Grey scheme have been made by a number of experimentalists and this work has been reviewed by Anderson, Andres and Fenn (3) and by others.

In an attempt to combine the high intensity of nozzle sources with increased particle energy and source temperature, Skinner (4) substituted a shock tube for the reservoir of cold gas. By operating the shock tube under tailored interface conditions he was

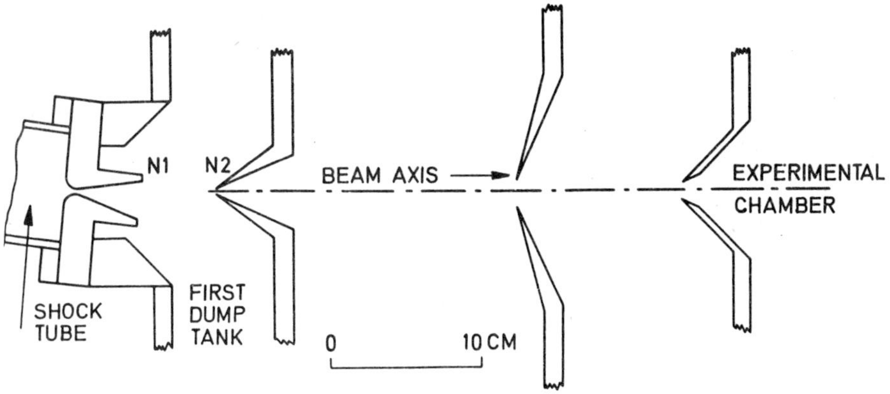

Fig. 5-2-1 Nozzles and collimators required to form shock heated beam

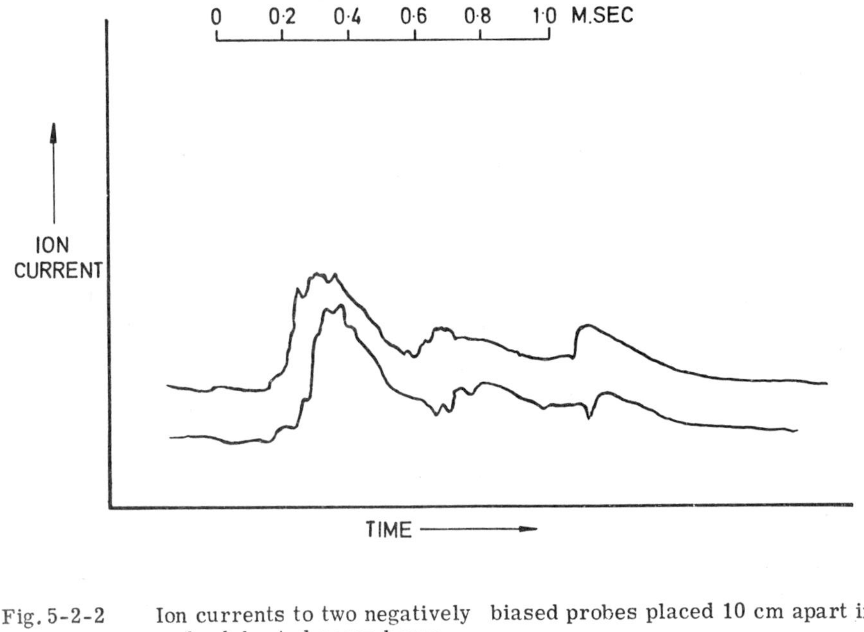

Fig. 5-2-2 Ion currents to two negatively biased probes placed 10 cm apart in a shock heated argon beam

able to sustain a beam of nitrogen for more than 10^{-3} second. In more recent experiments Skinner and Moyzis (5) obtained a nitrogen beam with particle flux close to 2.4×10^{18} molecules cm^{-2} s^{-1}, particle energy 1.2 eV, semi-angle of divergence 1° and flow duration greater than 10^{-3} s.

Jones (6) and the authors have each built an apparatus based on essentially similar principles to that of Skinner and Moyzis (17). In this paper we shall briefly describe the apparatus and report the results of measurements of particle velocity and dissociation in shock-heated molecular beams. A detailed account of this work is given elsewhere (7).

Apparatus

Gases were heated in a pressure-driven shock tube of total length 20 ft. and internal diameter $2\frac{3}{4}$". Stationary gas at the end of the tube, which had been heated first by an incident shock and then by a normally-reflected shock, expanded through a nozzle N1 (Fig. 5-2-1). The gas was then formed into a molecular beam by nozzles and collimators that are identical to those developed by Skinner and Moyzis. The largest dump tank in the present apparatus had a diameter of only 3 ft which is substantially smaller than the tanks used by Skinner and Jones. The smaller apparatus was, however, able to sustain a steady beam for about 10^{-3} second.

Thin film resistance thermometers were used to make a number of measurements of particle flux in beams of nitrogen, oxygen and argon. The results confirmed the work of Skinner and Moyzis (5) and Jones (6).

Particle Velocity

Particle velocities were measured in shock-heated beams of argon, oxygen and in beams consisting of mixtures of argon with either oxygen or hydrogen. This was done by placing two negatively-charged wire probes, spaced 10 cm apart, in the path of the beam. Small amounts of ionization caused currents to flow to the probes and clearly-defined fluctuations were observed in these currents.

Similar fluctuations were found in the currents to both probes but the downstream probe responded later than the upstream probe. From this delay, and the spacing between the probes, a velocity could be deduced which consistently agreed quite closely with theoretical predictions. An example of probe currents measured in an argon beam is shown by figure 5-2-2.

It might be argued that the probe potentials would accelerate ions and so give false values of particle velocity. The potentials are, however, limited to a region which is characterised by the Debye length (8) in the ionized gas flow. This length is given by:

$$\Delta = 6.9\,(T/n)^{\frac{1}{2}}\ \text{cm} \qquad \text{(Eq. 5-2-1)}$$

For a beam with only 0.1 percent ionization n is of order 10^{10} cm^{-3} and T cannot exceed the electron temperature in the shock tube ($< 10^4$ °K). Thus the Debye length might typically be of order 10^{-2} cm which is far less than the distance between the probes.

The energy of particles in a beam expanded to high Mach numbers from a source at temperature T°K is,

$$\tfrac{1}{2}\, mv^2 = \left\{\frac{\gamma}{\gamma - 1}\right\} kT \qquad \text{(Eq. 5-2-2)}$$

If the expanding gas is a binary mixture of light and heavy particles the energy of

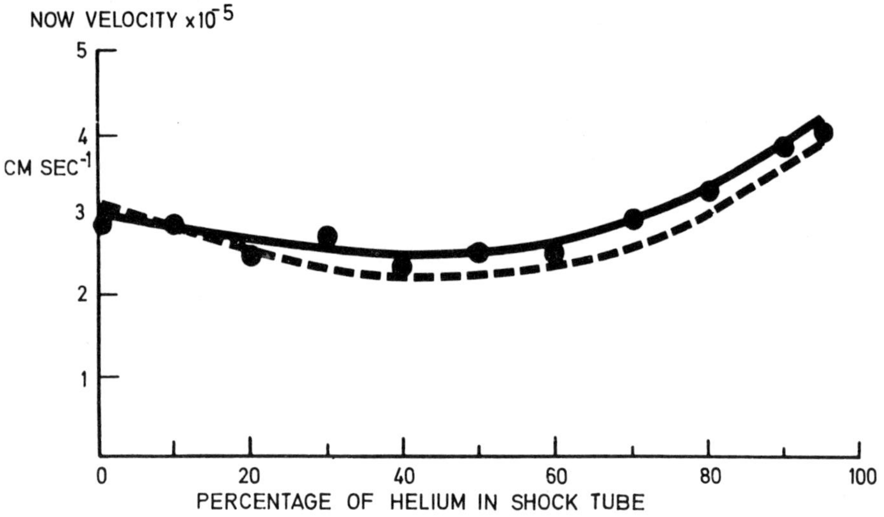

Fig. 5-2-3 Velocities in beams obtained from shock heated mixtures of helium and argon.

The dotted curve shows calculated velocities and the continuous curve locates experimental points.

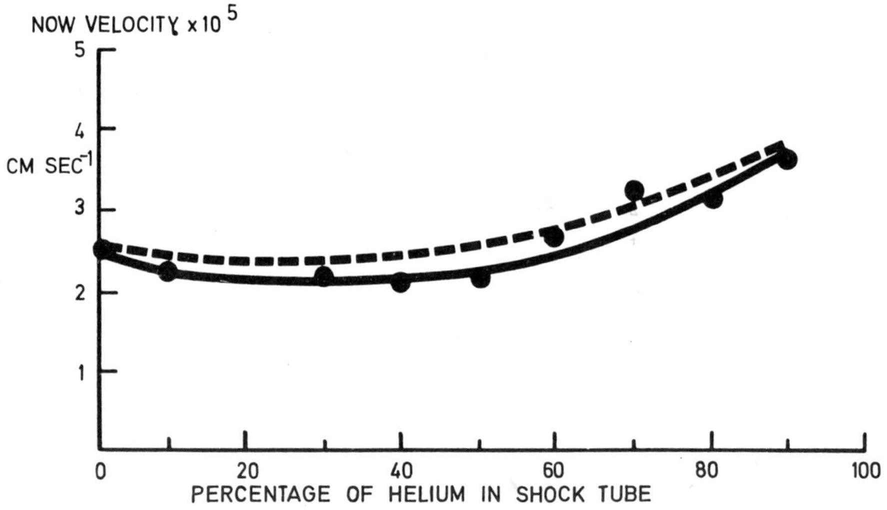

Fig. 5-2-4 Velocities in beams obtained from shock heated mixtures of helium and oxygen.

The dotted curve shows calculated velocities and the continuous curve locates experimental points.

the heavier particles is:

$$\frac{1}{2} m_h v^2 = \frac{m_h}{m_m} \frac{\gamma_m}{\gamma_m - 1} kT \qquad \text{(Eq. 5-2-3)}$$

In the expansion of a dense binary mixture all particles will attain the same velocity so that the more massive particles will have the greater energy. Anderson, Andres and Fenn (9) have employed this principle to obtain energetic ($\lesssim 4$ eV) argon atoms by expanding mixtures of argon and hydrogen in a nozzle-beam source.

The continuous curve of figure 5-2-3 shows velocity measurements in beams formed from shock-heated mixtures of argon and helium. In these experiments the proportion of argon was varied but the total pressure (P_1) of the un-shocked mixture was always 10 torr. The pressure ratio across the shock tube diaphragm was kept constant ($P_4/P_1 = 530$) so that the equilibrium temperature (T_5) of the shock-heated mixture varied with composition.

The particle velocity could also be obtained by calculation from equation 5-2-3. These theoretical values are shown by the dotted curve of figure 5-2-3. The temperature in equation 5-2-3 was set equal to the calculated (14) equilibrium temperature (T_5) behind the reflected shock. This was obtained from the measured incident shock Mach number.

Similar results for mixtures of oxygen and helium are shown by figure 5-2-4. It will be seen that the results of both experiments agree quite well with theory.

The highest velocity measured was 4.0×10^5 cm s^{-1} in a mixture containing 5 per-cent argon in helium. This corresponds to an argon energy of 3.6 eV. Higher energies could have been produced by using stronger shocks, by using hydrogen instead of helium as the carrier gas or by using smaller proportions of argon. No attempt was made in these experiments to separate heavy and light molecules in the beams.

Measurements of the Dissociation of Oxygen

Dissociation in molecular beams of oxygen was investigated by bombarding the beams with 200 eV electrons. In partially dissociated oxygen the following electron impact ionization processes are possible:

$$O_2 + e \rightarrow O_2^+ + 2e \qquad \text{(Eq. 5-2-4)}$$

$$O_2 + e \rightarrow O^+ + \ldots\ldots \qquad \text{(Eq. 5-2-5)}$$

$$O + e \rightarrow O^+ + 2e \qquad \text{(Eq. 5-2-6)}$$

The first and third reactions represent the simple ionization of molecules and atoms respectively. The second reaction denotes the production of O^+ by all forms of dissociative ionization.

Atomic and molecular ions were extracted at right angles from the molecular beam and focussed and accelerated by an ion optical system similar to that described by Giese (10). Before collection ions were separated in the sector field of a mass spectrometer magnet.

Care was taken to ensure that the collection efficiency of atomic and molecular ions was the same. It can then be shown that the ratio of collected currents of atomic

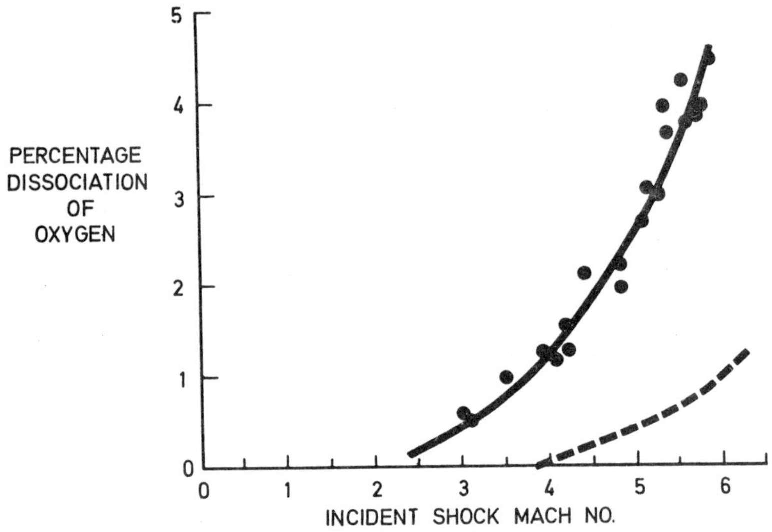

Fig. 5-2-5 Dissociation in shock heated beams of oxygen

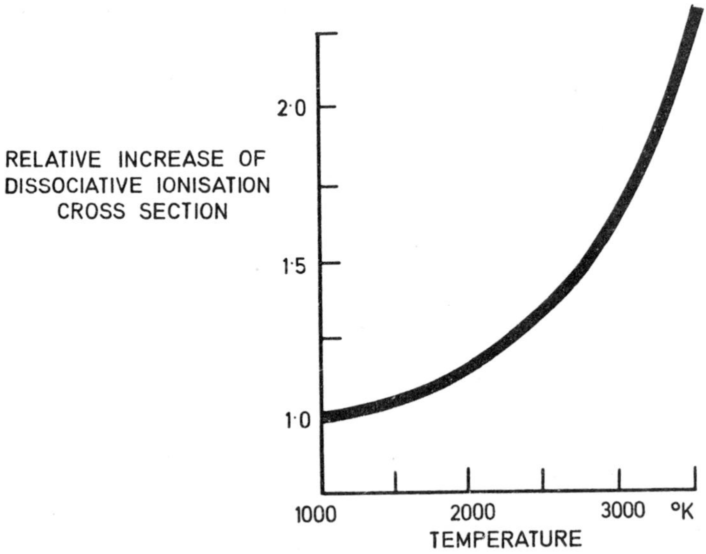

Fig. 5-2-6 Increase in dissociative ionisation cross section of oxygen necessary to reconcile theory with measured dissociation in a shock heated oxygen

and molecular ions is:

$$\frac{I_A}{I_m} = \frac{Q_D \ (1 - \alpha) + 2\, Q_A \ \alpha}{(1 - \alpha)\, Q_M} \qquad \text{(Eq. 5-2-7)}$$

where Q_M, Q_D and Q_A are the cross sections for reactions 4, 5 and 6. Values of the ionisation cross sections were used that have been measured by Fite and Brackmann (11) and by Rapp and Englander-Golden (12).

Equation 5-2-7 was then used to calculate α and the results are shown by the continuous curve of figure 5-2-5 as a function of the incident shock Mach number.

It must be pointed out that values of Q_M and Q_D have been used which were measured for oxygen at or near room temperature. However, in the present experiment the oxygen is heated to temperatures up to 3,500°K. Rapp, Sharp and Briglia have predicted that dissociative ionization is likely to depend quite sensitively upon the vibrational excitation of a molecule and this may account for the disparity between experiment and theory.

Theoretical values of dissociation are shown by the dotted curve in figure 5-2-5. These calculations, which were performed independently by Bernstein (14) and the authors, give the equilibrium dissociation behind a normally reflected shock. According to Hall and Russo (15) one might expect gas to pass from a shock tube to the beam without appreciable recombination.

It is plausible that the disparity between theory and experiment is due to an enhanced dissociative ionization cross section of vibrationally-excited oxygen. It is, for example, well known (16) that dissociative recombination,

$$e + O_2^{+} \rightarrow O + O$$

depends sensitively upon the molecular excitation of O_2^+. It is only necessary to assume the relative increase of Q_D with temperature shown by figure 5-2-6 to bring the present experiment into line with theory. In preparing figure 5-2-6 it was assumed that the excitation of oxygen in the beam corresponded to the calculated equilibrium temperature (14) in the shock tube.

Dissociation in Shock-Heated Mixtures

The experiments described in the previous section were performed for a range of shock strengths. Optimum flow duration is, however, only obtained if the shock tube is operated under tailored interface conditions. For hydrogen driving oxygen this implies an incident shock Mach number of about 6 if both gases are initially at room temperature.

To obtain completely dissociated oxygen, whilst simultaneously preserving tailoring, shocks were generated in mixtures of argon and oxygen. Argon is a dense monatomic gas which can easily be raised to temperatures that are sufficient to dissociate small proportions of oxygen mixed with it. Figure 5-2-7 shows the measured dissociation of oxygen in various mixtures of oxygen and argon. These experiments were performed with a shock speed $\approx 2.0 \times 10^5$ cm.sec^{-1}, which approximates closely to the speeds required for tailoring and a total initial gas pressure (P_1) of 10 torr. The dissociation was calculated from the measured ion currents by the same methods used for the results in figure 5-2-5. The true dissociation of oxygen was observed and, since this was signified by the complete absence of the O_2^+ ion, we conclude that dissociation was certainly complete.

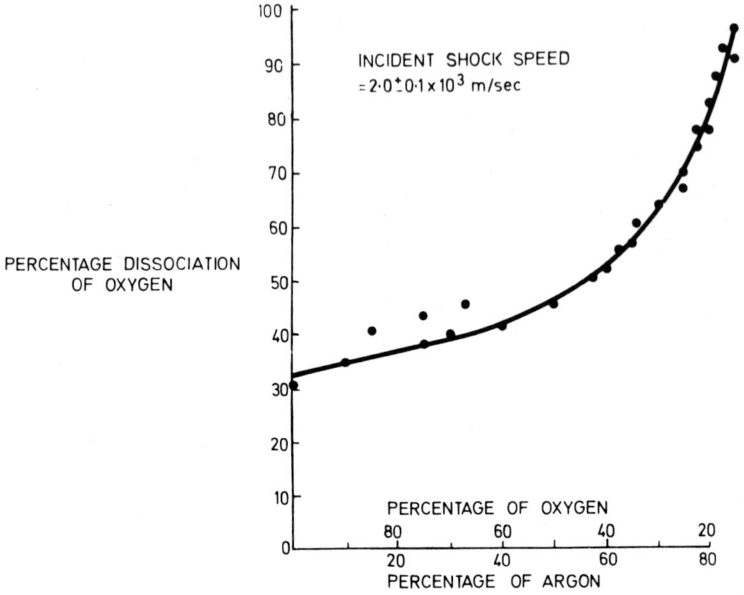

Fig. 5-2-7 Dissociation in shock heated beams of argon and oxygen mixtures

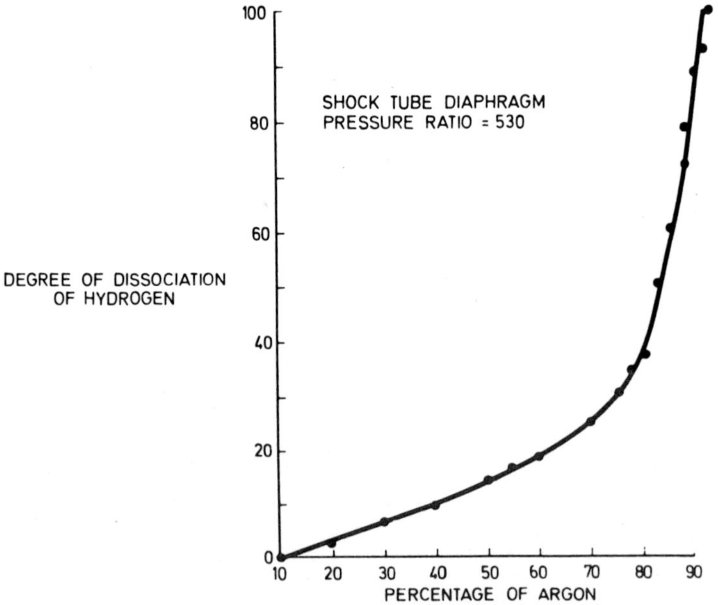

Fig. 5-2-8 **Dissociation of oxygen in shock heated mixtures of hydrogen and argon**

Similar results for argon hydrogen mixtures are shown by figure 5-2-8. However, in these experiments the shock tube diaphragm pressure ratio was constant (P_4/P_1 = 530) so that the shock speed varied with composition. Under tailoring conditions complete dissociation of hydrogen was observed for mixtures containing about 10 percent of hydrogen.

In some of the experiments described above hydrogen was used to generate shocks in gas consisting partly or wholly of oxygen. Combustion at the contact surface prevented tailoring and so shortened the duration of steady beam flow from 10^{-3} s to a few hundred microseconds. This combustion was prevented by fitting an additional thin diaphragm to the shock tube at a point about 10 " downstream from the main diaphragm. The space between these diaphragms was filled with argon which effectively separated hydrogen from the oxygen. Steady flow, lasting for 10^{-3} s was then obtained for beams containing oxygen.

Acknowledgements

We wish to acknowledge support from the Ministry of Technology and the assistance of K. Crozier in constructing parts of the apparatus.

References

1. **Kantrowitz, A., Grey, J.,** Rev. Sci. Inst., 22, 328, 1951.

2. **Stern, O.,** Zeit. f. Physik, 2, 49, 1920.

3. **Anderson, J.B., Andres, R.P., Fenn, J.B.,** Advances in Chemical Physics, 10, 275, 1966.

4. **Skinner, G.T.,** Phys. Fluids 4, 1172, 1961.

5. **Skinner, G.T., Moyzis, J.,** Phys. Fluids, 8, 452, 1965.

6. **Jones, T.V.,** Ph.D. Thesis, University of Oxford, 1966.

7. **Spence, D.,** Ph.D. Thesis, University of Newcastle upon Tyne, 1967.

8. **Spitzer, L.,** 'Physics of Fully Ionized Gases', Interscience, New York, 1956.

9. **Anderson, J.B., Andres, R.P., Fenn, J.B.,** AGARD Conf., Oslo, 1965.

10. **Giese, C.F.,** Rev. Sci. Inst. 30, 260, 1959.

11. **Fite, W., Brackmann, R.T.,** Phys. Rev. 113, 815, 1959.

12. **Rapp, D., Englander-Golden, P.,** J. Chem. Phys. 43, 1464, 1965.

13. **Rapp, D., Sharp, T.E., Briglia, D.D.,** Lockheed Rep. LMSC 6-74-64-45.

14. **Bernstein, L.,** Min. Aviation, Aero.Res.Council Current Paper, CP626, 1963.

15. **Hall, J.G., Russo, A.L.,** 'Proc. First. Conf. Kinetics, Equilibria and Performance of High-Temperature Systems', Butterworths, London, 1959.

16. **Whitten, R.C., Poppoff, I.G.,** 'Physics of the Lower Ionosphere', Prentice-Hall, N.J., 1965.

List of Mathematical Symbols

I_A	current of atomic ions
I_m	current of molecular ions
k	Boltzmann's constant
m	mass of particle
n	electron density (cm^{-3})
Q	cross section of reactions
T	electron temperature (°K)
v	particle velocity
α	degree of dissociation
γ	ratio of specific heats
Δ	Debye length

Subscripts

$_h$	refers to properties of heavier gas
$_m$	refers to mean properties of mixture

Commentary on Chapter 5-2

R.M. Hobson

Experimental work such as that reported here by Dr. Dolder provides us with further insight into the problems of forming molecular beams of a required species distribution, and also into the problems associated with using the beam system as a 'frozen flow' sampling device for determining reactions and reaction rates associated with inelastic collision processes in the reflected shock heated gas.

In some situations, such as crossed particle beam collision experiments, the necessity of good resolution imposes a requirement of averaging data over several runs each of which must be of the maximum running time possible and be stable during this period. In this work the probe current profiles are employed to measure velocity, apparently giving good agreement with theoretical values. However, do these profiles give any indication of a stable beam during the total running time which appears to be about 1.2×10^{-3} sec. from the traces of his figure 5-2-2? This profile could be interpreted to imply that the beam initially maximises to a condition which is reasonably stable for a period of about 200 μsec, then falls to a lower level during another 200 μsec and fluctuates about this level for the remainder of the running time. Can the latter region be associated with shock reflections in the source region due to imperfect tailoring?

However the very significant feature of this work is in the fact that the mass spectrometer detector shows a seven fold increase in O concentration above what would be expected in the reflected shock heated gas. Spence and Dolder have developed a plausible expanation of this large increase in O concentration and they attribute it to a large collision cross section for dissociative ionization which may be associated with the vibrationally excited O_2. Unfortunately these collision cross sections have not, as far as I know, yet been measured experimentally. Thus it is difficult to assess whether or not the variation in cross section with temperature, shown in his figure 5-2-6 and matching the experimental results given here, does in fact overestimate the influence of vibrational excitation in dissociative ionization.

An alternative explanation may be sought if the flow is not completely frozen between the nozzle and the mass spectrometer source region. If shocks exist in the jet downstream of the nozzle or if there is severe skimmer-jet interaction leading to stand off shocks at the skimmer nose, then excited but undissociated O_2 in the beam might undergo dissociation in its path from nozzle to detector. However it does not appear to be likely that this could account for the 7 fold increase in O concentration. Nevertheless a recent communication from Dr. Skinner states that at higher stagnation temperatures in the shock tube he finds that losses occur at the first nozzle. He has now replaced this nozzle with a cone fabricated from brass shim and with a 35° half angle. This modification is shown in figure A5-2-1 which is taken from Skinner's drawing. The overall result has been to make the beam intensity steadier than it ever was before. In the work reported by Dolder the earlier arrangement, in which the first skimmer is treated as a free molecular flow device, is used.

Finally, the role of the detector should be considered in these measurements of degree of dissociation. This is obtained by relating the measured ratios of O^+ and O_2^+ formed by electron bombardment in the mass spectrometer source to the expression containing the collision cross sections of the various reactions leading to atomic and molecular ions. Assuming that the mass spectrometer shows no mass discrimination in analysing different ions from the same source region, then it only remains to look in detail at the ionizing region. The ion current at a fixed electron energy may be related to the source parameters as follows:

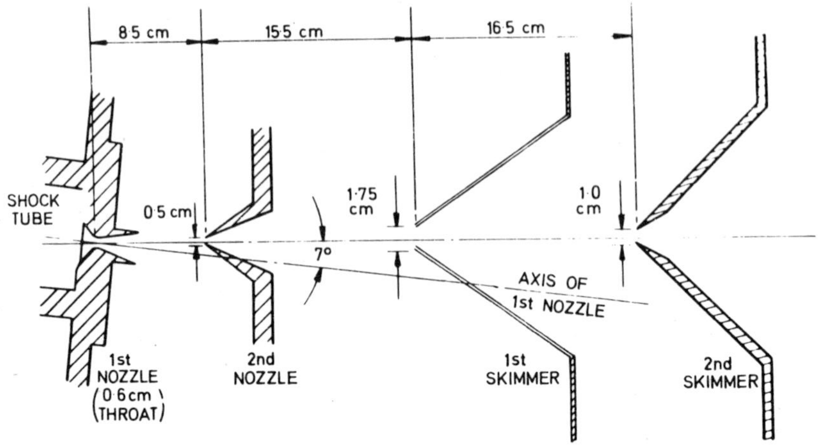

Fig. A5-2-1 Modified skimmer introduced by Skinner

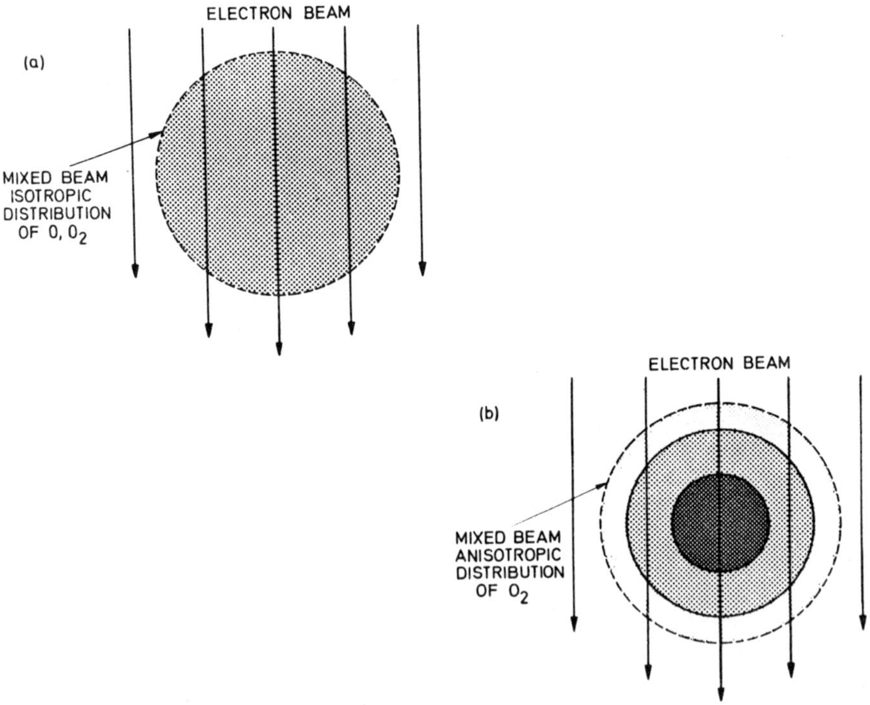

Fig. A5-2-2 Diagrams comparing the influence of an anisotropic distribution of heavy and light components (b) with those of an isotropic distribution (a) on ionization processes.

$$I_+ = \Phi A \ell Q_+ [N]$$

where Φ is the electron flux, A is the cross sectional intersection area defined by the intersecting electron and neutral beams, ℓ is the electron path length through the neutral gas, Q_+ the cross section for ionization and [N] the neutral gas density. Thus the cross sections Q'_+ and Q''_+ can be compared if Φ, A and ℓ are constant in both ionizing reactions leading to I''_+. This is the situation shown in figure C5-2-2(a) in which the electron beam sweeps a volume of O and O_2 in which the atoms and molecules are isotropically distributed. However in mixed beam flows the heavier component, in this case the O_2, is preferentially focussed on the beam axis and the consequent anisotropic distribution of O_2 across the neutral beam cross section is sketched in figure C5-2-2(b). In such a case the values of A and ℓ corresponding separately to the atom and molecule ionization processes will differ and the I'_+/I''_+ relationship cannot be directly related to the cross sections unless the radial distributions are known. Such could be the case in the experiment reported here and the O^+/O_2^+ ratio reported may be characteristic of the mass spectrometer source rather than of the degree of dissociation in the neutral beam.

It would be well worth while to resolve this problem. If Dolder's explanation is correct, then it implies that mass spectrometer detectors may not prove satisfactory in determining relative concentrations of neutral species in mixed beams, unless the degree of excitation and corresponding ionization cross sections are well known, and this is certainly not the case at present.

Part 6

Measurements using Ionization

6-1

Measurements of Dissociative Recombination Rates in Partially Ionized Gases in the Temperature Range 1000 to 3000°K

J. N. FOX AND R. M. HOBSON
Department of Pure and Applied Physics, Queen's University,
Belfast, N. Ireland.

Summary

A technique for the measurement of recombination rates at elevated gas temperatures is described. A shock tube is used to provide a high temperature gas environment, at a precise and predetermined temperature, into which a limited volume of ionization is introduced by means of a microwave or RF pulse. The subsequent decay of ionization with time is related to the dissociative recombination coefficient and its variation with gas temperature determined in the case of argon. Problems of measurement, associated with metastable atom concentrations, are emphasised and the possible effect of excited states of molecular ions on the temperature dependence in argon (found to be $T^{-3/2}$) is discussed.

Sommaire

On décrit une technique expérimentale permettant l'étude des vitesses de recombinaison à température élevée. On utilise un tube à choc pour fournir le gaz à haute température à une température bien définie et choisie à l'avance et on ionise par micro-onde ou impulsion HF un petit volume de gaz. L'amortissement de l'ionisation avec le temps est lié au coefficient de recombinaison dissociative et sa variation en fonction de la température est déterminée pour l'argon. Les problèmes de mesure liés aux concentrations d'atomes métastables sont examinés et l'effet possible des états excités des ions moléculaires stables sur la variation en fonction de la température pour de l'argon (évolution en $T^{-3/2}$) sont discutés.

Introduction

Experimental measurements of the rate of recombination over a temperature range between room temperature and a few thousand degrees are required. Such temperatures are found in the ionosphere where, for example, in the F_2 layer, the electron, ion and gas temperatures are about 1400°K. In the ionosphere, molecular ions are present and dissociative recombination can occur even though the pressure in this region is very low (10^{-5} - 10^{-6} torr). Biondi (1) has obtained values for the recombination of ions that are characteristic of this region, such as N_2^+, O_2^+ and NO^+, and these values are about 10^{-7} cm^3/sec at 300°K, which typifies dissociative recombination as the dominant process. However, such measurements have been carried out generally at temperatures around 300°K, rather than at the much higher temperatures associated with the ionosphere. This is particularly significant, as the temperature dependence of the recombination process still remains uncertain from the theoretical point of view, and there is little experimental evidence on which to base with confidence an extrapolation of the coefficient from room temperature values to about 1500°K. The afterglow of flames provides suitable high-temperature experimental situations, but it is difficult to vary the temperatures over a wide range without substantially changing the composition of the flame, and hence the recombining ions. Calcote *et al* (2) have, however, observed no change in the recombination coefficient in flames between 1700 - 2100°K. They suggested that the electron loss process was dissociative recombination by the reaction:

$$H_3O^+ + e \rightarrow H_3O^* \rightarrow H + H_2O ,$$

since H_3O^+ has been shown to be the dominant ion in the afterglow, and the high coefficient that Calcote *et al* measured supports this. Wilson and Evans (3), on the other hand, have observed recombination behind shock waves in argon containing dilute lean mixtures of hydro-carbons and oxygen, in which the above process is said to dominate, and they quoted a temperature dependence of $T^{-1.98}$ between 2500° and 5000°K. They also detected the independence of the coefficient to pressure, as expected.

Stein *et al* (4) also used a shock tube technique to study recombination in air, where the proposed reaction is:

$$NO^+ + e \rightarrow NO^* \rightarrow N + O$$

Their result, at 2900° K, is in reasonable agreement with that of Lin *et al* (5) at 5000°K extrapolated by a $T^{-3/2}$ dependence. Lin *et al* obtained their value of recombination coefficient by measuring the forward-going ionization process behind shock waves, and deducing the electron loss mechanism from the way in which the equilibrium ionization level was obtained. Other experiments have used the afterglow quenching technique, in which the electrons are selectively heated to a known temperature by microwaves, while the ions and neutral gas remain at room temperature. Chen, Leiby and Goldstein (6) found a temperature dependence in helium of $T_e^{-3/2}$ between 300° and 1500°K. However, it appears unlikely that the process measured in that case was dissociative recombination, as was pointed out in a later analysis by Ferguson *et al* (7) and Oskam and Mittelstadt (8) who have shown that the dominant process in helium at room temperature is probably collisional radiative recombination. In neon, Hess (9) reports a $T_e^{-0.25}$ dependence between 300° and 600° K and $T_e^{-0.4}$ between 900° and 2400° K. Frommhold and Biondi (10) report a best fit of $T_e^{-0.25}$ to measurements of recombination involving Ne_2^+ and N_2^+ in neon and neon-nitrogen mixtures. On the other hand Nygaard (11) found that $T_e^{-1.4 \pm 0.2}$ was the best fit to his measurements of dissociative recombination in neon afterglows.

It is not obvious why these measurements under apparently similar conditions should produce such different results. It is possible, however, that the state of vibrational excitation of the ionic species involved may be an important parameter and this could well differ from experiment to experiment depending, among other parameters, on the magnitude and duration of the electron heating pulse employed.

The work reported below provides a method of measuring recombination over a wide range in temperature, with the electron, ion and neutral temperatures being equal, and in gases that cannot be investigated using the technique of Wilson and Evans (3) in which the decay of electrons produced in the chemi-ionization processes behind a shock wave can be related to recombination with the molecular ions known to be formed in the chemi-ionization reactions. The results are compared with other measurements at room temperature, and the measured dependence is discussed in the light of the model of dissociative recombination of Bates and Dalgarno (12).

Apparatus

The gas being studied was heated in the shock tube by a shock wave travelling at between Mach 3 and Mach 5. The temperature of the shocked gas was between 1000° and 3000°K and was at a pressure of 30 to 100 torr for a typical initial pressure of 3 torr in the shock tube. Under these conditions there was no measurable thermal ionization in the gas, which acted purely as a heat bath in which charged particle experiments could be carried out. A volume of gas was ionized by a short pulse of microwave power immediately ahead of the shock front. This ionized gas was subsequently swept up into the shocked gas forming a well defined slice of ionization which moved with time from the shock front to the contact surface region. The charged particles constituting this ionized slice quickly reached the characteristic temperature of the shocked gas and, under these conditions of pressure, ambipolar diffusion was not a dominant process. The subsequent decay in number density was due to recombination reactions at the temperature of the shock heated gas, and could be measured from the levels of ionization detected at stations along the shock tube.

The 1 cm internal diameter shock tube is shown schematically in figure 6-1-1. The vacuum envelope was constructed from Q.V.F. glass sections, and vacuum seals between these were achieved with Dowty bonded seals. The diaphragm material was Melinex polyester film and, for the range of shock velocities covered, thicknesses between 0.00025 and 0.001 inches were used. Before each measurement, the shock tube was evacuated to about 3×10^{-7} torr and the combined leak and outgassing rate under closed off conditions was 3×10^{-6} torr/min. The experiment was completed within one minute of closure and, as the 'spectroscopic grade' gas used was rated at an impurity level of 4 parts per million, the overall impurity level during the experiment was less than 7 p.p.m. In order to obtain slow shocks of about Mach 3.0, nitrogen was used as the driver gas, and for faster shocks of about Mach 5.5, hydrogen was used. Intermediate velocities were obtained by mixing these two gases in suitable proportions. Small changes in shock velocity were obtained by using different thicknesses of Melinex. The method of rupturing the diaphragm was to increase the pressure in the high pressure section until the diaphragm burst, and the velocities of the resulting shocks were found to be reproducible within 0.1 of a Mach number.

Downstream of the diaphragm, a 1 cm i.d. precision bore glass tube formed the shock tube proper. This tube was separated from the vacuum envelope so that it could pass through a section of waveguide as shown in figure 6-1-2. In this region ionization was achieved by a pulse of microwave power, 0.5 microseconds long and 2.5 kW in magnitude delivered from a pulsed magnetron. The magnetron was triggered from a signal produced by the passage of the shock front across a platinum film thermal transducer, stationed upstream of this breakdown section. By using suitable time delays, ionization could be created at any instant relative to the shock front, and usually the gas was ionized just ahead of the shock front. Downstream of this section, further platinum films monitored the shock front velocity, which determined the temperature of the shocked gas, while double probes at stations along the tube measured the decay in ion density. These probes each consisted of two platinum wires 0.010 inch diameter stretched diametrically across the tube. Glass sleeves insulated the probes so that only a 0.5 cm length of probe in the centre of the tube was exposed. This restricted ion collection to just the shocked gas, so that no contributions were made by the boundary layer.

A sufficiently large voltage was applied between the probes to ensure that the current flowing round the circuit was equal to the saturated ion current to one probe. Although conditions were such that in a static plasma ion-neutral collisions would take place in the electrostatic region of influence or sheath around the probe - and thereby would be expected to reduce the flux of charged particles to the probe to a level well below that of the random ion current through an equivalent area in the plasma - it has been shown (13) (14) that because of the flow, the current collected i_p could be related to the ion density n by the expression:

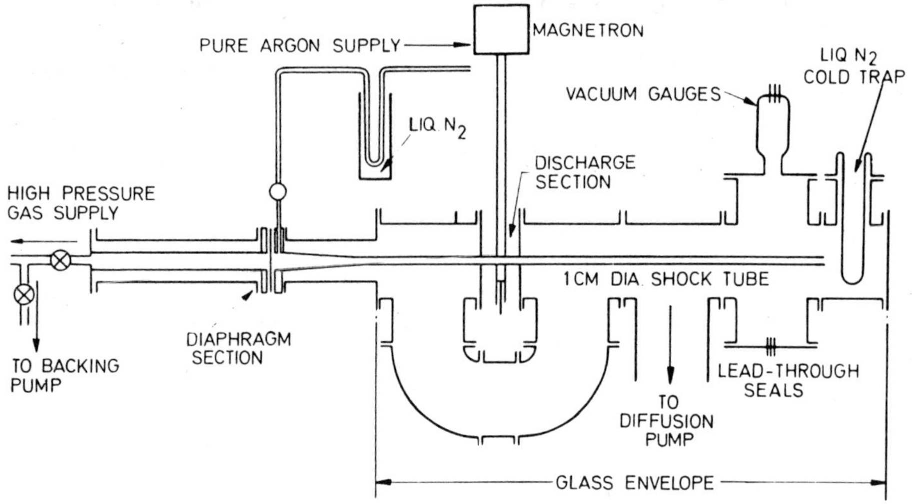

Fig. 6-1-1 1 cm diameter shock tube

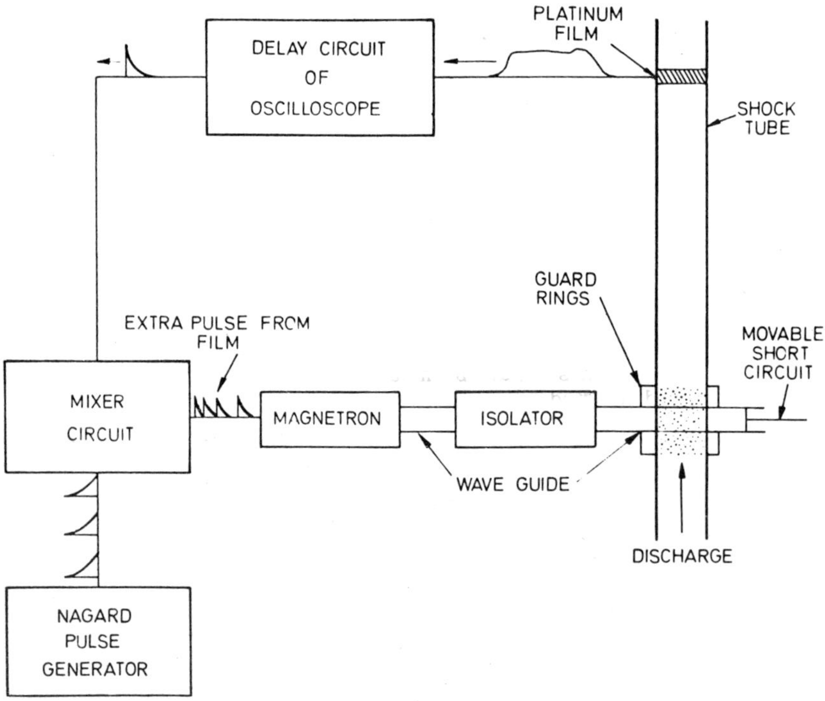

Fig. 6-1-2 Block diagram of breakdown system

$$i_p = K \frac{1}{4}\left[n e \sqrt{\frac{8kT}{\pi M}}\right] A_p \qquad \text{(Eq. 6-1-1)}$$

where T is the ion temperature, M the ion mass and A_p is the probe surface area. K is a proportionality constant, which has been found experimentally to be approximately 3 in the case of high Mach number shock waves. However, as a direct calibration is not at present available for the lower Mach number shocks used in the experiments reported here, it will be assumed, for the purpose of processing the raw probe data, that K = 1.

Results

It has been shown (15) that the ionized gas does in fact remain as a well defined slice during its passage from the shock front towards the contact surface, and that the path taken by the slice is the particle path predicted by Mirels (16), In the initial discharge, high temperature electrons and, in inert gases, atomic ions, will be created. When dissociative recombination is measured, it is assumed that the electrons have cooled so that electron, gas and ion temperatures are equal and that all atomic ions have formed molecular ions by the three-body process:

$$2A + A^+ \longrightarrow A_2^+ + A$$

At sufficiently high pressures and collision frequencies this reaction forming the molecular ion will be rapid. Phelps and Brown (17) and Holt *et al* (18) have shown that the frequency of conversion of helium ions to He_2^+ is about 50 p^2 where p is the pressure in torr. The time t, after which molecular ions dominate, is given by:

$$t = \frac{1}{\beta} \log \frac{\beta}{\alpha A_o}$$

where β is the frequency of conversion, α is the recombination coefficient and A_o is the initial number density of atomic ions. If β for the gas under test is approximately the same as for helium, then it is reasonable to suppose that for a temperature T°K:

$$\beta = 50 \; p^2 \left[\frac{T}{300}\right]^{\frac{1}{2}}$$

since β will be proportional to the ion collision frequency which is a function of $T^{\frac{1}{2}}$. Therefore, for 1000°K and 30 torr, β is 8 x 10^4 and t is 26 microseconds. The time allowed before measurement commenced in the work reported here - namely 150 microseconds - appears to have been sufficiently long to have ensured that molecular ions predominated.

Electrons cool quickly to the first excitation potential of the atom by inelastic collisions and then more slowly by elastic collisions with neutral atoms. Biondi and Brown (19) allowed 50 microseconds for the electrons to cool, and the linearity of their $^1/n$ versus time plot confirms that this time delay was sufficient. At the higher temperatures involved in our experiment, it would be expected that the collision frequency would have been greater so that the electrons would cool in a shorter time. In the 150 microseconds available, it was reasonable to assume that the electrons were indeed cooled to the gas temperature before measurements were made.

The first measurements were made in argon, and typical signals received from the probes and recorded on Tektronix 555 oscilloscopes are shown in figure 6-1-3. Traces 2 to 4 are the signals from the first four probes. Each trace started 80

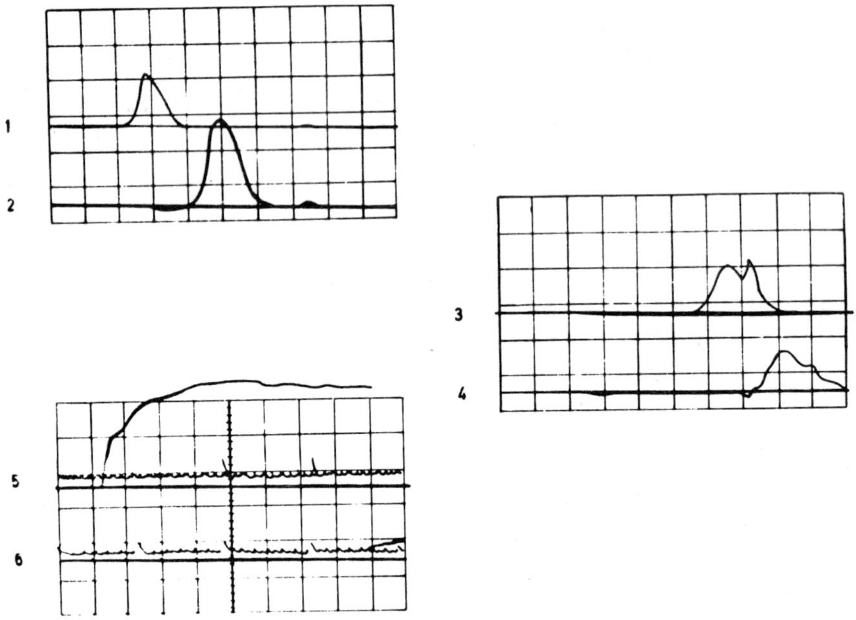

Fig. 6-1-3 Responses of double probes

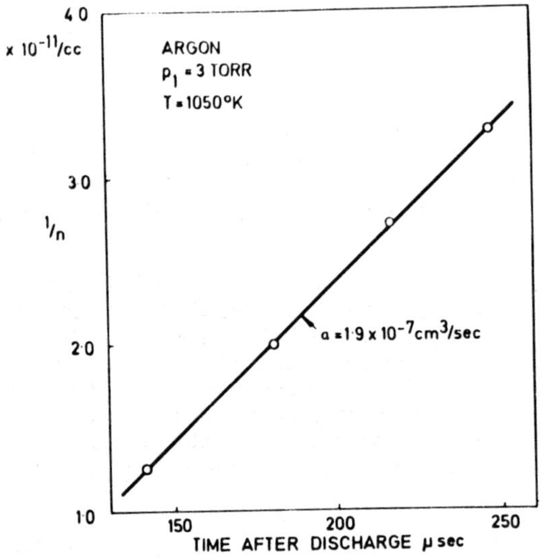

Fig. 6-1-4 Typical 1/n versus time plot

microseconds after the ionizing pulse, and the sweep speed was 20 microseconds per large division. The vertical scales are 5 microamps per large division for the first probe and 2 microamps per large division for the remaining three probes. The signal of trace 3 showed an extra peak after the main peak. This was caused by turbulence near the contact surface. Traces 5 and 6 showed the responses of the platinum films on the same time scale as the probe signals, together with the calibrating signal from the Tektronix 180 Time Mark generator.

For this shot the driver gas was nitrogen, the diaphragm 0.0005 inch Melinex, and the initial pressure 3 torr. The velocity of the shock was Mach 2.9 and the shocked gas temperature 1050°K. The ion density was measured from the maximum of the signals, and the raw probe data were related to ion number density using equation 6-1-1. A plot of $1/n$ against time is shown in fig. 6-1-4 and the slope of the straight line is 1.8×10^{-7} cm^3/sec, which is the recombination coefficient. The results of measurements made between 1000° and 3000°K are shown in fig. 6-1-5 The variation of the coefficient clearly follows a $T^{-3/2}$ dependence over the range covered. Extrapolation of the line to 300°K gives a value of 1.2×10^{-6} cm^3/sec to be compared with the value of 6.7×10^{-7} cm^3/sec obtained by Oskam and Mittelstadt (8) using the cavity technique of Biondi and Brown (19). This value, within a factor of two of the very accurate room temperature result of Oskam and Mittelstadt indicates that the proportionality constant K in equation 6-1-1 should have had the value 1/2.

The results obtained when krypton was used as the test gas have, up to the present, been less conclusive. During the afterglow, the ion density first began to increase and then continued to decrease as shown in figure 6-1-6. The prominence of this change in the character of the decay can be seen to be dependent on pressure. The discontinuity appeared to be dependent also, not on the time interval after the cessation of the discharge, but on the time duration of the slice in the shocked gas. The only explanation that seems to fit these observations is that the increase in ionization was due to the action of metastables, possibly:

$$Kr^m + Kr^* \longrightarrow Kr^+ + Kr$$

At higher pressures the rate of this process would be quicker, hence the greater effect at these pressures. The rate would also be quicker at high temperatures, hence the dependence on the time duration of the slice in the shocked gas.

Since microwave discharges are known to produce a high metastable population, it was decided to ionize the gas using an R.F. system operating at about 10 Mc/s. The decay observed from this, although the number density was too low for recombination to dominate, did not show the secondary ionization process seen with the microwave system. Also the rate of decay as indicated in figure 6-1-6 was much greater than the decay after the microwave discharge, indicating that the ionization process responsible for the increase was affecting the whole of the decay.

Conclusions

Bates (20) has shown that the recombination coefficients of about 5×10^{-7} in the noble gases, other than helium, are consistent with the process being dissociative recombination described as follows:

$$A_2^+ + e \longrightarrow A_2^* \longrightarrow A^* + A$$

Bates and Dalgarno (12) state that if the process $A_2^* \longrightarrow A^* + A$ were the controlling rate for the overall reaction, then a temperature dependence of $T^{-3/2}$ would be most likely. In this case auto-ionization via $A_2^* \longrightarrow A_2^+ + e$ would be more probable than stabilization via $A_2^+ \longrightarrow A^* + A$.

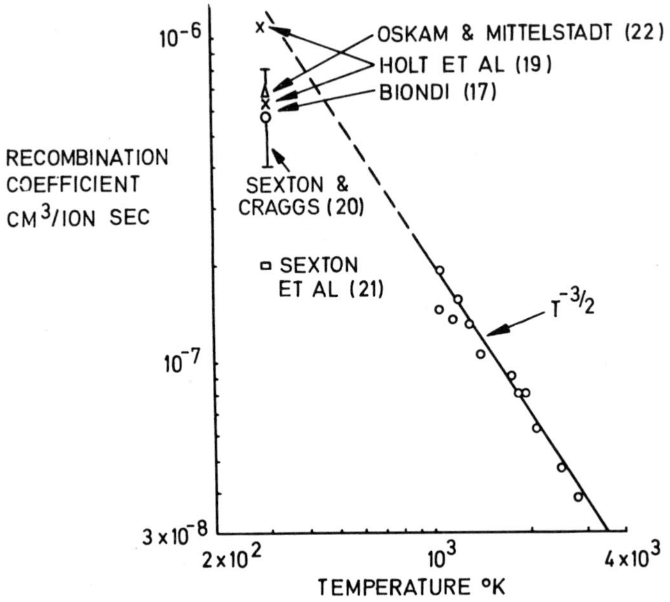

Fig. 6-1-5 Recombination coefficient as a function of temperature

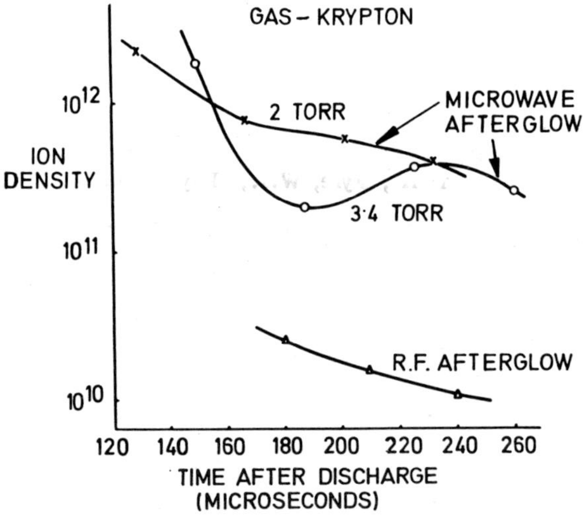

Fig. 6-1-6 Number density decay in krypton

If, however, the rate limiting process were $A_2^+ + e \longrightarrow A_2^*$ then at high temperatures a $T^{-3/2}$ law might be anticipated after a more complex initial dependence.

The results given here show that the $T^{-3/2}$ dependence is followed in the case of argon, and thus imply that autoionization is more probable than stabilization. The somewhat empirical theoretical model of the process, involving potential energy curve crossings, would suggest that the transitions from Ar_2^+ may be to an energy level of Ar_2^* which corresponds to a high-vibrational energy state of the molecule taking part in the recombination process. This is a significant parameter and may well explain why the preferential electron heating experiments in other noble gases have given inconsistent results. However, the $T^{-3/2}$ dependence measured in the present work in argon should not, for the same reason, be automatically applied to other noble gases, in which excited states differing from the argon excited states may be present in the recombination process.

It is anticipated that current work on the mechanism of collection, and work using the R.F. breakdown modification will soon permit data on the recombination coefficients for krypton and other inert gases to be obtained.

Acknowledgments

This work has been supported by the Ministry of Aviation under agreement PD/58/05. One of the authors (J.N.F.) is grateful to the Ministry of Aviation and Clarke Chapman and Company Limited, Gateshead, for their joint financial support during the course of this work.

We are also grateful to Mr. A. Cunningham for his assistance in the experimental work.

References

1. **Biondi, M.A.,** Ann. de Geophys. 20, 34, 1964.

2. **Calcote, H.F., Kurzins, S.C., Miller, W.J.,** Symp. Combust. 10th Cambridge Univ. 1964, 605, 1965.

3. **Wilson, L.N., Evans, E.W.,** Chem. Phys. 46, 859, 1967.

4. **Stein, R.P., Scheibe, M., Syverson, M.W., Shaw, T.M., Gunton, R.C.,** Phys. Fluids, 7, 10, 1641, 1964.

5. **Lin, S.C., Neal, R.A., Fyfe, W.I.,** Phys. Fluids, 5, 1633, 1962.

6. **Chen, C.L., Leiby, C.C., Goldstein, L.,** Phys. Rev. 121, 1391, 1961.

7. **Ferguson, E.E., Fehenfeld, F.C., Schmeltekopf, A.L.,** Phys. Rev. 138, A381, 1965.

8. **Oskam, H.J., Mittelstadt, V.R.,** Phys. Rev. 132, 1445, 1963.

9. **Hess, W.,** Z. Naturforsch., 20, 451, 1965.

10. **Frommhold, L., Biondi, M.A.,** Gaseous Electronics Conf. Georgia 1966, unpublished.

11. **Nygaard, K.J.,** Phys. Letters 22, 56, 1966.

12. **Bates, D.R., Dalgarno, A.,** 'Atomic and Molecular Processes', Acad. Press Inc., New York, 1962, p. 264.

13. **Fox, J.N.,** Thesis, Q.U.B., 1966.

14. Chap. 6-4.

15. **Fox, J.N., Hobson, R.M.,** 'Proc. VII Int. Conf. Ionization Phenomena in Gases, Belgrade. 1965.

16. **Mirels, H.,** Aerospace Corp. Rep. No. TDR-169, (3230-12) TN-5, 1962.

17. **Phelps, A.V., Brown, S.C.,** Phys. Rev. 86, 102, 1952.

18. **Johnson, R.A., McClure, B.T., Holt, R.B.,** Phys. Rev. 80, 376, 1950.

19. **Biondi, M.A., Brown, S.C.,** Phys. Rev. 76, 1697, 1949.

20. **Bates, D.R.,** Phys. Rev. 77, 718, 1950.

Commentary on Chapter 6-1

K.W. Michel

As pointed out by Prof. Hobson, these investigations are of relevance to upper atmosphere physics. They contribute, however, also to the understanding of some astrophysical processes. Shock waves, responsible for the heating of the corona of the sun, also travel through pre-ionized gases (3). Hence this technique, a variation of Dr. Wray's (1) and Dr. Hartunian's (2) methods, deserves additional merit as a small-scale simulation of such phenomena.

As opposed to other plasma-shock techniques, it should provide nearly stationary conditions. Deviations from ideal shock behavior, however, might result even from small heat sources in the wake of the shock. The slice of pre-ionized gas has a somewhat different enthalpy, which might give rise to flow instabilities more significant even than boundary layer phenomena alone (4). Being aware of the disturbing influence of small heat sources behind the incident shocks, one should perhaps control, in this case, the amount of energy that is transferred into both ionization and electronic excitation by the discharge.

Concerning the interpretation of the data, it appears that for more exact evaluation, the dissociation equilibrium of the dimeric ionized species must be considered. Their bond dissociation energy may be less than 1 eV (ne) (5) and appreciable decomposition is calculated at these temperatures and pressures.

Prof. Hobson has shown that the probe technique is a useful tool in measuring over-all electron concentrations. Naturally, supplementation of these data by direct observation of individual ion concentrations perhaps by spectroscopic methods, would be desirable for confirming the detailed kinetics.

1. **Wray, K.,** J. Chem. Phys. 44, 62, 1966.

2. **Hartunian, R., Thompson, W., Hewitt, E.,** J. Chem. Phys. 44, 1765. 1966.

3. **Biermann, L.,** Naturwissenschaften 33, 118. 1946.

4. **Stenlow, R.A., Cohen, A.,** Phys. Fluids 5, 97, 1962.

5. **Biondi, M.A.,** Ann. de Geophys. 20, 34, 1964.

6-2

Measurement of Radical Concentrations in Reacting Gases using Electron Spin Resonance

K. HOYERMANN and J. WOLFRUM
Institut für Physikalische Chemie, Universität Göttingen, Germany.

Summary

To measure quantitatively atom concentrations with ESR, the influence of microwave power, saturation, and modulation amplitude on line width and first moment are studied. The results are applied mainly to elementary reactions ($O + H_2O + C_2H_2$, $O + NH_3$, $H + C_2H_2$) whose rate constants and temperature dependence are determined with high accuracy. The flow system and sampling device employed are described in detail.

Sommaire

Pour mesurer quantitativement des concentrations atomiques au moyen de la résonance du spin électronique, on étudie l'influence de la puissance des micro-ondes, de la saturation et de la modulation d'amplitude sur l'épaisseur des raies et le premier moment. Les résultats sont appliqués surtout aux réactions élémentaires ($O+H_2O + C_2H_2$, $O + NH_3$, $H + C_2H_2$) dont on détermine avec précision les constantes de réaction ainsi que leur variation avec la température. On décrit en détail le montage utilisé pour l'écoulement et le prélèvement des échantillons.

Introduction

For a detailed understanding of reaction kinetics it is necessary to know the absolute rate constants of elementary reactions. In most cases these are free radical or atom reactions, so that the measurements required are the concentrations not only of stable reactants and products but also of radicals. While precise determination of the concentrations of stable species is possible using gas chromatography and mass spectrometry, this is much more complicated for atoms and free radicals. Wrede gauges, catalytic probe detectors, chemical exchange and titration techniques, photometric devices and rapid sampling mass spectrometers have all been used for this purpose, but their individual shortcomings,not discussed here, must be kept in mind. Perhaps the most general, but not necessarily the most accurate device is the mass spectrometer with molecular beam sampling. With this, low electron energies are necessary if radicals areto be resolved from dissociative ionisation products, with the result that the effective sensitivity and accuracy are considerably reduced. Difficulties arise also from mass separation effects for very light components such as H-atoms.

To these methods of measuring radical concentrations a new tool has recently been added: the electron spin resonance spectrometer, which measures the characteristic magnetic moments of atoms and radicals. The method is highly specific and sensitive, allows essentially unambiguous identification with little interference, has considerable generality, and gives information about paramagnetic ground and excited states. While the study of free radicals by electron resonance methods in liquids, matrices of frozen molecules, and solids has become one of the most im-

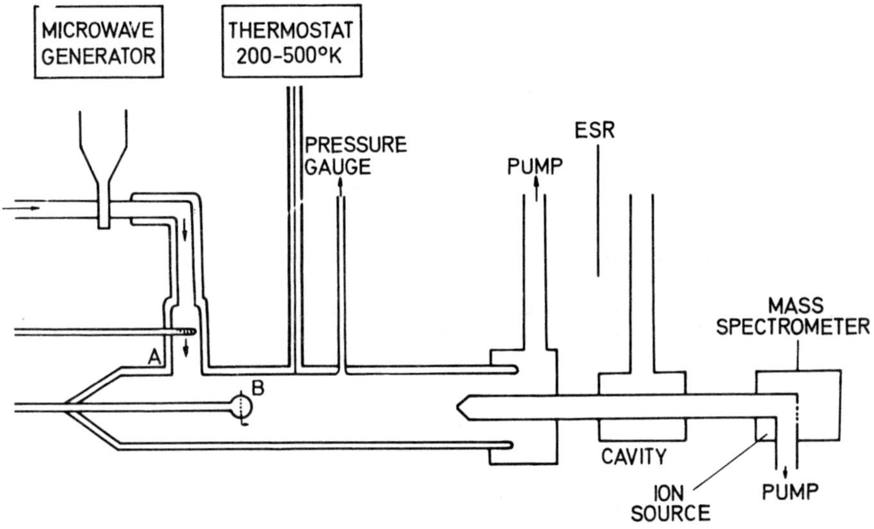

Fig. 6-2-1 Schematic diagram of the apparatus

portant areas of physical chemistry, much less work has been done in the gas phase. Nevertheless, since the first electron resonance studies of the paramagnetic gases O_2 NO, and NO_2 were carried out by Beringer and Castle (1) (2), many more atoms (for example H, O, N, P, I, F, Cl, Br, S) (2 through 11) and free radicals (OH, NF_2, SH, SeH, TeH, SO, ClO, BrO, SeO, NS etc) (12 through 17) have been detected in the gas phase.

There are two main difficulties with ESR techniques. The relationship between the measured ESR signal intensity and a radical concentration is theoretically clear, but complex and dependent on many parameters. The second problem is that of getting the labile species into the microwave cavity in the gap of the magnet. Solution of this second problem depends on the special features of the radical source, andtwo alternatives exist.

Nalbandyan *et al* (18 through 21) have arranged for the whole of the diffusion flame to be surrounded by the ESR resonator, and in this way have determined the dependence of the overall concentrations of atoms and radicals (H, O, OH, S, SO etc) on the composition of the mixture in rarefied flames. Obviously as it was not possible to measure conditions at any particular point, interpretation of the results was complicated by the influence of the high gas temperatures on the ESR signal lines, and accurate radical concentrations were unobtainable.

Carrington *et al* (17) designed a cylindrical quartz mixing cell (46 mm i.d.), in which radicals could be produced by adding a second gas to the products of a high frequency discharge, for example SO from O + OCS. By this technique it was possible to obtain valuable information about a number of properties of diatomic radicals (bond lengths, nuclear hyperfine constants, Λ-doubling intervals etc), but no kinetic investigations were carried out.

In practice, however, the source of free radicals is often situated outside the cavity, for example in burners, engines, flow systems etc. In these cases samples have to be introduced into the resonator by a suitable sampling device. It is advisable to have a low pressure in the resonator so as to reduce radical recombination; however, a molecular beam would not be convenient because its density would be too low. Attention must therefore be paid to possible losses of radicals at the wall and in the gas phase. However the pressure reduction makes possible the measurement of local concentrations. By the expansion, the gradients in the sampling tube become small, so that the conditions in the cavity are practically constant.

Although many theoretical papers have been published about the determination of absolute concentrations of radicals in the gas phase, reported experience in analysing reacting gas mixtures is scarce. More information is needed about the influence of saturation, 'Modulation broadening', calibration factors, etc.

Our experimental work designed to circumvent these problems was based on the measurements of Westenberg and de Haas (22).

Experimental Method

A schematic diagram of the apparatus we used, which consisted of a glass gas-flow system and a sampling probe designed to permit simultaneous application of ESR and mass spectrometry is shown in figure 6-2-1.

Flow System

The carrier gas (Ar, He) was mixed with a few percent of the gas to be dissociated (N_2, H_2) and passed into a quartz tube (12 mm i.d.) through a microwave discharge cavity, operating at 2450 Mc/sec supplied by a Raytheon PGM-IOXI generator. Ground state oxygen atoms (O^3P) free of molecular oxygen were generated by dissociation of N_2 and subsequent titration with NO (23) at injector A. From the discharge the flow passed through a multiple bend (not shown on figure 6-2-1) Pyrex tube into the main flow tube (25-35 mm i.d.). This led to reactivation of electronically and vibrationally excited species and avoided photolytic effects at A (24)(25). The whole system was double walled and could be thermostatically controlled between 200° and 500°K, gas temperature being uniform within 1°C over the 15 cm test section, monitored by a Pyrex coated Pt-resistance thermometer.

Reactant gases (H_2, C_2H_2, NH_3) were added through a movable injection tube. For obtaining rapid mixing of the reactants with the main flow, a sphere (B, 10 mm o.d.) with a ring of small holes was located at the downstream end of the tube. As could be monitored by the mass spectrometer, mixing was in fact complete within less than one millisecond.

The flow rates of the gases were controlled by stainless steel needle valves, and measured either by calibrated capillary flow meters or - if the flow rates were below 1.0 cm^3 NTP/sec - by timing a measured pressure rise in a known vacuum. All pressures were measured with an absolute (diaphragm-type) pressure gauge (MCT I, MAT Bremen).

At low pressures (1-2 torr), surface reactions and back diffusion become significant with the fast flow technique. In most cases, therefore, we worked with higher pressures (5-15 torr) and large flow velocities (10-20 m/s), so that the longitudinal diffusion corrections and the pressure drop were negligible. The mean time required for diffusion from the wall to the axis, where the probes were positioned, was then less than the reaction time: the time scale being well-defined by 'plug-flow' conditions existing since the test section was too short for a laminar flow profile to develop.

The high sensitivity of the ESR spectrometer (to 10^{11} spins/cm^3) and the mass spectrometer, to a mole fraction of 10^{-7} enabled us to work at very low atom concentrations, below 10^{-3} mole fraction, so that the influence of third-order combination reactions could be neglected even at high pressures.

Sampling Tube

The sampling tube was made of quartz (o.d. 10 mm) with a hole specially constructed so that contact of the gas with it at high pressure would be minimal and so that mass separation of the gases would not occur. Its behaviour was monitored with the mass spectrometer.

Before each measurement, the whole flow system including sampling tube was cleaned with a 5-8 percent solution of HF, and dried at 200°C. Neither treatment with boric acid etc.(26) nor the use of electrical discharge (27) was considered suitable. As the magnet and the cavity of the ESR could be moved along the sampling tube, the atom losses could be determined to the 1st order, less than 10% at a distance of 50 cm) so that an extrapolation to the probe tip appeared to be feasible. Any influence of the sampling tube on the kinetics to be measured could be seen from the consistency of the ESR and mass spectrometric measurements, each giving independent data.

Relation Between ESR Signal and Atom Concentration

The principle of ESR spectroscopy is as follows. In bringing a paramagnetic species, characterised by the total magnetic moment (μ) and the total angular momentum (J), into a strong magnetic field (H_o, 3-10 k Gauss) the degeneracy is removed. A ladder of 2J + 1 energy levels appears. By applying a microwave field (H_1 frequency 9×10^3 MHz) to this system, transitions between these levels are induced. Magnetic resonance occurs at the field strength H_o. The energy absorbed by this system is proportional to the employed power (P) of the H_1- field, the 'line-strength' (Q'_{at}) and the number of absorbing particles (N). The absorption line is modulated by a 100 kHz-field with an amplitude (MA) for better sensitivity. This signal (S) is correlated to the first moment (I), a measure of the absorbed energy, by the integral $\int^{\infty} (H-H_o)\, S\, dH$, where H_o is the magnetic field at the resonant condition. The theoretical relationship between the number of atoms (N) and the ESR-signal has been studied by several authors (28-31) (22) and is given by:

$$N = Q'_{at} \cdot p^{-1/2} \cdot MA^{-1} \cdot I \qquad \text{(Eq. 6-2-1)}$$

Saturation

At low energy, the first moment increases as $I \sim p^{1/2}$. At higher power the saturation occurs as predicted by theory: the line is strongly broadened while peak height decreases. In order to correlate atom concentrations to a standard - here O_2 in the state K=1, J=2, M=0 $\longrightarrow$ 1 called F-line(32) - no saturation must be present. In their parent gases this is the case for N, H, O atom at microwave powers of 0.1, 0.2, 100 m watt, respectively. This means that the decisive spin-lattice relaxation time T_1 for H-atoms, for instance, is of the order of 5×10^{-6} sec. This time is essentially short compared to the velocity of the gas flowing through the cavity.

Broadening by Amplitude Modulation

In order to increase the sensitivity of the ESR spectrometer, the H_o-field is modulated by a 100 kHz field, having an amplitude of MA. At low values of MA the absorption signal is differentiated by this 100 kHz field. Upon increasing the MA the signal height goes up until a maximum is reached at $(MA)_{max}$. It falls off again at higher values. This is demonstrated for H-atoms in figure 6-2-2. Theory shows that the first moment is proportional to $w^2.h$ if MA << $(MA)_{max}$:

$$I \sim w^2.h \text{ if } MA \ll (MA)_{max} \qquad \text{(Eq. 6-2-2)}$$

This is realized experimentally, numerical integration is not better. This moment is proportional to w. h at MA = $(MA)_{max}$:

$$I \sim w.h \text{ if } MA = (MA)_{max} \qquad \text{(Eq. 6-2-3)}$$

But the measured values of peak height and width show that this maximum is not very sharply defined, i.e. with a variation in peak height of 10 percent the width varies between 1.3-3.4 (arbitrary units) - more than a factor of 2.5. An evaluation using equation 6-2-1 and 3 has a great uncertainty if measurements are performed at MA = $(MA)_{max}$. The advantage of measuring around $(MA)_{max}$ is the higher sensitivity. One can measure at $(MA)_{max}$ if the same modulation amplitude is used when measuring the calibration factor of the atom relative to the standard. An additional pressure broadening is not to be considered if one always maintains the same pressure. For exact measurements of the atom concentrations one needs the calibration factors Q'_{at} for the different atoms if all other parameters have the same values. In this way one can avoid major errors. (The values of $(MA)_{max}$ for N, H, O, O_2 are: 0.01, 0.06, 2.6, 1.3 Gauss).

Although there are already two papers (22) (23) available a determination, especially for the H-atoms, has to be repeated as shown below.

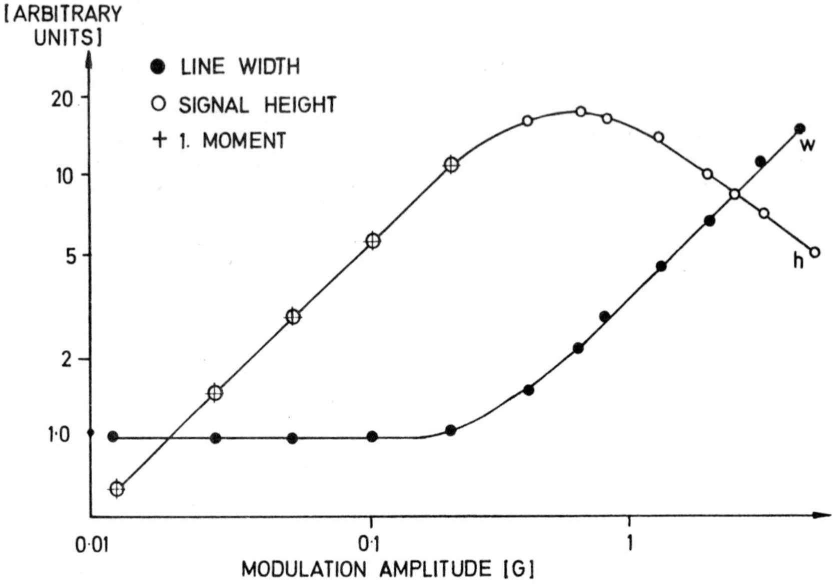

Fig. 6-2-2 The ESR signal plotted as a function of the modulation amplitude

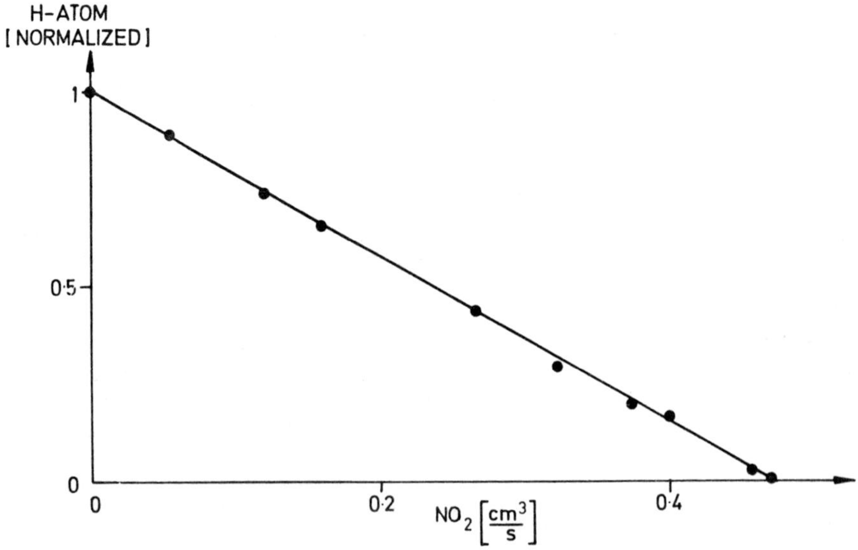

Fig. 6-2-3 Disappearance of H-atoms during NO_2 titration

Calibration Factors

Relating a ESR-signal to a certain quantity of atoms the number (N) of atoms must be known exactly. For the atoms H, O, N titration reactions are used.

H-atoms can be determined by titration with NO_2. But besides the 'modulation broadening', two difficulties arise when titrating the H-atoms with NO_2: (a) the titration curve-H-atoms plotted against added NO_2- is not a straight line as expected. Small H-atom concentrations, at large amounts of NO_2 consumed, give a relatively high signal. (b) There is a disturbing large uncertainty of 1-1.5 in the value of NO_2/H[34,35]. As we suspected that saturation might have occured in the previous experiments, we added 1 percent O_2. In this manner we obtain the expected titration curve (Fig. 6-2-3).

By using the mass spectrometer at the same time we are able to determine that the stoichiometry of NO_2/H is 1.49. This supports the earlier assumed mechanism:

$$H + NO_2 \longrightarrow NO + OH$$

$$OH + OH \longrightarrow H_2O + O$$

$$O + OH \longrightarrow O_2 + H$$

(The NO produced was probably responsible for the confusing behaviour of the titration curve of Westenberg, de Haas (33).

Thus we get for the calibration factor Q_H of H-atoms (1 line) relative to O_2 (F line) (Q_H is defined as Q'_H/Q'_{O_2}) a value of $Q_H = 6.3 \times 10^{-3}$.

The determination of Q for the O- and N-atoms is relatively simply done with our experimental apparatus on the basis of the equation:

$$N(^4S) + NO(^2\pi) \longrightarrow N_2(^1\Sigma) + O(^3P)$$

The NO consumed at the titration point controlled by ESR is a quantitative measure of the N-atom and the O-atom concentration. So the Q_o for the six-line-composite of the O-atom relative to the O_2 (F-line) is $Q_o = 3.7 \times 10^{-3}$.

Examples

In order to demonstrate briefly the usefulness of the method we will discuss some elementary reactions.

The Reaction of H_2 with O

This important reaction in hydrogen and hydrocarbon oxidation has often been studied (36) (37), however, the accuracy was not very high because of the complexity of the reactions when using O_2 and H_2 as primary reactants.

As described in 2.1, we measure with the ESR simultaneously the decrease in O-atom and the increase in H-atom concentration as a function of the distance between the sampling hole and the H_2 inlet (B). (Simultaneous measurement of the H-atom concentration was not possible with the photometric method of Clyne and Thrush (35) on account of the overlapping of some spectral lines and in the ESR studies of Westenberg and de Haas (38) due to great atom losses at the wall). Our finding is:

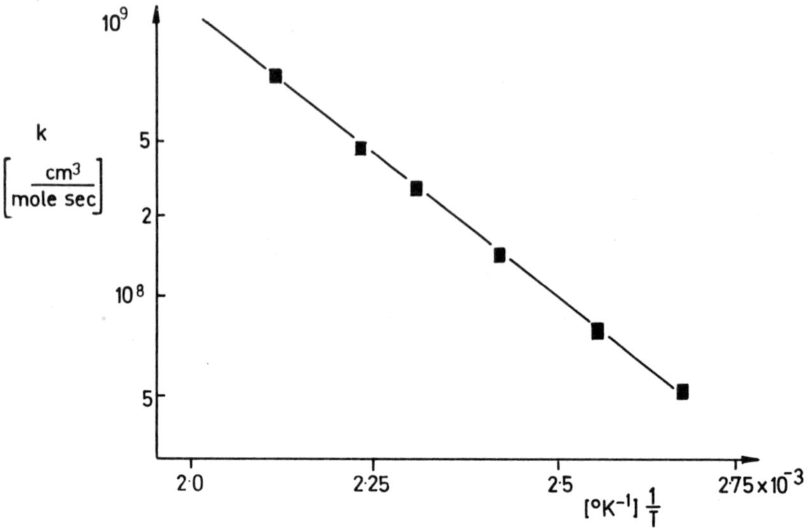

Fig. 6-2-4 The rate constant of the reaction $O + H_2$ as a function of temperature

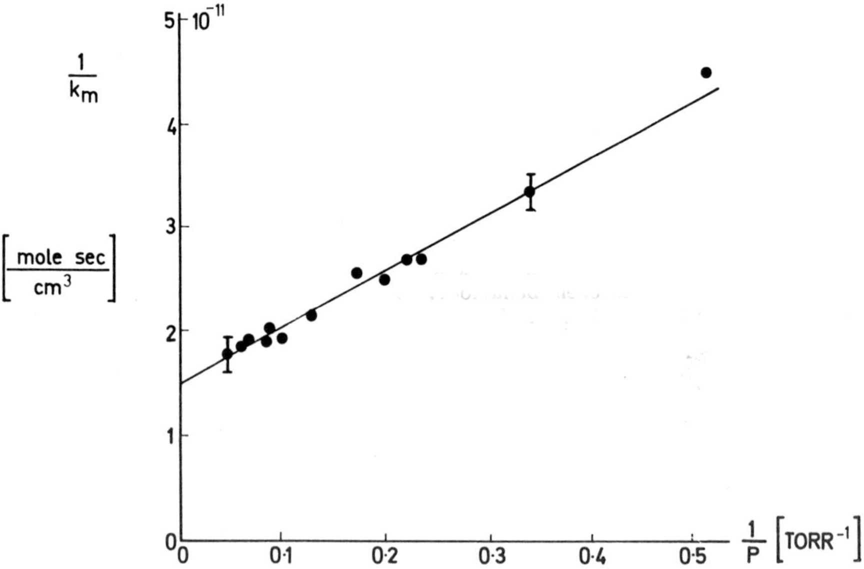

Fig. 6-2-5 Variation of the measured rate constant k_m with pressure in the reaction $H + C_2H_2$

$$-\frac{d[O]}{dt} = \frac{d[H]}{dt}$$

which is consistent only with the mechanism:

1) $O + H_2 \longrightarrow OH + H$

2) $O + OH \longrightarrow O_2 + H$

This can be shown independently with the mass spectrometer, for in addition to O_2 and H_2 no other products, such as water etc., are detectable

The rate constants of this reaction (1), measured at several temperatures, fit the equation $k_1 = 1.30 \times 10^{13} \exp - \frac{9400}{RT} \frac{cm^3}{mole\ sec.}$ in good agreement with Clyne and Thrush (38) (39) who used a quite different method. The temperature dependence is shown in Arrhenius form in figure 6-2-4.

The reaction vessel is easily replaced by a flat burner. Measurements on a hydrogen-oxygen flame yield a rate constant at 1000°K, which is in agreement within a factor of 2, if extrapolated with the given expression for k_1. Furthermore the independent findings with the mass spectrometer and the ESR and the agreement with Clyne and Thrush indicate that the concentration measurements with this sampling system are correct.

Reaction of C_2H_2 with O

Many studies have been made of the reactions of oxygen and hydrogen atoms with hydrocarbons. Photochemical methods lead usually to relative rate constants. Absolute rate constants can be obtained in discharge-flow system, but further complications arise here as in many other cases from consecutive reactions. Since under our conditions the reaction $O + OH \longrightarrow O_2 + H$(see above)is very fast, simultaneous measurement of O- and H-atom concentration by ESR gives valuable information about mechanisms.

The reaction of acetylene with atomic oxygen was extensively studied by Arrington *et al* (40) in a flow system similar to our using mass-spectrometric and photometric techniques. They found a stationary H atom concentration, which was independent of initial acetylene concentration. Using the same conditions of pressure, concentrations, flow speed as Arrington *et al* and measuring O- and H-atom concentrations with the ESR and acetylene concentration with the mass spectrometer we found a different behaviour. By lowering the initial acetylene to oxygen atom ratio from 10 : 1 to 0.1 : 1 increasing amounts of H atoms were observed (41). Comparison with the decrease of acetylene concentration shows that two H atoms were formed for every acetylene molecule reacted. This indicates that the major pathway of the mechanism of Arrington *et al*, a delicate balance between formation and removal of H atoms, is questionable. From these results, it is clearly shown that the measurements with the ESR give essential information about mechanisms. An unequivocal reaction scheme, however, cannot be established from this observation alone. Further investigations including a rapid sampling mass-spectrometer are therefore underway.

Reaction of C_2H_2 with H

This reaction may be interesting from a theoretical point of view in investigating chemicaly activated complexes, and of practical value in explaining the behaviour of acetylene in photolysis, radiolysis, pyrolysis and low pressure flames. The measurements on this reaction show two distinct advantages of the ESR method:

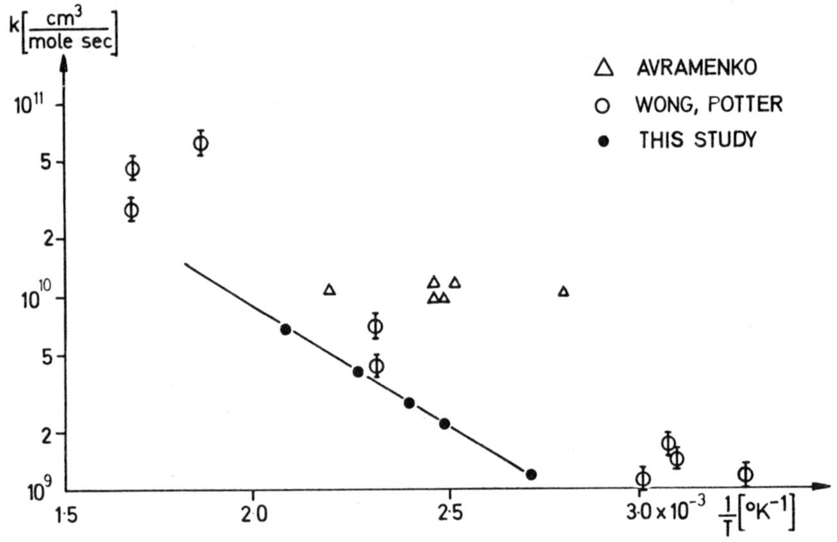

Fig. 6-2-6 The rate constant of the reaction $O + NH_3$ as a function of temperature

first, it yields high precision; second, it allows great variation of pressure. The experimental results are: H-atoms are consumed in a second order reaction:

$$-\frac{d[H]}{dt} = k_m [H] [C_2H_2]$$

$1/k_m$ varies linearly with $1/p$, the C_2H_2 concentration remains nearly constant, and only small amounts of ethylene can be detected with the mass spectrometer (using D instead of H atoms for better detection). These facts are consistent with the following mechanism.

$$1)\ C_2H_2 + H \rightleftharpoons C_2H_3^*$$

$$2)\ C_2H_3^* + M \longrightarrow C_2H_3 + M$$

$$3)\ C_2H_3 + H \longrightarrow C_2H_2 + H_2$$

(The asterisks denote species which are vibrationally excited above the dissociation limit). This gives an expression for k_m :

$$\frac{1}{k_m} = \frac{1}{2k_1}\left(1 + \frac{\overleftarrow{k_1}}{k_2}[M]\right)$$

This formula shows that k_1 may be obtained by measuring k_m over a wide range of pressure and extrapolating to $p = \infty$. In varying pressure from 1.9 .... 22 mm Hg our k_m changes only by a factor of three. The results, shown in figure 6-2-5, clearly indicate the high accuracy of H-atom measurement which can be obtained with the ESR and give $k_1 = 3\times5\cdot10^{10}$ cm^3/mole sec at 303°K.

In experiments with a catalytic probe detector the measured points scatter by almost a factor of two, so that no pressure dependence between 1 and 15 mm Hg could be found (42) (43).

Michael and Weston (44) have used Lyman -α-photometry for following the hydrogen atom concentration in the pressure region of 0.5 - 17 mm Hg, but their observed rate constant ($k_1 = 3.6\times10^{11}$ cm^3/mole sec) has a peculiar dependence on total pressure.

The investigation of Volpi and Zocchi (45) is limited from the principal disadvantages of using a mass spectrometer for determination of atomic hydrogen concentrations, and the small pressure variation (1.35 - 3.0 mm Hg) which they used.

Reaction of NH_3 with O

This reaction may be a good example for the usefulness of the ESR method in systems where atomic oxygen can be difficult to monitor with the mass spectrometer because its signal is superimposed by other species. We measure the O-atoms with ESR and NH_3 with the mass spectrometer. Using a large excess of ammonia the rate constant for the reaction:

$$NH_3 + O \Rightarrow \text{products}$$

could be calculated from the expression:

$$\ln \frac{[O]}{[O]_o} = -k\ [NH_3]\ t$$

The subscript o refers to the concentrations at t = o. The experimental k values are plotted against 1/T. They form a good straight line, which can be represented by:

$$k = 1.5 \times 10^{12} \exp(-5100/RT) \frac{cm^3}{mole\ sec}$$

The data of Avramenko *et al* (46) and Wong and Potter (47) are also included in figure 6-2-6. No temperature dependence could be found by Avramenko *et al* The data of Wong and Potter show a large scatter because their difficulties in monitoring O atom concentration mass spectrometrically at m/e=8 eV since NH_3 produces a strong peak at m/e = 16. Nevertheless, our values showed no difference by repeating the measurements several months later.

Conclusion

It was the purpose of this review to show that ESR studies of reacting gases give considerable information on reaction schemes. The method described is not restricted to reactions of O- and H-atoms. The concentrations of many other atoms and radicals can be determined with high accuracy using a similar calibration procedure. Also information about various properties of radicals can be obtained. But it must be kept in mind that there are limitations due to the complexity of this method. The full advantage of the ESR becomes apparent in combination with mass spectrometers, UV-, IR-spectrometers, etc. Then the ESR seems to be a powerful addition to the methods previously available.

References

1. **Beringer, R., Castle, J.G.,** Phys. Rev., Vol.78, 1950, p. 581.
2. **Beringer, R., Castle, J.G.,** Phys. Rev., Vol.80, 1950, p. 228.
3. **Beringer, R., Rawson, E.B.,** Phys. Rev., Vol.87, 1952, p. 228.
4. **Rawson, E.B., Beringer, R.,** Phys. Rev., Vol.88, 1952, p. 677.
5. **Heald, M.A., Beringer, R.,** Phys. Rev., Vol.96, 1954, p. 645.
6. **Dehmelt, H.G.,** Phys. Rev., Vol.99, 1955, p. 527.
7. **Brown, R.L.,** J. Chem. Phys., Vol.44, 1966, p. 2827.
8. **Bowers, K.D., Kamper, R.A., Lustig, G.D.,** Proc. Phys. Soc. (London), B, Vol. 70, 1957, p. 1176.
9. **Radford, H.E., Hughes, V.W., Beltran-Lopez, V.,** Phys. Rev., Vol.123, 1961, p. 153.
10. **Beltran-Lopez, V., Robinson, H.G.,** Phys. Rev., Vol.123, 1961, p. 1161.
11. **Vanderkooi, N., MacKenzie, J.S.,** Advan. Chem. Ser., Vol.36, 1962, p. 98.
12. **Radford, H.E.,** Phys. Rev., Vol.122, 1961, p. 114.
13. **Piette, L.H., Johnson, F.A., Booman, K.A., Coburn, C.B.,** J. Chem. Phys., Vol.35, 1961, p. 1481.
14. **Radford, H.E.,** J. Chem. Phys., Vol.40, 1964, p. 2732.

15. **McDonald, C.C.,** J. Chem. Phys., Vol.39, 1963, p. 2587.

16. **Daniels, J.M., Dorain, P.B.,** J. Chem. Phys., Vol.40, 1964, p. 1160.

17. **Carrington, A., Levy, D.H.,** J. Phys. Chem., Vol. 71, 1967, p. 2.

18. **Balakhin, V.P., Gershenzon, Y.M., Kondrat'ev, V.N., Nalbanyan, A.B.,** Doklady Akad. Nauk SSSR, Vol.154, 1964, p. 883.

19. **Balakhin, V.P., Gershenzon, Y.M., Kondrat'ev, V.N., Nalbanyan, A.B.,** Doklady Akad. Nauk SSSR, Vol.154, 1964, p. 1142.

20. **Azatyan, W.W., Nalbanyan, A.B., Sarkisyan, E.N.,** Doklady Akad. Nauk SSSR, Vol. 158, 1964, p. 179.

21. **Balakhin, W.P., Gershenzon, Y.M., Nalbanyan, A.B.,** Doklady Akad. Nauk SSSR, Vol.172, 1967, p. 375.

22. **Westenberg, A.A., de Haas, N.,** J. Chem. Phys., Vol.40, 1964, p. 3087.

23. **Kistiakowsky, G.B., Volpi, G.G.,** J. Chem. Phys., Vol.27, 1957, p. 1141.

24. **Morgan, J.E., Schiff, H.J.,** Can. J. Chem., Vol.41, 1963, p. 903.

25. **Gesser, H., Hussain, S.,** Nature, Vol.201, 1964, p. 290.

26. **Greaves, J.C., Linnet, J.W.,** Trans. Farad. Soc., Vol.55, 1959, p. 1346.

27. **Kaufman, F.,** 'Progress in Reaction Kinetics,' Vol.1, Pergamon Press, Oxford, 1961, p. 1.

28. **Townes, C.H., Schalow, A.L.,** 'Microwave Spectroscopy,' McGraw-Hill, New York, p. 342.

29. **Ingram, D.J.E.,** 'Free Radicals as Studied by Electron Spin Resonance,' Butterworths, London 1958, p. 28.

30. **Krongelb, S., Strandberg, M.W.P.,** J. Chem. Phys., Vol.31, 1959, p. 1196.

31. **Halbach, K.,** Phys. Rev., Vol.119, 1960, p. 1230.

32. **Tinkham, M., Strandberg, M.W.,** Phys. Rev., Vol.97, 1955, p. 951.

33. **Westenberg, A.A., de Haas, N.,** J. Chem. Phys., Vol.43, 1965, p. 1550.

34. **Philipps, L.F., Schiff, H.J.,** J. Chem. Phys. Vol.37, 1962, p. 1233.

35. **Clyne, M.A.A., Thrush, B.A.,** Proc. Roy. Soc. (London) A, Vol.275, 1963, p. 544.

36. **Jost, W.,** (Ed.) 'Low Temperature Oxidation', Gordon and Breach, New York, 1965, p. 44.

37. **Fristrom, R.M., Westenberg, A.A.,** 'Flame Structure', McGraw-Hill, New York, 1965, p. 367.

38. **Westenberg, A.A., de Haas, N.,** J. Chem. Phys., Vol.46, 1967, p. 490.

39. **Hoyermann, K., Wagner, H.G., Wolfrum, J.,** Ber. Bunsengesellschaft Physik. Chem. 1967, p. 599.

40. **Arrington, C.A., Brennen, W., Glass, G.P., Michael, J.V., Niki, H.,** J. Chem. Phys., Vol.43, 1965, p. 525.

41. **Hoyermann, K., Wagner, H.G., Wolfrum, J.,** Z. Phys. Chem. NF, Vol.55, 1967, p. 72; also Ber. Bunsengesellschaft Physik. Chem., 1967.

42. **Tollefson, J.L., LeRoy, D.J.,** J. Chem. Phys., Vol.16, 1948, p. 1057.

43. **Dingle, J.R., LeRoy, D.J.,** J. Chem. Phys., Vol.18, 1950, p. 1632.

44. **Michael, J.V., Weston, R.E.,** J. Chem. Phys., Vol.45, 1966, p. 3622.

45. **Volpi, G.G., Zocchi, F.,** J. Chem. Phys., Vol.44, 1966, p. 4010.

46. **Avramenko, L.I., Kolesnikova, R.V., Kuznetsova, N.L.,** Izv. Akad, Nauk SSSR Otd. Khim Nauk, Vol. 6, 1962, p. 983.

47. **Wong, E.L., Potter, A.E.,** J. Chem. Phys., Vol.43, 1965, p. 3371.

List of Mathematical Symbols

h	line height of the ESR-signal
H_o	resonant magnetic field, Gauss
H_1	microwave magnetic field
I	first moment, see p. 461.
J	total angular momentum
k	rate constant, cm^3 $mole^{-1}$ sec^{-1}
MA	modulation amplitude of the 100 k Hz field.
N	number of the absorbing particles
P	employed microwave power, mW
p	pressure, mm Hg
Q'_{at}	calibration factor of atom at
Q_{at}	calibration factor relative to O_2 molecule
R	gas constant
S	ESR-signal
T	temperature, °K
T_1	spin-lattice relaxation time, sec

t	reaction time
w	line width of the ESR-signal
μ	magnetic moment
[ ]	concentration

6-3

Low Pressure Discharge Flow Systems and their Application to the Chemical Aspects of Propulsion

H.F. FOUNTAIN AND E. HUTTON
Department of Chemistry, Manchester University, England.

Summary

The discharge-flow technique is described in detail, and novel methods that have been developed for the measurement of atom and radical concentrations are reviewed. The particular advantage of the technique is that it enables fast reactions to be studied in a controlled environment, uncomplicated by the many simultaneous processes that occur in most other fast reaction techniques.

A summary is made of results relevant to combustion research, and some of the most important of these, together with their application to propulsion systems, are discussed in detail.

The advantages and disadvantages of the technique are emphasised, and the precautions that must be taken in the application of results to practical systems are pointed out.

Sommaire

L'emploi d'une décharge électrique dans un écoulement gazeux permet de déterminer la concentration en atomes et en radicaux libres. Cette méthode et ses perfectionnements sont décrits. Elle a le grand avantage de permettre l'étude de réactions rapides dans un milieu de composition connue, sans la complication des processus parallèles rendant difficile l'emploi des autres méthodes.

Les résultats obtenus pour l'étude de la combustion sont résumés et les plus importants pour les applications propulsives sont analysés.

On souligne notamment les avantages et les inconvénients de cette technique et les précautions à prendre dans l'application des résultats aux emplois réels.

Introduction

The combustion processes in rocket engines may often be simulated for research purposes by burning the fuel-oxidant mixture as a conventional flame at the pressure range of interest. This technique has led recently to a better understanding of combustion processes (1) (2). Knowledge has now increased to the point where distinction may be drawn between the two previously disputed mechanisms of flame propagation (3), i.e. the radical-chain mechanism, and the mechanism of thermal diffusion of products upstream against the gas flow. However, the processes are still not well established in detail.

In hydrocarbon flames this is due to the inherent complexity of the combustion process. A fast burning flame at one atmosphere pressure, for example, may have a reaction-zone of about 0.002 cm in thickness corresponding to a reaction time of

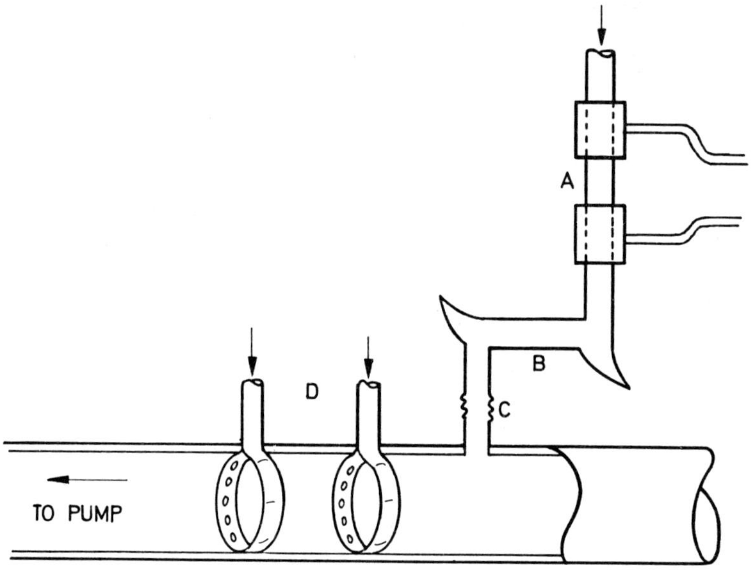

Fig. 6-3-1 Schematic Diagram of the Apparatus.

ten microseconds during which the gas is heated by as much as 3,000 degrees centigrade (3). This results in a temperature gradient of several million degrees per second within the reaction zone. Consequently, non-equilibrium conditions exist in this region and the overall process becomes prohibitively complicated to analyse. Some simplification may be attained by burning flames at reduced pressures, in which case the reaction zone thickness is increased (4). Even under these conditions, however, non-equilibrium processes are still observable: radicals and intermediates, which constitute important species in the reaction path, are short-lived and formed in such small stationary state concentrations that they cannot be readily detected by any techniques at present available.

Observations on the reaction zone and the post reaction zone gases have, nevertheless, yielded very useful results which, when combined with information obtained by other techniques e.g. the flow discharge technique, give some indication of the nature of the reaction sequence (1) (2).

The discharge-flow technique has the advantage over conventional flame studies that reactions may be investigated in carefully controlled environments under very much simplified conditions. The number of simultaneous and consecutive reactions can be minimised and non-equilibrium effects often completely eliminated. In addition, atom and radical concentrations, rate constants and activation energies may be reliably determined and the reactions uniquely ascribed (5) (6) (7).

Description of the Experimental Method

It has long been known that when an electrical discharge is passed through a gas at low pressure, atoms are formed by dissociation of gaseous molecules. Strutt (8) investigated some properties of the gaseous products of discharges, and significant contributions were made later by Wood (9), Bonhoeffer (10), Harteck (11) and others. However it was not until the late 1950's that innovations were introduced which gave the technique the accuracy and scope that it enjoys at the present time (12).

A significant advance was the introduction of radiofrequency and microwave discharges in which there is no contact between the electrodes and the gas, thus eliminating the contamination of the discharge products by electrode materials. Radiofrequency power is coupled to the gas externally through two thin metallic foil electrodes wrapped around the discharge tube in which the atoms are formed. Microwave power is supplied to the discharge by means of a resonant cavity. With either of these methods several percent of the total gas may be dissociated. The working pressure range for the radiofrequency discharge is from 0 to 5 torr, and for the microwave discharge from about 0.6 to 20 torr (13).

A schematic diagram of the apparatus is shown in figure 6-3-1. Two radiofrequency electrodes are wrapped around the discharge tube (A), which is constructed of quartz to withstand the high temperatures attained in this region, and is connected to the reaction cell via Rayleigh horns (B) and a quartz-pyrex seal (C). Ions formed in the gas as it flows through the discharge are completely removed within a few centimetres, and within this distance the gas is cooled to room temperature by contact with the tube walls. The reaction cell is normally from one to five centimetres in diameter, and as the discharged gas flows at a constant rate into the cell it traverses equal distances in equal periods of time. At given positions in the cell, fixed jets (D) allow reactants to be added, and these jets are multi-perforated to ensure rapid and efficient mixing. Measurements of pressure and the total input flow-rate of the gases enable the velocity of the gas within the tube to be found, and the reaction times calculated.

Removal of atoms on the surface of the tube does occur, but this can be minimised by coating the walls with substances that inhibit heterogeneous recombination: acid coatings have been found effective (14) (15). For most atoms, the wall recombination coefficient γ for walls thus treated falls to about 10^{-4} to 10^{-5}, so that these effects become small and may be allowed for in the calculation of the rates of homogeneous gas phase reactions.

Minute quantities of impurity in the gas that is passed through the discharge may have a significant effect on the properties of the discharged gas. Addition of small amounts of water vapour to nitrogen have been found to increase the nitrogen atom yield by up to a factor of ten (16), whilst the presence of hydrogeneous impurities in discharged oxygen accelerate the recombination of oxygen atoms (17).

The concentrations of reactants that react exothermally must be small enough to ensure that the rate of heat release is not too rapid, so that the reaction takes place under approximately isothermal conditions.

Accurate data can be obtained only if allowance is made for the facts: that flow systems possess a finite concentration gradient in the direction of flow; that high atom concentration gradients may lead to an appreciable atom concentration diffusion effect; and that any non-uniformity of atom concentration across the tube diameter may introduce errors. The magnitudes of these errors have been detailed (18). To minimise them for a pressure range of 1 to 5 torr and a tube diameter of 1 to 4 cm, the velocity of the gas is generally restricted to between 100 and 500 cm per second. Under these conditions only small corrections are necessary and can be taken into account in the analysis of experimental results.

Measurement of Atom and Radical Concentration

The atom concentrations obtainable from discharges are sufficiently high to enable direct and highly specific methods of detection to be used. Some of the most generally used methods are summarised here.

The Wrede-Harteck Gauge (19) (20) - The gauge consists of a tube, through the wall of which are one or more small holes having diameters less than one-tenth of the mean free path of the gaseous molecules. Behind the holes is a small closed volume containing an efficient catalyst for the recombination of the atoms. Molecules and atoms effuse through the hole, the atoms are recombined inside the cavity and hence only molecules effuse out.

A mass balance shows that a pressure difference (ΔP) will develop across the hole, and when sufficient time has been allowed for the system to reach equilibrium, the fractional percentage of atoms in the gas (α) may be obtained from:

$$\alpha = \frac{3.41\ \Delta P}{P} \times 100$$

where P is the pressure of the gas, and the molecules are all assumed to be diatomic.

The sensitivity of the method depends on the accurate measurement of small pressure differences: as these can be determined accurately to better than 10^{-4} torr (21), atom concentrations of 10^{-11} moles cm^{-3} can be measured quantitatively.

Mean free paths decrease with increase in pressure, so that at higher pressures smaller diameter holes must be used, and the time allowed for equilibration must be increased. This time factor sets an upper limit of about 0.7 torr on the pressure to which the method is applicable. Fritted discs, containing large numbers of fine capillaries, have been used to extend this pressure range (22), but objections have been raised as to the validity of some of the results obtained by this method (23).

The Wrede-Harteck gauge was the first to be developed to give absolute values of atom concentrations. It is insensitive to the presence of excited species. For simple recombination reactions at low pressures, it is good, but since it is non-specific and gives the total atom concentration irrespective of atom type, it is of no use when mixtures of atoms are present. This factor, together with the length of time required to ensure equilibrium, and the necessity of measuring very accurately small changes of pressure, has led to the adoption of alternative methods of detection in most current kinetic studies.

Thermal Methods - If a thermocouple probe, coated with a suitable catalyst, is placed in the gas flow, the atoms will recombine on the surface with release of heat. Recombination reactions are strongly exothermic, and a small atom concentration will therefore give rise to a relatively large temperature increase on the probe. This temperature rise may be directly related to the atom concentration (24).

To make the thermocouple method quantitative it must be calibrated against one of the absolute methods. Other disadvantages are: that the recombination efficiency of the catalyst may change with temperature; that excited states contribute to the heat release; that it is non-specific to atom type; and that it is difficult to estimate heat losses from the probe. One advantage is that it removes only a small fraction of the total atoms and hence causes very little disturbance to the system.

The Isothermal probe is another thermal method, also depending on the measurement of the heat released by recombining atoms (25) (26). The probe consists of several turns of platinum wire coated with an efficient catalytic surface and of sufficient length to ensure complete recombination of all atoms present in the gas. The wire is heated electrically by passage of a current, and with the discharge switched off, the power dissipation (W_o) and the resistance (R_o) of the wire are measured. The discharge is then switched on, atoms recombine on the wire with release of heat, and the electrical power is decreased (to W_1) until the measured resistance of the wire is again R_o. The temperature of the probe is then the same in the presence and in the absence of atoms, and so the necessity of making a correction for heat loss is eliminated. The atom flowrate (F) can be calculated from the difference in energy input by use of the expression:

$$F = \frac{W_o - W_1}{\Delta H}$$

where ΔH = the heat of atom recombination.

The isothermal probe method is absolute but it is both non-specific and sensitive to excited molecules, properties which may be utilised to follow reactions of excited states. For atoms formed in the absence of excited states and where only one type of atom is present e.g. hydrogen atom recombination reactions, it is a very reliable method (27).

Photometric Methods - The most widely used methods for the measurement of atom concentration depend on quantitative measurements of the intensities of the characteristic glows emitted from mixtures of atoms and radicals.

Estimation of Atomic Oxygen - When a small quantity of nitric oxide is added to atomic oxygen a greenish-yellow glow is emitted, the intensity of which is proportional to the product of the nitric oxide and oxygen atom concentrations (12). The nature of the emission has been the subject of some controversy (7) but certainly is due to emission from excited nitrogen dioxide, probably formed by reactions:

$$O + NO + M \rightarrow NO_2^* + M \quad \text{(Reaction 6-3-1)}$$

$$NO_2^* + M \rightarrow NO_2 + M \quad \text{(Reaction 6-3-2)}$$

$$NO_2^* \rightarrow NO_2 + h\nu \quad \text{(Reaction 6-3-3)}$$

Reaction 6-3-1 and the reaction 6-3-4:

$$O + NO + M \rightarrow NO_2 + M \quad \text{(Reaction 6-3-4)}$$

are so slow under the experimental conditions, compared to other oxygen atom reactions, that their effect on the rate of removal of oxygen atoms can be ignored.

The nitrogen dioxide, formed in reactions 6-3-1 and 6-3-4, rapidly regenerates nitric oxide in the fast reaction:

$$O + NO_2 \rightarrow NO + O_2 \quad \text{(Reaction 6-3-5)}$$

so that the nitric oxide concentration remains unchanged throughout the region of observation. Thus for a given nitric oxide concentration the intensity of the glow is proportional to the oxygen atom concentration. The intensity of the glow at several positions along the reaction cell is measured and gives the relative atom concentration as a function of distance or reaction time.

Relative atom concentrations can be converted into absolute values if the absolute atom concentration at any one position in the tube is known. This value is obtained at one fixed position by addition of nitrogen dioxide which reacts rapidly with atomic oxygen according to reaction 6-3-5. If oxygen atoms are in excess, the NO-O glow persists along the tube, but if the nitrogen dioxide flow rate is equal to or in excess of the oxygen atom flow rate, then the oxygen atoms are removed at the nitrogen dioxide injection point and the glow is extinguished. The minimum flow rate of nitrogen dioxide necessary to completely extinguish the glow is equal to the flow rate of the oxygen atoms at the position of nitrogen dioxide addition.

Similar specific photometric methods have been devised to measure the concentrations of other types of atoms.

Hydrogen - A red glow is observed when nitric oxide is added to atomic hydrogen and the processes responsible have been shown to be (27):

$$H + NO + M \rightarrow HNO^* + M \quad \text{(Reaction 6-3-6)}$$

$$M + HNO^* \rightarrow HNO + M \quad \text{(Reaction 6-3-7)}$$

$$HNO^* \rightarrow HNO + h\nu \quad \text{(Reaction 6-3-8)}$$

$$H + HNO \rightarrow H_2 + NO \quad \text{(Reaction 6-3-9)}$$

Reaction 6-3-6 is very slow and removes a negligible quantity of hydrogen atoms, Reaction 6-3-9 is fast and quickly regenerates nitric oxide. The nitric oxide concentration is thus maintained constant along the length of the tube and the intensity of the glow is directly proportional to the hydrogen atom concentration. Relative values were converted into absolute values by isothermal probe measurements.

Nitrogen - Nitrogen atom concentrations have been photometrically obtained from measurements of the intensity of the yellow emission associated with active nitrogen (28). Nitrogen atoms recombine according to a third order mechanism and produce excited nitrogen molecules that are responsible for the emission:

$$N + N + M \rightarrow N_2' + M \qquad \text{(Reaction 6-3-10)}$$

$$N_2' \rightarrow N_2 + h\nu \qquad \text{(Reaction 6-3-11)}$$

The square root of the intensity of the yellow glow is proportional to the atomic nitrogen concentration. Absolute concentrations may be obtained by use of the fast 'titration' reaction:

$$N + NO \rightarrow N_2 + O \qquad \text{(Reaction 6-3-12)}$$

Chlorine - The products of a discharge through chlorine emit an orange-red glow, the intensity of which can be related to the chlorine atom concentration (29) (30). The processes responsible for the emission are more complex than was first envisaged (31), but if a detector is used which is sensitive only to light of wavelength less than 5, 800A°, the emission intensity is proportional to the square of the chlorine atom concentration. Absolute values can be determined by the fast 'titration' reaction:

$$Cl + NOCl \rightarrow NO + Cl_2 \qquad \text{(Reaction 6-3-13)}$$

Hydroxyl Radicals - The hydroxyl radical possesses an absorption spectrum in the near ultraviolet, with a band head at 3064A°. Del Greco and Kaufman (32) monitored hydroxyl radical concentrations by quantitative measurements of the absorption by the radical in this wavelength region. A discharge through water vapour was found to be an unreliable method of producing hydroxyl radicals for kinetic studies, and the reaction of atomic hydrogen with nitrogen dioxide was used instead in:

$$H + NO_2 \rightarrow OH + NO \qquad \text{(Reaction 6-3-14)}$$

If the atomic hydrogen is in excess, the flow-rate of the nitrogen dioxide is equal to the flow-rate of hydroxyl radicals at the position of nitrogen dioxide addition.

Chlorine Monoxide Radicals - Clyne and Coxon (33) measured chlorine monoxide concentrations by an absorption method. They determined the spectral extinction coefficients of the radical, thus enabling absolute values to be obtained.

The advantages of the photometric measurements are that they are specific, absolute, fast, reliable and are not costly to set up, since the detector system consists only of a photomultiplier power supply and simple recording equipment.

Detection by E.S.R. - All atoms and free radicals contain unpaired electrons and can in principle be detected by E.S.R. methods. Useful results obtained by this method are now beginning to be published (34) (35) (36). The advantages of this technique are: that it is an absolute method; that excited species do not interfere with the measurements; and that different atomic species can be simultaneously measured. The disadvantages are: that the apparatus is extremely expensive; that limitations of cavity design dictate that measurements be made over such an extended length of the reaction cell that the time resolution is not as good as with other methods; and that measurement of absolute concentration involves lengthy and complicated integrations.

Mass Spectrometric Methods - Mass spectrometric detection can be made an absolute method, but requires expensive apparatus. It is particularly useful for the

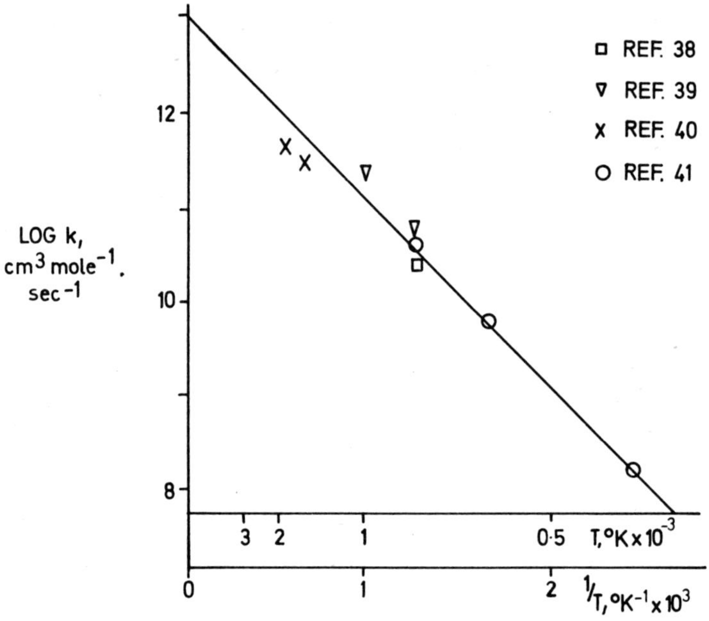

Fig. 6-3-2 Experimentally Determined Rate Constants for the reaction $O + H_2 \rightarrow OH + H$.

identification of small concentrations of transients that cannot be readily observed by other methods. Since it is specific it can be used to study reactions that occur with the simultaneous production of many types of atoms, radicals and excited states (37).

Application of the Flow Discharge Technique to Some Combustion and Propulsion Problems

The high atom concentrations obtainable, and the accuracy of the measuring techniques, allow some of the important elementary reactions that occur in the complex combustion process to be investigated with great precision, but at temperatures much less than those encountered in flames. However the simple Arrhenius plot of log k versus the reciprocal of the absolute temperature gives, to a first approximation, a straight line, the slope of which is related to the activation energy for the reaction. In the region of high flame temperatures, measurements taken over a large temperature range are bunched in a confined region of the graph, and accurate values of activation energies are hard to obtain. In contrast, a variation of only a few degrees in the vicinity of room temperature covers a large portion of this plot and accurate activation energies can be calculated from results obtained in this region. Thus rate constants at elevated temperatures may be confidently extrapolated from the lower temperature data, as illustrated in figure 6-3-2.

The Hydrogen-Oxygen System

Reference is made to the combustion literature for review of the 'best values' of reaction rate data for the hydrogen-oxygen system, (42) and it can be seen that the discharge-flow technique is perhaps the most successful for obtaining reliable basic data for this system over a wide range of conditions.

The CO + OH Reaction

The exothermic oxidation of carbon monoxide to carbon dioxide occurs via the reaction:

$$CO + OH \rightarrow CO_2 + H \qquad \text{(Reaction 6-3-15)}$$

Probe sampling measurements of flame in the temperature range 1, 200 to 2, 000°K indicated an activation energy of 7.5K cals $mole^{-1}$ for this reaction (43) (44). Dixon-Lewis, Wilson and Westenberg (45), using a flow method, found this value was too high to explain their room-temperature results which indicated an activation energy of 0.6K cals $mole^{-1}$, in agreement with other measurements (46). The revised value gives an entirely different connotation to the role of reaction 6-3-15 in cool flames and in the secondary combustion that occurs in the cooler regions of rocket exhausts, where fuel rich gases from the combustion chamber intermix with the surrounding air.

Hydrogen-Oxygen Reactions

Atomic oxygen-hydrocarbon reactions have been studied in discharge flow systems (36) (47) (48) (49), but complications occur because the products of the primary reactions are susceptible to subsequent attack by atomic and molecular oxygen. The system may be simplified by producing oxygen atoms in the absence of molecular oxygen by passing nitrogen through the discharge and reacting the nitrogen atoms with nitric oxide according to equation 12 in (49). If the flow rate of nitrogen atoms is equal to the flow rate of nitric oxide, then atomic oxygen is formed quantitatively according to this equation in the absence of molecular oxygen and in the presence of chemically inert molecular nitrogen. A recent investigation of the rate of hydrocarbon-oxygen reactions with simultaneous measurements of the hydrogen atom concentrations produced gives an indication not only of the identity of the initial reaction but also the subsequent fate of some of the primary reaction products (36).

Flame Ionisation Processes

Within the reaction zones of hydrocarbon flames positive ions and electrons are produced in great excess above their thermal equilibrium concentrations (50). The ionisation in this region is thought to arise predominantly from highly energetic radical and atom reactions (50) (51) but attempts to identify these have been inconclusive and in some cases contradictory (51) (52). Reactions observed in low pressure discharge-flow experiments between atomic oxygen and acetylene produce the same type of chemi-excited species as are observed in flame reaction zones (53), indicative of the fact that similar processes are occuring in both systems. On this basis the chemi-ionisation reactions are now being studied by several investigators using the discharge-flow method (37) (54) (55) (56). Results, again, do sometimes appear contradictory, but the mechanism is obviously complex and much work has still to be done.

Flame Inhibition

Halogens constitute the most common type of conventional flame inhibitors, and probably act through a radical chain inhibition mechanism (2). The discharge-flow technique is ideally suited to the investigation of such problems, although as yet, with the exception of some work on chlorine-oxygen reactions (33), little has been done in this field.

Iron carbonyl is found to be a very effective flame inhibitor (57) and addition of iron carbonyl to atomic flames produces solid particles and gives rise to some spectacular effects (53). The results were qualitative and a detailed re-investigation with the more quantitative present-day techniques would seem to be worthwhile.

Limitations of the Technique

As previously mentioned, these experiments are normally performed in the region of room temperature. In general, however, the experimental scatter on the Arrhenius plot is sufficiently small to enable accurate extrapolations to be made, as evidenced by the coincidence of this data with such direct flame measurements as are available as shown previously.

Discharge-flow studies are normally carried out in a pressure range of 1 to 5 torr. The role of third order reactions become more important at high pressures and a reaction that is second order at low pressures may proceed by an entirely different mechanism at high pressures. For example the recombination reaction of hydroxyl radicals is second order under discharge-flow conditions (32), but predominantly third order at 200 torr and above (58). Data obtained from low pressure measurements may be applied to high pressure systems only if it can be established that the mechanism of the reaction does not change with pressure. Hydrogen atom recombination is a third order process even at 1 torr (59), and this mechanism is still operative at higher pressures.

Strongly endothermic reactions and those with high energies of activation do not proceed at a measurable rate at room temperature, but may be very rapid at elevated temperatures. Their rate constants cannot be directly measured in flow systems but can be obtained by measuring the rate of the reverse reaction and applying thermodynamic arguments. For example, one of the most important radical branching reactions in flames is:

$$H + O_2 \rightarrow O + OH \qquad \text{(Reaction 6-3-16)}$$

which is 16.5k cal. $mole^{-1}$ endothermic. Its rate can be calculated from the rate of the reverse reaction, together with standard thermodynamic data (60).

Significant simplification of complex processes is generally obtainable by use of the discharge flow technique, but when attempts to do this are made for some flame reactions the course of the reaction and the nature of the final products may be changed. Acetylene-atomic oxygen mixtures undergo a fast reaction with emission spectra similar to those observed in flames, but the methane-atomic oxygen reaction is not accompanied by any emission of comparable intensity or spectral distribution to that observed in combustion (61). It may be argued that the methane system has been simplified to such an extent that the rate of primary attack on methane can be studied directly, whilst the acetylene system is still too complex and the secondary reactions so fast that such an investigation is difficult. On the other hand the methane system has been so much simplified that all but the initial reactions proceed by different mechanisms from those operating at higher temperatures. With acetylene these reactions do still occur within a more easily controlled environment.

Conclusion

No attempt has been made to review the subject comprehensively. The technique has been described and the application in a few selected fields discussed in order to illustrate some of its uses and limitations in the study of combustion and propulsion problems. The general applicability of the method to a wide range of chemical kinetic problems ensures a rapid and continued future development.

References

1. **Fristrom, R.M., Westenberg, A.A.,** 'Flame Structure', McGraw Hill, New York, 1965.

2. **Fenimore, C.P.,** 'Chemistry in Premixed Flames', Pergamon, Oxford, 1964.

3. **Gaydon, A.G., Wolfhard, H.G.,** 'Flames, Their Structure, Radiation and Chemistry', Chapman and Hall, London, 1953.

4. **Gaydon, A.G., Wolfhard, H.G.,** Proc. Roy. Soc. A, 194, 169, 1948.

5. See International Combustion Symposia Volumes 7 to 11, The Combustion Inst.

6. 'Chemical Reactions in the Lower and Upper Atmosphere', Interscience, New York, 1961.

7. 'Chemical Reactions in the Atmosphere', Disc. Faraday Soc., 37, 1964.

8. **Strutt, R.J.,** Proc. Roy. Soc. A, 85, 219, 1911.

9. **Wood, R.W.,** Proc. Roy. Soc. A, 97, 455, 1920.

10. **Bonhoeffer, K.F.,** Z. Phys. Chem., 113, 199, 1924.

11. **Harteck, P.,** Trans. Faraday Soc., 30, 134, 1934.

12. **Kaufman, F.,** Proc. Roy. Soc. A, 247, 123, 1958.

13. **Greaves, J.C., Linnet, J.W.,** Trans. Faraday Soc., 55, 1355, 1959.

14. **Harteck, P., Reeves, R.R., Mannella, G.,** J. Chem. Physics, 29, 1333, 1958.

15. **Ogryzlo, E.A.,** Canad. J. Chem., 39, 2556, 1961.

16. **Kaufman, F., Kelso, J.R.,** J. Chem. Physics, 32, 301, 1960.

17. **Kaufman, F., Kelso, J.R.,** Disc. Faraday Soc., 37, 26, 1964.

18. **Kaufman, F.,** 'Progress in Reaction Kinetics', Vol. 1, Pergamon Press, Oxford, 1961.

19. **Wrede, E.,** Z. Instrumentenkunde, 48, 201, 1928.

20. **Harteck, P.,** Z. Phys. Chem., 139, 98, 1928.

21. **Groth, W., Warneck, P.,** Z. Physik. Chem. 10, 323, 1957.

22. **Beckney, H.D., Warneck, P.,** Z. Naturforsch, 10a, 62, 1955.

23. **Greaves, J.C., Linnet, J.W.,** Trans. Faraday Soc., 55, 1338, 1958.

24. **Linnet, J.W., Marsden, D.G.H.,** Proc. Roy. Soc. A, 234, 489, 1956.

25. **Tollefson, E.L., Le Roy, D.J.,** J. Chem. Physics, 16, 1057, 1948.

26. **Elias, L., Ogryzlo, E.A., Schiff, H.I.,** Canad. J. Chem., 37, 1690, 1958.

27. **Clyne, M.A.A., Thrush, B.A.,** Trans. Faraday Soc., 57, 1305, 1961.

28. **Berkwitz, J., Chupka, W.A., Kistiakowsky, G.B.,** J. Chem. Physics, 25, 457, 1956.

29. **Bader, L.W., Ogryzlo, E.A.,** J. Chem. Physics, 41, 2926, 1964.

30. **Hutton, E., Wright, M.,** Trans. Faraday Soc., 61, 78, 1965.

31. **Clyne, M.A.A., Stedman, D.H.,** Private communication, 1967.

32. **Del Greco, F.P., Kaufman, F.,** Disc. Faraday Soc., 33, 128, 1962.

33. **Clyne, M.A.A., Coxon, J.A.,** Trans. Faraday Soc., 62, 1175, 1966.

34. **Westenberg, A.A.,** J. Chem. Physics, 43, 1544, 1965.

35. **Westenberg, A.A., De Haas, N.,** J. Chem. Physics, 43, 1550, 1965.

36. **Brown, J.M., Thrush, B.A.,** Trans. Faraday Soc., 63, 630, 1967.

37. **Arrington, C.A., Brennen, W.R., Glass, G.P., Michael, J.V., Niki, H.,** J. Chem. Physics, 43, 525, 1965.

38. **Baldwin, R.R.,** Trans. Faraday Soc., 52, 1344, 1956.

39. **Azatian, V.V., Voevodskii, V.V., Nalbandian, A.B.,** 'Kinetika i Kataliz,' 2, 340, 1961.

40. **Fenimore, C.P., Jones, G.W.,** J. Chem. Physics, 65, 993, 1961.

41. **Clyne, M.A.A., Thrush, B.A.,** Proc. Roy. Soc. A, 275, 544, 1963.

42. **Bascombe, K.N.,** Explosives Res. and Dev. Est., Survey No. 1/5/65, Waltham Abbey, Essex, 1965.

43. **Westenberg, A.A., Fristrom, R.M.,** J. Phys. Chem. 65, 591, 1961.

44. **Fenimore, C.P., Jones, G.W.,** J. Phys. Chem., 62, 1578, 1958.

45. **Dixon-Lewis, G., Wilson, W.E., Westenberg, A.A.,** J. Chem. Physics, 44, 2877, 1966.

46. **Dixon-Lewis, G., Sutton, M.M., Williams, A.,** Trans. Faraday Soc., 61, 255, 1965.

47. **Avramenko, L.I., Kolesnikova, R.V.,** Adv. Photochem., 2, 25, 1964.

48. **Elias, L., Schiff, H.I.,** Canad. J. Chem., 37, 1075. 1959; 38, 1657, 1960.

49. **Elias, L., Schiff, H.I.,** Canad. J. Chem. 36, 1159, 1958.

50. **Calcote, H.F.,** Eighth Symp., on Combustion, Williams and Wilkins, 1962.

51. **Green, J.A., Sugden, T.M.,** Ninth Symp. on Combustion, Academic Press, New York, 1963, p. 60.

52. **Cummings, G.A. McD., Hutton, E.,** Combustion and Flame, 10, 195, 1966.

53. **Gaydon, A.G., Wolfhard, H.G.,** Proc. Roy. Soc. A., 213, 366, 1952.

54. **Fontijn, A., Baughman, G.L.,** J. Chem. Physics, 38, 1784, 1963.

55. **Fontijn, A., Miller, W.J., Hogan, J.M.,** Tenth Symp. on Combustion, Combustion Inst. 1965.

56. **Arrington, C.A., Brennen, W., Glass, G.P., Michael, J.V., Niki, H.,** J. Chem. Physics, 43, 1489, 1965.

57. **Lask, G.W., Wagner, H.G.,** Eighth Symp. on Combustion, Williams and Wilkins, 1962.

58. **Caldwell, R., Back, R.A.,** Trans. Faraday Soc., 61, 1939, 1965.

59. **Larkin, F.S., Thrush, B.A.,** Disc. Faraday Soc., 27, 112, 1964.

60. **Clyne, M.A.A.,** Ninth Symp. on Combustion, Academic Press, New York, 1964.

61. **Geib, K.H., Vaidya, W.M.,** Proc. Roy. Soc. A, 178, 351, 1941.

6-4

Probe Measurements of Charged Particle Concentrations and Ionization Relaxation Phenomena in Low Density Supersonic Gas Flows

S. W. RADCLIFFE, T. I. MCLAREN and R. M. HOBSON
Department of Pure and Applied Physics, The Queen's University of Belfast, Belfast 7, Northern Ireland

Summary

The limitations on measurements conducted with probes in partially ionised low density supersonic gas flows are discussed in relation to the geometry of the probe. Experiments are described which relate some features of the probe response to interactions between the supersonic gas flow and the probe surface geometry. A preliminary theoretical formulation of the mechanism of charged particle collection in relation to the free stream ion/electron concentration is put forward and compared with experiment.

Measurements of thermal ionization rates in shock heated monotomic gases are given, and the initial reactions leading to ionisation are determined and their cross sections measured.

Sommaire

On analyse les limitations des possibilités d'emploi de sondes dans des écoulements supersoniques de gaz partiellement ionisés et à basse densité, en relation avec la géométrie de la sonde. Des expériences sont décrites qui montrent l'effet sur l'allure de la réponse des interactions entre l'écoulement supersonique et la géométrie de la sonde. Une analyse théorique préalable du mécanisme de collection de particules chargées en relation avec la concentration ion/électron dans un jet libre est comparée aux résultats expérimentaux.

On présente les vitesses d'ionisation thermique d'un gaz monoatomique chauffé par choc, on détermine les réactions initiales permettant l'ionisation du gaz, ainsi que les sections efficaces correspondantes.

Introduction

Processes leading to ionization in flames and detonations have potentially powerful technological applications and knowledge of ionization processes is required particularly in connexion with the re-entry of space vehicles. Much remains to be learned about the detailed reactions that lead to ionization in a gas taken from room temperature to a temperature of several thousand degrees K in short time. In collision physics, recent measurements of the behaviour of noble gases (1 through 4) have indicated the potentiality of precise measurements of the rate of ionization for obtaining inelastic collision cross sections that are otherwise experimentally difficult to measure, especially where multiple collisions and collective phenomena play a major part in the chain of reactions.

Successful techniques of measurement of ionization in shock tubes include optical spectroscopy (1) (5), optical interferometry (6) and microwave interferometry (2) (3). Optical diagnostics provide adequate resolution for studies of shock wave processes where levels of ionization can change appreciably in time intervals of a few microseconds. However they are generally applicable only to relatively high electron concentrations (say, above 10^{15} electrons cm^{-3}).

Microwave techniques are suitable for use at lower electron concentrations, but their spatial resolution is characteristic of a guide wavelength, so that they are at a disadvantage in shock tubes where charged particle concentrations can change rapidly within a few millimetres.

The electrostatic probe, a familiar diagnostic technique in static plasma systems, is attractive as a measuring device for charged particles in shock tube flows since it is capable of operating over a broad range of number densities and has a spatial resolution characteristic of its dimensions. Several types of probe system have been proposed and developed (7) (8) (9) for use in determining the degree of ionization in supersonic plasma flows, but few of these are suitable as a diagnostic in cases where the ionization level varies with time. For example, the stagnation probe geometry of Talbot (7) and of Pollin (8) is not suitable for the present purpose because several microseconds can elapse during the flow of the gas from the free stream to the probe surface. Further, the charged particle number density at the probe surface may be many times different from that in the free stream.

For the present work it was decided to continue using two wires stretched parallel to one another on a cross section of the shock tube. This arrangement has satisfactorily yielded information on events in an electrically driven shock tube (10). With a wire diameter of 0.03 cm, transit times from the free stream to the probe should be less than one microsecond, and as there are no substantial probe supports to destroy the flow of shocked gas, the ionization profile may be checked at various situations along the shock tube.

We have used transverse cylindrical double probes also to measure rates of ionization behind shock waves moving into spectroscopically pure argon. These experiments are currently being extended to provide measurements of the reaction coefficients for other noble gases. With this completed a range of gases will be available for use as heat baths in which studies of the collision parameters and reaction constants for other molecular systems can be conducted.

Electrostatic Probe Phenomena

Collection Mechanisms

The double probe system was first proposed by Johnson and Malter (11) and the technique has been extended with some success to the examination of ionized gases in shock tubes (12) (13). In the present mode of operation the differential voltage, V_d, between the probes is sufficiently large for the current, i_d, round the circuit to be the ion current to the more negative probe arm. The free stream ion number density N^+ can thus be deduced if a precise relationship between it and this ion current can be found. Experience in the electrically driven shock tube has shown that if it is assumed, as in the free molecule approximation, that the ion current falling on the probe is characteristic of the random thermal current in the gas, the estimates of N^+ that are then obtained are satisfactory. Thus:

$$i_d = \left[\frac{1}{4} N^+ e \sqrt{\frac{8kT}{\pi M}}\right] Ap \qquad \text{(Eq. 6-4-1)}$$

Support for this model theory was ostensibly provided by a microwave cavity calibration experiment (13), performed under typical conditions in the pressure driven shock tube. These conditions assume 'free fall' collection during which no collisions take place in the sheath region. However careful examination of the mean free paths of ions and neutrals in the shocked gas leads to a conclusion that such a mechanism would be unlikely. Thus for the typical case of a M10 shock wave travelling into argon at a pressure of 1 torr, the neutral-neutral mean free path is 0.002 cm, while the ion-neutral mean free path is about 0.0003 cm because of electric polarization effects. Comparison of these lengths with the probe diameter, 0.03 cm, suggests that the conditions are closer to the continuum regime and that the motion of the ions is mobility controlled.

Two theories are available for the mobility controlled collection of positive ions by cylindrical probes in stationary plasmas; by Schulz and Brown (14), and Zakharova *et al* (15) respectively. Both theories, which assume that the potential difference between the probe and the space occurs across a finite sheath, were checked experimentally by the authors and found to be satisfactory. The applicability of these theories to the present situation was tested in the following experiment. For fixed initial conditions of argon pressure and shock Mach number the values of i_d at a fixed position in the shocked gas were measured for a range of values of V_d. This meant that i_d could be related to the potential difference between probe and plasma, V_p. The graph for the case of $A_p = 0.21\ cm^2$ and N^+ - 5 μsecs laboratory time behind the shock wave is shown in figure 6-4-1.

Two important features are evident from this investigation. First, the values of N^+ deduced according to the mobility controlled theories are orders of magnitude greater than that given by equation 6-4-1: for example, with $V_p = 10V$ and $i_d = 0.77$ mA., equation 6-4-1 yields a value for $N^+ = 2.9 \times 10^{11}$ ions cm^{-3}, while (14) would give $N^+ = 6.4 \times 10^{12}$ ions cm^{-3}, and (15) would give $N^+ = 1.4 \times 10^{14}$ ions cm^{-3}. The two mobility theories agree that the sheath thickness is about 1.5 probe radii. Secondly it is seen that:

$$i_d \propto V_p^{0.64} \qquad \text{(Eq. 6-4-2)}$$

In stationary conditions where the characteristic parameters of the ionized gas are similar to those in the high velocity shocked gas, the mobility theories both lead to a relationship:

$$i_d \propto V_p^{0.25} \qquad \text{(Eq. 6-4-3)}$$

It thus appears that the effects of the high velocity flow are sufficiently dominant to invalidate the direct application of these theories to the present situation.

The aerodynamic flow in the vicinity of the probe is complex. The axial flow Mach number of the ionized gas varies non-uniformly because of the build up of the wall boundary layer (16); at the shock front the value is 1.29 for an incident shock Mach number of 10, increasing rapidly mostly in the early part of the shocked gas, to reach 1.76 at the contact surface. From the shocked gas viscosity calculated according to the tables in (17) Reynolds numbers were in the range 900 - 1400 cm^{-1} for all the conditions in these experiments. According to the classification of Schaaf and Chambre (18) the flow here belongs to the slip regime where continuum effects, such as the formation of bow shock waves, may still be expected to occur. Mathematical analysis of collection under these conditions is difficult.

However to take some account of the gas flow relative to the probe it was proposed that an impact model similar to that outlined by Davis and Harris (19) for highly ionized gas flows was relevant. Thus referring to figure 6-4-2 it was suggested that all ion paths that crossed an area S in the free stream would be diverted sufficiently under the influence of the electric field to intercept the probe surface, and that intuitively S would be similar in size to the impact area of the probe. Thus the equation:

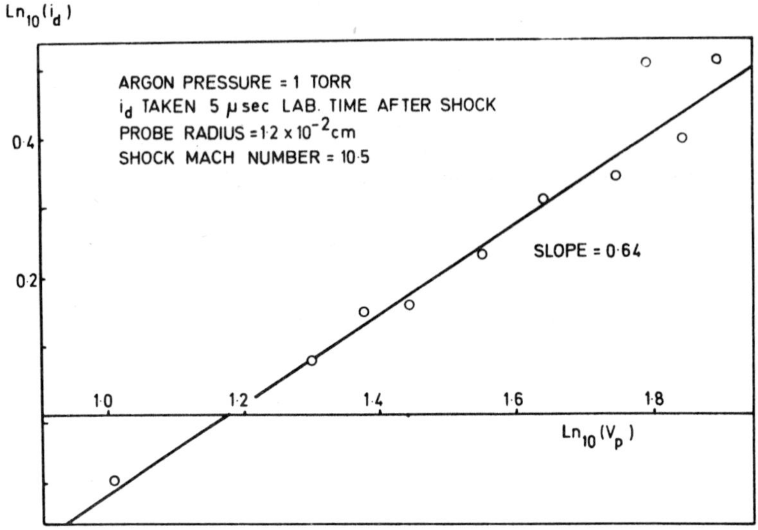

Fig. 6-4-1 Current-voltage characteristic.

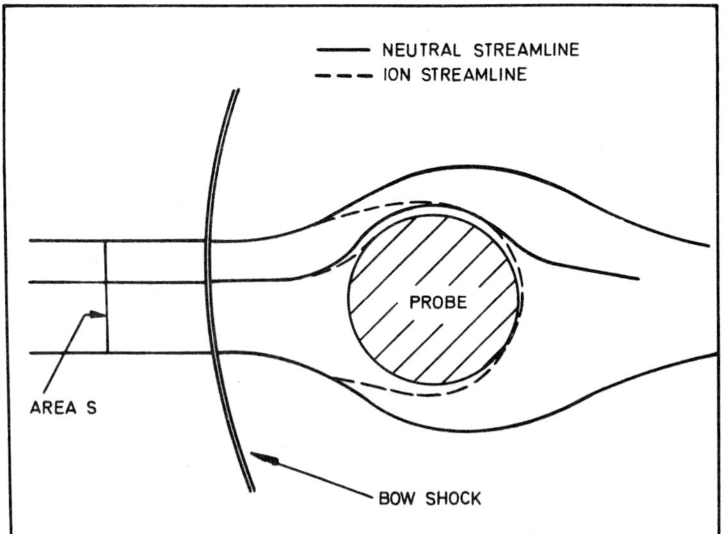

Fig. 6-4-2 Impact collection model.

$$i_d = N^+ueS \qquad \text{(Eq. 6-4-4)}$$

where u is the drift velocity of the shocked gas, relating the probe current to the free stream ion density. The dependence in equation 6-4-2 could thus be explained by a variation of S with V_p. Although for transverse cylindrical probes equation 6-4-4 gave values of N^+ which agreed with those indicated by the cavity experiment, its concepts were shown to be invalid by an experiment comparing the currents to two probes with the same A_p, but with values of S differing by a factor of 20: the currents collected were similar. Further evidence that i_d is characteristic of A_p rather than S was found by comparing the currents to six ion collectors with the same S and various A_p. The probes were constructed from needles aligned parallel to the shock tube axis and were differentiated by the thickness of the thin film of epoxy resin with which each was sheathed, so as to provide a range of collecting areas. It was observed that there was proportionality between i_d and A_p, and that the magnitude of i_d was again typical of collection according to equation 6-4-1.

The most likely collection mechanism is, therefore, that ions are carried to within a short distance of the probe surface, mainly by the motion of the neutral particles, while for the last few mean free paths the progress of the ions is determined by a combination of mobility and free fall effects. If ions are brought to the vicinity of the probe surface at a rate greater than, or comparable to, that at which they may be removed by free-fall collection, then the collection may be expected to be typical of the product of surface area and random thermal current. Since, in the shocked gases of these experiments, the drift velocity was about 1.5 times greater than the mean thermal speed, it was reasonable to expect that the volume in the proximity of the probe surface would be replenished sufficiently rapidly with charged particles. This problem is complex involving the combined actions of non-isotropic, aerodynamic and electrostatic influences in determining the paths of the ions to the probe. However until a better description of the phenomenon is available, advantage may be taken of this probe diagnostic by calibrating it against an independent means of finding the charged particle number density.

Probe Calibration

The apparatus for using a microwave cavity to measure electron concentrations in a shock heated gas, described in (13) (20), is shown schematically in figure 6-4-3. The X-band klystron is preset to operate at a frequency, f, greater than the resonant frequency, f_o, of the empty cavity. When the weakly ionized shock heated gas passes through the cavity, a resonance will be detected at the frequency f when the electron concentration in the cavity is that corresponding to the frequency shift $(f - f_o)$. The time at which this resonance occurs can be related to the shock front, whose path has been recorded using ancillary diagnostics such as thin film detectors and photomultipliers.

Improvements were made in this technique.

(a) Because the ionisation profile in a gas is sensitive to variation of the shock velocity, double probe systems were situated as closely as possible upstream and downstream of the cavity, and their individual responses were averaged and compared with that of the cavity. We found that the responses of similar upstream and downstream probes agreed to within 30 percent.

(b) Using a Tektronix Type 180A Time Mark generator to calibrate all oscilloscope traces and inter-scope delays, the instant of resonance was located to within a microsecond.

(c) The periods during which the shock front was contained in the cavity, or when the highly ionized region close to the contact surface was about to enter the

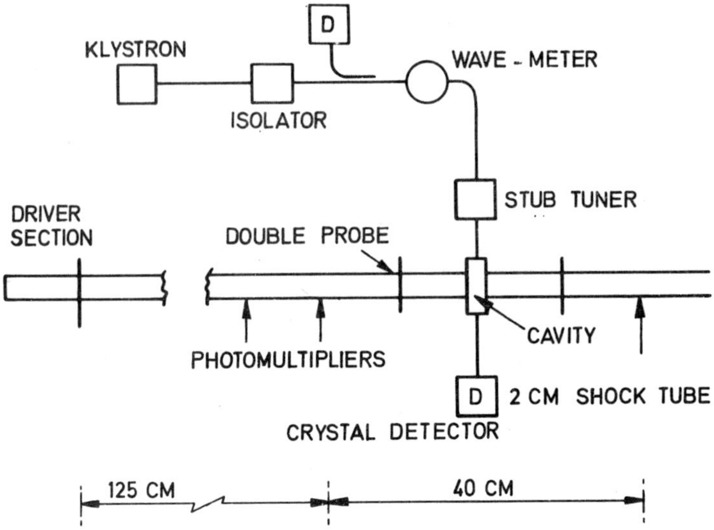

Fig. 6-4-3 Schematic of calibration arrangement.

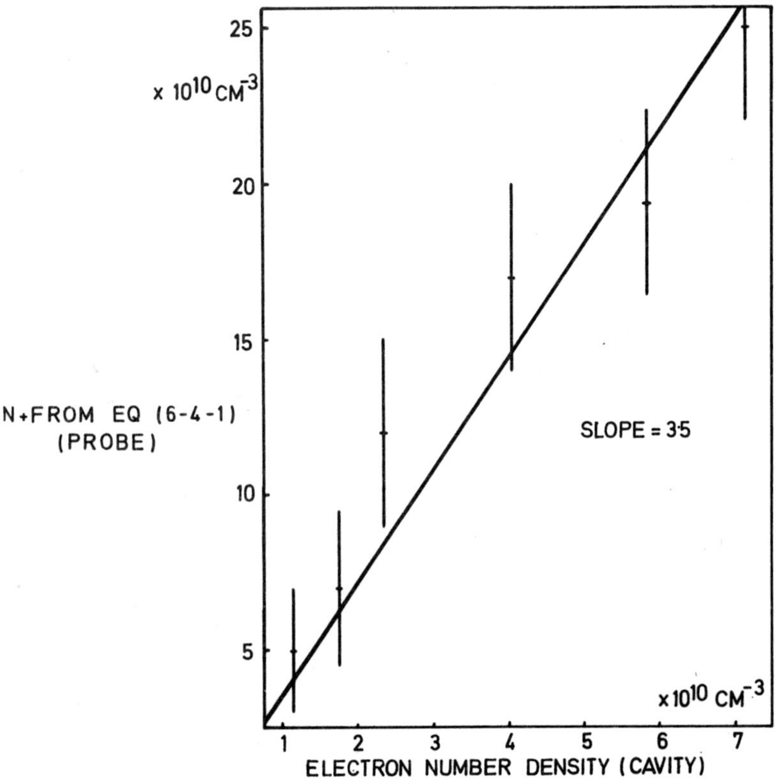

Fig. 6-4-4 Comparison of probe and cavity results.

cavity, were not recorded since the electron density profile in the cavity would then have been rapidly changing.

At the instant of resonance, the value of electron density as calculated from $(f - f_o)$ was assumed to be equal to N^+ at the mid-point of the cavity (when a cavity is filled with a slowly increasing ionization profile, this is a reasonable approximation). The experiment was carried out using spectroscopic grade argon in a 2-cm i.d. shock tube with a downstream pressure of 2 torr and with a shock Mach number of 9.2. The probes were similar to those used in the work of the following section, and a differential voltage of 5V was applied.

Figure 6-4-4 shows points obtained during this experiment; the ordinate is the value N^+ obtained from the probe responses using equation 6-4-1 and the abscissa is the value of electron number density found from the cavity frequency shift. The error bars on the points indicate the probable accuracy of determining the instant of cavity resonance. The ion density N^+ calculated from equation 6-4-1 is about 3.5 times larger than that obtained from the cavity estimate, assuming an equality of the ion and electron densities, indicating that the value of K for that experimental arrangement was about 3.5. This factor may represent an enhancement of the current actually collected by the probe as a result of either the electric field or the ionized gas flow. Possibly relaxation phenomena occurring in the gas behind the bow shock wave were responsible for part of the factor, but the self-consistency of the measurements reported in the following section, and their close agreement with measurements in other laboratories (2) (4) indicate that they did not dominate the probe response in pure argon. This conclusion must not be generalised at present to include similar measurements in pure gases of lower excitation and ionization energies, or to gases containing impurities of low ionization potential.

This preliminary calibration indicates that the probe data may be reduced to free stream ion number densities by means of the following expression:

$$i_d = K \left[\frac{1}{4} N^+ e \sqrt{\frac{8\,KT}{\pi\, m}}\right] Ap \qquad \text{(Eq. 6-4-5)}$$

in which K is an experimentally determined constant equal to about 3.5 for the present experimental arrangement. Experiments are currently being extended to determine K to a higher precision.

Initial Ionization Processes in Argon

Introduction

It is now well established (1) (2) (4) (6) that when a shock wave raises the temperature of pure argon from about 300°K to 10,000°K, thermal ionization proceeds in distinct stages, described recently (6) as Regimes I, II and III. In Regime I atom-atom inelastic collisions are responsible for a relatively slow rate of ionization. In Regime II electron-atom collisions predominate and the rate of ionization is significantly greater than in I. In Regime III the level of ionization approaches and finally reaches an equilibrium state and electron loss mechanisms such as recombination become increasingly important. The measurements reported here were concerned only with the processes taking place in Regime I. The evidence available indicates that here the ionization process is a two step mechanism involving the first excited state of the neutral atom and that the reactions are as follows:

$$A + A \xrightarrow{k^*} A^* + A \qquad \text{(Eq. 6-4-6)}$$

$$A^* + A \xrightarrow{k^+} A^+ + A + e \qquad \text{(Eq. 6-4-7)}$$

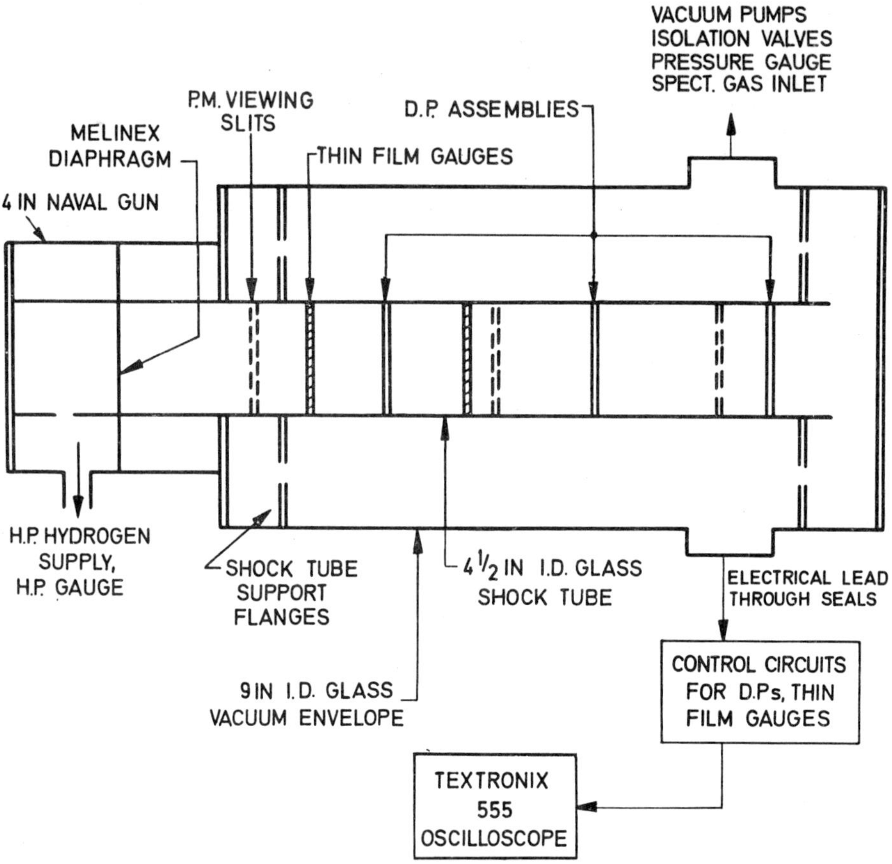

Fig. 6-4-5 Schematic of $4\frac{1}{2}$" Shock Tube.

The results showed that the degree of excitation (N^*/N_a) was between 10^{-6} and 10^{-8}, and hence that the reaction $A^* + A^* \rightarrow A^+ + A + e$ may be neglected.

The inelastic collision cross sections σ^* (E), σ^+ (E), and rate constants k^* (T), and k^+ (T) corresponding to reactions described by equations 6-4-6 and 6-4-7 were measured experimentally in the work reported below.

Apparatus

Experiments were carried out in the precision bore, 4.5 in internal diameter, glass shock tube, described in (21) and shown schematically in figure 6-4-5. These have confirmed and extended earlier work in a 4 cm i.d. shock tube of similar construction. The driver end of the 4.5 in tube was the modified breech of a 4" naval gun, the rifled barrel of which was bored out to 4.5 inches. The breech loading mechanism was adapted to facilitate quick diaphragm changing and to minimise the recycling time. The barrel length was 4.3 metres and all observations were made in the precision bore glass extension section downstream of the muzzle. The overall length was 12 metres. A cold hydrogen driver was used with Melinex (Mylar) diaphragms burst by overpressurisation, and shocks in the range Mach 8 to Mach 12 were driven into argon at initial pressures p_1 between 0.2 and 3.0 torr. The variation in level of ionization with time behind the shock front was measured using the floating double probe technique discussed in the previous section. Further diagnostic techniques used in these experiments included thin platinum film gauges and collimated photomultipliers. These enabled accurate measurements to be made of the shock velocity, and of the location of both the shock front and the contact surface (12). Such measurements were carried out at stations spaced 20 cms apart along the shock tube, and in each instance a detailed x - t plot of the shock front was constructed for at least the 200 cm section spanning the probe detector position. In this way the shock velocity at the probe location was determined to an accuracy better than ± 1 percent at Mach 10 and this represents a maximum uncertainty in shocked gas temperature of ± 200°K. Using Melinex diaphragms for all the work reported here, the shot-to-shot consistency for common downstream conditions and driver pressure was high and the diagnostic responses were completely reproducible.

The ultimate vacuum attained before a shot was better than 5×10^{-7} torr, and the combined leak and outgassing rate for the shock tube gave a pressure rise of less than 2×10^{-6} torr/minute when the vacuum isolation valves were closed and a lN_2 cold trap was used. During any experiment the usual time delay between isolation and the bursting of the diaphragm was less than 1 minute, so that the maximum impurity level in the gas at 1 torr initial pressure was about 7 ppm including the 4 ppm inherent in the spectroscopic grade gas used.

In this work a typical double probe consisted of two 0.03 cm diameter tungsten or steel wires held firmly under tension, positioned diametrically in the shock tube, parallel to one another and about 1 cm apart. The ends of the wires passed through 0.065 cm diameter holes drilled cleanly in the glass wall of the tube and were secured on mountings fixed to the outside wall of the shock tube (21). Electrical connections from the probes passed through the vacuum envelope via multiple lead-through seals to the probe voltage control circuit situated outside the envelope, as shown in figure 6-4-5. In general, several double probes were installed spaced about 40 cms apart. Electrical leads from the platinum thin film detectors located on the inside wall of the shock tube, were taken out in a similar way.

By means of the external circuit a differential voltage V_d could be applied between the wires forming the double probe, and the differential probe current flowing in the circuit was recorded using a Tektronix P6016 current probe coupled via a Type 131 amplifier to a Tektronix 555 double beam oscilloscope. The time constant of

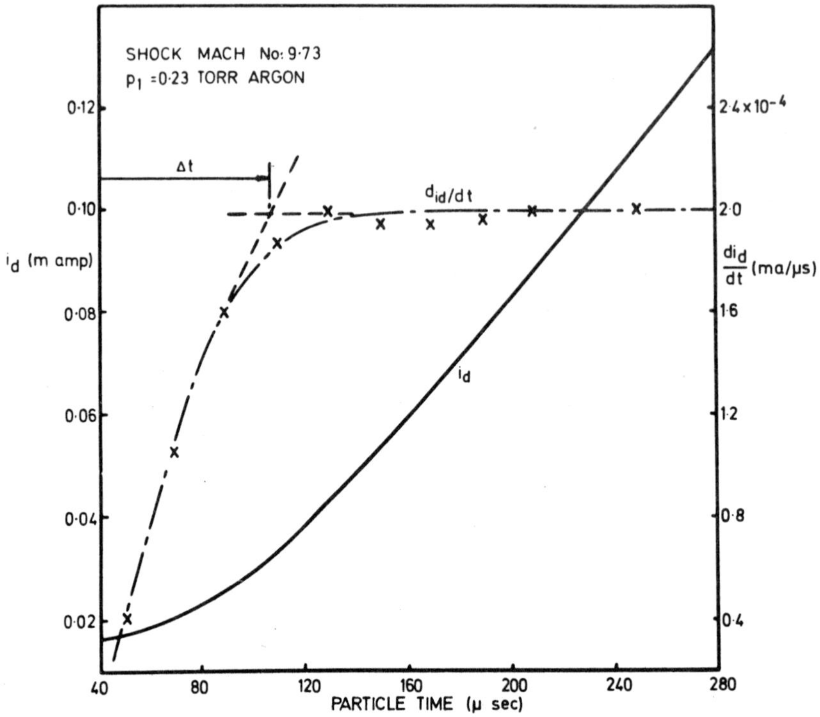

Fig. 6-4-6 Analysis of double probe response.

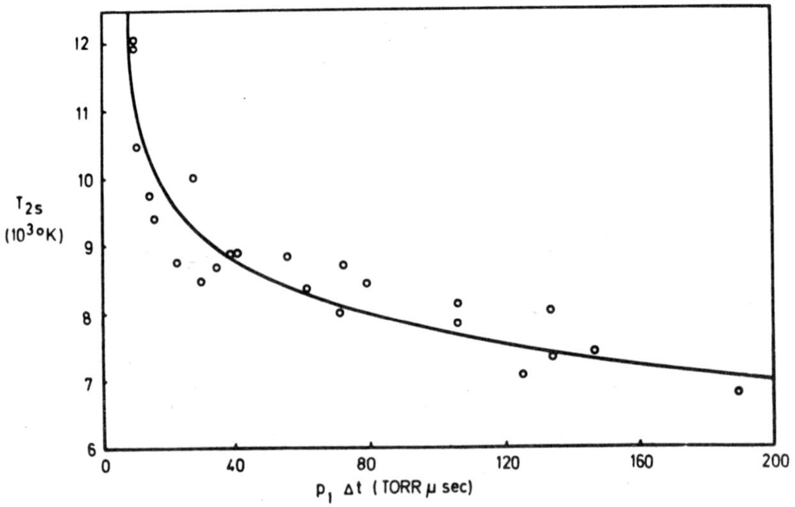

Fig. 6-4-7 Ionization incubation Time.

the complete circuit was about 0.1 μ sec.

Theory

Theoretical expressions relating ionization rates, cross sections and reaction rates for the processes described in equations 6-4-6 and 6-4-7 have been adequately formulated by Harwell and Jahn (2) and Morgan and Morrison (4) and are summarised below:

a. In a binary collision process, such as described by equation 6-4-6 the rate at which excited particles A^* are produced can be described by:

$$\frac{dN^*}{dt} = 4 S_1^* (\pi m)^{-1/2} N_a^2 (kT)^{3/2} \left(\frac{E_a}{2kT} + 1\right) \left(\exp\left(-\frac{E_a^*}{kT}\right)\right)$$

$$= k^* (T) N_a^2 \qquad \text{(Eq. 6-4-8)}$$

In this expression it is assumed that the cross section σ^* for the binary collision leading to the excited state A^* is linear with energy above threshold so that $\sigma^*(E) = S_1^* (E - E_a^*)$ for $E > E_a^*$. E_a^* is the excitation energy involved in the transition $A \rightarrow A^*$, E is the relative energy of the colliding particles, and N_a is the number density of ground state atoms of mass m. Similarly the rate at which A^+ are produced by $A^* \rightarrow A^+$ collisions can be expressed as:

$$\frac{dN^+}{dt} = 8 S^+ (\pi m)^{-1/2} N_a N^* (kT)^{3/2} \left(\frac{E_a'}{2kT} + 1\right) \exp\left(\frac{-E_a}{kT}\right)$$

$$= k^* (T) N_a N^* \qquad \text{(Eq. 6-4-9)}$$

Here E_a' corresponds to the minimum energy required to ionize an excited atom.

b. Excited argon atoms are rapidly ionized by the mechanism described in equation 6-4-7 so that the overall ionization rate is controlled by the rate of excitation.

c. Ionization through atom-atom collisions is the dominant process until in Regime II the rate of excitation of argon atoms through electron-atom collisions exceeds the atom-atom excitation rate.

d. In Regime I, if collisional and radiative de-excitation are neglected, ionization is negligible behind the shock front for a time given by $\Delta t = (k^+ N_a)^{-1}$ and the level of excited atoms builds up at a steady rate. At times greater than $(k^+ N_a)^{-1}$ ionization proceeds at a constant rate with $\frac{dN^+}{dt} = \frac{dN^*}{dt}$.

e. After the incubation period, the electron density will increase linearly with time behind the shock front until electron-atom collisions become predominant in Regime II when the electron density will increase exponentially.

f. Equation 6-4-8 can be differentiated with respect to $(\frac{1}{kT})$ and if $kT \ll E_a^*$ then:

$$\frac{d(\ln \dot{N}^+)}{d(kT)^{-1}} = -E_a^* \qquad \text{(Eq. 6-4-10)}$$

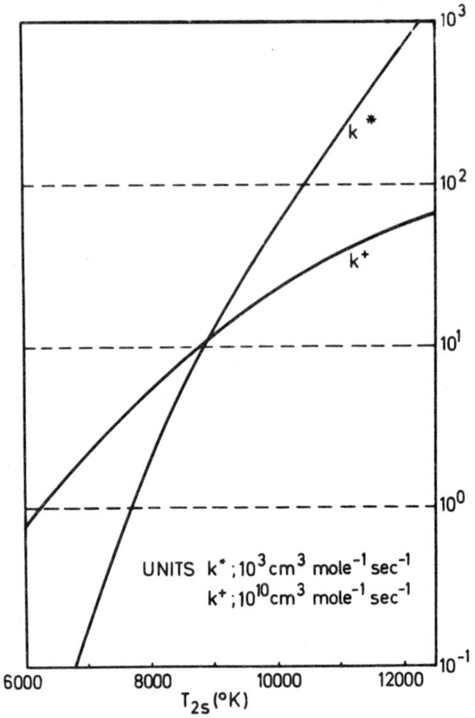

Fig. 6-4-8 Rate constants k*, k+.

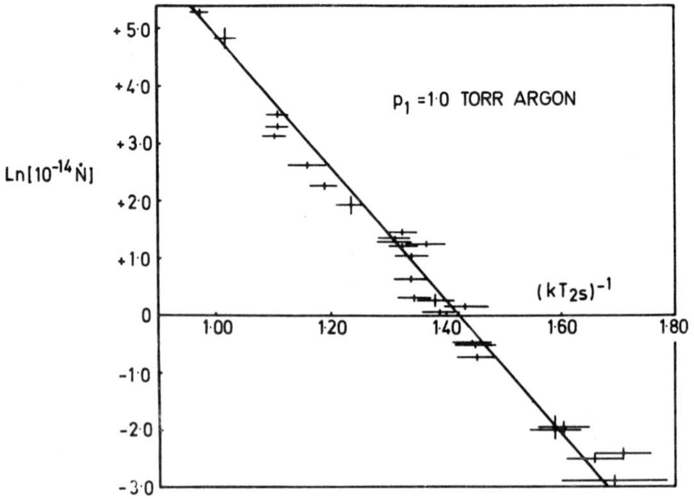

Fig. 6-4-9 Arrhenius plot of ionization rate.

This is the Arrhenius equation from which the characteristic activation energy E_a^* may be calculated, if the derivative expressions are first obtained by experiment.

Flow Corrections

The effects on the flow due to the boundary layer in low pressure shock tubes have been theoretically treated by Mirels (16). The implications relating to kinetic rate measurements in shock tubes have been discussed elsewhere (22) and the non linear time correction curve that relates laboratory to particle time is shown. A further significant correction to the measured rates is consequent upon variations in shocked gas number density and temperature through the shocked gas length. This has been developed by Mirels (16) and evaluated for the conditions of this experimental work (22) (23).

Analysis of Data

The probe response recorded gave the variation of ion number density with laboratory time behind the shock front. All oscilloscope traces and time delays between responses were time calibrated using a Tektronix 180A Time Mark Generator. The laboratory time scales were then converted to true particle time scales using the curves shown in (22) and (23) and the corrected probe current profile was plotted as in figure 6-4-6. The local slope of the profile so obtained was measured for small increments of time, and each value of $\frac{di_d}{dt}$ (proportional to $\frac{dN^+}{dt}$) was corrected for the variation in density and temperature through the shocked gas. These corrected values also are plotted against time in figure 6-4-6, and as predicted for a process described by equations 6-4-6 and 6-4-7 the ionization rate is seen to increase during an 'incubation' time, Δt, and then reach a constant value corresponding to the linear portion of the curve. The constant value of $\frac{di_d}{dt}$ is converted to an equivalent $\frac{dN^+}{dt}$ using equation 6-4-1.

Results

Figure 6-4-7 shows a plot of the incubation time normalised for density against shocked gas temperature. From this k^+ is evaluated as a function of temperature and this is shown in figure 6-4-8.

An Arrhenius plot of ln $(\dot{N}^+)/(kT_2)^{-1}$ for which the rates of ionization are obtained from current profiles (Fig. 6-4-6) is shown in figure 6-4-9. All data shown here were obtained in Argon at $0.2 \leq p_1 \leq 3.0$ torr, in the 4.5 in shock tube and were then quadratically normalised to a number density corresponding to $p_1 = 1$ torr. The activation energy as measured by the slope of the Arrhenius plot was 11.4 ± 0.3 e.v., the line drawn in on figure 6-4-9 having the slope 11.6 e.v. The maximum degree of ionization obtained during these experiments was of the order of 10^{-6}. For

$$\text{time} > (k^+ N_a)^{-1}, \quad \frac{dN^+}{dt} = \frac{dN^*}{dt} = k^* N_a^2 \qquad \text{(Eq. 6-4-11)}$$

Thus the variation of k^* (T) with temperature was obtained and this also is plotted in figure 6-4-8.

The steady state degree of excitation can now be established since:

$$\frac{N^*}{N_a} = \frac{k^*(T)}{k^+(T)}$$

The variation of $\frac{N^*}{N_a}$ with temperature is tabulated in Table 6-4-1 where it is seen to be in the range 10^{-6} - 10^{-8} for the conditions of these experiments.

Using equations 6-4-8 and 6-4-9 the cross section proportionality constants S_1^*, for the reaction A + A → A* and S^+ for the reaction A* + A → A^+ + A + e can be related as follows:

$$\frac{S^+}{S_1^*} = \frac{N_a}{2N^*}\left[\frac{E_a^* + 2kT_2}{E_a' + 2kT_2}\right] \exp\left[-\frac{E_a^*}{kT_2} + \frac{E_a'}{kT_2}\right] \qquad \text{(Eq. 6-4-12)}$$

Results for equation 6-4-12 are tabulated in Table 6-4-1, showing the cross section for the second step of the ionizing process to be about 10^3 greater than that for the original excitation collision. No previous estimates of S^+ for argon are available, although Johnston and Kornegay (24) have estimated $\frac{S^+}{S_1^*}$ to be about 10^3 for Xenon. However, they had made no allowance for the variation of cross section with energy.

The value of $S_1^* = (3.3 \pm 0.6)\ 10^{-20}$ cms^2/e.v. is smaller than the value of 7 x 10^{-20} cms^2/e.v. found by Morgan and Morrison to give agreement with the results of Petschek and Byron for Regimes II and III.

Discussion

The factor of two between the result obtained here for S_1^* and the value reported by Morgan and Morrison is more than contained in the additional correction terms we used to allow for the flow non uniformities resulting from the boundary layer effects. The activation energy measured by the probes for the process of ionization was close to that of the first excited state of argon, and this agrees with the conclusions of our earlier work. The transverse cylindrical double probe does therefore seem to provide a self-consistent diagnostic technique for use when measuring free stream ionization rates behind incident shock waves in pure argon. The validity of applying a quadratic density normalisation factor as appropriate to a binary collision process such as one described by equation 6-4-6 to the measured ionization rates is confirmed by the Arrhenius plot of figure 6-4-9. The normalisation factor here varied over some two orders of magnitude, corresponding to the initial pressure range of 0.2 to 3.0 torr argon. Furthermore the results are entirely self-consistent over a broad range of temperature, from below 7000°K to 12000°K.

This measure of agreement has been reached on the assumption that the probes were recording the number density of free stream ions just as if they were collecting ions under free fall conditions; in a stationary plasma this would imply particle mean free paths greater than the probe diameter. The collection mechanism is certainly not dominated by the ion mobility which -drawing the analogy with mobility controlled collection in a stationary plasma - would lead to free stream ion densities perhaps two orders of magnitude greater than those obtained in the previous section. In fact the calibration experiment indicated that, if anything, the probes were measuring about 3.5 times the free stream ion density detected by the cavity. This is opposite to what would have been measured had mobility considerations been limiting the current collected at the probe surface.

It appears that the collection process is in fact dominated by the gas flow. The experiments with probes of different collecting surface areas show that the probe current measured was proportional to the collector surface area. These and the other experiments reported briefly previously led to a collection model in which the incident gas flow replenished ions and electrons within about one mean free path of the probe surface, at a rate at least equal to that at which they were collected from the flow under free fall conditions. Further experiments are being conducted to determine the mechanism in more detail and to enable absolute measurements of free stream ion density to be obtained more precisely than the factor of 3.5 that has been measured in the preliminary calibration.

Conclusions

The electrostatic probe is seen to be a potentially powerful diagnostic technique for the measurement of ionization rates in supersonic flows. Certainly in argon the good temporal resolution of which it is capable has enabled reaction rates and the cross section proportionality constant S^+ to be measured, in contrast to the limitations in resolution and sensitivity of other diagnostic techniques for such measurements. However, care should be taken in applying the technique to other gases before calibration experiments are conducted in these gases, or until a dependable model of the collection process is available. This is particularly relevant to cases in which the characteristic relaxation times for free stream ionization processes may differ substantially from those of argon and thereby lead to secondary ionization processes arising out of the interaction of the probe with the incident flow.

Acknowledgements

This work has been supported by the Ministry of Technology (Ministry of Aviation).

One of us (T.I.McL.) is grateful for a Research Fellowship firstly from the Ministry of Aviation and latterly from Science Research Council. Another (S.W.R.) is grateful to Ministry of Education, Northern Ireland, for a Research Studentship held during part of this work.

References

1. **Petschek, H.E., Byron, S.,** Annals of Physics, 1, 270 1957.

2. **Harwell, K.E., Jahn, R.G.,** Phys. Fluids, 7, 2, 214, 1964.

3. **Kelly, A.J.,** J. Chem. Phys., 45, 5, 1723 1966.

4. **Morgan, E.J., Morrison, R.D.,** Phys. Fluids, 8, 9, 1608, 1965.

5. **Coates, P.B., Gaydon, A.G.,** Proc. Roy. Soc., A293, 452, 1966.

6. **Wong, H., Bershader, D.,** J. Fl. Mech. 26, 459, 1966.

7. **Talbot, L.,** Phys. Fluids, 3, 2, 289, 1960.

8. **Pollin, I.,** Phys. Fluids, 7, 9, 1433, 1964.

9. **Chung, P.M., Blankenship, V.D.,** A.I.A.A. J., 4, 3, 442, 1966.

10. **McLaren, T.I., Fox, J.N., Hobson, R.M.,** Nature 198, 1264, 1963.

11. **Johnson, E.O., Malter, L.,** Phys. Rev. 80, 58, 1950.

12. **McLaren, T.I., Hobson, R.M.,** Proc. VIIth Int. Conf. on Phenomena in Ionized Gases, Vol. 2, 759, 1965.

13. **Fox, J.N., Hobson, R.M.,** Proc. VIIth Int. Conf. on Phenomena in Ionized Gases, Vol. 2, 763, 1965.

14. **Schulz, G.A., Brown, S.C.,** Phys. Rev., 98, 1942, 1955.

15. **Zakharova, V.M., Kagan, Y.M., Mustafin, K.S., Perel, V.I.,** Z. Tech. Fiziki, 30, 422, 1960.

16. **Mirels, H.,** Phys. Fluids 9, 10, 1907, 1966.

17. **Hirschfelder, J.O., Curtiss, C.F., Bird, R.B., Spotz, E.L.,** High Speed Aerodynamics and Jet Propulsion, Princeton Univ. Press Vol.1, Section D. 1958.

18. **Schaaf, S.A., Chambre, P.L.,** High Speed Aerodynamics and Jet Propulsion, Princeton Univ. Press Vol.1, Section H. 1958.

19. **Davis, A.H., Harris, I.,** Advances in Applied Mechanics, Supp. 1. Academic Press, 1961.

20. **Fox, J.N., McLaren, T.I., Hobson, R.M.,** Phys. Fluids 9, 12, 2345. 1966.

21. **Fox, J.N., McLaren, T.I., Radcliffe, S.W., Hobson, R.M.,** To be published.

22. **McLaren, T.I., Hobson, R.M.,** To be published.

23. **McLaren, T.I., Hobson, R.M.,** VIIIth Int. Conf. on Phenomena in Ionized Gases, Vienna, Section 4.5, 1967.

24. **Johnston, H.S., Kornegay, W.,** Trans. Faraday Soc., 57, 1563, 1961.

Table 6-4-1

T_{2s} (°K)	$\frac{N^*}{N_a} = \frac{k^*}{k^+}$ (10^{-7})	$\frac{S^+}{S^*}$ (10^3)	S^* $(10^{-20}cm^2\ e.v.^{-1})$	S^+ $(10^{-17}cm^2\ e.v.^{-1})$
7000	0.09	0.56	3.8	2.1
8000	0.39	0.59	3.4	2.0
9000	1.04	0.69	3.2	2.2
10000	2.41	0.76	3.1	2.3
11000	5.27	0.79	2.9	2.3
12000	11.06	0.71	2.7	1.9

Mathematical List of Symbols

A_p	Probe surface area.
E_a^*, E_a'	Activation energies for processes 6 and 7.
f	Klystron frequency.
f_o	Resonant frequency of empty cavity.
$k^*(T)$, $k^+(T)$	Temperature dependent rate constants for processes 6 and 7
k	Boltzmann's constant.
K	Probe calibration constant.
m_e	Mass of electron
M	Mass of ion
N_a, N^*, N^+	Number densities of neutral atoms excited atoms and ions respectively.
$\frac{dN^*}{dt}$	Rate of ionization through process 6
$\frac{dN^+}{dt} = \dot{N}+$	Rate of ionization through process 7
p_1	Initial shock tube gas pressure.
S	Probe impact area.
T	Ion temperature.
T_2	Atom temperature in shocked gas.
T_{2S}	Atom temperature immediately behind shock front.
u	Shocked gas flow velocity.
V_d, i_d	Differential probe voltage and current respectively.
V_p	Probe to plasma potential.
$\sigma^*(E)$, $\sigma^+(E)$	Energy dependent collision cross sections for processes 6 and 7 respectively.
S_1^*	Excitation cross section coefficient defined by $\sigma^* = S_1^* (E - E_a^*)$ where E is the relative energy of the collecting atoms.
S^+	Ionization cross section coefficient for process 7 defined by $\sigma^+ = S^+ (E - E_a')$.
Δt	Ionization Incubation Time.

Part 7

Plasma Flow Measurements

7-1

Plasma Diagnostic Techniques

A. E. FUHS
Naval Postgraduate School, Monterey, Calif., U.S.A.

Summary

Active magnetic probes permit transducers to be designed which avoid problems due to heating from the plasma. Active probes gain information about plasma properties through eddy currents induced in the ionized gas. A family of instruments based on the magnetic-field-plasma interaction is discussed.

Low density plasmas can be diagnosed by observing radiation excited by a high energy electron beam. Scattered electrons also yield information. The conditions that are required for the application of this technique are examined. The types of data to be expected are discussed.

One experiment which is simple to execute but complex to interpret involves electrostatic probes in a static plasma. Use of probes in a flowing plasma adds a new degree of complexity and increases the interpretation difficulties. A survey of current theories is presented and regimes are defined where the various theories are adequate.

Interferometry leads to knowledge concerning the index of refraction, n, of a gas. If the gas is a plasma the concentration of electrical charges has an influence on n. Holographic interferometry, which is discussed in considerable detail, offers a new method for measuring n in two dimensions. In addition, there is hope of unraveling a three dimensional variation in n; three-dimensional interferograms are obtained.

Giant pulse lasers are a source suitable for other plasma visualization techniques, namely schlieren and shadowgraph. Scattered radiation yields information on electron density. The experiment involving scattered light is in contrast to electrostatic probes; the apparatus is quite complex whereas the data reduction is relatively straightforward.

A new technique using microwaves to determine electron density profiles is discussed; the technique is in the conceptual stage. A diagnostic method based on excitation of electroacoustic waves is also discussed.

Sommaire

Des sondes magnétiques actives permettent la réalisation de montages évitant les difficultés dues au chauffage par le plasma. Les sondes actives tirent leurs informations sur les propriétés des plasmas à partir des courants induits dans le gaz ionisé. On discute des caractéristiques d'une famille de sondes basées sur l'interaction champ magnétique-plasma.

Les diagnostics de plasmas à faible densité s'obtiennent par observation du rayonn-
ff* ement excité par un faisceau d'électrons à haute intensité. La dispersion des

électrons donne aussi quelques renseignements. On examine les conditions dans lesquelles cette technique peut être utilisée. On discute les types de données que l'on peut obtenir.

Une expérience facile à faire, mais difficile à interpréter, est l'emploi de sondes électrostatiques dans un plasma au repos. L'emploi de ces sondes dans un écoulement de plasma ajoute un degré de complexité supplémentaire et accroît les difficultés d'interprétation. On présente une revue des théories actuelles et on définit les régimes ou les diverses theories s'appliquent.

L'interférométrie permet d'obtenir des renseignements sur l'indice de réfraction n d'un gaz. Si ce gaz est un plasma, la densité de charges électriques a une influence sur n. Les hologrammes interférométriques, technique analysée en detail, donnent un moyen de mesurer n dans deux dimensions. Il y a également un certain espoir de pouvoir connaître n dans la troisième dimension : des interférogrammes à trois dimensions ayant été obtenus.

Des lasers géants à fonctionnement par impulsion constituent une source intéressante pour d'autres visualisations dans les plasmas, notamment méthode des ombres et strioscopie. Le rayonnement diffus donne des indications sur la densité électronique. Cette dernière technique s'oppose à celle des sondes électrostatiques : le montage est compliqué, mais l'interprétation des résultats est relativement facile.

On analyse une nouvelle technique de détermination de la densité électronique au moyen d'ondes ultra-courtes. La technique est encore en cours de développement. Une méthode basée sur l'excitation d'ondes électroacoustiques est également discutée.

Introduction

With the advent of hypersonic flight, real gas effects became a new area of study for aerodynamicists. Real gas effects include ionization and the production of plasma, and aerodynamicists and physicists with an applied or engineering outlook put together plasma and fluid motion with the goal of MHD power generation. Rocket exhausts have significant levels of ionization; consequently, electrical engineers found difficulty with telemetry to rocket boosted vehicles. Need for an understanding of plasma phenomena arose in many engineering problems. Accompanying the need for information there have been new requirements for instrumentation, and new instruments have been developed. Not all the instrumentation techniques have been borrowed by plasma engineers from fusion physicists. Some widely-used wind tunnel techniques also have found their way into the plasma research laboratory, and Schlieren and shadowgraphs of θ-pinches have appeared recently in the literature.

Plasmas associated with fusion research usually have higher T_e and much lower mass density ρ when compared to many engineering plasmas. Fusion research plasmas usually are fully ionized, while many engineering devices have plasmas in which the degree of ionization is very slight. Low density is encountered in plasma propulsion, plasma diodes, and MHD wind tunnels. Hence, when a technique is borrowed from fusion research, modification may be necessary. For example, Langmuir probe data from fusion research can usually be interpreted in terms of a collision-free model, but when a Langmuir probe is used in an MHD generator, the density is usually so high that a continuum model must be used.

Magnetic Probes

Magnetic probes can be classified as active or passive. When the transducer provides the magnetic field for the measurement, it is an active probe; otherwise, it is passive. Passive probes are discussed extensively in the literature; for an excellent summary the reader is referred to Lovberg's article (1) in the book edited by Huddlestone and Leonard.

A subclassification for magnetic probes applies to an active transducer which may be with or without electrodes. A thorough discussion of electrode-type probes appears in the monograph by Shercliff (2). Electrodeless probes are discussed in the monograph by Fuhs (3).

Electrodeless instruments depend on the interaction between the applied magnetic field and the plasma. Eddy currents arise from the interaction due to a time varying magnetic field or due to plasma motion through the field. A parameter that characterises the origin of the eddy currents is $\omega L/U$. Instruments have been used where this parameter is small, unity, and large.

Another parameter for classification of electrodeless active probes in the magnetic Reynolds number which is expressed in the form:

$$R_m = \mu \sigma U L$$

for probes with $\omega L/U \ll 1$. An alternative form for probes with $\omega L/U \gg 1$ is:

$$R_m = \mu \sigma \omega L^2$$

which is also related to the customary expression for skin depth.

RF Bridge

One of the first applications of the RF bridge to laboratory plasma diagnostics was by Koritz and Keck (4) who used the device to measure the electrical conductivity of the wake of a projectile in a ballistic range. Subsequently Grabowsky and Durran (5) designed a RF bridge for use with an MHD accelerator or wind tunnel. Fuhs, Grabowsky, and Gibb (6) studied the feasibility of using the RF bridge to measure the electrical conductivity of the plasma sheath surrounding a reentry vehicle (RV). A self-balancing RF bridge was developed by Fuhs and Gibb (7) for use on board a RV. It should be emphasized that although the motivation for developing the self-balancing bridge was for use on board a RV, it is a valuable addition to any laboratory instrumentation based on bridges.

A performance map for a RF bridge is shown in figure 7-1-1. Details for determining a performance map are given in (6). The lower limit for measuring electron density is established by a signal too small to detect. An upper bound on n_e is imposed by the skin depth of the applied magnetic field. This is indicated by a line with a value $S/\delta = 10$ where S is the skin depth and δ is the thickness of plasma layer to be measured; for this case $\delta = 0.03$ m. Another aspect that sets an upper bound on electron density is that the signal becomes no longer linearly related to plasma conductivity. This is indicated by $\Delta R/Z = 0.03$ where ΔR is the change of coil resistance due to plasma and Z is the complex impedance of one leg of bridge. At low density the conductivity becomes a tensor in the presence of a magnetic field. The line drawn for $\omega_c/\nu = 1$ is based on B = 0.5 gauss (Earth's field). Plotted in figure 7-1-1 are two curves labeled $\mathcal{R} = 10$ and $\mathcal{R} = -10$. The quantity $\mathcal{R}$ is the ratio of the in-phase component of plasma current to the reactive current. For $\mathcal{R} = +10$ the coil is losing power as radiated electromagnetic waves. For $\mathcal{R} = -10$ inertia of the electrons has caused a phase shift of the current relative to the electric field. Figure 7-1-1 shows a wide operating range in ambient gas density and electron density: the desired operating region is the region labeled 'linear region' between signal and σ.

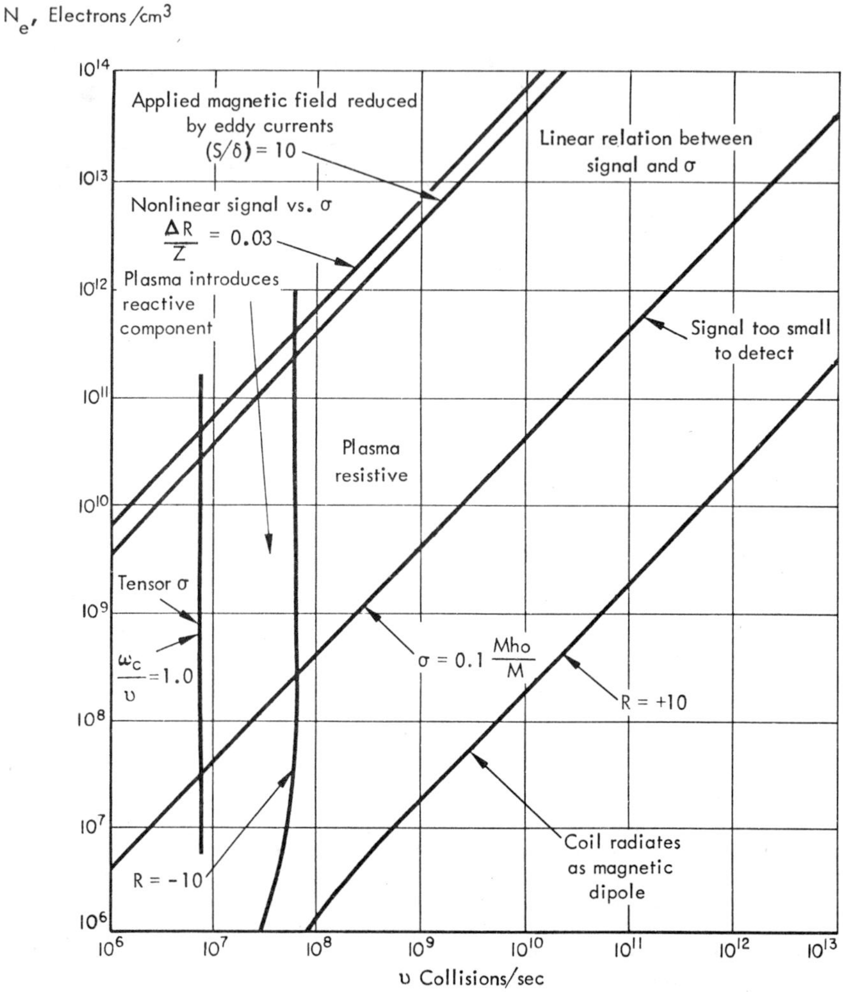

Fig. 7-1-1 Performance map for RF electrical conductivity bridge

Figure 7-1-1 was drawn for a bridge frequency, f = 1 megahertz; the coil radius was $r_c = 0.03$m; the coil inductance, 10^{-5} henry, and the distance from the coil to the plasma, a = 0.003m.

A block diagram of a self-balancing bridge is given in figure 7-1-2, and the bridge circuit is shown in figure 7-1-3. When biased correctly, a field effect transistor FET acts as a variable resistor. The resistance from the source to the drain is a function of the gate voltage. For example, one FET 2N2498 was found to have source-to-drain resistances of 600, 820, and 1140 ohms for gate voltages of -200, 0, and 200 millivolts.

The time constants for the self-balancing networks are controlled by $R_1 C_1$ and $R_2 C_2$; see figure 7-1-3. There are two feedback networks. One controls unbalance in phase with the reference voltage, and the other controls the quadrature component of unbalance. For a detailed description of the self-balancing bridge, see (7).

RF electrical conductivity bridges operate with the parameter $\omega L/U >> 1$. One can usually choose ω to make $\mu \sigma \omega L^2$ less than unity. If this is done, data interpretation is considerably less complex.

Instruments with Small $\omega L/U$

For this class of instruments the eddy currents in the plasma are the result of plasma motion through the magnetic field. The quantity that is measured is the product of electrical conductivity and plasma velocity, σU (3). These instruments measure magnetic Reynolds number since the variable quantities in the expression for R_m are σ and U. The usual item of interest is electron density M_e. Knowing the plasma velocity, the electrical conductivity can be determined. The electrical conductivity, for a magnetic field small enough to have $\omega_c \tau$ much less than unity, is proportional to the ratio of n_e to the electron collision frequency.

When there is a spatial variation in σU, due to variation either in σ or U or both, the instruments measure a weighted average. Instruments giving an average $\langle \sigma U \rangle$ have been built to operate with permanent magnets or with a frequency at the low end of the audio range.

An instrument to determine the spatial profile of σU has been used. The basis for its design is the fact that spatial averages are obtained for a given transducer geometry. By varying the geometry a series of averages is obtained and these can be related to the profile of σU.

Other Electrical Conductivity Meters

The presence of plasma near a coil carrying an alternating current manifests itself by changing the resistance and inductance of the coil, thereby altering phase angles, shifting the resonant frequency, and increasing the power dissipation. Each of these phenomena has been the basis for an instrument design.

Change in resonant frequency was used by Blackman (8) and by Savic and Boult (9) to measure electrical conductivity. Change in coil resistance - or more precisely, apparent change in resistance - due to the plasma was used by Moulin and Massê (10) to determine plasma conductivity. Olson and Lary (11) measured power dissipation with a probe immersed in the plasma. Crapo, Hill, Plummer, and Marshal (12-14) put the coil in series with a fixed impedance; because changes in coil impedance alter the current, the voltage measured across the coils was then related to conductivity. A determination of phase angle was used by Yosim, Grantham, Luchsinger, and Wike (15) to measure conductivity. Hollister (16) used a single

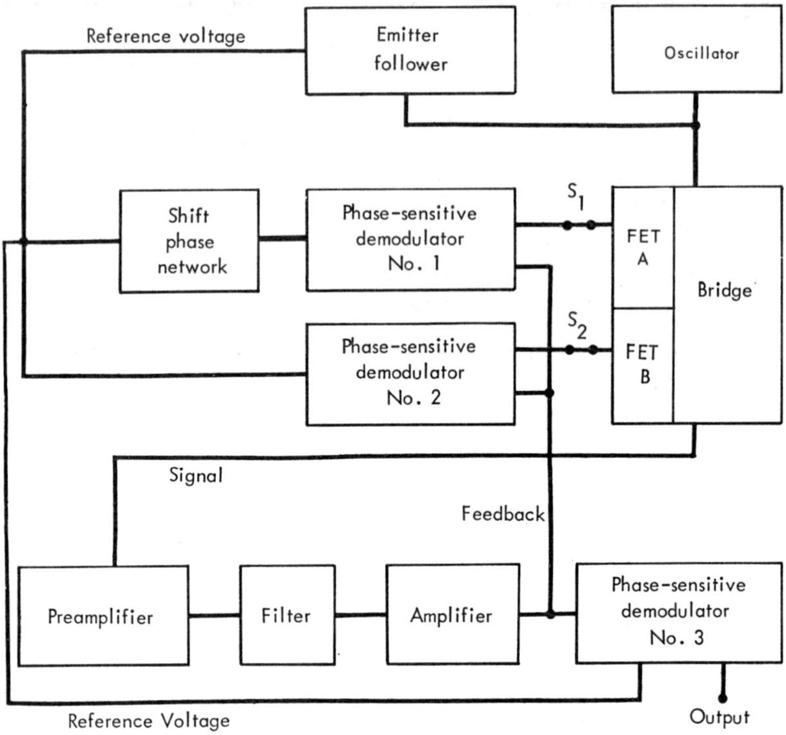

Fig. 7-1-2   Block diagram of the self-balancing bridge

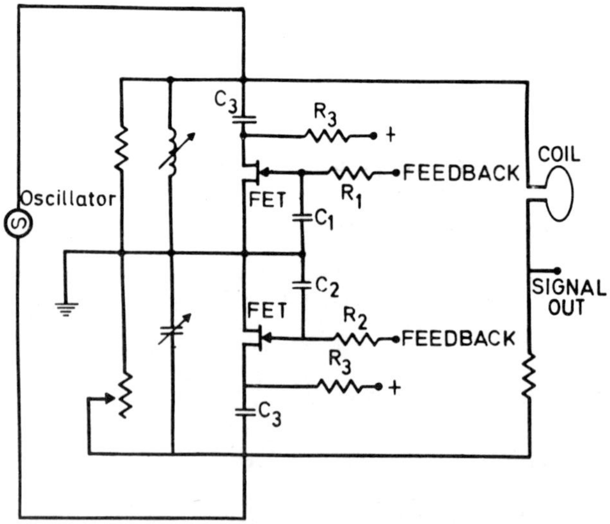

Fig. 7-1-3   Bridge circuit

turn coil within a large solenoid. Inside the coil was a coaxial plasma, so that the coupling between the solenoid and single coil was altered by the plasma, a fact that was related to σ. Gourdine (17) used a coil system with $\omega L/U$ near unity, and with this both velocity and electrical conductivity could be measured.

Velocity and Turbulence

It has been observed in arc jets that a flowing plasma may have inhomogeneities that can be transported by convection. Among these may be regions of higher or lower emitted radiance, and by observing the intensity with two phototubes and correlating the signals, the velocity of the plasma can be determined. Non-uniform plasma velocity U or electrical conductivity σ also may be transported in this way, hence, using two σU transducers, the convection of fluctuations in σU can be followed and the velocity can be determined (18).

Consider a plasma flowing through a constant magnetic field and suppose that a coil adjacent to the plasma is linked with the applied magnetic field. If the flow of plasma is turbulent, there will be fluctuating eddy currents that results in a time varying magnetic field at the coil. If the flow is laminar, there will be no voltage at the coil terminals. This simple transducer, described in (3) and (19) thus becomes a turbulence detector.

The voltage induced in the coil is:

$$e(t) = -\frac{N\mu_o}{4\pi}\frac{\partial}{\partial t}\int_A \int_V \frac{\sigma(E + U \times B) \times r}{r^3} \cdot n \, dA \, dV. \qquad \text{(Eq. 7-1-1)}$$

To solve for $\sigma(\vec{r}, t)$ and $U(\vec{r}, t)$ knowing $e(t)$, it is also necessary to simultaneously solve for the potential leading to $\vec{E}$. The equation for this potential is:

$$\sigma\nabla^2\varphi + \nabla\sigma \cdot \nabla\varphi = \nabla \cdot \sigma(U \times B).. \qquad \text{(Eq. 7-1-2)}$$

The normal vector can be related to a Cartesian coordinate system $\vec{n} = \vec{e}_i n_i$ where n_i is the component of $\vec{n}$ in the i-th direction. Identifying e_1 as the voltage for the case $n_1 = 1$, $n_2 = n_3 = 0$, and using three coils, e_1, e_2, e_3 are obtained. Correlations of the voltages, e.g.:

$$E_{ij} = \frac{\int e_i e_j \, dt}{\sqrt{\int e_i^2 \, dt} \sqrt{\int e_j^2 dt}} \qquad \text{(Eq. 7-1-3)}$$

could, perhaps, provide new insight into the turbulent flow of plasmas.

Flow Direction

Suppose that a magnetic field detector (e.g. a flux gate magnetometer) is placed between the poles of an U-shaped permanent magnet, the end-faces of the poles of which form a plane. Suppose then that the detector is oriented with its axis normal to this plane: the detector will indicate zero magnetic flux midway between the poles. Let N and S designate the north and south poles, and D, the detector. Plasma flow with streamlines passing sequentially over NDS will cause, say, a positive response from the detector. Hence, if the flow is reversed, i.e., SDN, the response will be negative and if the flow is normal to a line containing SDN, the response will be zero. This transducer then is sensitive to flow angle (20). It is one form of σU transducer.

Applications

Active magnetic probes find applications in connection with missile wakes, rocket exhausts, reentry plasma sheaths, ballistic ranges, shock tubes, MHD power generation, plasma propulsion, MHD wind tunnels, and arcjets. The minimum electrical conductivity that can be measured depends on the transducer size and the volume of plasma. An electrical conductivity of 0.1 mho/m can be measured using a RF bridge, and a value of 10^4 (mho/m)(m/sec) can be measured with the σU transducer.

Electron Beams

Electron beams have been used rather extensively to determine gas density and temperature in the laboratory (21 through 30). An inelastic collision of a high energy electron from the beam with a diatomic molecule causes an electronic transition, the rotational and vibrational states of the molecule being preserved. The radiation spectrum of the molecule contains information about the vibrational and rotational temperatures. One molecule that has been investigated thoroughly, and which can be used effectively with electron beams, is nitrogen, the predominant molecule in air. However, some plasmas used in propulsion and energy conversion apparatus may not contain nitrogen molecules: the use of electron beams with other molecules is at present being studied.

Measured Quantities

The ratio of two vibrational bands can be related to vibrational temperature; in the case of nitrogen, the [0, 0] and [1, 0] bands of N_2^+(1 -) system are commonly used. To obtain the rotational temperature T_r, the intensity of individual rotational lines can be measured, and the ratio of intensity can be related to temperature. An alternate method, which is less demanding on the experimental apparatus, is to measure the intensity of a complete vibrational band, e.g., N_2^+(1 -) [0, 0], and the intensity of one part of that band: the ratio of these two intensities can then be calibrated in terms of T_r which is also a measure of T_t, the translational temperature.

Gas density can be measured in two ways: (a) for a given chemical composition, beam attenuation is a function of path length and gas density, hence the comparison of the beam current leaving the electron gun with the beam current transmitted is related to the density of the gas; (b) observations of electrons scattered by the gas can be adequately related to the density of the gas, using Rutherford scattering as a model for correlating data.

If the local beam current, beam energy, excitation cross section, transition probability for various radiative processes, and cross sections for collisional de-excitation are known, a measurement of the absolute intensity of a vibrational band can be related to the partial density of the radiating species. An absolute intensity measurement is required however which is considerably more complicated than a relative intensity measurement.

Experimental Problems

In an electron beam experiment, optics define the column of hot plasma, that radiates energy to the optical sensors, and all the radiation that the optical sensors detect come from within the small region traversed by the beam. In these circumstances the background radiation should be carefully investigated to ensure that the signal-to-noise ratio is high.

The beam voltage is usually selected so that scattering and beam divergence are small within the plasma. The beam is then a narrow pencil, and as the column as defined by the optics also has a very small solid angle, the alignment of the beam and optics is critical. The presence of changing magnetic fields alters the position of the beam.

For most plasmas encountered in engineering devices, excitation of plasma waves is not a problem, mainly because n_e is small. However, for more dense plasmas, a significant fraction of the beam energy can be absorbed by waves. The electrons perturb the local electric field and these perturbations propagate from the charge (31). A wave pattern analogous to shock waves in supersonic flow carries away energy and causes a drag on the charge.

Once a molecule has been excited by the beam, there are several mechanisms that compete during the process of de-excitation. Quenching of excited states may therefore cause errors in the interpretation of the data, and the need for quenching may fix the upper pressure limit at which the experiment is feasible.

In the case of nitrogen, the molecules to be observed are ionized by the electron beam. The secondary electrons may have sufficient energy to cause additional ionization and excited N_2^+: consequently there may be errors in the interpretation of the data as a result of the radiation due to this secondary electron excitation.

An electron gun must be designed with high perveance to prevent the beam spreading due to the formation of virtual cathodes (see Pierce (32)). The gun must operate with the acceleration chamber in a hard vacuum to prevent arcing; a quick acting valve or differential pumping is therefore required. The optics of an electron gun should be shielded from stray magnetic fields. Electrons leave the cathode with small energy and even a weak magnetic field can distort the electron optics severely. Photomultiplier tubes are also sensitive to magnetic fields and mumetal should be used generously.

An electron gun can be pulsed to provide a periodic signal, and synchronous detection can then be used to enhance the signal against the background radiation. Filters or monochromators, used to select the desired spectral region, may have secondary transmission peaks. Radiation at these secondary transmission peaks may contribute to the background without aiding the signal.

One method of measuring gas density is to count scattered electrons using a scintillation counter. Care must be taken to block out visible radiation, and since the energy of the electrons is relatively low, a thin screen must be used: there is some danger that this might contain pinholes. Blocks containing holes are frequently used to define solid angles; electrons scattered into such a solid angle being counted. However it may be possible to charge such an aperature and the voltage on the block would then distort the trajectories of these electrons.

Self-absorption is not a problem with nitrogen since the radiation is not to the ground state. However if, with other molecular systems, the observed radiation does arise from transition to the ground-state, self-absorption could be a problem. In an MHD generator, or in a wind tunnel testing, ablation products can form carbon soot, which attenuates the radiation to be measured.

Choice of beam voltage is dictated in part by beam divergence and attenuation: the greater the gas density, the higher the voltage required. Beam divergence is due to multiple scattering of electrons by nuclei of ambient plasma.

Summary on Electron Beams

Electron beams yield considerable detailed information about a slightly-ionized plasma. Each measurement must be carefully designed for the conditions of the particular experiment. When the pressure is above a small fraction of an atmosphere, the electron beam technique may not be feasible.

Electrostatic Probes

An electrostatic probe, commonly known as a Langmuir probe, consists of an electrode exposed to the plasma. A second electrode has contact with the plasma so that the plasma forms one part of a circuit. By applying different potentials to the probe and measuring the currents, a characteristic curve is obtained. Information concerning electron density, electron temperature, and possibly ion temperature can be deduced from the current as a function of potential.

Due to the experimental simplicity and the satisfactory measurements obtainable under suitable conditions, Langmuir probes are used extensively. However, considerable information about the plasma is needed and interpretation of data is quite complicated, formulas for practical use being limited to special cases.

Stationary Plasma (33 through 51)

Several probe geometries have been studied in detail including flat plate (49), cylindrical (49), and spherical probes (46) (50) (51). The flat plate probe is a one dimensional problem and has the least complicated geometry. When sheath thickness d is much less than the electron or ion mean free path λ in the plasma, the charged particle motion is free fall in the sheath. The plasma is assumed to be neutral to the edge of the sheath, with zero electric field at that point. A space charge limited diode is used to describe spatial variation of potential and charge density within the sheath. The sheath thickness is the order of a Debye length h.

When collisions occur in the sheath, the free fall model is replaced by the high density space charge limited diode. The particle velocity is taken as being proportional to the local electric field using mobility as a proportionality factor.

These simple models lead to an understanding of the gross features of a characteristic curve, i.e., voltage vs current. Important facets have been omitted; for example, there is a presheath or transition region (39). Within this region the potential varies from plasma potential to a value differing in magnitude by an amount $kT_e/2$ for the case of electrons that are hot relative to ions.

One problem associated with other probe geometries is that of determining the sheath thickness and the effective collecting area of the probe. A model similar to that for a flat plate can be used, with appropriate space charge equations, for cylinders or spheres. This model is reasonable when sheath thickness and probe size are much smaller than the mean free path λ. When the probe or sheath is large compared to λ, a more detailed and accurate analysis is required. With a thick sheath the charged particles can experience orbits around the probe.

One important quantity is the flux due to random thermal motion of the charged particles. It is:

$$j = \frac{n\bar{v}}{4} \qquad \text{(Eq. 7-1-4)}$$

and it assumes a Maxwellian distribution. If the current density to a probe exceeds this value, caution needs to be observed in the interpretation of data, because it is likely that the plasma is disturbed by the probe.

Moving Plasma (52 through 75)

When the plasma is moving relative to the probe, additional phenomena are introduced, to account for which the conservation equations of gas dynamics are used. Additional parameters enter the problem along with added boundary conditions. A probe may be extended into the plasma stream at the end of a sting, or it may be flush mounted on the wall of a duct or body. French (52), Muntz, Harris, and Kaegi (53), and Scharfman (61 through 64) have studied probes immersed in the plasma stream. Chung (60) (67 through 70) (75), Talbot (65) (66), and Scharfman, Bredfeldt, Guthart, and Morita (63) have investigated flush mounted probes.

In a static plasma the mass density ρ may be uniform or nearly so. In a moving plasma there may be large variations in density along with gradients in the electron and ion concentration, the gradients being a consequence of plasma motion (61 through 72). Diffusion becomes important: outside the region of the sheath the electron charge is nearly balanced by the ion charge, and when $n_e \cong n_i$ ambipolar diffusion occurs. Electrons diffuse much faster than ions and tend to run ahead of ions. When this charge separation happens, an electric field builds up, which retards the electrons and drags the ions along behind the electrons. An ambipolar diffusion coefficient D_a can be derived in terms of electron and ion diffusion coefficients and mobilities. It is roughly twice as large as the ion diffusion coefficient, $D_a \cong 2D_i$.

At the wall a boundary layer forms. Variation of flow properties in the boundary layer is determined in part by plasma viscosity μ and thermal conductivity k. There are three quantities that are transported by gradients and which lead to three dimensionless parameters. The Schmidt number, $Sc = \mu/D\rho$, is the ratio of diffusion of momentum to diffusion of species. The Prandtl number, $Pr = \mu C_p/k$, is the ratio of the effects due to momentum transport, to the conduction of energy. A third parameter is the Lewis number, $Le = \rho C_p D/k$, which is the ratio of diffusion of mass to diffusion of energy. The Lewis number is the ratio of Pr to Sc.

For many applications the degree of ionization is low, and electron-neutral and ion-neutral collisions dominate. The transport coefficients μ and k can be evaluated for the neutral gas. For the diffusion of ions and electrons it is necessary to use the appropriate D_i, D_e, or D_a. There being a Schmidt number based on each of these. A low degree of ionization also decouples the fluxes of electrons and ions from the motion of the fluid. The plasma flow can be solved independently of current density. For example, laminar flow over a flat plate leads to a Blasius profile; knowing this profile, the fluxes of charged particles can be superimposed.

There are four characteristic lengths, namely the Debye length h, the boundary layer thickness δ, the mean free path λ, and sheath thickness d. From these four lengths three dimensionless ratios can be formed, h/δ, h/λ, and h/d. For a continuous plasma h/d may range from 0.1 to 0.001 or so, whereas for a rarified plasma h/d is the order of unity. The ratios h/δ and h/λ can be varied by changes in electron density, electron temperature, gas density, and gas temperature.

As in the case of static probes the ratio T_e/T_i remains an important variable because potentials and saturation currents are functions of the temperature ratio. Relaxation times for electrons can be much larger than those for ions and in an expanding gas T_e/T_i may be larger than unity since T_e has a tendency to lag.

The boundary conditions introduce new parameters. For low wall temperatures there is a flux of charged particles towards the wall but none leaving it, since recombination readily occurs on a cold wall. A cold wall will not emit electrons, hence thermionic emission is not important. When the probe becomes hot, however, a flux of thermionic electrons must be accounted for. Emitting probes have been studied by Hall, Kemp, and Sellen (73 through 74). A probe can be floating which means it has no net current, i.e., $j_i = j_e$; alternatively it can be drawing either a positive or a negative current. At the edge of the boundary layer the values for

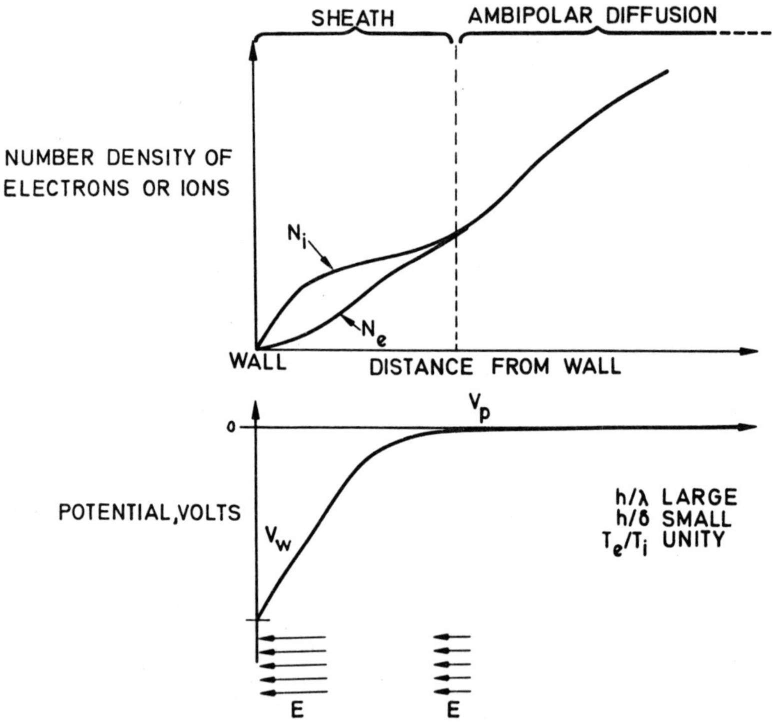

Fig. 7-1-4 Qualitative variation of electron and ion number density, potential and electric field near a catalytic, floating, wall

n_e, T_e, etc., must be specified. The sheath edge is a boundary where, in many analyses, the ambipolar diffusion solution is matched to a sheath (Fig. 7-1-4).

Chemical reactions, ionization, and recombination may occur. Obviously the problem is much less complex is the flow is frozen. For many problems this is a valid assumption since flow times are short relative to time for recombination.

Interferometry

An interferometric measurement gives an integral of the index of refraction n along the optical path. For a neutral gas, n is related to the mass density ρ hence a direct measurement of density is obtained. For an ionized gas the index of refraction depends on the dispersion relation for the plasma which has contributions from both free electrons and bound electrons. Under suitable conditions an interferogram yields data on electron density n_e.

Classical Interferometry (76 through 83)

Classical interferometry, as we use the term here, refers to techniques employed before the advent of the laser.

The fringe shift N due to an optical path L is:

$$N = \frac{1}{\lambda} \int_0^L (n_2 - n_1)\, dx \qquad \text{(Eq. 7-1-5)}$$

where λ is the wavelength of light. The index of refraction for the gas being measured is n_2, and n_1 is a reference index. The fact that this is written as an integral emphasizes the fact that an average index along the path is measured. For two dimensional distributions, where $n \neq n(x)$, the average n equals the actual n. For axisymmetric flows the average n can be related to variation of n along the path.

A quantity known as the refractivity X is introduced. This is defined as $X = n-1$. A plasma will have:

$$X_{plasma} = X_A + X_M + X_1 + X_e \qquad \text{(Eq. 7-1-6)}$$

where A, M, I, and e refer to atoms, molecules, ions, and electrons. The refractivity is the sum of the refractivity of each component as long as intermolecular forces can be neglected. The refractivity of atoms, molecules, and ions is:

$$X_r = 2\pi n_r \alpha_r \qquad \text{(Eq. 7-1-7)}$$

where n is the number density of species r and α_r is the polarizability. Polarizability:

$$\alpha_r = \frac{\text{induced dipole moment}}{\text{electric field}} \qquad \text{(Eq. 7-1-8)}$$

Ions also exert an influence due to the fact there is a net charge. This feature will be discussed when electrons are considered. The refractivity, which is dimensionless, has a value in the range 10^{-4} to 10^{-3}.

The index of refraction for electrons is a rather long expression which reduces to:

$$n \cong 1 - \frac{1}{2}\left(\frac{\omega_p}{\omega}\right)^2 \qquad \text{(Eq. 7-1-9)}$$

for the special case $\nu/\omega << 1$ and $\omega_p/\omega << 1$. Here ν is the momentum-exchange collision frequency for electrons and ω_p is the plasma frequency; if n has the units cm^{-3}, $\omega_p = 5.6 \times 10^4 \sqrt{n_e}$. The refractivity for electrons is:

$$\chi_e = -\frac{\omega_p^2}{2\omega} \qquad \text{(Eq. 7-1-10)}$$

Introducing numerical values for constants and changing ω to a corresponding wavelength λ yields:

$$\chi = -4.46 \times 10^{-14} \lambda^2 n_e \qquad \text{(Eq. 7-1-11)}$$

As used in equation 7-1-11 the units of λ are cm. Combining equations 7-1-5 through 7-1-10 gives:

$$N = \frac{L}{\lambda} \Sigma\, 2\pi n_r \alpha_r\; 4.46 \times 10^{-14} L\lambda n_e - \frac{X_o L}{\lambda} \qquad \text{(Eq. 7-1-12)}$$

where the summation includes all particles excluding electrons. A reference refractivity is χ_o.

For ions an expression similar to equation 7-1-9 can be written substituting the ion plasma frequency Ω_p for ω_p. Due to the fact that $\Omega_p << \omega_p$, the refractivity contribution due to ion charge is neglected in equation 7-1-12.

For the usual low-temperature gas the first term in equation 7-1-12 dominates. For a low density plasma the second term dominates. Conditions can occur where both the bound (first term in equation 7-1-12) and free (second term in equation 7-1-12) charges are important. A single number for N does not provide enough information to obtain n_r or n_e from equation 7-1-12. Note that the first term varies as $1/\lambda$ and has a wavelength dependence in α_r. Polarizability and refractivity do not have a strong dependence on λ; for example, for argon at STP (λ in microns):

$$\chi_A = 2.79 \times 10^{-4} \left(1 + \frac{5.59 \times 10^{-3}}{\lambda^2}\right) \qquad \text{(Eq. 7-1-13)}$$

The second term in equation 7-1-12 varies as λ. Consequently measurements at different λ permit separation of the effects due to n_r and n_e.

The need for different wavelengths suggests a need for the use of white light. Incidentally, white-light interferometry is one technique for which a laser is unsuitable. It is necessary to distinguish between the index of refraction for phase velocity n_p and the index of refraction for group velocity n_g. Shifts of an individual fringe, i.e., a fringe due to light in a narrow range of wavelengths are determined by n_p. For monochromatic interferometry it is n_p that is significant, and the preceding equations 7-1-5 to 7-1-13 imply n_p. Shift of the center of contrast is determined by n_g. A center of contrast is the locus where all of the monochromatic fringe patterns have the same phase. For neutral gases $n_p \cong n_g$. As a result the center of contrast and the individual fringes are shifted in the same direction by nearly the same amount.

The group refractivity for electrons is given by equation 7-1-9 and 7-1-11 along with the relation between phase and group indices, which is:

$$n_g = n_p - \lambda \frac{dn_p}{d\lambda} \qquad \text{(Eq. 7-1-14)}$$

The result is for electrons:

$$n_g = 1 + \tfrac{1}{2} \left(\frac{\omega_p}{\omega}\right)^2 \qquad \text{(Eq. 7-1-15)}$$

The phase refractivity is equal and opposite to the group refractivity. The group velocity is less than the speed of light.

As an illustration of the interferogram resulting from equation 7-1-12 consider a strong shock wave in a monatomic gas. Initially the gas density increases, which in turn increases the refractivity and shifts the fringes upward as shown in figure 7-1-5. This is the region where the first term of equation 7-1-12 is important. As the gas begins to ionize, the negative second term in equation 7-1-12 becomes important. The fringes move in the direction of decreasing refractivity. The negative net fringe shift shown in figure 7-1-5 indicates that the second term is larger than the first term of equation 7-1-12. Interferograms similar to the sketch of figure 7-1-5 have been observed by Alpher and White (78).

Since the complete flow field is sampled by the usual interferometer, the interferogram provides a method for testing for conical flow, as was done by Giese, Bennett and Bergdolt (83). Conical flow is one example of self similar flow in spatial coordinates. There are numerous self similar flows both steady and nonsteady; nonsteady flow has one spatial and one time coordinate. Blast waves and the waves driven by a piston in uniform motion are examples of nonsteady self similar flow. It is suggested that interferometric techniques provide a general method for testing any experimentally observed flow for self similarity.

Interferometry Using Modulation of Laser Light (84 through 86)

It has been observed that when the light from a laser is reflected back into the laser cavity, a modulation of output occurs. (Fig. 7-1-6a). If the mirror is moved, the laser output is modulated at doppler frequency. Ashby and Jephcott (84) used this fact to construct an interferometer (Fig. 7-1-6b). The laser beam passes through the plasma and is reflected by the mirror. Changes in electron density cause changes in refractivity and in optical path length.

Laser intensity can be monitored with a photomultiplier tube. The product of the electron density and the geometric path length L giving one fringe shift is readily obtained from equation 7-1-12. For many low density plasmas the first term is negligible. Recognizing that the beam makes two passes through the plasma and setting N = 1 gives:

$$n_e L = \frac{2.24 \times 10^{13}}{2\lambda} \qquad \text{(Eq. 7-1-16)}$$

For a helium-neon gas laser λ is 3.39μ. One fringe shift is obtained if $n_e L = 3.3 \times 10^{16}$ cm/cm^3. It is believed that this is the minimum value that can be successfully measured. Fractional fringe shifts are difficult to interpret since laser output is not a simple function of path length.

It is at the point where a value of λ is required for evaluation of equation 7-1-16

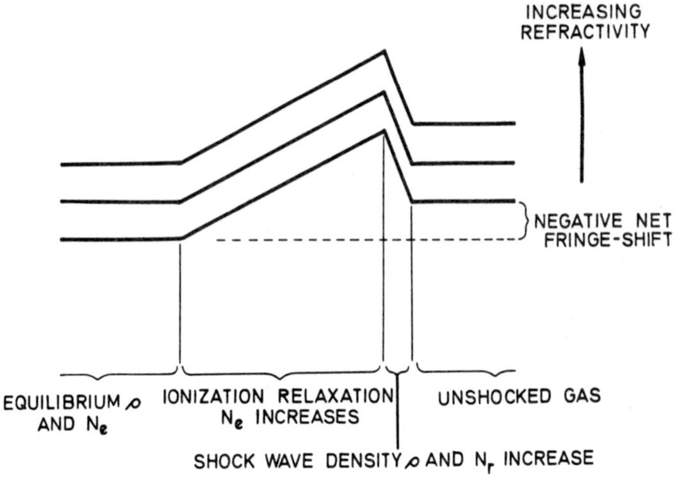

Fig. 7-1-5 Qualitative interferogram of shock wave structure in a monatomic gas with ionization relaxation

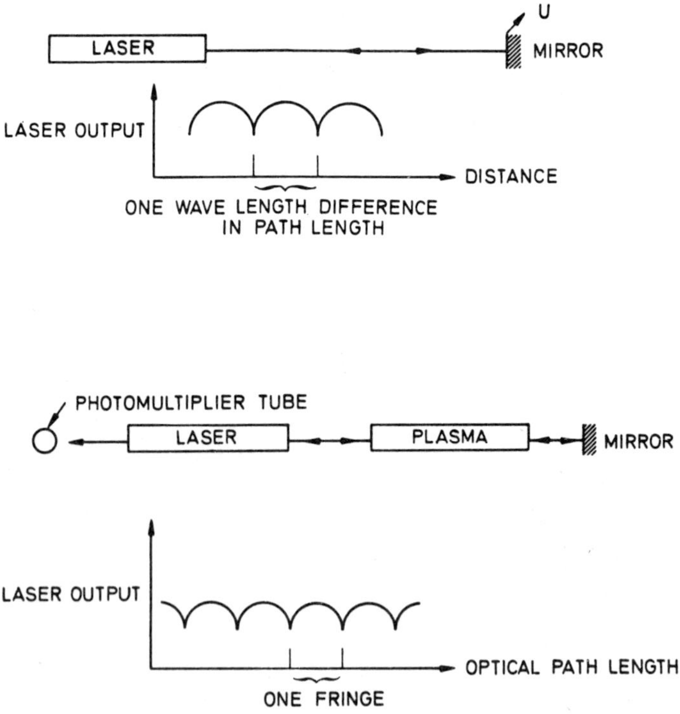

Fig. 7-1-6 Interferometry using modulation of laser

that one advantage of this technique becomes apparent. The red line at $.63\mu$ is coupled to the infrared line at 3.39μ in a HeNe gas laser. When changes in path length of 3.39μ radiation occur, the laser output is modulated at both $.63\mu$ radiation and 3.39μ; but the results can be observed in the visible. The optics can be aligned for 3.39μ using visible light at $.63\mu$.

The time response of the interferometer is an important consideration when transient plasmas are being observed. Response is controlled by processes occurring in the laser, such as excitation and depopulation of various states (85). Above 10 kHz the modulation of the $.63\mu$ radiation begins to droop. Modulation can be detected up to 3 MHz. The maximum rate of change that can be followed is:

$$\frac{d(n_e L)}{dt} = 10^{23} \frac{\text{cm}}{\text{cm}^3\text{sec}}$$

For example a plasma that changed from zero to 10^{15}cm/cm^3 in 10 nanosec could be followed. Measurements in the infrared region show that the 3.39μ modulation is flat to 1 MHz. Some fast IR detectors can be used especially as the radiation intensity is high.

Fluctuations of n_e in a steady flow plasma device can be determined using the laser modulation technique. When fringe shifts are due to variations in n_e, the equation is:

$$N = \frac{L}{2\lambda_o} \left(\frac{\omega_{p1}}{\omega}\right)^2 - \frac{\omega_{p2}}{\omega}^2 \qquad \text{(Eq. 7-1-17)}$$

Determination of N gives the difference in n_e at two different times 1 and 2. To obtain the absolute value of electron density another measurement is needed. Boornard, Nicastro, and Bollmer (86) measure the relative intensity of continuum radiation for which it can be shown that:

$$\left(\frac{\omega_{p2}}{\omega_{p1}}\right)^2 - \frac{I_2}{I_1}^{\frac{1}{2}} \qquad \text{(Eq. 7-1-18)}$$

The two measurements combined yield absolute n_e.

There are several advantages to the technique which Ashby, Jephcott, Malein, and Raynor (85) enumerate: the number of components is few; optical adjustments are simple; an infrared measurement can be aligned and monitored in the visible range, and the detection of interference fringes is straightforward.

Holographic Interferometry (87 through 100)

Gabor (87 through 89) introduced the word hologram to identify the photographic plate that records the wave pattern incident upon it. In his work he recorded a Fresnel diffraction pattern. Gabor's motivation for his research was to improve the resolution of electron microscopes. Holograms made from Fresnel diffraction in Gabor's method have a twin image. When one image is in focus, the other is present out of focus. Numerous papers have been published dealing with means for eliminating the twin image.

Leith and Upatnieks (92 through 94) used communication theory to gain insight. Within this framework the twin image leads to a signal-to-noise ratio of unity.

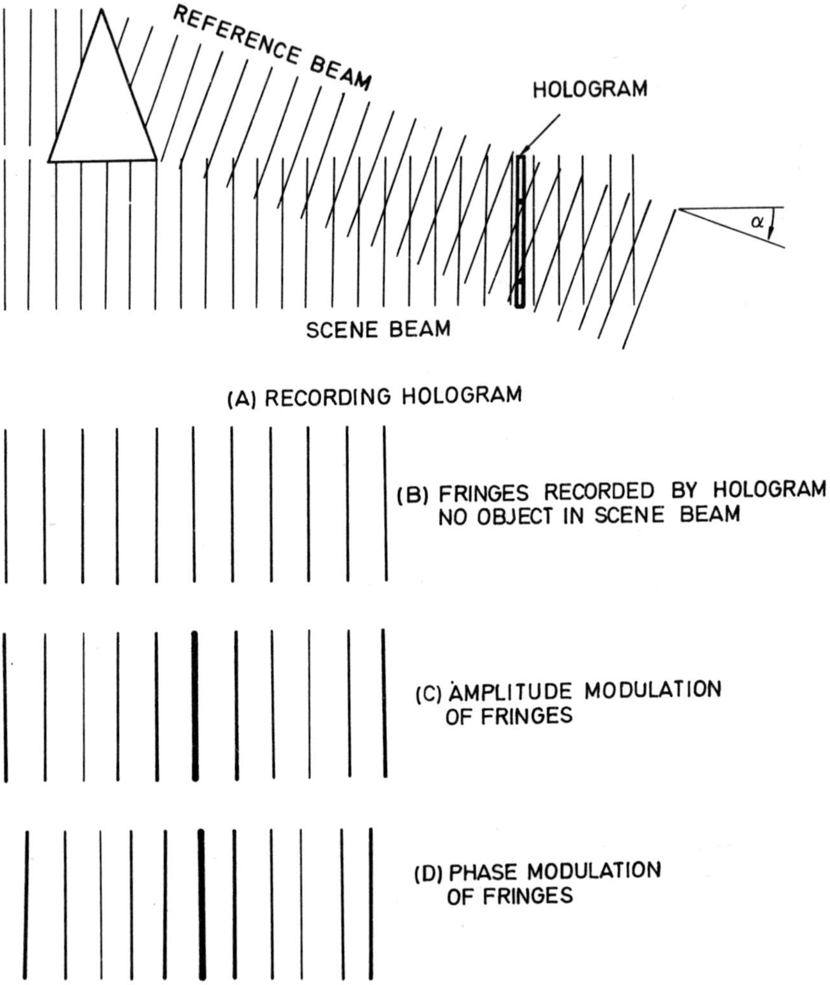

Fig. 7-1-7 Two beam holography

Three processes were identified in holography: a) defocusing or spatial-frequency dispersion of image, b) hologram recording process, and c) focusing or compression which occurs in reconstruction. A two beam method was devised to eliminate the twin image and the intermodulation distortion term. The idea of spatial filtering was introduced.

A hologram formed by recording the Fresnel diffraction pattern is limited to the special case where most of the incident beam is transmitted without scattering. In the photographic recording, phase information is lost. The two beam method records both phase and amplitude information as can be explained with the help of figure 7-1-7. The scene beam and reference beam constructively interfere to cause a darkening of the film. Where destructive interference occurs, the emulsion is not exposed. Since the light is coherent, the locations of constructive interference are fixed and fringes are formed (Fig. 7-1-7b). If the light is not coherent, the regions of constructive interference will meander over the film causing uniform darkening of exposure. The spacing of the fringes s is:

$$s = \frac{\lambda}{\tan \alpha} \qquad \text{(Eq. 7-1-19)}$$

Equation 7-1-19 can be used to estimate the resolution requirements for film. For He-Ne laser λ is 6.3×10^{-4} mm. If α is 20°, then s is 1.73×10^{-3} mm. For the hologram, film with at least 1/.00173 = 580 lines/mm resolution is required.

Returning to figure 7-1-7 consider the influence of an object placed in the scene beam. The diffraction pattern causes amplitude modulation (Fig. 7-1-7c). Changes in optical path lengths alter the phase of wavefronts in the scene beam relative to wavefronts in the reference beam. Phase modulation of the fringes then results (Fig. 7-1-7d), both phase and amplitude information being recorded. The fringes serve as a carrier using the communication analogy.

If a hologram is made (Fig. 7-1-7a), it has the properties of a diffraction grating. Sunlight is diffracted into its spectrum. The hologram serves as a transmission and reflection grating. First and second orders can be seen.

Using the two beam technique Leith and Upatnieks (93) demonstrated ability to reconstruct transparent lettering on a dark background and continuous tone objects. A laser was introduced as the source. In a subsequent paper (94) the use of diffuse illumination was described. Coherent radiation passing through frosted glass gives a random amplitude and phase from point to point so that a wavefront is not clearly defined; however, the amplitude and phase at any given point is independent of time. Frosted glass makes it possible to see the virtual image through the hologram as if it were a window. A three dimensional scene appears in three dimensions with parallax.

There are numerous advantages to diffuse illumination. It is possible to see both the real and virtual images. Flaws due to dust particles are eliminated. Each part of a hologram records the complete scene. Poor quality optics can be used. Increased dynamic range of recording is possible; bright regions are spread out over the entire hologram. Intensity variations of 10^4 to 10^5 have been recorded. Leith and Upatnieks (94) did not give a theoretical formulation for diffuse illumination.

In addition to Fresnel diffraction holograms experiments were conducted with Fraunhofer diffraction holograms. Multicolor wavefront reconstruction was demonstrated (94). Transient phenomena, using both multispike and Q-switch laser

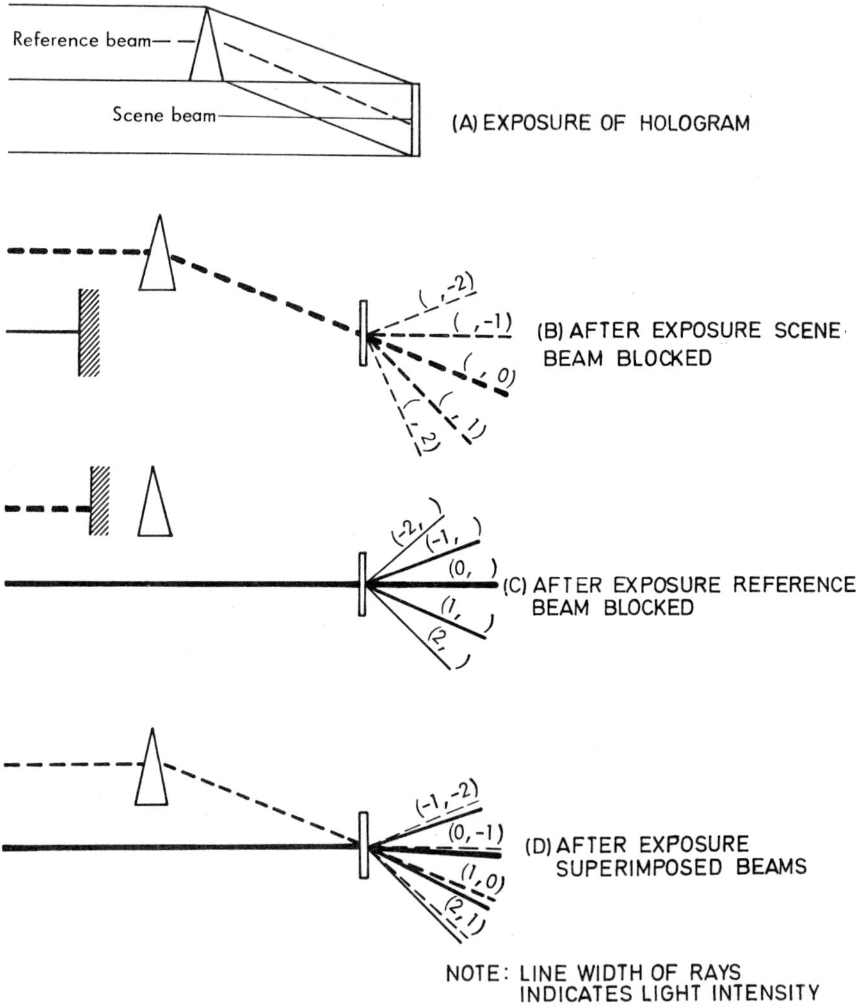

Fig. 7-1-8 Interferometry with a holographically reconstructed comparison beam

operation, have been recorded by holography (95).

The two beam method can be extended to provide an interferometer as was done by Brooks, Heflinger, and Wuerker (96). A hologram (Fig. 7-1-8) is made using two beams as shown in Fig. 7-1-8a. There will be fringes (Fig. 7-1-7b). If the hologram is replaced in the same position after the development as it occupied during exposure, the result is an interferometer. When the scene beam is blocked, (Fig. 7-1-8b) the reference beam can be seen to be diffracted into the various orders. Likewise when the reference beam is blocked, the scene beam is diffracted by the hologram (Fig. 7-1-8c). When both beams illuminate the hologram, the diffracted beams overlap. If an object is placed in the scene beam, there will be interference between (0,) beam and (, -1) beam. Fringes are produced. Downstream of the hologram the (0,) beam has the wave pattern of scene beam with object, and the (, -1) beam has holographically reconstructed wavefronts of the scene beam without an object. Any of the superimposed beams (-1, -2), (0, -1), (1, 0), or (2, 1) display interference patterns. It is only in the (0, -1) set of beams that the object can be seen.

Typical interferograms that can be obtained by this technique are shown in figures 7-1-9 and 7-1-10 taken by Holds (99) at the Naval Postgraduate School. Figure 7-1-9 shows both infinite fringe, also identified by some scientists as single fringe, and finite fringe interferograms. The dots represent prediction of fringe location. It should be emphasized that there are no adjustable parameters in the prediction which involves only the following data: Mach number, cone angle, wind tunnel free-stream density, index of refraction of air, and ratio of heat capacities of air. The density distribution along the optical path was obtained from a Busemann apple curve which is a graphical solution of conical flow.

Figure 7-1-10 shows four dark field interferograms of four randomly chosen subjects. The adjectives 'dark field' mean that diffuse illumination of the hologram was not used. A camera focused on the test object is required to obtain a picture. The heated rod was oriented horizontally and parallel to the scene beam. The heated flat plate was oriented vertically. A burning match is shown.

Heflinger, Wuerker, and Brooks (91) have demonstrated that three dimensional interferograms are obtained when diffuse illumination is used. When the test object is viewed from different positions, the fringe pattern changes. In principle one should be able to use the three dimensional distribution of intensity of light to determine a three dimensional distribution of refractivity, $\chi(x, y, z)$. Knox (97) obtained holograms of microscopic dynamics phenomena using a pulsed laser. All objects within a 1000 cm^3 volume were recorded, and objects could be selected for study by proper focus of a microscope. Holds and Fuhs (98) performed extensive numerical calculations of the flow field surrounding a projectile flying in a krypton atmosphere. These calculations were used to refine the fringe prediction of an interferogram taken by Heflinger, Wuerker, and Brooks (91).

An alternate method for obtaining a holographic interferogram is to use double exposure (91). The Q-switch of a giant pulse laser is made to switch twice. During one pulse the scene beam without object is recorded, and during the second pulse the scene beam with object is recorded. Transient phenomena require short duration pulses. Short duration pulses imply limited coherence length. Limited coherence length necessitates careful attention to optics. Single exposure interferograms can be obtained using the technique developed by De and Sévigny (100).

For plasma measurements one would combine the holographic interferometer technique with the plasma flow such as discussed in connection with equation 7-1-12 and figure 7-1-5.

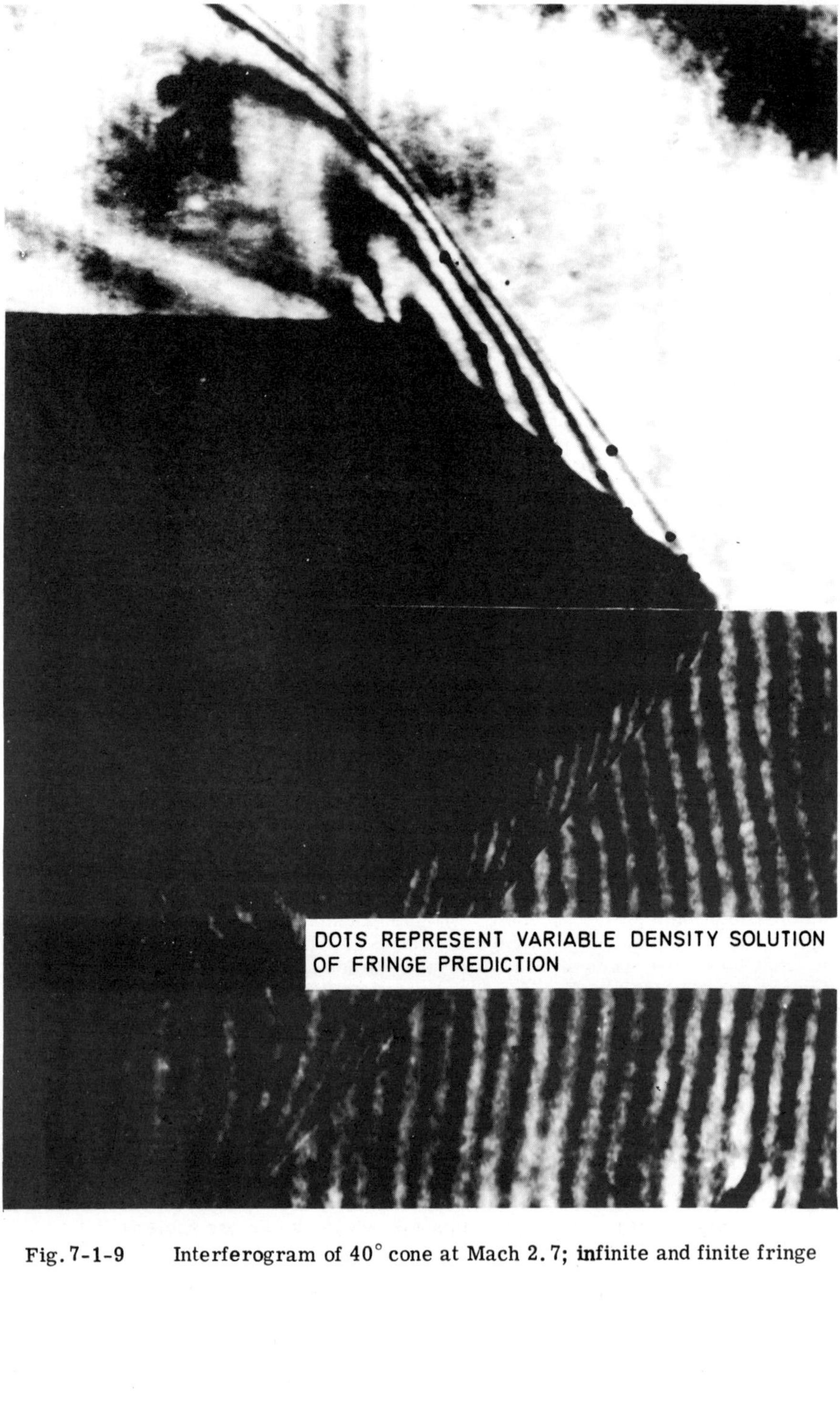

Fig. 7-1-9 Interferogram of 40° cone at Mach 2.7; infinite and finite fringe

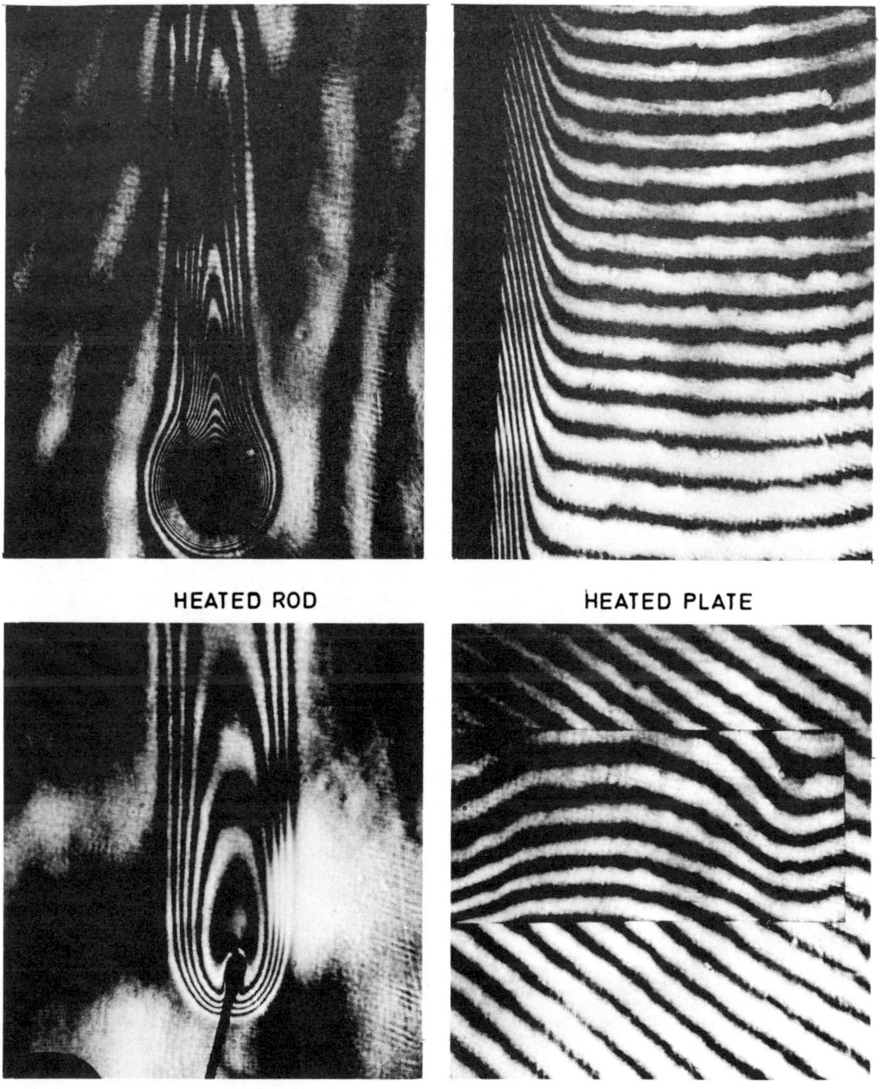

Fig. 7-1-10   Dark field interferograms of various subjects

Visualization Techniques

Considerable information can be gained from photographs of a self luminous plasma. The previous section discussed one visualization technique, interferometry. There are two other techniques widely used in aerodynamics and more recently applied in plasma physics; these are schlieren and shadowgraph.

Photography

For steady state or slowly varying phenomena, as may occur in MHD generators, rocket exhausts, and hypersonic tunnels, conventional photography using single exposure and movies both in black and white and color are useful. For transient phenomena more complicated photographic systems are required. One commonly used technique is streak photography. Time resolution of the order of a microsecond can be achieved. Rotating mirror and rotating drum cameras permit framing rates up to 10^6 frames/sec. Rotating mirror cameras have been built where the mirror is spun and suspended magnetically. Using magnetic levitation in a vacuum, rotational speeds limited only by tensile strength of the mirror can be achieved.

Several fast shutters have been developed for photography. These shutters have been used to Q-switch lasers. Kerr cell shutters, which are based on the Kerr electro-optic effect, are widely used as shutters and Q-switches. One disadvantage of a Kerr cell as a Q-switch is the influence of Raman scattering. A Faraday-switched ruby laser has been developed (101). Magneto-optic shutters, which are based on the Faraday effect, have been used for photography. Fast shutters have been developed which use strong electromagnetic forces to move small masses. For example, aluminum foil can be moved in a few microseconds when a capacitor is discharged through an adjacent coil.

The image converter camera consists of an electronic image converter or image-intensifier with associated optics and electronics. A continuously swept deflection voltage makes streak photography possible. Writing speeds of a few tens of cm/microsec are possible. By stepping voltages a few photographs with exposure times of about 5 nanosec and framing speeds of 2×10^7 frames/sec can be obtained.

Schlieren

Schlieren systems are widely used in aerodynamics. Liepmann and Roshko (102) have an excellent discussion of schlieren systems. For neutral gases over a wide temperature range there is no need to distinguish between mass density, index of refraction, and refractivity. These three quantities can be simply related to each other. For a plasma, however, it is necessary to interpret schlieren photographs in terms of optical refractivity. The important quantity in regard to schlieren is the angular deflection of a ray caused by varying optical pathlength of test gas. This angular deflection, commonly denoted by ϵ, is given by:

$$\epsilon = \int_0^L \left(\frac{1}{n} \frac{\partial n}{\partial y} \right) dx \qquad \text{(Eq. 7-1-20)}$$

where x is distance along the ray and y is normal to the ray. The appropriate expression for n is obtained from equations 7-1-6, 7-1-7 and 7-1-11 and the definition, $n = 1 + X$. If X_e dominates, a schlieren photograph gives direct information about n_e. In fact, an equation for lateral gradients of n_e can be derived using equation 7-1-11 and equation 7-1-20 above. Schlieren photographs have been obtained in plasmas (103-104).

In times past schlieren techniques have not been used for quantitative results as much as for qualitative insight into a complicated flow. There have been recent papers devoted to questions of calibration and quantitative data from schlieren systems (105-107).

Shadowgraph

As in the case of schlieren, shadowgraph techniques have been used for many years by aerodynamicists. More recently plasma physicists (108-109) have obtained shadowgraphs. The variation of illumination at the viewing screen is determined by the second derivative of index of refraction. It can be shown (102) that the fractional change of illumination is:

$$\frac{\Delta E}{E} \sim \int_0^L \frac{1}{n}\left(\frac{\partial^2 n}{\partial y^2} + \frac{\partial^2 n}{\partial z^2}\right) dx \qquad \text{(Eq. 7-1-21)}$$

In plasma fusion and plasma propulsion devices the electron refractivity usually dominates. An arc jet, MHD generator, and rocket exhaust would probably have the refractivity of neutrals as the dominant effect with smaller contributions from electrons.

Light Scattering by Electrons (111 through 115)

Before the discovery of the laser there were no light sources that could be used successfully for light scattering on electrons. The cross section for scattering of light by electrons is extremely small being the order of the classical electron radius (112), which is $r_o^2 = (e^2/mc^2)^2 = 7.9 \times 10^{-26}cm^2$.

Several advantages result from the use of Thomson scattering as a diagnostic technique. The time resolution is the same as the pulse duration of the laser (10^{-8} sec). A laser beam is a few mm in diameter. The solid angle subtended by the detector can be made sufficiently small so that only a few cubic millimeters of plasma are examined. The lower limit of electron density that can be measured is a function of laser power, electron temperature, detector capability, and other factors to be discussed shortly.

There are two limiting cases (116 through 118) which are identified in terms of a parameter α defined as:

$$\alpha = \frac{\lambda}{\sin \theta/2} \sqrt{\frac{n_e e^2}{4\pi k T_e}} \qquad \text{(Eq. 7-1-22)}$$

The two limiting cases are illustrated (Fig. 7-1-11). When $\alpha << 1$, there is scattering by individual electrons. Besides the ability to determine n_e from absolute measurements of incident and scattered laser radiation, this case also makes it possible to determine electron temperature. By measuring the line profile of the scattered radiation the line half width $\Delta\lambda$ (Å) can be related to electron temperature T_e (°K) (115). If the scattering angle is 90°, if the electrons have a Maxwellian velocity distribution, and if a ruby red line is used:

$$\Delta\lambda = 0.3\sqrt{T_e} \qquad \text{(Eq. 7-1-23)}$$

The amount of scattered radiation is small which complicates line profile measurements.

The scattered power P_s is:

$$P_s = r_o^2 \, n_e \, \frac{P_o}{A} \, \Delta\Omega \, \sin^2 \beta \qquad \text{(Eq. 7-1-24)}$$

where P_o is the incident laser power in a beam of cross sectional area A, with scattering into a solid angle $\Delta\Omega$. The angle between the electric vector of the incident radiation and the direction of observation is β. Plane polarized light is required, if not from a theoretical point of view, by practical considerations of experimental technique and data reduction.

The other limiting case is $\alpha \gg 1$ which occurs for low electron temperature and large electron density. Scattering is due to clusters of electrons around ions; it is due to fluctuations in electron density. The relation between P_s and P_o remains that given by Eq. 7-1-24. The profile of the scattered radiation is more complex for large α than for small α. Several papers (116 through 120) have examined the profiles for large α. The shape of the profile depends on the ratio T_e/T_i and consists roughly of a central peak at the incident laser frequency ω_o with two sidebands at $\omega_o \pm \omega_p$. The width of the center peak is determined largely by ion temperature.

When α is near unity, a measurement of the spectrum of the scattered radiation yields both electron temperature and density. It is not necessary to measure P_o and P_s absolutely. Consequently $\alpha \cong 1$ is considered to be the most convenient condition experimentally.

In making profile measurements, the minimum measurable n_e depends on the experimental details of the spectrometer, the detector, the laser energy, and the number of points used to define the profile. Briefly there is a minimum scattered radiant energy that can be measured. Since the photon count can be integrated over the pulse width, it is energy and not power.

There is also a maximum value of n_e that can be measured. This is established by the ratio of background radiation to scattered radiation. Most of the background has been attributed to bremsstrahlung. Another upper bound is established by absorption of the laser beam by the plasma. If this absorption does occur, a measurement of an undisturbed plasma has not been accomplished.

Figure 7-1-12 is a plot adopted from Malyshev (111), of minimum and maximum n_e and α. Values of n_e and T_e to be encountered in plasma propulsion and energy conversion devices are tabulated in table 7-1-1. Values from table 7-1-1 can be plotted in figure 7-1-12. It should be emphasized that the curves for minimum and maximum n_e (Fig. 7-1-12) can be shifted by assumed capability of the experimental apparatus (111). Even so there appear to be engineering plasma which are amenable to diagnosis by light scattering from electrons.

Microwave Plasma Probes

Microwaves have been used extensively to probe plasmas (126) (127). The first plasma to be investigated by microwave methods was the ionosphere (128) (129). Microwaves are used in connection with arc jets (130), ballistic ranges, rocket exhausts (131), thermonuclear fusion, reentry plasma sheath, MHD accelerators, shock tubes, and most any other plasma occuring in nature or the laboratory. No attempt will be made to discuss this rather voluminous subject. One particular topic will be selected for discussion.

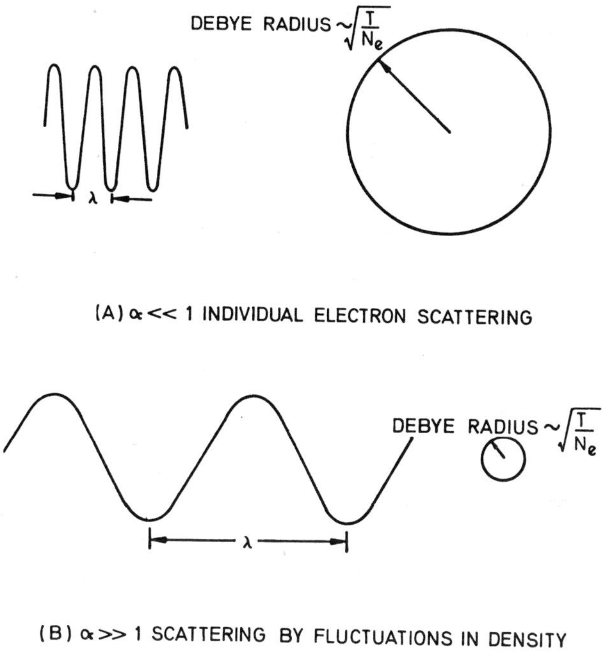

Fig. 7-1-11 Two limiting cases for light scattering by electrons

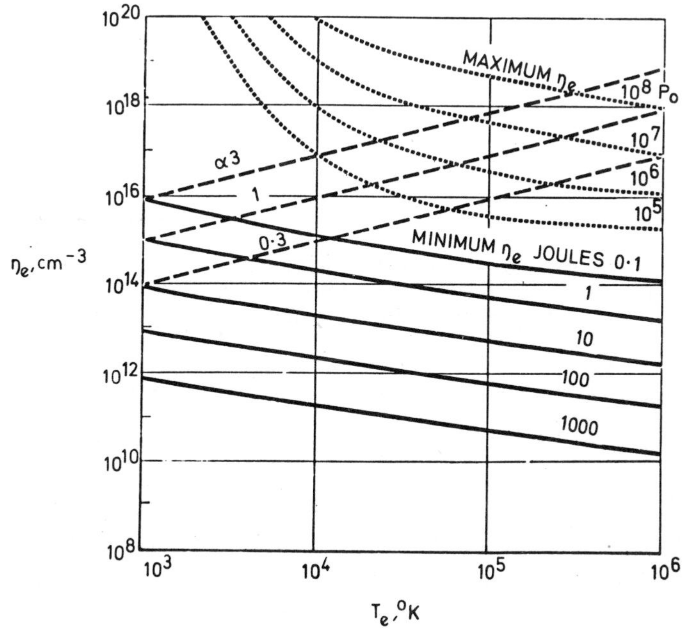

Fig. 7-1-12 Limits for electron scattering

A problem that occurs repeatedly is the determination of the spatial distribution of electron density, and the σU profile meter is one solution developed by the author (132). A microwave technique for obtaining profiles of n_e has been used by Anisimov, Vinogradov, Goland, and Konstantinow (133) and extended by other scientists (134-135). The method, one advantage of which is the fast time response, employs several microwave frequencies all of which are less than the maximum ω_p in the plasma. The phase angle of the reflected wave is measured relative to the incident wave. The phase angle is:

$$\Phi(\omega) = \frac{2\omega}{c} \int_{o}^{x(\omega)} n(x, \omega)dx - \frac{\pi}{2} \qquad \text{(Eq. 7-1-25)}$$

where ω is the microwave frequency; n, the index of refraction; and $x(\omega)$, the critical point at which reflection occurs. This formula assumes the validity of the geometric-optics approximation, which requires:

$$\frac{\lambda}{n_e} \nabla n_e << 1 \qquad \text{(Eq. 7-1-26)}$$

The fractional variation in electron density over the distance of a wavelength should be small. Short wavelengths tend to alleviate this problem; however, there must be sufficient n_e to cause reflection. For a given wavelength λ (cm) the required electron density cm^{-3} is:

$$n_e = \frac{1.12 \times 10^{12}}{\lambda^2} \qquad \text{(Eq. 7-1-27)}$$

A plane wave is assumed in the analysis leading to equation 7-1-25. Diffraction effects are not important if:

$$\frac{\lambda}{2\pi} << L \qquad \text{(Eq. 7-1-28)}$$

where L is a characteristic plasma dimension.

In order to invert equation 7-1-25 several frequencies are necessary. Each frequency requires a separate set of antennas, oscillators, detectors, amplifiers, and other microwave plumbing.

The following suggested method of determining spatial variations in n_e by using one frequency only is appropriate where fast time response is not desired. Time response is determined by the rapidity with which a current can be established in the field coil.

An applied magnetic field B is provided by a coil (Fig. 7-1-13) large enough to give a nearly uniform field. Corresponding to this field is an electron cyclotron frequency ω_c = eB/m = 17.6 mHz/gauss. With no applied magnetic field the extraordinary wave penetrates to x_b where $\omega_p(x_b) = \omega$. The phase angle Φ_b is measured. When a field is turned on such that $\omega_c/\omega = 0.5$, the reflection plane occurs at x_a and a phase angle Φ_a is measured. By measuring a series of Φ for cyclotron frequencies in the range $o < \omega < \omega_c$ sufficient information is obtained to invert the integral formula:

Table 7-1 Representative electron density and temperature for various devices

	n_e, cm^{-3}	T_e, ^0K
Arc jet	10^{15} - 10^{17}	5000
Shock tube	10^{16} - 10^{18}	7000
MHD generator	10^{13} - 10^{15}	2000
Rocket exhaust	10^{11} - 10^{14}	1500
Reentry plasma sheath, sharp nose vehicle	10^{6} - 10^{10}	4000
Reentry plasma sheath, blunt nose	10^{6} - 10^{15}	8000
Ballistic range	10^{7} - 10^{9}	1500
MHD accelerators	10^{14} - 10^{16}	6000

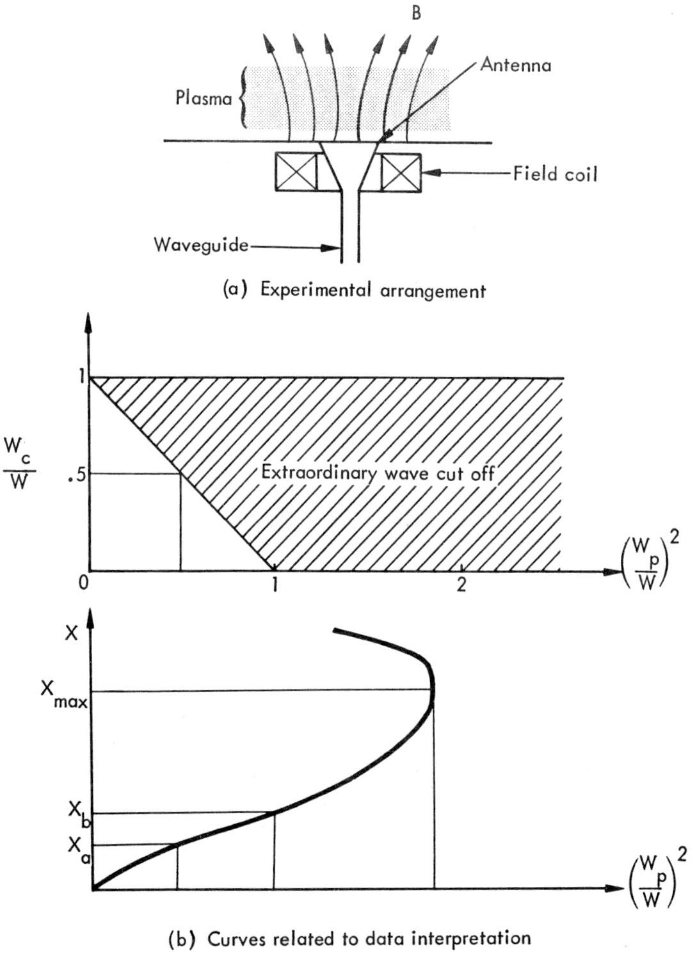

Fig. 7-1-13 Plasma probing using an extraordinary wave

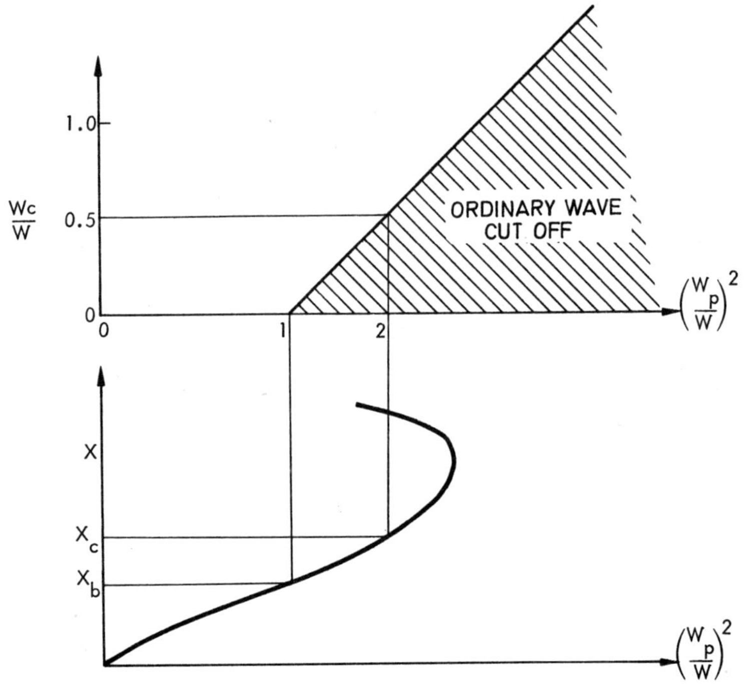

Fig. 7-1-14 Plasma probing using an ordinary wave

$$\Phi(\omega_c) = \frac{2\omega}{c} \int_0^{x(\omega_c)} \sqrt{\left(1 - \frac{\omega_p^2(x)}{\omega^2} \frac{1}{1-\omega_c/\omega}\right)}\, dx - \frac{\pi}{2} \qquad \text{(Eq. 7-1-29)}$$

In equation 7-1-29 $\omega_p^2(x)/\omega^2$ and $x(\omega_c)$ are unknown. For the profile (Fig. 7-1-13) only part of the spatial distribution of n_e has been obtained. The ordinary wave can be employed to determine the remainder of the profile up to the peak (Fig. 7-1-14). When ω_c/ω equals 0.5, the ordinary wave penetrates to x_c and a phase angle Φ_c is measured. As before Φ is measured as a function of ω_c/ω. The integral formula:

$$\Phi(\omega_c) = \frac{2\omega}{c} \int_0^{x(\omega_c)} \sqrt{\left(1 - \frac{\omega_p^2}{\omega^2} \frac{1}{1+\omega_c/\omega}\right)}\, dx - \frac{\pi}{2} \qquad \text{(Eq. 7-1-30)}$$

is inverted to find $\omega_p^2(x)/\omega^2$ and $x(\omega_c)$. The solution of the integral equation will be the subject of a subsequent article.

Assume a microwave frequency $f = 10^{11}$ Hz is used; the wavelength λ being 0.3cm. Equation 7-1-27 required an electron density not less than about $1.3 \times 10^{13}\text{cm}^{-3}$. If 0.1 is assumed to be much less than 1 in equation 7-1-26, then the maximum gradient in n_e is $4 \times 10^{12}\text{cm}^{-4}$. A distance of about 3 cm is required for x_{max} (Fig. 7-1-14). The positive gradient of the plasma is about 10 free space wavelengths in thickness. Equation 7-1-28 is not a particularly difficult condition to satisfy. A magnetic field of 5700 gauss is required for the extraordinary wave. This numerical example makes it apparent that the technique is suitable for large scale plasma devices with large n_e, but that it is not appropriate for probing their boundary layers with small n_e.

Plasma Waves

Spitzer (112) groups plasma waves into three categories: electro-magnetic waves, hydromagnetic waves, and electrostatic waves. Plasma diagnostics by means of electromagnetic waves is a well developed technique as discussed in the previous section. Two key problems in plasma diagnostics are the determination of n_e and spatial distribution of n_e as might be characterized by the gradient of n_e. Using electromagnetic waves as a probe, the anticipated n_e determines the frequency. There exists an awkward range of frequencies or wavelengths ranging from millimeter waves to near infrared. Wave speed for electromagnetic waves in a plasma is somewhere near the speed of light. Given a frequency, as determined by n_e, a length is established. The gradient of n_e establishes another length. These two may not be compatible.

Electromagnetic waves are only one phenomena which are dependent on the items to be measured, n_e and dn_e/dx. Electrostatic waves or MHD waves may have desirable features for plasma diagnostics. Use of these waves may require careful tailoring of the probe to the plasma. Techniques so developed may not be very general in range of applicability. All phenomena in which n_e or dn_e/dx occurs as a variable should be screened for applicability. As examples of this philosophy consider the following successful examples.

Trivelpiece and Gould (136) measured phase relationships of space charge waves in cylindrical plasma columns. Knowing the geometry of the plasma and surroundings and the magnetic field, the electron density could be measured. Nickel, Parker, and Gould (137) account for various Tonks-Dattner resonances in terms of spatial distribution of electrons in a nonuniform plasma column. The column was divided into regions where waves could propagate and regions where waves could not. The reflection of electroacoustic waves with different ω provides a means for measuring dn_e/dx. Dattner (138) relates plasma density to column geometry. Schmitt (139) could correlate Tonks-Dattner resonances with plasma frequency.

Electroacoustic waves propagate at a wave speed slow relative to electromagnetic waves. Consequently wave lengths are short. These waves may be appropriate for probing small scale plasmas.

Conclusions

A variety of plasma diagnostic techniques has been discussed covering the range from proven methods to new concepts. Both active and passive magnetic probes are well developed. The family of active probes include methods for measuring electrical conductivity, flow velocity, and flow angle. Electron beams yield information about rotational and vibrational temperature, mass density, and mass fraction of radiating species. Use of electron beams is limited, at least at present, to low density and to plasmas containing certain molecules. In the area of electrostatic probes more research is needed for a continuum, flowing plasma. The laser offers many experimental advantages; certainly holography and interferometry provide many new and fruitful avenues of research. Thomson scattering requires complicated equipment and careful attention to experimental detail. It has potential of becoming a universally applied technique although at present results are meager. Some scientists doubt the validity of data obtained in this manner.

Microwaves continue as a standard diagnostic technique. At wavelengths shorter than a few millimeters there is a void of equipment and experience. A concept for a method to determine spatial variation of n_e was presented. There are specialized measurement needs which require specialized techniques; electroacoustic and other plasma waves may be useful.

References

1. **Lovberg, R.H.,** 'Magnetic Probes', Plasma Diagnostic Techniques, Ch. 3. R.H. Huddlestone and S.L. Leonard, Ed., New York, Academic Press, 1965.

2. **Shercliff, J.A.,** 'The Theory of Electromagnetic Flow-Measurement', Cambridge Univ. Press, 1962.

3. **Fuhs, A.E.,** 'Instrumentation for High Speed Plasma Flow', AGARDograph 96, New York: Gordon and Breach Science Pub. 1965.

4. **Koritz, H.E., Keck, J.C.,** 'A Technique for Measuring the Electrical Conductivity of Wakes of Projectiles at Hypersonic Speed', Res. Rep. 167, AVCO Everett Res. Lab. Sept., 1963. Also Rev. Sci. Inst., 35, Feb. 1964.

5. **Durran, D.A., Grabowsky, W.R.,** 'Development of a Hot Shot Gas Source for Pulse MHD Wind Tunnel Study', Proc. 6th Symp. on Engineering Aspects of MHD, U. Pittsburgh, April 1964.

6. **Fuhs, A.E., Grabowsky, W.R., Gibb, O.L.,** 'Use of Radio Frequency Bridge for Re-Entry Plasma Diagnostics', J. Spacecraft and Rockets, 4,

No. 3, March 1967.

7. **Fuhs, A. E., Gibb, O. L.,** 'A Self-Balancing Radio Frequency Bridge', SSD-TR-67-44, Aerospace Corp., El Segundo, California, May 1967.

8. **Blackman, V. H.,** 'Magnetohydrodynamic Flow Experiments of a Steady State Nature', American Rocket Soc., Preprint 1007-59.

9. **Savic, P., Boult, G. T.,** 'A Frequency Modulation Circuit for the Measurement of Gas Conductivity and Boundary Layer Thickness in a Shock Tube', Nat. Res. Council of Canada, NRC No. 6349, May, 1961. Also J. Sci. Inst., 39. June 1962.

10. **Moulin, T., Massé, J.,** 'Sur la Mesure par Sonde Haute Fréquence de la Conductivité d'un Plasma Faiblement Ionisé'. Symp. Int. sur la Production MHD D'Energie Electrique, Paris, July 1964.

11. **Olson, R. A., Lary, E. C.,** 'Electrodeless Plasma Conductivity Probe Apparatus', Rev. Sci. Inst., 33. Dec. 1962.

12. **Marshall, T., Hill, L. L.,** 'An RF Technique for Determining the Electrical Conductivity of an Ionized Gas in a Flow Field', U.S. Naval Ord. Lab. Rep. 6878, July, 1961.

13. **Marshall, T., Hill, L. L., Crapo, B. J.,** 'Radio-Frequency Probes for Ionized Wake Studies', U.S. Naval Ord. Lab. Rep. 62-64, June 1962.

14. **Hill, L. L., Marshall, T., Crapo, B. J.,** 'A Probe for Electrical Conductivity Measurements in Ionized Gases', U.S. Naval Ord. Lab. Rep. 62-186, Dec. 1962.

15. **Yosim, S. J., Grantham, L. F., Luchsinger, E. B., Wike, R.,** 'Electrodeless Determination of Electrical Conductivities of Melts at Elevated Temperatures', Rev. Sci. Inst., 34, Sept. 1963.

16. **Hollister, D. D.,** 'A Technique for the Experimental Determination of the Electrical Conductivity of Plasmas', AIAA J. 2, Sept., 1964.

17. **Gourdine, M. C.,** 'A Technique for Making Local Measurements of the Conductivity and Velocity of Plasma jet', Rep. N. PLR-71, Plasmadyne Corp., Santa Ana, Calif., June 1960.

18. **Fuhs, A. E.,** 'Study of Turbulence in an Arc Plasma Jet Using Electromagnetic Induction', AIAA Paper No. 64-375, June 1964. AIAA J. 3, 1965.

19. **Fuhs, A. E.,** 'Direct Determination of Transition to Turbulence During Reentry', AIAA Entry Tech. Conf., Williamsburg, Va., Oct. 1964.

20. **Fuhs, A. E., Gibb, O. L.,** 'A Magnetohydrodynamic Flow Angle Indicator', TDR-169 (3153) TN-4, Aerospace Corp., El Segundo, Calif., Nov. 1962. Also Fourth Symp. Engineering Aspects of Magnetohydrodynamics, Univ. of Calif., April, 1963. Rev. Sci. Inst., 35. Nov. 1964.

21. **Schumacher, B. W., Gadamer, E. O.,** 'Electron Beam Fluorescence Probe for Measuring the Local Gas Density in a Wide Field of Observation', Can. J. Phys., Vol. 36, 1958.

22. **Grun, A. E., Schopper, E., Schumacher, B.,** 'Electron Shadowgraphs and

Afterglow Pictures of Gas Jets at Low Densities', J. Appl. Phy., Vol.24, 1953.

23. **Schumacher, B.W.,** 'Some Aspects of Flow Visualization and Local Density Measurements in Rarefied Gases by Means of Electron Beam Probes', Ontario Res. Foundation Rep. PRR 6604. 1966.

24. **Falckenberg, H.R., Schumacher, B.W.,** 'The Possibilities for Gas Density Gauging by Single Scattering of Electrons', Can. J. Phys., Vol.40, 1962.

25. **Schumacher, B.W.,** 'Gauging Gas Density with Fast Charged Particles', Nucleonics, Vol.18, 1960.

26. **Grün, A.E.,** 'On the Fluorescence of Air, Excited by Fast Electrons: Light Yield as a Function of Pressure', Can. J. Phys., Vol.36, 1958.

27. **Muntz, E.P.,** 'Measurement of Rotational Temperature, Vibrational Temperature, and Molecule Concentration, In Non-Radiating Flows of Low Density Nitrogen', Univ. of Toronto, Inst. Aerophysics, Rep. 71, April 1961.

28. **Muntz, E.P., Abel, S.J., Maguire, B.L.,** 'The Electron Beam Fluorescence Probe in Experimental Gas Dynamics', Supp. to IEEE Trans. on Aerospace, 1965.

29. **Muntz, E.P., Abel, S.,** 'The Direct Measurement of Static Temperatures in Shock Tunnel Flows', Gen. Electric, Space Sci. Lab., R64SD25, 1964.

30. **Zempel, R., Mallin, J., Muntz, E.,** 'An Electron Beam Excitation Technique for the Measurement of Low Densities in Hypersonic Shock Tunnel Flows', Gen. Electric, Space Sci. Lab., R63SD21, 1963.

31. **Denisse, J.F., Decroix, J.L.,** 'Plasma Waves,' Interscience Publ., New York. 1963.

32. **Pierce, J.R.,** Theory and Design of Electron Beams, Van Nostrand, New York, 1954.

33. **Tonks, L., Langmuir, I.,** 'A General Theory of the Plasma of an Arc', Phys. Rev., 34, 1929.

34. **Bohm, D., Burhop, E.H., Massey, H.S.,** Characteristics of Electrical Discharges in Magnetic Fields (A Guthrie, R.K. Wakerling, Ed., New York: McGraw-Hill, Ch.2. 1949.

35. **Allen, J.E., Boyd, R.L., Reynolds, P.,** 'Collection of Positive Ions by a Probe Immersed in a Plasma', Proc. Phys. Soc., London, B70. 1957.

36. **Bohm, D.,** Characteristics of Electrical Discharges in Magnetic Fields, A. Guthrie, R.K. Wakerling, Ed., New York: McGraw-Hill, Ch.3, 1949.

37. **Bernstein, I.B., Rabinowitz, I.N.,** 'Theory of Electrostatic Probes in a Low Density Plasma', Phys. Fluids, 2. 1959.

38. **Langmuir, I., Mott-Smith, H.,** 'Studies of Electrical Discharges in Gases at Low Pressures', Gen. Elec. Rev. 27, Pt.1, 1924.

39. **Allen, J.E., Thonemann, P.C.,** 'Current Limitation in a Low-Pressure Mercury Arc', Proc. Phys. Soc., London, B67. 1954.

40. **Wehner, G., Medicus, G.,** 'Reliability of Probe Measurements in Hot Cathode Gas Diodes', J. Appl. Phys., 23. 1952.

41. **Schultz, G.J., Brown, S.C.,** 'Microwave Study of Positive Ion Collection by Probes', Phys. Rev., 98. 1955.

42. **Boyd, R.L.,** 'Collection of Positive Ions by a Probe in an Electrical Discharge', Proc. Roy. Soc., London, A201. 1950.

43. **Johnson, E.O., Malter, L.,** 'A Double-Probe Method for Measurements in Steady and Time-Varying Discharges', Phys. Rev., 77. 1950.

44. **Chapman, R.A.,** 'Measurement of Plasma Electron Temperature by an External Langmuir-Type Probe', MHD Symp. Int. sur la Production MHD D'Energie Electrique, Paris, July 1964.

45. **Sturtevant, B.,** 'Diffusion in a Slightly Ionized Gas with Application to Effusion from a Shock Tube', Phys. Fluids, 4. 1961.

46. **Su, C.H., Lam, S.H.,** 'Continuum Theory of Spherical Electrostatic Probes', Phys. Fluids, 6. 1963.

47. **Cohen, I.M.,** Asymptotic Theory of Spherical Electrostatic Probe in a Slightly Ionized, Collision-Dominated Gas', Phys. Fluids, 6. 1963.

48. **Loeb, L.B.,** 'Basic Processes of Gaseous Electronics', Berkeley: Univ. of Calif., 1955.

49. **Chen, F.F.,** 'Electric Probes , Ch. 4., Plasma Diagnostic Techniques, R.H. Huddlestone, S.L. Leonard, Ed., New York: Academic Press, 1965.

50. **Guderley, K.G.,** 'Limiting Cases in the Theory of Spherical Plasma Probes', USAF, OAR, Aerospace Res. Lab., ARL 66-0113, 1966.

51. **Guderley, K.G., Valentine, E.M.,** 'Theory of Spherical Probes in Monoenergetic Plasmas', USAF, OAR, Aerospace Res. Lab. ARL-66-0024, 1966.

52. **French, J.B.,** 'Langmuir Probes in a Flowing Low Density Plasma', Rep. No. 79, Inst. Aerophysics, Univ. of Toronto, Aug. 1961.

53. **Muntz, E.P., Harris, C.J., Kaegi, E.M.,** 'Techniques for Experimental Investigation of the Properties of Electrically Conducting Hypersonic Flow Fields', Rep. R63SD27, Space Sci. Lab. Missile and Space Div., Gen. Elec. Co., Pennsylvania, March, 1963.

54. **Kaegi, E.M., Muntz, E.P.,** 'Driver-Driven Gas Mixing and Its Effect on Shock Tunnel Test Time', Third Hypervelocity Tech. Symp., Denver Res. Inst., Univ. of Denver, Colorado, March, 1964.

55. **Bourdeau, R.E., Donley, J.L., Serbu, G.P., Whipple, E.C.,** 'Measurements of Sheath Currents and Equilibrium Potential on the Explorer VIII Satellite 1960'. NASA TND-1064.

56. **Basu, S., Fay, J.A.,** 'Ionization in Detonation Waves', Seventh Symp. Int. on Combustion. London: Butterworths, 1958.

57. **Lam, S.H.,** 'A General Theory for the Flow of Weakly Ionized Gases', AIAA Paper 63-459, 1963.

58. **Su, C.H.,** 'Compressible Plasma Flow over a Biased Body', AIAA J. 3, 1965.

59. **Wasserstrom, E., Su, C., Probstein, R.,** 'Kinetic Theory Approach to Electrostatic Probes', Mass. Inst., of Tech. Fluid Mech. Lab. Publ. 64-5, 1964.

60. **Chung, P.M., Blankenship, V.D.,** 'Theory of Electrostatic Double Probe Comprised of Two Parallel Plates', AIAA J. 4, 1966.

61. **Weissman, D.E., Scharfman, W.E., Guthart, H.,** 'Electrostatic Probe Measurements in a Short Mean Free Path Plasma', Stanford Res. Inst., Tech. Rep. No.1, Contr. AF 04(694)-931, July, 1966.

62. **Scharfman, W.E.,** 'The Use of Langmuir Probes to Determine the Electron Density Surrounding Reentry Vehicles', Stanford Res. Inst., Final Rep. Contr. NAS 1-3942, June, 1965.

63. **Scharfman, W.E., Bredfeldt, H., Buthart, H., Morita, T.,** 'The Use of Langmuir Probes in Diagnosing Ionized Flow Fields', Stanford Res. Inst., ARPA Order 281-62, Project Code 7400.

64. **Bredfeldt, H.R., Scharfman, W.E., Guthart, H., Morita, T.,** 'The Use of Ion Probes in Reentry Physics', Stanford Res. Inst., Tech. Rep. 26, May, 1965.

65. **Talbot, L.,** 'Theory of the Stagnation-Point Langmuir Probe', Phys. Fluids, 3. 1960.

66. **Talbot, L.,** 'Note on the Stagnation-Point Langmuir Probe', Phys. Fluids, 5. 1962.

67. **Chung, P.M.,** 'Electrical Characteristics of Couette and Stagnation Boundary-Layer Flows of Weakly Ionized Gases', Phys. Fluids, 7. 1964.

68. **Chung, P.M.,** 'Langmuir Potential Associated with Couette Flow of Viscous Plasma', Phys. Fluids, 5. 1962.

69. **Chung, P.M., Mullen, J.F.,** 'Nonequilibrium Electron Temperature Effects in Weakly Ionized Stagnation Boundary Layers', AIAA Preprint 63-161, 1963.

70. **Chung, P.M.,** 'Diagnostic Equations of Electrostatic Double Probes for Continuum Plasmas', Univ. of Illinois at Chicago Circle, Dept. of Energy Engr., TR-E5, Dec., 1966.

71. **Dix, D.M.,** 'Energy Transfer Processes in a Partially Ionized, Two-Temperature Gas', AIAA J. 2, 1964.

72. **Camac, M., Kemp, N.H.,** 'A Multitemperature Boundary Layer', AIAA Preprint 63-640, Aug. 1963.

73. **Hall, D.F., Kemp, R.F., Sellen, J.M.,** 'Plasma-Vehicle Interaction in a Plasma Stream', AIAA J. 2, 1964.

74. **Kemp, R.F., Sellen, J.M.,** 'Plasma Potential Measurements by Electron Emissive Probes', Rev. Sci. Inst., 37. 1966.

75. **Chung, P.M., Blankenship, V.D.,** 'Approximate Analysis of an Electrostatic Probe for Electron Density Measurements on Re-entry Vehicles', J. Spacecraft and Rockets, 3. 1966.

76. **Klein, A.F.,** 'A Survey of Optical Interferometry as Applied to Plasma Diagnostics', Chap. 12. Physico-Chemical Diagnostics of Plasmas. T.F. Anderson, *et al*. Ed. Northwestern Univ. Press, 1963.

77. **Alpher, R.A., White, D.R.,** 'Optical Refractivity of High-Temperature Gases. I. Effects Resulting from Dissociation of Diatomic Gases', Phys. Fluids, 2. 1959.

78. **Alpher, R.A., White, D.R.,** 'Optical Refractivity of High-Temperature Gases. II. Effects Resulting from Ionization of Monatomic Gases', Phys. Fluids, 2. 1959.

79. **Alpher, R.A., White, D.R.,** 'Interferometric Measurement of Electron Concentrations in Plasmas', Phys. Fluids, 1. 1958.

80. **Shukhtin, A.M.,** 'Interferometric Method for the Determination of the Gas Density and Electron Concentration in a Plasma', Optics and Spectroscopy, 10. 1961.

81. **Candler, A.C.,** Modern Interferometers, Hilger and Watts, London, 1951.

82. **Bennett, F.D., Carter, W.C., Bergdolt, V.E.,** 'Interferometric Analysis of Airflow about Projectiles in Free Flight', J. Appl. Phys., 23. 1952.

83. **Giese, J.H., Bennett, F.D., Bergdolt, V.E.,** 'A Simple Interferometric Test for Conical Flow', J. Appl. Phys., 21. 1950.

84. **Ashby, D.E., Jephcott, D.F.,** 'Measurement of Plasma Density Using a Gas Laser as an Infrared Interferometer', Appl. Physics Letters, 3. 1963

85. **Ashby, D.E., Jephcott, D.F., Malein, A., Raynor, F.A.,** 'Performance of the HeNe Gas Laser as an Interferometer for Measuring Plasma Density', J. Appl. Physics, 36. 1965.

86. **Boornard, A., Nicastro, L.J., Vollmer, J.,** 'Determination of Plasma Density by Laser Interferometric and Continuum Radiation Intensity Measurements', Appl. Physics Letters, 7. 1965.

87. **Gabor, D.,** 'A New Microscopic Principle', Nature, 161. May 1948.

88. **Gabor, D.,** 'Microscopy by Reconstructed Wave Fronts', Proc. Roy. Soc., A197. 1949.

89. **Gabor, D.,** 'Microscopy by Reconstructed Wave Fronts: II', Proc. of Phys. Soc., 64. 1951.

90. **Stroke, G.W.,** 'An Introduction to Coherent Optics and Holography,' New York: Academic Press, 1966.

91. **Heflinger, L.O., Wuerker, R.F., Brooks, R.E.,** 'Holographic Interferometry', J. Appl. Phys., 37. 1966.

92. **Leith, E.N., Upatnieks, J.,** 'Reconstructed Wavefronts and Communication Theory', J. Opt. Soc. of Amer., 52. 1962.

93. **Leith, E.N., Upatnieks, J.,** 'Wavefront Reconstruction with Continuous-Tone Objects', J. Opt. Soc. of Amer., 53. 1963.

94. **Leith, E.N., Upatnieks, J.,** 'Wavefront Reconstruction with Diffused Illumination and Three-Dimensional Objects', J. Opt. Soc. of Amer., 54. 1964.

95. **Brooks, R.E., Heflinger, L.O., Wuerker, R.F., Briones, R.A.,** 'Holographic Photography of High-Speed Phenomena with Conventional and Q-Switched Ruby Lasers', Appl. Physics Letters, 7. 1965.

96. **Brooks, R.E., Heflinger, L.O., Wuerker, R.F.,** 'Interferometry with a Holographically Reconstructed Comparison Beam', Appl. Physics Letters, 7. 1965.

97. **Knox, C.,** 'Holographic Microscopy as a Technique for Recording Dynamic Microscopic Subjects', Science, 153. 1966.

98. **Holds, J.H., Fuhs, A.E.,** 'A Refined Analysis of a Holographic Interferogram', J. Appl. Phys. 1967.

99. **Holds, J.H.,** 'Aeronautical Applications of Holographic Interferometry', M.S. Thesis, Dep. of Aeronautics Naval Postgrad. School, Monterey, Calif. June, 1967.

100. **De, M., Sévigny, L.,** 'Three-Beam Holography', Appl. Physics Letters, 10. 1967.

101. **Ascoli-Bartoli, U., Benedetti-Michelangeli, G., Lovisetto, L.,** 'A Faraday-Switched Ruby Laser', Appl. Physics Letters, 8. 1966.

102. **Liepmann, H.W., Roshko, A.,** Elements of Gas Dynamics, Wiley, New York, 1957.

103. **Ascoli-Bartoli, U., Martellucci, S., Mazzucato, E.,** Proc. of 6th Int. Conf. Ionization Phenomena in Gases, Paris, 1963. Vol. 4, North-Holland, Amsterdam, 1964.

104. **Lovberg, R.H.,** 'The Measurement of Plasma Density in a Rail Acceleration by Means of Schlieren Photography', IEEE Trans. on Nucl. Sci., 11. 1963.

105. **Meyer-Arendt, J.R., Shettle, E.P.,** 'Calibration of Schlieren Systems', Appl. Optics, 4. 1965.

106. **Reisman, E., Sutton, P.M.,** 'Measurement of Air Temperature Distributions with the Schlieren Interferometer', Appl. Optics, 4. 1965.

107. **Brackenridge, J.B., Peterka, J.,** 'Criteria for Quantitative Schlieren Interferometry', Appl. Optics, 6. 1967.

108. **Levine, M.A., Haggerty, J.C.,** 'Shadowgraph of Self-Luminous Objects',

Appl. Optics, 2. 1963.

109. **Ascoli-Bartoli, U., Martellucci, S., Mazzucato, E.,** Proc. of 6th Int. Conf. Ionization Phenomena in Gases, Paris, 1963, Vol. 4, North-Holland, Amsterdam, 1964.

110. **Johoda, F.C., Quinn,** W.E., **Ribe, F.L.,** 'Plasma Experiments with a 570-Kj Theta-Pinch', J. Applied Physics, 35. 1964.

111. **Malyshev, G.M.,** 'Plasma Diagnostics by Light Scattering on Electrons', Soviet Physics - Tech. Physics, 10. 1966.

112. **Spitzer, L.,** Physics of Fully Ionized Gases, New York: Intersci. Publ., 1956.

113. **Wessel, G.K., Rothenberg, H.C., Zendle,** B., 'Ruby Optical Masers for In-Flight Measurement of Hypersonic Flow Field Parameters', Vol. II, Advances in the Astronautical Sciences, Amer. Astronaut, Soc., 1963.

114. **Hughes, T.P.,** 'A New Method for the Determination of Plasma Electron Temperature and Density from Thomson Scattering of an Optical Maser Beam', Nature, 194. 1962.

115. **Chandrasekhar, S.,** Radiative Transfer. Oxford Univ. Press, 1950.

116. **Salpeter, E.E.,** 'Electron Density Fluctuations in a Plasma', Physical Rev., 120. 1960.

117. **Daugherty, J.P., Farley, D.T.,** 'A Theory of Incoherent Scattering of Radio Waves by a Plasma', Proc. Roy. Soc. (London), A259. 1960.

118. **Fejer, J.A.,** 'Scattering of Radio Waves by an Ionized Gas in Thermal Equilibrium', Can. J. Phys., 38. 1960.

119. **Rosenbluth, M.N., Rostoker, N.,** 'Scattering of Electromagnetic Waves by a Nonequilibrium Plasma', Phys. Fluids, 5. 1962.

120. **Shafranov, V.D.,** Reviews of Plasma Physics, Vol. 2, New York: Consultants Bureau, 1963.

121. **Stehle, P.,** 'High Intensity Limit of Thomson Scattering', J. Opt. Soc. Am., 53. 1963.

122. **Brown, L.S., Kibble, T.W.,** 'Interaction of Intense Laser Beams with Electrons', Phys. Rev., 133A. 1963.

123. **Goldman, I.I.,** 'Intensity Effects in Compton Scattering', Soviet Physics - JETP, 19. 1964.

124. **Mandel, L.,** 'Thomson Scattering of Intense Light Beams', J. Opt. Soc. Am., 54. 1964.

125. **Fiocco, G., Thompson, E.,** 'Thomson Scattering of Optical Radiation from an Electron Beam', Phys. Rev. Letters, 10. 1963.

126. **Wharton, C.B.,** 'Microwave Techniques', Ch. 11, Plasma Diagnostic Techniques - Academic Press, 1966.

127. **Heald, M.A., Wharton, C.B.,** 'Plasma Diagnostics with Microwaves,' Wiley, New York, 1965.

128. **Ratcliffe, J.A.,** 'Magneto-Ionic Theory,' Cambridge Univ. Press, London. 1959.

129. **Mitra, S.K.,** 'The Upper Atmosphere,' Asiatic Soc., Calcutta, 1952.

130. **Warder, R., Nigham, W.L., Brodwin, M., Cambel, A.B.,** 'Microwave Diagnostics of Arc-Heated Plasmas', Dynamics of Manned Lifting Planetary Entry, S.M. Scala, A.C. Harrison, M. Rogers, Ed., New York: Wiley, 1963.

131. **Balwanz, W.W., Weston, J.P.,** 'The Prediction of Rocket Exhaust Interference with Radio Signals', ARS Preprint 2591-62, Oct., 1962.

132. **Fuhs, A.E.,** 'A Technique for Obtaining the Electrical Conductivity Velocity Profile', Ch. 19, Electromagnetic Aspects of Hypersonic Flight, W. Rotman, H. Moore, and R. Papa, Ed; Baltimore: Spartan Books, 1964.

133. **Anisimov, A.I., Vinogradov, N.I., Golant, V.E., Konstantinov, B.P.,** 'Method of Investigating Electron Spatial Distribution in a Plasma', Soviet Physics, Tech. Physics, 9. 1961.

134. **Raizer, M.D., Strelkov, P.S., Frank, A.G.,** 'Localization of a Linear High-Frequency Current in a Gas by Means of a Quadrapole Magnetic Field', Soviet Physics-Technical Physics, 9. 1964.

135. **Dnestrovskii, Y.N., Kostomarov, D.P.,** 'Plasma Probing by an Electromagnetic Field', Soviet Physics-Tech. Physics, II, 1966.

136. **Trivelpiece, A.W., Gould, R.W.,** 'Space Charge Waves in Cylindrical Plasma Columns,' J. Applied Physics, 30. 1959.

137. **Nickel, J.C., Parker, J.V., Gould, R.W.,** 'Resonance Oscillations in a Hot Nonuniform Plasma Column', Phys. Rev. Letters, 11. 1963.

138. **Dattner, A.,** 'Resonance Densities in a Cylindrical Plasma Column', Phys. Rev. Letters, 10. 1963.

139. **Schmitt, H.J.,** 'Acoustic Resonances in Afterglow Plasmas', Appl. Physics Letters, 4. 1964.

List of Mathematical Symbols

a	distance, coil to plasma
A	area
B	magnetic field
c	speed of light
C	capacitance
d	sheath thickness
D_a	ambipolar diffusion coefficient
D_e	electron diffusion coefficient
D_i	ion diffusion coefficient
e	voltage; electron charge
E	electric field
E_{ij}	voltage correlation function
f	cyclic frequency cycles/sec
h	Debye length
I	intensity of radiation
j	current density
k	Boltzmann constant; thermal conductivity
L	characteristic length
Le	Lewis number
m	electron mass
n	index of refraction; normal vector
n_e	electron density
n_g	group index of refraction
n_i	ion density
n_p	phase index of refraction
N	coil turns; fringe shift
P_0	incident power
Pr	Prandtl number
P_s	scattered power
r_c	coil radius
r_o	classical electron radius
$\mathcal{R}$	ratio of resistive component of current to reactive component
R	resistance
R_m	magnetic Reynolds number
s	fringe spacing
S	skin depth
Sc	Schmidt number

t	time
T_e	electron temperature
T_r	rotational temperature
T_t	translational temperature
U	plasma velocity
$\overline{v}$	average thermal velocity
V	volume
Z	complex impedance
α	polarizability; angle between beams in two beam holography; scattering parameter
β	angle between electric vector of incident radiation and direction of observation
δ	thickness of plasma layer; boundary layer thickness
ϵ	angular deflection of a ray
θ	scattering angle
λ	mean free path; wavelength
μ	magnetic permeability; viscosity coefficient
ν	electron collision frequency
ρ	density
σ	electrical conductivity
τ	mean time between collisions
φ	potential
Φ	phase angle
χ	refractivity
ω	angular frequency radians/sec
Ω	solid angle
ΔR	change of coil resistance
ω_c	cyclotron frequency
ω_p	electron plasma frequency
Ω_p	ion plasma frequency

Commentary A on Chapter 7-1

A.B. Cambel.

The author is well known for his contribution to plasma diagnostics both in the laboratory and in flight. This chapter differs from his previous publications in that it addresses itself primarily to the instrumentation rather than to the phenomena to be studied or the properties to be measured.

Whereas in the past a researcher could be either an experimentalist or a theoretician, for plasma diagnostics he must excel in both. Even though the major contribution of a plasma diagnostician may be in determining data experimentally, he must be well versed in theoretical aspects of electrodynamics, gas dynamics, statistical mechanics and quantum theory. Those problems that are amenable to theoretical treatment are as a rule difficult to simulate in the laboratory and the converse is also true.

It would have been helpful if the author had differentiated between the diagnostics of static plasmas and those of dynamic plasmas. It would have been similarly helpful had he provided a map of temperature versus electron density with the various diagnostic instruments described, but Table 7-1-1 is a fine substitute for this. It is interesting to note however that the figures he gives differ from those cited in other references. The AGARD community might well consider establishing standards of plasma diagnostics reminiscent of the AGARD/ONERA standard circular flame-holder that researchers in flame stabilisation found so useful in the late 1950's.

In the interesting account of optical measurements insufficient attention has perhaps been paid to the problem of optical thickness which is of such great importance. Although visualisation techniques may not give a detailed account of plasma characteristics they do provide useful 'ball park' impressions. The ramifications of microwave diagnostics, and the possible sources of error, are quoted in insufficient detail and other omissions are plasma spectroscopy, Raman scattering and the possible applications of cavity measurements, and the Zeeman effect.

The content of the chapter is wide in scope and the approach is refreshing. The author must be congratulated for undertaking this most demanding review.

Commentary B on Chapter 7-1

T.H. Peters.

This chapter gives an excellent summary of a wide variety of plasma diagnostic methods.

Magnetic probes for measuring the product of conductivity and flow velocity, that is to say the magnetic Reynolds Number, and the electrical conductivity usually measure only an average of the plasma dynamics inside a coil. The author has shown how spatially distributed values may be obtained by means of different coil geometries and frequencies. I would add that where electric currents and magnetic fields are present in the plasma, the use of simple Hall probes can be advantageous.

The author has paid considerable attention to interferometry which indicates index of refraction and hence, in a plasma, electron density. Holographic interferometry, which is now being developed in many institutes may become a powerful method for determining spatially distributed quantities. The scattering of laser light has been shown to be an excellent method of measuring electron densities and electron and ion temperatures.

These new methods interlock with the well known spectrographic diagnostic techniques that have been applied by gas discharge physicists and astrophysicists since the end of the last century.

7-2

Internal Physical Measurements in MHD Plasma Accelerators

R.H. LOVBERG
University of California, La Jolla, California, USA.

Summary

A restricted set of correlated measurements is described, in which the electromagnetic field structure and the electron density distribution are determined, and inferences concerning the efficiency of acceleration and partitioning of electron and ion currents are made. The apparatus employed for this research is an inductive "theta pinch". Electric and magnetic fields are measured by the use of probes, and the electron density distribution is determined through the use of a modified schlieren photographic technique. Current partitioning is inferred through insertion of the experimental data into the generalized Ohm's Law.

Sommaire

On décrit un ensemble de mesures liées, dans lequel on a déterminé la structure du champ électromagnétique et la distribution électronique auxquelles on a lié le rendement de l'accélération et la répartition de courant électronique et ionique. L'appareillage utilisé est un "theta pinch" à induction. Les champs électrique et magnétique sont mesurés au moyen de sondes et la distribution de densité électronoque au moyen d'un montage strioscopique modifié. On déduit la répartition de courant en introduisant les données expérimentales dans la loi d'Ohm généralisée.

Introduction

Electromagnetic plasma accelerators have the unique advantage for space propulsion that the velocity of the exhaust need not be limited by the greatest temperature that can be produced or withstood in an inner chamber. A large velocity, i.e. a large energy in one coordinate direction, can be imparted to an ionized gaseous propellant (plasma) without first going through a phase of thermal heating where the energy is partitioned equally in all degrees of freedom.

Energy may be transferred to the propellant either by direct application of an electric field to ionized atoms (ion engines), or by exploiting the electrical conductivity of the plasma. In a device employing the latter principle, an electric current $\vec{J}$ is passed through the plasma and allowed to interact with a superimposed magnetic field $\vec{B}$, producing a force $\vec{F} = \vec{J} \times \vec{B}$ per unit volume of plasma, in a direction normal to both current and field. Such a device is usually called a plasma engine, or an MHD engine. There are many varieties being studied, ranging from steadily operating systems (MHD arcs, Hall accelerators) to plasma guns, in which accelerating forces and powers of great magnitude are applied as short duration impulses. This paper is concerned mainly with experiments related to the plasma gun type, although most of the experimental techniques are applicable, with modifications, to the other classes of electrical engines.

Problems of Measurement

Research into magnetoplasmadynamic systems poses problems of unusual difficulty because it is necessary to determine the properties of the propellant gas as well as the structure of the electromagnetic field in which it is immersed. The presence of strong, sometimes complete ionization introduces phenomena characteristic of plasma, e.g. oscillations and instabilities of several kinds, most of which are detrimental to the achievement of efficient thrustors.

The coupling between the field and the plasma is a function of the temperature and density of the electrons and ions, which, in a typical situation, can vary by orders of magnitude over a very small part of the system. The physical processes governing this coupling between field and plasma are not well understood, electric propulsion technology being still in the basic research phase rather than the development phase. They are probably quite complicated, however, because 'reasonable' assumptions about the field-plasma system have proved inadequate for practical development. All attempts to construct prototype plasma engines of various kinds have failed. While it has usually been easy to produce a plasma stream having approximately the correct velocity for an intended application, the best attainable efficiency has been discouragingly low because of obscure instabilities or mysterious plasma losses or an anomalous plasma resistance.

To obtain a complete mapping of the structure of the electric and magnetic fields it is not necessary to measure the complete distribution function of all components of the plasma: some unmeasured parameters may be inferred through the use of theoretical equations such as Boltzmann's and Maxwell's equations. Unfortunately experimental techniques are not yet sufficiently good in either variety or precision to make a complete specification of most plasma systems possible, even with the aid of theory.

In plasma guns especially, the entire acceleration may occur in less than a microsecond, so that experimental analysis requires instruments having the sharpest possible time resolution and the widest possible frequency response.

The fundamental work on the physics of electric propulsion is usually done on experimental devices that in configuration are quite unlike actual linear accelerators. Real plasma guns, for instance, may have complicated geometry, preventing the plasma from having uniformity along any coordinate direction, unlike the theory being verified or employed, which is usually one-dimensional. Furthermore the reduction of experimental data on field structures and plasma distributions into such inferred quantities as current density and momentum density is a reasonable task only when the system geometry is simple. These considerations commend the use of a pinch or an inverse pinch discharge, both of which produce plasma-field systems that locally can be made nearly one-dimensional (1).

Geometry of Present Experiments

In our experiments we used an induction pinch or theta pinch device to determine the means by which momentum is transferred to a non-ionized propellant gas by a rapidly moving magnetic shock such as that produced in any highly impulsive plasma accelerator. In an induction pinch device, the usual sequence of events is: (i) a neutral gas is introduced at low pressure (typically ~ 0.1 torr) into the region between a pair of electrodes; (ii) a high voltage is switched across the electrodes, causing electrical breakdown in the gas; (iii) because the rate of current rise in the plasma is sufficiently great, the electromagnetic 'skin effect' forces nearly all of the current to flow in a thin layer of gas at that edge of the electrodes into which the current flows; (iv) the interaction between this sheet of current and the magnetic field arising from its flow causes the sheet to accelerate along the electrodes, and finally to emerge as a plume into the space beyond the electrode region. In the common case of a coaxial gun, the sheet is, ideally, an azimuthally uniform layer of radial current, highly localized in the axial direction and propagated along the system axis.

The interaction between this magnetic shock and the propellant gas may produce (a) acceleration of the gas to a speed less than that of the current sheet, (b) stagnation, or entrainment of the gas within the sheet, the so-called 'snow plow' effect, or (c) specular reflection of the gas from the sheet, which should occur in the limiting case when the propellant is fully pre-ionized and essentially collisionless. Whichever of these modes actually occurs in any particular system has a strong effect upon its overall efficiency. For instance the 'snow plow' entrainment of propellant divides the energy of the encounter equally(?) between translational kinetic energy and internal thermal energy of the accelerated gas, whereas the specular reflection of plasma from a current sheet is a perfectly efficient process, converting all of the work done by the magnetic field upon the gas into translational kinetic energy.

The theta pinch is an induction discharge that has no electrodes at all. Instead an induction coil, usually a single turn having an axial length somewhat greater than its diameter, is fitted closely around a tubular vacuum vessel into which the propellant under study is introduced at low pressure. A very rapidly rising surge of current in sent around this coil in the azimuthal direction. Obeying Faraday's law of induction, a strong azimuthal electric field is generated within the test vessel and this causes breakdown with consequent flow of current in a thin cylindrical shell within the gas. A strong inward radial force upon the plasma is produced through the $\vec{j} \times \vec{B}$ coupling.

This current sheet is the same in its internal microscopic behaviour and local macroscopic structure as that found in a plasma gun provided that the current and gas densities are the same. However, it is much more uniform structurally along its surface than the layer in a probable gun and, to the extent that its thickness is substantially less than the shell radius, it is close to being a one-dimensional structure.

The theta pinch permits easy experimental access: optical windows may be put in the ends of the plasma tube, giving a direct view of the pinch cylinder along its axis. High speed photography may be used to record the collapsing plasma shell, using the intrinsic plasma luminosity, and time-resolved spectroscopy may be used to obtain information on the detailed state of the gas.

Apparatus Parameters and Diagnostics

The theta pinch used in our work had a coil length of 25 cm, and the inner diameter of the vacuum vessel was 10 cm. Energy was stored in a 15 microfarad capacitor charged to 16 kV. The coil current reached about 2×10^5 amperes in a time of 1.6×10^{-6} seconds after closing the spark gap switch.

Measurements were made in nitrogen at a pressure of about 0.05 torr, which is typical of values used in propulsion-oriented plasma gun experiments. Since we employed no pre-ionization, the coupling between the applied fields and plasma was not strong until the second half-cycle of the oscillating current in the coil. It was usually observed that only a low level of ionization occurred during the first half-cycle of current flow in these devices, but this was sufficient to allow the formation of a highly conducting sheet when the driving current was reversed (2).

A result of this breakdown delay, which is typical of induction pinch devices, was that a portion of the magnetic flux within the coil during the first half-cycle was trapped within the cylinder of azimuthal plasma current that began to flow just prior to the reversal of coil current. Thus when this plasma sheet began to contract, it acted as a separation layer between an inner core of 'forward' flux and an outer region of 'reverse'flux. This characteristic will be apparent in the field data which follow.

Measurements of the magnetic field strength were made with conventional miniature magnetic probes, whose outputs were integrated and displayed on an oscilloscope (3). The oscilloscope sweep was triggered by a signal derived from the current flow in the main coil, and the sweep circuitry was set to give a delay equal to the duration of the first half-cycle, about 3 microseconds, followed by a sweep time of two microseconds. At each sampling point two or three sweeps from as many discharges of the pinch were superimposed photographically to provide evidence of reproducibility of the system from one pulse to the next. After many initial measurements to verify that this plasma was quite uniform along the azimuthal (θ) and axial (z) coordinates during the first implosion, it was only necessary to map the field structure along a radius, in order to specify it in all dimensions: we mapped the B_z distribution in 5 mm intervals from one wall to the other.

The electric field strength within the plasma was mapped using a differential electric probe (3). In this technique, a pair of closely spaced electrodes is exposed to the plasma, and the difference in their floating potentials is measured. In the absence of extremely large electron temperature gradients, it can be assumed that this difference is equal to the difference in local plasma potential, and thus, that this difference divided by the electrode spacing is the average electric field along a line connecting them. The procedure for acquiring E-field data was the same as that used for measuring B, except that the reproducibility from shot to shot of the machine was always poorer, and more traces had to be overlaid photographically at each field point in order to get adequate statistics.

Finally, the electron density distribution was inferred in these experiments through the use of schlieren photography: we employed a double-traversal system aligned along the coil axis. The plasma conditions were sufficiently uniform along z (up to within about 1.5 cm of the coil end during compression) that one could infer the local density with reasonable accuracy simply by dividing the measured integral of the density along z, by the coil length. A Q-switched ruby laser supplied the illumination, the effective exposure time being about 2×10^{-8}s.

A satisfactory representation of the general distribution of the plasma was given by simple schlieren photographs obtained from a double knife-edge system. These represented transverse gradients in the refractive index of the test medium and the refractive index of well-ionized gas was related to the free electron density by the equation:

$$\mu = \sqrt{\left(1 - \frac{r_e}{\pi} \lambda^2 \eta_e\right)}$$

where μ is the refractive index, r_e is the classical electron radius ($r_e = 2.8 \times 10^{-13}$ cm), λ is the transmitted wavelength, and n_e is electron density. We thus related refractivity directly to electron density (4).

The usual schlieren picture is dark or light according to the value of refraction gradients in the field. A quantitative interpretation would therefore require a sensinometric measurement, and calibration of the materials used - a difficult task. More useful for quantitative work is a second representation, which, while it has a superficial resemblance to an oscilloscope trace, is actually a curve of refractive index gradient versus position at a given instant. Details of this optical system, known as the double-inclined-slit, or DIS system, are described elsewhere (5), (6): it is quite simple to calibrate. A single integration of its representation gives electron density directly, provided that the viewing geometry is sufficiently simple, as it was in our experiments.

Theoretical Equations

The measurements made of $\vec{B}$, $\vec{E}$, and n_e were used to derive other plasma properties from theoretical equations, principally the equation of motion in a plasma and the generalised Ohm's law defined by the Boltzmann equation. Certain other variables were eliminated by justifiable assumptions.

We have inferred from certain linear plasma acceleration studies that ion current can be important when the current sheet itself is the ionizing agent (7). Bodin and Newton (8) have reported that the bulk rotation that would be associated with ion current in a theta pinch is present in certain experiments.

The generalised Ohm's law, which sums the various contributing sources of electric field in plasma, may be written:

$$\vec{E} + \vec{v} \times \vec{B} - \frac{1}{n_e e}\left[\vec{j} \times \vec{B} - \nabla \rho_e\right] - \eta \vec{j} = 0 \qquad \text{(Eq. 7-2-1)}$$

Here, $\vec{v}$ is the centre-of-mass plasma velocity, nearly equal to that of the ions, $\vec{j}$ is the total current density, and η is the plasma resistivity. In the cylindrical coordinates of our theta pinch, it is convenient to resolve this equation into radial and azimuthal components:

$$E_r + v_\theta B_z - \frac{1}{n_e e}\left[j_\theta B_z - \frac{\partial p_e}{\partial r}\right] = 0 \qquad \text{(Eq. 7-2-2)}$$

$$E_\theta - v_r B_z - \eta j_\theta = 0 \qquad \text{(Eq. 7-2-3)}$$

We assumed that $\vec{j}$ was entirely azimuthal, as it must be in a symmetrical system.

Consider equation 7-2-2. The second and third terms are in the ratio of the ion component of the azimuthal current, to the total current. Suppose we assume that $j_{\theta i} \ll j_{\theta e}$, and also that the radial electron pressure gradient is much less than $|\vec{j} \times \vec{B}|$, as it usually is in a plasma gun. We then have only:

$$E_r = \frac{j_\theta B_z}{n_e\, e} \qquad \text{(Eq. 7-2-4)}$$

This approximation has a simple physical interpretation: if electrons carry all the current, they alone experience directly the radial $j_\theta\, B_z$ force but they are prevented from running inward by the space charge field E_r that is set up when they attempt to leave the ions. This field E_r acts directly upon the ions, and accelerates them. Since the ions have nearly all the mass of the plasma, however, the force density on the ions ($n_e e E_r$) must equal the total $j_\theta B_z$, hence equation 7-2-4. We have assumed single ionization so that $n_i = n_e$, but a correction to a higher average z_i is simple.

Equation 7-2-4 then, offered us a simple means of estimating n_e if we could measure E and B distributions directly. $\vec{j}$ could be derived from $\vec{B}$ through Maxwell's equation.

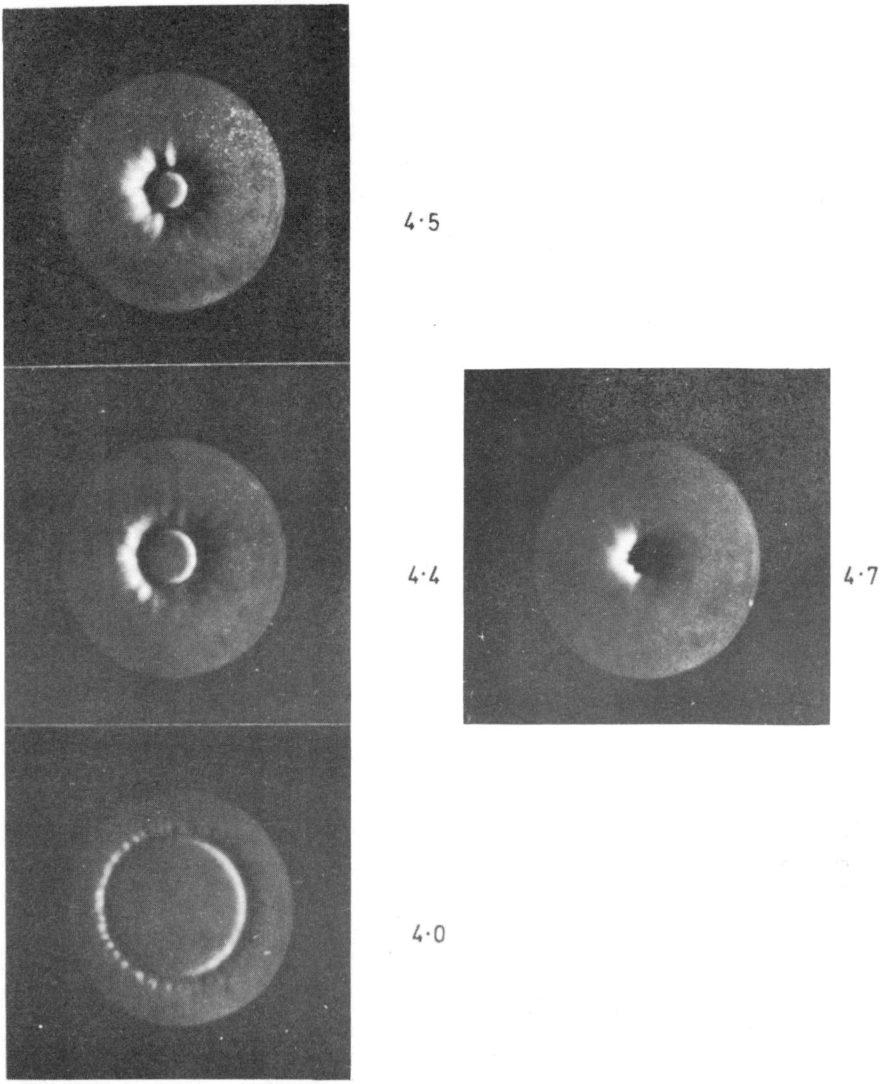

Fig. 7-2-1 Schlieren photographs of theta-pinch discharge in nitrogen. Times indicated are in micro-seconds after the beginning of first half-cycle current flow.

Being interested here in the partitioning of the current between the ions and the electrons, we could not neglect the second term in Eq. 7-2-2. We also needed separate data on the distribution of n_e in order to be able to solve for v_θ, whence:

$$v_\theta = \frac{j_\theta}{n_e e} - \frac{E_r}{B_z} \qquad \text{(Eq. 7-2-5)}$$

still assuming that:

$$\frac{\partial p_e}{\partial r} \ll j_\theta B_z$$

Experimental Results

To give an overall view of the behaviour of the theta pinch as a radial accelerator of plasma, figure 7-2-1 shows a set of four schlieren photographs of a contracting current sheet in nitrogen. As mentioned before, the view was along the coil axis and through about 25 cm of plasma. The visual field diameter was 10 cm. It is evident that except for minor flutes on the outside of the plasma shell, the geometrical regularity of the discharge was good. Even at 4.5 microseconds, the time after the trigger signal at the start of the first half-cycle, the view through the cylinder, then about 1.5 cm in diameter, was unobstructed by any instabilities or irregularities.

Figure 7-2-2 shows both kinds of schlieren presentation used in this work. The instants selected show (i) the initial implosion of the current sheet, while it is still relatively thin and discrete, and (ii) an instant just after peak plasma compression at the axis; the plasma has begun to diffuse outward through the confining field. By interpreting the full-field photograph as a side-illuminated three-dimensional model of the electron density, it may be seen that the displacement of the DIS schlieren was indeed proportional to the gradient of the electron density.

The DIS schlieren was analysed quantitatively by graphical integration of line tracings, as shown in figure 7-2-3 (points in parentheses are uncertain due to faintness of the rapidly rising part of the DIS curve). A tracing across the whole tube diameter is shown, and under it are the curves obtained by integrating both branches of the gradient curve from the wall inward. Both branches agree very well as to the peak density, but because of the faintness of the interior points, do not return toward zero together. Indeed it is not obvious from these data that we should expect the central density to be zero, although later and better data than these do seem to integrate to a value very near to zero.

Figure 7-2-4 is a map of the axial magnetic field B_z and the radial electric field E_r as obtained using the probing procedures described earlier. The trapping of flux from the first half-cycle is evident. Each point plotted represents the amplitude of the oscilloscope trace for one shot of the machine. Where only one point is plotted, as in the B_z data, it signifies that the two discharges made at that position gave coincident results.

It is significant that $E_r(r)$ and $B_z(r)$ cross zero exactly together, as would be the case for the simple situation for which equation 7-2-4 holds.

In order to test this hypothesis, we have computed, using equation 7-2-4, a distribution $n_e(r)$, and have compared it with $n_e(r)$ as derived from the schlieren DIS traces taken for the same instant. Both plots are shown in fig. 7-2-5 together with the E and B profiles that were used. It is clear that excellent agreement has been

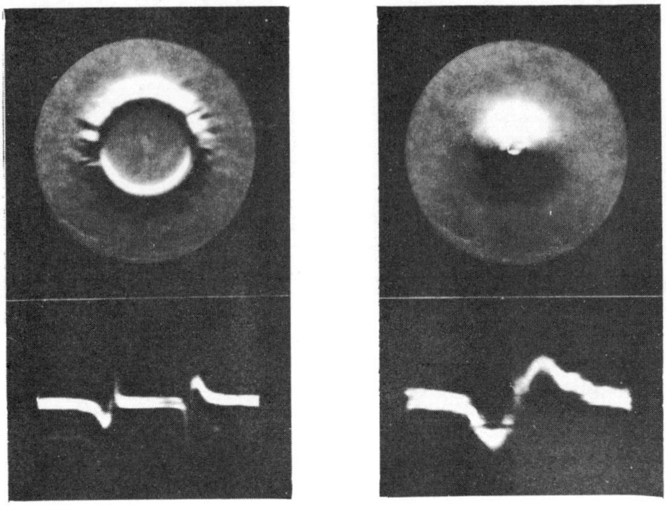

Fig. 7-2-2 Schlieren and double-inclined-slit (DIS) representations of the nitrogen theta pinch for times before and after the maximum axial compression

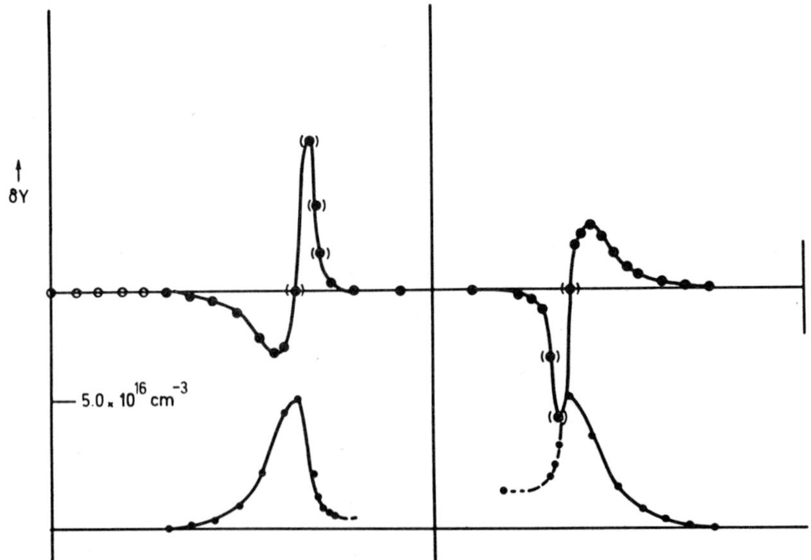

Fig. 7-2-3 Tracing of DIS record and integral curves which represent the electron density distribution. Points in parentheses are relatively uncertain due to faintness of original trace.

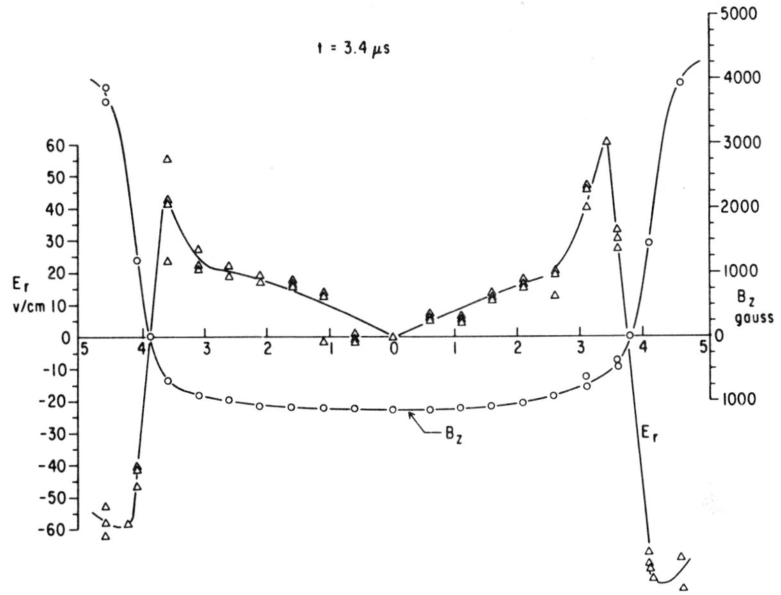

Fig. 7-2-4 Distributions (as a function of radius) of the axial magnetic and radial electric fields in nitrogen discharge.

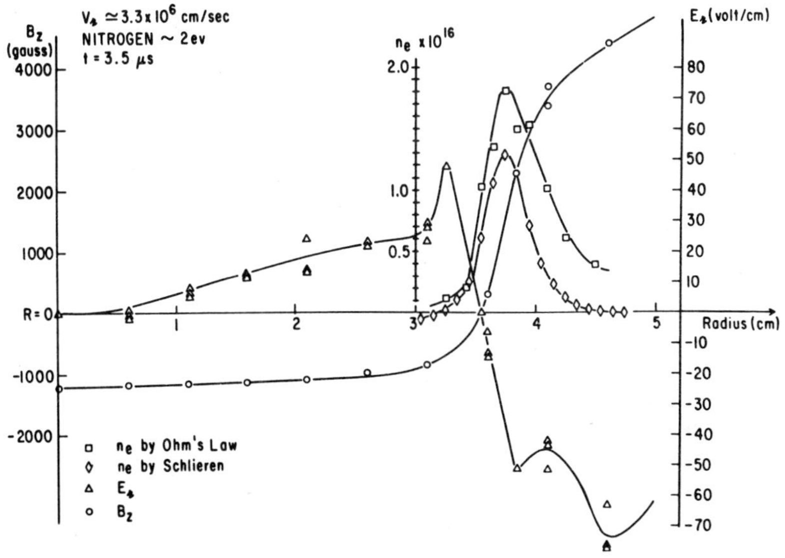

Fig. 7-2-5 Electron density distributions as obtained from the given electric and magnetic field plots (squares) and from schlieren records (diamonds).

obtained as both n_e curves are plotted on an absolute scale, and have involved completely different equipment calibrations. The schlieren distribution has been calculated on the assumption that the plasma was uniform over the entire coil length. If we were to insert the observed fact that the plasma uniformity ceased at about 2 cm from the coil ends i.e. to assume that it was only 20 cm long or so, the two curves would come closer together.

From these data we conclude that in this discharge $v_\theta = \sim 0$, because equation 7-2-4, which assumes the ion current to be zero, gives results that correlate well with the measured electron distribution.

For equation 7-2-4 it was also assumed that the electron pressure gradient was negligible. Given the values of n_e in fig. 7-2-5 it can be shown that the electron pressure gradient will be negligible if T_e is no greater than 1 or 2 ev. It is in fact possible to make an independent estimate of T_e in this experiment by tabulating the rate of change of total flux trapped inside the plasma radii. From Faraday's law, this rate of flux loss is known to be equal to the integral of E_θ around the current sheet. We may thus obtain an approximate value for the local E_θ in the plasma by involving the symmetry of the system. j_θ, however, is just the radial derivative of B_z, so we can estimate the plasma resistivity and hence, through the Spitzer relationship (9), the electron temperature. The electron temperature is then found to be about 2 ev, thus fully justifying the use of equation 7-2-4 for this system.

There is also good agreement between the measured total population of plasma electrons in the current sheet and the number that would be predicted assuming complete ionization and entrainment of the gas through which the current sheet moved.

The simple 'snow plow' sheet model appears to be applicable to this system, and presumably also to linear plasma accelerators which operate in this regime of field strengths and gas and current densities.

Conclusions

These experiments demonstrate how the association of a number of experimental techniques can be used to provide still further information about the properties of a plasma system. These experiments too can be improved by the employment of additional diagnostics, and work is now in process to obtain a detailed space and time distribution of the electron temperature through the scattering of laser light. We hope to obtain experimentally, *inter alia*, the value of the numerical factor relating electrical resistivity and $T_e^{-3/2}$.

Acknowledgements

Portions of this work were supported by the Advanced Research Projects Agency (Project Defender) and was monitored by the U.S. Army Research Office - Durham, under Contract DA-31-124-ARO-D-257.

Also supporting that part of this work which was performed at the General Atomic division of General Dynamics Corporation was the Texas Atomic Energy Research Foundation.

References

1. **Jahn, R.G., von Jaskowsky, W.,** A.I.A.A. J. 1, 1809. 1963.

2. **Lovberg, R.,** Bull. Am. Phys. Soc. 10, 216. 1965.

3. **Huddlestone, R.H., Leonard, S.L.,** (ed.) 'Plasma Diagnostic Techniques' Academic Press, New York, 1965.

4. **Thompson, W.B.,** 'An Introduction to Plasma Physics', Pergamon Press, London 1962.

5. **Beans, J.W.,** 'Physical Measurements in Gas Dynamics', Princeton Univ. Press, Princeton, N.J., 1954.

6. **Lovberg, R.H.,** A.I.A.A. J. 4, 1215. 1966.

7. **Lovberg, R.H.,** 'Acceleration of Plasma by Displacement Currents Resulting from Ionization', Proc. VI Int. Conf. on Ionization Phenomena in Gases, Paris, Vol. IV, 1963.

8. **Bodin, H.A., Newton, A.A.,** Phys. Fluids 6, 1338-1345. 1963.

9. **Spitzer, L.,** 'The Physics of Fully Ionized Gases', 2nd ed., Interscience, New York, 1962.

List of Mathematical Symbols

$\vec{B}$	superimposed magnetic field
e	electron charge
E	electrostatic field
$\vec{F}$	force
$\vec{J}$	electric current
$\vec{j}$	total current density
n_e	electron density
r	radial coordinate
r_e	electron radius $= 2.8 \times 10^{-13}$ cm
$\vec{v}$	centre of mass plasma velocity (nearly equal to that of the ions)
Z	axial coordinate
h	plasma resistivity
θ	azimuthal coordinate
μ	refractive index
λ	transmitted wavelength
p_e	electronic pressure

Commentary on Chapter 7-2

P. Ricateau

Le canon à plasma, même de type coaxial, forme un plasma de géométrie compliquée qui se prête mal aux mesures et tout spécialement aux mesures optiques. Il est dès lors très naturel d'essayer de reproduire l'accélération magnétique du plasma avec le même gaz et dans le même domaine de pression à l'aide d'un "théta pinch" cylindrique où l'observation est beaucoup plus aisée. L'analogie entre ces deux types de machine est cependant loin d'être complète:

a) Les électrodes peuvent jouer un rôle important dans le canon à plasma.

b) L'effet de compression du flux magnétique du "théta pinch" n'existe pas dans le canon à plasma. Les phénomènes cessent donc d'être comparables quand le rapport des inductions externes et internes à la couche de plasma approche de l'unité.

c) Enfin, la forme de la surface de plasma étant très différente, tout spécialement le signe de sa concavité, des instabilités hydromagnétiques peuvent apparaître dans un cas et non dans l'autre.

Néanmoins, l'étude présentée donne une image exceptionnellement fine de la densité électronique dans la couche de plasma permettant de concevoir un modèle de mécanisme d'accélération. Il s'agira ensuite de le tester sur les accélérateurs du type canon à plasma.

Les quelques observations que je désire présenter ou les questions que je poserai concernent:

le choix de la nature et de la pression du gaz,

l'étendue de la zône préionisée et le modèle d'accélération,

l'existence possible d'une rotation du plasma et son effet sur la mesure de densité par l'intermédiaire de la loi d'Ohm,

l'explication microscopique du courant ionique.

Nature et pression du gaz:

Le gaz est de l'azote à une pression de 50 microns de mercure. Dans une étude antérieure, l'auteur a experimenté sur l'hydrogène. Essayant de mesurer la densité électronique par la relation approchée $E_r = j_\theta B_z/n_e e$, il a constaté que le champ E_r ne s'annulait pas en même temps que j_θ et B_z, tout au moins au début de l'accélération. S'agit-il de l'effet du terme de pression électronique? L'auteur a-t-il de nouvelles mesures de température électronique dans ce cas? Il est probable en effet que le refroidissement radiatif des électrons est bien plus efficace dans l'azote que dans l'hydrogène.

En ce qui concerne l'azote, dans quelles limites de pression les mesures de densité électronique par les deux méthodes employées sontelles possibles et cohérentes?

Etendue de la zône préionisée et modèle d'accélération:

On sait que la première demi-alternance du champ qui dure dans cette expérience 3 microsecondes ne produit pas un effet de pinch notable car le gaz, initialement non ionisé, ne s'oppose pas immédiatement à la pénétration du champ magnétique.

Dans cette première phase, l'ionisation est initiée dans la couche de gaz voisine de la paroi, mais elle se propage très vraisemblablement vers le centre du tube par un choc hydromagnétique imparfait, de sorte que l'ensemble du gaz est au moins partiellement ionisé. Ceci expliquerait pourquoi les courbes de densite de la fig.7-2-3 n'aboutissent pas à zéro a l'intérieur de la surface du choc. Etant donné la densité initiale du gaz, l'ionisation complète donnerait une densité de charge de 3.10^{15} cm^{-3} compatible avec les résultats de mesure. Dans ces conditions, le modèle 'boule de neige' ne serait pas entièrement justifié.

En ce qui concerne l' "imperméabilité" de la zône de choc, il serait très intéressant de comparer en fonction du temps le nombre total de charge contenu dans cette zône, connu par la mesure du profil de n_e, au nombre de particules de gaz balayé par la zône depuis l'instant initial de la contraction.

Rotation possible du plasma et effet sur la mesure de densité:

En observant certains détails de la zône de choc, on a parfois mesuré une rotation importante de la zône lumineuse (référence 8 du chapitre) dans un théta pinch sujet à une forte instabilité de type flûte et où par conséquent la symétrie de révolution a disparu.

Dans une structure symétrique, la loi d'Ohm en projection radiale s'écrit:

$$E_r + v_\theta \cdot B_z - \frac{1}{n_e e} j_\theta B_z = - \frac{1}{n_e e} \frac{\partial pe}{\partial r}$$

Si on néglige le terme de pression électronique, la mesure de n_e est facile à condition que le terme v_θ B_z puisse être également négligé. v_θ est la vitesse des masses et non la vitesse ionique.

Si le modèle de 'boule de neige' est retenu, il ne peut pas y avoir transfert de quantité de mouvement de la zône de choc à la partie centrale. Si donc on néglige l'effet des courants de déplacement radiaux, on trouve que le moment de rotation moyen de la zône est un invariant dans le mouvement. Par conséquent, le moment de rotation de la zône ne pourrait provenir que d'une impulsion initiale due à une réflexion non spéculaire des ions avant que la zône se détache de la paroi du tube.

Mécanisme du courant ionique:

En l'absence de moment de rotation des masses, le courant ionique moyen dans la zône est très faible puisque $\frac{(j_e)}{(j_i)} = \frac{m_e}{m_i}$ mais il pourraît être important localement étant donne les échanges de particules (électrons principalement) entre un point et un autre de la zône situés sur des rayons différents.

Microscopiquement, on pourrait s'attendre à un mouvement de dérive local des ions de l'ordre de E_r/B_z, c'est-à-dire comparable en valeur absolue à la vitesse des électrons. Ceci soulève une critique qui semble assez grave en ce qui concerne la mesure détaillée du profil de densité par l'intermédiaire de la loi d'Ohm.

Je n'ai pas compris le mécanisme différent proposé par l'auteur dans (6) et (7) pour rendre compte du déplacement des ions. Il repose sur une étude de trajectoire de l'ion et de l'électron après l'instant d'ionisation. Le neutre est au repos dans le système du laboratoire et la champ électrique (c'est-à-dire $- \nabla U + \frac{\partial \vec{A}}{\partial t}$) est, semble-t-il, supposé nul. Dans ces conditions, l'ion et l'électron formés devraient rester immobiles. Transposé dans le référentiel mobile à la vitesse v_o cela se traduit par le fait que l'accélération magnétique $- \vec{v}_o \times \vec{B}$ est exactement opposée au champ $\vec{v}_o \times \vec{B}$ induit par le déplacement du référentiel.

7-3

Magnetogasdynamic Properties of Stationary Supersonic Plasma Jets

J. HÜGEL
Deutsche Versuchsanstalt für Luft und Raumfarht C.V.
Stuttgart - Vaihingen, Germany.

Summary

A device, similar to an arc-jet engine, has been operated stably within a current range of 500 to 3000 amps at an ambient pressure of 0.5 mm Hg. With argon and nitrogen as working gases mass flow rates of 0.075 to 0.6 g/s were applied. The velocity produced resulted from two mechanisms namely aerodynamic acceleration due to heating and expansion processes, and electro-magnetic acceleration due to self-magnetic body forces. The latter originate from the interaction of the arc current with its self-induced magnetic field. The relative importance of both effects is found to depend on arc current, mass flow rate and electrode and nozzle configuration. The exit Mach number of the plasma flow increased with increasing current. A mean magnetic Reynolds number calculated from the experimental data exhibited a stronger than linear dependence on arc current and reached values of the order of one at high currents. The magnetic pressure number (Cowling number) of the flow inside the nozzle is a marked function of arc current and mass flow rate. With increasing current this quantity increases to attain constant values for flow conditions with predominant self-magnetic velocity production.

Sommaire

On utilise un montage semblable à un arc-jet dans le domaine d'intensité allant de 500 à 3000 ampères, à la pression de 0,5 torr. Le gaz actif est l'argon où l'azote et les débits-masse varient de 0,075 à 6 g/s. L'accélération est due à deux processus, l'accélération aérodynamique obtenue par chauffage et détente et l'accélération électro-magnétique sous l'effet des forces magnétiques. Ce dernier effet provient de l'interaction du courant de l'arc avec son propre champ magnétique. L'importance relative des deux effets dépend de l'intensité de l'arc, du débit-masse et de la configuration des électrodes et de la tuyère. Le nombre de Mach de sortie de l'écoulement de plasma croît avec l'intensité. On détermine un nombre de Reynolds magnétique moyen à partir des données expérimentales. Il croît plus rapidement que linéairement avec l'intensité et devient de l'ordre de l'unité aux courants élevés. Le coefficient de pression magnétique (nombre de Cowling) de l'écoulement dans la tuyère est une fonction rapidement variable du courant d'arc et du débit-masse. Lorsque l'intensité croît, cette grandeur croît aussi pour atteindre une valeur constante lorsque les conditions de l'écoulement correspondent à l'effet prédominant du champ magnétique induit dans le processus d'accélération.

Introduction

In recent years plasma jets of high velocity have become of increasing interest in many fields of experimental research. The design of plasma accelerators is closely associated with the development of continuously working arc jet engines, for by utilizing electromagnetic body forces the exhaust velocity of arc jet engines can be increased substantially at a simultaneous increase of overall efficiency. Apart

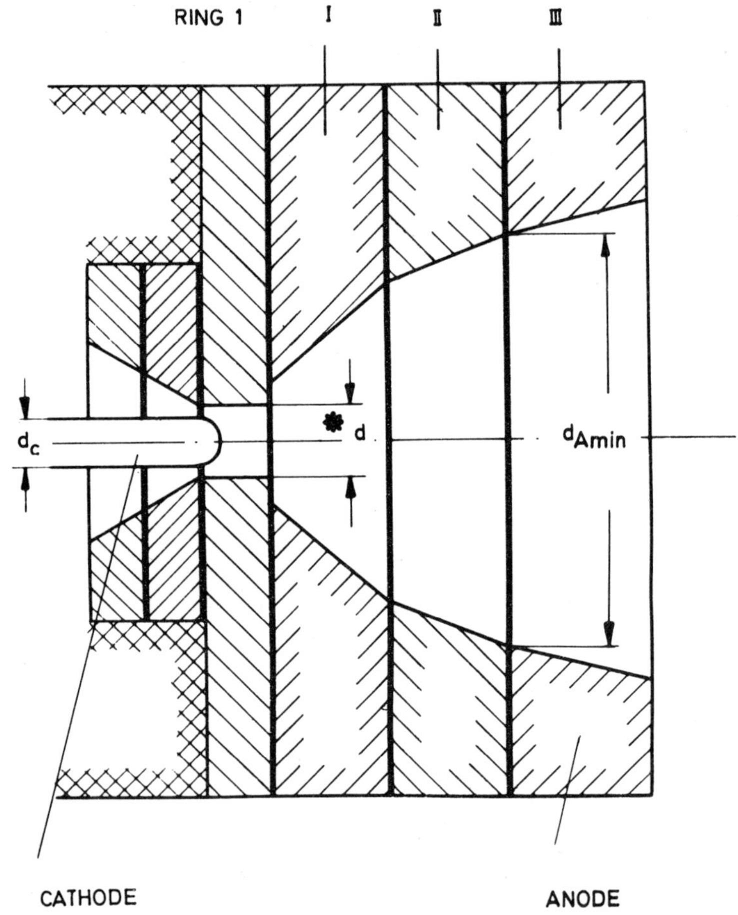

Fig. 7-3-1 Schematic presentation of apparatus

from their application as thrustors and wind tunnel facilities these devices have been used recently in magneto-gasdynamic experiments which until then had exclusively employed shock tubes and condensor discharges (1).

The so-called selfmagnetic effect was investigated experimentally and theoretically as early as 1955 for the case of the high-current carbon arc (2), and this effect later became of great interest when higher currents were applied in arc jet engines. The first performance data reported (3) (4) in this text probably cannot be considered as a realization of selfmagnetic velocity production in arc jet engines because external magnetic fields were applied to avoid anode destruction. Subsequent experiments however indicated substantial increases of thrust even with field strengths as low as 200 to 500 Gauss (5). We decided to design and operate a device without an external magnetic field, in order to obtain evidence about the self-magnetic effect in arc jet engines. In the course of this work it became evident that the device might serve as a plasma accelerator for magneto-gasdynamic experiments. In particular, the properties of the plasma jet would meet the requirements for experimental verification of the interaction effects between plasma flows and axisymmetric magnetic fields as theoretically treated in (6).

Experimental Apparatus and Procedure

A schematic representation of the axisymmetric device is given in figure 7-3-1. The convergent-divergent nozzle is constructed of individual water-cooled copper segments which are insulated against each other: this prevents the electric current entering the nozzle wall in an uncontrolled manner. The two larger segments (ring II and III) may be connected alternately to anode potential. The region of the anode attachment is thus relatively well known, permitting meaningful comparison of experimental data with theoretical expressions. In the experiments reported here, the anode is represented by ring III which had an exit diameter of 80 mm. All other segments were electrically floating. The nozzle throat consisted of a 12 mm long cylindrical portion with $d^* = 12$ or 16 mm. The rod-shaped cathode of 50 mm length was made of thoriated tungsten. Its axial position was variable with respect to the throat. The cathode diameter d_c was either 8 or 10 mm for nozzle throat diameters $d^* = 12$ or 16 mm, respectively.

With argon and nitrogen as working gases the device was operated in a current range of 500 to 3000 A. The ambient pressure in the vacuum tank was kept constant at $p_\infty = 0.5$ mm Hg at all power levels and mass flow rates. The following quantities were measured: thrust; arc current and voltage; mass flow rate; chamber and ambient pressures; and heat losses to the electrodes, arc chamber and nozzle walls. The evaluation of the experimental data and the calculation of additional quantities of interest were performed with the aid of a computer.

For a set of operating conditions, with $d^* = 12$ mm, local Pitot pressures and Mach numbers were obtained using wedge-type and Pitot probes. The water-cooled flat-nosed cylindrical probe with an inner diameter of 7 mm and an outer diameter of 20 mm was 90 mm long.

Operation Mode and Flow Phenomena

The mode of operation of the device is determined by the low pressure level at which the arc discharge is maintained. Chamber pressures of $p_o \approx 15$ to 100 mm Hg according to gas type, mass flow rate, arc current and nozzle throat diameter, cause relatively large and spotless arc-onsets at the cathode (7). By cascading the nozzle, the anode attachment can be achieved in regions of very low pressure (ring III in figure 7-3-1). Under this condition the current transport to the anode occurs in a diffuse mode. This is explained by the different temperatures of electrons and heavy particles (8). For conditions similar to those discussed here the

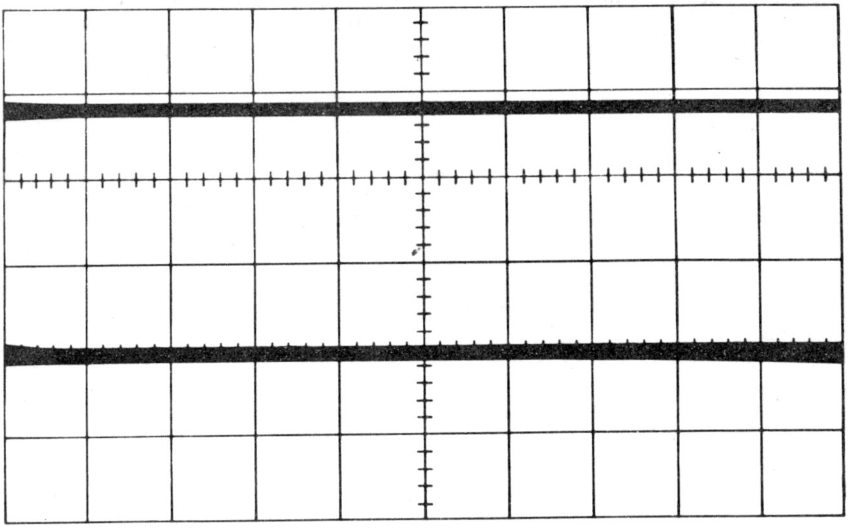

Fig. 7-3-2 Argon: Oscillogram of arc voltage at J = 1200 A and $\dot{m}$ = 0.15 g/s. Horizontal scale 1 ms/div, vertical scale 10 V/div

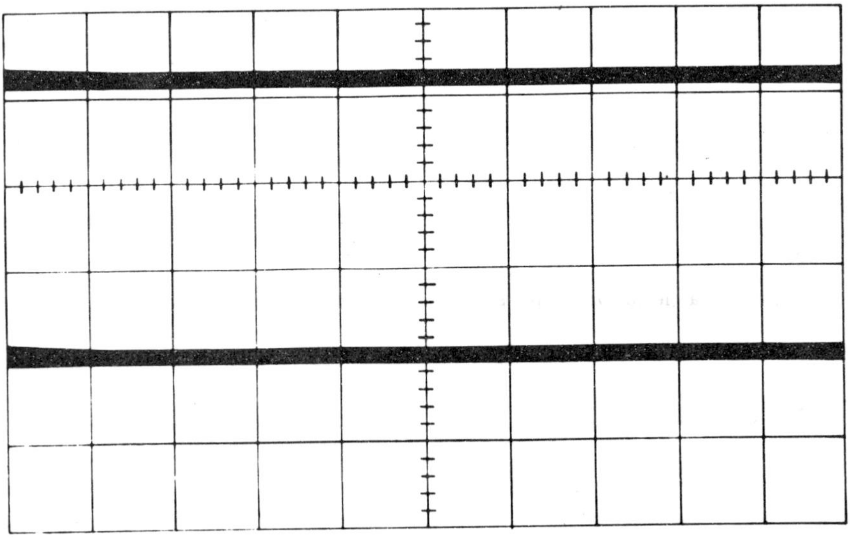

Fig. 7-3-3 Nitrogen: Oscillogram of arc voltage at J = 1200 A amd $\dot{m}$ = 0.15 g/s Horizontal scale 1 ms/div, vertical scale 10 V/div

electron temperature may exceed the ion temperature by a factor of 3 (9). In spite of the low gas temperature in the vicinity of the cooled walls the electrical conductivity governed by the electron temperature is sufficiently high. Thus a formation of one or several spots, representing channels of high conductivity leading from the arc column to the wall, is not as necessary unlike the case at high pressure where particle temperatures departing less from an uniform value.

Visual observation of the anode walls, as well as of the exhaust jet, found no evidence of the presence of spots. This was supported by the oscillograms of the arc voltage, which showed no sort of fluctuations, as can be seen on the records of the arc voltage at $J = 1200$ A and $\dot{m} = 0.15$ g/s for argon and nitrogen Figs. 7-3-2 and 7-3-3), the vertical scale of which is 10 V/div. Only at very high amplification were fluctuations observed and this corresponded exactly to the frequencies originating from the power supply. From these facts it may be concluded that the device operated in a completely stable mode.

The appearance of the plasma flow was markedly affected by the gas type. With nitrogen as working gas a diffuse flow was present without significant radial variation of brightness. Figure 7-3-4 shows the typical appearance of a nitrogen plasma jet which was relatively independent of arc current and flow rate. A somewhat similar feature was observed with argon at low mass flow rates ($\dot{m} < 0.1$ g/s). With moderate and high $\dot{m}$, however, a distinct bright core surrounded by dark boundaries was characteristic of argon flow, (Fig. 7-3-5). Moreover, a shock pattern in the jet became visible. This shock originated from the discontinuous area increase behind the nozzle throat (compare Fig. 7-3-1), and was by no means an indication of flow separation. The supersonic character of the entire plasma jet is evident from figure 7-3-5.

Discussion of Experimental Results

Velocity Production

Velocity production in the device discussed here was based on two mechanisms: thermodynamic processes like heating and expansion of a working fluid and electromagnetic body forces. In the absence of an external magnetic field the latter originated from the interaction of the arc current with its self-induced magnetic field. A determination of the selfmagnetic thrust for conditions such as those encountered in our experiments is given by Peters and Ragaller (10) and the following discussion is based upon this.

With electromagnetic forces present the momentum equation may be written:

$$\rho\ (\vec{w}\nabla)\ \vec{w}\ =\ -\nabla p\ +\ (\vec{j}\times\vec{B}) \qquad \text{(Eq. 7-3-1)}$$

It is convenient to consider the two thrust producing mechanisms separately. Considering electromagnetic acceleration only, integration of equation 7-3-1 over a control volume τ yields the electromagnetic thrust:

$$\int\rho\ (\vec{w}\Delta)\ \vec{w}\ d\tau\ =\ \oint p_m\ d\vec{A}\ +\ \int(\vec{j}\times\vec{B})\ d\tau \qquad \text{(Eq. 7-3-2)}$$

where p_m denotes the pressure caused by $\vec{j}\times\vec{B}$ forces. The self-induced magnetic field of the arc is related to the current density by Maxwell's equations:

$$\vec{j}\ =\ \frac{1}{\mu_o}\ (\nabla\times\vec{B}) \qquad \text{(Eq. 7-3-3)}$$

$$(\nabla\quad\vec{B})\ =\ 0 \qquad \text{(Eq. 7-3-4)}$$

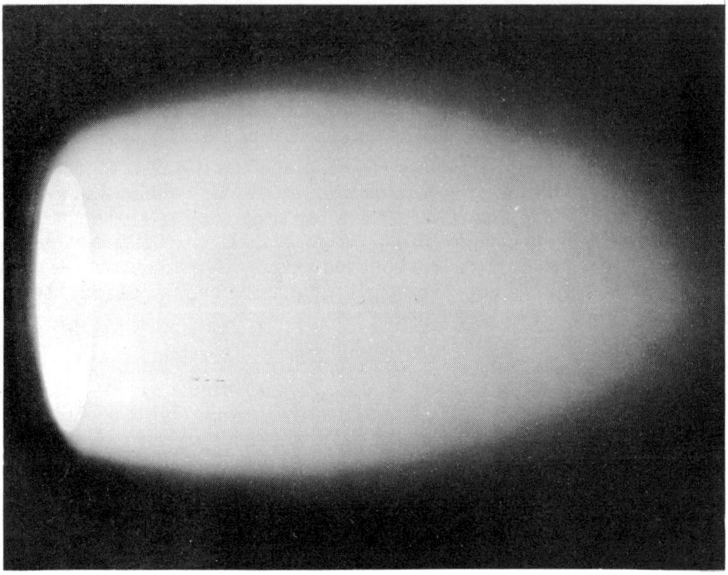

Fig. 7-3-4 Nitrogen plasma jet at J = 1200 A, $\dot{m}$ = 0.15 g/s and p_{∞} = 0.5 mm H

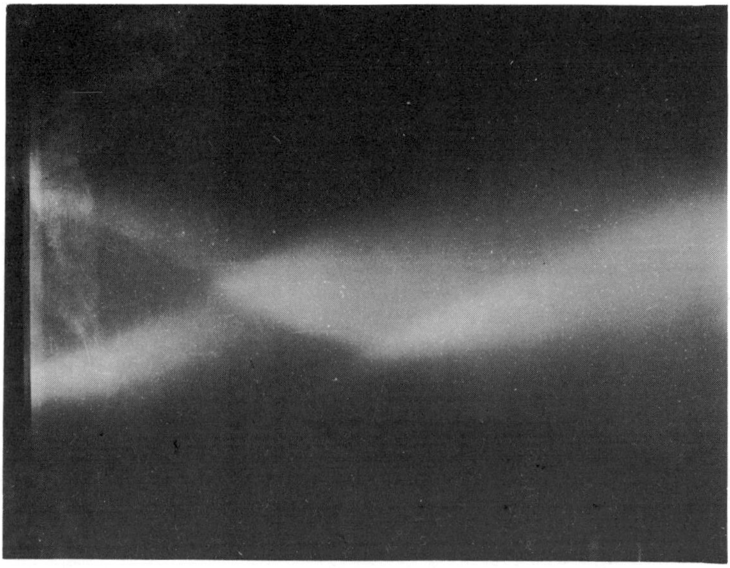

Fig. 7-3-5 Argon plasma jet at J = 1200 A, $\dot{m}$ = 0.5 g/s and p_{∞} = 0.5 mm Hg with wedge probe

Assuming one-dimensional flow and current density distributions, according to (10), for an arc column diverging from a radius r_1 at the cathode to a radius r_2 at the anode:

$$\rho_2 W_2^2 A_2 - \rho_1 W_1^2 A_1 = \frac{\mu . J^2}{8\pi} + \frac{\mu . J^2}{4\pi}\left(\frac{1}{4} + \ln \frac{r_2}{r_1}\right) \quad \text{(Eq. 7-3-5)}$$

or:

$$\dot{m}\,(W_2 - W_1) = \frac{\mu . J^2}{4\pi}\left(\frac{3}{4} + \ln \frac{r_2}{r_1}\right) \quad \text{(Eq. 7-3-6)}$$

Equation 7-3-6 represents the selfmagnetic thrust causing a theoretical velocity gain of:

$$\Delta W_m = \frac{1}{\dot{m}} \frac{\mu . J^2}{4\pi} \ln 2.12 \frac{r_2}{r_1} \quad \text{(Eq. 7-3-7)}$$

It is evident from equation 7-3-7 that high currents, low mass flow rates and large values of $\frac{r_2}{r_1}$ should be applied for proper utilization of the selfmagnetic effect. A determination of the ratio $\frac{r_2}{r_1}$ appears generally rather difficult. However, with the anode being represented by ring III, and the other segments electrically floating, r_2 cannot become smaller than $\frac{d_{A\,min}}{2}$. On the other hand, the radial dimension of the cathode spot r_1 is limited by the cathode $\frac{d_c}{2}$. Thus a minimum value of $\frac{r_2}{r_1} = \frac{d_{A\,min}}{d_c}$ is attained independent of operating conditions.

The aerodynamic thrust component is usually expressed in the form:

$$F_{ae} = c_F\, p_o\, A^* \quad \text{(Eq. 7-3-8)}$$

For one-dimensional perfect gas flow the thrust coefficient c_F is determined theoretically as a function of $\frac{A_e}{A^*}$, γ and n. Its value is estimated to be of the order of 1.6 for our experimental conditions.

Hence the theoretical mean velocity is:

$$W = \frac{1}{\dot{m}}\left(\frac{\mu . J^2}{4\pi} \ln 2,12 \frac{d_{A\,min}}{d_c} + c_F\, p_o\, A^*\right) \quad \text{(Eq. 7-3-9)}$$

With fixed geometry and constant mass flow rate the theoretical velocity according to equation 7-3-9 is a function of arc current only, since $p_o = f(J)$ as well, and c_F remains approximately constant. With known variations of $w = f(J)$ and $p_o = f(J)$ one is in a position to classify the measured thrust (velocity) data with respect to their origin. Evidence obtained in this way about selfmagnetic acceleration in arc jet engines was recently published by the author (11). However, the flow being by no means one-dimensional (substantial radial variation of most properties being inherent in plasma flows), and with considerable frictional effects being involved, the real "mean" velocities are far below the values predicted by simple one-dimensional theory.

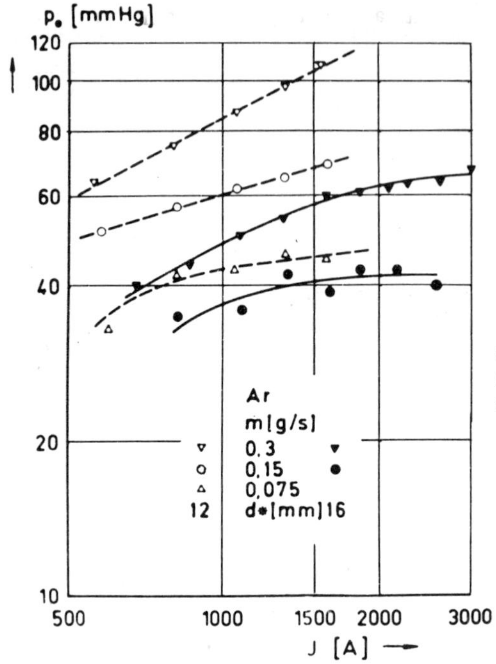

Fig. 7-3-6 Argon: Variation of chamber pressure with arc current, mass flow rate and nozzle throat diameter

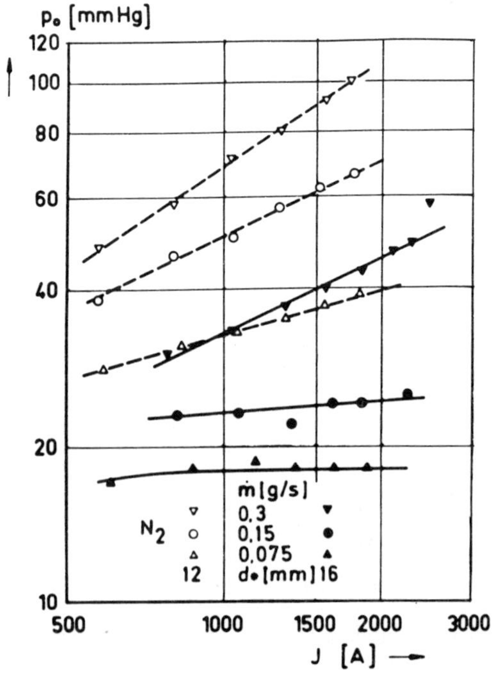

Fig. 7-3-7 Nitrogen: Variation of chamber pressure with arc current, mass flow rate and nozzle flow diameter

The variation of chamber pressure p_o with arc current is shown in figures 7-3-6 and 7-3-7 for argon and nitrogen, respectively. At lower mass flow rates the dependence of p_o on J is less strong than with higher $\dot{m}$. A substantial decrease of the pressure level is caused by the larger nozzle throat of $d^* = 16$ mm. A tendency is evident of the chamber pressure increasing only in a certain current range and acquiring constant values at high currents.

Assuming $p_e = p_\infty$ values of the mean velocity may be calculated from the measured thrust by:

$$w = \frac{F}{\dot{m}} \qquad \text{(Eq. 7-3-10)}$$

Of course, this quantity is mainly of qualitative interest. In Figs. 7-3-8 and 9, w is plotted vs. J with $\dot{m}$ and d^* as parameters. An inspection of the curves reveals a typical feature: at all mass flow rates with increasing current the slopes of the curves increase, approaching a value of $\frac{d \lg w}{d \lg J} = 2$ which is characteristic of the selfmagnetic effect as given by equation 7-3-7. The lower the $\dot{m}$ the lower the current at which the proportionality of w to J^2 is attained. With the lowest value of $\dot{m}$ the velocity is practically determined within the entire current range by selfmagnetic acceleration (Fig. 7-3-9). On the other hand, at $\dot{m} = 0.3$ gs^{-1}, approximately 2200 A with nitrogen and 2600 A with argon must be applied to obtain significant evidence of this effect (Figs. 7-3-6 and 7-3-7). Increase in pressure is not effective in markedly increasing the velocity with increasing current. The considerable velocity increase obtained at approximately constant values of p_o i.e. with constant aerodynamic thrust component, implies that the velocity production there is predominantly of an electromagnetic nature.

Comparison of the performance data of the two working gases indicates that somewhat higher velocities are obtained with argon when the current is less than approximately 1400 A. This might be due partly to the higher chamber pressures and thus higher aerodynamic thrust components obtained with argon. On the other hand, it is certainly due to some extent to the fact that nitrogen - a molecular gas - has to be dissociated. The dissociation energy must be considered lost, since generally no recombination occurs within the nozzle of an arc jet engine (12). The different physical behaviour of the two gases can be demonstrated by the kinetic efficiency, defined as the ratio of kinetic energy to the total enthalpy of the jet, i.e.

$$\eta_{kin} = \frac{w^2/2}{h} \qquad \text{(Eq. 7-3-11)}$$

Figures 7-3-10 and 7-3-11, showing the variation of η_{kin} with current, give clear evidence of the dissociation effect. In a region of predominantly thermal velocity production the values of η_{kin} for nitrogen are considerably lower than those for argon as working fluid.

The strong increase of kinetic efficiency with arc current is noteworthy. With the slope of the curves approaching values of $\frac{d \lg \eta_{kin}}{d \lg J} \approx 2$ confirmatory evidence of the selfmagnetic effect is presented.

Figures 7-3-8 to 7-3-11 show that when the device is equipped with the smaller nozzle throat, higher velocities are produced at higher efficiencies. Although the chamber pressures are higher with $d^* = 12$ mm at all corresponding conditions of operation except at high nitrogen flow rates, the aerodynamic thrust component

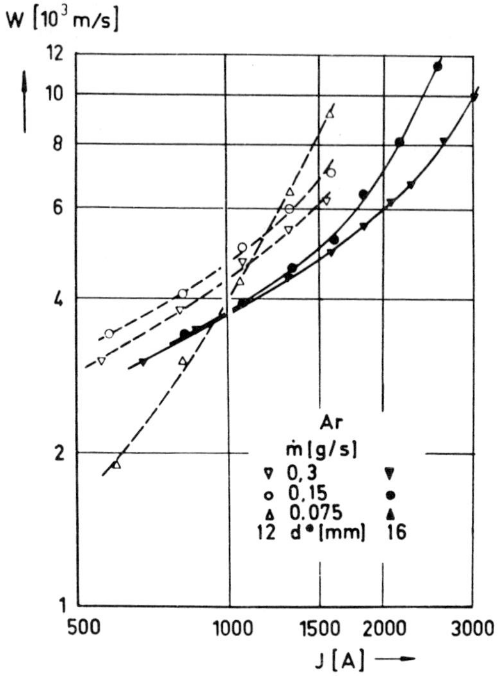

Fig. 7-3-8 Argon: Variation of mean velocity with current, mass flow rate and nozzle throat diameter

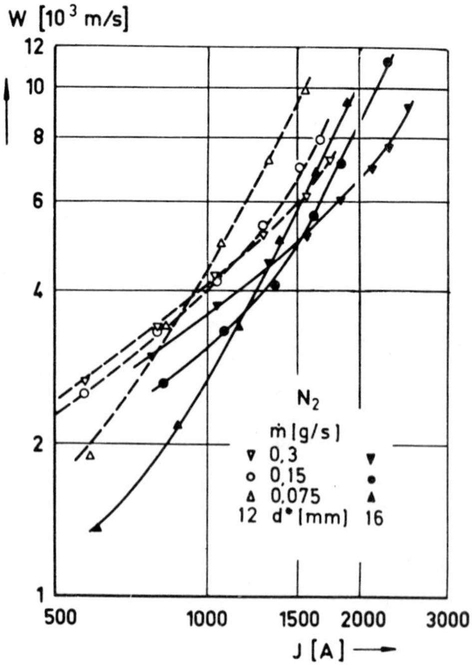

Fig. 7-3-9 Nitrogen: Variation of mean velocity with current, mass flow rate and nozzle throat diameter

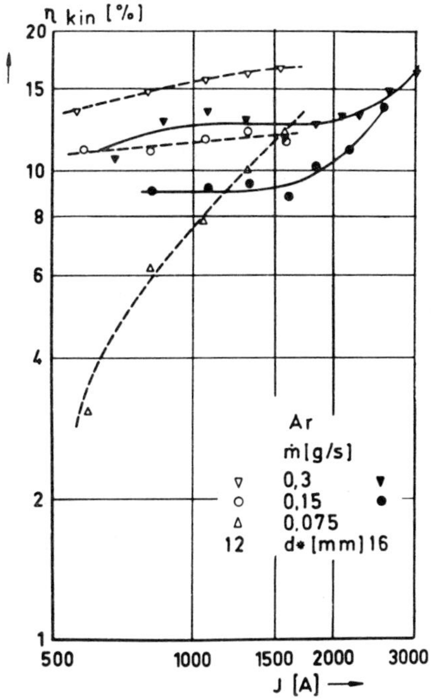

Fig. 7-3-10 Argon: Variation of kinetic efficiency with current, mass flow rate and nozzle throat diameter

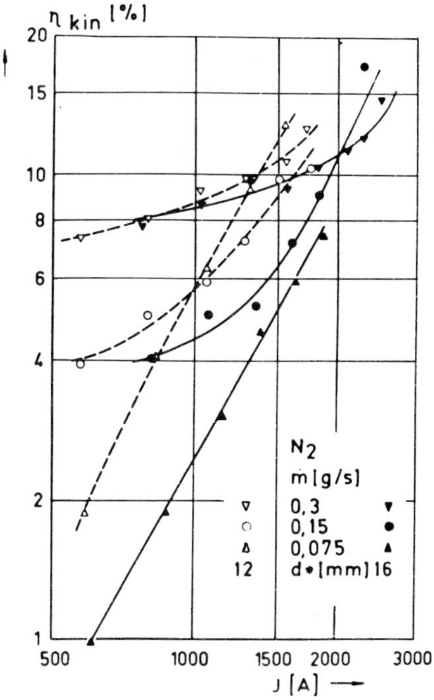

Fig. 7-3-11 Nitrogen: Variation of kinetic efficiency with current, mass flow rate and nozzle throat diameter

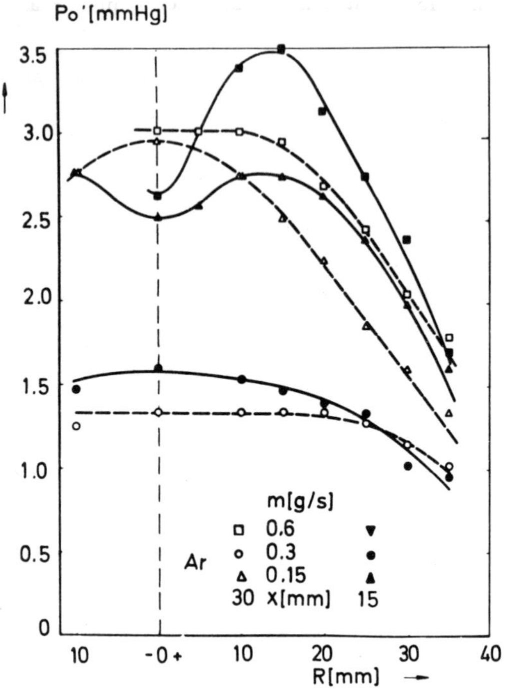

Fig. 7-3-12 Argon: Radial Pitot pressure distributions at J = 800 A for various mass flow rates

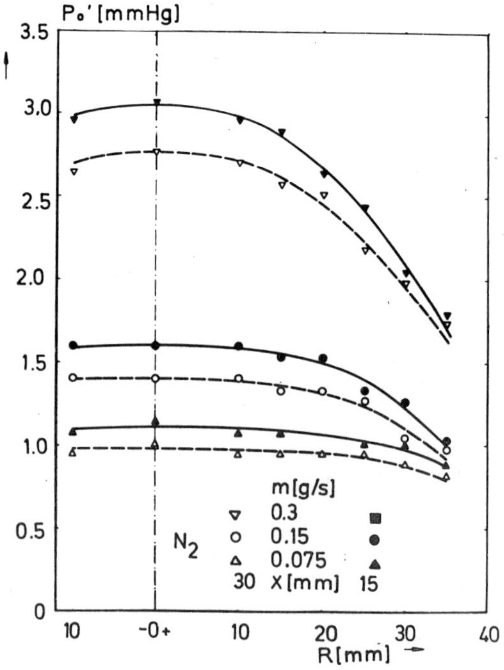

Fig. 7-3-13 Nitrogen: Radial Pitot pressure distributions at J = 800 A for various mass flow rates

$c_F\ p_oA^*$ is practically equal for both nozzle throat diameters, which implies that a different phenomenon causes the thrust to decrease with larger throat dimension. As a matter of fact, in arc jet engines the radial mass flux distribution exhibits a distinct minimum at the jet axis and a maximum in the cooler boundary (13). At low pressure levels in the arc chamber and along the nozzle throat, i.e. at large throat diameters, relatively thick boundary layers are built up. Moreover, at the larger d^* the plasma flow is in contact with the cold walls over larger areas. Therefore a higher fraction of $\dot{m}$ might flow in the more voluminous boundary layer, thus causing lower thrust.

Mach Number of the Plasma Jet

Mach number is especially useful because it can be related to the kinetic efficiency (cf. (10)), usually it is determined either by optical methods or by Pitot pressure measurements. In plasma jets of not too low density, the density increase across a shock is sufficient to produce radiation, but as mentioned earlier, we had to use mass flow rates, $\dot{m} \gtrsim 0.3$ g/s in our experiments in order to obtain any visible shock pattern.

A water-cooled wedge probe was traversed across the plasma flow with its plane surface parallel to the jet axis, see figure 7-3-5. The weak shock originating from the leading edge of the probe was then taken to be a Mach line and the Mach number approximately determined from the angle between shock and jet axis according to:

$$\sin \alpha_M = \frac{1}{M} \qquad \text{(Eq. 7-3-12)}$$

The form of the slightly curved shock is affected by both a radial Mach number gradient and the presence of a relatively thick boundary layer on the probe surface. Since both effects are complex, this method is appropriate only for regions of insignificant Mach number gradient and with the probe positioned at a sufficient distance away from that region. Values of the Mach number have therefore been evaluated only at the jet axis by this method.

However, to obtain an idea of the radial Mach number distribution several Pitot pressure traverses have been performed 15 and 30mm away from the nozzle exit. With the flow conditions encountered in our experiments the pressure measurements have to be interpreted very carefully. Because of the low density and high temperature of the flow, viscous, slip and relaxation effects must be taken into account (14). The significance of these phenomena and their influence on the measurements for plasma flows similar to those above has been discussed by Krause (15) who concluded that the measured pressures can be identified with the Pitot pressures in good approximation.

Local Pitot pressures p_o' at constant arc current of J = 800 A plotted vs. radial position are shown in figures 7-3-12 and 7-3-13. With low flow rates of argon and nitrogen, the distribution of pressure is found to have values of p_o' approximately constant across a substantial portion of the jet. At the larger distance from the nozzle exit, the pressures are lower and the profiles even flatter, as might generally be expected of free jets. At larger $\dot{m}$ the pressures in the flow core are substantially higher than in the boundary regions. The profiles obtained with high flow rates of argon at x = 15 mm show a distinct indent at the axis. This phenomenon is caused by the shock running out of the nozzle (cf. Fig. 7-3-5). When traversing at this axial position, the probe crosses a region of lower static pressure enclosed by the shock. Moreover, in this region the pressure loss due to the normal shock in front of the probe is higher than in regions farther downstream.

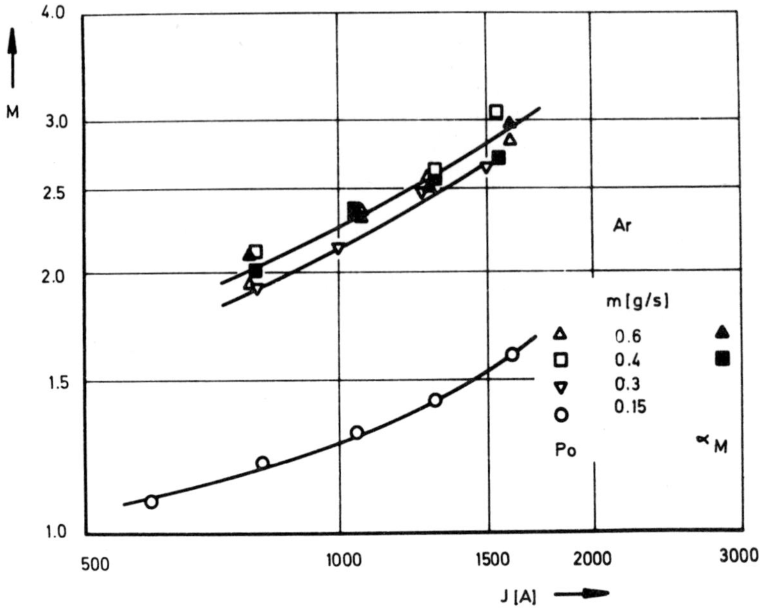

Fig. 7-3-14 Argon: Variation of Mach number at the jet axis with arc current. Mach numbers evaluated from Pitot pressures and measurements of Mach angle, resp.

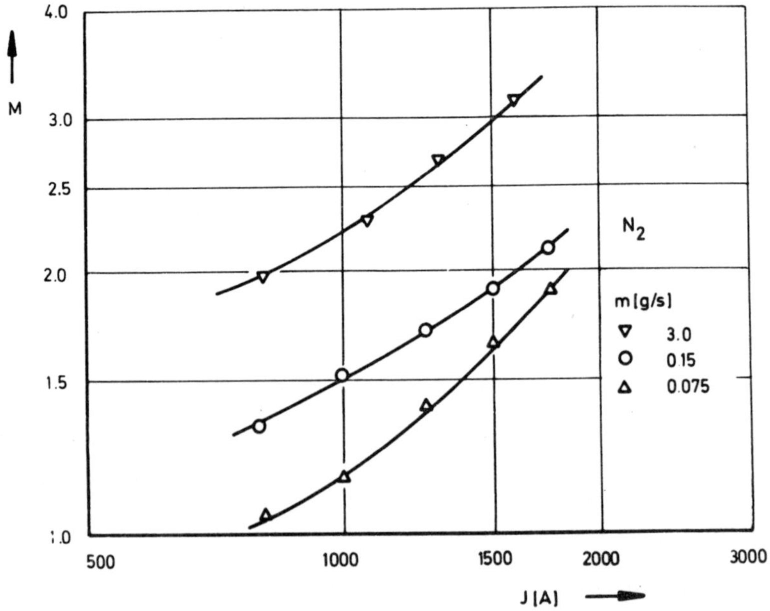

Fig. 7-3-15 Nitrogen: Variation of Mach number at the jet axis with arc current. Mach numbers evaluated from Pitot pressures.

The derivation of the Mach numbers from the measured Pitot pressure can be performed with sufficient accuracy from the Rayleigh formula (13) (15) for which purpose the value of γ for a perfect gas may be used. The static pressure is assumed to be constant within the jet, which it is definitely not at high $\dot{m}$, and equal to the ambient pressure. Values of the pressure at R = O and x = 30 mm have been evaluated to demonstrate the variation of Mach number with arc current. From figures 7-3-14 and 7-3-15 a substantial increase of M with increasing J is evident. Larger mass flow rates cause higher values of M. Of course, if the variation of some "mean" values of Mach number were to be considered, the curves corresponding to high $\dot{m}$ would be shifted towards lower values of M. Anyhow, a distinct rise of η_{kin} with increasing current and somewhat higher kinetic efficiencies at larger $\dot{m}$ might be expected from the curves in figures 7-3-14 and 7-3-15. In fact, for the current range investigated with the Pitot probe this is confirmed by the tendency shown in the curves in figures 7-3-10 and 7-3-11.

Values of the Mach number obtained by measuring the Mach angle α_M, shown for two flow rates of argon, are included in figure 7-3-14. The surprisingly good agreement of the results seems to justify the methods of measurement and various approximations used.

Magnetic Reynolds Number

In characterizing particular problems of magneto-gasdynamics it has been found valuable to define various dimensionless parameters (16) (17). The interaction between the flow field of an electrically conducting medium and a magnetic field may be described by the so-called magnetic Reynolds number.

Starting from Ohm's law for a neutral plasma with constant conductivity:

$$\vec{j} = \sigma[\vec{E} + (\vec{w} \times \vec{B})] \qquad \text{(Eq. 7-3-13)}$$

and assuming stationary currents, the magneto-fluidmechanic diffusion equation is obtained by applying several vector operations (16), i.e.:

$$\frac{\partial \vec{B}}{\partial t} = \nabla \times [\vec{w} \times \vec{B}) - \frac{\vec{j}}{\sigma}] \qquad \text{(Eq. 7-3-14)}$$

With $\sigma \to \infty$ the second term on the right-hand side of equation 7-3-14 vanishes, and the case of the so-called frozen-in fields i.e. with the magnetic lines locked in the fluid is obtained.

However, in all magneto-gasdynamic experiments and applications, the conductivity is of finite magnitude. Therefore it appears useful to define the scalar ratio of the two terms on the right-hand side of equation 7-3-14 as a criterion for the relative motion between magnetic field and flow field, i.e.:

$$Rm = \frac{wB}{j/\sigma} = \frac{wB\sigma}{j} \qquad \text{(Eq. 7-3-15)}$$

This expression is called magnetic Reynolds number and may be interpreted as the ratio of induced current density to total current density. If $Rm \gg 1$, the magnetic field will stay with the flow and will be greatly affected by the motion of the fluid. If $Rm \ll 1$, the magnetic field is not noticeably influenced by the flow.

In our experiments, without an applied magnetic field, the relation between magnetic field and current density was as given by equation 7-3-3. Substitution for j in equation 7-3-15 yields:

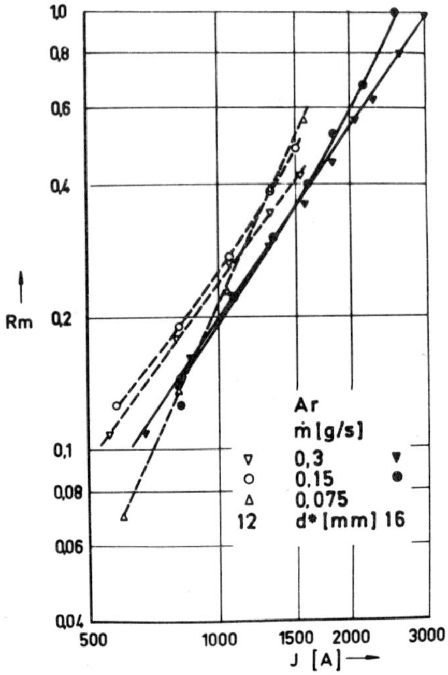

Fig. 7-3-16 Argon: Variation of magnetic Reynolds number with arc current, mass flow rate and nozzle throat diameter

Fig. 7-3-17 Nitrogen: Variation of magnetic Reynolds number with arc current, mass flow rate and nozzle throat diameter

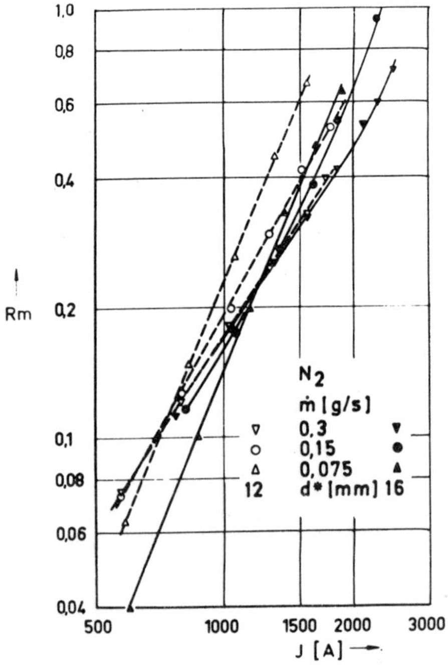

$$Rm = \frac{wB\sigma}{B/\mu_o l} = \mu_o \sigma w l \qquad \text{(Eq. 7-3-16)}$$

with l denoting a characteristic length. The electrical conductivity is approximately calculated from experimental data as follows:

$$\sigma = \frac{j}{E} = \frac{J}{A} \cdot \frac{1}{U/l} = \frac{Jl}{AU} \qquad \text{(Eq. 7-3-17)}$$

The characteristic length is assumed to be the axial distance from cathode tip to the anode ring. For the geometry of our device $l \approx d_{A\,min}$ is taken. Hence:

$$Rm = \mu_o \frac{4}{\pi} \frac{J}{U} w \qquad \text{(Eq. 7-3-18)}$$

The results of these calculations are presented in figures 7-3-16 and 7-3-17. The values of Rm rise from Rm $\ll$ 1 at low currents up to the order of unity at arc currents of 2000 to 3000 A. The variation of magnetic Reynolds number appears to have been relatively independent of gas type, mass flow rate and nozzle throat geometry. Over the current range investigated, the slope $\frac{d \lg Rm}{d \lg J}$ of the curves has a value of about 2. Thus an approximate dependence of Rm on J^2 might be postulated for our experiments.

Extrapolation of the magnetic Reynolds number towards even higher currents indicates that values of Rm $\approx$ 10 should be achieved with arc currents of about 6000 A (provided adequate experimental verification). This might be of considerable interest for magneto-gasdynamic experiments where Rm $\gg$ 1 is required.

Cowling Number (Magnetic Pressure Number)

Of the various definitions of the Cowling number (16) (17), we are interested in its physical interpretation as "magnetic pressure number". According to (17) the Cowling number is defined by the ratio of magnetic pressure to dynamic pressure, i.e.:

$$Co = \frac{B^2/2\mu_o}{\rho w^2/2} \qquad \text{(Eq. 7-3-19)}$$

If Co $\ll$ 1, the gasdynamic motion is practically not affected by the magnetic field. If Co reaches the order of unity or becomes larger, the flow field is considerably influenced by the magnetic field.

In the following an expression of the Cowling number is derived which contains only experimental measurable terms. Herein the flow region within the anode ring is considered. The dynamic pressure may be approximated by:

$$\tfrac{1}{2}\rho w^2 \approx \frac{1}{2} \frac{\dot{m}\, w}{d^2_{A\,min}\pi/4} \approx \frac{1}{2} \frac{F}{d^2_{A\,min}\pi/4} \qquad \text{(Eq. 7-3-20)}$$

With the assumption of constant current density over the cross section the selfinduced magnetic field is expressed as function of the radius:

$$B(r) = \frac{\mu_o}{2} j\, r \qquad \text{(Eq. 7-3-21)}$$

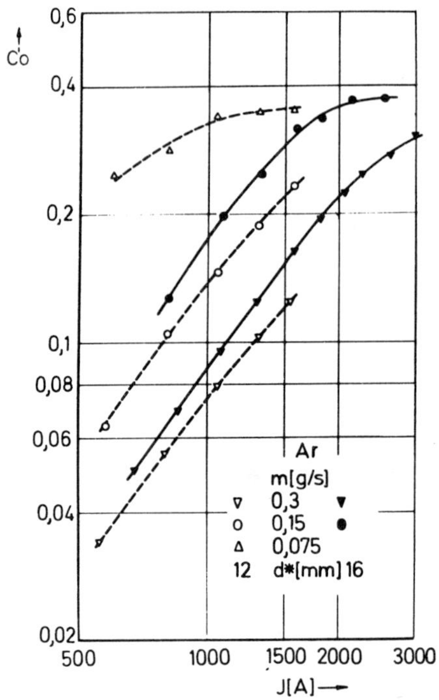

Fig. 7-3-18 Argon: Variation of Cowling number (magnetic pressure number) with arc current, mass flow rate and nozzle throat diameter

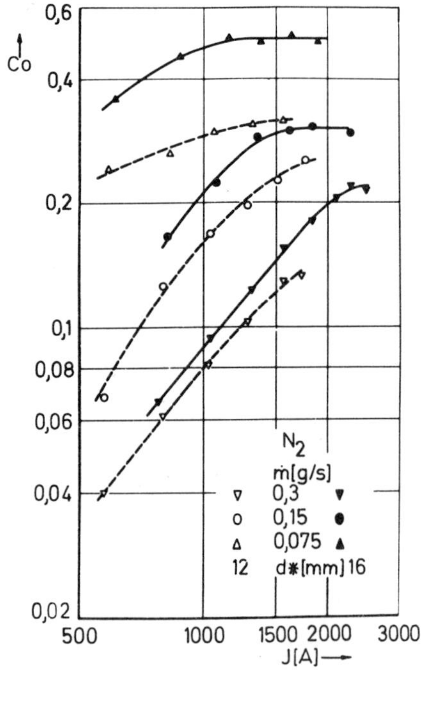

Fig. 7-3-19 Nitrogen: Variation of Cowling number (magnetic pressure number) with arc current, mass flow rate and nozzle throat diameter

The proceeding calculations are performed for $B = B\left(\frac{d_{Amin}}{2}\right)$ so that equation 7-3-21 becomes:

$$B = \frac{\mu_o}{2}\,\frac{J}{d^2_{Amin}\pi/4}\,\frac{d_{Amin}}{2} \qquad \text{(Eq. 7-3-22)}$$

The magnetic pressure at this particular radius amounts to:

$$\frac{B^2}{2\mu_o} = \frac{\mu_o}{2}\,\frac{J^2}{d^2_{Amin}\pi^2} \qquad \text{(Eq. 7-3-23)}$$

Finally one obtains for the Cowling number:

$$Co = \frac{\mu_o J^2}{4\pi F} \qquad \text{(Eq. 7-3-24)}$$

A radial dimension does not appear in equation 7-3-24 since both the mass flux and the current density have been determined by applying the same cross section. If some 'mean' value of B instead of $B = B\left(\frac{d_{Amin}}{2}\right)$ were introduced into equation 7-3-19, this would result in a numerical factor lower than $\frac{1}{4\pi}$ in equation 7-3-24.

A closer inspection of equation 7-3-24 indicates that the Cowling number might serve as a criterion for the relative magnitude of the two velocity producing mechanisms. According to equation 7-3-9 the total thrust of the device consists of an aerodynamic and an electromagnetic component, so that equation 7-3-24 can be rewritten:

$$Co = \frac{\mu_o}{4\pi}\,\frac{J^2}{F_{ae} + F_m} = \frac{\mu_o}{4\pi}\,\frac{J^2}{F_m\,(1 + F_{ae}/F_m)} \qquad \text{(Eq. 7-3-25)}$$

Considering operating conditions with predominant selfmagnetic velocity production, i.e. $F_m \gg F_{ae}$ or $F_{ae}/F_m \rightarrow 0$ one arrives at the approximate expression:

$$Co \approx \frac{\mu_o}{4\pi}\,\frac{J^2}{\mu_o/4\pi \cdot J^2 \ln 2.12\,(d_{Amin}/d_c)} \;. \qquad \text{(Eq. 7-3-26)}$$

Thus, with increasing selfmagnetic velocity production the Cowling number approaches a limiting value of:

$$Co_1 = \frac{1}{\ln 2.12\,\frac{d_{Amin}}{d_c}} \qquad \text{(Eq. 7-3-27)}$$

which is determined exclusively by electrode dimensions.

Figures 7-3-18 and 7-3-19 show the variation of Cowling number with arc current. With increasing J the curves rise approaching constant values of Co. In fact, the operation conditions where $\frac{d \lg Co}{d \lg J} \rightarrow 0$ correspond well with those of predominant selfmagnetic thrust production as discussed above. With the smaller d^*, lower Cowling numbers are obtained because of the higher aerodynamic thrust components involved. Evaluation of equation 7-3-27 for the electrode dimensions used in the

kk

present investigations yields values of $Co_1 \approx 0.35$ and 0.38 depending on the cathode diameter used. Comparison with the 'real' limiting Cowling numbers of the order of 0.3 to 0.5 shows sufficient agreement.

Acknowledgment

The author is indepted to Dr.Th. Peters for many valuable discussions. The help of Mr. W. Hennig in the experimental work is gratefully acknowledged.

This work was sponsored by the Bundesministerium für Wissenschaftliche Forschung, German Federal Republic, by contract No. RFF 3005.

References

1. **Ragaller, K.,** 'Experimental study of a plasma jet in an axisymmetric magnetic field', Paper pres. at VIIIth Int. Conf. Phen. Ionized Gases, Vienna, 1967.

2. **Maecker, H.,** 'Plasmaströmungen in Lichtbögen infolge eigenmagnetischer Kompression', Z. Phys. 141, 198-216. 1955.

3. **Ducati, A.C., Muehlberger, E., Giannini, G.,** 'High specific impulse thermo-ionic accelerator', AIAA Paper No. 64-668. 1964.

4. **John, R.R., Bennett, S., Connors, J.F.,** 'Experimental performance of a high specific impulse arc jet engine', AIAA Paper No. 64-669. 1964.

5. **Krülle, G., Hügel, H.,** 'Zur elektromagnetischen Geschwindigkeitserzeugung in Lichtbogentriebwerken', Z. Flugwiss. 14, 323-333. 1966.

6. **Ragaller, K.,** 'Über die Wechselwirkung eines Plasmastrahls mit einem achsensymmetrischen Magnetfeld', Z. Naturforschg. 21a, 725-732. 1966.

7. Hugel, H., **Krülle, G., Peters, T.,** 'Investigations on plasma thrusters with thermal and self-magnetic acceleration', AIAA J. 5, 551-558. 1967.

8. **McKee, H.B., Dean, R.C., Pytte, A.,** 'On cooled anodes in contact with a laminar arc-heated flow', IEEE Trans. Nuclear Science NS 11, 66-91. 1964.

9. **Bohn, W.L., Beth, M.-U., Nedder, G.,** 'On spectroscopic measurements of velocity profiles and non-equilibrium radial temperatures in an argon plasma jet', J. Quant. Spectrosc. Radiat. Transfer 7. 1967.

10. **Peters, T., Ragaller, K.,** 'Erzeugung hoher Plasmageschwindigkeiten durch Heizung und eigenmagnetische Beschleunigung in Lichtbögen', DLR FB 64-28. 1964.

11. **Hügel, H.,** 'Self-magnetic acceleration in arc-jet engines', Paper pres. at VIIIth Int. Conf. Phen. Ionized Gases, Vienna, 1967.

12. **John, R.R., Bennett, S., Connors, J.F.,** 'Arcjet engine performance: experiment and theory', AIAA J. 1, 2517-2525. 1963.

13. **Krause, S.,** 'Messungen mit gasdynamischen Sonden an einem Überschallplasmastrahl', DLR FB 66-49. 1966.

14. **Potter, J.L., Baily, A.B.,** Pressures in the stagnation regions of blunt

bodies in the viscous-layer to merged-layer regimes of rarefied flow, AEDC-TDR-63-168. 1963.

15. **Krause, S.**, 'Measurements with aerodynamic probes in plasma jets produced by electrothermal and Hall current accelerators', Paper pres. at 30th AGARD Propulsion and Energetics Panel Meeting, Munich, 1967.

16. **Cambel, A.B.**, 'Plasma physics and magnetofluidmechanics', McGraw-Hill Book Comp. Inc., New York, 1963.

17. **Pai, Shih-J.**, 'Magnetogasdynamics and plasma dynamics', Springer-Verlag, Wien, 1962.

List of Mathematical Symbols

A	area
B	superimposed magnetic field
C_F	thrust coefficient
C_O	cowling number
d	diameter
e	electron charge
E	electrostatic field
F	force
h	total enthalpy
j	total current density
J	electric current
$\dot{m}$	mass flow
p	pressure
r	radial coordinate
Rm	magnetic Reynolds number
t	time
U	arc voltage
w	velocity
x	distance from nozzle exit
α_M	mach angle
γ	ratio of specific heats
η_{kin}	kinetic efficiency

μ_o	permeability
ρ	density
σ	electrical conductivity
τ	control volume

Subscripts and Superscripts

ae	aerodynamic
c	of cathode
e	exit
l	limiting
m	electromagnetic
o	in arc chamber
1	at cathode
2	at anode
∞	ambient
*	nozzle throat

7-4

Analysis of the Current Flow of a Condenser Discharge of an Electrodeless MHD Motor

A. MEERT
Deutsche Versuchsanstalt für Luft und Raumfahrt e.V. Stuttgart - Vaihingen, Germany.

Summary

An analysis of the time dependence of the current due to a high-current condenser discharge permits the determination of the magnitude of losses in the discharge circuit as well as of the energy input into the plasma. The problems connected with the measurement of the discharge current are known and some of them are mentioned in this paper. Difficulties arise in the correct evaluation of the curves of the discharge current. A procedure based on a mathematical abstraction permits a very precise analysis of the current to be made. Some procedures for the measurement of the discharge current are discussed.

Sommaire

L'analyse de la variation en fonction du temps de l'intensité obtenue par décharge d'un condensateur à intensité elevée, permet la détermination des pertes dans le circuit ainsi que l'énergie mise dans le plasma. Les problèmes liés à la mesure du courant sont connus, on en rappelle quelques-uns. Les difficultés apparaissent lorsqu'on veut déterminer de façon précise la variation de l'intensité. On présente deux procédés de calcul. Le premier est une approximation basée sur l'introduction et la détermination d'un coefficient d'amortissement et d'une pulsation fonction du temps. Le second, purement mathématique, permet une analyse précise de l'intensité. On analyse quelques méthodes de mesure de l'intensité.

Introduction

High current capacitor discharges permit the generation of short duration power outputs in the megawatt range. This type of discharge is therefore of great interest, both for pure plasma research and for application to such things as MHD generators and accelerators. One important requirement is that the electrical energy stored in the capacitor should be transferred into the plasma with the highest possible efficiency. Knowledge of this efficiency would permit the calculation of the losses in the discharge circuit and of the energy absorbed by the plasma.

The efficiency can be derived from the time rate of decay of the discharge current which, to a first approximation, may be represented by a damped sine function. The suggestion by Preukschat (1) that values of the damping rate and the angular frequency could be measured at different points of the current function and averaged, was shown to be insufficiently accurate by Wichmann (2) who also demonstrated this experimentally. The variation of damping rate with time is not primarily due to measurement scattering, but to physical phenomena like the heating or acceleration of the plasma.

Physically, the inductance and resistance of the circuit are unknown functions of
kk* time, while capacitance may be assumed to be constant. Solution of the relevant

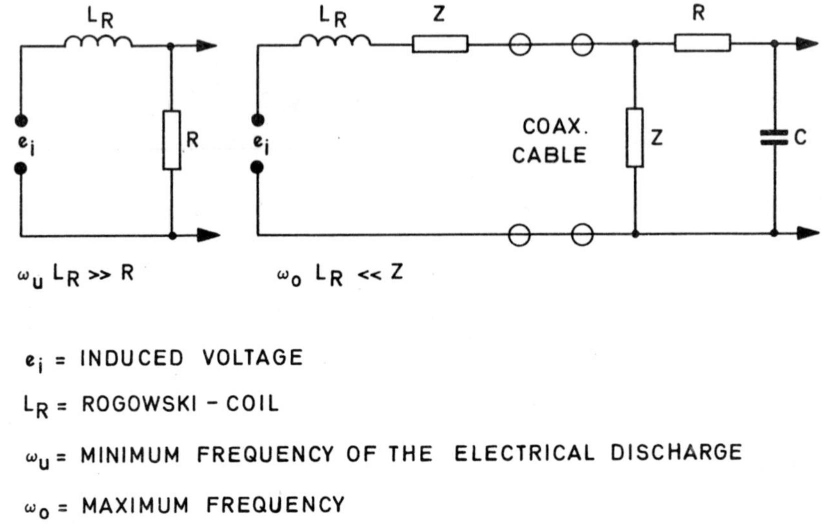

Fig. 7-4-1 Electronic methods of integration

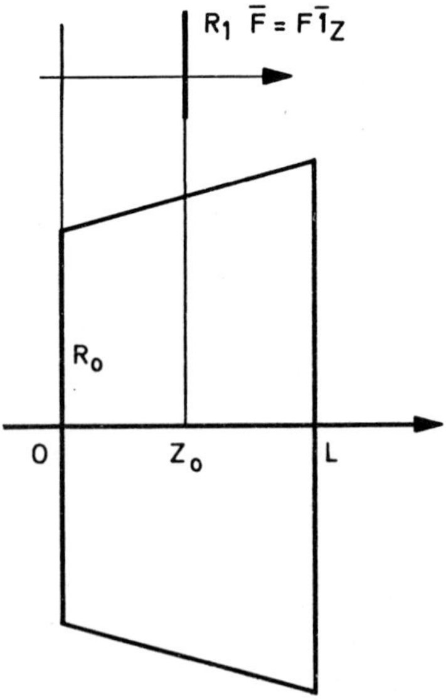

Fig. 7-4-2 Position and coordinates of the pick up coil

equations therefore gives information about the motion of the plasma, in addition to the efficiency of the discharge.

Current Measurement

For measurement of the extremely high currents associated with capacitance discharges, procedures based on Ohm's Law are unsuitable. Instead it is necessary to use some form of transformer to measure the current in the leads from the capacitor to the plasma. In practice the so-called Rogowski transformer, which consists of an annular coil with evenly spaced toroidal windings, is unsuitable because it greatly increases the inductance of the leads.

Schreitmüller (3) described a current transformer designed specially to overcome this difficulty. In the capacitor banks used in our experiments, band-shaped conductors with a gap of less than 1 mm were used as leads in order to keep the inductance low. The current transformer consisted of a very flat, elongated coil which was inserted perpendicularly into this gap. This facilitated precise measurement of current but caused virtually no increase in the inductance of the leads. Moreover, interference fields could be screened very easily, so that large signal-to-noise ratios were obtainable.

The magnetic flux across the coil for the Rogowski coil is:

$$\phi = \mu_o nAi \qquad \text{(Eq. 7-4-1)}$$

and the induced voltage across the coil is:

$$e_i = -\frac{d\phi}{dt} = -\mu_o nA\frac{di}{dt} \qquad \text{(Eq. 7-4-2)}$$

to obtain i from which, measurement of e_i is necessary, and the integration may be obtained electronically as shown in figures 7-4-1. For the Schreitmüller coil it is necessary to measure the voltage $e_s(t)$ across the coil as well as the voltage $e_{cc}(t)$ across a calibrated coil:

$$e_s(t) = L_s\frac{di}{dt} \qquad \text{(Eq. 7-4-3)}$$

$$e_{cc}(t) = L_{cc}\frac{di}{dt} \qquad \text{(Eq. 7-4-4)}$$

and:

$$L_s = \frac{e_s(t)}{e_{cc}(t)} L_{cc} \qquad \text{(Eq. 7-4-5)}$$

As an alternative to the above mentioned method for measuring the current, a pick-up coil may be used. The induced voltage in a pick-up coil is:

$$e_i(t) = -\frac{d\phi(t)}{dt} \qquad \text{(Eq. 7-4-6)}$$

where $\phi(t)$ is the flow through the surface of the loop, i.e.:

$$\phi(t) = \int\int \bar{B} \cdot \overline{dA} \qquad \text{(Eq. 7-4-7)}$$

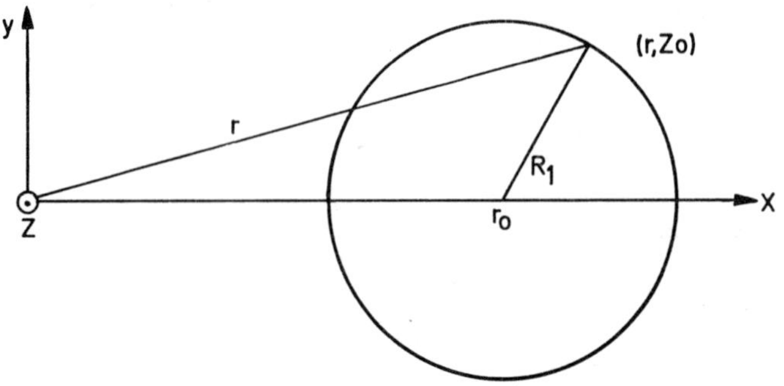

Fig. 7-4-3 Transformation of cylindrical into cartesian coordinates

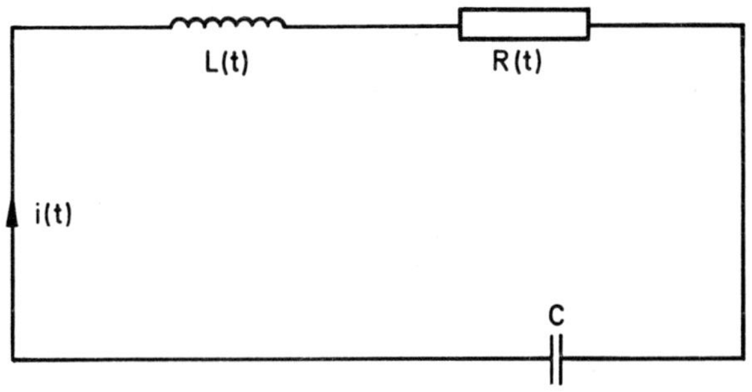

Fig. 7-4-4 Diagram of the equivalent oscillating circuit

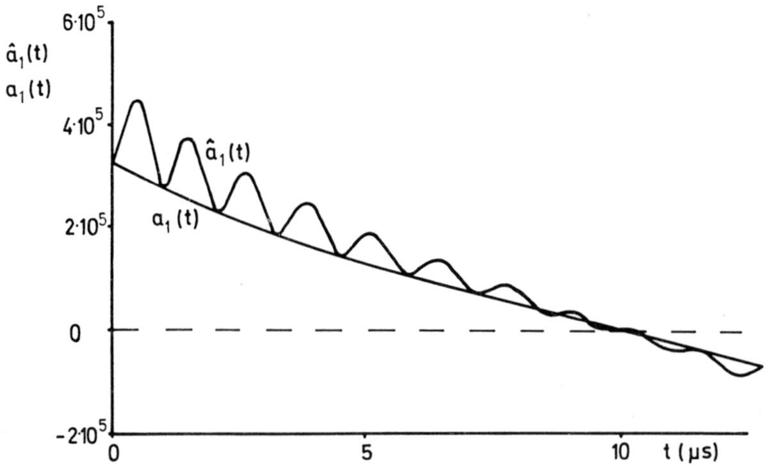

Fig. 7-4-5 Comparison between the calculated a_1 (t) and the given function a_1 (t)

In principle this method can be used for quantitative current measurement also, but knowledge of the distribution of the magnetic field is then required. The field can be described by:

$$\bar{B} = \bar{B}\,(t, r, z) = i(t)\,b(r, z)\,\bar{1}_B\,(r, z) \qquad \text{(Eq. 7-4-8)}$$

Given:

$$d\bar{A} = dA(r, z)\,\bar{1}_A\,(r, z) \qquad \text{(Eq. 7-4-9)}$$

and after integration with respect to time:

$$i(t) = \frac{\int_0^t e_i(t)\,dt}{\int\int b(r, z)\,\bar{1}_B \cdot \bar{1}_A\,dA(r, z)} \qquad \text{(Eq. 7-4-10)}$$

We obtained this integration electronically and computed the surface integral mathematically. The pick-up coil was positioned so that only the axial component of the magnetic field contributed to the flow (Fig. 7-4-2). The surface integration was then not dependent on z, because $z = z_0$ for the whole loop surface.

The axial H-field of the discharge coil is given by (4), viz

$$\bar{H}_z(t, r, z) = \frac{i(t)}{R_o}\;\Xi\;(r, z)\,\bar{1}_z \qquad \text{(Eq. 7-4-11)}$$

where Ξ (r, z) is the function of the normalised field distribution, containing the coil's directive function and the corresponding current distribution function as parameters. It follows (Fig. 7-4-2):

$$b(r, z) = \frac{\mu_o}{R_o}\;\Xi\;(r, z) \qquad \text{(Eq. 7-4-12)}$$

$$\bar{1}_A \cdot \bar{1}_B = 1 \qquad \text{(Eq. 7-4-13)}$$

and the surface integral can be simplified:

$$i(t) = -\frac{R_o \int_0^t e_i(t)\,dt}{\mu_o \int\int \Xi\;(r, z_o)\,dA(r)} \qquad \text{(Eq. 7-4-14)}$$

As there is no axial symmetry, it is easier to express this equation (given in cylindrical coordinates) in Cartesian coordinates (Fig. 7-4-3).

With:

$$r = (R_1^2 + 2\,xr_o - r_o^{\,2})^{\frac{1}{2}} \qquad \text{(Eq. 7-4-15)}$$

one gets:

$$\int\int \Xi\;(r, z_o)\,dA(r) = 2\int_{r_o - R_1}^{r_o + R_1} \Xi\;(r, z_o)\left[R_1^2 - (x-r_o)^2\right]^{\frac{1}{2}} dx \qquad \text{(Eq. 7-4-16)}$$

which can be programmed easily.

The influence of the magnetic field caused by the current flow in the plasma is not accounted for in the function Ξ (r, z); this method is therefore correct only if there is no gas in the discharge tube. When a plasma is generated, the use of pick-up coils is suitable only for quick orientations.

Oscillating Circuit with Time-Dependent Elements

Figure 7-4-4 shows a circuit diagram equivalent to the discharge path. C is the given capacitance; L(t) and R(t) are mathematically equivalent quantities causing the same current i(t) as in the actual circuit.

Mathematically the behavior of this oscillating circuit has to be described by a homogeneous linear differential equation of 2nd order with variable coefficients. First there is:

$$\frac{de_i}{dt} + \frac{de_r}{dt} + \frac{de_c}{dt} = 0 \qquad \text{(Eq. 7-4-17)}$$

with:

$$\frac{de_i}{dt} = \frac{d^2L(t)}{dt^2}\, i(t) + 2\,\frac{dL(t)}{dt}\,\frac{di(t)}{dt} + L(t)\,\frac{d^2i(t)}{dt^2} \qquad \text{(Eq. 7-4-18)}$$

$$\frac{de_r}{dt} = \frac{dR(t)}{dt}\, i(t) + R(t)\,\frac{di(t)}{dt} \qquad \text{(Eq. 7-4-19)}$$

$$\frac{de_c}{dt} = \frac{i(t)}{C} \qquad \text{(Eq. 7-4-20)}$$

From this results:

$$\frac{d^2i(t)}{dt^2} + a_1(t)\,\frac{di(t)}{dt} + a_o(t)\, i(t) = 0 \qquad \text{(Eq. 7-4-21)}$$

with:

$$a_1(t) = \left[R(t) + 2\,\frac{dL(t)}{dt}\right]\frac{1}{L(t)} \qquad \text{(Eq. 7-4-22)}$$

$$a_o(t) = \left[\frac{1}{C} + \frac{dR(t)}{dt} + \frac{d^2L(t)}{dt^2}\right]\frac{1}{L(t)} \qquad \text{(Eq. 7-4-23)}$$

As a rule, the differential equation 7-4-21 has to be solved for a given pair $a_1(t)$ and $a_o(t)$, to obtain values for the function i(t) and its derivatives. In our case the problem is reversed.

Calculation of the Pair

In principle, the problem may be solved in two different ways. The first method is a logical extension of the procedure used in (2). At any instant in time, certain values of the frequency and the damping rate for the oscillating circuit exist. An approximate determination of these values is possible from the graphical representation of the current and permits a calculation of the pair. This procedure is very instructive because of its physical background, but it is not precise and is very laborious. More exact results can be obtained by the second method, which follows.

If function $a_1(t)$ is given, the other function $a_o(t)$ can be determined from equation 7-4-21. For each value of $a_1(t)$ there exists a unique value of $a_o(t)$ but also an infinite number of solutions. The problem is to find a criterion that leads to the determination of the pair appropriate to the system under consideration.

If all the functions are continuous, equation 7-4-21 gives:

$$\frac{d^3i}{dt^3} + a_1(t)\frac{d^2i}{dt^2} + a_o(t)\frac{di}{dt} + \frac{da_1(t)}{dt}\cdot\frac{di}{dt} + \frac{da_o(t)}{dt}\, i = 0 \qquad \text{(Eq. 7-4-24)}$$

Equation 7-4-21 and equation 7-4-24 can be resolved in $a_o(t)$ and $a_1(t)$:

$$a_o(t) = \frac{\left(\frac{d^2i}{dt^2}\right)^2 - \frac{di}{dt}\left(\frac{d^3i}{dt^3} + \frac{da_1(t)}{dt}\cdot\frac{di}{dt} + \frac{da_o(t)}{dt}\cdot i\right)}{\left(\frac{di}{dt}\right)^2 - i\,\frac{d^2i}{dt^2}} \qquad \text{(Eq. 7-4-25)}$$

$$a_1(t) = \frac{-\frac{di}{dt}\cdot\frac{d^2i}{dt^2} + i\left(\frac{d^3i}{dt^3} + \frac{da_1(t)}{dt}\cdot\frac{di}{dt} + \frac{da_o(t)}{dt}\, i\right)}{\left(\frac{di}{dt}\right)^2 - i\,\frac{d^2i}{dt^2}} \qquad \text{(Eq. 7-4-26)}$$

The expressions for $a_o(t)$ and $a_1(t)$ contain a term that disappears when $di/dt = 0$ in the case of $a_o(t)$, and when $i = 0$ in the case of $a_1(t)$. This feature permits discrete series of exact values to be obtained for the functions $a_o(t)$ and $a_1(t)$ if a sufficient number of periods of the function $i(t)$ is available.

In the particular case where the functions $a_o(t)$ and $a_1(t)$ have a small variation during a convenient time interval, we can retain as approximate values:

$$\hat{a}_o(t) = \frac{\left(\frac{d^2i}{dt^2}\right)^2 - \frac{di}{dt}\cdot\frac{d^3i}{dt^3}}{\left(\frac{di}{dt}\right)^2 - i\,\frac{d^2i}{dt^2}} \qquad \text{(Eq. 7-4-27)}$$

$$\hat{a}_1(t) = \frac{-\frac{di}{dt}\cdot\frac{d^2i}{dt^2} + i\,\frac{d^3i}{dt^3}}{\left(\frac{di}{dt}\right)^2 - i\,\frac{d^2i}{dt^2}} \qquad \text{(Eq. 7-4-28)}$$

The current $i(t)$ and its derivatives can be determined by electronic integration and differentiation respectively by the methods mentioned in section 2. Figure 7-4-6 shows the shape of $\hat{a}_1(t)$ calculated by the last equations compared with function $a_1(t)$ calculated by equation 7-4-22. The waviness of $\hat{a}_1(t)$ is due to the fact that the gradients of $a_o(t)$ and $a_1(t)$ could not be taken into account since these were not available at that time.

Calculation of the Inductance, Resistance and Efficiency

In the following the time dependence of the inductance and resistance may be calculated using the known values for the pair. Solving equations 7-4-22 and 7-4-23 for R and L, one gets:

$$\ddot{L} - a_1 \dot{L} + L(a_o - \dot{a}_1) = \frac{1}{C} \qquad \text{(Eq. 7-4-29)}$$

$$R = a_1 L - 2\dot{L} \qquad \text{(Eq. 7-4-30)}$$

This form of the equation is not suitable for numerical calculation, because $\dot{a}_1$ has to be determined. However, by introducing an auxiliary quantity H, the system can be put into a form suitable for numerical calculation by the Runge-Kutta method:

$$\dot{L} = a_1 L - H \qquad \text{(Eq. 7-4-31)}$$

$$\dot{H} = a_o L - \frac{1}{C} \qquad \text{(Eq. 7-4-32)}$$

$$R = a_1 L - 2\dot{L} \qquad \text{(Eq. 7-4-33)}$$

The advantage of this presentation is that calculation of L and R is now possible without differentiating a_1. To obtain L from equation 7-4-31 and equation 7-4-32 values of the initial conditions $L(t=0)$ and $H(t=0)$ are required. $H(0)$ is calculated by:

$$H(0) = \frac{1}{2}\left[a_1(0)\,L(0) + R(0)\right] \qquad \text{(Eq. 7-4-34)}$$

so that a single-valued solution is obtainable by specification of the initial inductance and resistance. ($\dot{L}(0)$ cannot be given as there exists no stated nor measured value).

As the principal component of the circuit inductance is the inductance of the coil, this value can be used for $L(0)$ with sufficient accuracy. $R(0)$, however, cannot be specified from the outset because spark gaps are used and these produce a discontinuity in the time dependence of the resistance: at the beginning of the discharge the resistance is infinite, but an instant later the resistance of the equivalent oscillating circuit becomes finite. $R(0)$ has to be understood therefore as $\lim_{t\to 0} R(t)$. As this value cannot be measured, a secondary condition has to be introduced, from which $R(0)$ can be calculated.

The entire available energy E_t is:

$$E_t = \frac{1}{2}\, C\, e_{co}^2 \qquad \text{(Eq. 7-4-35)}$$

This energy E_t is consumed through the equivalent resistance $R(t)$. From this, we have the relation:

$$\int_0^{\infty} R(t)\, i^2(t)dt = E_t \qquad \text{(Eq. 7-4-36)}$$

If R(0) is assumed to be a parameter for the calculation, a family of curves for the single parameter R(t) results from the system of differential equations 7-4-30, 7-4-31, 7-4-32 and 7-4-33 for any given value of L(0). Only one of these curves satisfies equation 7-4-35. By using this condition, R(t) can be calculated exactly.

Two initial values $R_a(0)$ and $R_b(0)$ corresponding to the functions $R_a(t)$ and $R_b(t)$ are required to satisfy:

$$\int_0^{\infty} R_a(t) i^2(t)dt < E_t < \int_0^{\infty} R_b(t) i^2(t)dt \qquad \text{(Eq. 7-4-37)}$$

The correct initial value R(0) can be calculated using the regula falsi procedure. From this value, R(t) and L(t) may be determined by equations 7-4-31, 7-4-32 and 7-4-33. Since the current function i(t) is damped and converges to zero for large values of t, only a sufficiently high value of the upper limit is required for the numerical integration of 7-4-36. This value can be easily determined by consideration of the convergency: for instance, the boundary values may be increased gradually until the variation between two succeeding integral values is below a reasonable limit of accuracy.

With R(t) known, the efficiency of energy transfer can be calculated. This quantity is defined by:

$$\eta(t) = \frac{E_{pl}(t)}{E(t)} \qquad \text{(Eq. 7-4-38)}$$

where $E_{pl}(t)$ is the energy absorbed by the plasma, and E(t) the total energy expended up to this time. E(t) is given by:

$$E(t) = \int_0^{t} R(\tau)\, i^2(\tau)\, d\tau \qquad \text{(Eq. 7-4-39)}$$

To determine $E_{pl}(t)$ a fictitious resistance $R_{pl}(t)$ is introduced by the relation:

$$E_{pl}(t) = \int_0^{t} R_{pl}(\tau)\, i^2(\tau) d\tau \qquad \text{(Eq. 7-4-40)}$$

From the difference between the equivalent resistance without gas discharge R^* (circuit losses) and the equivalent resistance with gas discharge R(t) the resistance of the plasma can be obtained with sufficient accuracy:

$$R_{pl}(t) = R(t) - R^* \qquad \text{(Eq. 7-4-41)}$$

The overall efficiency of energy transfer then results from:

$$\eta_f = \lim_{t \to \infty} \eta(t) \qquad \text{(Eq. 7-4-42)}$$

References

1. **Preukschat, A.W.**, 'Elektrodenlose Ringentladung als Mittel zur Schub-erzeugung. Bestimmung des Energieübertragungswirkungsgrades und Abschätzung der kinetischen Energie des Gases', Jahrbuch der WGLR, S. 377-381. 1963.

2. **Wichmann, H.G., Meert, A., Savary, M.M.**, 'Über den Einfluß der Primärkreiskomponenten auf die Beschleunigung des Plasmas bei elektrod-enlosen MHD-Triebwerken', Jahrbuch der WGLR, S. 416-421. 1964.

3. **Schreitmüller, K.**, 'Impulsaufheizung eines vorionisierten Wasserstoff-kanals', Dissertation Techn. Hochschule Stuttgart, 1967.

4. **Meert, A.**, 'Analyse der Feldkonfiguration einer rotationssymetrischen Spule mit beliebiger Mantellinie', ZAMP. Vol. 17, No. 1, S. 157-168. 1966.

List of Mathematical Symbols

A	Coil area
a_1, a_o	Pair of coefficients of the diff. equation
B	Induction
C	Capacitance of the oscillating circuit
E(t)	Energy
E_t	Total available energy
e_c	Voltage across capacitor
e_{co}	e_c at t=0
e_{cc}	Voltage across the calibrated coil
e_i	Induced voltage across the coil
e_1	Voltage across inductor
e_r	Voltage across resistor
e_s	Voltage across the Schreitmüller coil
H = H(t)	Auxiliary quantity
$\bar{H}_z$	Axial magnetic vector field
i = i(t)	Current
L = L(t)	Inductance of the oscillating circuit
L_{cc}	Inductance of calibrated coil
L_s	Inductance of Schreitmüller coil
n	number of windings / unit length
R_o	Radius at the discharge coil at z=0 (Fig. 7-4-2)
R_1	Radius of the pick-up coil (Figs. 7-4-2 and 7-4-3)
R^*	Resistance of the circuit without gas discharge
R = R(t)	Resistance of the oscillating circuit with gas discharge
R_a, R_b	Auxiliary resistance values
$R_{p1}(t)$	Fictitious plasma resistance
r	Cylindrical coordinate
r_o	Distance: centre pick-up coil - z-axis
t	Time
x	Cartesian coordinate
z	Cylindrical coordinate
α	Damping rate
η(t)	Efficiency
η_f	Overall efficiency
μ_o	Permeability in air
Ξ (r, z)	Function of the normalized axial field distribution
ω	Angular frequency

$\bar{1}$ Unit vector in the relevant direction

NB. Use of the MKSA system is assumed.

Part 8

Gas Chromatography

8-1

Gas Chromatography

J. L. LAUER
Sun Oil Company, Marcus Hook, Pennsylvania, U.S.A.

Summary

When a sample of a mixture of gases is injected into a (carrier) gas stream about to enter and pass through a tube either wholly filled or impregnated on the inner surface by a liquid and/or a solid or solid powder, the different components of the mixture emerge from the tube after different (retention) times, if their solubilities in the liquid or absorptivities at the solid surfaces are sufficiently different. This separation of gases is analogous to that of dyes in a liquid solution when the latter is percolated through an adsorption column - hence the name gas chromatography. However, a physical property detector or transducer is always necessary downstream to indicate the flowby of a particular component. Preferably the value of this property is very different for the sample and the carrier gas. Usually the components, even when fully separated by the column, follow in rapid succession and their amounts are very small, making a full identification often difficult. When all the possible components in the mixture and their relative retention times in a particular column are known, a qualitative and quantitative analysis can already be achieved from the position and strength of the signal emitted by a thermal conductivity (hot wire) or other non-specific detector and recorded on a strip chart. Otherwise a fast-scanning spectrophotometer or another specific detector should be used. Particular advances in this area have recently been made: e.g., electron capture and affinity detectors specifically for halogen-containing compounds, flame ionization detectors, and even small and fast chemical reaction detectors are now available. Conversely, the retention times of a particular column are characteristic of various physicochemical properties of the material in the column and can be used for their convenient determination. Gas chromatography can be fully automated and made very fast and sensitive in the parts-per-million-and-below range.

Sommaire

Lorsqu'un échantillon de mélange de gaz est injecté dans un écoulement (porteur), et pénètre dans un tube soit rempli, soit imprégné sur sa surface interne d'un liquide et/ou d'un solide ou d'une poudre les divers composants du mélange ressortent du tube après des temps (de reténtion) différents, si leurs solubilites dans le liquide ou leurs adsorptivités sur les surfaces solides sont suffisamment différentes. Cette séparation de gaz est analogue aux colorants dans une solution liquide lorsque cette dernière traverse une colonne d'adsorption, d'où provient le nom de chromatographie gazeuse. Il est cependant nécessaire d'utiliser un détecteur de propriétés physiques ou un capteur placé à l'aval du tube pour indiquer le composant particulier dans l'écoulement. Il est avantageux que la valeur de cette propriété soit differente pour l'échantillon et le gaz porteur. Généralement, les composants, lorsqu' ils sont complètement séparés par la colonne, se succèdent rapidement et leurs quantités sont très faibles, ce qui rend difficile une identification complète. Lors-

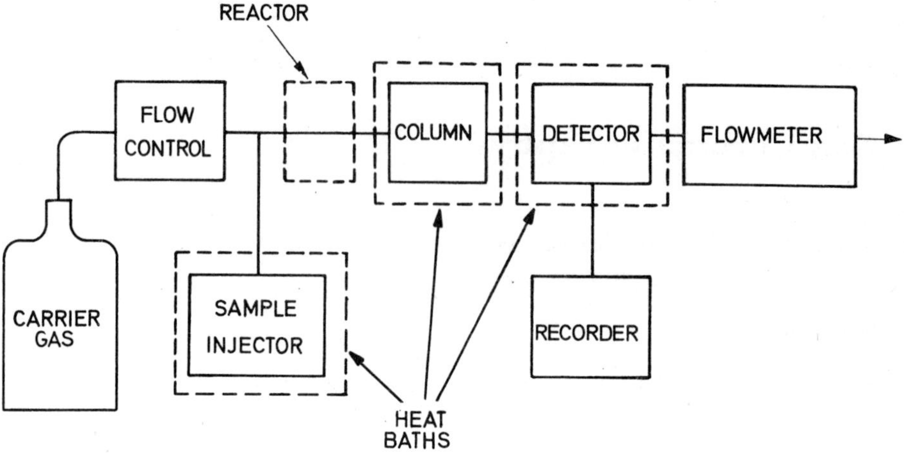

Fig. 8-1-1 Block diagram of an elementary gas chromatograph

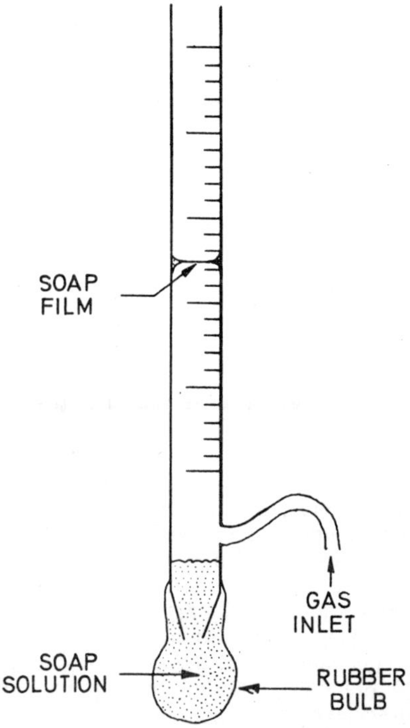

Fig. 8-1-2 Soap bubble flow meter

que tous les composants du mélange et leurs temps de rétention respectifs sont connus, une analyse qualitative et quantitative est possible à partir de la position et de l'intensité du signal émis par conductivité thermique (fil chaud) ou tout autre émetteur non spécifique et enregistré sur bande. On peut utiliser aussi un spectrophotomètre à commutation rapide ou tout autre détecteur adapté. Des progrès récents ont été obtenus dans ce domaine: capture d'électrons et détecteurs d' affinité réalisés spécialement pour les composés halogénés, détecteurs d'ionisation dans les flammes, détecteurs de petites dimensions pour réactions chimiques rapides sont maintenant disponibles. D'autre part, les temps de rétention d'une colonne donnée sont caractéristiques des propriétés physico-chimiques du matériau utilise et peuvent être utilisés commodément pour leur détermination. La chromatographie gazeuse peut être rendue automatique, rapide et sensible pour le domaine de la particule par million et en-dessous.

Introduction

Chromatography is a physical method for separating components in a mixture. It depends on differences in the distribution coefficients of the components between a fixed and a mobile phase. The name chromatography derives from experiments carried out by Tswett (1) in 1906: plant pigments were placed on the top of a glass tube containing powdered chalk and observed to spread out into zones of different colors when washed down the 'column' ('eluted') by means of light petroleum. Eventually these zones, travelling at different speeds, could become completely separated and yield solutions of but a single solute emerging from the lower end of the column. Separation was facilitated because the substances were colored, but the method in no way depends on color and is applicable to mixtures of gases as well as to mixtures of liquids. Instruments are then required to take the place of the eye. Gas chromatography in particular came into being in 1952 (2) when electronics took off on its first large postwar jump.

Basic Apparatus

The instrumentation of gas chromatography is inherently much simpler and usually considerably less expensive than that of other recent instrumental techniques of chemical analysis. The basic parts, shown in figure 8-1-1 are: (i) the tank of carrier gas used for eluting the gas mixture to be separated, (ii) the pressure regulator and flow meter, (iii) the sample injection system, (iv) the separation column, (v) the heating and temperature control system, (vi) the detection system, and (vii) the recorder or other device indicating the output of the detector. Note that some of the components may have to be kept at different temperatures and that a reactor may be inserted to convert the sample to compounds more suitable for separation than those present originally. For example, a typical 'student chromatograph' might use helium at 2-3 atmospheres of pressure and a flow rate of 70 ml/min. as the carrier gas, a soap bubble flow meter made out of a pipette (Fig. 8-1-2), a gas syringe and a self-healing rubber septum (Fig. 8-1-3) for introducing the unknown gas mixture, e.g. air, a few feet of coiled 5 mm diameter copper tubing filled with so-called molecular sieve, a kind of clay, as the separating column, and a hot wire bridge connected to an amplifier and strip chart recorder as the detection system. A simple hot-wire bridge comparing the resistances of two identical filaments, one exposed to the carrier gas and the other to the carrier gas containing the unknown mixture, is shown in figure 8-1-4. The column may be wrapped in electric heating tape and kept at about 50°C by means of a Variac-type transformer. A typical record obtained for air with such a system is represented in figure 8-1-5. The physical quantity recorded is thermal conductivity. Only two peaks are apparent: one occurring at a time, from the moment of injection, that is characteristic of oxygen; the second at a time that is characteristic of nitrogen. These characteristic "retention times" are established by calibration with pure materials. The areas under the peaks are indicative of the re-

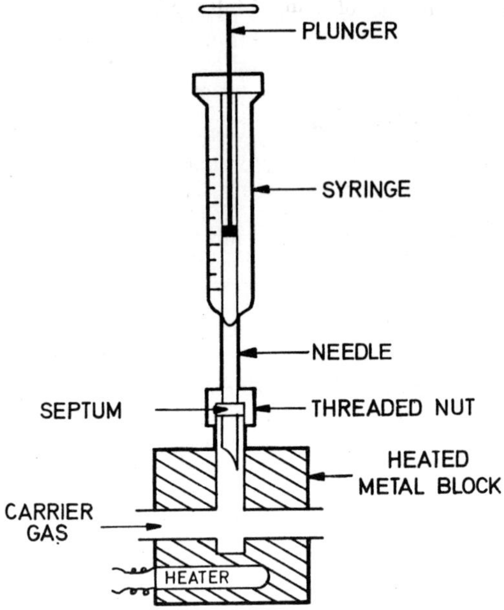

Fig. 8-1-3
Syringe and rubber septum injection system

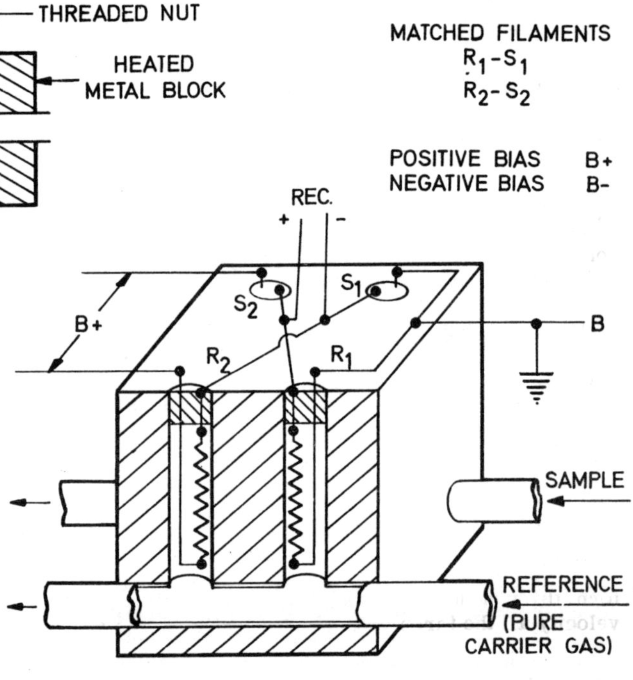

Fig. 8-1-4
Thermal conductivity detector. The electrical circuitry is built around a Wheatstone bridge design. (Courtesy Gow-Mac Instrument Co.)

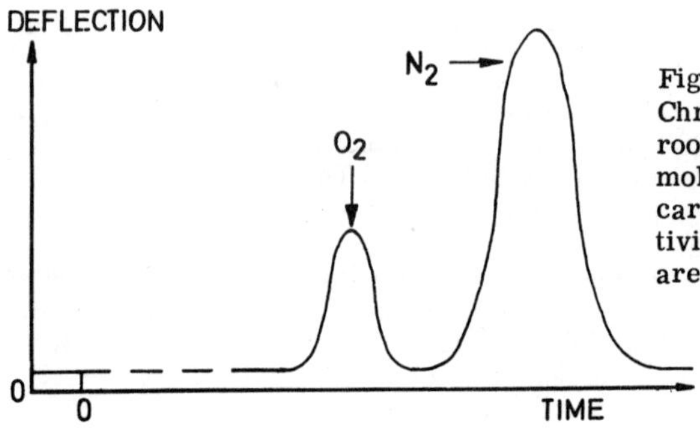

Fig. 8-1-5
Chromatogram of air obtained near room temperature with a Linde 5A molecular sieve column, helium carrier gas, and a thermal conductivity detector. Argon and oxygen are superimposed.

lative proportions of the two gases and of the size of sample injected.

Applications

Already from this elementary example some of the advantages and limitations of gas chromatography as a means of chemical analysis emerge. Clearly the method can be made very sensitive, and is simple, rapid and inexpensive. Disadvantages are its limitation to gases, the need for calibration, and the sequential detection of the components. It will be shown later that prior knowledge of the nature of the sample is essential for the selection of the separation column and the detector. Gas chromatography is not as general a method as infrared spectroscopy. However, all these shortcomings can usually be overcome. Furthermore, gas chromatography can be adapted to measurements of such physico-chemical parameters as adsorption isotherms, diffusivities in porous materials, heats of adsorption, and surface areas. It can also, in combination with a flow reactor, be helpful in determinations of molecular structure and in the elucidation of reaction mechanisms.

Principles

General

When the carrier gas is in contact with a stationary liquid phase (e.g. a coated capillary or liquid-impregnated solid support), the separation procedure is referred to as 'gas-liquid chromatography' (GLC). When it is in contact with a solid phase only, the procedure is called'gas-solid chromatography' (GSC). The basis for the separation of the components of a slug of gas mixture carried through a column in a constant flow of inert carrier gas is their different solubilities in a stationary liquid phase, or adsorptivities to a stationary solid phase.

Because of solubility or adsorption in the stationary phase, a volume of carrier gas V_R greater than the column volume V_m is required to sweep out a given vapor component of a gas mixture. For a 'linear' relation between concentrations of solute i.e., the separated vapor in the carrier gas and in the liquid or adsorbed phases, this excess volume is proportional to the volume of stationary phase in the column V_s; hence, V_R or the 'retention volume' of this particular vapor is given by:

$$V_R^\circ = V_m^\circ + KV_s \qquad \text{(Eq. 8-1-1)}$$

Because the carrier gas experiences a pressure drop in flowing through the column, a correction (indicated by superscript $^\circ$) is applied to make the flow velocity uniform throughout. V_m° is known as the 'hold-up volume'. Since for uniform flow velocity all the three volumes of equation 8-1-1 are proportional to time intervals:

$$t_R = t_m + t_r \qquad \text{(Eq. 8-1-2)}$$

Relative retention times t_r are the quantities most easily compared for different substances under a given set of conditions. Hold-up times t_m are generally determined in one of two ways: (a) by injecting a quantity of insoluble vapor ($K = 0$) with a slug of vapor sample to be analyzed (air is usually injected more or less unintentionally) or (b) by measuring the retention times of a homologous series of compounds, and plotting them against molecular weight (3). The latter method has to be used when the detector is insensitive to air. t_r is the total recovery time.

Retention times are always referred to band peaks. Littlewood (4) has listed many variables for which the measured retention times should be corrected before they may be compared with tabulated values for identification of the unknown components, or before publication. However, even so, the experience in the author's laboratory has been that a component in an unknown mixture cannot be identified on the basis of

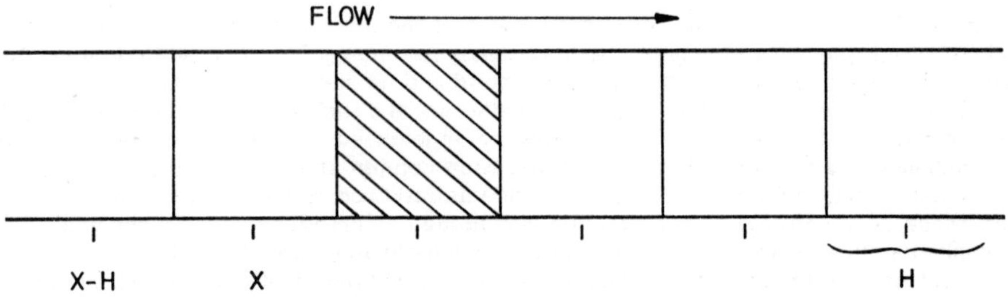

Fig. 8-1-6 Division of column into theoretical plates
(after Gidding (5), p. 25)

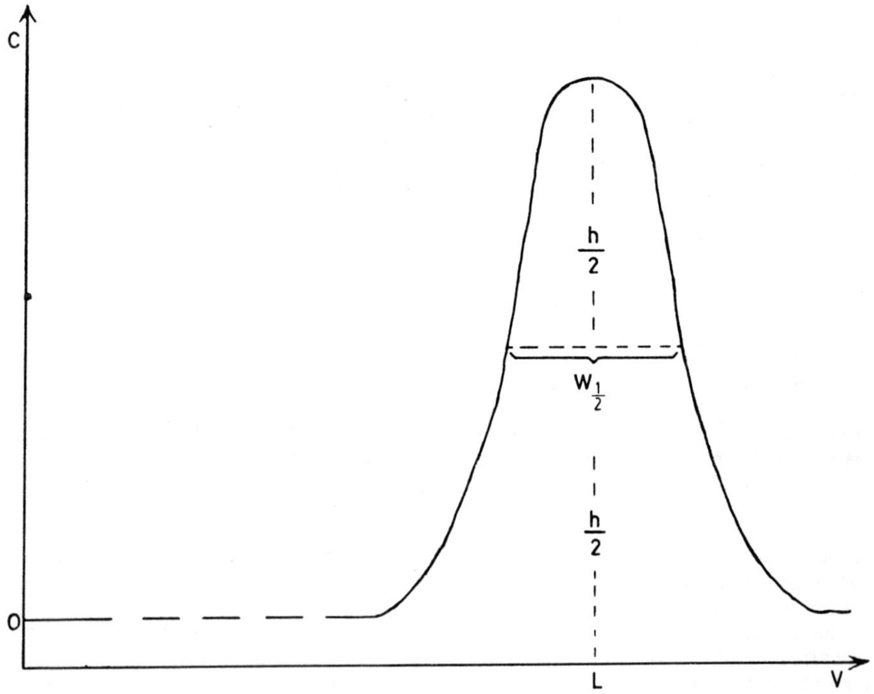

Fig. 8-1-7 Estimation of plate-number from retention and peak width at half-height
(after Littlewood (4), p. 131)

published retention times. Either the material whose presence is suspected - if available - is sent through the apparatus under identical conditions and its retention time compared with that of the unknown, or - as will be discussed below, the separated components emerging from the column are analyzed by a specific detector such as a spectrometer. The latter is then, in general, used only for identification, i.e., qualitative not quantitative analysis.

Quantitative Analysis

With the usual non-discriminating detector, such as the hot wire bridge (thermal conductivity detector) illustrated in figure 8-1-4 only one peak can be obtained for every component of a gaseous mixture. In contrast, spectroscopic methods of analysis generally provide many peaks for every component. This peak is usually recorded on a strip chart recorder so that the location of the peak in time, with respect to the time of sample introduction, can be used for qualitative identification, as indicated before (cf. also Fig. 8-1-5). For simple and routine quantitative analysis: (i) the output of the detector-recorder system must be linear with concentration; (ii) the flow rate must be constant so that the abscissa may be converted to volume of carrier gas; and (iii) the shape of the peak should be narrow and symmetrical. The last remark refers in particular to differentiating detectors such as the hot-wire bridge. Under these conditions peak area can be used as a function of the total amount of component present in the unknown gas mixture. Peak area can be obtained by planimetry on the recorder charts, for example. But for routine applications we have found it preferable to have these integrations computed automatically from digital records.

Column Performance

The heart of the chromatograph is the column - packed or capillary - in which the separation is carried out. The following discussion of its behaviour will lay the foundation for the practical applications of chromatography.

Dynamics of Zone Spreading (Kinetic Factors)

In the preceding section it was shown that a gas or vapor partially adsorbed on, or soluble in a stationary phase is swept out of a column by a volume of carrier gas that is greater than the dead volume of the column. To achieve separation of a slug of gas (vapor or 'solute') injected into a carrier gas stream: (a) the slug must decompose into "zones" corresponding to the different components of the sample, every zone traveling at a different velocity; but at the same time (b) every zone will spread continuously over an increasingly larger volume because of various diffusion effects. Fortunately it has been found that mechanism (a) is faster than mechanism (b), for the reasons set out below (following Giddings (5)).

Basis of Chromatographic Separation - By analogy with distillation columns (6), chromatographic columns are assumed to be divided into theoretical plates of height H, there being equilibrium between the phases in every plate (Fig. 8-1-6). The effective transport or diffusion coefficient for solute molecules in the gas phase is then one-half, since flow in one direction only is considered, the product of drift velocity v_d times H. But since transport of solute is taking place in the mobile phase only, the drift velocity is given by the product of average velocity of the mobile phase fluid, v, and the fraction of solute at equilibrium in the mobile phase, i.e.:

$$v_d = v \, . \, V_m/(V_m + KV_s) \equiv Rv \qquad \text{(Eq. 8-1-3)}$$

The appropriate convective diffusion equation:

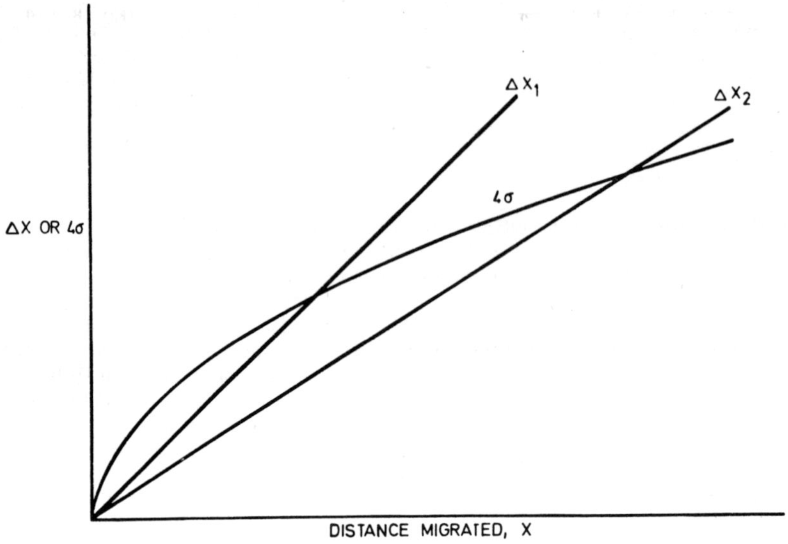

Fig. 8-1-8 The increase of the separation gas $\Delta\psi$ and the increase of zone width, 4σ, as a function of the distance migrated.

The $\Delta\psi$ and 4σ lines must intersect before separation is achieved. The component corresponding to $\Delta\psi_1$ is eluted before that corresponding to $\Delta\psi_2$

(after Giddings, (5), p. 33)

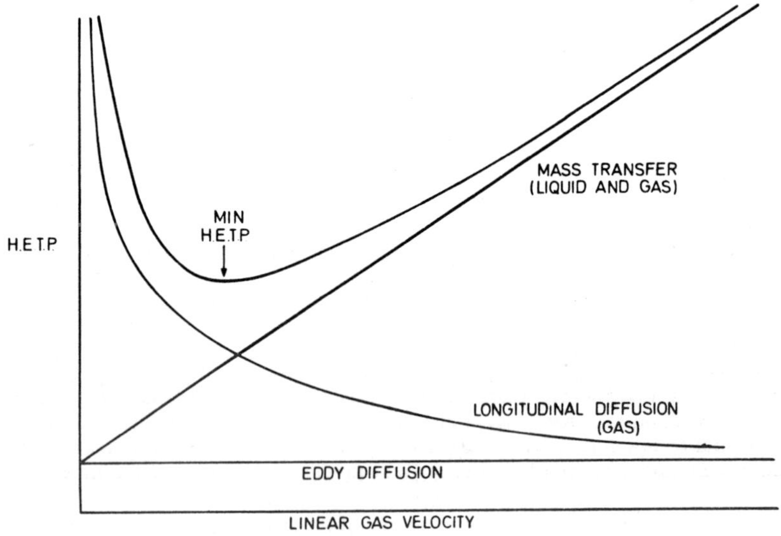

Fig. 8-1-9 Curve of H. E. T. P. against gas velocity

The experimental curve is resolved into three contributing functions, (a) eddy diffusion, A, (b) longitudinal diffusion in the gas phase, B/ , and (c) mass transfer in both phases, C and ν.

$$\frac{\partial c}{\partial t} = -v_d \frac{\partial c}{\partial \psi} + D \frac{\partial^2 c}{\partial \psi^2} \qquad \text{(Eq. 8-1-4)}$$

leads to:

$$\frac{\partial c}{\partial t} = - Rv \frac{\partial c}{\partial \psi} + \frac{RvH}{\alpha} \frac{\partial^2 c}{\partial \psi^2} \qquad \text{(Eq. 8-1-5)}$$

and thus to a standard deviation of:

$$\sigma = \sqrt{(RvtH)} = \sqrt{(v_d tH)} = \sqrt{(V_R^\circ H)} \qquad \text{(Eq. 8-1-6)}$$

Hence, since the number of theoretical plates is defined by

$$N = V_R^\circ / H \qquad \text{(Eq. 8-1-7)}$$

the practically useful result

$$N = V_R^{\circ 2} / \sigma^2 \qquad \text{(Eq. 8-1-8)}$$

is obtained. For the variance, σ^2, can be determined in the customary fashion from the peak width $\omega_{\frac{1}{2}}$ half-way up a peak ($^2 = 5.54\,\sigma^2$, (cf. Fig. 8-1-7) on a chromatogram. And although the above equilibrium picture does not tell us anything about the causes for the diffusion, it does permit comparisons to be made between columns in terms of their numbers of theoretical plates, and it allows for the resolution of the variance, σ^2, into variances contributed by the physical parameters involved. The basis for chromatographic separation can be explained if the variance σ^2 is considered as resulting from a random walk of solute molecules after n steps of step length l_s, i.e.

$$\sigma^2 = l_s^2 n \qquad \text{(Eq. 8-1-9)}$$

It is immediately apparent that the width of a zone or concentration profile, 4σ, is proportional to $\sqrt{n}$ and thus to $\sqrt{x}$, i.e., to the square root of the distance traveled. On the other hand, since every zone travels at uniform velocity, the gap or distance of separation between any two zone centers (corresponding to two different components in the sample injected into the carrier gas stream) increases in direct, α proportion to ψ ($\Delta\psi \, \alpha \, \psi$). Thus, the gap eventually outdistances the spreading influences (Fig. 8-1-8), making chromatographic separation possible.

Broadening Factors - The following broadening factors are thought to be of primary influence on column performance: (i) longitudinal molecular diffusion; (ii) mass transfer; and (iii) eddy diffusion. More or less empirical equations have been derived for the 'height, equivalent to a theoretical plate' contributed by these factors. It is now generally assumed that eddy diffusion is independent of flow rate; longitudinal diffusion inversely proportional to it, and mass transfer directly proportional to it; and it has become customary to lump them together in the so-called van Deemter equation (actually first given by Keulemans and Kwantes (7)), viz:

$$H = A + B/v + Cv \qquad \text{(Eq. 8-1-10)}$$

Here:

$$A \simeq 2\lambda d_p \qquad \text{(Eq. 8-1-11)}$$

where λ is an empirical parameter:

$$B = 2\gamma_m \overline{D}_m + 2\gamma_s \overline{D}_s (1 - R)/R \qquad \text{(Eq. 8-1-12)}$$

and:

$$C = 2R(1 - R)\, t_d \qquad \text{(Eq. 8-1-13)}$$

where t_d is the desorption time for solid adsorbents, or:

$$t_d = d^2/2\overline{D}_s \qquad \text{(Eq. 8-1-14)}$$

for diffusion-controlled kinetics, where d is the effective diffusion depth.

By referring to the factors lumped into equation 8-1-10, (compare equations 8-1-11 through 8-1-14), it is clear that for small plate heights: (a) particle diameters should be small; on the other hand, fór practical reasons pressure drops must not be excessive; (b) the carrier gas should be of high molecular weight (as $\overline{D}_m \propto M^{-\frac{1}{2}}$); (c) a liquid stationary phase should be spread out thinly (d should be small) and made up of a viscous liquid ($\overline{D}_s \propto (\text{viscosity})^{-1}$), and (d) particle size should be uniform to reduce eddy effects. The role of temperature is not very clear from the kinetic equations; for reasons discussed later it should be as low as possible.

Figure 8-1-9 is a typical van Deemter plot (Eq. 8-1-10) showing how the three contributing terms are found.

An average packed column containing particle sizes in the range of 30-120 mesh and about 3 meters long would normally contain 1000-5000 theoretical plates. However, capillary columns may contain millions of theoretical plates in normal operation.

Thermodynamic or Equilibrium Factors

Because of the subsequent discussion of applications of gas-chromatography to physical properties determinations, a few comments on the nature of the distribution or partition coefficients are in order. These are equilibrium factors, distinct from the kinetic factors discussed in the preceding section.

Liquid Stationary Phases - The appropriate thermodynamic relation is Raoult's law. First, the distribution coefficient K is redefined by:

$$K = \rho_s \beta \qquad \text{(Eq. 8-1-15)}$$

Then, for dilute solutions:

$$\beta = R' T_c / M_1 \gamma' p_2^\circ \qquad \text{(Eq. 8-1-16)}$$

In general, solutes will be eluted from a solution column in the order of their vapor pressures, those having high vapor pressure - and usually low boiling point - first. Solutes with small activity coefficients, forming non-ideal solutions, will have larger β's and thus be eluted later than normal. For high column efficiency, the molecular weight of the liquid stationary phase should be high, provided the liquid remains thermally stable and the variation of γ with M_1 does not cancel out M_1. p_2° is also a measure of the free energy difference between gas and liquid phases, for:

$$\Delta G = RT \ln (\gamma' p_2^\circ / P^*) \qquad \text{(Eq. 8-1-17)}$$

The variation of vapor pressure with temperature is described by the Clausius - Clapeyron equation:

$$\frac{d\ln p_o^2}{dT} = \frac{\Delta H_v}{R^1T^2} \qquad \text{(Eq. 8-1-18)}$$

The dependence of activity coefficients on the molecular structure of solute and solvent remains largely unsolved (8). Polarity and molecular weight are probably the most important parameters (9) (10) (11). As it turns out, gas chromatography is the most powerful technique for the determination of activity coefficients at infinite dilution (12).

Solid Adsorbents - In early chromatography solid adsorbents were thought to be of little use because their isotherms are non-linear. For example, the Langmuir isotherm:

$$c_m = \frac{a'c_s}{1+b'c_s} \qquad \text{(Eq. 8-1-19)}$$

is linear only for small concentrations. It was necessary to improve the sensitivity of detectors before concentrations that were sufficiently small could be used. Figure 8-1-10 shows peak shapes obtained as a function of isotherms. With linear isotherms the peaks are Gaussian ($c_s' \propto c_m$). The retention time is independent of concentration, as shown by the dotted curves. Peaks with a sharp front and a tailing rear boundary are caused by Langmuir type isotherms. The peak shape is caused by the main portion of the solute band eluting more rapidly than the leading front edge because only a limited number of sites is available for adsorption. Retention time is generally a function of sample size. The anti-Langmuir isotherm (Fig. 8-1-10) is often obtained where the solute has low solubility in the stationary liquid phase, e.g. alcohols in hydrocarbons. The liquid phase becomes modified by the solute molecules, causing the main portion of the band to elute more slowly than the extremities. Retention time is also a function of sample size.

It should be noted that the 'tailing' observed with Langmuir type isotherms is frequently removed by increasing the sample size. This phenomenon is closely associated with the existence of another kind of site, the tailing site, whose heat of adsorption exceeds that of the normal sites (13). Increasing the sample size may saturate these sites. A so-called tail reducer or poison may do the same and is often better.

It has been noticed by some chromatographers (14) that a liquid stationary phase will occasionally continue to function at temperatures well below its melting point. Has it become a solid adsorbent or absorbent? Barrer has written a review (15) on the specificity of physical adsorbents which may be relevant. Plastic packings quite often behave like liquid stationary phases (see below), though they may even be quite crystalline according to X-ray diffraction. I think that many gaseous separations that are presumed to have resulted from differential solubility in a liquid have really resulted from differential adsorptivities at a colloid/gas interface. Possible ways of applying colloidal stationary phases have been proposed(16).

Packed Columns vs. Capillary Columns

Packed columns with liquid stationary phases require 'inert' solid supports to provide a large liquid surface area of low depth. Most of the supports used are diatomaceous earths, but even though these are generally pretreated or poisoned by material chemically similar to the compounds to be analysed, they are still never absolutely inert and may cause tailing of the Langmuir type. Much has been written

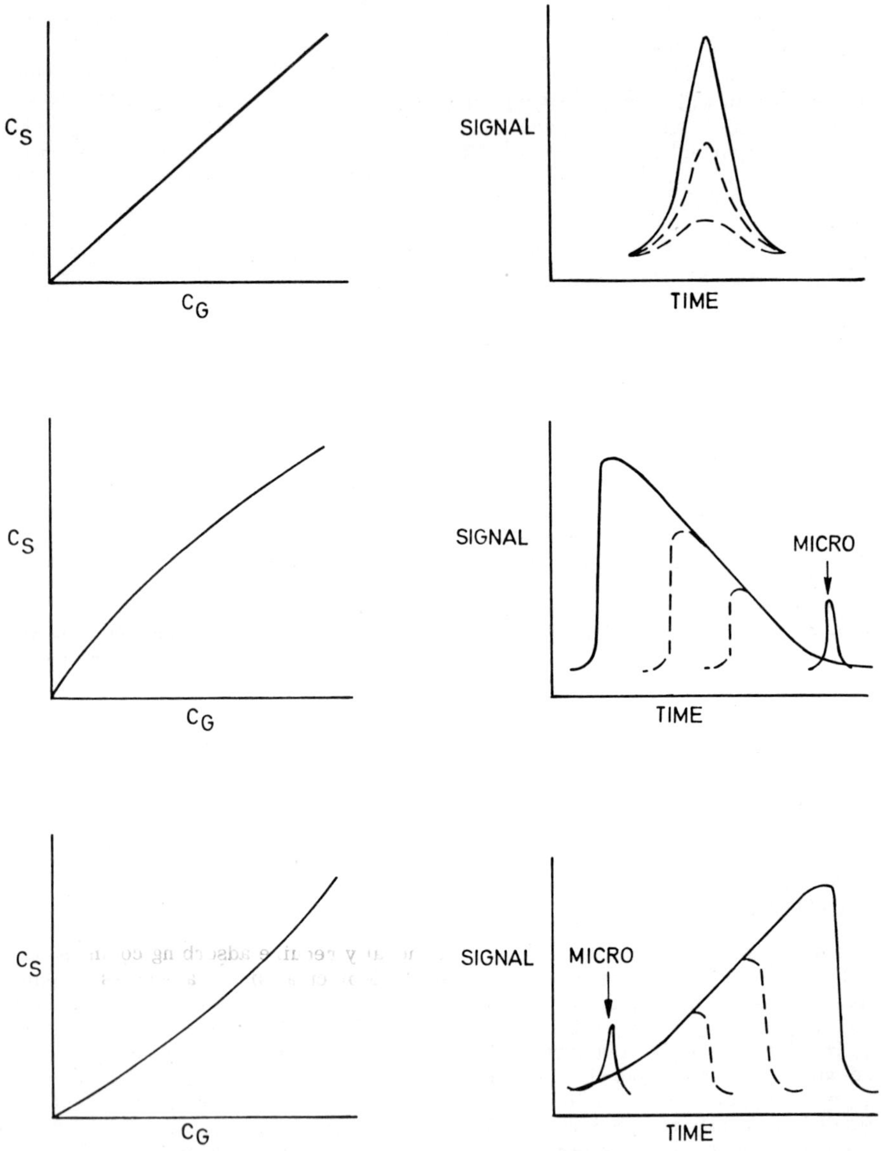

Fig. 8-1-10 Peak shapes as a function of isotherms

Only for the linear isotherm is the retention time of the peak independent of sample size. The peak called "Micro" is located at a position determined by extrapolation of peak retention time to zero sample size about the art of deactivating supports (17). Very inert materials, like teflon, are not porous and require very long tube lengths.

Early in the history of gas chromatography it was realized that fine capillary columns, with very thin films ($<0.1\mu$) of stationary phase, would be faster and more efficient than packed columns and would not require supports. Since a high number of theoretical plates can be obtained far more easily (18) with capillary columns than with packed columns ($\sim 10^6$), the first gas-chromatographic applications essentially involved the separation of multi-component hydrocarbon mixtures (19). Subsequent studies (20) dealt with the possibility of carrying out rapid analyses by means of appropriate selection of the column parameters and operating conditions. Today, with the availability of the extremely sensitive detectors required to handle the very small samples that these columns can pass without being overloaded, capillary (or Golay) chromatography has become indispensable in automatic laboratory analytical instrumentation.

Essential features of these columns are low pressure drop, uniformity, lack of a support and hence of eddy diffusivity, and compactness (it is easy to coil a hundred meters of 0.5 mm outside 0.1 mm inside diameter capillary tubing).

Because of their high-speed performance, samples of reformate that would take 25 minutes to separate and analyze by packed columns, have been analyzed in 7 seconds (13). As will be shown, capillary chromatography has also been applied as a research tool in catalyst studies (21) and in research on rapid reactions such as occur in engine combustion (14).

Selection of Stationary Phases

When the mixture is very complex, it is generally impossible to separate every component in a single column. Because of their speed, capillary columns are capable of covering wide molecular weight ranges of complex materials such as the lower boiling petroleum fractions. However, packed columns can often be designed to greater specificity.

In general, in selecting a liquid stationary phase, the principle of ***similia similibus solvuntur*** is applicable; e.g., hydrocarbon solutes will be retained, ***ceteris paribus***, more in hydrocarbon stationary phases and polar solutes, such as alcohols, in polar stationary phases, and so on. This generalization shows how desirable it is to effect preliminary separation of complex samples into groups. The most common stationary liquid phases are listed in Table 8-1-1.

Permanent gases and low-boiling vapors generally require adsorbing columns. Common adsorbents are listed in Table 8-1-2. Molecular sieves are types of clay whose microcapillaries can differentiate ('sift') molecules by their 'apparent' size, 'apparent', because some investigators believe electrostatic polarizability rather than molecular geometry to be the criterion (22). 'Poropak' is a trade name for polymers consisting of styrene cross-linked with divinylbenzene (23). These polymers are highly porous, yet inert, and serve the functions of both the liquid phase and the solid support. Apparently the solute molecules partition directly from the gas phase into the amorphous polymer. The elimination of the conventional diatomaceous solid support removes the adsorptive sites which may cause tailing even of highly polar molecules like water. The porous polymer beads can, however, also be coated with conventional liquid phases. Thus Poropak Q with polyethylenimine has been used to separate amines.

The basic tubing for all columns should be stable, flexible and heat resistant. Most columns in the author's laboratory are made of stainless steel.

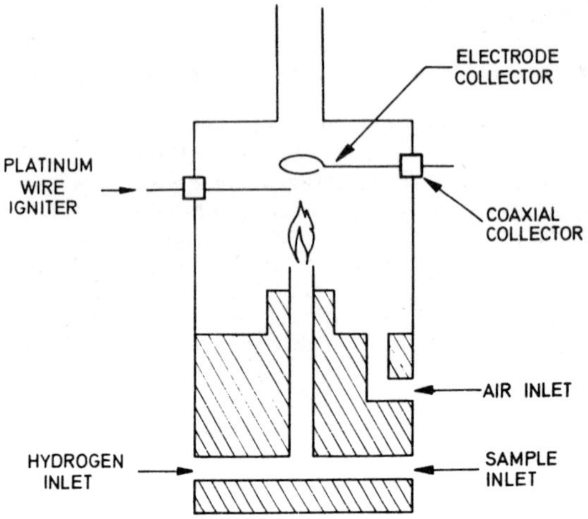

Fig. 8-1-11
Hydrogen flame ionization detector (schematic)

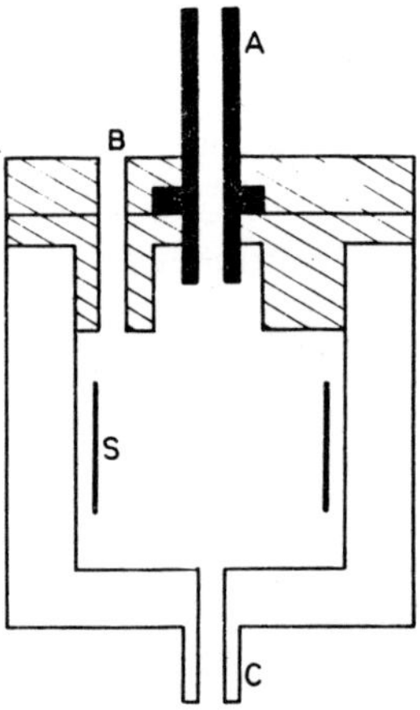

Fig. 8-1-12
Helium or argon ionization detector (schematic)

A - Carrier gas inlet, B - gas outlet,
C - scavenge (pure carrier) gas inlet,
S - source of ionizing radiation

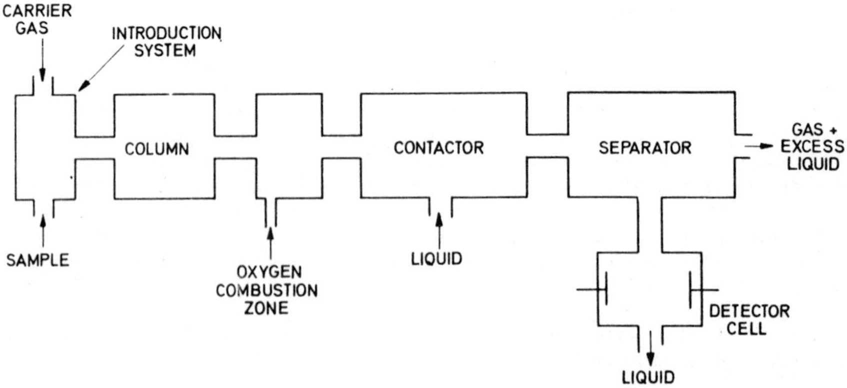

Fig. 8-1-13 Block diagram of an electrolytic conductivity detector (after Coulson (31))

Temperature Programming

When mixtures of components with a wide range of boiling points are analyzed on isothermal columns, the components with low boiling points give sharp peaks in short succession, while the components with high boiling points are eluted quite late, spaced widely apart and often lost because of their broadness. By 'programming' the column temperature (i.e. , linearly increasing it during elution) it is possible to obtain even spacings and sharp peaks throughout (24).

Injection Systems

An extensive literature has been published about experimental methods. The nature of the carrier gas is, of course, often crucial. Vacuum methods have been employed for unstable materials, like lithium and aluminum alkyls. Care is usually taken to inject samples into the carrier gas stream as small slugs of vapour. Periodic injections are possible, using rotating valves, for continuous monitoring of streams. In some applications, see below, a chemical reaction precedes injection.

Detectors

Non-specific Detectors

Almost any physical property can be made the basis of detection. Next to the thermal conductivity cell (26), the most common non-specific detectors are ionization detectors such as the flame ionization detector (Fig. 1-8-11, (27)) and the helium and argon ionization detectors (Fig. 8-1-12, (28)). The helium ionization detector (29) is perhaps the most sensitive detector in general use, applicable in the parts per billion (i.e. 10^{-9}) range for permanent gases.

Class-Specific Detectors

Electron Capture Detectors - Complex molecules - particularly when they contain electronegative elements such as halogens, or groups with large electron affinity such as hydroxyls - will capture slow electrons to form negative ions. Any inert carrier gas can be used. The greater the electron affinity, the larger is the field in which the electrons may be captured. These negative ions cause a decrease in the ionization current in an ionization chamber irradiated by radioactive elements, and this can be measured.

In those analytical problems where a small quantity of strongly absorbing compound is present in an excess of weakly adsorbing material, the method has its greatest potentialities (30). An example is the analysis of oxidation products in hydrocarbons.

Electrolytic Conductivity Detection - The electrolytic detector cell is a small, gas-liquid contactor fitted with electrolytic conductivity electrodes (31). It performs the functions of scrubbing the electrolytes out of the gas and into the water, and of detecting these electrolytes in the liquid phase. All organic compounds are first burnt quantitatively with oxygen as shown in figure 8-1-13 and, for example, this will convert sulfur compounds to SO_2.

By substituting catalytic hydrogenation for the oxidation, halogens and sulfur can be converted to HCl and H_2S and nitrogen to NH_3. It is clear that many interesting separations can thus be achieved.

Catalytic Combustion Ionization Detectors - These detectors should find application in differentiating between branched chain and straight chain hydrocarbons present as small traces. A platinum filament catalyst is heated to above 500°C in an oxidizing atmosphere containing the hydrocarbon. Ions produced can be collected, say by a platinum ion collector, downstream with respect to the filament. The number of ions produced by the hydrocarbons is an exceedingly small fraction of the number of carbon atoms oxidized and - in contrast to flame ionization - varies widely depending on the molecular structure of the hydrocarbon and the catalyst temperature. The production of ions is an order of magnitude greater for branched hydrocarbons than for straight-chain hydrocarbons. The mechanism involved is most likely one of chemi-ionization (32) (33) (34) (35).

Alkali Flame Detector - An alkali flame detector which gives sensitivities and selectivities for phosphorus exceeding those possible with other detectors by as much as 100 times has been described by Aue (36). Because the detector is specific for phosphorus, Aue found that he could detect phosphorus compounds with sensitivities in the picogram rather than the nonagram range; moreover, the technique could be used to measure phosphate derivatives with high selectivity and sensitivity. The Aue detector is a hydrogen flame detector that is modified by adding an alkali-salt source. Aue used a variety of compounds as alkali sources. The alkali salt is brought into contact with the flame using platinum helices, ceramic beads or pressed pellets. The electrodes are spirals, cylinders or single loops, that are either movable or stationary.

Specific Detectors - Identification of Components

These detectors are mostly used for qualitative identification in conjunction with a non-specific detector. They are spectrophotometers or mass spectrometers with a scanning rate capable of producing a complete spectrum in less than the time between successive chromatographic zones.

The Beckman # 102 Infrared Spectrophotometer with which I am familiar, achieves spectral resolution by rotating a disc with three peripherally arranged interference filters of continuously varying thickness in front of the detector. The rotational speed of the disc is fed into the amplification system of the detector. Infrared spectra extending from 2-14 microns can be recorded in a few seconds by means of an oscillograph. As different elution zones pass by a thermal conductivity detector, the individual components are picked up also by the spectrophotometer and can be identified from their spectra.

Even faster and more sensitive is the Block Interferometer Spectrometer which is capable of carrying out a complete 1-14 scan in a second. Various mass spectrometers are even faster and readily adaptable to gas chromatography. One use for them is to check for chemical reaction between the sample and the stationary phase.

Some Selected Analytical Applications

Gas-chromatographic analysis permits fast and sensitive analyses to be obtained cheaply. In process control, timed sample injection valves may be used to get periodic records. Systems with a capillary column and ionization detectors have been found suitable for repetitive analyses made at intervals of only a few seconds.

For research, it is usually desirable to have sufficient basic components on hand to put a system together for a particular application. Since 'Swagelok' or other ready-made fittings are now available to connect tubing in a leakproof manner, it is easy to change from one system to another. Sometimes several columns have to be placed in parallel, or in series, to analyze different groups of compounds, and the same is often done with non-destructive detectors. The number of permutations possible can be extremely great.

It should be pointed out that most of the standard columns are now commerically available, with methods for regeneration of the stationary phase and pretreatments stated in cookbook fashion. Similarly valve and injection systems, thermostatically controlled ovens for columns and detectors, and a great many detectors, amplifiers and other components are now stock items with many laboratory supply houses.

Likewise, the selection of columns and detectors for common problems is now frequently a computer exercise, retention times, or volumes on different columns, and and all other pertinent information having been tabulated for thousands of substances (37) (38).

Analysis of Permanent Gases

A very sensitive method of determining mixtures of H_2, Ar, O_2, N_2, and CH_4 employs a 20 foot long molecular sieve column at room temperature and a helium ionization detector (29). This length of the column is necessary to separate Ar from O_2; the other gases are eluted far enough apart. For all these gases the sensitivity is better than 3 parts-per-billion (ppb).

If H_2O and CO_2 are included in the above mixture, a separate sample should be analyzed with the same detector but using a Poropak Z (29) column at 200°C instead of the molecular sieve column. The other gases will hardly be separated now, but CO_2 and H_2O will follow CH_4 after fairly long intervals. The sensitivity for these gases is only about 20 ppb. because of tailing.

Indirect Analysis by Reaction Products

A distinction can be made between 'Gas Chromatographic Analysis by Reaction Products' and 'Reaction Gas Chromatography', depending on whether the reaction to which the injected substances have been subjected is completed before the reaction products enter the columns or whether it is completed in the column. The former is primarily a procedure of quantitative chemical analysis. In the latter, the principal information sought is usually related either to quantitative identification, to the kinetics of a particular reaction or series of reactions, or to catalytic activity. Very often, however, the distinction is semantic only.

Prior reaction of a mixture to be analyzed quantitatively is usually required when (a) the mixture is so complicated that one is satisfied with an analysis according to type of compound, e.g. the amount of alcohol or aldehyde present, or (b) the material to be analyzed is not volatile or too unstable even for injection with a solvent in a heated column. To some extent a type analysis can be carried out by a class-specific detector, often in conjunction with prior reaction. However, reaction prior to introduction into the column, and use of a non-specific detector, represents a much more versatile procedure.

The following comments follow Beroza and Coad (39) to a degree.

Subtraction - In subtraction two parallel columns and detectors are used. One of the columns will pass a class of compounds whose presence is suspected in the sample. The other will not. The determination may be made from the peaks caused by the non-absorbed gas sample. The principal subtracting agents used in petroleum analysis are molecular sieves (normal paraffins), sulfuric acid (olefins), sulfuric acid and silver sulfate (olefins as well as aromatics), maleic anhydride (conjugated diolefins), magnesium perchlorate (water), and boric acid (primary and secondary alcohols).

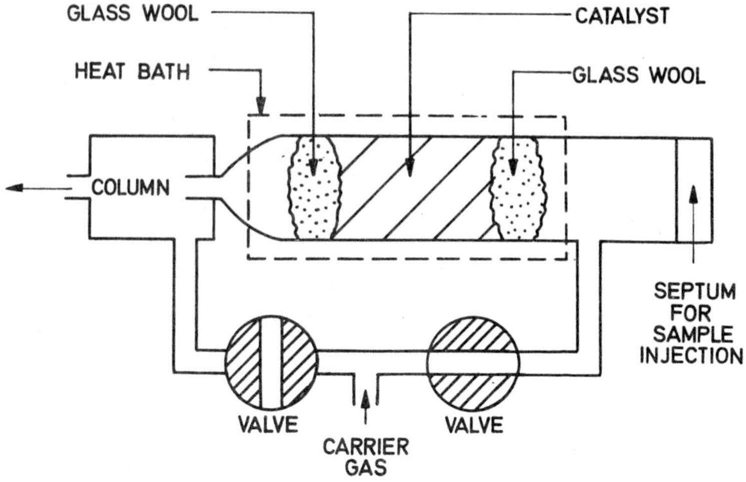

Fig. 8-1-14 Reactor inlet system

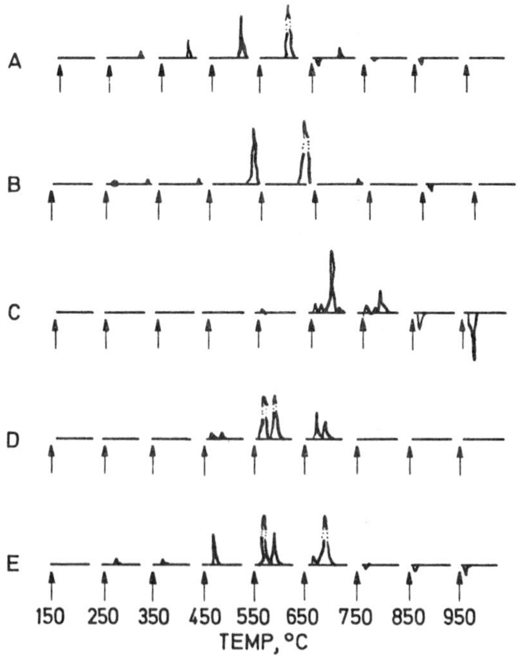

Fig. 8-1-15
Temperature series pyrograms of polymers
A
poly (methyl methacrylate)
B
poly (ethyl methacrylate)
C
poly (methyl acrylate)
D
50% methyl methacrylate-50% styrene copolymer
E
50% methyl methacrylate-50% styrene mixture
(after Lehrle and Robb (41))

Pyrolysis - The pyrolysis technique, first suggested in 1954 (40), has now become a standard procedure in the analysis of non-volatile materials and apparatus is commercially available. The substance is cracked quickly and without catalytic decomposition, and the volatile degradation fragments are carried directly into the gas chromatograph for analysis (Fig. 8-1-14). The 500-800°C range is most useful for polymers, but other temperature ranges have been used. The technique can be made to give repeatable results and is useful for identification or quantitative analysis. In many instances, quite a number of decomposition products are formed whose peaks form a 'spectrum' similar to an infrared spectrum (Fig. 8-1-15). Lehrle and Robb (41) recently reviewed the techniques used. Many investigators obtained fast heating by discharging a condenser between electrodes. Simon and coworkers (42) obtained nearly ideal heating conditions by coating a ferromagnetic conductor with the sample and warming it to constant temperature in 20-30 milliseconds using high frequency induction heating. Owing to the drastic change in the magnetic permeability of the magnetic conductor at the Curie point, the energy input drops at that temperature.

The pyrograms of Fig. 8-1-15 (41) require a few comments. They were obtained by mounting a polymer sample on a filament, raising the temperature of the filament in steps of 100 or 150°C at a rate of about ten seconds per step, and recording the chromatogram after every step. Fig. 8-1-15 (a) and 8-1-15 (b) show only 'upzipping', but Fig. 8-1-15 (c) contains peaks due to complicating side reactions. Mixtures and copolymers are clearly differentiated (Fig. 8-1-15 (d) and (e)). Peak reversal at elevated temperatures is caused by the detector.

Clearly this procedure can be extended to the determination of polymer degradation kinetics and, hence, can become useful in ablation studies. For this purpose, Lehrle and Robb state, (i) the temperature of the sample must be uniform and accurately known, (ii) the sample must be heated as rapidly as possible to the required temperature, and then cooled as quickly as possible at the end of the degradation period in order to define closely the duration of the degradation, and (iii) the sample thickness must be such that the reaction is not controlled by the rate of diffusion of products through the molten polymer film. By improved temperature measurement techniques and by varying heating rates and sample thicknesses it is possible to obtain meaningful Arrhenius-type plots from which reaction rates and activation energies can be determined.

An example is the degradation of polymethylmethacrylate. This polymer on degradation produces an almost quantitative yield of monomer. This is best interpreted in terms of a mechanism whereby the polymer chain is initially ruptured, followed by depolymerization to yield the monomer by a chain reaction ('upzipping'). There are four possible variants: (a) end initiation and upzipping to the end of the chain, (b) end initiation and bimolecular termination, (c) random scission initiation - upzip to the end of the chain, and (d) random scission initiation - bimolecular termination. These four possibilities all have a first order dependence of rate on the weight of polymer remaining at a given time. If k_{obs} is the apparent specific rate constant, where:

$$\frac{dm}{dt} = k_{obs} \cdot W \qquad \text{(Eq. 8-1-20)}$$

or, in integrated form:

$$-\ln(1 - X) = k_{obs}\, t \qquad \text{(Eq. 8-1-21)}$$

mm* it is clear that k_{obs} is independent of the degree of polymerization D for (a) and (d),

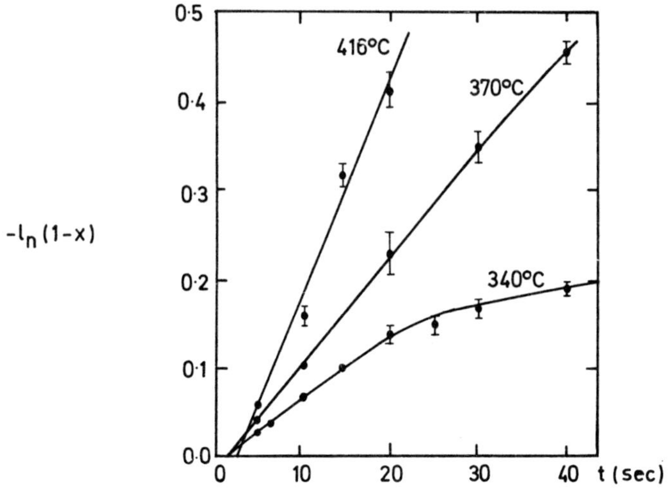

Fig. 8-1-16 Arrhenius plots for polymethylmethacrylate pyrolyses (after Lehrle and Robb (40))

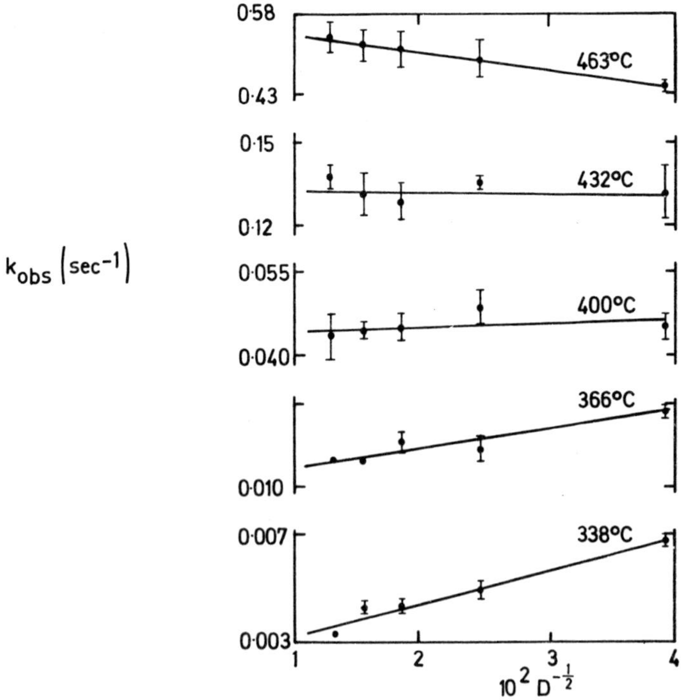

Fig. 8-1-17 Polymethylmethacrylate degradation. Dependence of specific rate on molecular weight (after Lehrle and Robb (40))

proportional to $1\sqrt{D^{\frac{1}{2}}}$ for (b), and proportional to D for (c). A study was made of the degradation of a series of fractionated polymethylmethacrylate specimens at a series of temperatures. The results in terms of Arrhenius plots and plots of rate constant versus $1\sqrt{D^{\frac{1}{2}}}$ are shown in figures 8-1-16 and 8-1-17.

These results lead to the following interpretation. At the lower temperatures, chains are initiated at a terminal link, and termination is by bimolecular interaction. The interpretation is also supported by the fall-off in the rate with time at the lower temperature, which can be due to the loss of weak end links,probably the unsaturated terminal bond produced by disproportionation. At the highest temperature the mechanism is one of random initiation and complete unzipping. The two mechanisms overlap at intermediate temperatures.

Hydrogenolysis - Here the reactor usually consists of a tube, filled with palladium catalyst and heated in an electric furnace. A common application of hydrogenolysis in the petroleum laboratory is to organic sulfur compounds, which are quantitatively converted to hydrocarbons and hydrogen sulfide (43).

Elemental Analysis - When pyrolysis is carried to the extreme, e.g., by contact with a catalyst and a source of oxygen at very high temperatures, organic compounds break down to CO_2, water, and small molecules with other elements. The exposure of organic molecules to hydrogen under similar drastic conditions produces methane, water and small molecules containing other elements. These molecules are then analyzed by gas chromatography. Such procedures have speeded up elemental analyses (CHN analyses take 10 minutes) and made them possible on as little as 0.2 mg of compound.

Elemental analysis by gas chromatography has been fully described by Francis (44).

Functional Group Analysis - Many of the standard reagents for functional groups have been used to convert the original samples into volatile derivatives. They are then separated in a column, and can be recovered, e.g., by flash decomposition. Thus, lower aldehyde and ketone constituents of very volatile natural substances have been converted into 2, 4-dinitrophenylhydrazones and then 'flash exchanged' back to the free carbonyl compounds (45).

Reaction Gas Chromatography

By combining gas chromatography with reactors, generally tube reactors, in such a way that the reaction is not carried to completion, it is possible to determine kinetic parameters. Thus, in some instances, the information obtained in this way in pyrolytic studies has been complementary to shock tube studies. For example, Cramers and Keulemans (46) carried out pyrolyses on samples of 50 micrograms and were able to determine the influence of varying temperature, time, and concentration on reaction rate, product distribution, and recombination reactions. Furthermore, the chromatograms of the pyrolyses products themselves resemble mass-spectrograms and can help elucidate reaction mechanisms. Specifically, the kinetics of the decomposition of ethyl acetate and cyclopropane at 762°K and 873°K have been reported by these authors.

This technique is probably the simplest and fastest for getting kinetic data, and a great many reactions have already been studied by it. Since this work is representative of many investigations that are currently being carried out, it seems appropriate to discuss it in more detail.

Figure 8-1-18 is a schematic diagram of the reactor, the multiport coupling valve, and the detector system. The reactor consists of a metal tube (gold is best, for it allows deposits to be burned off) of 1 m length and 1 mm inner diameter and

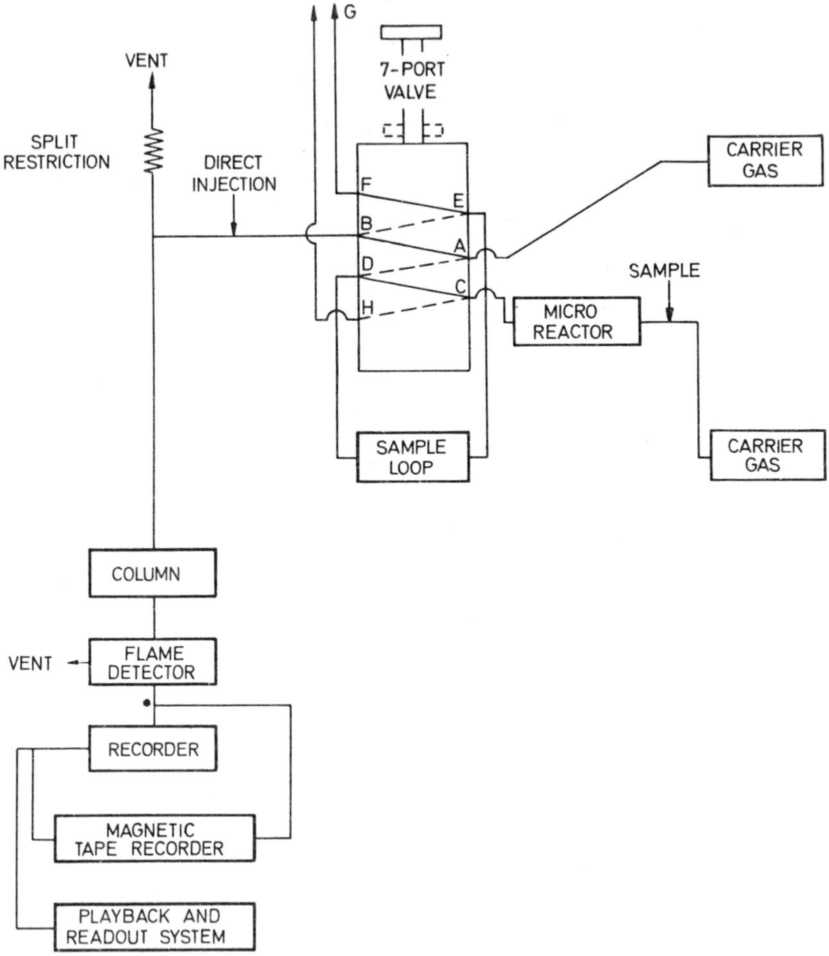

Fig. 8-1-18 Apparatus used in gas pyrolysis studies of kinetics and decomposition products.
Upper position (full line) - sample loop is connected to reactor; lower position (dotted line) - contents of sample loop to chromatograph
(after Cramers and Keulemans (46))

appropriate heaters. The carrier gas is preheated to reaction temperature. Isothermal conditions can be achieved easily for endothermic degradation reactions. The seven-port coupling valve makes possible (a) independent flow control of the reactor and the chromatograph, with the advantage of great ease in measuring reaction times, and (b) both pulse injection and continuous introduction of samples into the chromatograph, thus creating stationary reaction conditions.

Gas leaving the reactor may follow two paths: it may go to the vent directly (CH), or it may first pass through a sample loop and then go to vent (CDEFG). When the reactants of a pulse-injected sample have reached the sample loop, the seven-port valve is actuated and the carrier gas of the column entrains the contents of the sample loop. The reactants then meet a stream splitter device and about 1 percent of the gas goes into the analyzer. It is obvious that the system can be adapted to a continuous flow reactor with intermittent sampling.

The carrier gas stream to the reactor may follow two paths:

(a) It may go directly to the reactor. In the meantime the vapor space of the sample bottle can be purged with a controlled flow of argon to remove oxygen from the system, as oxygen has a accelerating effect on the decomposition rate of many hydrocarbons, for example.

(b) The reactor carrier gas may pass the vapor space above the sample on its way to the reactor. By controlling the temperature of the sample bottle, any convenient concentration of sample in the reactor gas can be obtained.

In the reduction of the data, first order kinetics are assumed. Allowance is made for the increase in gas volume resulting from decomposition.

The principal application of this method has been to product studies. Heterogeneous reactions can be analyzed similarly, after substitution of an appropriate reactor.

In the case of reactions of type $A \rightleftharpoons B + C$ over a heterogeneous catalyst and in which the reactant is introduced as a pulse, the chromatographic properties of the bed may serve to separate the products of reaction from each other and so drive the reaction further toward completion than would be possible in an equilibrium system. Roginskii, *et al*., (47) and Matsen (48) have studied the dehydrogenation of cyclohexane over platinum and shown that substantially complete conversion to benzene can be achieved as a result of the chromatographic separation of the product band in the catalyst bed. See (49), for further examples of gas-chromatographic studies in catalytic reactions.

Reactions on Chromatographic Columns

If the adsorbent used in the packing of a chromatographic column has catalytic properties, adsorbates may be chemically transformed while they are being eluted along the column. The virtue of this technique is that the resulting reaction rates are expressed in terms of concentrations on the surface of the catalyst rather than in the gas phase. Habgood (49) has shown how it is possible to derive rate constants and activation energies for first order reaction sequences. The product of reaction is formed along the whole length of the column, and the retention volume of the product peak is spread between its normal value and the retention volume of the reactant, depending on whether the product was formed near the inlet or near the exit of the column. Expressions to describe the shapes of the peaks corresponding to various values of the retention volumes, and of the first order rate constant, have been described by various authors (50) (51) (52) (53).

However, while the expression of the rate of a catalytic reaction in terms of surface concentrations rather than gas phase concentrations does remove the possibility of confusion resulting from changes in the adsorptivity of a catalyst, it does not necessarily afford any additional insight into the mechanism of the reaction. For example, migration can occur from nonreactive adsorption sites to catalytically active sites, the catalytic sites being only a tiny fraction of the whole surface.

Physico-Chemical Parameters

From the theoretical discussion given earlier, it is apparent that many physico-chemical parameters can be determined from the retention times and the shapes of peaks in a chromatogram. Molecular weights (54), free energies and heats of evaporation and adsorption, or desorption, diffusivities, surface areas, and isotherms have been so determined. Often data reduction requires the use of the van Deemter equation (Eq. 8-1-10).

Of particular value is the determination of surface area of solid catalysts according to Nelsen and Eggertsen (55). The classical procedure (B.E.T. method) is exceedingly tedious and, in particular, requires good vacuum techniques. The new method requires adsorption of the gas by the solid from a stream of non-adsorbed helium, cooling the adsorbent, and allowing it to warm up. A peak is produced on the chart which is in the reverse direction to the adsorption peak. The potentiometer plots curves for both adsorption and desorption and the area under either one of these curves is a measure of the gas adsorbed. A calibration is, however, usually necessary with a sample of known surface area.

Leffler (56) and Davis and Scott (57) determined effective diffusivities of catalysts from the mass transfer coefficient (C) of the van Deemter equation. By analyzing the moments of chromatographic curves, Schneider and Smith (58) were able to evaluate intraparticle effective diffusion coefficients, from which surface diffusivities could be extracted.

The heat of adsorption may be determined from a plot of log retention volume, corrected to 0°C,vs. reciprocal of the absolute temperature (59).

In regard to isotherm determination of adsorbents from the shape of the elution peak (Fig. 8-1-10) it should be remarked that the greatest usefulness of the method lies in cases where the high temperature or the reactivity of the adsorbate on the solid makes other methods extremely difficult. The application of this analysis is limited by diffusional and other band-broadening effects. It is probably better to work at relatively low flow rates and to correct for diffusional band broadening than to use higher rates where the rate of desorption may give significant broadening at the rear edge, not balanced by a similar broadening at the leading edge (60) (61).

However, the chromatographic method of determining adsorption under reaction conditions developed by Tamaru (62) and his coworkers is particularly valuable in studies of heterogeneous catalysis. He passed a steady stream of reactants, or of a reaction mixture through a catalytic column and used this stream as the carrier gas for the chromatographic elution of small samples of one of the reactants or products injected ahead of the column.

Continuous Gas Chromatography

Although it has often been stated that one of the drawbacks of gas chromatography lies in its sequential nature, i.e., samples cannot be analyzed continuously, a number of devices have been described in the literature for doing just that. So far, none of them has been proved practical, but a significant breakthrough can be expected soon.

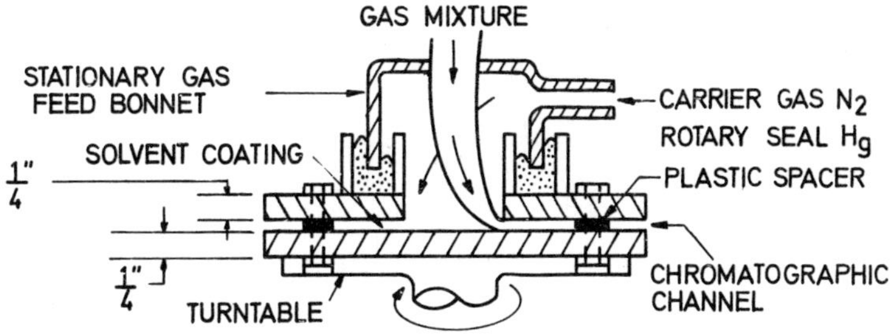

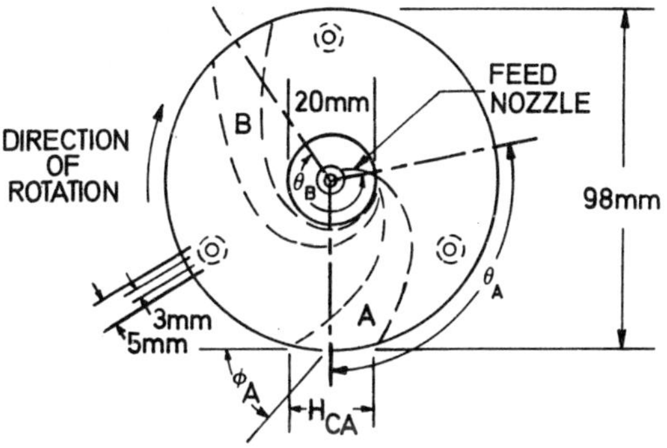

Fig. 8-1-19
Continuous disc chromatograph.
(Top) Vertical section.
(Bottom) Discs showing central supply orifice and paths of hypothetical components A and B. H_{CA} is the length of discharge zone for component A.
(after Sussman and Huang (63))

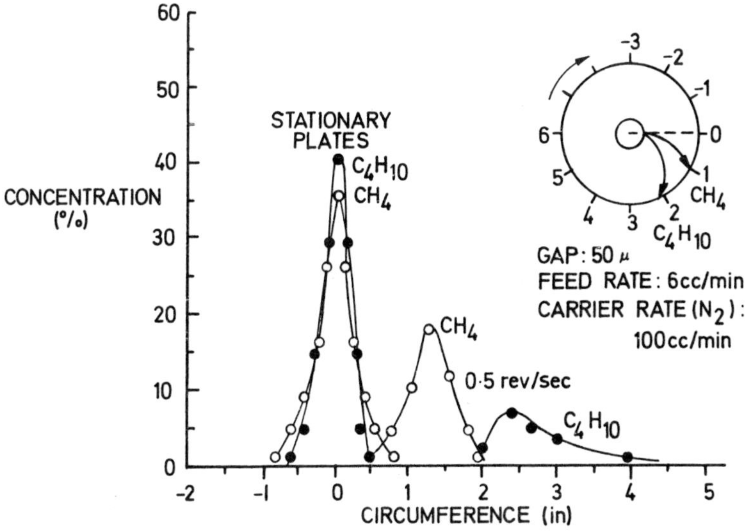

Fig. 8-1-20 Effluent concentration as a function of peripheral position Mixture of methane and butane (37:63) ; 6 cm^3/min.

The device of Sussman and Huang (63) appears to offer some intriguing possibilities, largely because it does not depend on packing (Fig. 8-1-19). It might properly be described as a continuous capillary gas chromatograph. The previous continuous gas chromatographs were all impractical because of the disadvantages arising from a non-uniform flow resistance of a packed bed. The new device achieves chromatographic separation with a radial-flow chromatographic channel formed between two closely spaced (50-75 microns) disc surfaces, optically flat and solvent coated, rotating at one-half or one revolution per second. This technique provides high capacity and immediate response in a fraction of a second. In effect, the device is a kind of mass spectrometer in which the pole pieces are formed by coated metal surfaces. Mixtures of methane and butane have been separated at flow rates of 6-30 cc/min. with 100-150 cc/min. of nitrogen carrier gas in a chromatographic channel only 39 millimeters long. Continuous monitoring of reactions should thus be possible (Fig. 8-1-20).

References

1. **Tswett, M.,** Ber. deut. bot. Ges., 24, 235. 1906.

2. **James, A.T., Martin, A.J.P.,** Biochem. J., 50, 679. 1952.

3. **Peterson, M.L., Hirsch, J.,** J. Lipid Research, 1, 132. 1959.

4. **Littlewood, A.B.,** 'Gas Chromatography,' New York, Academic Press, 1962

5. **Giddings, J.C.,** 'Dynamics of Chromatography, Part 1, Principles and Theory,' New York, Marcel Dekker, Inc., 1965.

6. **Martin, A.J.P., Synge, R.L.M.,** Biochem. J. 35, 1358. 1941.

7. **Keulemans, A.I.M., Kwantes, A.,** in 'Vapor Phase Chromatography' Proc. First Symp. London, May, 1956 (D.H. Desty, ed.), New York, Academic Press, 1957.

8. **Van Ness, H.C.,** 'Classical Thermodynamics of Non-electrolyte Solutions,' Oxford, Pergamon Press, 1964.

9. **Pierotti, G.J., Deal, C.H., Derr, E.L., Porter, P.E.,** J. Am. Chem. Soc. 78, 2989. 1956.

10. **Perotti, G.J., Derr, E.L., Deal, C.H.,** Ind. Eng. Chem. 51, 95. 1959.

11. **Hildebrand, J.H., Scott, R.L.,** 'The Solubility of Non-electrolytes,' 3rd Edn., New York, Dover Publ., Inc., 1964.

12. **Kwantes, A., Rijnders, C.W.S.,** 'Gas Chromatography,' 1958, 125 (D.H. Desty, ed.), London, Butterworths, 1958.

13. **Giddings, J.C.,** J. Chem. Phys., 26, 169. 1957.

14. **Desty, D.H., Goldup, A., Swanton, W.T.,** 'Gas Chromatography 1961,' New York, Academic Press, 1962.

15. **Barrer, R.M.,** J. Coll. 'Interface Science' 21, 415-34. 1966.

16. **Schwartz, R.D., Brasseaux, D.J., Shoemake, G.R.,** Analytical Chemistry, 35, 496-505. 1957.

17. **Offenstein, D.M.**; 'The Chromatographic Support,' Advances in Chromatography, Vol.3 (J.C. Giddings and R.A. Keller, eds.), New York, Marcel Dekker, Inc., 1966.

18. **Golay, M.I.E.**, 'Gas Chromatography'(V.J. Coates, ed.), New York, Academic Press, 1958.

19. **Desty, D.H., Goldup, A., Whyman, B.M.F.**, J. Inst. Petrol. 45, 287. 1959.

20. **Purnell, J.H., Quinn, C.P.**, 'Gas Chromatography,' 1960, (R.P.W. Scott, ed.), London, Butterworths, 1960.

21. **Pitkethly, R.C., Goble, A.G.**, Symp. on Instrumental Techniques in Study of Catalyst Mechanisms, A.C.S. Spring Meeting, Boston, Mass., April 1959.

22. **Benson, S.W., King, J.W.**, Science, 150, 1710-13. 1965.

23. **Hollis, O.L.**, Anal. Chem. 38, 309, 1966.

24. a. **Said, A.S.**, 'The Theory of Programmed Temperature Gas Chromatography,' Chapter 6, Gas Chromatography, N. Brenner *et al* eds, New York, Academic Press, 1962.

b. **Mikkelsen, L.**, 'Advances in Programmed Temperature Chromatography,' Advances in Chromatography, Vol. 2, J.C. Giddings and R.A. Keller, eds., New York, Marcel Dekker, Chp.6, 1966.

25. **Longi, P., Mazzocchi, R.**, Chem. Ind. (Milan) 48 7, 718. 1966.

26. **Lawson, A.E., Miller, J.M.**, 'Thermal Conductivity Detectors in Gas Chromatography'(Review) J. Gas Chromatography 273. 1966.

27. **Sternberg, J.C., Gallaway, W.S., Jones, D.T.L.**, 'The Mechanism of Response of Flame Ionization Detectors', Gas Chromatography (N. Brenner, ed.) New York, Academic Press, Ch.6, 1962.

28. **Karmen, A.**, 'Ionization Detectors for Gas Chromatography, Advances in Chromatography, Vol.2. J.C. Giddings and R.A. Keller, eds., New York, Marcel Dekker, Ch.6, 1966.

29. **Hartmann, C.H., Thompson, K.**, Aerograph Res. Notes, Spring 1967. Varian Assoc., Walnut Creek, Cal., 1, 1967.

30. **Lovelock, J.E., Lipsky, S.R.**, J. Am. Chem. Soc. 82, 431. 1960.

31. **Coulson, D.M.**, 'Electrolytic Conductivity Detection in Gas Chromatography,' Advances in Chromatography, Vol.3, J.C. Giddings and R.A. Keller, eds., New York, Marcel Dekker, 1966.

32. **Woods, F.J., Umstead, M.E., Johnson, J.E.**, U.S. Naval Res. Lab. Rep. 6316, Oct. 1, 1965.

33. **Folmer, O.F., Yang, K., Perkins, G.**, Anal. Chem. 35, 454. 1963.

34. **Perkins, G., Folmer, O.F.**, Gas Chromatography, Fowler, L., ed., New York, Academic Press, 1963.

35. **Perkins, G., Yang, K., Folmer, O.F.,** Nature 198, 198. 1963.

36. **Aue, W.A.,** Sci. Res. 2, 16, 1967.

37. Preston Gas Chromatography Abstract Cards, 909 Pitner Ave., Evanston, Ill. 60206.

38. **Knapman, C.E.H.,** ed., Gas Chromatography Abstracts, 1958-1962, London, Butterworths, 1962.

39. **Beroza, M., Coad, R.A.,** Reaction Gas Chromatography (Review), J. Gas Chromatography, 1966.

40. **Davison, W.H.T., Slaney, S., Wragg, A.L.,** Chem. Ind. (London), 1354. 1954.

41. **Lehrle, R.S., Robb, J.C.,** J. Gas Chromatography, 88. Feb. 1967.

42. **Simon, W., Kriemler, P., Voellmin, J.A., Steiner, H.,** J. Gas Chromatography, 53. Feb. 1967.

43. **Thompson, C.J., Coleman, H.J., Hopkins, R.L., Rall, H.T.,** J. Gas Chromatography, Jan. 1967.

44. **Francis, H.J.,** Anal. Chem. 36, 31A-47A. 1964.

45. **Ralls, R.W.,** Anal. Chem. 32, 1960, 332-6

46. **Cramers, C.A.M.G., Keulemans, A.I.M.,** J. Gas Chromatography. 58. Feb. 1967.

47. **Roginskii, S.Z.,** *et al*. Kinetika i Kataliz, 3, 529. 1962.

48. **Matsen, J.M.,** *et al*. J. Phys. Chem. 69, 522. 1965.

49. **Habgood, H.W.,** 'Chromatography,' The Solid-Gas Interface, Vol. 2, New York, Marcel Dekker, 631. 1967.

50. **Kallen, J., Heilbronner, E.,** Helo. Chem. Acta., 43, 489. 1960.

51. **Boreskova, E.G.,** *et al*., Kinetika i Kataliz, 5, 903. 1964.

52. **Roginskii, S.Z., Rozental, A.L.,** ibid. 5, 104. 1964.

53. **Gaziev, G.A.,** *et al*., ibid. 4, 688. 1963.

54. **Martin, D.E., Purnell, J.H.,** Trans. Far. Soc. 62, 710. 1966.

55. **Nelsen, R., Eggertsen, F.T.,** Anal. Chem. 30, 1387. 1958.

56. **Leffler, A.J.,** J. of Catalysis 5. 1966, 22-26

57. **Davis, B.R., Scott, D.S.,** Preprint 48D, Symp. on Fundamentals of Heat and Mass Transfer, 58th Ann. Meeting Philadelphia, Pa., 1965, American Inst. Chem. Eng., New York, 1966.

58. **Schneider, P., Smith, J.M.,** Chem. Engineering, in press.

59. **Toth, J., Graf, L.,** Magyar Kem. Folyirat, 66, 1960, 123-8.

60. **Cramer, E.,** Monatsh. Chem. 92, 1961, 112-5.

61. **Habgood, H.W.,** 'Gas Chromatography'(Review) Annual Review of Physical Chemistry, 13, 1962, 259.

62. **Tamaru, K.,** Nature, 183, 1959, 319.

63. **Sussman, M.V., Huang, C.C.,** Science 156, 1967, 974.

Acknowledgement

The author wishes to thank Sun Oil Company for providing him with some time and the clerical help needed for the preparation of this manuscript. Further he wishes to express his gratitude to Mr. D.L. Camin of the Sun Oil Company for introducing him to gas chromatography and to Mr. P.A. Antal for helping him with many analyses and with the drafting of the figures.

a'	coefficient of Langmuir isotherm
A	eddy diffusion term in van Deemter equation
b'	coefficient of Langmuir isotherm
B	longitudinal diffusion term in van Deemter equation
c	concentration (solute per unit volume of column)
c_m	concentration of vapor in mobile phase
c_s	concentration of vapor in stationary phase
C	mass transfer term in van Deemter equation
d	effective diffusion depth
d_p	particle diameter
D	degree of polymerization
$\overline{D_m}$	diffusion coefficient in the mobile phase
$\overline{D_s}$	diffusion coefficient in the stationary phase
H	height of a theoretical plate
ΔH_v	heat of vaporization of pure liquid solute
k_{obs}	apparent specific rate constant
K	partition coefficient $= p_s\beta$
1_s	step length in a random walk

m	mass
M	molecular weight
M_1	molecular weight of the solvent
n	number of steps in random walk
N	number of theoretical plates in a column
p_c	vapor pressure at temperature T
p_2°	vapor pressure of pure solute at temperature Te
p^*	a standard pressure
R	retention ratio
R'	universal gas constant
t	time
t_d	desorption time for solid absorbents
t_r t_R	relative and total retention times
t_m	hold-up time
T	temperature
T_c	column temperature
v	average mobile phase fluid velocity
v_d	drift velocity
V_m	volume of mobile phase
V_m°	hold-up volume corrected to constant flow
V_R°	retention volume corrected to constant flow
V_s	volume of stationary phase
$w_{1/2}$	peak width at half height
W	weight of polymer
X	fractional conversion in time t
β	specific retention volume at temperature Tc
γ_m	obstructive factor or tortuosity in the mobile phase
γ_s	obstructive factor or tortuosity in the stationary phase

γ^1	activity coefficient
λ	eddy diffusion coefficient
ρ_s	density of stationary phase
σ^2	variance of elution zone
ψ	location along column

Table 8-1-1. Common Liquid Stationary Phases

Liquid	Max. Temp. (°C)	Used for	Polarity
Squalane	125	Hydrocarbons, ethers ketones, halides	Non-polar
Apiezon grease	300	Above, higher b.p.	Non-polar
Silicon grease	300	Same	Non-polar
Tricresyl phosphate	170	Oxygenated Hydrocarbons	Slightly polar
Dimethylsulfolane	20	Low boiling hydrocarbons and olefins	Polar

Table 8-1-2. Common Absorbents

Absorbent	For Separation of
Activated carbon	H_2, $O_2 + N_2$, CO, NO, CH_4
Silica gel	CH_4, C_2H_6, CO_2, C_2H_4
Alumina	C_2H_6, C_2H_4, C_3H_8, C_3H_6, Cyclopropane
Linde "Mol. Sieve 5A"	H_2, O_2, N_2, CH_4, CO, C_2H_6
Porous polymer beads	H_2O, nitrocompounds, CO_2, C_2H_6, CH_3C_1 $C_2H_5C_1$, $C_2H_4C_1$, C_1 - C_5 alcohols, C_1 - C_3 acids, C_1 - C_3 glycols and oxides, air (N_2, O_2, Ar), N_2O

Index

On a donné les traductions ci-dessous quand il y a beaucoup de différence entre les mots propres des deux langues. Quand les différences essentiales sont principalement de l'orthographe, les termes anglais ou français sont donnés tout seuls. Fréquemment, nombreuses des chiffres qui suivent un titre en-tête, dequoi les premiers mots se trouve en langue français, sont références aux textes français

D

E

F

K

L

M

N

O

P

Q

R

S

T

U

V

W

X

Z